Springer Water

Series Editor

Andrey G. Kostianoy, Russian Academy of Sciences, P.P. Shirshov Institute of Oceanology, Moscow, Russia

AF895781

The book series Springer Water comprises a broad portfolio of multi- and interdisciplinary scientific books, aiming at researchers, students, and everyone interested in water-related science. The series includes peer-reviewed monographs, edited volumes, textbooks, and conference proceedings. Its volumes combine all kinds of water-related research areas, such as: the movement, distribution and quality of freshwater; water resources; the quality and pollution of water and its influence on health; the water industry including drinking water, wastewater, and desalination services and technologies; water history; as well as water management and the governmental, political, developmental, and ethical aspects of water.

More information about this series at http://www.springer.com/series/13419

El-Sayed Ewis Omran · Abdelazim M. Negm
Editors

Climate Change Impacts on Agriculture and Food Security in Egypt

Land and Water Resources—Smart Farming—Livestock, Fishery, and Aquaculture

Springer

Editors
El-Sayed Ewis Omran
Faculty of Agriculture
Suez Canal University
Ismailia, Egypt

Abdelazim M. Negm
Faculty of Engineering
Zagazig University
Zagazig, Egypt

ISSN 2364-6934 ISSN 2364-8198 (electronic)
Springer Water
ISBN 978-3-030-41631-7 ISBN 978-3-030-41629-4 (eBook)
https://doi.org/10.1007/978-3-030-41629-4

© Springer Nature Switzerland AG 2020
This work is subject to copyright. All rights are reserved by the Publisher, whether the whole or part of the material is concerned, specifically the rights of translation, reprinting, reuse of illustrations, recitation, broadcasting, reproduction on microfilms or in any other physical way, and transmission or information storage and retrieval, electronic adaptation, computer software, or by similar or dissimilar methodology now known or hereafter developed.
The use of general descriptive names, registered names, trademarks, service marks, etc. in this publication does not imply, even in the absence of a specific statement, that such names are exempt from the relevant protective laws and regulations and therefore free for general use.
The publisher, the authors and the editors are safe to assume that the advice and information in this book are believed to be true and accurate at the date of publication. Neither the publisher nor the authors or the editors give a warranty, expressed or implied, with respect to the material contained herein or for any errors or omissions that may have been made. The publisher remains neutral with regard to jurisdictional claims in published maps and institutional affiliations.

This Springer imprint is published by the registered company Springer Nature Switzerland AG
The registered company address is: Gewerbestrasse 11, 6330 Cham, Switzerland

Preface

With the writing of this preface, a long journey comes to an end, offering you glimpses of beautiful views. At the end of 2017, the idea of writing this book was conceived. One major motivation to write this book is the fact that the climate change impact is increasingly attracting the world's attention. Does this really happen? If so, how much of it is because of human activity? How far will climate change adaptation be possible? What action can or should we take to fight it? How much is that going to cost? Or is it too late to take useful action already? This book aims to provide answers to all these questions by providing the best and latest available information.

The current generation's primary concern is the undesirable changes taking place in global warming and global climate and environment. There is nothing permanent. We all know that it is imminent to change. Why are we so alarmed by the global climate change? Climate change writing often falls into one of two opposite traps, both of which could be are avoided by applying the ideas of this book. This is not a story of gloom and doom, of the inevitable catastrophe of the climate. On the contrary, it outlines, in more detail than usual, what can and should be done to prevent real disaster risks. It is also not, however, a story of complacent congratulations on "win-win" policies, cautiously incremental steps, and "green" policy choices. It calls us to an endeavor worthy of the twenty-first century's resources and ingenuity, to bold initiatives with large costs—and far greater benefits. Climate change has become, as well as one of the most dynamic, the defining environmental legal and political challenge of the twenty-first century.

This book is written to help us make sense of the climate change discussion, especially on the question of whether this problem can be solved. It is now widely accepted that our climate has changed and will continue to change due to our economy based on fossil fuel, our transformation of the surface of the planet, and the increasing number of people and their growing wealth. Nevertheless, there is increasing confusion about the solutions. Some people think the only way is to drastically change our way of life. Give up our cars; give up our heating, no air travel anymore. Return to the Middle Ages, some people think that in the near future, technology will give us plenty of CO_2-free energy at low cost. Others believe that nuclear power is the only solution, as renewable energy and energy efficiency will never significantly reduce CO_2 emissions. Others believe that nuclear power is the only solution, as renewable energy and energy efficiency will never significantly reduce CO_2 emissions. Therefore, the key message of this book is intended to deserve widespread attention to the idea that "The climate and development crises will be solved together, or not at all. Moreover, the faster we begin, the less painful—and more possible—the solutions will be." The fact that we can control climate change is what our current knowledge tells us. We cannot avoid further changes and further negative impacts altogether, but we can avoid the most severe impacts of climate change, so that things remain manageable. We want people to understand that and see that this is only possible if strong and decisive action is now being taken.

This book's goal is to introduce the role of ecosystems in the climate system and the changing Earth system. The book consists of 25 chapters and contributed by more than 35 scientists, specialists, and researchers from Egypt. Keeping in mind the philosophy of "The Art of Climate Change," chapter "Introduction to 'Climate Change Impacts on Agriculture and Food Security in Egypt'" gives an overview of climate change impacts on agriculture and food security in Egypt as it presents the main technical issues of the chapters. The book begins with an overview of our knowledge of the climate system, the changes taking place, the opportunities for further climate change, and the impact on human and natural systems. Briefly, it gives the rationale for the rest of the book on how to control climate change and limit it to manageable proportions.

The 23 main chapters of the book are roughly organized in five main parts, which sequentially observe the climate change impacts on agriculture and food security in Egypt. A brief description of each of the part is as follows:

The first part is organized to present a comprehensive overview of climate change in Egypt. This part consists of three chapters. Two ways were identified to give us an overall impression of climate change in Egypt. The first way is linked to governance and institutional structure of climate change in Egypt. The second way is the role of science, technology, and innovation in addressing climate change challenges in Egypt.

The second part, which consists of five chapters, delves deeply into the land and water resources concepts. This part covers five potential impacts, which were identified to study the effect of climate change on land and water resources in Egypt. First is the effect of climate change on agriculture, which has specifically become a global issue due to the ever-growing need to provide food security and end hunger. Second are the impacts of climate change on microbial activity in agricultural Egypt's soil. The third influence is through the mapping of Egypt vulnerability to sea level rise. Fourth inspiration is soil itself as driver and victim of climate change. The final control is through soil carbon sequestration for climate change mitigation.

The third part, which involves three chapters, reviews the various smart farmings. This topic is covered in detail in seven approaches, which were identified acknowledged for using smart farming as a way to mitigate climate change impacts on agriculture. The first is assessing vulnerability and enhancing adaptive capacity and resilience to climate change; second, the pollination of flowering plants is one of the most important ecological services in natural and agricultural ecosystems; third, change of agricultural calendar as a response to climate variability; fourth, projected crop coefficients under climate change in Egypt; fifth, rice production in Egypt: the challenges of climate change and water deficiency; six, smart farming approach is using nanotechnology for real-time control of the red palm weevil under climate change; and finally, how to face climate change: urban gardening and sustainable agriculture are discussed.

The fourth part, which contains five chapters, discusses the different methods and approaches for livestock, fishery, and aquaculture. Five different ways are recognized for potential climate change impacts on livestock, fishery, and aquaculture. First, potential climate change impacts on livestock and food security nexus in Egypt is presented; second, the influence of climate changes on animal feed production; the third way is through algae and fish; the fourth way is climate change impact on immune status and productivity of poultry as well as the quality of meat and egg products; and the final way in this book is climatic change and chicken immunity.

The final part, which consists of three chapters, provides a comprehensive overview of the socioeconomic impacts and green sustainability. Three scenarios have been identified to study the socioeconomic impacts of climate change. The first scenario deals with agriculture and rural communities' vulnerability in the Nile Delta. The second scenario is related to water–climate–food nexus for green sustainability. The third scenario is an efficiency analysis approach.

Lastly, the final chapter briefly summarizes the most significant findings and recommendations of the book. The concluding chapter highlights major challenges to achieving equitable and sustainable water security in Egypt and offers cautious prospects for the future.

At this point, this book on the impacts of climate change on agriculture and food security in Egypt emerged. There are several books written around the world on the topic; however, this book aims to highlight and focus on recent advances in the impacts of climate change on agriculture and food security in Egypt. A group of experts has come together to write this book, which aims to pass on the new knowledge and information available to the readers. Depending on your interests and preferences, one can start the journey from anywhere. Readers and beneficiaries vary from academics, professionals, and scientists to students from undergraduate to graduate.

This book is intended to be of interest to all stakeholders in the climate change sector: financing institutions, users' associations, planners, designers, training, and research institutions. I wish it would act as a guide book on climate change for those interested readers. We believe the information presented here will be of the greatest help policy-makers, managers, and researchers interested in a broad perspective of urgent climate change issues in Egypt and hope that the book will contribute to some real if modest, progress toward the beneficial management of climate change.

Without the great efforts of all the authors, advances in this book would not have been possible, and I am sure that their valuable contributions will increase the significance of the book. To meet Springer's high-quality standards, it would not have been possible to produce this book and make it a reality without their patience and effort in writing and revising the various versions. All appreciation and gratitude must be extended to include all members of the Springer team who have worked long and hard to make this volume a reality for researchers, graduate students, and scientists worldwide. I have to thank all the experts who contributed to the chapter review processes. I hope it has widely read. If we want to avoid past misunderstandings, then we need to change direction and start taking advantage of the knowledge base of scientists. We did not have that opportunity a decade ago. Now it is the right time.

The editors strongly welcome to receive feedback and comments from the audiences to improve the future editions.

Ismailia, Egypt
Zagazig, Egypt
May 2019

El-Sayed Ewis Omran
Abdelazim M. Negm

Contents

An Overview of Climate Change in Egypt

Introduction to "Climate Change Impacts on Agriculture and Food Security in Egypt" 3
El-Sayed Ewis Omran and Abdelazim M. Negm

An Overview of Paleo-Climate Evidence in Egypt 21
Hassan Khozyem

Governance and Institutional Structure of Climate Change in Egypt 45
Mohamed A. S. Abdel Monem and Phoebe Lewis

Role of Science, Technology and Innovation in Addressing Climate Change Challenges in Egypt 59
Mohamed A. S. Abdel Monem and Ismail A. El Ghandour

Land and Water Resources

Impact of Climate Change on Plant-Associated Fungi 83
Anindya Chanda, Heba Maghrawy, Hossam Sayour, Phani M. Gummadidala and Ola M. Gomaa

Impacts of Climate Change on Microbial Activity in Agricultural Egyptian Soils 97
Samy Abd El-Malik Mohamed Abd El-Azeem

Soils as Driver and Victim of Climate Change in Egypt 115
Reda Ragab Shahin

Soil Carbon Sequestration for Climate Change Mitigation: Some Implications to Egypt 151
Heba Elbasiouny and Fathy Elbehiry

Mapping Egypt Vulnerability to Sea Level Rise Scenarios 183
Aymen Solyman and Tarek Abdel Monem

Smart Farming

Agricultural Production in Egypt: Assessing Vulnerability and Enhancing Adaptive Capacity and Resilience to Climate Change 205
Mohamed A. S. Abdel Monem and Biljana Radojevic

Vulnerability of Crop Pollination Ecosystem Services to Climate Change 223
Mohamed A. M. Osman and Mohamed A. Shebl

Exploring Changes in the Agricultural Calendar as a Response to Climate Variability in Egypt 249
El-Sayed Ewis Omran

Projected Crop Coefficients Under Climate Change in Egypt 273
Samiha Ouda

Rice Production in Egypt: The Challenges of Climate Change and Water Deficiency 295
Heba Elbasiouny and Fathy Elbehiry

Nano-technology for Real-Time Control of the Red Palm Weevil Under Climate Change 321
El-Sayed Ewis Omran

Facing Climate Change: Urban Gardening and Sustainable Agriculture 345
Eid M. Koriesh and Islam H. Abo-Soud

Livestock, Fishery, and Aquaculture

Potential Climate Change Impacts on Livestock and Food Security Nexus in Egypt 423
Tarik S. K. M. Rabie

Influence of Climate Changes on Animal Feed Production, the Problems and the Suggested Solutions 451
Salah A. Attia-Ismail

Algae and Fishes: Benefits and Hazards 465
Nermin Adel Hussein El Semary

Climate Change Impact on Immune Status and Productivity of Poultry as Well as the Quality of Meat and Egg Products 481
Ahmed Soliman and Assem Mohamed Safwat

Climatic Change and Chicken Immunity 499
Akrum M. M. Hamdy

Socioeconomic Impacts and Green Sustainability

Climate Change, Agriculture and Rural Communities' Vulnerability in the Nile Delta 525
Dalia M. Gouda

Water Climate Food Nexus for Green Sustainability 577
Sherien A. Zahran

Climate Impact on Egyptian Agriculture: An Efficiency Analysis Approach 603
Fadi Abdelradi and Dalia Yassin

Conclusions

Update, Conclusions, and Recommendations to "Climate Change Impacts on Agriculture and Food Security in Egypt" 627
El-Sayed Ewis Omran and Abdelazim M. Negm

An Overview of Climate Change in Egypt

Introduction to "Climate Change Impacts on Agriculture and Food Security in Egypt"

El-Sayed Ewis Omran and Abdelazim M. Negm

Abstract This chapter introduces briefly the main technical components of the chapter presented in the book. The technical contents of the chapters are presented theme-wise arrangement. The book contains 25 chapters arranged under five themes to covers different topics associated with climate change impacts on agriculture and food security in Egypt.

Keywords Climate change · Egypt · Climate variability · Land and water resources · Smart farming · Livestock · Fishery · Aquaculture · Socioeconomic impacts

1 Background/Overview

The current generation's primary concern is the undesirable changes taking place in global warming and global climate and environment. There is nothing permanent. We all know that it is imminent to change. Yet why are we so alarmed by the global climate change? Climate change writing often falls into one of two opposite traps, both of which could be are avoided by applying the ideas of this book. This is not a story of gloom and doom, of the inevitable catastrophe of the climate. On the contrary, it outlines, in more detail than usual, what can and should be done to prevent real disaster risks. It is also not, however, a story of complacent congratulations on "win–win" policies, cautiously incremental steps, and "green" policy choices. It calls us to an endeavor worthy of the twenty-first century's resources and ingenuity, to bold initiatives with large costs—and far greater benefits. Climate change has become,

---

E.-S. E. Omran
Department of Soil and Water, Faculty of Agriculture, Suez Canal University, Ismailia 41522, Egypt
e-mail: ee.omran@gmail.com

A. M. Negm (✉)
Department of Water and Water Structures Engineering, Faculty of Engineering, Zagazig University, Zagazig 44519, Egypt
e-mail: amnegm@zu.edu.eg; amnegm85@yahoo.com

© Springer Nature Switzerland AG 2020
E.-S. E. Omran and A. M. Negm (eds.), *Climate Change Impacts on Agriculture and Food Security in Egypt*, Springer Water, https://doi.org/10.1007/978-3-030-41629-4_1

as well as one of the most dynamic, the defining environmental legal and political challenge of the 21st century.

This book is written to help us make sense of the climate change discussion. Especially on the question of whether this problem can be solved. It is now widely accepted that our climate has changed and will continue to change due to our economy based on fossil fuel, our transformation of the surface of the planet, and the increasing number of people and their growing wealth. Nevertheless, there is increasing confusion about the solutions. Some people think the only way is to drastically change our way of life. Give up our cars; give up our heating, no air travel anymore. Return to the Middle Ages, some people think that in the near future, technology will give us plenty of CO_2-free energy at low cost. Others believe that nuclear power is the only solution, as renewable energy and energy efficiency will never significantly reduce CO_2 emissions. Therefore, the key message of this book is intended to deserve widespread attention to the idea that "The climate and development crises will be solved together, or not at all. Moreover, the faster we begin, the less painful—and more possible—the solutions will be." The fact that we can control climate change is what our current knowledge tells us. We cannot avoid further changes and further negative impacts altogether, but we can avoid the most severe impacts of climate change so that things remain manageable. We want people to understand that and see that this is only possible if strong and decisive action is now being taken. The recommendations given under the different themes of the book in chapter "Update, Conclusions, and Recommendations to "Climate Change Impacts on Agriculture and Food Security in Egypt" can guide taking such decisions.

2 Themes of the Book

Such climate change will affect Egypt's natural resources, particularly the major resources characterized by the relative scarcity of Egypt's land and water resources. This leads to a direct and far-reaching impact on the agricultural sector and thus on the food security of Egypt. The four dimensions of food security include climate change: food availability, access, capacity, and stability. So, the objective of the book is to address the following main theme.

- An overview of climate change in Egypt.
- Land and water resources.
- Smart farming.
- Livestock, fishery, and aquaculture.
- Socioeconomic impacts.

The next sections present a summary of each chapter presented in the main body of the book under its related theme.

3 Chapter's Summary

The following subsection presents the salient features of the chapters presented in the book under each theme of the book.

3.1 An Overview of Climate Change in Egypt

This section is covered in four chapters from 1 to 4. Chapter two is titled "An Overview of Paleo-climate Evidence in Egypt". It explains how can the knowledge that we can get from the past can help us to understand the consequence of the current warming. The earth system is a closed system has its energy from the solar and cosmic rays and formed of four interconnected spheres (lithosphere, atmosphere, hydrosphere, and biosphere). The change in chemical or physical conditions in each one of these sphere is monitor directly for the others. Several internal and external processes have a strong effect on modifying the earth climate such as extraterrestrial impacts, volcanism, and injection of greenhouse gaseous into the atmosphere, these processes are recorded in the geologic history of the earth as a catastrophic and/or hyperthermal events. The ecosystem has three different ways to respond to the climate change; extinction and it is the worst response and is recorded several time in earth history five of them were major mass extinctions; evolution, in which the Genera are adapting themselves to the new climatic condition; or rising of new genera. In several cases, these three scenarios are recorded together in other only one or two of these scenarios can be observed. To sum up, the study of the paleoclimatic changes that prevailed in the earth's history is an important issue to understand the current climatic changes. Some extinction and hyperthermal events can be used as an analog for the current changes and help to predict what the future hide for the humankind.

While chapter three is about "Governance and Institutional Structure of Climate Change in Egypt" focuses on the effective implementation of climate change policies and strategies in Egypt which requires strong governance and institutional structure. More coordination between the different stallholders and more effective involvement of the private sector and civil society organizations in the climate change implementation plans are required.

Successful climate change governance systems should ensure the implementation of climate policy integration at the national level. This chapter address the governance and institutional structure of climate change in Egypt listing the different institutions that are partners to the Egyptian Environmental Affairs Agency (EEAA) handling the issues related to climate change in different ministries and sectors in Egypt. The chapter also covers the environmental policies and legislation in Egypt starting from the Law 4/1994 as the first environmental protection law in Egypt, listing all related laws as well as the already implemented projects. The chapter also discusses the important role of civil society organizations and their active engagement as has been highlighted in the Third Communication of Egypt (2016). Also, the challenges facing the private sector have been discussed. The chapter also reflected on the issue of gender mainstreaming in the climate change policies in Egypt heightening the importance of the development of the National Strategy for Mainstreaming Gender in Climate Change in Egypt. The chapter also covers the engagement of Egypt in the regional and global climate change governance focusing on being the president of the African Ministerial Conference on the Environment (AMCEN) and Coordinator of the Committee of the African Heads of State and Government on Climate Change (CAHOSCC) for the period 2015–2017. Also, the chapter discussed the main challenges facing climate change governance in Egypt highlighting the need for better coordination between different stockholders.

The second way is linked to governance and institutional structure of climate change in Egypt. There is still a marked lack of support for the growth of institutional capacity and the frameworks needed to address climate change at local [1, 2]. A cross-cutting problem, such as climate change, can not be handled by a single department of government or ministry. The inclusion of climate change in policy decisions requires consideration of the causes and consequences of climate change in non-climate sectors [3]. "Integrated strategies are a relatively novel approach to govern highly complex issues that involve several sectors and levels of governance. These are a key tool for fostering not only sustainable development, but also convergence or ecological modernization of environmental policy" [1]. Both concepts promote the integration of environmental concerns into sectors with significant environmental implications long before sustainable development dominated the discourse on the environment [4]. In Egypt, despite a stated commitment to grassroots involvement in climate change adaptation processes and actions, the actual focus in national adaptation plans is on technical and infrastructure options for adaptation, with little attention given to the social and institutional context in which these options operate [2]. In order to better serve the needs and interests of the most vulnerable populations, future national-level efforts to develop adaptation plans need to consider the role of local institutions more centrally.

Chapter four elaborates on the "Role of Science, Technology, and Innovation in Addressing Climate Change Challenges in Egypt" because climate change becoming a serious man-made environmental challenge with its serious impact on the livelihood. It provides them with the tools and technologies needed to address the impact of climate change, which should be the main task of international communities [5]. Therefore, science, innovations, and technologies are being to be more important than ever. They have been focused by agricultural research initiatives that contribute to the climate change adaptation efforts and enable the farmers and policy-makers to draw their strategies and actions to face climate change impacts. Egypt vision 2030 stated that "Knowledge, innovation, and scientific research contribute to achieving the goals of sustainable development and to cope with the impact of climate change on agricultural sectors, new technologies and innovations have been tested and verified by agricultural research initiatives that contribute to the climate change adaptation efforts and enable the farmers and policy-makers to draw their strategies and actions to face climate change impacts. The Egyptian research system is making great effort to overcome challenges related to innovation, and scientific research and already making big progress with regards to assessment of the climate change impact on Egyptian agriculture as well as suggesting scientific climate change adaptation measures Because of the cost of research, funding for scientific research and development needs long-term commitments which are not affected by rapid policy changes. There is also a need for more international cooperation.

3.2 Land and Water Resources

This theme is covered into five chapters. The chapter titled "Impact of Climate Change on Plant-Associated Fungi" discusses an important aspect related to climate change which is its impact plant-associated fungi. The chapter features the relationship between climate change and the diversity of soil fungi such as decomposers, mutalists, and pathogens. Since climate change directly affects abiotic stressors such as temperature, humidity, drought, salinity and nutrient limitations, the relationship between plants and mycorrhizal fungi is discussed and linked to the increase in agricultural productivity. The increase in temperature and carbon dioxide directly affects the activity of saprophytic and pathogenic fungi, consequently leading to change the soil fungal profile. The effect includes fungal enzymatic activity as well as secondary metabolites and eventually have an overall effect on soil mycobiomes. In order to reach precision farming, the state of the art technologies has to be used. Modern tools such as Global Positioning Systems (GPS), Geographic Information Systems (GIS) and Wireless Sensory Networks (WSN) offer precise monitoring for key biomarkers. WSN is gaining attention because it can be coupled with corrective action providing a fair opportunity to mitigate harsh climatic changes.

While the chapter titled "Impacts of Climate Change on Microbial Activity in Agricultural Egyptian Soils" is devoted how climate change affects the microbial processes and activity of enzymes in Egyptian soils and how they are used as an indicator of their quality and are likely to respond rapidly to climate change. Egyptian soils have a great diversity of microorganisms such as bacteria, actinobacteria, fungi, and arbuscular mycorrhizal fungi that play a vital role in nutrient cyclings, the breakdown of soil organic matter, and increasing soil fertility. However, we still have no general framework or know remarkably little about their response to climate change. Moreover, the soil is a complex habitat for microbial growth, and the structure and function of microorganisms are tremendously complex in the soil. These complexations lead the difficult to predict the effects of climate change on the activity of Egyptian soil microorganisms. In this chapter, we review the currently available researches regarding the impact of climate change on soil microbial activities such as microbial populations, microbial biomass, enzymes activity, soil beneficial microorganisms in Egyptian soils.

Climate change is possible to change the composition, function, and abundance of soil microorganisms, as well as plant-microbe interactions that together affect the quality of soils. Microorganisms inhabiting Egyptian soils have been exposed to high temperature and accumulating of carbonates and soluble salts in desert soils, and the other inhabiting in alluvial soils in the Nile Delta and the Qattara Depression of the Western desert. Soil microbial activity will affect by climate change directly or indirectly. These effects include increasing temperature, elevated or increasing the concentration of CO_2, rise changing soil moisture content, increasing of soil salinity, and drought. The activity of soil microorganism and/or enzyme is significant as a sensitive indicator of soil biological quality. These activities are informative to determine changes in soil biochemical properties that are affected by environmental stress from natural phenomena or anthropogenic activities. Bacteria and fungi living in the soil control the breakdown of organic matter in the soil and its release into the atmosphere as carbon dioxide [6]. Climate warming is expected to increase soil microbe development, promote more CO_2 release, as well as positive climate change feedback. A better understanding of microbial processes is likely to improve climate change predictions [7]. Recent research has therefore attempted to quantify the impact of warming on soil microbes and the carbon cycle processes they control and to explain this using efficiency metrics such as microbial carbon [6].

The third chapter under the current theme is titled "Soils as Driver and Victim of Climate Change in Egypt". This chapter explains how the soil can act as a driver and Victim of climate change. Agricultural soil is an important driver of GHG emissions. Emissions from clay soils are higher compared to sandy soils, especially in the presence of high moisture and organic matter contents and high pH that encourage CO_2 and N_2O flux. Global warming increases CO_2 emissions with an exponential relationship by increasing the temperature of the surface layer of agricultural soils. Nitrogen fertilizers are a major source of N_2O emissions both in dry soils due to nitrification process or denitrification in waterlogged soils as well as methane emissions in rice fields. In Egypt, about 25 million tons of rice husk and sugarcane wastes are burned in open field, which contributes strongly to climate change.

On the other hand, agricultural land is a victim of climate change as the global warming accelerates the breakdown of soil organic matter, which reduces soil fertility, water storage, and microbial activity, destroys soil structure and increases their susceptibility to soil compaction. The surface layer of soil is expected to lose about 55 million tons of carbon by the year 2050. Egypt's agricultural land is experiencing increased drought throughout the country. Northern coasts of the Delta are intruded by rising sea level, and salt-affected soils cover about 10% of the Delta area.

On the other hand, the chapter titled "Soil Carbon Sequestration for Climate Change Mitigation: Some Implications to Egypt" show how in carefully managed croplands, soil C sequestration can be substantial and represents a potentially constructive portion for mitigating the increased levels of atmospheric CO_2. The soil is the largest terrestrial carbon (C) stock, and those factors that affect C retention and release also influence on atmospheric CO_2 levels. Soil C sequestration represents about 90% of the total mitigation practices of climate change and about 10% of emission reduction. There is a great concern of soil carbon (C) sequestration and its role in absorbing atmospheric CO_2 not only because of its impacts on climate change mitigation but also because of its positive impacts on the sustainability of crop productivity, soil fertility, and soil quality. Cultivation has resulted in considerable loss of soil C due to chemical and biological decomposition of soil organic carbon (SOC), as well as erosion by wind and water. There is a general agreement that many agricultural ecosystems have a huge potential to sequester carbon in the soil, which could decrease CO_2 concentrations in the air and mitigate its global emissions. Egyptian soils are low in their C content. Thus its potential to sequester C is high. Therefore, good management practices should be considered for enhancing soil C sequestration in Egyptian soils, especially in degraded and desert soil. Climate change is indicated as an increasing trend in atmospheric temperature and global changes in weather conditions [8].

Climate change's effect on global SOC stocks has recently been widely recognized [8].

Moving from the soil to the sea, the chapter titled "Mapping Egypt Vulnerability to Sea Level Rise Scenarios" is presented to highlight the threat of sea level rise on the Nile Delta and present a methodology to develop a web vulnerability mapping tool using Geographic Information Systems to study the different sea level rise scenarios and to identify and map the areas that are most vulnerable to SLR on the Nile Delta coast. Four scenarios are implemented to assess the impact of SLR, for 25, 50, 75, and 100 cm.

The Nile Delta shoreline extends from Alexandria in the west to Port Said in the east, with a total length of about 285 km. The Nile Delta region in the Mediterranean coastal zone represents the major industrial, agricultural, and economic resource of the country. The region is characterized by relatively low land elevation, which leaves it severely exposed to rising sea levels. In addition, it suffers from local land subsidence, compounding the effects of rising seas (https://www.stimson.org/sites/default/files/file-attachments/Mohamed_1.pdf). This area is subject to shoreline changes resulting from erosion and accretion, subsidence, and sea level rise resulting from climate change.

In order to investigate and analyze such environmental problems and their impacts, we must have a tool (SLRS) to understand these changes. The SLRS is a web mapping tool based on the assessment of the IPCC definition of vulnerability. National and subnational geographic and socio-economic data have been gathered for the area of study from different sources including Shuttle Radar Topography Mission (SRTM), land cover, urban areas, industrial zones, protected areas, power plants, and population data.

3.3 Smart Farming

The smart farming theme is covered in 7 chapters. The chapter titled "Agricultural Production in Egypt: Assessing Vulnerability and Enhancing Adaptive Capacity and Resilience to Climate Change" address the challenges facing the agricultural production in Egypt due to climate change in the context of the impact of climate change on the agricultural and water sectors globally and at the regional levels. The international community has endorsed five principles of sustainable food and agriculture in order to implement the 2030 Agenda, which provide a framework for policy dialog and the development of appropriate policies, strategies, regulations and incentives [9]. More than 40% of the Arab population has already been exposed to drought and other climate disasters [10]. The chapter also discusses the definition of the term vulnerability in general, analyze the vulnerability of the Near East and North Africa Region to climate change due to its limited natural resources of water scarcity and land degradation, that will be even worse by climate change providing detailed information on the impact of climate change on different farming system in the Arb Region and the regional efforts including the " Regional Initiative for the Assessment of the Impact of Climate Change on Water Resources and Socio-Economic Vulnerability in the Arab Region' (RICCAR)". The chapter also provides a summary of the farming system in Egypt and its vulnerability to climate change focusing on the water resources (the Nile, rainfall and the groundwater) as well as the impact of climate change due to expected sea level rise and changes in the ecosystems of the lakes in the Nile Delta. The chapter also discusses the vulnerability of both crop and livestock production systems in Egypt as well as the fishery production system providing an estimation of the change in the crop, livestock and fishery production and the impact of that on the small scale farmers in Egypt. The chapter also discusses the strategies that have been developed by the Egyptian government to enhance the adaptive capacity and resilience to climate change including "Egypt vision 2030" and the climate change adaptation efforts undertaken by the government.

The second chapter under the smart farming theme is titled "Vulnerability of Crop Pollination Ecosystem Services to Climate Change". The chapter explains how climate change threatens the ecosystem services, especially crop pollination, as a result of its impact on the biodiversity of pollinators, reduces crop productivity, and negatively impacts global food security. In Egypt, over the last 20 years, due to the increase of temperature, a lot of impacts have affected the agriculture ecosystem, especially to flora and fauna, including wild and domestic bees. This led to the disappearance of some species, and some other species have become more superior for dominating in the ecosystem. In the future, it is expected that there will be increasing impacts. Beside the impact of climate change in Egypt, several bee species are threatened by nesting fragmentation, habitat elimination, degradation, and urbanization. Climate change is also expected to have some more changes in the bee generations and their intercorrelations with their foraging resources. There is evidence that many parasitic mites, diseases, and pathogens have a great impact on the managed honey bee colonies across Egypt, which probably the factors are leading to Colony Collapse Disorder (CCD). Pesticides also are a key factor behind colony collapse disorder (CCD), directly and in tandem with two leading co-factors, pathogens, and a shortage of natural resources. Climate change and human interference such as miss management of honey bee races have a great influence on the spreading and weakness of honeybees. Recently, there is a lot of efforts for protecting and conserving plant pollinators with a particular interest with honey and solitary bees. In Egypt, many efforts are ongoing for protecting the habitats of some solitary bees present in Egypt. These attempts have already succeeded in re-nesting and establishing three bee species.

Several efforts must be taken into consideration such as enhancing the health of pollinators "honeybees and solitary bees," avoid extensive using of chemical pesticides, encourage local farmers for cultivating bee-friendly plants and protecting the natural nesting habitats of solitary bees. Promote the ecological intensification strategy such as intercropping, crop rotations, farm-level diversification, and reduced agrochemical use for promoting biodiversity beneficial to agricultural production. Therefore, it is strongly recommended to mitigate the potential adverse impacts of climate change on the diversity and efficiency of insect pollinators, especially, managed honeybees and other solitary bee species.

The third chapter in this section is titled "Change of Agricultural Calendar as a Response to Climate Variability". It demonstrates the need for farmers to adjust their agricultural calendar and switch to farming practices that make better use of natural resources. Climate change often has an indirect impact on agricultural production by affecting the development and distribution of crop pests and animal diseases, increasing the rate and spread of harmful diseases, weather conditions, decreasing water supply and irrigation, and increasing soil erosion severity. Important agricultural adaptation methods include new crop varieties and animal species suitable for drier situations, irrigation, crop divarication, implementation of mixed crop and livestock farming systems, and changes in the dates of agricultural activity. In response to short-term climate variability, some of these methods (e.g., changes in agricultural activity dates) are classified as coping responses. Many factors and critical information, including plant calendars, are considered [11]. In its growth cycle, the sequential series of various phenological stages of a crop determines the so-called crop calendar [12].

Crop calendars provide the timing of crop sowing, growing, and harvesting periods. Different sources of information were used on crop calendars. There is no research focusing on how farmers in the face of climate change are adapting their entire agricultural calendar. Thus, as a response to climate variability in Egypt, this chapter aims to explore observed changes in the agricultural calendar. It also aims to illustrate the factors affecting the decision of farmers to adjust their agricultural calendar as a climate variability coping mechanism.

Land surface phenology (LSP) metrics were used as a default for crop calendars and criteria such as the beginning and end of the season were applied to identify the pixel-level growth period of active agricultural vegetation.

The chapter titled "Projected Crop Coefficients Under Climate Change in Egypt" addresses the quantification of how climate change will affect the values of crop coefficient (Kc) for several important crops in Egypt in 2030 as this is very important in the management of water resources in the future. Projection of Kc values for the cultivated crop is important for the future planning of water resources. The available weather data in 2030 contained only solar radiation and temperature (maximum and minimum), which not enough to calculate evapotranspiration (ETo) using the Penman-Monteith equation (P-M). To overcome that, the monthly values of ETo in 2016 were calculated using the Penman-Monteith equation (P-M) and Hargreaves-Samani equation (H-S). Then, the monthly ETo (H-S) values were regressed on the monthly ETo (P-M) values, and prediction equations were developed for each agro-climatic zone of Egypt. These equations were used to project ETo values under climate change in 2030 using RCP6.0 climate change scenario resulted from MIROC5 climate change model. The developed ETo values were used to run BISm model and to calculate Kc values for 14 field crops, 7 fruit crops, and 13 vegetable crops, where the date and the value of each Kc growth stage, as well as the water consumptive use of each crop in 2030 were calculated. Comparison between Kc values in 2016 and 2030 for field and vegetable crops revealed that the values of Kcini were higher in 2016, compared to its counterpart values in 2030. The values of Kcmed and Kcend were similar or lower in 2016, compared to its counterpart values in 2030. Whereas, there was no change in the values of Kc for fruit crops between 2016 and 2030.

A practical example is given by the chapter titled "Rice Production in Egypt: The Challenges of Climate Change and Water Deficiency". This chapter show how the changing in rice management practice; such as decreasing ploughing, creating another alternative to rice straw burning and balanced fertilizer application would lead to mitigating of greenhouse gas emission from rice cultivation and improving soil organic matter (SOM) stocks, subsequently soil quality and productivity. Climate change has been one of the major global environmental problems of the 21st century. Rice is the main cereal crop for over 50% of the world's population. Rice cultivation is known as an important emitter of greenhouse gas emission, especially methane due to rice management practices and burning of rice straw after harvesting. However, many studies confirmed that rice soils accumulate carbon higher than other crops such as wheat and corn. The cultivated area of rice in Egypt is approximately 650,000 ha from the whole cultivated area in Egypt; approximately 3.3 million ha; i.e. around 20% of the cultivated area in Egypt. Egypt relies on the Nile for 97% of its water requirements. The expected scenario of water deficiency in the Nasser Lake due to the Grand Ethiopian Renaissance Dam construction, with pulling of deficiency from Dam Lake; is emphasizing on wasting approximately 1.7 million ha of Egypt's cultivated area. As well, the expected high scenario of a relative sea level rise in Egypt; especially Nile Delta increases the amount of land that lying under risk from inundation in the north Nile Delta by 300 km^2, which estimated by one-fifth of the total agricultural land in the northeast Nile Delta only. Also, all crops are projected to have a decrease in yields and an increase in irrigation needs. Thus; all these challenges will increase the stresses on rice production and decrease soil C storage in Egypt as a result of climate change and water shortage due to establishing GERD.

On the other hand, the chapter titled "Nano-technology for Real-Time Control of the Red Palm Weevil Under Climate Change" explains how to utilize the nano-techniques using acoustic and thermal sensors, to detect the infection of RPW at early stages and consequently protect the Red Palm Weevil. In fact, the current measures used to control the insect are not effective enough to succeed in eliminating the insect because of the great difficulty in early detection of infection and reaching all life stages inside the trunk. The first objective of this study is to detect the presence of living stages of RPW, which are hidden in the palm tree. Nano-sensor system using acoustic and thermal sensors was developed for significant recognition of RPW in an earlier phase of the infestation.

The second objective was to propose Nano-natural (plants and minerals) extract for RPW treatment and cure. Nano-minerals were used and prepared in natural leaves extract to control RPW, which is a cost-effective and environmentally friendly method. The nanoparticle minerals were used and prepared in *Moringa oleifera* leaves extract to control RPW. Nano-minerals extract (NME) drastically decreased developmental stages of *R. ferrugineus*. The compound is a 100% natural solution. It is derived from natural plants and minerals that can significantly aid in controlling RPW. Also, it is safe to use on all plants and crops including natives.

The third objective was to propose a Nano-ultrasonic system to prevent the date palm field from infected RPW. Ultra-sonic sensors were used to prevent the RPW to attack the date palm trees. Unlike traditional methods, which are time-consuming, and labor intensive, this system offers the advantages of keeping the palm trees intact; reduce costs, as well as saving time and money in the process of pest's infestation detection.

The smart farming theme ends with the chapter titled "Facing Climate Change: Urban Gardening and Sustainable Agriculture" to demonstrate how every person or every official in the municipality or community can play a role in mitigating climate change. Either by reducing emissions from his activities or by reducing the levels of incoming emissions through sustainable gardening. Managers must follow the right scientific methods for choosing the appropriate garden plants or following the agricultural practices that reduce emission levels. Measures to combat climate change encompasses two primary titles: removal of the maximum greenhouse effective gases and decreasing causes of greenhouse gas emissions. The direct greenhouse gases are carbon dioxide, nitrous oxide, and methane. Nitrous oxide is approximately 300 and methane around 30 times than carbon dioxide at trapping heat in the atmosphere. Climate change affects and is affected by all communities, but its treatment must begin with the actions of individuals. Trees in the urban area strongly reduce pedestrian level heat stress by absorbing and reflecting solar irradiance.

Vegetation in gardens is one of the most important components affecting climate change. Urban gardening, mainly consist of trees resource, is a valuable asset. Trees' benefits to the human being were most pronounced in their contribution to environmental benefits. Thus, plants in gardens were found to provide a particularly important function in mitigating climate change and maintaining the environmental quality of communities.

Gardeners can help lessen the global warming pollutants associated with waste disposal by turning leaves, grass, woody garden clippings, and dead garden waste into mulch or compost, then using it in the garden. Recycling these wastes will not only reduce methane emissions from landfills but also improve the garden's soil and help it store carbon. One of the innovative methods that reduce greenhouse gas emissions is to make and use biochar. Because nitrous oxides are an important greenhouse gas, better management of nitrogen fertilizers can reduce its emissions. The four main management factors that help reduce nitrous oxide emissions from applied nitrogen fertilizer are commonly known as the 4R's: right application rate; right formulation (fertilizer type); right timing of application; right placement at the plant's root zone as possible.

3.4 Livestock, Fishery, and Aquaculture

This theme is covered in 5 chapters. The chapter titled "Potential Climate Change Impacts on Livestock and Food Security Nexus in Egypt" illustrates the interactions between and among the energy, water, land, and climate systems that take place within a social and economic context. Regardless of the way that the phenomenon of worldwide atmosphere changes, and the local effects, it is normal that Egypt would be one of the countries most influenced by the impacts of climate change. These impacts are reflected in high temperatures, changing rainfall pattern, rising sea levels, and the expanding recurrence of atmosphere related disasters, which posture dangers to farming, agriculture land, water supply, and food security. Agriculture land is a constrained that ought to be utilized effectively and in a way that keeps up its esteem with the goal that it can produce an adequate amount of food, fuel, and forage in both the short and long term. To enhance rural life and enhance the food security in the district, it will be fundamental, subsequently, to enhance the efficiency of this agricultural framework through presenting water and soil preservation procedures, providing quality seeds, and empowering the utilization of manure. From the viewpoint of water administration, intercessions to increment agricultural profitability ought to incorporate projects to extend irrigated areas, enhance the water-maintenance properties of soil, and enhance profitability and water-utilize efficiency. To enhance production, and accordingly, farmers, population, and food security are the two primary components prompting the expansion and the requirement for more developed agribusiness in both rainfed and irrigated farming.

Added to the above, the chapter titled "Influence of Climate Changes on Animal Feed Production, the problems, and The Suggested Solutions" presents the efforts done by some selected countries across the world to get the benefit of how the climate change influenced the animal feed production. Climate change is a long-term change in the earth's climate, especially a change due to an increase in the average atmospheric temperature. Climate change demonstrates itself through the noticeable variation in the weather, including temperature, rain fall, and wind. Agriculture in the arid land is vulnerable to climate changes since it depends on water resources and prevailing atmospheric conditions. The influences and consequences of these variables on agriculture might operate negatively or positively. There is a reciprocal effect between climate change and animal production. The production of livestock contributes about 18% of global greenhouse gas emissions from all human activities. The farm animals are prone to the adverse impacts of the changes in climate.

The changes in climate will impact both of the quantity and quality of forage production as well as their reliability. Major impacts of climatic changes on feed crops and grazing systems would be changes in herbage growth, changes in the composition of pastures, changes in herbage quality, and the offset of biomass yield increases. Since pastures depend on rainfall, any changes in rainfall patterns will affect the plants on pasture.

Climate Change Convention in Paris and declared an agreement to "hold the increase in the global average temperature to well below 2 °C above pre-industrial levels… recognizing that this would significantly reduce the risks and impacts of climate change", pursue efforts to limit the temperature increase even further to 1.5 °C and undertake and communicate ambitious efforts to contribute to the global response to climate change by strengthening the mobility of countries to deal with the impacts of climate change.

Moreover, the chapter titled "Algae and Fishes: Benefits and Hazards" show how algae play a major role as primary producers and form the base of the food chain as they are the major diet for many aquatic animals. They are either microscopic or macroscopic, and both can be cultured for mass production of fish and crustaceans feed. They can also cause the favourable pigmentation of several animals. The iron fertilization experiments aimed at increasing algal proliferation but their impact is still under debate. Nonetheless, algae have their harmful side as they can be toxin producers and biofoulers. Aquatic animals such as fish and crustaceans provide human beings with high-quality protein diets. Meals of aquatic animals are based partially or even entirely on algae. Marine algae can be grown as maricultures in marine environments in various forms. Also, algae can be used for fish diets in inland aquacultures. However, the types of algae vary according to the type of aquaculture. Algae can also be grown not only for their nutritional value but also for their pigments that can affect the colour and health of some aquatic animals. Recently the iron limitation in marine environments attracted the attention of scientists as the deficiency in this element caused a reduction in phytoplanktons. Iron fertilization in the ocean was performed in order to enrich some marine environments with iron, which in turn would lead to an increase of the phytoplankton. Scientists propose that this would cause mitigation of global warming as algae take in carbon dioxide, the main factor responsible for global warming, and release oxygen during photosynthesis. The number of aquatic animals is also expected to increase. However, the short and long term overall effects of those experiments are yet to be evaluated.

Toxic algae can also adversely affect aquatic animals and can cause a massive die off. The two main algal groups responsible for toxins production are cyanobacteria and dinoflagellates.

They can spread from one water body to another through ballast water or biofouling aquatic vessels such as ships and boats. This, in turn, would be harmful and even lethal for aquatic animals and would jeopardize food security. Thereby, surveillance of aquatic vessels must be performed.

On the other hand, the chapter titled "Climate Change Impact on Immune Status and Productivity of Poultry as well as the Quality of Meat and Egg Products" is presented to discuss the impacts of climate change on immune status and productivity of poultry. Also, it explains how climate change affects the quality of meant and egg products. Additionally, the chapter focuses on the key principles (nutritional or managerial practices) to alleviate the adverse impacts of heat stress.

Current poultry production systems comprise large numbers of birds that are housed together. This results in heat stress, which affects the poultry systems making them more susceptible to heat stress. Heat stress causes (i) inconvenience and high mortality rate for birds, and (ii) lower or lost production, which therefore reduces the profitability. Both production performance and feed conversion ratio are affected by heat stress conditions, which affect the production rate. Other effects that are connected to heat stress include immunity reduction and weak immune response to vaccines that decrease the resistance of birds to many infectious diseases. In laying hens, the production is significantly decreased and does not reach to the peak a decrease in the egg quality (e.g., thin and breakable eggshell) in addition to lower egg weight with small size.

The last chapter in this section is titled "Climatic Change and Chicken Immunity ". The chapter is presented to compile the current knowledge about the importance and impact of climatic change, warming the global, on chicken production, focusing on chicken immunity. Chickens have very strong built-in defenses (immunity) against diseases that are caused by overrunning of the body by various microorganisms and toxins. Under stress conditions, avian blood tolerates a change from acid-base balance to alkaline balance. There is a decline in the plasma, a reduced level of vitamin C in the adrenal cortex, a reduction in lymphocytes, and a depression of the immune response. As the temperature rises, the birds undergo many changes—increased water consumption, respiration rate, body temperature, inferior egg quality, and susceptibility to diseases.

The highly specific adaptive immune mechanisms are affected by heat stress. In more specific, heat stress deteriorates the cell-mediated immune responses. As a result of heat conditioning, biochemical and physiological mechanisms were induced to cope with heat stress; this induction may have delayed production of additional acute phase proteins to protect the cells from damage. The stress hormones—cytokine interactions are responsible for altered immune functions during heat stress. Modern-day molecular biology tools can help in understanding various cellular and molecular mechanisms involved in the production, physiological and immunological aspects of the poultry birds, which in turn can help in the development of breeds more adapted to the climate changes.

3.5 Socioeconomic Impacts

The socioeconomic impacts of climate change are covered in three chapters. The chapter titled "Climate Change, Agriculture, and Rural Communities' Vulnerability in the Nile Delta" discusses the severe impacts of climate change on agricultural production and rural communities in the Nile Delta in Egypt. However, immediate challenges, such as population growth, land fragmentation, and urban expansion on agricultural land if not timely curbed will further exacerbate future climate change impact. Agriculture in Egypt in general and in the Nile Delta in particular, is currently facing other intense challenges resulting from the rapid population growth that is exceeding 2.5% annually [13], thus augmenting agricultural land and irrigation water scarcities.

Farmers in Egypt are expected to suffer from climate change impacts, and the Nile Delta in the northern part of Egypt—comprising about 50% of the agricultural land—is identified as one of the world's three "extreme" vulnerability hotspots.

The Nile Delta is central for the production of field crops' and the provision of related agricultural employment in Egypt. However, there are present socio-economic (e.g., population growth, land fragmentation and urbanization of agricultural land) and natural resources (e.g., water and soil) challenges that negatively affect agricultural production, farmers' livelihoods, and national food security.

The chapter explores the change in agriculture in the Nile Delta governorates, in terms of field crops' agricultural land area, production, net value, and employment, stemming from socio-economic challenges, and those caused by the climate change stimuli, here, temperature increase and sea-level rise by 2030.

Population growth and land fragmentation are expected to constrain farmers' resilience to averse the negative impacts of climate change, while urbanization on agricultural land would increase the effects of climate change on the local and national levels contributing to the eruption of social tensions.

Though all Nile Delta governorates are suffering from the presence of high illiteracy and poverty rates, and the absence of strong associative structures, qualified extension information and target group specific support, some governorates are to be affected more by climate change than others.

While the chapter titled "Water Climate Food Nexus for Green Sustainability" deals with Water-Climate-Food Security (WCF) Nexus approach as an important tool to reduce future challenges due to climate hazards. Whatever applying this approach need to devote more time and efforts for developing a coordination mechanism horizontally between different sectors and vertically between many stakeholders and authorities. Implementing for such mechanisms, it could be used the natural resources with optimum manner.

Green Sustainable Development (GSD) is that meets the needs of the present without compromising the ability of future generations to meet their own needs. At the core of sustainable development is the need to consider "three pillars" together: society, the economy, and the environment. No matter the context, the basic idea remains the same—people, habitats, and economic systems are inter-related.

The interrelationship between many Sustainable Development Goals (17 SDGs) supporting the nexus approach. Among the SDGs 17 goals; Achieving SDG 13 on climate Action required a close nexus with other key SDGs on the social agenda; especially goal 1 reducing poverty, SDG 2 for food security, SDG 3 improving health, SDG5 Gender equity, SDG6 for water Access, SDG 10 for inequality and SDG 15 for land and Ecosystems, and finally SDG 17 for Partnerships.

The last chapter in this section is titled "Climate Impact on Egyptian Agriculture: An Efficiency Analysis Approach" is presented to assess the impact of average maximum temperature, humidity and solar radiation through the technical efficient governorates using panel frontier models. Climate Smart Agriculture (CSA) is an approach that deals with restructuring agricultural systems to cope with climate change through three main objectives, which are sustainability increasing productivity and incomes, mitigating GHG emissions, and adaptation and building resilience. The Intergovernmental Panel on Climate Change (IPCC) has released a special report on 2018 on the impact of global warming of 1.5 °C and GHG emission pathways.

Additionally, the report has shown that most climate models showed robust results of regional climate characteristics differences [14].

Both agriculture and climate change impact each other. Climate change affects agriculture in different ways through changing the intensity, the frequency, and occurrence of events (rainfall, heat waves, changes in pests and diseases, changes in carbon dioxide, droughts) that are not common in the location´s history. Egypt is considered one of the countries that are vulnerable to climate change impact because it is located in a region that falls between semi-arid and arid conditions, where 94.5% of Egypt total area is desert. The 5.5% of the area is populated with 95 million inhabitants that where the River Nile and the coastal area. The objective of this study is to assess the climate impact using different indicators (humidity, average maximum temperature) on cereals production and technical efficiency in five governorates (Behera, Gharbia, Kafr-Elsheikh, Dakahlia, and Sharkia) from the year 2000 to the year 2009.

The graphical analysis of the mean Vegetation Condition Index (VHI) exhibits a positive trend over time for the analyzed governorates and the Agricultural Stress Index (ASI) indicator exhibits high fluctuations for season 2 compared with season 1 considering the higher the index value the more likelihood of facing drought. The increase in the ASI was more intense for Behera, Gharbia, Sharkia, which represent the old lands and New Valley that represent the new lands. Results of the frontier analysis, production elasticity estimates indicate that, for all governorates, the cultivated area has the largest positive effect on the quantity produced. In the case of the rice model, cultivated land, labor and fertilizers are found to have statistically significant influence; lagged rice production has an adverse impact on current production because of the government policy to minimize water resources consumption. Also the results of the analysis show that the average maximum temperature and humidity are found to increase the technical inefficiency of the cereals producers.

The book ends with the conclusions and recommendations in last chapter.

Acknowledgements The editors who wrote this chapter would like to acknowledge the authors of the chapters for their efforts during the different phases of the book including their inputs in this chapter.

References

1. Mubaya C, Mafongoya P (2017) The role of institutions in managing local level climate change adaptation in semi-arid Zimbabwe. Climate Risk Manag 16:93–105
2. United Nations (2018) Implementation of the framework for capacity-building in developing countries. Framework convention on climate change subsidiary body for implementation forty-eighth session (Bonn, 30 April to 10 May 2018)

3. UNEP (2018) Decision adopted by the conference of the parties to the convention on biological diversity. Convention on biological diversity, United Nations, Treaty Series, Registration No. I-54113 (Fourteenth meeting. Sharm El-Sheikh, Egypt, 17–29 November 2018, Agenda item 21)
4. Kleinschmit D, Arts B, Giurca A, Mustalaht I, Sergent A, Pülzl H (2017) Environmental concerns in political bioeconomy discourses. Int For Rev 19(1):41
5. Hiwasaki L, Hill L (2018) Building resilience through socially equitable climate action. International Development Research Centre
6. Walker WN, Kaiser C, Strasser F, Herbold C, Leblans N, Woebken D, Janssens I, Sigurdsson B, Richter A (2018) Microbial temperature sensitivity and biomass change explain soil carbon loss with warming. Nature Climate Change 8(10). https://doi.org/10.1038/s41558-018-0259-x
7. Bradford MA (2016) Managing uncertainty in soil carbon feedbacks to climate change. Nat Clim Change 6:751–758
8. Soleimani A, Hosseini SM, Bavani ARM, Jafari M, Francaviglia R (2017) Simulating soil organic carbon stock as affected by land cover change and climate change, Hyrcanian forests (northern Iran). Sci Total Environ 599–600:1646–1657
9. FAO (2017) The state of food security and nutrition in the world. Food and Agriculture Organization of the United Nations, Rome, Italy
10. Koday K (2017) The International Policy Centre for inclusive growth. Soc Prot Arab Spring 14(3)
11. Meroni M, Rembold F, Urbano F, Csak G, Lemoine G, Kerdiles H, Perez-Hoyoz A (2018) The warning classification scheme of ASAP—Anomaly hot Spots of Agricultural Production, vol 2.1. https://doi.org/10.2760/798528
12. Patel JH, Oza MP (2014) Deriving crop calendar using NDVI time-series. https://doi.org/10.5194/isprsarchives-XL-8-869-2014
13. Capmas (2017) Egypt 2018 statistical abstract. Capmas, Cairo, Egypt
14. IPCC (2018) Summary for policymakers. In: global warming of 15 °C An IPCC special Report on the impacts of global warming of 15 °C above pre-industrial levels and related global greenhouse gas emission pathways, in the context of strengthening the global response to the threat of climate change, sustainable development, and efforts to eradicate World Meteorological Organization, Geneva, Switzerland, 32 pp

An Overview of Paleo-Climate Evidence in Egypt

Hassan Khozyem

Abstract In this chapter, we discuss the global paleoclimatic and paleoenvironmental changes focusing on their causes and consequences. Two main processes can yield the climatic changes; the earth's internal processes and the extraterrestrial impacts. Both of them have a strong effect on the earth's system. The paleoclimatic change is well preserved in the earth's sedimentary record and can be reviled by using multidisciplinary studies including mineralogy, geochemistry, and the fossil contents. Egypt is a key area of one of the most pronounced climatic changes that occurred in the earth's geologic history; the Paleocene Eocene thermal maximum (PETM) that used recently as analog for the current warming.

Keywords Paleoclimate · Paleoenvironment · Mass extinction events · Hyperthermal events · PETM · Current warming

1 Introduction

> Earth is superorganism, the oceans are the heart of Earth's global system, and the forests are the lungs …. James Hutton (1785)

The earth's climate is defined between two climatic end members; polar climate and tropical wet climate. Between icy poles and the steamy tropics, wide variety of climatic conditions are occurring and contribute to Earth's geologic history and consequently effect on biodiversity and geologic heritage. In 1985; James Lovelock; proposed the Gaia Hypothesis to explain the environmental changes that occurred on the Earth's surface. The Gaia Hypothesis discuss five main aspects, which are: (1) Earth is an Organism; (2) Life significantly affects Earth's environment; (3) Different life sorts modify the environment to reach to its betterment of life; (4) the deliberately and/or consciously of life can control the global environment; and (5) Studying of the Gaia hypothesis needs to activate the interdisciplinary thinking.

H. Khozyem (✉)
Geology Department, Faculty of Sciences, Aswan University, Aswan 81528, Egypt
e-mail: H.M.Khozyem@aswu.edu.eg

© Springer Nature Switzerland AG 2020
E.-S. E. Omran and A. M. Negm (eds.), *Climate Change Impacts on Agriculture and Food Security in Egypt*, Springer Water, https://doi.org/10.1007/978-3-030-41629-4_2

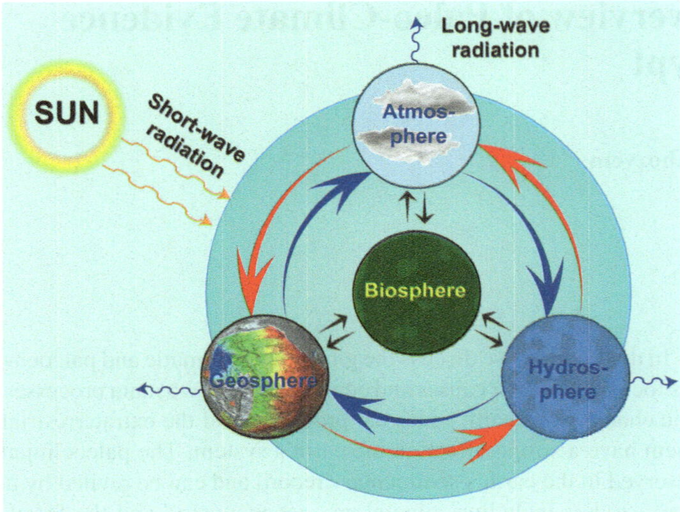

Fig. 1 The interaction between the earth system components

Earth is a complex system of interacting physical, chemical, and biological processes. The Earth system is comprised of four interconnected spheres of components and processes: atmosphere, geosphere, hydrosphere (incl. cryosphere) and biosphere (Fig. 1). Changes in any of these components have a direct and/or indirect effect on the other components. Consequently participates in the redistribution of climatic belts. Recently we can also add the human impact as an effective factor has a strong impact on the Earth's climate. In reality, no component of the Earth system can be considered in isolation from any other component. The Earth system operates within and is strongly influenced by, the exosphere (sun, solar system, and galaxy).

The term Climatic Changes is proposed to explain the changes in the distribution of weather patterns that are lasting and extends for a certain period. Several proxies are proposed to be used as paleoclimatic tracers (i.e., oxygen—$\delta^{18}O$, sulfur—$\delta^{33}S$ and organic and inorganic carbon—$\delta^{13}C$) as well as the sedimentological data. These proxies together increase our vital knowledge into the composition of atmosphere–ocean-biosphere system. For example, the low $\delta^{18}O$ suggesting increasing of greenhouse phenomena due to the high level of the greenhouse gaseous (CO_2, CH_4 as well as H_2S). In contrast, the high $\delta^{18}O$ indicates the proliferation of phytoplankton in the oceans [1].

2 Palaeoclimate Through Geological Time

The dramatic instability of the Earth's climate is a point of debate. Many hypotheses have been proposed to explain the climate variations through the geologic time such as changes in solar radiance due to the instability of earth's Orbital and Milankovitch cycles, change in the greenhouse gases (carbon dioxide or methane) level in the atmosphere [2], or change in the Oceanic thermohaline circulation. The climate can be affected by two different major processes; internal processes such as changes in the atmosphere, ice, vegetation, and changes in the ocean and land surface; and external processes like changes in plate tectonics, change in the Earth's orbit, and changes in the intensity of the Sun [3]. The main control on the paleoclimate was the paleoposition due to the continental drift (Fig. 2). Through the geologic time, the Earth's climate varies in cyclical rhythm over a number of time-scales fluctuating between hot to cold weather as well as dry and wet climate (Fig. 3).

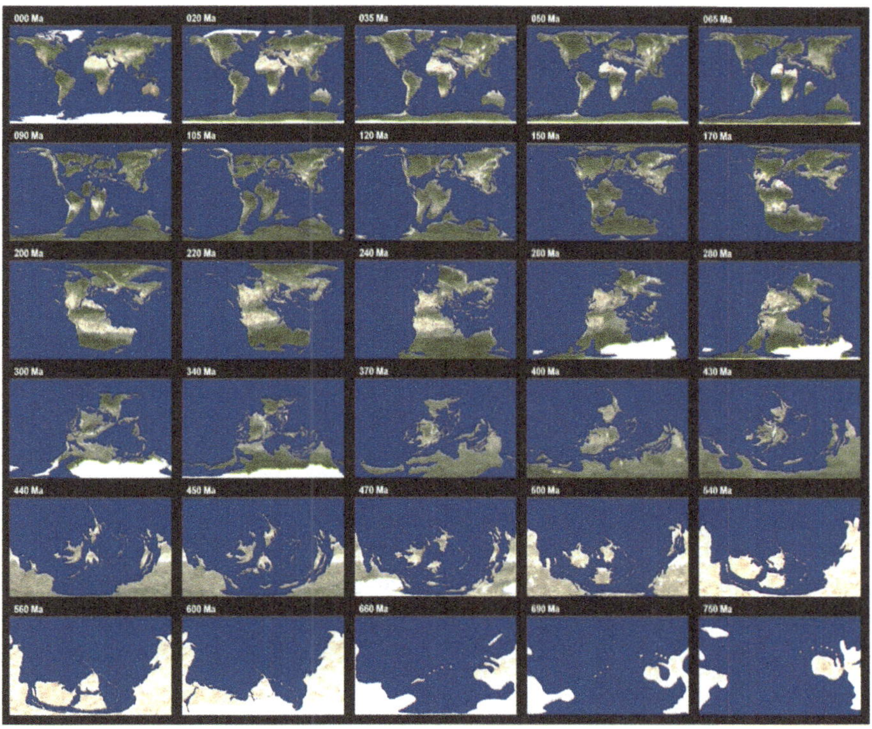

Fig. 2 Sequence of paleomap reconstruction showing the evolution of the continents through the last 750 Ma. (The visible paleo-earth project, planetary habitability laboratory @ UPR Arecibo—phl.uora.edu)

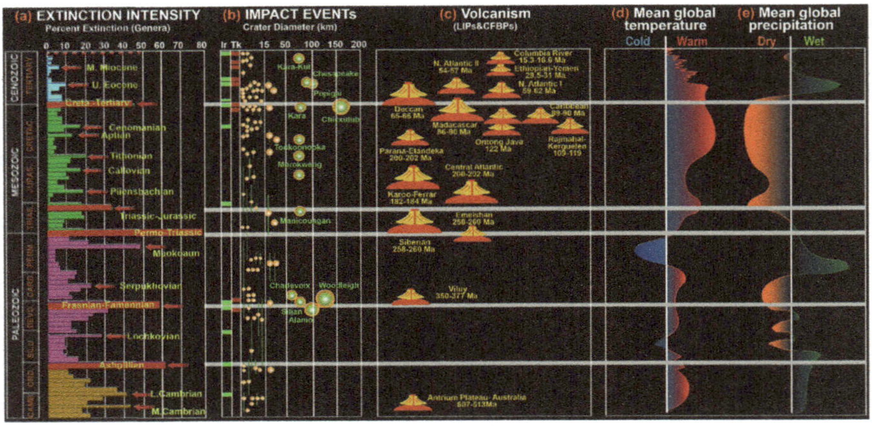

Fig. 3 Mass extinctions, impacts, and large igneous provinces during the Phanerozoic. Stratigraphic subdivisions and numerical ages from the 2004 International Stratigraphy Chart (ICS) of [4], Genera compilation from [5–7]; impact database from [8], LIPS and CFBP database from [9]. Note that the Chicxulub impact predates the K-T boundary by 300 ky. Modified after [10]. Together with climate variations in Earth history (modified after [3])

The climatic evolution led to the evolution of life through the Phanerozoic era and helps to divide the Phanerozoic into three parts: Palaeozoic (~542–251), Mesozoic (~251–65 Ma) and Cenozoic (65–0 Ma) [11]. The Phanerozoic geochemical climatic model of [12] depends mainly on the carbon, oxygen and sulphur cycles and these cycles have a very closed consistent relationship to the Geological and paleontological studies [13, 14].

The relationships between the carbon cycle, oxygen cycle, and biological productivity can be summarized as that the period of low atmospheric CO_2 it always coincided with periods of high atmospheric O_2 and the high biological productivity is occurring. During cold periods a large amount of atmospheric CO_2 is trapped by ocean cold waters. The atmospheric CO_2 drawdown via Photosynthesis. The decayed organic matter from land plants and phytoplankton are removed from the atmosphere-biosphere system by buried and release the oxygen to the atmosphere. During high-CO_2 level the ocean acidify due to the less dissolved oxygen contents, led to the stress conditions inhibit the phytoplankton from oxygen production, while on land, aridity and/or wildfire effects negatively on the plant photosynthesis [1].

3 Causes of and Consequences of Global Paleoenvironmental and Climatic Changes

A vast number of studies have emphasized the initial causes of the climatic and palaeoenvironmental changes that took places during the Earth's history. The proposed scenarios depend on multidisciplinary studies includes paleontological, sedimentary, volcanic, asteroid impact, and paleo-CO_2 and paleo-temperature trends. The causes of climatic and palaeoenvironmental changes can be summarized as one or even more of the following:

Extraterrestrial impacts (Fig. 3) that led to a sequence of processes includes explosion, fragmentation, cratering, melting and vaporization of the immediately surrounding rocks [15–17].

(a) The results of any asteroid impact depend mainly on the crater size and includes earthquakes, tsunami waves, high level of aerosols input into the atmosphere such as sulfur and carbon dioxides, mercury, dust. The consequences of the impact are increasing the acid rain due to the high greenhouse gases level in the atmosphere changing the climate to hot humid climate, increase in ocean acidity [18]. The long-lasting of the greenhouse gases in the atmosphere leads to prolongation of the period of high global temperature that affect on both biosphere and hydrosphere [19].

(b) The large volcanic eruptions, The Large Igneous Provinces (LIPs) thought to play a significant role in global climatic changes that can be observed in Paleo-records [20]. Several efforts were made to link the main mass extinctions and oceanic anoxia events resulted from climatic and paleoenvironmental changes to the major volcanic eruptions even (Fig. 3), though it's still in debate [21, 22, 23, 24, 9, 25, 20, 26].

(c) Several Maas extinctions and hyperthermal events show a good correlation with the LPIs occurred during their time and their relation is well established such as Late Ordovician mass extinction [27] Middle Permian and the Permo-Triassic mass extinction event [21, 28, 29, 30], end-Pliensbachian extinction and Toarcian [31, 32] end-Triassic mass extinction [33], early Cretaceous [34]; Cretaceous-Tertiary Mass Extinction events [21, 35, 36, 9, 37, 38, 39] end of Cretaceous [37] and recently PETM [40, 41]. The large volcanic eruption has the same effect as the extraterrestrial impacts.

(d) Huge CO_2 release, several sources of the CO_2 were proposed explaining several climatic changes in the Earth's history. Wildfires, extensive burning of peat and coal deposits linked with the arid period [42]; (2) thermogenic methane linked to hydrothermal injection in organic-rich sediments [43–45]. (3) the drying of isolated epicontinental seas which led to rapid oxidation of organic matter [46], (4) melting of the methane-rich permafrost [47] and (5) Catastrophic methane release from the continental sheilf [48, 49]. All these sources can be led to increasing the greenhouse conditions and forces the climatic changes.

The biosphere is very sensitive to the climatic and environmental change and has three different types of response to these changes; 1—Mass extinction, 2—Evolution, and 3—Rising of new genera. Five major mass extinction events took places through the geologic history, as well as several hyperthermal events led to the evolution and/or introducing new creatures to the Earth's record.

During the Mass extinction events, and based on the High-resolution palaeoecological and geochemical studies indicates extensive ecological stress conditions led to turnover and recovery of different marine genera extended through intervals lasting millions of years and extends upheavals affecting terrestrial vegetation.

4 Major Mass Extinction Events

4.1 The Ordovician-Silurian Mass Extinction Event (Fig. 3)

A global extinction event is occurring during the Hirnantian Age (445.2–443.8 Ma) of the Ordovician Period and the subsequent Rhuddanian Age (443.8–440.8 Ma) [50] of the Silurian Period. Eighty-five percent of life species were extinct. In which one-third of all brachiopod and bryozoan families were disappeared, as well as numerous groups of conodonts, trilobites, and graptolites [51]. Two main reasons led to the end-Ordovician extinction; the first wave of extinction may be linked to rapid cooling occurred at the end of the Ordovician [5, 52, 53, 10] and the second is caused by the sea-level fall associated with the glaciation. The drop in sea level would have drained the large epicontinental seas and reduced the available habitat for organisms that favored those settings. The CO_2 level is decreased from 5000 to 3000 ppm during the glaciation, whereas the sea level is lowered by about present level [54]. The combination of the sea-level drop, and reducing ecospace on continental shelves due to the glaciation, are likely driving agents for the Ordovician mass extinction.

4.2 Late Devonian (Frasnian-Fammenian)

A global extinction event is occurring during 375–360 Ma close to the Devonian–Carboniferous transition and lasted about 20 Ma [10]. At the end of the Frasnian Age (Fig. 3). About 19% of all families, 50% of all genera and at least 70% of all species were extinct [55]. The end of Devonian was marked by a large multi extraterrestrial impacts (Bolide also called fir bale; Fig. 3) Woodleigh, Alamo, Charlevoix, Warburton and Siljan with diameters of 120, 100, 54, 52 400 km respectively [16, 17, 56]. Late Devonian mass extinction was a series of extinction events [54]. The first was around 387 Ma extinction in which about 30% of Genera was extinct. The second, around 374 Ma and about 58% of the pelagic Genera were extinct including several species of Ammonoids, Cricoconaids, Placoderms, Conodonts, Agnathans, as well as some of the benthic groups such as Rugose corals, Trilobites, Ostracods and Brachiopods [10]. The last around 359 Maand includes extinction of 30% of Genera including several species of affected Placoderms, ammonoids, conodonts, stromatoporoids, rugose corals, trilobites and ostracods. The CO_2 levels decreased from ranges of 3200–5200 ppm to below 500 ppm [57]. In the terrestrial realm, the development in the plant megaphyll leaves containing high stomata density allowed vegetation to adapt to the cool low-CO_2 conditions of the fore coming cold periods [14, 58]. The land had been colonized by plants and insects. In particular, the rapid rising of *Archaeopteris* forests during the closing stages of the Devonian [59]. The sedimentary record includes evidence reveal several environmental changes Such as exists of widespread anoxia in oceanic bottom waters; the high rate of carbon burial [59], tropical reef and benthic organisms were devastated. The most important evidence is the cyclic sea-level changes around the Frasnian–Famennian Kellwasser event, in which the sea level rise is associated with the onset of anoxic deposits [60].

4.3 Permo-Triassic Mass Extinction

The Great Dying, occurred between 251.941 ± 0.037 and 251.880 ± 0.031 Ma ago, and lasting for the duration of 60 ± 48 ka [61]. The Earth's largest mass extinction event that killed 57% of all families, 83% of all genera and 90–96% of all species [62]. These percentage includes about 96% of all marine species and about 70% of terrestrial species and insects [63]. This event is associated with a major eruption of Siberian volcanism [64, 39, 65]; Fig. 3) with a large asteroid impact (Araguinha, Brazil, with a diameter of 40 km). The atmospheric CO_2 level is increased to about 3400 ppm [66, 67]. The geochemical investigation across the Permo-Triassic boundary [68] based on isotopic $d^{13}C$ and $d^{34}S$ indicates that increasing the aridity that led to dropping in the rate of organic burial starting from late Permian. The aridity led to a decrease in on land biomass from forests to grassland. An abrupt drop in oxygen level and increasing both pyrite and organic carbon led to rising of euxinic marine conditions. The consequence of the previous conditions is to forsees the extinction of vertebrates and loss of giant insects and amphibians, as well as increase in ocean acidity, melting of polar ice, change in ocean circulation speed that result in rising anoxia producing a huge amount of H_2S that released to the atmosphere and accelerate the sea and land mass extinction [19].

4.4 Triassic- Jurassic Mass Extinction

Took place around 201.3 Ma characterized by the extinction of 70–75% of all species [69] including Most archosaurs, therapsids, and the large amphibians were vanished, only dinosaurs were resisted with few terrestrial fauna. The aquatic ecosystems were dominant archosaurs, whereas in the marine environment were dominated the non-archosaurian diapsids. This event was coincident with the opening the central Atlantic magmatic province [9, 70, 71] as well as the large Manicouagan impact (Fig. 3). Its characterized by perturbation in the carbon cycle with increasing of CO_2 level reached about 1300–2200 ppm that causes the rising extreme warming conditions.

4.5 *Cretaceous-Tertiary Mass Extinction*

The KPg boundary is placed around 66 Ma and is considered as the 2nd largest mass extinction event in the earth's history. About 17% of all families, 50% of all genera and 75% of all species became extinct [72]. this event is characterized by: a- presence of extraterritorial Iridium anomaly associated with red clay layer; b-Mass Extinction of planktic foraminifera at the KPB followed by evolutionary trend of Danian species; c- negative shift in $\delta^{13}C$ due to increased atmospheric CO_2 level [73–76]. Chicxulub impact was the acceptable scenario that proposed to explain this event; [77–86]. Recently several studies focused on the relation between Deccan volcanism eruption and the KPB Mass Extinction event [87, 88, 37, 89, 90, 91, 92, 38, 93, 94, 95].

5 Hyperthermal Events

During the last 65 million years, the Earth has undergone many episodes of natural global warming and cooling due to various causes. The sedimentary rocks hold a good record of the environmental changes during this part of Earth's history. Most of these environmental changes are linked to carbon cycle perturbations that resulted in the increase or decrease in Earth's temperature linked to variations between extremes of expansive warmth with ice-free poles to extremes of cold with massive continental ice-sheets and polar ice caps. Such extreme climatic changes are controlled on a global scale by Earth's orbital geometry and/or plate tectonics [96].

The Eocene (55.8 $\pm$ 0.2 to 33.9 $\pm$ 0.1 Ma. [11], is a period in Earth history characterized by climatic extremes varying from the greenhouse to icehouse conditions (Fig. 4). The warmest period of the Cenozoic era coincides with an extreme warming event at the Paleocene-Eocene transition known as Paleocene-Eocene Thermal Maximum (ETM1 or PETM) [97]. The climate warmed steadily until the Early Eocene Climatic Optimum (EECO) ~53.5 Ma when temperatures and greenhouse gas concentrations peaked at their maxima for ~1.5 myr [98]. These two major events are followed by several abrupt and short hyperthermal events including H2 (53.6 Ma), I1 (53.3 Ma), I2 (53.2 Ma) and ETM-3 (52.8 Ma) [98]. A major negative $\delta^{13}C_{org}$ excursion marks each of these hyperthermal events (Fig. 4)

The most recognizable amongst these warming events is the Paleocene-Eocene Thermal Maximum (PETM), which was first recognized [99] in the Ocean Drilling Program (ODP) site 690B cores, and associated with a temperature rise of 5–9 °C

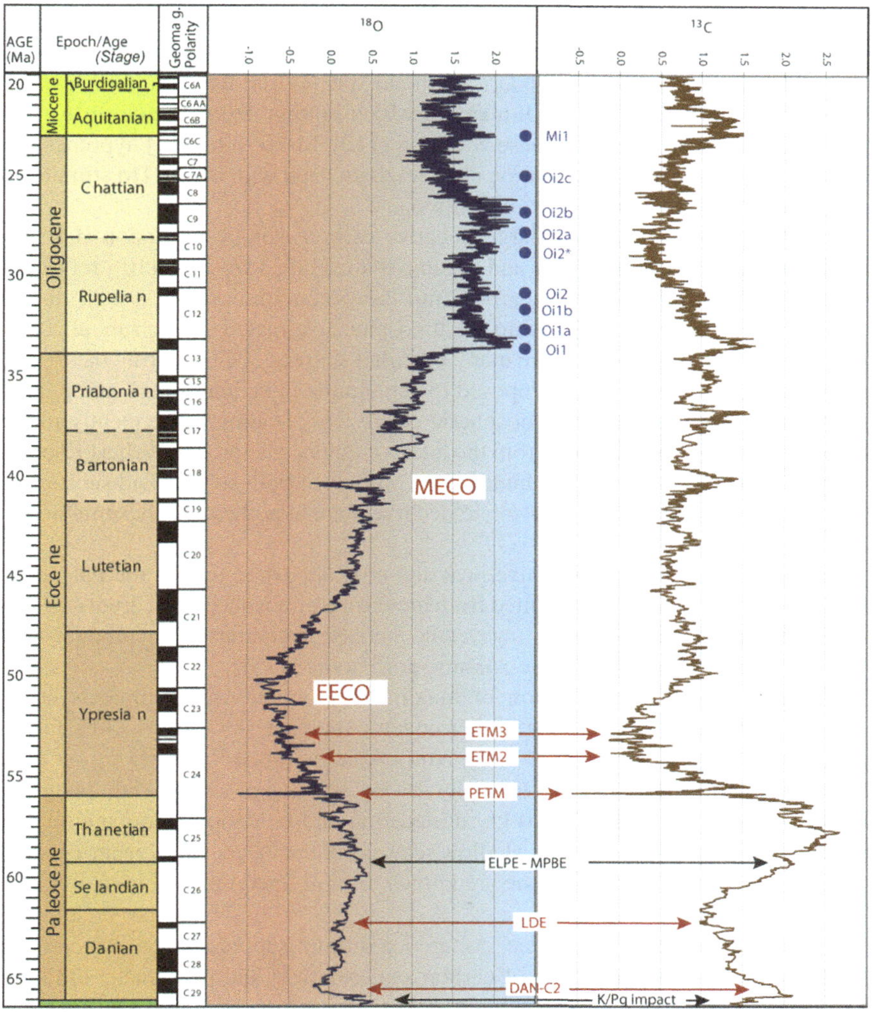

Fig. 4 Position of main Paleogene climatic events against the generalized oxygen and carbon isotope curves for the Paleogene [11]

over about 10,000 years (Fig. 4). Subsequently, this thermal maximum was shown to gradually decrease to pre-excursion $\delta^{13}C_{org}$ values over about 150,000–220,000 years [45]. The PETM is characterized by a decline in oxygen isotope ratios indicating warming of 3–4 °C of surface water and 6 °C in deep water. This increase in temperature is associated by a negative shift in carbon isotope of nearly −2‰ for the benthic foraminifera and −4‰ for planktic foraminifera. The PETM events, led to major species extinctions in deep marine benthic foraminifera [100–102] but diversification in planktic foraminifera with subtropical affinities [103–105]. Mineralogically, the

PETM is marked by increased detrital input and kaolinite contents due to intense on-land weathering during the hot humid climate [106, 107]. In the terrestrial realm, it led to the diversification of modern mammal species and their migration across the northern continents [108–110] many of which immigrated to the North American continent during the greenhouse warming [110]. Moreover, [111] hypothesized that the temperature and precipitation changes associated with the PETM stimulated rapid shifts in plant community composition.

The PETM is marked by a large negative carbon isotope excursion (CIE) in terrestrial and marine carbonates and organic matter [99, 112, 113, 110], reflecting a rapid release of ^{13}C-depleted carbon into the ocean-atmosphere system due to injection of huge amount of CO_2 into the atmosphere estimated to reach about 2000–2600 Gt over 10 ka [96]. There are many potential sources of CO_2 during the PETM and many scenarios have been proposed to explain the injection of high amounts of CO_2 [96] in both marine and atmospheric ecosystems leading subsequently to the perturbation of the carbon cycle from the latest of Paleocene into the earliest Eocene. More than twenty-five years and hundreds of studies of both surface and subsurface sections covering the PETM interval yielded five main hypotheses were summarized [114].

Wildfires: Burning of the extensive peat and coal deposited during the Paleocene (δ^{13}C of ~−22‰) could have resulted from increasing atmospheric O_2, dryer climate, and/or uplift of coal basins [42]. However, no increase in combustion byproducts was observed in cores from neither the Atlantic nor Pacific [115].

Thermogenic methane: Injection of magma into organic-rich sediments could have caused the explosive release of thermogenic methane (δ^{13}C of ~−30‰) from Cretaceous-Paleocene mudstones in the North Atlantic [43, 44, 116, 45].

Drying epicontinental seas: Tectonically driven isolation of an epicontinental seaway could have led to rapid (<20 ka) desiccation and oxidation of organic matter (δ^{13}C of ~−22‰) [46]. However, shallow seaways of the Paleocene-Eocene covered vast areas of central Asia, and none are known to have dried up coincident with the PETM [117]

Permafrost: During the Paleogene, Antarctica did not support a large ice cap and may have stored huge quantities of carbon as permafrost and peat that could have been rapidly defrosted and oxidized, releasing carbon (δ^{13}C of ~−30‰) [47].

Catastrophic methane release: the most recognized scenario explaining the PETM event is the catastrophic methane release from hydrates (clathrates) [48, 49].

This scenario invokes methane released from the continental margin that could have led to carbon dioxide input estimated at 2000×10^9 metric tons over 10,000 years [118], which was suggested as potentially the main cause for the PETM [119, 120]. Clathrates are stable in deep-sea sediments but can be destabilized by increasing temperature caused by changes in ocean circulation [48, 121] by decreasing pressure resulting from slope failure [97] or sea level changes. The clathrate release could also result from late Paleocene volcanic activities linked to the opening of the North Atlantic Ocean [122, 9]. Whatever the source of CO_2 released during the

PETM, global temperature increased about 5–9 °C during a period of about 200 ka [123–125].

The termination of the PETM was a response to an increase in the intensity of the marine biological pump (productivity feedback) resulting in the drawdown of atmospheric CO_2 and subsequent carbon sequestration in the ocean [126, 127]. The export production did not facilitate the rapid removal of excess carbon from the atmosphere [127]. Thus, the most likely mechanism for atmospheric CO_2 removal appears to be increased silicate weathering, which occurred at much faster rates than previously assumed [128, 129].

The PETM is the best analog resemblances to the human-caused climate change unfolding today. Most notably, the culprit behind the PETM was a massive injection of heat-trapping greenhouse gases into the atmosphere and oceans, comparable in volume to today's persistent burning of fossil fuels. Knowledge causes and mechanisms of the PETM could help us foresee climate warming in our near future and design possible ways to mitigate its potentially adverse impact. The PETM is believed to have been caused by greenhouse gas emissions, similarly to today's anthropogenic warming and therefore can serve as analog to the actual warming. However, the current warming is estimated to be 10 times faster than the PETM.

6 Paleocene-Eocene Thermal Maximum in Egypt

Throughout the geologic history of Egypt affected by the global climatic changes and subjected to several hyperthermal events, especially during Paleocene and Eocene. The trend of the carbon isotope shift that represents the global carbon cycle perturbation and consequently the climatic change is identical to those obtained from several places in Egypt such as Red Sea (El-Beida section) (Fig. 5). Amongst the hyperthermal recorded in the Egyptian section, we will discuss the Paleocene Eocene Thermal Maximum event in its type locality.

In 2003 the International Commission for Stratigraphy (ICS) designated the Dababiya Quarry in southern Egypt (Dababiya village, Luxor) as the Global Stratigraphic Section and Point (GSSP) for the Paleocene-Eocene (PE) boundary as marked by the PETM event. The PE boundary is placed in the basal part of the Esna Formation [130, 131].

Based on the following criteria: (1) the negative organic carbon isotope excursion (CIE), (2) the extinction of deep water benthic foraminifera (*Stensioina beccariiformis*), (3) the transient occurrence of planktonic foraminifera (*Acarinina africana, A. sibaiyaensis, Morozovella allisonensis*) during the $\delta^{13}C$ excursion, (4) the transient occurrence of the *Rhomboaster* spp.—*Discoaster araneus* (RD) nanofossils assemblage, and (5) an acme of the dinoflagellate *Apectodiniu.*

Based on these criteria the Dababiya section was considered as one of the most complete and expanded Upper Paleocene to Lower Eocene sequences and representative of this boundary event globally [130, 131]. The age estimate of the CIE onset has been recently determined to be 56.011 Ma using radiometric data obtained from

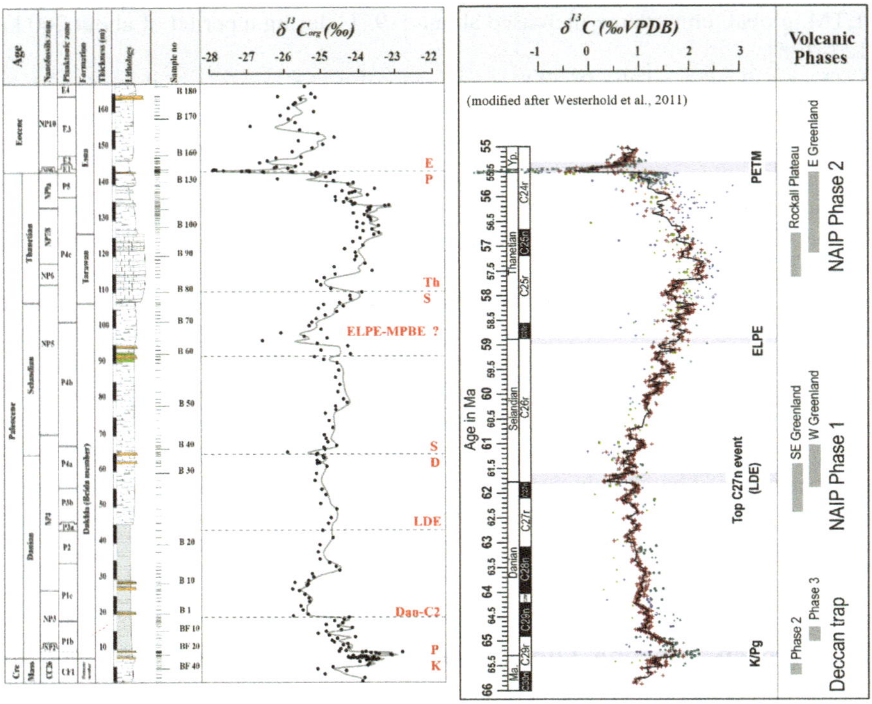

Fig. 5 Global correlation between El-Beida section $\delta^{13}C_{org}$ record and the international standard curve of [116] with major LIPs occurred during the time interval from the upper cretaceous to early Eocene (Khozyem et al., in prep)

a marine ash layer [116], whereas dating based on the zircon crystals collected in a bentonite located in the upper part of the CIE gave age of 56.09 ± 0.03 Ma [132]. Astronomically calibrated cyclostratigraphy timescale indicates a duration of 150–220 ka for the CIE [133, 134], whereas calculations based on the extra-terrestrial ^{3}He fluxes suggest a duration between 120 and 220 ka [135].

The time interval between the pre-CIE $\delta^{13}C$ value and the CIE-minimum has been estimated to be 10–25 ka [136, 137, 138, 118] the accuracy of this determination depends on the presence of the uppermost Paleocene dissolution levels. The Dababiya GSSP (Luxor, Egypt; Fig. 6) provides important clues to understand the warming mechanism before, during, and after the PETM events. Before the PETM event (topmost Paleocene), gradual decreases in $\delta^{13}C_{Carb}$ and $\delta^{13}C_{org}$ are linked to gradual warming (Fig. 7), which may have resulted from the late Paleocene volcanism (North Atlantic magmatic province).

This warming reaches its maximum (at the P/E boundary) leading to the abrupt melting of the methane hydrates stored on the continental shelf. Released methane hydrate is associated with some specific criteria such as (1) Ocean acidification as observed by dissolution of carbonate minerals in Dababiya and elsewhere, (2)

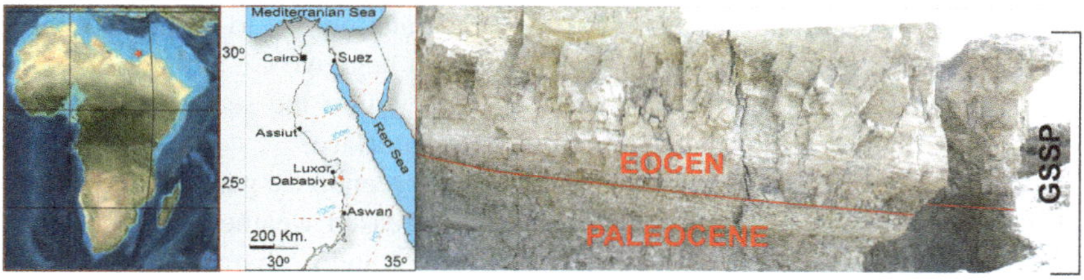

Fig. 6 Location map and field photograph of Dababiya GSSP, Loxur, Egypt. Modified after [136]

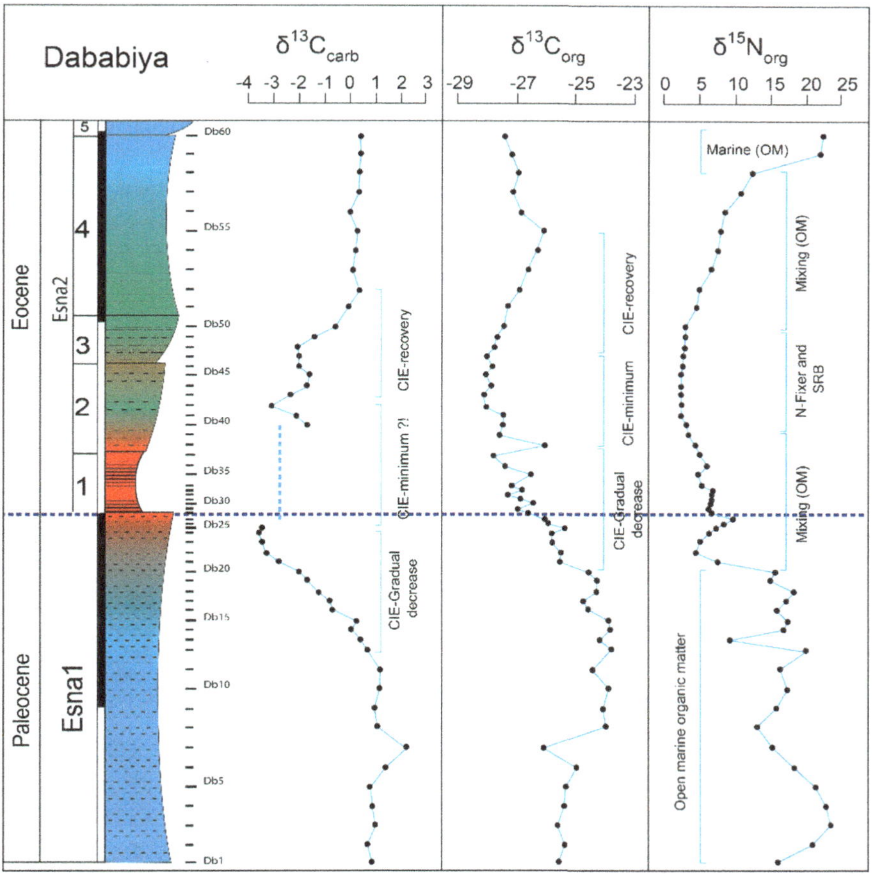

Fig. 7 $\delta^{13}C_{Carb}$, $\delta^{13}C_{org}$, and $\delta^{15}N$ patterns in front of the Dababiya GSSP illustration. Note the gradual decrease in carbon isotopic compositions and the abrupt shift in $\delta^{15}N$. Modified after [138]

anoxic condition that could be linked to oxygen exhalation due to partial oxidation of methane, this anoxia is marked in Dababiya by a negative Ce-anomaly and the precipitation of individual euhedral pyrite crystals within the sediments, and (3) methane oxidation to CO_2 at the water/atmosphere interface.

The excess of CO_2 in the atmosphere is acting as greenhouse gases and increases the weathering of silicate rocks. This period of increased weathering is observed at the Dababiya GSSP 60 cm above the P/E boundary and is characterized by high phyllosilicate contents including kaolinite, high CIA and CIW, high Ti, Cr and K input, $\delta^{13}C_{Carb}$ minimum value (Fig. 8). The top of that interval is characterized by anoxic conditions marked by (1) a negative Ce-anomaly, (2) enrichment of redox-sensitive trace elements and their ratios (V, U, V/Cr, Ni/Co, V/V + Ni), (3) abundant pyrite framboids (>2 μm in diameter), which indicate anoxic conditions within the water column, (4) increased productivity sensitive trace elements (Pb, Zn, Cu, Co and Mo) that could only be enriched under anoxic conditions associated with high organic matter accumulation leading to the development of anoxic conditions, (5) a strong and persistent decrease in $\delta^{15}N_{org}$ reflecting significant bacterial activity such as sulfur reducing and cyanobacteria activities as indicated by low N isotope and the presence of EPS-like structures within the PETM interval; and (6) $\delta^{13}C_{org}$ minimum recorded value.

The PETM recovery phase starts at the top of the interval marked by intense weathering (top of the upper anoxic level). The high nutrients input delivered into the ocean reactivates the biological pump leading to the re-establishment of more

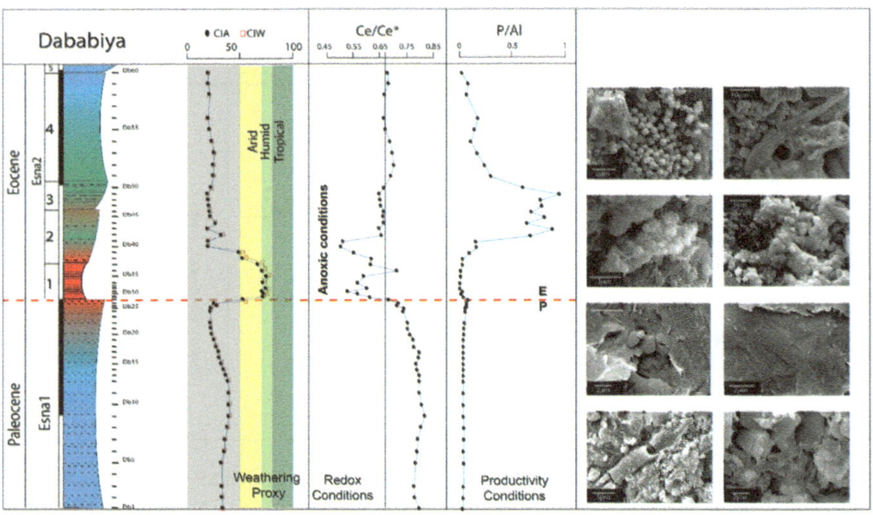

Fig. 8 CIA and CIW curve indicates the intensive weathering during the PETM, Ce- anomaly and the redox parameters showing the different anoxic intervals recorded at the Dababiya GSSP, and phosphorus as an indicator of Paleoproductivity at the recovery interval. The SEM pictures show the observed features during the anoxic interval of the PETM in Dababiya GSSP. Modified after [138]

oxic and suitable conditions as indicated by the increase of productivity indicators such as Ba, P and both $\delta^{13}C_{Carb}$ and $\delta^{13}C_{org}$, which gradually return to the background values observed below the PETM interval.

The PETM at Dababiya GSSP can be therefore divided into four main phases (Fig. 9): (1) pre-PETM gradual warming during the latest Paleocene), (2) PETM main phase with methane hydrate release, (3) environmental response (weathering and anoxia), and (4) finally, recovery phase.

7 Anthropocene Warming

In 2016, both Anthropocene Working Group (AWG) of the Subcommission on Quaternary Stratigraphy (SQS) of the International Commission on Stratigraphy (ICS), defined the Anthropocene as the period of the commencement of significant human impact on the Earth's geology and ecosystems [139–142]. Human activity has had dramatic impacts on landscape, the subsurface and Earth systems, and driving significant atmospheric, chemical, physical and biological changes. The Anthropocene climatic changes are determined to start from the beginning of the Agricultural Revolution 12,000–15,000 years ago. The ratification process continues, and thus a date remains to be decided definitively. The most notable environmental change occurring on earth has a direct link with the beginning of the industrial revolution. Whereas some recent studies linked the beginning of the Anthropocene era to begin approximately 8000 years ago with the development of agriculture and beginning of stable humans societies. It started when the ancient farmers cleared forests to grow crops [143, 144].

The main factor forcing the Anthropocene climate change is increasing the greenhouse gaseous due to the burning of fossil fuel. The burning of fossil fuel due to the increased human activities push the atmospheric carbon dioxide (CO_2) content to higher levels. In 2013, the CO_2 emissions increased atmospheric CO_2 concentration by 100 ppm to reach approximately 400 ppm from 280 ppm that recorded from Holocene or pre-industrial [145], these values continue increasing through 2015–2016 [146] showing a rising trend above 400 ppm. This significant increases in CO_2 has a strong impact on the Earth's climate system it is occurring much faster, with a greater extent, than any recorded warming events during the geologic history of the earth [147]. Most of this increase is due to the combustion of fossil fuels such as coal, oil, and gas, although smaller fractions result from cement production and from land-use changes (such as deforestation).

Geologically, this climate warming is commonly compared with the rapid short-term ~5 °C warming of the PETM except that during the PETM major diversification in marine and terrestrial life and significant species extinctions occurred only in deep-water benthic foraminifera [148]. Therefor to get a better understanding of the Anthropocene climatic changes we have to compare it with the different rapid warming and mass extinction events such as Cretaceous-Paleogene mass extinction as well as the Paleocene Eocene climatic optimum to gain insights into potential

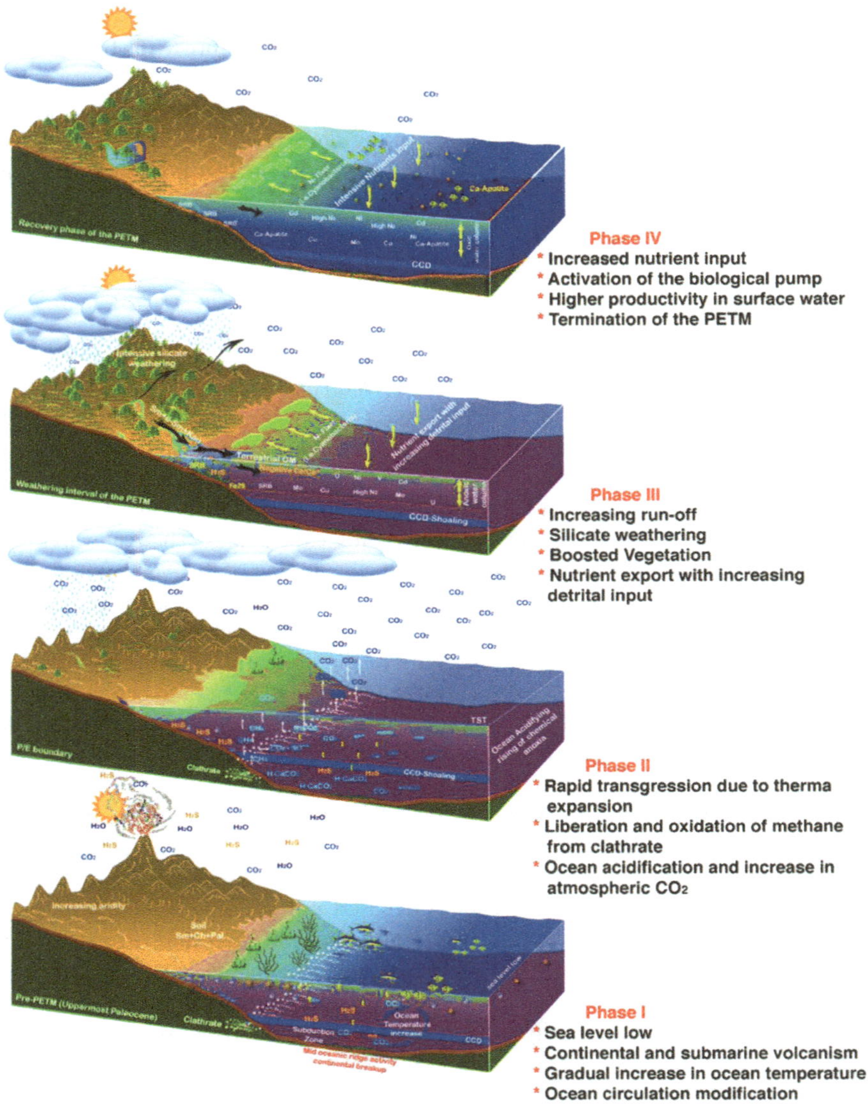

Fig. 9 Block diagrams representing the four phases of the PETM event: Phase 1—gradual warming during late Paleocene, Phase 2—PEB interval, Phase 3—intensive silicate weathering phase, and phase 4—marine recovery and drawdown of atmospheric CO_2. Note that this scenario is partly based on [122, 138]

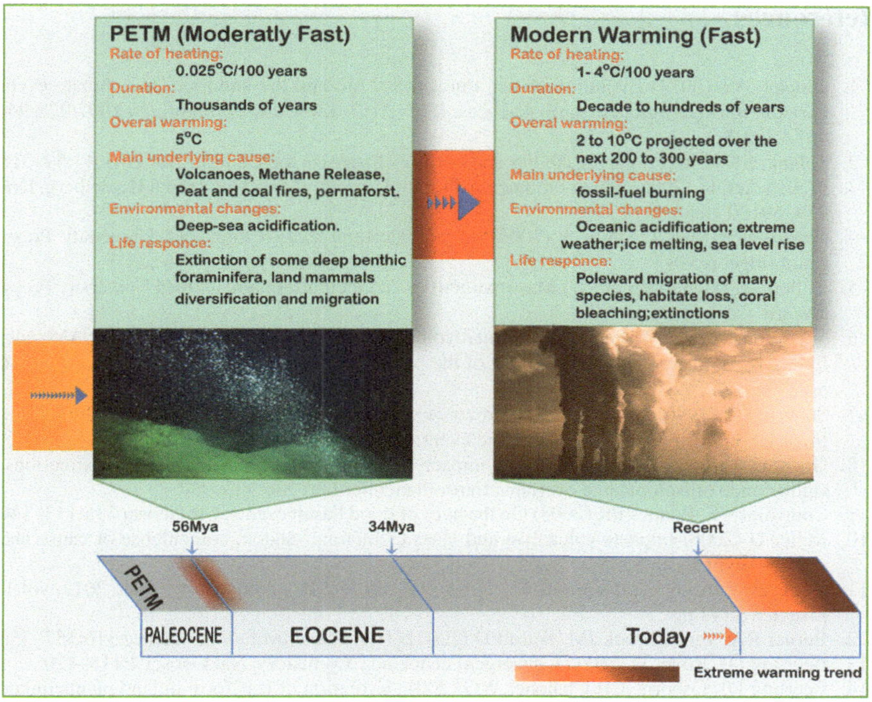

Fig. 10 Comparison between the PETM and the current warming that reflect the importance of the PETM study, and how it can help to understand the future climatic changes and the effect of the temperature increase on various environments

Anthropocene scenarios (causes and consequences. The PETM extreme warming is a commonly used analog and predicted best-case scenario for the current rapid climate warming in the coming decades and centuries. We summarize the comparison between the PETM and the anthropogenic climate changes in Fig. 10.

8 Conclusion

Study of the different impacts of the climatic changes occurred in the Earth's geologic history can give a clue to predict the consequence of the current warming. The natural warming events took several hundreds of kilo years led to increase the earth temperatures by a few degrees, whereas the human impact accelerates the process by around 250 times. Therefor the human activities should be controlled by lows to stop and/or decrease the emission of the greenhouse gases to slow there steps toward the sex mass extinction event.

References

1. Glikson AY (2014) Evolution of the atmosphere, fire and the anthropocene climate event horizon. Springer briefs in earth sciences. ISSN 2191-5369, https://doi.org/10.1007/978-94-007-7332-5
2. Solanki SK (2002) Solar variability and climate change: is there a link? Sol Phys 43:59–513
3. Pfeilsticker, K. (2006). Paleo-climate. Institut für Umweltphysik, Universität Heidelberg, INF 229, 69120 Heidelberg
4. Gradstein F, Ogg J, Smith A (2004) A geologic time scale. Cambridge University Press, Cambridge, p 598
5. Hallam A, Wignall PB (1997) Mass extinctions and their aftermath. Oxford University Press, Oxford
6. MacLeod N (2003) The causes of Phanerozoic extinctions. In: Rothschild LJ, Lister AM (eds) Evolution on planet earth, the impact of the physical environment. Elsevier Books, Oxford, pp 253–278
7. Sepkoski JJ (1996) Patterns of Phanerozoic extinction: a perspective from global data bases. In: Walliser OH (ed) Global events and event stratigraphy. Springer, Berlin, pp 35–52
8. Glikson AY (2005) Asteroid/comet impact clusters, flood basalts and mass extinctions: significance of isotopic age overlaps. Earth Planet Sci Lett 236:933–937
9. Courtillot VE, Rennes PR (2003) On the ages of flood basalt events. CR Geosci 335:113–140
10. Keller G (2005) Impacts volcanism and mass extinction: random coincidence or cause and effect? Aust J Earth Sci 52:725–757
11. Gradstein MF, Ogg JG, Schmitz MD, Ogg GM (2012) The geologic time scale 2012, vol 1. Elsevier, 1144 pp
12. Berner RA, Vanderbrook JM, Ward PD (2007) Oxygen and evolution. Science 316:557–558
13. Beerling DJ, Royer D (2011) Convergent cenozoic CO_2 history. Nat Geosci 4:418–420
14. Beerling DJ, Osborne CP, Chaloner WG (2001) Evolution of leaf-form in land plants linked to atmospheric CO_2 decline in the Late Palaeozoic era. Nature 410:352–354
15. French BM (1998) Traces of catastrophe. Lunar Planet Inst 954:120
16. Glikson AY, Uysal IT (2013) Geophysical and structural criteria for the identification of buried impact structures, with reference to Australia. Earth Sci Rev 125:114–122
17. Glikson AY, Uysal IT, Fitz Gerald JD, Saygin E (2013) Geophysical anomalies and quartz microstructures, Eastern Warburton Basin, North-east South Australia: tectonic or impact shock metamorphic origin? Tectonophysics 589:57–76
18. Pope KO, Baines KH, Ocampo AC, Ivanov BA (1997) Energy volatile production and climatic effects of the Chicxulub Cretaceous/tertiary impact. J Geophys Res 102:21645–21664
19. Ward PD (2007) Under a green sky: global warming, the mass extinctions of the past, and what they can tell us about our future. Harper Collins, New York, p 242
20. Ernst RE, Youbi N (2017) How large igneous Provinces affect global climate, sometimes cause mass extinctions, and represent natural markers in the geological record. Palaeogeography, palaeoclimatology, palaeoecology
21. Bond DP, Wignall PB, Joachimski MM, Sun Y, Savov I, Grasby SE,... Blomeier DP (2015) An abrupt extinction in the Middle Permian (Capitanian) of the Boreal Realm (Spitsbergen) and its link to anoxia and acidification. Geol Soc Am Bull 127(9–10):1411–1421
22. Bond DPG, Grasby SE (2016) On the causes of mass extinctions. Palaeogeogr Palaeoclimatol Palaeoecol. https://doi.org/10.1016/j.palaeo.2016.11.005
23. Courtillot V, Fluteau F (2014) A review of the embedded time scales of flood basalt volcanism with special emphasis on dramatically short magmatic pulses. In: Keller G, Kerr A (eds) Volcanism, impacts, and mass extinctions: causes and effects. Geological Society of America Special Paper 505, pp 301–317
24. Courtillot V, Fluteau F, Besse J (2015) Evidence for volcanism triggering extinctions: a short history of IPGP contributions with emphasis on paleomagnetism. In: Schmidt A, Fristad KE, Elkins Tanton LT (eds) Volcanism and global environmental change. Cambridge University Press, Cambridge. ©

25. Ernst RE (2014) Large igneous Provinces. Cambridge University Press, Cambridge, 653p
26. Wignall PB (2005) The link between large igneous province eruptions and mass extinctions. Elements 1:293–297
27. Jones DS, Martini AM, Fike DA, Kaiho K (2017) A volcanic trigger for the Late Ordovician mass extinction? Mercury data from south China and Laurentia: Geology 45(7):631–634
28. Grasby SE, Beauchamp B, Knies J (2016) Early Triassic productivity crises delayed recovery from world's worst mass extinction. Geology 44(9):779–782. https://doi.org/10.1130/g38141.1
29. Grasby SE, Sanei H, Beauchamp B, Chen ZH (2013) Mercury deposition through the Permo-Triassic biotic crisis. Chem Geol 351:209–216
30. Sanei H, Grasby SE, Beauchamp B (2012) Latest Permian mercury anomalies. Geology 40:63–66
31. Percival LME, Witt MLI, Mather TA, Hermoso M, Jenkyns HC, Hesselbo SP, Al-Suwaidi AH, Storm MS, Xu W, Ruhl M (2015) Globally enhanced mercury deposition during the end-Pliensbachian extinction and Toarcian OAE: A link to the Karoo Ferrar large igneous Province. Earth Planet Sci Lett 428:267–280
32. Thibodeau AM, Bergquist BA (2017) Do mercury isotopes record the signature of massive volcanism in marine sedimentary records? Geology 45:95–96
33. Thibodeau AM, Ritterbush K, Yager JA, West J, Ibarra Y, Bottjer DJ, Berelson WM, Bergquist BA, Corsetti FA (2016) Mercury anomalies and the timing of biotic recovery following the end-Triassic mass extinction. Nat Commun 7(11147). https://doi.org/10.1038/ncomms11147
34. Charbonnier G, Morales C, Duchamp-Alphonse S, Westermann S, Adatte T, Föllmi KB (2017) Mercury enrichment indicates volcanic triggering of Valanginian environmental change. Sci Rep 7:40808. https://doi.org/10.1038/srep40808
35. Bond DP, Grasby SE (2017) On the causes of mass extinctions. Palaeogeogr Palaeoclimatol Palaeoecol 478:3–29
36. Courtillot V (2002) Evolutionary catastrophes: the science of mass extinction. Cambridge University Press, Cambridge
37. Font E, Adatte T, Sial AN, Lacerda LD, Keller G, Punekar J (2016) Mercury anomaly, Deccan volcanism and the end-cretaceous mass extinction. Geology 44:171–174
38. Punekar J, Keller G, Khozyem H, Hamming C, Adatte T, Tantawy AA, Spangenberg JE (2014) Late Maastrichtian-early Danian high-stress environments and delayed recovery linked to Deccan volcanism. Cretac Res 49:63–82
39. Wignall PB (2001) Large igneous provinces and mass extinctions. Earth Sci Rev 53:1–33
40. Khozyem H, Adatte T, Mbabi Bitchong A, Mahmoud A, Keller G (2017) The role of volcanism (North Atlantic Igneous province) In: The PETM events revealed by mercury anomalies. GSA Geological Society of America Abstracts with Programs, vol 49(6)
41. Khozyem H, Adatte T, Mbabi BA, Chevalier Y, Keller G (2016) Paleocene-Eocene thermal maximum triggered by volcanism? Evidence from mercury anomalies. GSA Annual Meeting in Denver, Colorado, USA—2016 Paper No. 106-2
42. Kurtz AC, KumP LR, Arthur MA, ZachoS JC, Paytan A (2003) Early Cenozoic decoupling of the global carbon and sulfur cycles: Paleoceanography 18:1090. https://doi.org/10.1029/2003pa000908
43. Svensen H, Planke S, Corfu F (2010) Zircon dating ties NE Atlantic sill emplacement to initial Eocene global warming. J Geol Soc Lond 167:433–436
44. Svensen H, Planke S, Malthe-Sorenssen A, Jamtveit B, Myklebust R et al (2004) Release of methane from a volcanic basin as a mechanism for initial Eocene global warming. Nature 429:542–545
45. Westerhold T, Röhl U, McCarren H, Zachos J (2009) Latest on the absolute age of the Paleocene-Eocene thermal maximum (PETM): new insights from exact stratigraphic position of key ash layers + 19 and—17. Earth Planet Sci Lett 287(3–4):412–419
46. Higgins JA, Schrag DP (2006) Beyond methane: towards a theory for the Paleocene-Eocene thermal maximum. Earth Planet Sci Lett 245:523–537

47. DeConto RM, Galeotti S, Pagani M, Tracy D, Schaefer K, Zhang T, Pollard D, Beerling DJ (2010) Past extreme warming events linked to massive carbon release from thawing permafrost. Nature 484:87–91
48. Dickens GR, O'Neil JR, Rea DK, Owen RM (1995) Dissociation of oceanic methane hydrate as a cause of the carbon isotope excursion at the end of the Paleocene. Paleoceanography 10:965–971
49. Kennett JP, Cannariato KG, Hendy IL, Behl RJ (2002) Methane Hydrates in quaternary climate change: the clathrate gun hypothesis: special publication, vol 54. AGU, Washington, D. C, p 224
50. Frakes LA, Francis JE, Syktus JI (1992) Climate modes of the phanerozoic. Cambridge University Press, Cambridge
51. Elewa A (2008) Late Ordovician mass extinction, p 252. ISBN 978-3-540-75915-7
52. Holland SM, Patzkowsky ME (2007) Gradient ecology of a biotic invasion: biofacies of the type Cincinnatian Series (Upper Ordovician), Cincinnati, Ohio region, USA. PALAIOS. SEPM Soc Sediment Geol 22(4):392–407. https://doi.org/10.2110/palo.2006.p 06-066
53. Holland SM, Patzkowsky ME (2015) The stratigraphy of mass extinction. Palaeontol Palaeontol Assoc 58(5):903–924. https://doi.org/10.1111/pala.12188
54. Kump LR, Arthur MA, Patzkowsky ME, Gibbs MT, Pinkus DS, Sheenan PM (1999) A weathering hypothesis for glaciation at high atmospheric pCO_2 during the Late Ordovician. Palaeoclimatol Palaeogeogr Palaeoecol 152:173–187
55. Briggs D, Crowther PR (2008) Palaeobiology II. Wiley, New York, p 223. ISBN 978-0-470-99928-8
56. Wang K, Attrep M, Orth CJ (2017) Global iridium anomaly, mass extinction, and redox change at the Devonian-Carboniferous boundary. Geology 21(12):1071–1074
57. Mora CI, Driese SG, Colarusso LA (1996) Middle to late Paleozoic atmospheric CO_2 levels from soil carbonate and organic matter. Science 271:1105–1107
58. Rothwell GW, Scheckler SE, Gillespie WH (1989) Elkinsia gen nov a Late Devonian gymnosperm with cupulate ovules. Bot Gaz 150:170–189
59. Algeo TJ, Scheckler SE (1998) Terrestrial-marine teleconnections in the Devonian: links between the evolution of land plants, weathering processes, and marine anoxic events. Philos Trans R Soc B: Biol Sci 353(1365):113–130. https://doi.org/10.1098/rstb.1998.0195
60. Bond DPG, Wignall PB (2008) The role of sea-level change and marine anoxia in the Frasnian-Famennian (Late Devonian) mass extinction. Palaeogeogr Palaeoclimatol Palaeoecol 263(3–4):107–118. https://doi.org/10.1016/j.palaeo.2008.02.015
61. Burgess SD (2014) High-precision timeline for Earth's most severe extinction. PNAS 111(9):3316–3321. Bibcode:2014PNAS..111.3316B. https://doi.org/10.1073/pnas.1317692111
62. Bergstrom Cl, Dugatkin TL (2012) Evolution. Norton, p 515. ISBN 978-0-393-92592-0
63. Brayard A, Krumenacker LJ, Botting JP, Jenks JF, Bylund KG, Vennin E, Olivier N, Goudemand N, Saucède T, Charbonnier S, Romano C, Doguzhaeva L, Thuy B, Hautmann M, Stephen DA, Thomazo C, Escarguel G (2017) Unexpected early triassic marine ecosystem and the rise of the Modern evolutionary fauna. Sci Adv 13(2):e1602159
64. Renne PR, Zhang Z, Richards MA, Black MT, Basu AR (1995) Synchrony and causal relations between Permian: triassic boundary crises and Siberian flood volcanism. Science 269:1413–1416
65. Wignall PB, Twitchett RJ (1996) Oceanic anoxia and the end Permian mass extinction. Science 272:1155–1158
66. McGhee GR (1996) The late Devonian mass extinction. Columbia University Press, New York
67. Royer DL (2006) CO_2-forced climate thresholds during the Phanerozoic. Geochim Cosmochim Acta 70:5665–5675
68. Berner RA (2005) The carbon and sulfur cycles and atmospheric oxygen from middle Permian to middle Triassic. Geochim Cosmochim Acta 69:3211–3217

69. Ryder G, Fastovsky DE, Gartner S (1996) The cretaceous-tertiary event and other catastrophes in earth history. Geological Society of America, p 19. ISBN 9780813723075
70. Hames W, McHone JG, Renne P, Ruppel C (2003) The Central Atlantic Magmatic Province: insights from fragments of Pangea. Geophys Monog Ser 136:267
71. Jourdan F, Marzoli A, Bertrand HS, Cirilli S, Tanner LH, Kontak DJ, McHone G, Renne PR, Bellieni G (2009) 40Ar/39Ar ages of CAMP in North America: implications for the Triassic–Jurassic boundary and the 40 K decay constant bias. Lithos 110:167–180
72. Raup D, Sepkoski J Jr (1982) Mass extinctions in the marine fossil record. Science 215(4539):1501–1503
73. Cowie JW, Ziegler W, Remane J (1989) Stratigraphic commission accelerates progress, 1984 to 1989. Episodes 12:79–83
74. Keller G (2001) The end-Cretaceous mass extinction in the marine realm: year 2000 assessment. Planet Space Sci 49(8):817–830
75. Keller G, Li L, Macleod N (1995) The Cretaceous/Tertiary boundary stratotype section at El Kef, Tunisia: how catastrophic was the mass extinction? Palaeogeogr Palaeoclimatol Palaeoecol 119:221–254
76. Remane J, Keller G, Adatte T, Pardo A (1999) Report on the international workshop on Cretaceous-Paleogene transitions. Episodes
77. Adatte T, Keller G, Burns S, Stoykova KH, Ivanov MI, Vangelov D, Kramar U, Stüben D (2002) Paleoenvironment across the Cretaceous-Tertiary transition in eastern Bulgaria. Geol Soc Am Spec Pap 356:231–251
78. Adatte T, Stinnesbeck W, Keller G (1996) Lithostratigraphic and mineralogic correlations of near K/T boundary clastic sediments in northeastern Mexico: implications for origin and nature of deposition. Geol Soc Am Spec Pap 307:211–226
79. Keller G (2011) Defining the Cretaceous-Tertiary boundary: a practical guide and return to first principles. In: Keller G, Adatte T (eds) The end-cretaceous mass extinction and the Chicxulub Impact in Texas. Society for sedimentary geology special publication 100, pp 23–42. ISBN: 978-1-56576-308-1
80. Keller G, Adatte T, Berner Z, Pardo A, Lopez-Oliva L (2009) New evidence concerning the age and biotic effects of the Chicxulub impact in Mexico. J Geol Soc Lond 166:393–411
81. Keller G, Adatte T (eds) (2011) The end-cretaceous mass extinction and the Chicxulub impact in Texas, pp 81–122. SEPM (Society for Sedimentary Geology)
82. Keller G, Khozyem HM, Adatte T, Malarkodi N, Spangenberg JE, Stinnesbeck W (2013) Chicxulub impact spherules in the North Atlantic and Caribbean: age constraints and Cretaceous-Tertiary boundary hiatus. Geol Mag 150:885–907
83. Keller G, Lopez-Oliva JG, Stinnesbeck W, Adatte T (1997) Age stratigraphy and deposition of near K/T siliciclastic deposits in Mexico: relation to bolide impact? Geol Soc Am Bull 109:410–428
84. Mateo P, Keller G, Adatte T, Spangenberg JE (2015) Mass wasting and hiatuses during the Cretaceous-Tertiary transition in the North Atlantic: relationship to the Chicxulub impact? Palaeogeogr Palaeoclimatol Palaeoecol 441:96–115
85. Schulte P, Alegret L, Arenillas I, Arz JA, Barton PJ, Bown PR, Collins GS (2010) The Chicxulub asteroid impact and mass extinction at the Cretaceous-Paleogene boundary. Science 327(5970):1214–1218
86. Sohl NF, Martinez RE, Salmeron-Urena P, Soto-Jaramillo F (1991) Upper cretaceous. The Gulf of Mexico basin: Boulder, Colorado, Geological Society of America, Geology of North America, v. J, pp 205–244
87. Caldeira K, Rampino MR (1990) Carbon dioxide emissions from Deccan volcanism and a K/T boundary greenhouse effect. Geophys Res Lett 17(9):1299–1302
88. Courtillot V (1990) Deccan volcanism at the Cretaceous-Tertiary boundary: past climatic crises as a key to the future? Global Planet Change 3(3):291–299
89. Font E, Carlut J, Rémazeilles C, Mather TA, Nédélec A, Mirão J, Casale S (2017) End-cretaceous akaganéite as a mineral marker of Deccan volcanism in the sedimentary record. Sci Rep 7

90. Keller G, Adatte T, Bhowmick PK, Upadhyay H, Dave A, Reddy AN, Jaiprakash BC (2012) Nature and timing of extinctions in Cretaceous-Tertiary planktic foraminifera preserved in Deccan intertrappean sediments of the Krishna-Godavari Basin, India. Earth Planet Sci Lett 341:211–221
91. Petersen SV, Dutton A, Lohmann KC (2016) End-Cretaceous extinction in Antarctica linked to both Deccan volcanism and meteorite impact via climate change. Nat Commun 7
92. Ponte JMN (2013) Magnetic mineralogy of Cretaceous-Tertiary sections (Tethys, Iran and India): links with the Deccan Phase-2 (Doctoral dissertation)
93. Renne P, Sprain C, Pande K, Richards M, Vanderkluysen L, Self S (2016) Tempo of the Deccan Traps eruptions in relation to events at the Cretaceous-Paleogene boundary. In EGU general assembly conference abstracts (18, 9496)
94. Renne P, Sprain CJ, Richards MA, Self S, Vanderkluysen L, Pande K (2015) State shift in Deccan volcanism at the Cretaceous-Paleogene boundary, possibly induced by impact. Science 350(6256):76–78
95. Robinson N, Ravizza G, Coccioni R, Peucker-Ehrenbrink B, Norris R (2009) A high-resolution marine 187 Os/188 Os record for the late Maastrichtian: Distinguishing the chemical fingerprints of Deccan volcanism and the KP impact event. Earth Planet Sci Lett 281(3):159–168
96. Zachos J, Pagani M, Sloan L, Thomas E, Billups K (2001) Trends, rhythms, and aberrations in global climate 65 Ma to present. Science 292:686–693
97. Katz ME, Pak DK, Dickens GR, Miller KG (1999) The source and fate of massive carbon input during the Latest Paleocene thermal maximum. Science 286:1531–1533
98. Zachos JC, Dickens GR, Zeebe RE (2008) An early Cenozoic perspective on greenhouse warming and carbon cycle dynamics. Nature 451:279–283
99. Kennett JP, Stott LD (1991) Abrupt deep-sea warming, palaeoceanographic changes and benthic extinctions at the end of the Palaeocene. Nature 353:225–229
100. Alegret L, Ortiz S, Molina E (2009) Extinction and recovery of benthic foraminifera across the Paleocene-Eocene thermal maximum at the Alamedilla section (Southern Spain). Palaeogeogr Palaeoclimatol Palaeoecol 279:186–200
101. Speijer R, Wagner T (2002) Sea-level changes and black shales associated with the late Paleocene thermal maximum: organic-geochemical and micropaleontologic evidence from the southern Tethyan margin (Egypt-Israel). Geol Soc Am Spec Paper 356:533–549
102. Thomas E (1989) Development of Cenozoic deep-sea benthic foraminiferal faunas in Antarctic waters. J Geol Soc Lond Spec Publ 47:283–296
103. Kelly DC, Bralower TJ, Zachos JC, Premoli-Silva I, Thomas E (1996) Rapid diversification of planktonic foraminifera in the tropical Pacific (ODP Site 865) during the late Paleocene thermal maximum. Geology 24:423–426
104. Lu G, Adatte T, Keller G, Ortiz N (1998) Abrupt climatic, oceanographic and ecologic changes near the Paleocene-Eocene transition in the deep Tethys basin: the Alamedilla section, southern Spain. Ecol Geol Helvatica 91:293–306
105. Luciani V, Giusberti L, Agnini C, Backman J, Fornaciari E, Rio D (2007) Paleocene-Eocene thermal maximum as recorded by Tethyan planktonic foraminifera in the Forada section (northern Italy). Mar Micropaleontol 64:189–214
106. Bolle MP, Adatte T (2001) Paleocene-early eocene climatic evolution in the Tethyan realm: clay mineral evidence. Clay Mineral 36:249–261
107. Bolle M-P, Tantawy AA, Pardo A, Adatte T, Burns SJ, Kassab A (2000) Climatic and environmental changes documented in the upper Paleocene to lower Eocene of Egypt. Eclogae Geol Helv 93:33–51
108. Bowen GJ, Beerling DJ, Koch PL, Zachos JC, Quattlebaum T (2004) A humid climate state during the Paleocene-Eocene thermal maximum. Nature 432:495–499
109. Bowen GJ, Clyde WC, Koch PL, Ting S, Alroy J, Tsubamoto T, Wang Y, Wang Y (2002) Mammalian dispersal at the Paleocene/Eocene boundary. Science 295:2062–2065
110. Wing SL, Alroy J, Hickey LJ (1995) Plant and mammal diversity in the Paleocene to early Eocene of the Bighorn Basin. Palaeogeogr Palaeoclimatol Palaeoecol 115:117–155

111. Smith T, Rana RS, Missiaen P, Rose KD, Sahni A, Singh H, Singh L (2007) Highest diversity of earliest bats in the Early Eocene of India. Naturwissenscehaften 94:1003–1009
112. Koch PL, Zachos JC, Dettman DL (1995) Stable isotope stratigraphy and paleoclimatology of the Paleogene Bighom Basin (Wyoming, USA). Palaeogeogr Palaeoclimatol Palaeoecol 115:61–89
113. Koch PL, Zachos JC, Gingerich PD (1992) Correlation between isotope records in marine and continental carbon reservoirs near the Paleocene- Eocene boundary. Nature 358:319–322
114. McInerney FA, Wing SL (2011) The Paleocene-Eocene thermal maximum: a perturbation of carbon cycle, climate, and biosphere with implications for the future. Ann Rev Earth Planet Sci 39:489–516
115. Moore EA, Kurtz AC (2008) Black carbon in Paleocene-Eocene boundary sediments: a test of biomass combustion as the PETM trigger. Palaeogeogr Palaeoclimatol Palaeoecol 267:147–152
116. Westerhold T, Röhl U, Donner B, McCarren HK, Zachos JC (2011) A complete high-resolution Paleocene benthic stable isotope record for the central Pacific (ODP Site 1209). Paleoceanography 26(2)
117. Gavrilov YO, Shcherbinina EA, Oberhänsli H (2003) Paleocene-Eocene boundary events in the northeastern Peri-Tethys. Geol Soc Am Spec Pap 369:147–168
118. Zachos JC, Röhl U, Schellenberg SA, Sluijs A, Hodell DA, Kelly DC, Thomas E, Nicolo M, Raffi I, Lourens LJ, McCarren H, Kroon D (2005) Rapid acidification of the ocean during the Paleocene-Eocene thermal maximum. Science 308:1161–1611
119. Dickens G (2011) Down the Rabbit Hole: toward appropriate discussion of methane release from gas hydrate systems during the Paleocene-Eocene thermal maximum and other past hyperthermal events. Clim Past 7:831–846
120. Katz M, Cramer B, Mountain G, Katz S, Miller K (2001) Uncorking the bottle: what triggered the paleocene/Eocene thermal maximum methane release? Paleoceanography 16(6):549–562
121. Dickens GR, Castillo MM, Walker JCG (1997) A blast of gas in the latest Paleocene: simulating first-order effects of massive dissociation of oceanic methane hydrate. Geology 25:259–262
122. Bowen GJ, Zachos JC (2010) Rapid carbon sequestration at the termination of the Palaeocene-Eocene thermal maximum. Nat Geosci 3:866–869
123. Sluijs A, Schouten S, Pagani M, Woltering M, Brinkhuis H, Sinninghe Damsté JS, Dickens GR, Huber M, Reichart GJ, Stein R (2006) Subtropical Arctic Ocean temperatures during the Palaeocene/Eocene thermal maximum. Nature 441:610–613
124. Weijers JWH, Schouten S, Sluijs A, Brinkhuis H, Sinninghe Damsté JS (2007) Warm Arctic continents during the Palaeocene-Eocene thermal maximum. Earth Planet Sci Lett 261:230–238
125. Zachos JC, Schouten S, Bohaty S, Quattlebaum T, Sluijs A, Brinkhuis H, Gibbs SJ, Bralower TJ (2006) Extreme warming of mid-latitude coastal ocean during the Paleocene-Eocene thermal maximum: inferences from TEX86 and isotope data. Geology 34:737–740
126. Bains S, Norris RD, Corfield RM, Faul KL (2000) Termination of global warmth at the Palaeocene/Eocene boundary through productivity feedback. Nature 407:171–174
127. Torfstein A, Winckler G, Tripati A (2010) Productivity feedback did not terminate the Paleocene Eocene thermal maximum (PETM). Clim Past 5:2391–2410
128. Kelly DC, Nielsen T, McCarren H, Zachos JC, Rohl U (2010) Spatiotemporal patterns of carbonate sedimentation in the South Atlantic: implications for carbon cycling during the Paleocene-Eocene thermal maximum. Palaeogeogr Palaeoclimatol Palaeoecol 293:30–40
129. Kelly DC, Zachos JC, Bralower TJ, Schellenberg SA (2005) Enhanced terrestrial weathering/runoff and surface ocean carbonate production during the recovery stages of the Paleocene Eocene thermal maximum. Paleoceanography 20:4023
130. Aubry M-P, Ouda K, Dupuis C, Berggren WA, Van Couvering JA, Members of the Working Group on the Paleocene/Eocene Boundary (2007) The global standard stratotype-section and point (GSSP) for the base of the Eocene Series in the Dababiya section (Egypt). Episodes 30(4):271–286

131. Dupuis C, Aubry M-P, Steurbaut E, Berggren WA, Ouda K, Magioncalda R, Cramer BS, Kent DV, Speijer RP, Heilmann-Clausen C (2003) The Dababiya Quarry section: lithostratigraphy, clay mineralogy, geochemistry and paleontology. In: Ouda K, Aubry M-P (eds) The upper Paleocene–Lower Eocene of the Upper Nile Valley: Part 1. Stratigraphy: micropaleontology, vol 49, pp 41–59
132. Jaramillo CA, Ochoa D, Contreras L, Pagani M, Carvajal-Ortiz H et al (2010) Effects of rapid global warming at the Paleocene-Eocene boundary on neotropical vegetation. Science 330:957–961
133. Aziz HA, Hilgen FJ, Luijk GM, Sluijs A, Kraus MJ (2008) Astronomical climate control on paleosol stacking patterns in the upper Paleocene-lower Eocene Willwood Formation, Bighorn Basin, Wyoming. Geology 36:531–534
134. Röhl U, Westerhold T, Bralower TJ, Zachos JC (2007) On the duration of the Paleocene-Eocene thermal maximum (PETM). Geochem Geophys Geosyst 8, Q12002 pp
135. Murphy BH, Farley KA, Zachos JC (2010) An extraterrestrial 3He-based timescale for the Paleocene-Eocene thermal maximum (PETM) from Walvis Ridge, IODP Site 1266. Geochim Cosmochim Acta 74:5098–5108
136. Khozyem H, Adatte T, Keller G, Tantawy AA, Spangenberg JE (2014) The paleocene-eocene GSSP at Dababiya, Egypt-Revisited. Episodes 37(2):78–86
137. Khozyem H, Adatte T, Spangenberg EJ, Tantawy A, Gerta Keller K (2013) Paleoenvironmental and climatic changes during the Paleocene-Eocene thermal maximum (PETM) at the Wadi Nukhul Section, Sinai, Egypt. J Geol Soc Lond 170:341–352
138. Khozyem H, Adatte T, Spangenberg JE, Keller G, Tantawy AA, Ulianov A (2015) New geochemical constraints on the Paleocene-Eocene thermal maximum: Dababiya GSSP, Egypt. Palaeogeogr Palaeoclimatol Palaeoecol 429:117–135
139. Castree N (2015) The Anthropocene: a primer for geographers. Geography 100 part 2: 66
140. Edwards LE (2015) What is the Anthropocene? Eos 96. https://doi.org/10.1029/2015eo040297
141. Waters CN, Zalasiewicz J, Summerhayes C, Fairchild IJ, Rose NL, Loader NJ, Shotyk W, Cearreta A, Head MJ, James PM, Syvitski, Williams M, Wagreich M, Barnosky AD, Zhisheng A, Leinfelder R, Jeandel C, Gałuszka A, Ivar do Sul JA, Gradstein F, Steffen W, McNeill JR, Wing S, Poirier C, Edgeworth M (2017) Global boundary stratotype section and point (GSSP) for the Anthropocene Series: where and how to look for potential candidates. https://doi.org/10.1016/j.earscirev.2017.12.016
142. Waters CN et al (2016) The Anthropocene is functionally and stratigraphically distinct from the Holocene. Science 351(6269). https://doi.org/10.1126/science.aad2622
143. Mason B (2003) Man has been changing climate for 8,000 years. Nature. https://doi.org/10.1038/news031208-7
144. Ruddiman WF (2003) The anthropogenic greenhouse era began thousands of years ago. Clim Change 61(3):261–293. https://doi.org/10.1023/b:clim.0000004577.17928.fa
145. ESRL Web Tea (2015) ESRL global monitoring division–global greenhouse gas reference network. noaa.gov
146. NASA (2016) Global climate change—vital signs of the planet—facts—Carbon Dioxide
147. Lewis SL, Maslin MA (2015) Defining the Anthropocene (PDF). Nature 519(7542):171–180. https://doi.org/10.1038/nature14258
148. Keller G, Mateo P, Punekar J, Khozyem H, Gertsch B, Spangenberg JE, Font E, Bitchong A, Adatte T (2018) Environmental changes across the Cretaceous-Paleogene mass extinction and paleocene-eocene thermal maximum: implications for the anthropocene. Gondwana Res 56:69–89. https://doi.org/10.1016/j.gr.2017.12.002

Governance and Institutional Structure of Climate Change in Egypt

Mohamed A. S. Abdel Monem and Phoebe Lewis

Abstract A cross-cutting issue like climate change, cannot be managed by one government department or single ministry. Assessing of the impact of climate change and effectively integrating adaptation and mitigation measures has to involve almost every sector in government with their administrative system mainstreaming climate change into their strategies and development policies and plans. Successful climate change governance systems should ensure the implementation of climate policy integration at the national level. All Egypt's economic sectors are linked in one way or another with climate change because they emit greenhouse gas (GHG) emissions or are themselves impacted by climate change. Notable industries include energy, transport, urban communities, water, agriculture, health and tourism. Environmental governance of these sectors started with the establishment of The Egyptian Environmental Affairs Agency (EEAA) in 1982 as the highest national authority in the country responsible for promoting and coordinating all efforts related to environmental protection. Achievements have been made at the strategic level with the engagement of relevant stakeholders to develop: (1) the National Environmental Action Plan of Egypt 2002–2017, (2) several rounds of Egypt's National Communication on Climate Change, (3) Intended Nationally Determined Contributions (INDC) to the Paris Agreement, (4) Egypt's National Strategy for Adaptation to Climate Change and Disaster Risk Reduction, (5) the country's National Energy Efficiency Strategy. Each of these developments has contributed to the mainstreaming of climate change in Egypt's institutional structures. However, while a framework for climate change adaptation exists in Egypt, local level policies and actions need to be strengthened to ensure effective implementation of adaptation and mitigation effects. There is also a need to formulate integrated climate change strategies, to enable access to climate finance mechanisms, and to develop and strengthen the policies, institutions, capacities, knowledge and transformative change required to build climate-resilient communities. The existing role of public and local civic institutions needs emphasis,

M. A. S. Abdel Monem (✉) · P. Lewis
Regional Office for the Near East and North Africa, Food and Agriculture Organization of the United Nations, FAO, P.O.Box: 2223, Dokki, Cairo, Egypt
e-mail: mohamed.abdelmonem@fao.org

M. A. S. Abdel Monem
Soil and Water Department, Egyptian Atomic Energy Authority, Cairo, Egypt

© Springer Nature Switzerland AG 2020
E.-S. E. Omran and A. M. Negm (eds.), *Climate Change Impacts on Agriculture and Food Security in Egypt*, Springer Water, https://doi.org/10.1007/978-3-030-41629-4_3

and new incentives are needed to promote the involvement of private organizations and institutions in facilitating climate change adaptation.

Keywords Capacity building · Civil society organizations · Legislation · Private sector

Acronyms

ACMEN	African Ministerial Conference on the Environment
AGN	Africa Group of Negotiators
AREI	Africa Renewable Energy Initiative
CAHOSCC	Committee of the African Heads of State and Government on Climate Change
CPI	Climate Policy Integration
CTRC	Carbon Trading Regulatory Commission
EEAA	Egyptian Environmental Affairs Agency
ETC	Egyptian Electricity Transmission Company
FEI	Federation of the Egyptian Industry
FiT	Feed-in Tariff
GHG	Greenhouse Gas
NGOs	Non-Governmental Organizations
PPA	Power Purchase Agreement
UNFCCC	United Nations Framework Convention on Climate Change

1 Introduction

Despite the growing body of literature dedicated to understanding adaptation, resilience and vulnerability to climate change at multiple scales, as well as the efforts of different development agencies within the donor community, there is still marked lack of support for developing the institutional capacity and the corresponding mechanisms necessary to address climate variation at the local level [1]. A cross-cutting issue like climate change, cannot be managed by one government department or single ministry. Assessing of the impact of climate change and effectively integrating adaptation and mitigation measures has to involve almost every sector in government with their administrative system mainstreaming climate change into their strategies and development policies and plans. Policy makers need to consider the governance of climate change as a complex, multi-actor process that is deeply embedded in local realities.

Mainstreaming climate change into policy decisions requires the consideration of climate change causes and consequences into non-climate sectors [2]. More specifically, climate policy integration (CPI) should incorporate the aims of climate change mitigation and adaptation into all the stages of policymaking in other policy sectors [3]. It should also be complemented by an attempt to aggregate the expected consequences mitigation and adaptation activities into an overall policy evaluation, and minimize contradictions between policies [4]. Integrated strategies are a relatively novel approach to govern highly complex issues that involve several sectors and levels of governance [4]. They are a key tool to foster not only sustainable development, but also environmental policy integration or ecological modernization. Both of which are concepts that promote the integration of environmental concerns into sectors with significant environmental ramifications long before sustainable development dominated the environmental discourse [1]. In fact, mainstreaming climate change helps to position climate-related issues as drivers for development rather than as an added burden for policy makers.

Successful integration of climate change into policies and institutions centers on two key components: horizontal coordination and vertical coordination. These govern the functional (i.e. who should do what) and procedural (i.e. how mainstreaming should happen) implications of integration. Horizontal coordination aims to integrate climate change policy within and across the different departments with a key objective of curbing the silo mentality (or departmentalism) that prevails in all institutions [2]. On the other hand, vertical coordination entails the coherent translation of climate policy design and implementation across the different tiers of government.

This integration applies—albeit in slightly different ways—across the various types of institutions. Effective local institutions are just as key to society's ability to respond to the impacts of climate change as larger ones. The role of the local institutions in shaping climate change adaptation measures and its outcomes is an essential governance mechanism. They are crucial as mediating bodies that connect households to local resources and link local populations to national policies and interventions [5]. This importance is often overlooked. For instance, in Egypt, despite a stated commitment to grassroots involvement in climate change adaptation processes and actions, the actual focus in national adaptation plans is on technical and infrastructure options for adaptation, with little attention given to social and institutional context in which these options operate [6]. Future national-level efforts to develop adaptation plans need to consider the role of local institutions more centrally if they seek to serve the needs and interests of the most vulnerable populations better.

2 Environmental Governance in Egypt

The Egyptian Environmental Affairs Agency (EEAA) was established 1982 as the highest national authority in Egypt responsible for promoting and coordinating all efforts related to environmental protection. The EEAA has the responsibility of implementing national environmental policies and setting up environmental standards for cases of conflicting interests. It has been highly instrumental in developing national capacity at the technical and institutional levels in the climate change related areas. The EEAA signed a protocol with the Federation of the Egyptian Industry (FEI) to promote cooperation on environmental protection in Egypt. The EEAA also ensures that the correct legislation related to the environment are in place as well as introducing amendments in the environmental law when needed. In addition to the EEAA, several other organizations are engaged in enhancing the national capacity to deal with the climate change challenges. These include environmental organizations, energy-related organizations, research centers, universities, governmental organizations, nongovernmental organizations, labor unions and political parties.

At the same time, the Ministry of Environment scaled up its "Climate Change Unit" to strengthen the climate change institutional framework at the national level and in so doing made it a "Central Department" in the Egyptian Environmental Affairs Agency in 2009.

Furthermore, efforts have been made at the sectoral level to strengthen the institutional framework of climate change. Two committees for climate change have been established in the Ministry of Agriculture and Land Reclamation and Ministry of Water Resources and Irrigation. Also, a "climate change information centre" has been initiated for the agriculture sector. Finally, climate change adaptation programme has been mainstreamed in the Agriculture Sustainable Development Strategy up to 2030.

The government's achievements have been made at the strategic level with the engagement of relevant stakeholders to develop:

- The National Environmental Action Plan of Egypt 2002/2017;
- First, second, and third Egypt's National Communication on Climate Change;
- Intended Nationally Determined Contributions (INDC);
- Egypt's National Strategy for Adaptation to Climate Change and Disaster Risk Reduction;
- National Energy Efficiency Strategy;
- National Strategy for Solid Waste Management.

In addition to the EEAA, several institutions played a leading role in integrating climate change policies in the national development plans and strategies agendas in Egypt. These institutions include:

- Energy-related organizations, such as the Organization for Energy Planning;
- The Egyptian Electricity Authority, the New and Renewable Energy Authority;
- Research Centres: such as the Agriculture Research Centre, and the National Research Centre;
- Universities: such as Cairo University, Alexandria University, Assiut University, Ain Shams University, Zagazig University, and other institutes of environmental studies;
- Governmental organizations and national laboratories, such as the General Organization for Industrialization, the Central Laboratory for Agriculture Climate, and the General Organization for Sanitary Drainage;

- Non-Governmental Organizations: It has been estimated that, in Egypt, there are more than 200 NGOs that work in environmental protection in general, some of them focus on climate change issues.

3 Environmental Policies and Legislation in Egypt

All Egypt's economic sectors are linked in one way or another with climate change because they emit greenhouse gas (GHG) emissions or are themselves impacted by climate change. Notable industries include energy, transport, urban communities, water, agriculture, health and tourism. Integration of the environmental policies to include the concerns of different sectors is a major challenge in the country. In response to this challenge, Egypt has paid much attention to environmental legislation. Examples of the most important laws include Law No. 48 of 1982 (better known as the Nile Protection Law) and Law No. 4 of 1994, which was amended in 1999 to include 104 articles regulating the various means of environmental work in different sectors. The latest amendments to the Egyptian constitution, passed in 2011, included the addition of Article 59, which states that environmental protection is a national duty and that the measures necessary to protect the environment shall be governed by law.

- Law 4/1994 is considered the first environmental protection law that describes the mandate of the Ministry of Environment and the Egyptian Environmental Affairs Agency (EEAA). The law concentrates on: Developing and following up on policies, plans and standards for environmental protection;
- Running pilot projects and promoting economic incentives, which respectively demonstrate and facilitate improved environmental performance;
- Conducting studies, monitoring, and supporting other bodies in their environmental monitoring duties;
- Preparing an environmental contingency plan for confronting environmental disasters and implementing a national system of Environmental Impact Assessment;
- Establishing natural protectorates, and managing bio-diversity within and outside these areas;
- Coordinating and following up on international environmental relations and obligations, preparing programs for environmental education.

The EEAA also introduced amendments to be included in a new environmental Law 9/2009 that address main gaps and challenges that were identified in Law 4/1994.

Moreover, the Prime Minister renewed the "National Committee for Climate Change" (established in 1997) by his Decree No. 272 in 2007. It is the Minister of Environment that heads the new Inter-Ministerial National Committee for Climate Change. The members represent a wide range of governmental, experts and non-governmental stakeholders, demonstrating a movement more towards improved integration. Furthermore, the Ministry of Environment scaled up the "Climate Change

Unit" to strengthen its climate change institutional framework on the national level, to be a Central Department in Egyptian Environmental Affairs Agency in 2009. Meanwhile, on the sectoral level, attempts to strengthen the institutional framework led to establishing two committees in Ministry of Agriculture and Land Reclamation and Ministry of Water Resources and Irrigation, in addition to establishing a climate change information centre for the agriculture sector, and conducting an adaptation program in Agriculture Sustainable.

Development Strategy up to 2030. The Committee was established to:

- Coordinate on a national level, regarding the participation of Egypt in the Framework-Convention for Climate Change;
- Develop an overall picture of the Egyptian policies and strategies for dealing with the issue of climate change;
- Review the National Action Plan for Climate Change;
- Follow up on the implementation of the Framework Convention for Climate Change.

Through this, a number of policies and measures have been implemented to decrease the GHG emissions and enhance the generation and use of renewable energy. We will now provide some examples of these. First, according to the decision of the Cabinet of Ministers in Egypt, the Ministry of Transport adopted a strategy that includes improving public transport, energy efficiency, and fuel switching. It also supports the development and use of new propulsion technologies and new methods for freight transport, including shifting from diesel to electrified railways and developing and using fuel cells technology.

Second, in October 2007, the Supreme Council for Energy adopted a strategy for energy supply and used, integrating the main policies and measures that could meet the longer-term challenges facing the national energy industry. The main barriers that currently prevent the industrial sector from achieving full energy conservation and considerable GHG emissions reduction include a lack of information related to GHG emission reduction opportunities in the sector in addition to financial barriers such as the lack of access to investment capital and or high-interest rate on investments.

Third, the Egypt Renewable Energy Law (Decree No. 203/2014) aims to encourage the private sector to produce electricity from renewable energy sources. The law provides several options for support to renewable energy projects and allows a gradual shift away from state-administered projects to privately financed projects. The key project structures envisaged are (Art. 2):

- Projects which are tendered and operated by New and Renewable Energy Authority (established in 1986);
- Projects which are tendered by the (state-owned) Egyptian Electricity Transmission Company (ETC) on a Build-Own-Operate basis, where the private investor enters into a long-term Power Purchase Agreement (PPA) with ETC.

Fourth, a series of projects have been established by private investors who sell the electricity generated by those projects to ETC on the basis of the feed-in tariff (FiT) enacted by Decree 1947/2014. The tariff provides 25 years guarantees for solar projects and 20 years for wind. These FiTs differentiate between project types and sizes. For solar projects, the FiT varies between USD 0.136 and USD 0.1434 for projects of 500 kW–20 MW and 20–50 MW, respectively. For wind projects, the FiT varies between USD 0.0957 and USD 0.1148 during the first five years of operation and then between USD 0.0460 and USD 0.1148 for the remaining 15 [7].

Another development in the private sector is commencement of public–private partnership (PPAs) with mostly large consumers who are being granted access to the grid. For example, the Presidential Decree No. 17/2015 (which amended Investment Law No. 8/1997) focuses on attracting energy investments in Egypt. This includes renewable energy, which is important since Egypt aims to produce 20% of electricity from renewable sources by 2020. The government is doing this by providing tax incentives through the Decree [7]. These include (1) reducing sales tax from 10 to 5%, and (2) setting low customs duties (around 2%) on equipment used for energy production. Additional non-tax incentives offered to energy producers include (1) refunding the expenses paid to extend infrastructure facilities to the project's land (e.g. grid connection), (2) subsidising training programs and social insurance for employees and (3) allocating government-owned land at discounted prices or free of charge [7].

In addition to these incentives, Egypt has made an important step towards achieving sustainable development by developing the "*Sustainable Development Strategy: Egypt's Vision 2030*" that serves as a roadmap for the country to achieve its desired sustainable development goals during the next 13 years. This strategy promotes the optimum use of available resources, enhancement of Egypt's competitiveness and revival of its historic leading role in the region. Moreover, such a strategy aims to fulfil the aspirations of the Egyptian people regarding their right to a decent standard of living. The goals outlined in the strategy are in line with the globally sustainable development goals (SDGs).

Finally, Egypt developed the "Egypt's National Strategy for Adaptation to Climate Change and Disaster Risk Reduction" in 2011. The objective of the strategy is to increase the resilience of Egyptian communities and build their capacity to address the risks and disasters resulting from climate change and its impact on various sectors and activities. These are notable steps, but to achieve the strategy, some important elements need to be present, including the political will to enforce, the provision of human, financial and natural resources, the reform and amendments of institutional frameworks, the amendment of legislation and laws, and lastly the strengthening of the national system of information dissemination.

4 Role of Civil Society Organizations and Private Sector

Egypt's National Strategy for Adaptation to Climate Change and Disaster Risk Reduction that was issued in 2011 emphasises the role of civil society organizations and community participation in developing, managing, monitoring, and evaluating climate change adaptation programmes and projects in close contact with the local communities focusing on the vulnerable groups. The strategy also highlighted, as an example, the significant role that "water users associations" can play in enhancing the sustainable use of water resources through awareness raising programmes and conveying the messages of water-saving techniques to the local communities.

The Third Communication of Egypt [8] highlighted the active role of non-governmental organizations (NGOs) in Egypt in formulating and implementing environmental conservation efforts. The Government of Egypt facilitates the growing responsibilities of the private sector through increasing the scope of incentives aimed at responding to investors' needs for being fully integrated into the Egyptian economy.

Labor unions and political parties are vital in facilitating better environmental management due, in part, to their experience in addressing industrial change, in protecting the workplace and related natural environments, and in promoting socially responsible economic development. The Egyptian Environmental Affairs Agency (EEAA) also encourages the efforts of the private sector to use cleaner technologies through environmentally-friendly industrial zones and processes aiming at increasing the efficiency of resource use, including reuse, recovery and recycling in order to reduce the amount of waste generated from production activities.

The private sector still faced administrative challenges in achieving their role in enhancing the environmental dimension of the development. However, the EEAA signed a protocol with the Federation of the Egyptian Industry (FEI) to promote cooperation for environmental protection in Egypt. Labour unions and political parties are vital in facilitating better environmental management due, in part, to their experience in addressing industrial change, and in promoting socially responsible economic development.

As part of this movement to implement industrial change, Egypt ratified the Kyoto Protocol on 12th January 2005, an act that when then followed by the establishment of the Egyptian Designated National Authority for Clean Development Mechanism "DNA-CDM". Investment costs of USD 1243 million were initially approved for 55 projects in 2009. These projects aimed to reduce GHG emissions by almost 8.3 million tons of CO_2-equivalent. They included a reduction of nitrous oxide emission from the fertilizer industry, renewable energy, fuel switching, methane capture and flaring from waste, and energy efficiency improvement.

5 Gender Mainstreaming in the Climate Change Policies in Egypt

Women are likely to be most negatively affected by climate change. This is due to a number of factors, including the roles women perform, their detachment from resources, and inability to access institutional power. To date, many efforts to mainstream gender have too often been confined to simplistic, ad-hoc and short-term technical interventions that have failed to challenge inequitable power structures and, therefore, also failed to ensure optimal implementation. Gender disparities remain among the deepest and most pervasive of all inequalities and hinder the best of development efforts profoundly [9].

Since climate change affects women and men differently, a gender perspective is essential to the formulation of effective policy development around mitigation and adaptation. The Egyptian government is aware of this angle and see options to address the challenge through, for example, the role of women in producers' organizations and cooperatives. This alignment is in fact necessary given that the Article 40 of the Egyptian Constitution mandates the equal rights of all citizens.

National Strategy for Mainstreaming Gender in Climate Change in Egypt was developed as a result of a partnership between different civil society organizations, key national institutions as well as regional and international agencies. The strategy acknowledges the need to bring the diverse roles, needs and contributions of women and men to the fore in the sustainable development agenda. Rather than merely adding women's participation onto existing strategies and programmes post facto, it aims to mainstream gender goals in order to transform unequal social and institutional structures by recognizing the promotion of gender equality as a central driving principle. The strategy is the result of a series of inputs, including interviews with policymakers, stakeholder consultations, research, and a stakeholder workshop. It was drafted on the basis on an analysis of the current national priorities concerning climate change that prioritized topics of key importance, which were water resources, agriculture and coastal zones [10].

Furthermore, gender was not considered in either the Initial National Communication (INC) or the Second National Communication (SNC) produced by the Government of Egypt for the United Nations Framework Convention on Climate Change (UNFCCC). The EEAA is, however, planning to include gender as a cross-cutting theme in the Third National Communication (TNC). In particular, the TNC recommended:

- The inclusion of gender perspectives into disaster reduction efforts at the local, regional and national levels, including in policies, strategies, action plans, and programmes. As well as, increase their participation and representation at all levels of the decision-making process;
- The analysis of climate change data such as drought, floods, or desertification from a gender-sensitive perspective, including the traditional knowledge and perspectives of women in the analysis and evaluation of the characteristics of key disaster risks;
- The assurance that women can be visible agents of change at all levels of disaster preparedness, including early warning systems, education, communication, information and advocacy;
- The building of the capacity of national and local women's groups and provide them with a platform to be heard;
- The inclusion of gender-specific indicators to monitor and track progress on gender equality targets.

6 Engagement in Regional and Global Climate Change Governance

At the regional level, Egypt has played a vital role in enhancing the climate change policies and strategies in Africa through its role as president of the African Ministerial Conference on the Environment (AMCEN) and Coordinator of the Committee of the African Heads of State and Government on Climate Change (CAHOSCC) for the period 2015 to 2017. The AMCEN presidency works very closely with the Africa Group of Negotiators (AGN), the African ministers, and the AMCEN bureau members to emphasize the African priorities during the global negotiations for the COP21 meetings in Paris. Efforts have been made by Egypt as president of AMCEN to maintain African unity and defend African interests.

Furthermore, AMCEN presidency spearheaded the development and promotion of the Africa Renewable Energy Initiative (AREI) that aims to scale of renewable energy deployment on the African continent as well as the promotion of the Africa Adaptation Initiative (AAI) that is considered a major initiative in mobilizing the needed support to significantly scale up adaptation actions in Africa. At the global level, Egypt ratified the UNFCCC in 1994 and the Kyoto Protocol in 2005. It submitted its First UNFCCC National Communication in July 1999 and the Second UNFCCC National Communication in June 2010 as well as signing the Paris Agreement in April 2016 [11].

7 Challenges Facing Climate Change Governance in Egypt

The most significant constraint to effective environmental policy making and implementation in Egypt is the lack of reliable and timely information indicating how various sectors of society impact the environment and whether development is becoming more sustainable or not. Various constraints related to the processes of environmental information collection, production and dissemination are evident in Egypt. These include uncoordinated institutional set-ups for monitoring activities, the absence of a common information system for monitoring organizations to feed data and findings into, the absence of comprehensive systematic methodologies for monitoring, the absence of valuation, and/or the undervaluation, of many natural resources, and the lack of financial resources for maintaining monitoring processes.

The coordination required to reach comprehensive and integrated environmental activities is significant because of the cross-sectoral nature of environmental issues. Although EEAA has the primary responsibility for coordination, numerous other entities are partners in environmental policies, playing roles in implementation and/or monitoring. In this respect, the Agency has to empower the already established inter-ministerial committees on each of the major relevant crosscutting environmental issues, such as water, energy, and climate change in order to coordinate between multiple competent authorities for the different specific environmental processes of concern. Inadequate capacity (technical, financial and institutional) remains one of the significant challenges affecting climate change adaptation in Egypt. Strengthening the capacity of major sectors aids government, local, and research sectors to fulfil its mandate to address climate change is very essential.

Third Egypt's National Communication on Climate Change [8] identified some critical policy and institutional issues that should be addressed to effectively formulate and implement related interventions for mitigating the impacts of climate change and variability. Policy and institutional reforms should strive to:

- Coordinate and harmonize climate change policies across several sectors.
- Embrace participatory management that gives more responsibility to users.
- Equitably share water resources and improve trans-boundary water management.
- Improve efficiency and sustainability of climate change interventions by building the capacity of various stakeholders.
- Enhance adaptively and applied research and information sharing to enhance efficient uses of resources and climate change adaptation.
- Promote public–private partnerships and increase investment in climate change and adaptation to climate change.

8 Recommendations for Institutional Strengthening

Egypt's Third National Communication on Climate Change [8] suggested clearly defined recommendations for both short and long-term institutional reforms to climate change governance in Egypt. The short-term recommendations include:

- The establishment of an Inter-Ministerial National Committee for Climate Change is to be chaired by H.E. the Prime Minister. Ministry of Environment should hold the technical secretariat and coordinate all other technical contents for the NCCC;
- The development of clear overall objectives and plans for climate finance in Egypt by the NCCC/EEAA, including the appointment of responsible managing agencies;
- The government should entrust the EEAA/Central Department of Climate Change with greater powers and functions and also improve the existing coordination mechanism among other governmental funding management agencies;
- The provision of clear arrangements by the Ministry of Finance on the funding of climate change programmes, including the establishment of designated budget items and sub-items for climate-related programs to be managed by the Ministry of Environment/EEAA;
- The establishment of an analysis and reporting mechanism by the NCCC and the Ministry of Finance for direct climate finance to cover the disbursement use and effectiveness of funding.
- The incorporation of carbon reduction and adaptation into climate change elements in the national development strategy and management system for international

cooperation and use of these elements as a key criterion for project selection in national and international investments;

- The elevation of climate change collaboration to the level of international relations strategy to facilitate climate funding into Egypt;
- The awareness raising by government among businesses and local authorities about the risks and vulnerabilities created by climate change to minimize impacts and help limit the cost of damages.

On the other hand, long-term recommendations for institutional reforms include:

- The endorsement legislation detailing the responsibilities and status of government entities involved in organizing and managing climate finance at various ministries;
- The establishment of an international development administration with responsibility for climate-related international cooperation (including finance) also under the direct supervision of the Ministry of International Cooperation;
- The establishment by the Ministry of Finance of an assessment and monitoring systems of climate finance and develop guidelines for assessing the performance of new climate funds;
- The establishment by the Ministry of Finance/Ministry of Environment of a Carbon Trading Regulatory Commission (CTRC);
- The provision to the EEAA of the full responsibility for the monitoring and management of a national carbon market, including, but not limited to allocation of emissions, development of appropriate standards, regulations and laws, establishment of appropriate bodies and mechanisms for the effective functioning of the market, oversight of development of derivative markets, and provision of guidance to concerned institutions;
- The development of new and improved climate finance mechanisms to leverage private sector capital in support of the public climate.

9 Conclusions and Recommendations

While a framework for climate change adaptation exists in Egypt, local level policies and actions need to be strengthened to ensure effective implementation of adaptation and mitigation effects. There is also a need to formulate integrated climate change strategies, to enable access to climate finance mechanisms, and to develop and strengthen the policies, institutions, capacities, knowledge and transformative change required to build climate-resilient communities. The existing role of public and local civic institutions needs emphasis, and new incentives are needed to promote the involvement of private organizations and institutions in facilitating climate change adaptation. Egypt's National Strategy for Adaptation to Climate Change and Disaster Risk Reduction [12] stressed the importance of reforming and modifying the institutional frameworks for climate change adaptation in Egypt. Such reformation would not necessarily entail establishing new institutions, but rather ensuring best practices in existing ones. Similarly important is the amendment of the prevailing legislation to be consistent with adaptation requirements. Egypt's Third National Communication on Climate Change [8] provided a clear analysis of not only the challenges facing the institutional set-up of the climate change management, but also opportunities for institutional strengthening. These can be harnessed to ensure the overall enforcement of effective institutional capacity building.

References

1. Osbahr H, Twyman C, Adger WN, Thomas DSG (2008) Effective livelihood adaptation to climate change disturbance: scale dimensions of practice in Mozambique. Geoforum 39:1951–1964
2. Mickwitz P, Kivimaa P (2007) Evaluating policy integration—the case of policies for environmentally friendlier technological innovations. Evaluation 13(1):68–86
3. Gamero Rus JM (2011) Climate policy integration: government programmes and strategies in Spain. https://www.um.edu.mt/__data/assets/pdf_file/0006/137778/gameroruspp t.pdf
4. Persson A (2004) Environmental policy integration: an introduction. Policy integration for sustainability background paper. Stockholm Environment Institute, Stockholm
5. Rayner J, Howlett M (2009) Introduction: understanding integrated policy strategies and their evolution. Policy Soc 28(2):99–109
6. Agrawal A (2008) The role of local institutions in adaptation to climate change, May 2008. https://doi.org/10.1596/28274
7. Grantham Institute (2017) Egypt—Grantham Research Institute on Climate Change and the Environment. London School of Economics and Political Science. www.lse.ac.uk/GranthamInstitute/country-profiles/egypt/
8. Egyptian Ministry of Environment (2016) Third National Communication of Egypt under the United Nations framework convention on climate change. https://unfccc.int/files/national_reports/non-annex_i_parties/biennial_update_reports/application/pdf/tnc_report.pdf
9. Lewis P, Abdel Monem MA, Impiglia A (2018) Impacts of climate change on farming systems and livelihoods in the Near East and North Africa—with a special focus on small-scale family farming. FAO, Cairo, p 92
10. IUCN (2011) The national strategy for mainstreaming gender in climate change in Egypt. Cairo, Egypt. https://genderandenvironment.org/resource/egypt-national-strategy-for-mainstreaming-gender-in-climate-change-in-egypt/
11. Egyptian Ministry of Environment (2010) Second National Communication of Egypt under the United Nations framework convention on climate change. https://unfccc.int/resource/docs/natc/egync2.pdf
12. Egypt National Strategy for Adaptation to Climate Change and Disaster Risk Reduction (2011) Cairo, Egypt. https://www.climasouth.eu/docs/Adaptation011%20StrategyEgypt.pdf

Role of Science, Technology and Innovation in Addressing Climate Change Challenges in Egypt

Mohamed A. S. Abdel Monem and Ismail A. El Ghandour

Abstract Science has been approaching atmosphere, change in global temperature, oceans and seas for a long time; however as the climate change becoming a serious man made environmental challenge with its serious impact on livelihood, the role of science and new innovations and technologies became even more important than ever. The Intergovernmental Panel on Climate Change (IPCC) is a scientific body for the assessment of climate change, established to provide the world with a clear scientific view on the current state of climate change and its potential environmental and socio-economic consequences. The IPCC reviews and assesses the most recent scientific, technical and socio-economic information produced worldwide relevant to the understanding of climate change. Thousands of scientists from all over the world contribute to the work of the IPCC on a voluntary basis. It has been well established scientifically that The Near East and North Africa (NENA) region is one of the most vulnerable regions to climatic change. The '*Regional Initiative for the Assessment of the Impact of Climate Change on Water Resources and Socio-Economic Vulnerability in the Arab Region*' RICCAR was initiated for a close scientific observation on the impact of climate change on the Region based on modeling and expected scenarios.To cope with the impact of climate change on agricultural sectors, new technologies and innovations have been tested and verified by agricultural research initiatives that contributes to the climate change adaptation efforts and enable the farmers and policy-makers to draw their strategies and actions to face climate change impacts. These efforts include integrating heat, drought, and salinity tolerance traits through genetic improvement into the crop and horticultural varieties. Similar breeding techniques have been used in the livestock sector. Crop simulation models can predict key crop characteristics over a wide range of climatic

M. A. S. Abdel Monem · I. A. El Ghandour (✉)
Food and Agriculture Organization of the United Nations (FAO), P.O. Box: 2223, Dokki, Cairo, Egypt
e-mail: ismailelghandour@yahoo.co.uk

M. A. S. Abdel Monem
e-mail: mohamed.abdelmonem@fao.org

M. A. S. Abdel Monem
Soil & Water Department, Nuclear Research Center, Egyptian Atomic Energy Authority, P.O. Box 13759, Cairo, Egypt

© Springer Nature Switzerland AG 2020
E.-S. E. Omran and A. M. Negm (eds.), *Climate Change Impacts on Agriculture and Food Security in Egypt*, Springer Water, https://doi.org/10.1007/978-3-030-41629-4_4

conditions, such as the timing of flowering and physiological maturity, through correct descriptions of phonological responses to temperature and day length. Egypt vision 2030 stated that "Knowledge, innovation, and scientific research contribute in achieving the goals of sustainable development in general, whether national or international, where the increase in the contribution of a large number of economic sectors in economic activity can be achieved by connecting scientific research outputs and innovation by the needs of these sectors". The Egyptian research system is making great effort to overcome challenges related to innovation, and scientific research and already making big progress with regards to assessment of the climate change impact on Egyptian agriculture as well as suggesting scientific climate change adaptation measures.

Keywords Climate change challenges · Science · Technology and innovation

Abbreviation

ARC	The Agricultural Research Center
EEAA	Egyptian Environmental Affairs Agency
ET	Evapotranspiration
FAO	Food and Agriculture Organization of the United Nation
GCMs	Global Circulation Models
GCM	A General Circulation Model
GIS	A Geographic Information System
IAEA	International Atomic Energy Agency
ICT	Information and Communication Technology
IPCC	The Intergovernmental Panel on Climate Change
MALR	Ministry of Agriculture and Land Reclamation
NARSS	The National Authority for Remote Sensing and Space Sciences
NENA	The Near East and North Africa region
NOAA	National Oceanic and Atmospheric Administration
QTL	Quantitative Trait Loci
RICCAR	The Regional Initiative for the Assessment of the Impact of Climate Change on Water Resources and Socio-Economic Vulnerability in the Arab Region
SDS	The Sustainable Development Strategy
SLR	Sea Level Rise
SMNP	The soil moisture neutron probe
SPOT	Satellite Pour l'Observation de la Terre
THI	Temperature Humidity Index
UNEP	The United Nations Environment Programme
VA	Vulnerability Assessment
WMO	World Meteorological Organization
WUF	Water Use Efficiency

1 Introduction

Although Climate change add more stress on the poor mainly in the developing countries and presents challenges, it also presents opportunities for development. Linking climate change and development became important to ensure increasing resilience of the poorest and most vulnerable communities, Providing them with the needed tools and technologies should be the main task of the international communities to approach impact of climate change [1].

Climate change phenomena are based originally on scientific facts where research tools are being used to confirm the data and information related to the cause of climate change as well as its impact. The strong evidence that significant global warming is occurring is based on measurement of air temperature and subsurface ocean temperature in addition to the increase in the sea level at the global scale, retreating glaciers as well as changes in many ecosystems retreating most of the warming in recent decades can be attributed to human activities [2]. Extensive scientific evidence made it clear that the main cause of the change in the climate of the past half century is human-induced increases in the amount of atmospheric greenhouse gases. This scientific opinion is expressed in technical reports and scientific articles by scientific national and international entities and by conducting surveys of opinion among climate scientists. According to Cook et al. [3] individual scientists, universities, and laboratories contributed to the overall scientific opinion via their peer-reviewed publications. Cook et al. [3], analyzed the evolution of the scientific consensus on the global warming in the peer-reviewed literature, examining 11,944 climate abstracts from 1991 to 2011, reaching the conclusion that 97.1% endorsed the consensus position that humans are causing global warming.

Complex, integrated mathematical models are being used for projecting future climate change. The model is essentially a representation of the many interactions within the climate which includes the atmosphere, oceans, land surface, and ice. Climate models are typically quantitative in nature and range from simple depictions of the Earth's climate to very complex ones. Climate models are based on well-established physical principles and have been demonstrated to reproduce observed features of recent climate and past climate changes. Results from a wide range of climate model simulations suggest that our planet's average temperature could be between 2 and 9.7 °F (1.1–5.4 °C) warmer in 2100 than it is today.

Global climate models (GCM) can simulate many important aspects of Earth's climate: large-scale patterns of temperature and precipitation, general characteristics of storm tracks and extratropical cyclones, and observed changes in global mean temperature and ocean heat content as a result of human emissions. Models can simulate the impacts at the local and global scale, e.g. crop yields, hydrological, and water quality models [4].

2 Role of Science in the Assessment of Climate Change Scenarios

2.1 At the Global Level

The science behind the fact that the atmosphere and oceans have warmed, sea levels have risen, and glaciers and ice sheets have decreased in size is supported by extensive studies based on main lines of evidence: (1) Physical principles established long time ago that carbon dioxide (CO_2), restrict the radiant flow of heat from Earth to space. This mechanism, known as the 'greenhouse effect', keeps Earth's surface and lower atmosphere considerably warmer than they would otherwise be. An increase in greenhouse gas concentrations raises the temperature of the surface. (2) Measurements from the recent past (the last 150 years) tell us that Earth's surface has warmed as atmospheric concentrations of greenhouse gases increased through human activities, and that this warming has led to other environmental changes. Although the climate varies from decade to decade, the overall upward trend of average global surface temperature over the last century is clear. (3) Climate models illustrate in scientific presentation better understanding of the causes of past climate changes, and to project the future climate change as well. Physical principles and knowledge of past models provide convincing evidences about the role of the increased in greenhouse gas concentration in the atmosphere in causing the recent climate change. It became obvious that, unless greenhouse gas emissions are reduced greatly, and greenhouse gas concentrations are stabilized, greenhouse warming will continue to increase. (www.science.org.au/.../climate-change-r.pdf, Feb. 2015).

The Intergovernmental Panel of Climate Change (IPCC) is a scientific body for the assessment of climate change, established by the United Nations Environment Programme (UNEP) and the World Meteorological Organization (WMO) to provide the world with a clear scientific view on the current state of climate change and its potential environmental and socio-economic consequences. The role of IPCC is to review and assess the most recent scientific, technical and socio-economic data and information relevant to the understanding of climate change that have been produced worldwide IPCC does not conduct any research nor does it monitor climate related data or parameters (http://www.ipcc.ch/organization/organization.shtml, 2018). Thousands of scientists from all over the world contribute to the work of the IPCC on a voluntary basis. The review is an essential part of the IPCC process, to ensure an objective and complete assessment of current information. Different viewpoints existing within the scientific community are reflected in the IPCC reports. Because of its scientific and intergovernmental nature, the IPCC embodies a unique opportunity to provide rigorous and balanced scientific information to decision makers.

IPCC report [2] indicated that human activities since 1750 until now are responsible "Golf the global atmospheric concentrations of carbon dioxide, methane and nitrous oxide and this increase far exceed pre-industrial values determined from ice cores spanning many thousands of years. The increases in the global carbon dioxide concentration are due primarily to fossil fuel use and land use change, while those of methane and nitrous oxide are primarily due to activities of the agricultural sector. The IPCC has very high confidence (representing at least a 9 out of 10 chance of being correct) in this conclusion."

"According to IPCC [2] warming of the global climate system became very obvious, as is now evident from observations of increases in global average air and ocean temperatures, melting of snow and ice at higher rate, and rising global average sea level. "Numerous long-term changes in climate have been observed at continental, regional and ocean basin scales. These include changes in arctic temperatures and ice, widespread changes in precipitation amounts, ocean salinity, wind patterns and aspects of extreme weather including droughts, heavy precipitation, heat waves and the intensity of tropical cyclones" [2]. The observed increase in global temperatures since the mid of the twentieth century is very likely can be attributed to the observed increase in anthropogenic greenhouse gas concentrations. Impact of human activities now extend to other aspects of climate, including ocean warming, continental-average temperatures, temperature extremes and wind patterns.

For the next two decades, even if the concentrations of all greenhouse gases and aerosols had been kept constant at the level of the year 2000, "a further warming of about 0.1 °C per decade would be expected. Continued greenhouse gas emissions at or above current rates would cause further warming causing further changes in the global climate system during the 21st century that would very likely be larger than those observed during the 20th century" [2].

The IPCC at (2018) "has identified far ranging and drastic consequences of climate change, to both humans and the environment including:

- Increasing species extinctions
- Reduction of coral reefs, mangrove forests, and tropical rainforests
- Threats to small island states in the Pacific as sea levels rise
- Increasing drought threats in Africa
- More severe flooding in densely populated river deltas in Asia
- More severe weather in hurricane prone zones". (http://www.ipcc.ch/organization/organization.shtml, 2018).

An early warning system for disasters and extreme climatic events requires a broad of different knowledge's base, depending on the substantial existing accurate scientific research in the geophysical, environmental and social science fields. Most of the data required for disaster management, has special components (usage of GPS data models, risk maps and scenarios), and also changes overtime. Therefore, the use of remote sensing and GIS has become essential in disaster management. Moreover, a range of tools (core prediction system models and tools, warning decision system tools) exist to help in providing warnings about events that could threaten the livelihood of the local communities. Remote sensing application and use of GIS has been successful to predict the occurrence of disastrous and to warn people on time.

Information and communications technology (ICT) is envisaged to be an important tool in the communication of climate change mitigation technologies, which are necessarily of low carbon footprint in order to reduce GHG emissions [5].

2.2 At the Regional Level

According to the latest IPCC assessment, the Near East and North Africa (NENA) region is one of the most vulnerable regions to climatic change. The Region is predicted to become even hotter and dryer than what it already is. A temperature rising up to 4 °C is expected by century's end. Rainfall may decline up to 40% by 2050 (and up to 60% in a 4 °C scenario), with an increase of severe drought and floods frequency and intensity. Sea level rise of about 0.1–0.3 m is also expected by the year 2050, augmenting seawater intrusion to coastal groundwater resources. It is anticipated that such a climate change scenario would reduce crop yield up to 30%. The livelihoods of a large segment of the region's population, particularly those depending on small scale rainfed agriculture, will be at great risk. The environment will be impacted as well due to the anticipated increase of desertification, land degradation and biodiversity losses. Ultimately, people migration will be accelerated.

The demand for a close observation on the impact of climate change on the Region has materialized through the establishment of the '*Regional Initiative for the Assessment of the Impact of Climate Change on Water Resources and Socio-Economic Vulnerability in the Arab Region*' [6].

Within the RICCAR project, future climate change data are generated through downscaling projections from three global circulation models, EC-EARTH, CNRM -CM5, and GFDL-ESM, exercised on the region using the regional climate model RCA4 to provide higher-resolution information (50 km). The two considered climate scenarios are based on selected Representative Concentration Pathways (RCP), specifically RCP 4.5 (moderate-case scenario) and RCP 8.5 (worst-case scenario). The projected climatic and hydrological variable considered (max and min temperature, and precipitation, runoff and evapotranspiration) are presented as the change between the twenty-year periods of 2046–2065 (mid-century) and 2081–2100 (end-century) with reference to the baseline period of 1986–2005. Here we highlight some relevant findings for temperature, precipitation and evapotranspiration.

Change in maximum temperatures: The change in maximum temperature to the middle of the century shows a varied increase between 0.3 and 2.4 °C in RCP 4.5 and from 1.1 to 3.4 °C under RCP 8.5. The higher increase is shown in the non-coastal areas, with the greatest increases projected in the Maghreb region, the south of Nile Valley, and the central and western parts of the Arab peninsula. Towards the end of the century, the change shows an increase between 1 and 3 °C in RCP 4.5 and from 2 to 5 °C with RCP 8.5. The areas showing higher increases are almost the same as in the scenarios of midcentury.

Change in minimum temperatures: The general change in annual average minimum temperatures to the middle of the century also shows a varied increase under RCP 4.5 that goes from less than 1 °C at the Mediterranean coasts to more than 2 °C in the Sahara desert, the Atlantic mountains and Euphrates-Tigris basin. Under RCP 8.5 the rise exceeds 1 °C almost everywhere in the Arab region and reaches 3–3.5 °C in some areas of the Sahara desert. The change of minimum temperatures is clearly more evident towards the end of the century increasing between 1 and 3 °C in RCP 4.5 and from 2 to 5 °C and more with RCP 8.5. The areas showing a higher increase are in the Sahara area in North Africa and East Africa, including Morocco and Mauritania. The increasing change in temperature becomes much more evident throughout the region with scenario RCP 8.5, when the increased temperature by more than 3 °C would cover most of the Arab region.

Change of precipitation: Decreasing trends can be seen in most of the Arab region towards the midcentury, with a reduction of about 90 mm in average annual precipitation for the Atlas Mountains when worst-case scenario is considered (RCP 8.5). For the end of the century, both scenarios show a reduction of the average annual precipitation reaching 90–120 mm/year in the coastal areas and mainly around the Atlas Mountains in the west and upper Euphrates and Tigris river basins in the East.

Changes in evapotranspiration (*ET*): The projected annual change in precipitations is compliant with projected changes in evapotranspiration. The most affected areas are spread out in the coastal zones of the Northern Africa (especially in the West: Morocco, Algeria and Tunisia), but also in Sahel, in the north of Horn of Africa as well as in the coasts of Arabian Peninsula (mostly in the south-west).In the most affected zones, ET would increase by 20–40 mm/year under the moderate-case (RCP 4.5), but it reaches 40–60 mm/year under the worse-case scenario (RCP 8.5) at mid-century. At end-century, ET would climb even higher and it would reach 100 mm/year in some areas (e.g. Morocco).

All projections were used to generate Vulnerability Assessment (VA) maps that had been used to identify Hotspots in the various sub-sector of agriculture. The climate projections, the vulnerability assessment and the hotspots where all used to evaluate the impact of climate change.

3 Innovations and Technologies for Climate Change Adaptation Measures

In addition to the traditional and indigenous knowledge, harnessing new technologies and innovations that are tested and verified by agricultural research initiatives will contribute to the climate change adaptation efforts and enable the farmers and policy makers as well to face the challenges of climate change.

3.1 Management of Water Resources

Water is one of the most affected natural resources by climate change. Options to improve water management in countries under water scarcity include; efforts to increase irrigation efficiency; enhancing saving water techniques and water harvesting—among others, to conserve the already limited water resources under different climate change scenarios. Often benefiting from the traditional knowledge of rural communities, water harvesting can also be enhanced through the application of satellite remote sensing.

The soil moisture neutron probe (SMNP) is portable equipment for measuring periodically soil water content at different depths. Data generated from this monitoring are used to calculate the soil water balance and estimate the total amount of soil water removed by both soil evaporation and plant transpiration [7–10]. WUE by crops as measured by the SMNP can be increased by up to 50% by changing irrigation technologies [11, 12] and/or management practices [13, 14], to improve groundcover and thus reduce evaporation from the soil surface. For example, approximately 25–50% of irrigation water can be saved by using drip irrigation over the traditional flood surface irrigation. Such savings also brought about other benefits including an increase in the efficiency of fertilizer applied to crops to the same extent (20–50%) and a reduction in nitrogen leaching losses beyond the plant rooting zone.

The measurement of natural variations in the abundance of stable isotopes of oxygen, hydrogen, carbon and nitrogen in soil, water and plant components can help to identify the sources of water and nutrients used by plants and to quantify water and nutrient fluxes through and beyond the plant rooting zone. It also an important tool in understanding different irrigation and land management practices and their impact on plant uptake of water and nutrients. The soil surface since the light isotopes (hydrogen-1 and oxygen-16) evaporate more readily than the heavy isotopes (hydrogen-2 {^{2}H} and oxygen-18 {^{18}O}). The natural isotopic ratios of hydrogen (^{2}H/^{1}H) and oxygen (^{18}O/^{16}O), which are often expressed as delta units (G^{2}H and G^{18}O) in soil water, water vapour within a plant canopy and plant leaves can provide estimates of soil evaporation and plant transpiration [15, 16]. Such information will enable irrigation and land management practices to be developed to minimize soil evaporation (the non-productive loss of water) and channel this water for crop production.

Use of nuclear and isotopic techniques has proved to be invaluable tools for improving WUE. Stable isotopes of water, ^{2}H and ^{18}O, at the natural abundance level; have provided useful tools for tracking and quantifying water flows within and beyond the plant rooting zone. These techniques show potential to partition evapotranspiration into soil evaporation and plant transpiration (the water component removed by plants for their growth). Data and information obtained from using the nuclear and isotopic techniques can be used: (i) To evaluate the efficacy of different irrigation and land management practices that minimize soil evaporation and optimize plant transpiration, (ii) To locate sources of water use by different plants so as to develop an integrate tree-crop system for sustainable food production particularly in drylands environments and (iii) To identify and develop management strategies that minimize the losses of water and associated fertilizers, pesticides, soils and animal manure from farmlands. The need to minimize such environmental impact from agricultural activities is increasingly important to enhance viable and sustainable agricultural systems. The carbon isotope discrimination has potential as a tool for screening and evaluating large samples of cultivars with increased WUE under water-limited conditions https://mirfali.wordpress.com/2013/08/09/page/58/.

The need to minimize such environmental impact from agricultural activities is increasingly important to enhance viable and sustainable agricultural systems. The carbon isotope discrimination has potential as a tool for screening and evaluating large samples of cultivars with increased WUE under water-limited conditions.

3.2 Breeding for Drought Tolerance

One promising approach to face the projected impact of climate change on crops is to integrate useful drought tolerance traits through genetic improvement into the cultivated varieties. Recently, some techniques and plant criteria have been widely used in screening for drought tolerance in plant breeding programs as the Drought susceptibility index that based on the field performance and is used to compare genotypes in terms of their ability to minimize the reduction in yield under drought-stressful compared with non-stressful environments. Chemical desiccation has been suggested as an indirect simple method to simulate post-anthesis drought for small-grain cereals [17]. Remote-sensing infrared thermometry (IR) was also proposed to meet the criteria of a rapid, instantaneous, nondestructive, and monitoring whole-plant response to drought in the field [18, 19]. The technique is an indirect method for assessing stomatal conductance of canopy leaves where canopies tend to warm up with stomatal closure.

To select and evaluate plant cultivars that can withstand drought, measurement of natural variations in the abundance of both light and heavy carbon stable isotopes (^{12}C and ^{13}C) in plant materials is increasingly being used by researches around the world. This technique obviates the need for measurements of the water budgets of a large number of plants during a large scale screening for WUE characteristics. Under drought, less carbon (in the form of carbon dioxide), particularly ^{13}C from the atmosphere is taken up by plants for growth because of plant stress, thus creating a major variation in the natural isotopic ratios of ^{13}C and ^{12}C in plant materials. A plant cultivar, which is resistant to water scarcity should display less depletion in ^{13}C compared with a susceptible cultivar. Such discrimination against ^{13}C (i.e., the difference between ^{13}C and ^{12}C, expressed as delta $G^{13}C$) in plant tissues (leaves and grains) has been successfully used in the selection of drought-resistant barley, wheat, rice and peanut (20) and (21) [20, 21]. Scientists have concluded that $G^{13}C$ in plant leaves and grain is negatively related to WUE. Carbon isotope discrimination (often abbreviated as (CID) measured in different plant parts at harvest can be used as a historical account on how water availability varied during the cropping season [22].

3.3 Breeding for Heat Tolerance

It is well established that increasing tolerance to temperature by conventional breeding is an obvious approach to reduce the adverse produced eventually. Thus, the selection of breeding lines for relatively hot regions is conducted under hot conditions [23]. Conventional breeding has also been used to intentionally develop new heat-tolerant crop genotypes. For example, a variety of broccoli has an improved head quality thanks to early maturation, because this trait prevents hot days later in the season to affect the heat-sensitive flower initiation developmental stage [24].

In addition, new varieties of cowpea showed higher average grain yield when grown under hot and long days during reproduction Ehlers and Hall [25]. And recurrent selection has also been successful for improving wheat yield using ancestor *T. tauschii* as a gene donor, leading to increased rates of grain filling and larger grains in BC1F6 plants [26]. Jha et al. [27] developed a list of quantitative trait loci (QTLs) associated with heat tolerance in various plants, including Arabidopsis, azuki bean, barley, brassica, cowpea, maize, potato, rice, sorghum, tomato and wheat. The authors showed several types of genetic markers linked to different traits of interest which spanned the various aspects of a plant's vulnerability to heat. This included QTLs for yield traits, such as fruit set or grain filling rate, under heat. Also, QTLs for several heat tolerance- related traits have been discovered, such as for lower canopy temperature during vegetative and reproductive stages and higher chlorophyll fluorescence in wheat [28–30]. Pinto and Reynolds [31] stated that, High chlorophyll fluorescence represents heat-tolerant photosynthesis, and lower canopy temperature reflects efficacious water uptake which has been associated with deep rooting. A major quantitative trait loci (QTL) for high temperature seed germination capacity in lettuce, Htg6.1, collocates with a temperature-sensitive gene encoding an abscisic acid biosynthesis enzyme (LsNCED4) [32, 33]. In potato, nine quantitative trait loci (QTLs) for internal heat necrosis in tubers were detected that each explain between 4.5 and 29.4% of the phenotypic variation [34]. Many studies have focused on the effect of high temperature on reproductive characteristics. This including pollen germinability, pollen tube growth, grain weight, days to heading, grain filling and post-anthesis leaf senescence, fruit set and quality traits such as white-back kernels in rice. In maize, five and six QTLs for pollen quality and tube growth have been identified with a high heritability of 0.64 and 0.68, respectively. In response to high-temperature stress, plants modulate the expression of a plethora of genes. These genes and their annotation could help to identify the processes that are induced or suppress such as those involved in adaptation and protection to heat stress [35].

3.4 Crop Simulation Models

Crop simulation models can predict several key crop characteristics over a wide range of climatic conditions, such as the timing of flowering and physiological maturity, through correct descriptions of phonological responses to temperature and day length. Furthermore, accumulation of yield needs to be predicted by accurately predicting the development and loss of leaf area and, therefore, a crop's ability to intercept radiation, accumulate biomass, and partition it to harvestable parts such as grain. Crop water use is also needs to be accurately predicted by correctly predicting evapotranspiration and the extraction of soil water by plant root [36]. The authors used the CropSyst model to study the effects of cropping management systems on crop productivity and the environment under different climate change scenarios. The CropSyst model allows using set of daily weather data spanning on a reasonable number of years to assess the impact of climate change on crops [36, 37] It is a multi-year, multi-crop, daily time step crop growth simulation model, developed with an emphasis on a user-friendly interface, and with a link to GIS software and a weather generator.

The currently developed FAO crop water productivity model AquaCrop. Steduto et al. [38], indicated that, the model has been developed to predict yield response to water for most field and vegetable crops under different range of irrigation and land management, practices that requires a range of data on transpiration and evaporation. These two components of evapotranspiration can be separated by using stable isotopes as outlined above.

4 Management of Sustainable Livestock Production Under Expected Harsh Environment

Giuseppe et al. [39], provided four elements to ensure low impact of environment on ruminant production: (a) feeding techniques that increase feed efficiency and can be adapted to different scenarios, (b) traditional and innovative methods of animal breeding, including new technologies of gene editing that enable the rational use of genetic resources, and (c) maintenance and enhancement of animal biodiversity.

Adaptation of ruminants to climate change: Temperature humidity index (THI), an index which combines temperature and humidity into a single value and is widely considered as a useful tool to predict the effects of environmental warming on farm animals [40]. Adaptation in the ruminant sector includes actions that may help to alleviate the direct or indirect negative effects of climate on animal health, welfare and productivity. The adaptation measures that can minimize the direct effects of climate change on animals are classified as structural interventions and management practices. Under intensive and semi-extensive production systems, the main structural interventions of adaptive significance include building orientation, insulation and reflectance, shading and ventilation, with or without the use of water [41]. Provision of shade may represent an effective adaptation measure under grazing or pastoral production systems as well. Management practices that may alleviate the negative impacts of climate change on ruminants include hormonal treatments to improve fertility [42], and nutritional and genetic approaches. Adoption of systems that increase water availability may be useful to limit the negative indirect effects of climate on ruminant production that may arise from reduced grassland or crop production, further more selecting plants with low water needs or high resistance to water shortages, and high resistance to emerging pathogens. Also, increasing surveillance and developing preventive tools (i.e. vaccines) for the control of emerging animal diseases would also be crucial to coping with climate change and its impact on farmed ruminants. Finally, adaptation options include developing and implementing meteorological warning systems and insurance systems, which are needed to reduce the impact of severe weather events and prevent loss of livestock [43].

In the field of Sustainable ruminant feeding there have been many scientific and technological advances in such as precision livestock feeding, identification and characterization of new feed resources, the reutilization of agricultural industry residues and biotechnologies for animal feed and animal feeding. Also a rediscovery and re-evaluation of agroforestry and silvopastoral systems, regarding food quality aspects and the maintenance of ecosystem services (e.g. carbon sequestration, land landscape maintenance, and biodiversity enhancement). The goals of precision feeding are to increase production efficiency and profitability, reduce environmental impact, improve product quality and safety, and improve animal health and well-being [44]. Regarding the animal breeding: In addition to meeting the growing demand for animal products, animal breeders will have to address climate change challenges to ensure the socioeconomic sustainability acceptance of ruminant farming. In particular, selection will have to (1) Improve animal health and welfare, (2) Increase tolerance to stressors, (3) Reduce livestock environmental impact and (4) Produce milk and meat having improved nutritional and functional properties.

Two strategies are possible to meet these objectives: (a) selection in cosmopolitan breeds, by integrating traditional and new breeding goals; and (b) exploitation of local breeds and their genes. Local breeds are the most adapted to low-input production systems and a range of environments, can produce food with unique properties, and can preserve cultural and social values. Local breeds also hold a reservoir of valuable genes that, once discovered, may be introgressed in cosmopolitan breeds by traditional or innovative methods Bewley [44].

5 Scientific Research and Innovations to Address Climate Changes in Egypt

The Agricultural Research Center, Colleges of Agriculture in the different Universities, Environmental institutes, The National Research Center, The Desert Research Center are the main agricultural institutions in Egypt that provide the science base to enhance the adaptive capacity of the country to cope with impacts of climate change. In addition to The National Water Research Center, National Institute of Oceanography and Fisheries and The Egyptian Atomic Energy Authority.

The Agricultural Research Center (ARC) was established in the early 1970s. With its more than 4000 researchers working in 16 Institutes, 13 Central Labs, 10 regional stations, and 4 researches, extension and training centers of excellence, ARC contributed significantly to the agricultural production in Egypt. ARC accomplishments include development of new varieties, improved agronomic practices, livestock development, maintenance of the national herds and better food processing techniques. The ultimate goal is to maximize the economic return per unit of land and water. However despite the numerous research institutions and the availability of a great number of distinguished agricultural researchers, it has not so far been possible to make appropriate use of this human resource wealth [45].

6 Monitoring, Assessment and Follow-Up of Climate Change Related Impact on Egypt

Because of its dependence on the River Nile and the agricultural activities that require access to climatic data that impact the natural flow of the Nile as well as seasonal rainfall, Egypt has a long history in dealing with natural meteorological observations. Since 1933 the General Authority for Meteorology has been operating through a network of 112 ground stations and air stations to monitor air pollution and cosmic radiation, 26 meteorological stations for monitoring agricultural climate are connected to international networks.

Egyptian Environmental Affairs Agency (EEAA) is operating 47 stations to monitor air pollution in Cairo, Alexandria and the capitals of the governorates of the Delta, Upper Egypt and Central Egypt as well as the major cities. No doubt Greenhouse gases are a part of the elements of pollution monitored by these stations, except ozone gas, which is monitored by a limited number of stations. Furthermore, the Institute for Higher Studies on the Environment, in the University of Alexandria, monitors the shores of the Mediterranean Sea, while the National Institute of Oceanography and Fisheries monitors the shores of the Red Sea. The Faculty of Science in Ain Shams University controls the quality of samples analyzed by both agencies, including the bacteriological and physical, chemical and trophozoite elements. The Environmental Affairs Agency publishes this data on a special website related to all elements including the temperature of the sea water surface.

The National Authority for Remote Sensing and Space Sciences (NARSS) has been monitoring the earth since 2007 through Egypt's Satellite "Egypt Sat 1", The National Authority for Remote Sensing has also set up a station to receive data coming from satellites of the U.S. National Oceanic and Atmospheric Administration (NOAA), the French earth observation satellites (SPOT), and others.

In the field of agro-climatic meteorology, the Ministry of Agriculture and Land Reclamation (MALR) oversees the operation of 26 agro-climatic stations as well as two environmental meteorological stations, but most of these stations need maintenance and calibration. As regards to wave height monitoring, the setup of relevant measurement devices began in the port of Alexandria in 1944, the mouth of Rasheed Branch in 1964, Lake Burullus in 1972, the mouth of the Damietta branch in 1990, the port of Abu Qir in 1992 and the port of Damietta in 1997. An electric power-generating station was established in 1998 in El-Arish. Additionally, a device was set up to measure the height of the waves on the Red Sea, and 11 automatic measuring devices were installed along the Suez Canal.

Egypt depends on obtaining data on the evolution of sea level rise on the Egyptian coast, temperature rise, rainfall, surface flow of the Nile waters and the amount of water that could reach the country as a natural flow of the River Nile. All of these activities require the constant monitoring of many aerial and ground variables, which may be within the country or abroad, particularly in the Nile Basin countries.

7 Climate Change Modeling in Egypt

Mathematical models have been used in Egypt to identify the impacts of climate change on water resources. Elshemy [46] used models and scenarios to study the impact of climate change on hydrodynamics and water quality of Lake Nubia. He used a proposed hydrodynamic and water quality model, which simulates three periods: I (2010–2039), II (2040–2069), and III (2070–2099)—with two emission scenarios, A2 and B1, for each period, including the average of 11 GCMs outputs. A theoretical process algorithm has been simplified and further developed to modify the initial conditions input file of dissolved oxygen due to GCC effects, the impacts of climate change on hydrodynamic and water quality characteristics of Lake Nubia were investigated. It is emphasized that the calibration and the scenarios cover only a short period of 2 weeks of simulation. This is due to the limited availability of historical data. Consequently the results and quantified effects do not reflect the variability of hydrodynamics and water quality extending over 1 year or even longer periods. The hydrodynamic characteristics studied include water surface levels, evaporative water losses, and reservoir thermal structure. The water quality parameters of the study include pH, dissolved oxygen, chlorophyll-a, orthophosphate, nitrate–nitrite, ammonium, total dissolved solids, and total suspended solids. The results showed that hydrodynamic and water quality characteristics of Lake Nubia would be significantly impacted by projected climate change.

Also identifies a critical need for developing a regional climate change, model for the Nile Basin and acquiring long-term records of hydrodynamic and water quality characteristics of the Aswan High Dam Reservoir for detailed investigation of climate change impacts [46].

The sensitivity of Nile flows to climate change is strongly affected by increasing the temperature due to climate change. Scenarios of the effect of climate change on the Nile flows using three Global Circulation Models (GCMs) to estimate future Nile flows. Strzepek et al. [47], developed ten different scenarios for Nile flows. Nine of the ten predict reductions in Nile flows from 10 to 90% by the year 2095. Even in the short term, by 2025 losses are estimated at 5–50% [48], Based on investigated hydro-meteorological impacts of climate change in the River Nile Basin using different scenarios he reported, temperature increases and possible precipitation decrease will significantly aggravate the existing water related problems.

Agrawala et al. [49], Focused in their studies of development and climate change in Egypt on coastal resources and the Nile adaptation measures on the supply-side include ways to improve rain-harvesting techniques, increasing extraction of ground water, water recycling, desalination, and improving water transportation recommending improved long-term forecasting as a necessity to enhance Egypt's ability to cope with prolonged drought.

Satellite images, GIS, mathematical models and scenarios have been instrumental in studying the impacts of climate change on the coastal area of the Nile Delta in Egypt through sea level rise (SLR) combined with geological and human factors, which is already subsiding at approximately 3–5 mm/year around the Nile delta [50]. Satellite maps of the Nile delta has been used to illustrate the potential impact of SLR on the available agricultural land in the Delta under different scenarios of the SLR that change with time as well as on the livelihood of the local communities. Simulation models were used in a study by Egypt's National Strategy for Adaptation to Climate Change And Disaster Risk Reduction (2011) on different agricultural areas over a period of 25–40 years to predict the statues of the major crops in Egypt and their water use under different climate change scenarios concluding that:

Simulation models were used in a study by Egypt's National Strategy for Adaptation to Climate Change and Disaster Risk Reduction (2011) on different agricultural areas over a period of 25–40 years to predict the statues of the major crops in Egypt and their water use under different climate change scenarios concluding that:

- Wheat: The productivity of wheat will fall by 9% if the temperature rises by 2 °C. The water consumption of this crop will rise by 6.2%, in comparison with the situation under the current weather conditions. A 4 °C rise will increase the deficit to 18%.
- Maize: Crop productivity will fall by 19% by the middle of this century if temperature increases by 3.5 °C, in comparison with the situation under the current weather conditions. Accordingly, water consumption will rise by 8%.
- Cotton: Climate change will have a positive impact on the productivity of cotton. Production will rise by 17% if the temperature rises by 2 °C. The water consumption of this crop will increase by 4.1–5.2% in comparison with the situation under the current weather conditions. If the temperature rises by 4 °C, production will increase by 31%. On the other hand, water consumption will rise by 10% in comparison with the situation under the current weather conditions.
- Rice: Productivity will fall by 11% in comparison with the situation under the current weather conditions. Water consumption will increase by 16%.
- Tomato: Tomato productivity will fall by 14% if the temperature increases by 2 °C. Water consumption of the crop will increase by 4.2–5.7% in comparison with the situation under the current weather conditions. The drop in productivity will reach 51% if the temperature rises by 3.5 °C.

- Sugar Cane: The results of studies on the climate change impact on the production of sugar from sugar cane indicate a 24.5% drop in productivity, and a 2.3% increase in water consumption. There will be a 25.6% drop in crop return, per unit of water.

8 Improved Crop Varieties and Livestock Breeds

Advances in crop breeding using different methodologies and techniques can produce varieties that are can resist temperature extremes, drought, and disease. Science can also be harnessed to breed hardy, more resilience livestock breeds that are capable of adapting to the harsh conditions that characterize dryland regions, particularly marginal areas.

Genetic engineering has been widely used in Egypt to produce drought, heat and salinity crop varieties. Scientists at the Agricultural Genetic Engineering Research Institute (AGERI) were able to produced drought-tolerant wheat by transferring a gene 'HVAI1' from barley into a local wheat variety. The new breed is consuming less irrigation water.

Aboul-Naga et al. [51], Studied the performance of crosses of Damascus goat with the local Barki raised by Bedouins at the arid coastal zone of Egypt. The study covered 159 breeders in the area from Borg-Arab to Barani in the North West Coast of Egypt. Results indicated that the Damascus crossbred are well adapted to the prevailing conditions, poor ranges and harsh management conditions in the arid area of the North West Coast of Egypt. The 25% Damascus crossbred does give more than double milk production than the native Barki, and their kids performed better at the market.

Management of forage crops for livestock in Egypt under heat, drought and salinity stress had been studied by El-Nahrawy [52]. He identified several forage crop species that can tolerate the harsh conditions of soil salinity as well as low water quality. He recommended several management options to alleviate degradation of the pasture lands in the Northwest Coast and North Sinai due to drought are to (a) Developing a tree seedling nursery capacity in the villages, and planting, in cooperation with local land users, of improved fodder trees and shrubs. (b) Enhance soil stabilization by the planting of windbreaks, using trees or shrubs with nutritive value. (c) Identifying useful local grassland species, developing seed collection and multiplication and over-seeding selected rangelands with seeds of good nutritive local grasses and legumes species.

9 Challenges of Innovation, Knowledge, and Scientific Research (SDS Egypt Vision 2030)

It was stated in the Egypt vision 2030 that "Knowledge, innovation, and scientific research contribute in achieving the goals of sustainable development strategy (SDS) in general, whether national or international, where the increase in the contribution of a large number of economic sectors in economic activity can be achieved by connecting scientific research outputs and innovation by the needs of these sectors". However challenges related to innovation, knowledge, and scientific research are divided into three main groups:

The first group: is characterized by its high impact and relative ease to control, hence it attracts the biggest share of attention as they could be encountered and overcome much easier compared with others. It includes:

- The legislative system being poor in motivating and protecting innovation: There is an absolute necessity to review legislation relevant to innovation protection, and develop an incentives structure, governmental expenditures, and protection of intellectual rights in order to achieve a motivating environment for innovation.
- Poor coordination between social needs and innovation: This can be observed in the low proportion of local content in several vital sectors as well as not taking advantage of scientific research outputs in facing the main challenges encountering Egyptian society.

The second group: of challenges could be overcome, yet it has a relative and limited impact, including:

- Inefficient sectorial planning: Sectorial planning must be focused on defining national priorities and linking them to the system of innovation, knowledge, and scientific research to achieve high levels of competitiveness and sustainability for industries and strategic sectors in Egypt.
- Lack of a comprehensive mechanism connecting knowledge with innovation: Due to the lack of a proper comprehensive mechanism to connect knowledge with innovation, the development of curricula and educational methods became irrelevant to beneficial social or economic innovation production, which in its turn led to significant imports of technology.
- Insufficient economic and financing incentives for innovation: Limited low expenditure allocated for scientific research and development has a negative effect on knowledge production as a result. Medium and small companies' inability to produce and market innovation: Due to the high costs of such activities, this consequently undermines their financial capacities.

The third group of challenges is characterized by its high impact and its long-term nature to be overcome:

- Poor innovation culture of society: This is one of the most difficult challenges facing the development of innovation, knowledge, and scientific research in Egypt, due to decades of inherited negligence, whether in terms of instilling the culture of

innovation in youth, or in fostering talents in schools, or in creative appreciation and motivation.
- Lack of awareness of the importance of intellectual property and its protection: In light of the limited implementation of Intellectual Property Laws in Egypt, researchers and innovators do not enjoy their intellectual rights.

10 Conclusions

Science policy interface is vital in approaching the climate change challenges globally and specifically in Egypt. Innovations and new technologies have the major role of developing new crops varieties that are resistant to possible impacts of climate change (heat, drought and salinity stress), disease, and producing higher yields. These technologies should be simplified to be close to the policymakers, who need to understand the nature and processes of scientific innovation and to enhance an environment in which it can flourish in responding to the emerging challenges. Water, food energy and climate change nexus can be the major concern of the scientific research plans, programs and projects in Egypt. Also, Policy timescales are generally shorter than scientific developments, so policy needs to demonstrate more continuity for fostering innovation. Because of the cost of research, Funding for scientific research and development needs long-term commitments which are not affected by rapid policy changes. There is also a need for more international cooperation.

11 Recommendations

This chapter highlights the following recommendations:

1. Guide the research agenda in Egypt to provide scientific facts to inform policy makers with information and knowledge related to climate change and its impact on agriculture sector.
2. Enhance the role of the scientific research entities in presenting the knowledge in a simple way for the decision makers and the general public.
3. Mainstream climate change assessment, impact and measures of adaptation in the research programs of the research institutions.
4. Establish strong linkages between the local civil society organizations and the agricultural research bodies in the country to empower these organization with the necessary tools and gaudiness to approach climate change challenges at the levels of local communities.

References

1. Lybbert T, Sumner D (2010) Agricultural technologies for climate change mitigation and adaptation in developing countries: policy options for innovation and technology diffusion issue brief no. 6. International Centre for Trade and Sustainable Development (ICTSD) & International Food & Agricultural Trade Policy Council (IPC)
2. IPCC (2007) Climate change 2007, impacts, adaptation and vulnerability. Contribution of working group II to the fourth assessment report of the intergovernmental panel on climate change, Cambridge University Press, Cambridge, UK, 2007. Retrieved from http://www.ipccdata.org/ddc_co2.html
3. Cook J, Nuccitelli D, Green SA, Richardson M, Winkler B, Painting R, Way R, Jacobs P, Skuce A (2013). Quantifying the consensus on anthropogenic global warming in the scientific literature. Environ Res Lett 8, 024024 (7 pp). Retrieved from https://doi.org/10.1088/1748-9326/8/2/024024
4. Pulatov B, Jönsson AM, Wilcke RAI, Linderson M-L, Hall K, Bärring L (2016) Evaluation of the phenological synchrony between potato crop and Colorado potato beetle under future climate in Europe. Agr Ecosyst Environ 224(2016):39–49. https://doi.org/10.1016/j.agee.2016.03.027
5. World Bank (2009) Development and climate change. The World Bank Group at Work. Washington DC 20433
6. RICCAR (2017) Regional initiative for the assessment of the impact of climate change on water resources and socio-economic vulnerability in the Arab Region. ESCWA, UN
7. Greacen El (1981) Soil water assessment by the neutron method (CSIRO, Adelaide S. Aust), 140 pp
8. International Atomic Energy Agency (2003) Neutron and gamma probes: their use in agronomy, 2nd edn. Training Course Series No 16, 73 pp
9. IAEA, FAO (2001) Use of isotope and radiation method in soil and water management and crop nutrition (Manual), Training Course Series No 14, 247 pp
10. IAEA (1996) Nuclear techniques to assess irrigation schedules for field crops, IAEA-TECDOC-888, 273 pp
11. Food and Agriculture Organization of the United Nations & International Atomic Energy Agency (2002). Deficit irrigation practices. Water Rep 22:102 pp. Retrieved from ftp://ftp.fao.org/agl/aglw/docs/wr22e.pdf
12. Kirda C, Moutonnet P, Hera C, Nielsen DR (eds) (1999) Crop yield response -deficit irrigation. Kluwer Academic Publishers, Dordrecht, The Netherlands
13. IAEA (2000) Comparison of soil water measurement using the neutron scattering, time domain reflectometry and capacitance methods, IAEA, TECDOC- 1137, 161 pp
14. IAEA (1998) Management of nutrients and water in rainfed arid and semi-arid areas, IAEA-TECDOC-1026
15. Williams DG, Cable W, Hultine K, Hoedjes JCB, Yepez EA, Simonneaux V, Er-Raki S, Boulet G, De Bruin HAR, Chehbouni A, Martogensis OK, Timouk F (2004) Evapotranspiration components determined by stable isotopes, sap flow and eddy covariance techniques. Agric For Meteor 125:241–258
16. Yepez EA, Williams DG, Scott RL, Lin G (2003) Partitioning overstory and understory evapotranspiration in a semiarid savanna woodland from the isotopic composition of water vapor. Agric For Meteor 119(2003):53–68
17. Saadalla MM (2001) Chemical desiccation as a screening method for post-anthesis drought tolerance in wheat. Bull Fac Agric Cairo Univ 52:385–400
18. Amani I, Fischer RA, Reynolds MP (1996) Canopy temperature depression association with yield of irrigated spring wheat cultivars in a hot climate. J Agron Crop Sci 176:119–126
19. Saadalla MM, Alderfasi AA (2000) Infrared-thermal sensing as a screening criterion for drought tolerance in wheat. Ann Agric Sci Ain Shams Univ 45(2):421–437
20. Hubick K, Farquhar G (1989) Carbon isotope discrimination and the ratio of carbon gains to water lost in barley cultivars. Plant Cell Environ 12:795–804

21. Rebetzk GJ, Condon AG, Richards RA, Farquhar GD (2002) Selection for reduced carbon isotope discrimination increases aerial biomass and grain yield of rainfed breadwheat. Crop Sci 42:739–745
22. Dercon G, Clymans E, Diels J, Merckx R, Deckers J (2006) Differential ^{13}Cisotopic discrimination in maize at varying water stress and at low to high nitrogen availability. Plant Soil 282:313–326
23. Mickelbart MV, Hasegawa PM, Bailey-Serres J (2015) Genetic mechanisms of abiotic stress tolerance that translate to crop yield stability. Nat Rev Genet 16:237–251. https://doi.org/10.1038/nrg3901
24. Farnham MW, Bjorkman T (2011) Breeding vegetables adapted to high temperatures: a case study with broccoli. Hort Sci 46:1093–1097
25. Ehlers JD, Hall AE (1998) Heat tolerance of contrasting cowpea lines in short and long days. Field Crop Res 55:11–21. Retrieved from https://doi.org/10.1016/s0378-4290(97)00055-5
26. Gororo NN, Eagles HA, Eastwood RF, Nicolas ME, Flood RG (2002) Use of Triticum tauschii to improve yield of wheat in lowyielding environments. Euphytica 123:241–254. https://doi.org/10.1023/A:1014910000128
27. Jha UC, Bohra A, Singh NP (2014) Heat stress in crop plants: its nature, impacts and integrated breeding strategies to improve heat tolerance. Plant Breed 133:679–701. Retrieved from https://doi.org/10.1111/pbr.12217
28. Lopes MS, Reynolds MP, McIntyre CL, Mathews KL, Jalal Kamali MR, Mossad M, Feltaous Y, Tahir ISA, Chatrath R, Ogbonnaya F, Baum M (2013) QTL for yield and associated traits in the Seri/Babax population grown across several environments in Mexico, in the West Asia, North Africa, and South Asia regions. Theor Appl Genet 126:971–984. Retrieved from https://doi.org/10.1007/s00122-012-2030-4
29. Pinto RS, Reynolds MP, Mathews KL, McIntyre CL, Olivares- Villegas J-J, Chapman SC (2010) Heat and drought adaptive QTL in a wheat population designed to minimize confounding agronomic effects. Theor Appl Genet 121:1001–1021. https://doi.org/10.1007/s0012-010-1351-4
30. Vijayalakshmi K, Fritz AK, Paulsen GM, Bai G, Pandravada S, Gill BS (2010) Modeling and mapping QTL for senescence-related traits in winter wheat under high temperature. Mol Breed 26:163–175. https://doi.org/10.1007/s11032-009-9366-8
31. Pinto RS, Reynolds MP (2015) Common genetic basis for canopy temperature depression under heat and drought stress associated with optimized root distribution in bread wheat. Theor Appl Genet 128:575–585. https://doi.org/10.1007/s00122-015-2453-9
32. Argyris J, Dahal P, Hayashi E, Still DW, Bradford KJ (2008) Genetic variation for lettuce seed thermos -inhibition is associated with temperature-sensitive expression of abscisic acid, gibberellin, and ethylene biosynthesis, metabolism, and response genes. Plant Physiol 148:926–947. Retrieved from https://doi.org/10.1104/pp.108.125807
33. Argyris J, Truco MJ, Ochoa O, McHale L, Dahal P, Van Deynze A, Michelmore RW, Bradford KJ (2011) A gene encoding an abscisic acid biosynthetic enzyme (LsNCED4) collocates with the high temperature germination locus Htg6.1 in lettuce (Lactuca sp.). Theor Appl Genet 122:95–108. Retrieved from https://doi.org/10.1007/s00122-010-1425-3
34. McCord PH, Sosinski BR, Haynes KG, Clough ME, Yencho GC (2011) QTL mapping of internal heat necrosis in tetraploid potato. Theor Appl Genet 122:129–142. Doi:10.1007/s00122-010-1429-z
35. Driedonks N, Rieu I, Vriezen WH (2016) Breeding from plant heat tolerance at vegetative and productive stages. Plant Reprod 29:67–79. doi 10.1007/s00497-016-0275-9
36. Reichter GM, Semenov MA (2005) Modeling impacts of climate change on wheat yields in England and Wales: assessing drought risks. Agric Syst 84(1):77–97
37. Stockle CO, Martin S, Campbell GS (1994) CropSyst, a cropping systems model: water/nitrogen budgets and crop yield. Agric Syst 46:335–359. https://doi.org/10.1016/0308-521X(94)90006-2
38. Steduto P, Fereres E, Hsiao TC, Raes D (2006) Yield response to water: the FAO revision framework and the crop-water productivity model AquaCrop. In: Proceedings of international

symposium on water and land management for sustainable irrigated agriculture, 4–8 Apr 2006. Cukurova University, Adana, Turkey
39. Giuseppe P, Francesconi AHD, Stefanon B, Sevi A, Calamari L, Lacetera N, Dell'Orto V, Pilla F, Ajmone Marsan P, Mele M, Rossi F, Bertoni G, Matteo Crovetto G, Ronchi B (2016) Sustainable ruminant production to help feed the planet. Ital J Anim Sci. Retrieved from http://dx.doi.org/10.1080/1828051X.2016.1260500
40. Bohmanova J, Misztal I, Cole JB (2007) Temperature-humidity indices as indicators of milk production losses due to heat stress. J Dairy Sci 90:1947–1956
41. Kuczynski T, Blanes-Vidal V, Li B, Gates RS, Nääs I, de Alencar Moura DJ, Berckmans D, Banhazi TM (2011) Impact of global climate change on the health, welfare and productivity of intensively housed livestock. Int J Agric Biol Eng 4:1–22
42. Chandra V, Sejian V, Sharma GT (2015) Strategies to improve livestock reproduction under the changing climate scenario. In: Sejian V, Gaughan J, Baumgard L, Prasad C (eds) Climate change impact on livestock: adaptation and mitigation. Springer, New Delhi (India), pp 425–439
43. Lacetera N, Segnalini M, Bernabucci U, Ronchi B, Vitali A, Tran A, Guis H, Caminade C, Calvete C, Morse A, Baylis M, Nardone A (2013) Climate induced effects on livestock population and productivity in the Mediterranean area. In: Navarra A, Tubiana L (eds) Regional assessment of climate change in the Mediterranean: volume 2: Agriculture, forests and ecosystem services and people. Advances in global change research 51. Dordrecht, Netherlands: Springer. Business Media; p. 135–156. https://doi.org/10.1007/978-94-007-5772-1_7, © Springer Science + Business Media Dordrecht 2013
44. Bewley J (2010) Precision dairy farming: advanced analysis solutions for future profitability. In: Proceedings of 1st North American conference precision dairy management; 2–5 March 2010; Toronto, Ontario, Canada, pp 1–8
45. MALR (2009) Sustainable agricultural development strategy "SADS", towards 2030. Ministry of Agriculture and Land Reclamation (MALR), Cairo, Egypt, 1p: 20, 26, 32, 43
46. Elshemy M (2013) Climate change impacts on water resources in semiarid regions: case study of Aswan High Dam Reservoir. In: Younos T, Grady CA (eds) Climate change and water resources, Hdb Env Chem (2013). 25, 69–98, Retrieved from https://doi.org/10.1007/698_2013_223,© Springer- Verlag Berlin Heidelberg 2013, Published online: 3 Apr 2013
47. Strzepek K, Yates D, Yohe G, Tol R, Mader N (2001) Constructing 'not implausible' climate and economic scenarios for Egypt. Integr Assess 2:139–157
48. Mohamed HI (2013) Impact of climate change on water resources in the Nile Basin and adaptation measures in Egypt. A review article presented in partial fulfillment of the requirement for promotion to the position of professor. Civil Engineering Department Faculty of Engineering, Assiut University
49. Agrawala S, Moehner A, El Raey M, Conway D, van Aalst M, Hagenstad M, Smith J (2004) Environment directorate environment policy committee. Working party on global and structural policies working party on development co-operation and environment. Development and climate change in EGYPT, focus on coastal resources and the Nile Organization for Economic Co-operation and Development
50. Gornitz V (2005) Natural Hazards. In: Schwartz ML (ed) Encyclopedia of coastal science, 678–684. Dordrecht, The Netherlands: Springer
51. Aboul-Naga A, Shaat I, Osman MA, Metawi HR, Hassan F (2008) Performance of crosses of Damascus goat with the local Barki raised by Bedouins at the arid coastal zone of Egypt. Sustainable development in drylands—meeting the challenge of global climate change. In: El-Beltagy A, Saxena MC (eds) Proceedings of the ninth international conference on development of dry lands 7–10 November 2008, Alexandria, Egypt, Editors:
52. El-Nahrawy MA (2008) Pasture and forage status in Egypt limitations and opportunities sustainable development in dry lands—meeting the challenge of global climate change. In: El-Beltagy A, Saxena MC (eds) Proceedings of the ninth international conference on development of dry lands 7–10 Nov 2008, Alexandria, Egypt

Land and Water Resources

Impact of Climate Change on Plant-Associated Fungi

Anindya Chanda, Heba Maghrawy, Hossam Sayour, Phani M. Gummadidala and Ola M. Gomaa

Abstract The global climate change can influence agricultural productivity by altering the plant-microbe interactions. Plant-associated fungi play important roles in these interactions by regulating nutrient transformation in soils, nutrient availability for plants and plant health and growth. The abiotic stressors that increase with the changing climate result in significant alterations in these processes. These alterations are either as a response to the changing biology of the plant or due to the direct effect of the stressors on the fungi. In this chapter we retrospect the current knowledge on the plant-associated fungi and discuss the effects of the changing climate on their interactions with their hosts. The goal of this review is to emphasize the need for more research on plant-fungal interactions that can increase the resilience of crops to climate change.

Keywords Plant-associated fungi · Agriculture · Fungal diversity · Fungal activity, and climate change · Molecular imprinted polymers · Wireless sensor networks · Precision farming

1 Introduction

Life as we know it has been transformed due to climate change. The heavy impact of climate change has taken its toll on everything. The rate and extent of damage are expected to become worse each year. Climate change impact has been the center of attention during the past few years. Policymakers have identified it as a problem that needs to be addressed by scientists. Climate action is the goal number 13 in the 2030

A. Chanda · P. M. Gummadidala
Department of Environmental Health Sciences, Arnold School of Public Health, University of South Carolina, Columbia, SC 29208, USA

H. Maghrawy · O. M. Gomaa (✉)
Radiation Microbiology Department, National Center for Radiation Research and Technology (NCRRT), Atomic Energy Authority, Cairo, Egypt

H. Sayour
Biomedical Chemistry Unit, Chemistry Department, Animal Health Research Institute Dokki, Cairo 12618, Egypt

© Springer Nature Switzerland AG 2020

E.-S. E. Omran and A. M. Negm (eds.), *Climate Change Impacts on Agriculture and Food Security in Egypt*, Springer Water, https://doi.org/10.1007/978-3-030-41629-4_5

agenda for sustainable development. This goal addresses the target action to combat climate change and its impact to improve the lives of people around the globe (www.un.org/sustainabledevelopment/climate-change-2/).

The estimated increase in temperature is expected to exceed 3 °C this century. An action plan was adopted upon Paris Agreement that took place in 2015. The implementation of the Paris Agreement was perceived as crucial step to fulfill the sustainable development goals by providing a roadmap to "reduce emissions and provide climate resilience". An Intergovernmental Panel on Climate Change [1] was formed back in 1988 upon the recommendation of the World Meteorological Organization (WMO) and United Nations Environmental Program (UNEP). IPCC has released until now, five assessment reports that cover the scientific, technical and socio-economic impacts of climate change. Currently, the IPCC is working on their sixth assessment report and three special reports to suggest ways for policymakers to mitigate the impact. Climate change impact and mitigation is not only a political decision, but it also requires the intervention of scientific research to study the phenomena, help draw a clearer picture and suggest innovative scientific solutions. Therefore, it has been placed among the top priority societal challenges in the Horizon 2020 competitive call for proposals addressed by the European Commission (EC www.ec.europa.eu/research/participants/portal/desktop/en/opportunities/h2020/).

The effect of climate change on agriculture has specifically become a global issue due to the ever-growing need to provide food security and end hunger. The impact is on the whole agro-ecosystem from crops and livestock to fisheries, which in turn can affect economic stability [2]. Implications of climate change on agriculture is not only due to effects on plants but also due to effects on biogeochemical properties of the soil on which they grow. An important consequence of changing climate is its impact on the microorganisms present in the agricultural soil. Not only does the climate-mediated changes in microbiota and their behaviors reduce food productivity but it extends to food safety, human health and economy when soil microbiota turn toxigenic under environmental influence.

Fungi are the primary decomposers in the soil. Hence it is implicative that changes in temperature and precipitation will change the diversity and functionality of fungal communities. Hence to find a solution, one has to step back and study the impact. From this standpoint, this chapter aims to review the current knowledge on the impacts of climate change on soil fungi diversity and activity from stress-inducing perspective. The review also addresses the recent initiative by the Egyptian Ministry of Higher Education and Scientific Research (MHESR) to encourage researchers to find solutions to overcome climate change impacts such as forecasting of future changes in diversity and food productivity, detecting potential natural disasters, changing building codes, protecting coasts and delta region. Climate change is at the top of environmental priorities stated in the National strategy for scientific research for Egypt.

2 Relation Between Climate Change and Diversity of Soil Fungi

Fungi are main constituents of global ecosystems. They convert the hard digested organic material into forms that other organisms can use. Fungal hyphae physically bind soil particles together, causing stable aggregates that increase water infiltration and soil water holding capacity. Soil fungi can be grouped into three general functional groups based on how they acquire their energy as follows (https://www.nrcs.usda.gov/wps/portal/nrcs/detailfull/?cid=nrcs142p2_053864):

(i) Decomposers (saprophytic fungi) convert the dead organic material into fungal biomass, carbon dioxide (CO_2), and small molecules, such as organic acids. These fungi generally consume complex substrates, such as the cellulose and lignin in wood and are essential in decomposing the carbon ring structures in some pollutants. They are important for immobilizing, or retaining, nutrients in the soil. Also many of the secondary metabolites of these fungi are organic acids, which increase the accumulation of humic-acid rich organic matters that are resistant to degradation and stay in the soil for hundreds of years.
(ii) Mutualists (mycorrhizal fungi) colonize plant roots, solubilize phosphorus and deliver phosphorus, nitrogen, micronutrients, and water to the plant. One major group of mycorrhizae, the *ectomycorrhizae* grow on the surface layers of the roots and are commonly associated with trees. The second major group of mycorrhizae is the *endomycorrhizae* that grow within the root cells and associated with grasses, row crops, vegetables, and shrubs. Arbuscular mycorrhizal (AMF) is a type of endomycorrhizal fungi.
(iii) Pathogens are parasites that reduce production and result in death when they colonize roots of plants such as *Verticillium*, *Pythium*, and *Rhizoctonia*, leading to major economic losses in agriculture each year [3].

While it is difficult to study the variations in fungal activity and growth within soil ecosystems as the mycelia are hidden below ground, it is easy to anticipate that arbuscular ectomycorrhizal, endomycorrhizal fungi and saprotroph fungi with different nutritional needs are affected differently by changing the climate. For example, saprotrophs may be directly affected by the abiotic change, but mycorrhizal fungi will be affected not only directly but also indirectly by effects on the host plants. As a result, modifications in the activity and growth period of fungal mycelia prompted by climate variation may have broad-scale changes on the carbon cycle of soil ecosystems.

3 Plants and Mycorrhizal Fungi: Effect of Climate Change on This Relationship

Abiotic stress is common in all agricultural environments as drought, temperature, salinity and nutrient deficiency or excess all have negative consequence on plant strength. However several lines of evidence suggest that plant-mycorrhizal fungi offer several beneficial effects to the host and increase their resistance to this stress. Approximately 6000 species of Glomeromycota, Ascomycota, and Basidiomycota have been stated as mycorrhizal fungi. Discussed here are a collection of such studies that demonstrate the importance of mycorrhizal fungi in protecting plants from the abiotic stressors and thereby are linked with increased agricultural productivity.

Arbuscular mycorrhizal fungi (AMF) symbiosis are obligate symbionts that colonize land plant root. These fungi receive carbohydrates from plant host and improve plant nutrient and water availability [4]. Elevated carbon dioxide levels usually enhance the colonization of these mycorrihizal fungi, which in turn promotes plant growth. However, in many ecosystems, plant growth promotion is limited by nutrient availability. Therefore elevation of carbon dioxide levels will not necessarily increase agricultural productivity under these field conditions [5]. The rise in temperature has a positive effect on arbuscular fungi colonization and hyphal length. It may also alter the structure of the AMF hyphal network and initiate a switch from vesicular hyphae responsible for storage in the cooler soil to extra mycorrihizal hyphal representing growth in warmer soils. This may be responsible for faster carbon distribution to the rhizosphere and improved respiration of the extra mycorrihizal hyphae at a high soil temperature. AMF may enhance plant growth and colonization for the majority of strains *Glomus intraradices* and *Glomus mossae* at a higher temperature. Drought stress can be a result of global warming. Studies show that drought improves arbuscule formation and hyphae enlargement of *Glomus* spp. (e.g. strain ZAC-19), but reduces its colonization (e.g. *Glomus fasciculatum*). Drought is responsible for the reduction of plant growth and both roots and aerial plant parts may be compressed. This may lead to changes in the distribution of photosynthates in the rhizosphere as well as in extra mycorrhizal mycelium.

Various plant species, most importantly forest trees, interact with ectomycorrhizal (ECM) fungi which in concert with AMF provide organic nitrogen, phosphate and other micronutrients to their plant hosts and also enhance seedling survival. Similar to AMF, elevated CO_2 concentrations may induce changes in ECM colonization and/or community structures. It has been reported that mycelial biomass production by *Hebeloma crustuliniforme* in *Pinus sylvestris* (L.) Karst. Seedlings increased three-fold under elevated CO_2, as compared to mycelial growth under ambient CO_2 conditions [6]. The mycorrhizal fungus *Pisolithus tinctorius*, which depended on plant assimilates of *Pinus silvestris* L., grew much faster in the presence of elevated CO_2, three fold higher mycorrhizal root clusters and two-fold higher biomass of extra mycorrhizal mycelia as compared to ambient CO_2 levels. These studies demonstrate that increased CO_2 can increase ECM colonization of host plants, although as observed in *Cenococcum geophilum and Suillus* sp. associated with Scots pine (*P. sylvestris* L.), fungal biomass does not increase under these conditions. Respiration of ECM strains of *C. geophilum, Suillus intermedius and Lactarius* cf. pubescens can be reduced under increasing temperature [5]. It has been reported that *C. geophilum* show decreased colonization in *Quercus myrsinaefolia* under increased temperature. Drought can significantly influence ECM colonization and community structures. Drought decreased mycorrhizal fungal colonization of some plant species such as Norway spruce trees, several Mediterranean shrubs, some *Pinus* spp. (*such as P. oaxacana* and *P. muricata* colonization with *Rhizopogon* sp.). But not of *Pinus taeda* seedlings. Moreover, a particular adaptation in colonization by some ECM species seems to occur, which may lead to community changes under decreased soil water availability. Hence, ECM strains bring beneficial effects to plants even under drought conditions and hence may find application in reforestation by relieving of drought stress in plants. Symbiosis with *C. geophilum* is stimulated under low water content as this fungus persists better under drought stress than others [5].

As mycorrhizal fungi that colonize plant roots grow into the rhizosphere, there are microsymbionts such as endophytic fungi of genera *Atkinsonella,Balansia,Balansiopsis,Echinodothis,Epichloë,Myriogenospora,Parepichloe* and *Neotyphodium* that exist in plant tissues and may grow within roots, stems and/or leaves, evolving to sporulate on plant or host-tissue. The ericoids, dark septate endophytes demonstrated increased colonization in ericaceous dwarf shrubs under increased atmospheric CO_2 concentrations [5]. In the case of *Neotyphodium coenophialum* and its host, tall fescue *Schedonorus phoenix*, endophyte infection stated to be higher under increased CO_2 compared with ambient CO_2. However, CO_2 enhancement did not affect the interfaces between host grasses and purpletop grass, *Tridens flavus* and their endophytic–fungal symbionts, *Acremonium lolii* and *Balansia epichloe*, respectively. The effect of CO_2 on the host plant and its endosymbionts may additionally change the plant carbohydrate content. This was confirmed with *N. lolii* and perennial ryegrass (*L. perenne*), where infected plants had higher carbohydrate contents than endophyte free plants, and was higher than under ambient CO_2 conditions. Elevated CO_2 may lead to the increased existence of endophyte infections resulting in overall effects on the ecosystem. Also under elevated CO_2 conditions associations between endophytic fungi and grasses defend plants against insect herbivores. Soil warming may influence favorable associations between plants and fungal endophytes. Temperature is a primary parameter that has a significant influence on the endophyte appearance in plant tissue. The endophyte infection of *N. coenophialum* of its host tall *fescue S. phoenix* was not obstructed by warming. Warming increased the density of different fungal endophyte genotypes within individual root sections of the plant without affecting the composition and the richness of the community. Any changes in endophyte communities seen with changes in temperature are attributed to the plant species diversity being affected by warming, leading to effects on endophytic fungal communities. Endophytic fungi provide a competitive advantage to their host plant by enhancing the resistance to environmental stresses. Some endophytes can enhance plant growth during drought change exposure. The fungal endophyte may increase water retention in the leaf sheath and therefore may protect the internal growing region from lethal dehydration.

4 Effects of Climate Change on Saprophytic Fungi

Many saprophytic fungi decompose carbon in soil efficiently more than mycorrhizal fungi. There is clear confirmation that fungal community varies vertically through the soil profile. Wood decay fungi, especially brown rot fungi and white rot basidiomycetes colonize on standing dead or fallen woody debris. Surface litter horizon is controlled by saprotrophic basidiomycetes and ascomycetes. Increase in temperatures upon global warming may increase plant biomass imbalance if high temperatures induce faster plant growth and soil nutrient mineralization. Warming may stimulate carbon losses from the soil by increasing rates of decomposition by the saprophytes. The saprophytic activity can be influenced directly through variations in temperature or precipitation or indirect effect as variations in soil moisture [7]. Increasing temperature will also increase the metabolic rates of saprophytes, thereby mobilizing the large carbon stocks in soil, and increasing soil respiration and decomposition.

5 Effects of Climate Change on the Pathogenic Fungi

Plant pathogens differ in the degree of host specificity and in the amount of physiological interactions they have with their plant hosts, depending on their mode of infection [8–11]. Necrotrophic pathogens that acquire nutrients from damaged host plants have restricted interactions with the active metabolism of host plants [12]. However, abiotic factors that lead or fasten tissue necrosis, such as increased O_3levels, may lead to infection by these types of pathogens [13]. On the other hand, biotrophic pathogens, as obligate parasites, have increased periods of physiological interaction with their hosts, as they develop nutrients from living cells [14, 15]. Therefore, the factors that affect plant growth, such as elevated levels of CO_2, may also affect the colonization of host plants by biotrophic pathogens by altering in host physiology [13–16]. Climate changes can also have direct effects on pathogens, in addition to host plants [17–22]. Pathogen existence in the absence of a host plant (e.g. over-wintering and over-summering) can be affected by temperature, the developments of tuber production and germination and host infection and are usually controlled by temperature and moisture conditions. Warming, rainfall events, or origination of infections earlier in the season can lead to more damaging action. After infections, temperatures and plant water capacity affect the rate of colonization of host tissues, the production of new inoculum and the appearance of symptoms by the host. These changes affect the pathogenicity of both biotrophic and necrotrophic pathogens, and the host defense responses [23]. Some pathogens as *Xylella fastidiosa* can cause disease on a wide range of host plants whereas other pathogens can only infect a narrow range of closely related plant species. Likewise, host defense responses can range from non-specific stress-induced responses to very specific reactions based on the presence of specific genes in both the host and pathogen.

Concentrations of CO_2 and O_3 have increased since the initiation of the industrial revolution, and they will continue to increase in the 21st century. It is well known that both elevated CO_2 and O_3 change plant function like photosynthetic capacity, water-use capacity and growth of the crop. These parameters are increasing with increasing CO_2, but are reduced by increasing O_3. These responses of plants to climate change also affect the host-pathogen interactions. Improved photosynthetic proficiency under increased CO_2provides additional carbohydrate that results in increasing starch and sugar levels in leaf tissue. Under elevated O_3, levels of starch and sugar decrease. Enrichment of sugar content of leaves of plants enhances the sugar-dependent pathogens. The negative effects of increased O_3 on plant physiology, growth and yield lead to decreased photosynthetic capacity that results in visible lesions, decreased leaf durability and premature leaf [23].

CO_2 concentrations affect several rust diseases of cereal crops. Infections by *Puccinia coronata* on oats, *Puccinia dispersa* on rye, and *Puccinia graminis* f.sp. tritici and *Puccinia recondita* f.sp. tritici on wheat were all enriched by concentrations of CO_2 in the 0.5–0.75% average [24]. Interestingly, soybean downy mildew, caused by *Peronospora manshurica*, another biotroph, decreased under increased CO_2 level than on those grown with ambient levels of CO_2 [23]. As this pathogen infects through stomata, increased CO_2 levels decreased the number of stomates per unit leaf area thereby providing less infection sites. Higher CO_2 levels also reduced the time that stomata were open, which lead to a reduction of the degree of infection. Increased CO_2 levels. However, lead to increased colony formation of the powdery mildew pathogen. These suggest that there are different mechanisms through which CO_2 levels regulate plant infections caused by biotrophic pathogens.

Soybean plants under elevated CO_2 grew faster than those in the ambient CO_2 and were infected by the necrotroph, *Septoria glycines*, the causal agent of the brown spot over the growing season [23]. Rice plants grown under increased CO_2 conditions showed an increased susceptibility of the leaf to the sheath blight disease. Wheat plants under elevated CO_2 levels demonstrated increased plant biomass as well as increased biomass of the fungal pathogen, *Fusarium pseudograminearum*, that causes crown rot in plants [13, 12]. These studies all support the idea that increased CO_2levels enhance greater plant growth, which in turn, lead to increased disease levels of necrotrophic pathogens.

Increased CO_2 level reduced leaf necrosis in potato (late blight of potato) caused by *Phytophthora infestans* infection [23]. This was attributed to increased β-1,3-glucanase activity under increased CO_2 levels, this enzyme plays a role in disease resistance. The greatest levels of resistance were observed when $CO_2$was significantly higher, and O_3 concentration was significantly lower than the ambient concentration. As O_3 levels increased to one or two times the ambient level, the positive effects of CO_2 decreased. Starch and soluble sugars increased in the leaves, and decreased in the tubers under an elevated level of O_3, indicating that changed nutrient quality of the leaves and translocation within the plant can influence disease development under these conditions [23].

6 Effect of Climate Change on Soil Mycobiomes

Direct exposure to the sun in Upper Egypt resulted in significant changes to mycobiomes. Thermophilic and thermotolerant fungi were significantly reduced in count and number. Fungi that exhibited changes in number under the tested conditions were *Alternaria alternata, Cochliobolus spicifer, Emericella nidulans* v. *dentata, E. nidulans* v. *lata, E. nidulans* v. *nidulans* and *E. rugulosa, Rhizopus stolinifer, Fusarium chlamydosporum, F. dimerum* and *Gibberella fujikuroi* var. *fujikuroi (anamorph), Cochliobolus sativus, Gliocladium roseum, Melanospora zamiae* and *Nectria heamatococca* (anamorph), *Aspergillus fumigatus, A. flavus* and *A. niger* [25]. Climate change affects not only the fungal diversity but also spore dispersal, sporocarp fruiting and inter-specific fungal interactions in the fungal community. All these determine the landscape of agricultural land [26]. Fungal community composition shifts according to the richness of soil and is dependent on the level of carbon in the soil. The richness of Ascomycota increases at a warmer temperature and dry weather, while the richness of Cryptomycota deceases under the same conditions. Changes in the community are also dependent on the moisture content in soil [27].

7 Direct Impact of Changing Temperatures on Fungi

Microbial interaction among themselves and with plants shapes the agricultural landscape and productivity [28]. Soil fungi comprise a huge portion of soil and play an important part in carbon breakdown. Plant-associated fungi produce the majority of lignin degrading enzymes such as ligninase and chitinase that are responsible for plant decomposition, that increases the availability of nutrients such as nitrogen and phosphorus in soil. Since fungi are sensitive to changes in air temperature and moisture, the alterations of these parameters upon climate change can alter their metabolic activity. Several reviews have focused on the impact of global environmental changes on plants and other terrestrial ecosystems. However, mycorrhizal fungi and their role in carbon sequestration in soils have been poorly studied despite their critical role [29]. An increase of 5 °C can result in a 2.75 fold increase of Mucoromycotina [30]. Increase in temperature also increases the activity of saprophytes thereby decreasing the availability of carbon and increasing carbon dioxide release in the soil. Increase in temperature also affects the extracellular enzyme activity (EEA). The main hydrolytic enzymes involved in labile to intermediate C decomposition are α-glucosidase (AG) that degrades starch, ß-glucosidase (BG) and cellobiohydrolase (CBH) that degrade cellulose and ß-xylosidase (BX) that degrades hemicellulose. The change in their

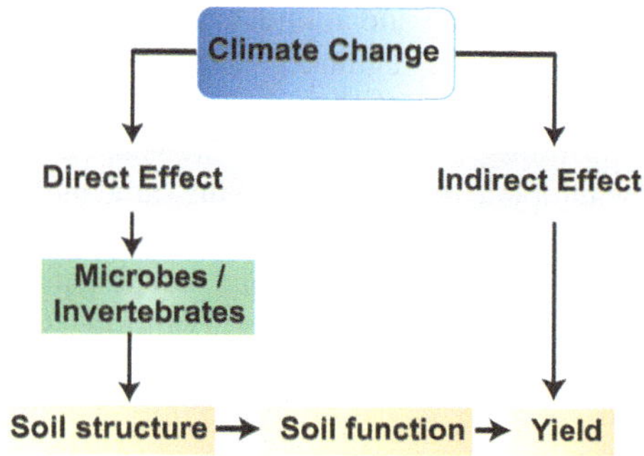

Fig. 1 Representation of possible scenarios for climate change impact on soil fungi

enzymatic activity depended on both temperature and precipitation [27]. While elevated temperatures influence soil fungi growth patterns, lower temperatures (below 30 °C) can increase their secondary metabolism, a process that results in an array of natural products. These metabolites, in turn, can have a significant influence on soil microbial communities and their chemical interactions. Figure 1 is a schematic representation of possible scenarios taking place under climate change.

8 The Need for Modern Tools for Monitoring Climate Changes and Some Closing Remarks

Screening and measurement of rhizosphere biomarkers due to microbial metabolites needs more attention and concern. The agriculture in Egypt needs more attention to Precision Farming (PF) which is an emerged management practice with the potential to increase profits by utilizing more precise information about agricultural resources through sensing and communication technology [31], a possible PF solution using Wireless Sensor Network (WSN) technology. Wireless sensor network (WSN) can perform an autonomous measurements of microbial metabolites in rizhosphere (root exudates) [32]. WSN technology is one of the modern used greenhouse tools, it is considered one of the main support in decisions. The obtained data could be remotely collected by sensor nodes networked to actuators for both alarming and corrective actions to be taken at any point of the greenhouse (point-of-care devices). Development of different sensors for both greenhouse environmental parameters or rhizosphere microbial biomarkers are successfully deploymented. Ease of locating sensor nodes according to needed monitored points is one of the main advantages of WSN technology, for more precise corrective actions compared to the cabled systems [33]. Moreoever, the local aquaponics industry still faces a lot of problems related to the degraded and uncontrolled water quality [34–39]. Real-time data can be accessed from the remote site and analyzed and a real-time reaction taken upon it. Climate change is not temporary but rather a ubiquitous continuous and unpredictable change.

Engineering and animal science researchers have to cooperate and activate directive 97-12 and regulation 178-2002 to monitor and control farms for food safety and security, public health, animal health, and welfare. Sensors based on molecular imprinting [40] creates template-shaped cavities in polymer matrices with memory of the template molecules with the same merits as of Lock and the Key theory/paradigm and are of great applications. Since the 90's, chemists succeeded in preparation of imprinted polymers (artificial receptor/plastic antibody) that imitating/mimicing biological natural molecular recognition explained by Lock and the key theory in the early 40's. In the last decade, the use of molecularly imprinted polymers (MIPs) in sensors have played high recognition ability, known as plastic antibodies, to extract target analytes from complicated sample matrices [16].

Analytes are inorganic elements up to macromolecular biomarkers of air, water and soil quality parameters of valuable insight into the status of both crops and fish in a highly localized way in integrated greenhouse farming like aquaponics strategy. Incorporation of MIP technology as advanced materials in sensor development and WSN leads to innovations in smart systems flourishing applications of precision agriculture Fig. 2.

The integrated system provided theoretical support for establishing data integration network and general framework solution of data collection for recirculation aquaculture via wireless sensor networks. The WSN is sensors-based connected to

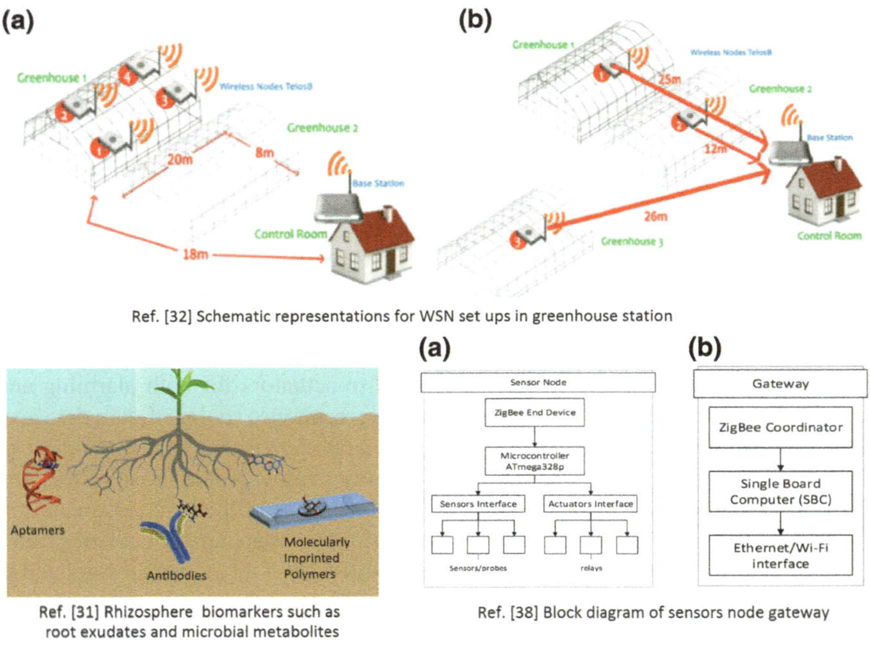

Fig. 2 WSN system in greenhouse for air, water and soil quality parameters control

Arduino microcontroller with built-in memory and RAM, and Zigbee is a specification for a suite of high-level communication protocols using small, low-power digital radios based on an IEEE 802 standard networks. Remote wireless sensor nodes through the integration with GSM were tested [34–36] and the high rate of return on investment (ROI) on WSN's control has attracted a large number of small to middle-level agricultural investors and stakeholders who tend to have a more scientific background than traditional farmers. WSN is currently being used as a means to address and raise awareness as to what governmental decisions to be taken on the National level to become self-reliant in producing its own food. Moreover evaluating economical potentials, in return would definitely provide the user with means to examine and assess the profitability of the system under many alternative scenarios [10].

Other technologies such as Global Positioning Systems (GPS) and Geographic Information Systems (GIS) are also used to compete for the picture by obtaining accurate maps and models for the agricultural field [41]. GIS uses an analytic framework to manage and integrate data, solve problems and understand a past problem or predict an upcoming one. Remote sensing and actuation have become a solution to rectify a problem. In conclusion, more effort is needed from scientists and policymakers to understand the impacts of climate change and mitigate its detrimental effects on agriculture and the quality of life. Multi-factorial ecological studies are essential for increasing this understanding.

9 Conclusions

Whether we like it or not, climate change will continue to occur in the future laying its heavy impact on every agricultural-related element. To overcome the effects of climate change and find solutions to reach food security, it has become very crucial to use Information Communication Technology (ICT) tools such as wireless sensor networks and actuators to predict the change by making important decisions ahead, these ICT tools are perceived now as indispensable tools for precision farming in Egypt. Decisions of where what and when to plant, as well as planning irrigation scheduling and fertilization, will soon be automated decisions required to reach the highest crop yield, and thus closing the food gap. On the other hand, finding solutions to shifts in soil microbial community and deterioration of plant produce under extreme climate conditions can be addressed using biotechnology tools such as genetic modification, stress tolerance induction, and synthetic biology to obtain new breeds that tolerate climate change increase productivity or enhance carbon sequestration. Molecular biology is expected to play another role which is developing new tools for solar energy conversion because a huge percentage of CO_2 emissions are attributed to agricultural emissions. Harnessing the energy from photosynthesis is also another biotechnology application that will add an ecological dimension through mitigating greenhouse gases produced from conventional energy production, this will consequently aid in preserving the environment and biodiversity.

10 Recommendations

- Integrate multidisciplinary sciences to solve climate change problem.
- Increase the awareness of farmers through media and directed workshops.
- Use Information and Communication Technology for precision farming and smart decision.
- Decision support should include local, regional, national and international levels.

References

1. IPCC: Climate Change (2014) Mitigation of climate change. In: Edenhofer O, Pichs-Madruga R, Sokona Y, Farahani E, Kadner S, Seyboth K, Adler A, Baum I, Brunner S, Eickemeier P, Kriemann B, Savolainen J, Schlömer S, von Stechow C, Zwickel T, Minx JC (eds) Contribution of working group III to the fifth assessment report of the intergovernmental panel on climate change. Cambridge University Press, Cambridge
2. FAO (2016) Climate change and food security: risks and responses
3. Kauserud H et al (2012) Warming-induced shift in European mushroom fruiting phenology. Proc Natl Acad Sci 109:14488–14493
4. Millar NS, Bennett AE (2016) Stressed out symbiotes: hypotheses for the influence of abiotic stress on arbuscular mycorrhizal fungi. Oecologia 182:625–641
5. Compant S, Van Der Heijden MG, Sessitsch A (2010) Climate change effects on beneficial plant–microorganism interactions. FEMS Microbiol Ecol 73:197–214
6. Fransson PM, Taylor AF, Finlay RD (2005) Mycelial production, spread and root colonisation by the ectomycorrhizal fungi *Hebeloma crustuliniforme* and *Paxillus involutus* under elevated atmospheric CO_2 Mycorrhiza 15:25–31
7. Allison SD, Treseder KK (2011) Climate change feedbacks to microbial decomposition in boreal soils. Fungal Ecol 4:362–374
8. Dickie IA et al (2017) The emerging science of linked plant-fungal invasions. New Phytol 215:1314–1332. https://doi.org/10.1111/nph.14657
9. Leonard S, Hommais F, Nasser W, Reverchon S (2017) Plant-phytopathogen interactions: bacterial responses to environmental and plant stimuli. Environ Microbiol 19:1689–1716. https://doi.org/10.1111/1462-2920.13611
10. Peyraud R, Dubiella U, Barbacci A, Genin S, Raffaele S, Roby D (2017) Advances on plant-pathogen interactions from molecular toward systems biology perspectives. Plant J 90:720–737. https://doi.org/10.1111/tpj.13429
11. Trebicki P, Dader B, Vassiliadis S, Fereres A (2017) Insect-plant-pathogen interactions as shaped by future climate: effects on biology, distribution, and implications for agriculture. Insect Sci 24:975–989. https://doi.org/10.1111/1744-7917.12531
12. McDonald MC, Solomon PS (2018) Just the surface: advances in the discovery and characterization of necrotrophic wheat effectors. Curr Opin Microbiol 46:14–18. https://doi.org/10.1016/j.mib.2018.01.019
13. Manning WJ, Tiedemann AV (1995) Climate change: potential effects of increased atmospheric carbon dioxide (CO_2), ozone (O_3), and ultraviolet-B (UV-B) radiation on plant diseases. Environ Pollut 88:219–245
14. Nickstadt A et al. (2004) The jasmonate-insensitive mutant jin1 shows increased resistance to biotrophic as well as necrotrophic pathogens. Mol Plant Pathol 5:425–434. https://doi.org/10.1111/j.1364-3703.2004.00242.x
15. Sandermann H Jr (2000) Ozone/biotic disease interactions: molecular biomarkers as a new experimental tool. Environ Pollut 108:327–332

16. von Tiedemann A, Weigel HJ, Jager HJ (1991) Effects of open-top chamber fumigations with ozone on three fungal leaf diseases of wheat and the mycoflora of the phyllosphere. Environ Pollut 72:205–224
17. Castillo JA, Plata G (2016) The expansion of brown rot disease throughout Bolivia: possible role of climate change. Can J Microbiol 62:442–448. https://doi.org/10.1139/cjm-2015-0665
18. Chowdhury FR, Nur Z, Hassan N, von Seidlein L, Dunachie S (2017) Pandemics, pathogenicity and changing molecular epidemiology of cholera in the era of global warming. Ann Clin Microbiol Antimicrob 16:10. https://doi.org/10.1186/s12941-017-0185-1
19. Clare FC et al (2016) Climate forcing of an emerging pathogenic fungus across a montane multi-host community. Philos Trans R Soc Lond B Biol Sci 371. https://doi.org/10.1098/rstb.2015.0454
20. Kriss AB, Paul PA, Madden LV (2012) Variability in Fusarium head blight epidemics in relation to global climate fluctuations as represented by the El Nino-Southern Oscillation and other atmospheric patterns. Phytopathology 102:55–64. https://doi.org/10.1094/PHYTO-04-11-0125
21. Lake JA, Wade RN (2009) Plant-pathogen interactions and elevated CO_2: morphological changes in favour of pathogens. J Exp Bot 60:3123–3131. https://doi.org/10.1093/jxb/erp147
22. Onaga G, Wydra K, Koopmann B, Chebotarov D, Sere Y, Von Tiedemann A (2017) High temperature effects on Pi54 conferred resistance to *Magnaporthe oryzae* in two genetic backgrounds of *Oryza sativa*. J Plant Physiol 212:80–93. https://doi.org/10.1016/j.jplph.2017.02.004
23. Eastburn D, McElrone A, Bilgin D (2011) Influence of atmospheric and climatic change on plant–pathogen interactions. Plant Pathol 60:54–69
24. von Tiedemann A, Firsching KH (2000) Interactive effects of elevated ozone and carbon dioxide on growth and yield of leaf rust-infected versus non-infected wheat. Environ Pollut 108:357–363
25. El-Shanawany AA, El-Ghamery AA, El-Sheikh HH, Bashandy AA (2004) Soil solarization and the composition of soil fungal community in upper Egypt. Ass Univ Bull Environ Res 7:137–152
26. Andrew C et al (2016) Climate impacts on fungal community and trait dynamics. Fungal Ecol 22:17–25
27. Looby CI, Treseder KK (2018) Shifts in soil fungi and extracellular enzyme activity with simulated climate change in a tropical montane cloud forest. Soil Biol Biochem 117:87–96
28. Classen AT et al (2015) Direct and indirect effects of climate change on soil microbial and soil microbial-plant interactions: what lies ahead? Ecosphere 6:1–21
29. Staddon P, Heinemeyer A, Fitter A (2002) Mycorrhizas and global environmental change: research at different scales. In: Diversity and integration in mycorrhizas. Springer, pp 253–261
30. Andrade-Linares DR, Veresoglou SD, Rillig MC (2016) Temperature priming and memory in soil filamentous fungi. Fungal Ecol 21:10–15
31. Manyika J, Chui M, Miremadi M, Bughin J, George K, Willmott P, Dewhurst M (2017) A future that works: automation, employment, and productivity. McKinsey Global Institute
32. Mastronardi E, Monreal C, DeRosa MC (2018) Personalized medicine for crops? Opportunities for the application of molecular recognition in agriculture. J Agric Food Chem 66:6457–6461. https://doi.org/10.1021/acs.jafc.7b03295
33. Ferentinos K, Katsoulas N, Tzounis A, Kittas C, Bartzanas T (2014) A climate control methodology based on wireless sensor networks in greenhouses. In: XXIX international horticultural congress on horticulture: sustaining lives, livelihoods and landscapes (IHC2014), vol 1107, pp 75–82
34. Adly I, Sayour HEM, Badawy NM, Ragai H (2013) Evaluation of real-time water quality monitoring system in Nile Tilapia–lettuce Aquaponic based on wireless sensor network. Paper presented at the 6th international conference of animal wealth, Hurgada, Egypt
35. Khan MI, Khan AM, Nouman A, Azhar MI, Saleem MK (2012) pH sensing materials for MEMS sensors and detection techniques. In: International conference on solid-state and integrated circuit

36. López M, Gómez J, Sabater J, Herms A (2010) IEEE 802.15. 4 based wireless monitoring of pH and temperature in a fish farm. In: MELECON 2010–2010 15th IEEE Mediterranean electrotechnical conference. IEEE, pp 575–580
37. Odema M, Adly I, Wahba A, Ragai H (2017) Smart aquaponics system for Industrial Internet of Things (IIoT). In: International conference on advanced intelligent systems and informatics. Springer, pp 844–854
38. Ragai H, Adly I (2012) Site-aware design of wireless sensor networks for O&G plants. In: Proceedings ubiquitous computing sensors and wireless sensor networks workshop
39. Ragai HF, Adly I, Sayour HEM, Wilson S (2017) Remote control and monitoring of fish farms using wireless sensor networks. Paper presented at the 12th international conference on computer engineering & systems
40. Sayour HEM, Tony RM, Rizk MMS (2012) Potentiometric biomimetic sensor based on a molecularly imprinted polymer for imidocarb determination in pharmaceutical formulation and veterinary drug residue in food. Paper presented at the 7th international conference on molecularly imprinted polymers—science and technology. AgroParisTech, 16 rue Claude Bernard, 75005, Paris
41. El-Kader SMA, El-Basioni BMM (2013) Precision farming solution in Egypt using the wireless sensor network technology. Egypt Inf J 14:221–233

Impacts of Climate Change on Microbial Activity in Agricultural Egyptian Soils

Samy Abd El-Malik Mohamed Abd El-Azeem

Abstract Egypt's agriculture sector is highly vulnerable to climate change and its exposed to environmental threats in different faces such as energy, water and food security. Specifically, climate change is possible to change plant species composition, abundance and function of soil community, and plant-microbe interactions that together affect the quality of agricultural soils. For microorganisms inhabiting Egyptian soils, some insight has been exposed to high temperature and accumulating of carbonates and soluble salts in Aridisols (desert soils), and the other inhabiting in alluvial soils (Entisols) in the Nile Delta and the Qattara Depression of the western desert. Egyptian soils have a great diversity of microorganisms such as bacteria, actinobacteria, fungi and arbuscular mycorrhizal fungi that play an important role in nutrient cyclings. Unfortunately, the soil is a complex habitat for microbial growth, and the structure and function of microorganisms are tremendously complex in the soil. These complexations lead the difficult to predict the effects of climate change on the activity of Egyptian soil microorganisms. Climate change will have direct and indirect effects on soil microbial activity. Climate changes include increasing temperature, elevated or increasing the concentration of CO_2, rise changing soil moisture content, increasing of soil salinity, and drought. These changes led to reversely impacted soil microbial communities that affect biogeochemical cycles of nutrients in agricultural soils. Therefore, understanding of microbial activity in the soil is essential for our ability to evaluate the necessity of biogeochemical cycles-climate feedbacks. Soil microbial activities play an important role in increasing soil fertility and recycling of nutrients within the soil. Activity in soil microorganism and/or enzyme is significant as a sensitive indicator of soil biological quality. These activities are informative to determine changes in soil biochemical properties that are affected by environmental stress from natural phenomena or anthropogenic activities. In this chapter, we review the currently available researches regarding the impact of climate change on soil microbial activity, especially in Egyptian soils. Soil microbial activity includes microbial populations, microbial biomass, enzymes activity, soil beneficial microorganisms and carbon sequestration in Egyptian soil.

S. A. El.-M. M. Abd El-Azeem (✉)
Soil and Water Department, Faculty of Agriculture, Suez Canal University, Ismailia 41522, Egypt
e-mail: samy_abdelmalik@agr.suez.edu.eg; sazeem18@gmail.com

© Springer Nature Switzerland AG 2020
E.-S. E. Omran and A. M. Negm (eds.), *Climate Change Impacts on Agriculture and Food Security in Egypt*, Springer Water, https://doi.org/10.1007/978-3-030-41629-4_6

Keywords Climate changes · Egypt · Enzyme activity · Microbial biomass · Microbial community

1 Introduction

Climate is the principal of all environmental aspects, it controls not only the growth and development of the plants but also their geographical distribution. Therefore, the main crops type and soil properties particularly biological features are correlated to the different climatic conditions [1]. Egypt is in the northeast of Africa extending approximately between 22° and 31°36′ N and 25° and 35° E with an area of approximately one million km^2. It is bounded by the Mediterranean Sea to the north and the Red Sea to the east. Egypt is generally homogeneous with low relief. The altitude varies gently from 133 m below the mean sea level (Qattara Depression) to nearly 2500 m (Southern Sinai), with an average of 306 m. Southern Sinai and the Red Sea hills are the most elevated areas, representing a topographic barrier for maritime influences from the Red Sea. The climate of Egypt is an arid or desert climate and generally characterized by moderate and rainy winters, hot and dry summers and a relatively high number of sunny hours (10 h day^{-1}). The annual average rainfall ranging from 60 to 190 mm along the Mediterranean coast to 25 to 60 mm in the Nile delta, and less than 25 mm in upper Egypt and adjacent areas. The rainy season extends only from September to April. The average maximum and minimum temperatures are 33.3 and 21.1 °C, respectively. January (July) is the coldest (warmest) month with an average temperature of 19.3 °C (33.9 °C). The maximum temperatures in summer commonly exceed 30 °C, while winter minimum temperatures rarely fall below 5 °C [2]. The general features of the soils of Egypt are very close to climate and vegetation in this region, for example, low organic matter and biological activity [1]. Changes in temperature and precipitation in Egypt were estimated using General Circulation Model (GCM). This model is accoupled gas-cycle/climate model that drives a spatial climate change scenario generator. The GCM estimates of temperature and precipitation changes in Egypt (Table 1).

Table 1 General circulation model estimates of temperature and precipitation changes in Egypt

Year	Temperature change (°C) mean (standard deviation)			Precipitation change (%) mean (standard deviation)		
	Annual	DJF[a]	JJA[b]	Annual	DJF	JJA
2030	1.0 (0.15)	0.8 (0.21)	1.1 (0.18)	−5.2 (7.93)	−8.9 (3.01)	10.7 (26.35)
2050	1.4 (0.22)	1.2 (0.30)	1.7 (0.26)	−7.6 (11.46)	−12.8 (4.35)	15.4 (38.07)
2100	2.4 (0.38)	2.1 (0.52)	2.9 (0.45)	−13.2 (19.95)	−22.3 (7.58)	26.9 (66.28)

Source Agrawala et al. [3]
[a]DJF is December, January, and February
[b]JJA is June, July, and August

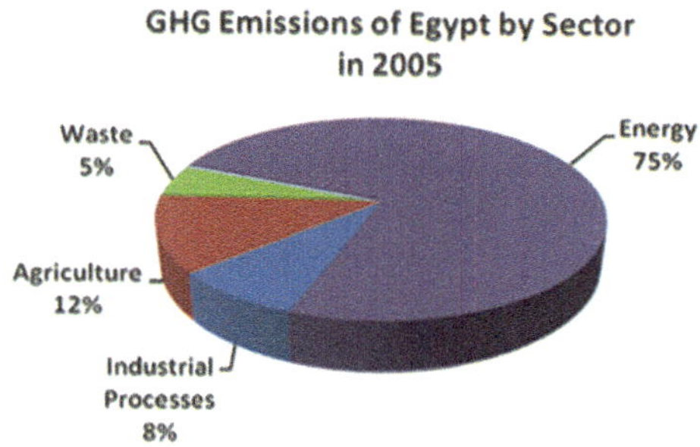

Fig. 1 Greenhouse gases (GHG) emission in Egypt by sector. *Source* El Massah and Omran [4]

Greenhouse gas (GHG) emissions of Egypt, carbon dioxide (CO_2), methane (CH_4) and nitrous oxide (N_2O), are considered minor and representing only 0.53% of the total world emissions in 2005. However these emissions have increased dramatically by around 82% in only 15 years. The main source of GHG emissions in Egypt is the energy sector and releasing about 75% of the total emissions in 2005 followed by the agricultural sector, industrial process, and waste (Fig. 1). Accordingly, the agricultural sector is the second largest emission sector in Egypt [4]. Therefore, changing of climate change especially increased the emission of CO_2 could affect soil biological activities.

Egyptian soils have a great diversity of microorganisms such as bacteria, actinobacteria, fungi and cyanobacteria as well as arbuscular mycorrhizal fungi [5, 6]. These microorganisms play a crucial role in agricultural ecosystems through their functions in nutrients cycling, soil-plant-microbe's interactions and transformations of soil organic matter. Additionally, numerous studies in Egypt indicated that the Egyptian soils contain beneficial plant growth promoting microorganisms (bacteria and fungi) that can be colonized plant roots and enhanced plant growth, yield, nutrients status in soil [7]. For instance, Hanna et al. [8] demonstrated that the considerable diversity of culturable endophytic diazotrophs bacteria in the plant-soil systems of north Sinai deserts. These results indicated that the possibility of these bacteria for future application as biofertilizers and biopesticides. Under greenhouse pot experiment, Abd El-Azeem et al. [9] found that the inoculation of eggplant seedlings with local plant growth promoting rhizobacterial strains lead to reduce the negative effects of salt stress using sandy soil from Ismailia, Egypt. Indeed, without or scarce of these microorganisms in agricultural soil ecosystems, the agriculture and food production are not existing. Unfortunately, climate change and soil mismanagement have the potential to affect the microbial activity in agricultural soils in different locations in Egypt. Climate change will have direct and indirect effects on soil microbial communities and their biogeochemical processes that they underpin.

Previous studies have been investigated the effect of climate changes on soil microbial activity. Some of the microbiological activities were assessed in these studies such as microbial populations or communities [10, 11], soil microbial biomass [12], soil microbial respiration [13, 14], the activity of soil enzymes [15, 16], soil nutrients transformations [17–19].

In Egypt, 93% of the cultivated soils are affected by salinization and waterlogging [20]. The distribution of water within a soil profile and its content in the soil are changed according to rainfall cycles during the season, soil temperature and irrigation periods. In summer particularly in arid and Mediterranean climates like Egypt ecosystems, surface soils have frequently exposed to dry long period followed by a relatively quick wetting. These conditions created two water stresses for microbial activity, the reducing water potential in drying soils and suddenly increase water potential after rewetting soils. Indeed, the moist and well-aerated soils are the optimum conditions for microbial activity, whereas the dry soils led to a decrease in the degradation of soil organic matter and respiration rates [20]. Additionally, Drying and rewetting soils may alter the microbial populations and compositions, indicating that some microbial species or groups are more vulnerable to the drying and rewetting stress than others. On the other hand, flooded soils may also decrease the decomposition rate of organic matter due to restricted aeration through decreasing the oxygen diffusion in water [14]. As there is no information available on the responses of Egyptian soil microorganisms to climate changes. In this chapter we will discuss the impact of climate change on belowground microbial community structure and function.

2 Soil Microbial and Enzyme Activities

Soil microbial activity plays an important role and the main source of the biochemical activities in the agricultural soils. The execration of root exudates in the rhizosphere lead to change the counts and community of soil microorganisms. Therefore, the type of cultivated plant species will affect the activity of soil microorganisms due to their different kinds of root exudates like low or high molecular weight organic compounds [21]. Biochemical processes or reactions affect the transformation and biogeochemical cycles of nutrients in soils. Soil microorganisms, plant roots, and soil animals have been mediated by biochemical processes. The mineralization of organic matter and nutrient cycles are conducted by a huge microbial community and involves a wide range of biological processes [22]. These processes are catalyzed by enzymes that produced by soil microorganism, plant and animal residues. Soil enzymes activities are classified into exoenzymes released from the living cell (free enzyme) and endoenzymes released from disintegrating cells, and enzymes bound to cell constituents. Soil exoenzyme is adsorbed on organic and mineral components or complexed with humic substances, or both and accumulated in soils [23]. The exoenzymes play a vital role in the degradation of organic macromolecules such as cellulose, hemicellulose and lignin, while endoenzymes mineralized smaller molecules like sugar and amino acids [24]. Therefore, several studies on soil enzyme activities have been conducted depending on measuring the rates of enzyme-catalyzed reactions in soil samples. The soil enzymes of β-glucosidase, ureases, phosphatases, and arylsulfatase are representatively being considered to trace the cycles of C, N, P and S in agricultural soils, respectively. The measured activity of the enzyme was

Table 2 Some substantial soil enzymes and their functions

Enzyme	Function	Reaction	Microbial activity indicator
Amidase	Hydrolysis of C–N bonds other than peptide bonds in linear amides	$R\text{-}CONH_2 + H_2O \rightarrow NH_3 + R\text{-}COOH$	N-cycling
Urease	Hydrolysis of urea into ammonium and carbon oxide	$Urea + H_2O \rightarrow CO_2 + 2NH_3$	N-cycling
Phosphatase	Hydrolysis of ester-phosphate bond and release of phosphate	$RNa_2PO_4 + H_2O \rightarrow R\text{-}OH + Na_2HPO_4$	P-cycling
Sulfatase (arylsulfatase)	Hydrolysis of organic-sulfate ester releasing inorganic sulfate	$R\text{-}O\text{-}SO_3^- + H_2O \rightarrow R\text{-}OH + H^+ + SO_4^{2-}$	S-cycling
β-glucosidase	Hydrolysis of cellulose	$Glucoside\text{-}R + H_2O \rightarrow glucose + R\text{-}OH$	C-cycling
Dehydrogenase	Electron transport system	$XH_2 + A \rightarrow X + AH_2$	Endoenzyme

Source Acosta-Martinez and Tabatabai [25], Gianfreda and Ruggiero [26], Skujins [27]

significantly correlated with some soil microbial parameter like microbial biomass and respiration [22]. Dehydrogenase is an endoenzyme and used as bioindicators and relating to soil fertility. The dominant soil enzymes that play a vital role in the decomposition of organic compounds and nutrient cycling in the soil are summarized in Table 2.

Soil microorganisms such as bacteria, fungi, actinobacteria, mycorrhizae and cyanobacteria have the enzyme of urease. This enzyme plays an essential role for degradation or hydrolysis of organic matter in agricultural soils. For example, El-Shinnawi [28] studied the activity of urease in different Egyptian soils treated with 0 and 5% farmyard manure and adjusted moisture at 50, 65 and 80% of the water holding capacity (WHC). He found that the activity of urease is very varied based on the soil type and the order of activity was the following: fertile, saline, alkali, calcareous, and sandy soil. Application of farmyard manure led to slightly increased the urease activity, and the optimum of moisture is 50% of WHC for the tested soils. These findings are in agreement with a study of Elsoury et al. [29], showing the addition of compost or farmyard manure significantly increased soil microbial biomass when compared to mineral fertilizers. The activity of urease, phosphatase and dehydrogenase significantly increased in wheat rhizospheric soil treated with organic amendments when compared to mineral fertilizer under field experiment in Burg El-Arab region, Egypt. The degradation of plant and animal organic nitrogen wastes in soils using urease-producing microorganisms leads to eliminate the vast wastes and soil pollution and provide the plant with available nitrogen simultaneously [30].

Plant growth promoting rhizobacteria, mycorrhizal fungi and phosphate solubilizing bacteria are applied as biofertilizers, biocontrol agents and/or phytostimulators in Egyptian agricultural system. These microorganisms can be colonized plant roots and promoted plant growth and yield as well as increased nutrients availability in soils especially N and P [31, 9, 7]. In soils with a low P bioavailability such as Egyptian soils, free-living P-solubilizing bacteria could release P ions from sparingly soluble inorganic and organic P compounds, and thereby contribute to an increase in the soil P pool that is available for the extraradical mycelium of the arbuscular mycorrhizal fungus and that can be transferred to the host [32, 33]. Unfortunately, soil enzyme activities and biochemical processes in agricultural soils will impact by climate change especially in arid climatic regions. Also, climate change affects the performance of plant growth promoting microorganisms. In the following section, we will describe the response of microbial activity to changes in temperature, CO_2 concentration, moisture content and salinity.

3 Response of Microbial Activity to Climate Change

Egypt is mostly located in the region that is highly vulnerable to climate change impact. Additionally, it is typically lack technological expertise and financial resources for adaptation to climate change or its mitigation. The impact of climate change on beneficial plant-microbe interactions has recently received considerable attention [34]. In the following sections, we discuss the effect of climate changes such as temperature, elevated CO_2, soil moisture content and soil salinity on Egyptian soil microbial activities like the population of soil microorganisms, soil microbial biomass and respiration, and soil microbe-plant interactions.

3.1 Effect of Temperature

Natural and/or anthropogenic activities increased atmospheric CO_2 levels, resulting in increased the temperature of the global surface about 1.8 and 3.6 (4–7) °C by the year 2100 [35]. For instance, increased temperature by 5 °C in a temperate region changed the relative abundances of soil bacteria and increased the bacterial to the fungal ratio (B/F ratio) of the community [36]. As the climate warms, the populations of soil microorganisms must adapt or die. In the soil microclimate, soil temperature can influence on enzyme pool sizes through their effect on the activity of enzymes, microbial efficiency and availability of substrate. Therefore, the overall and specific rate of enzyme production and the relative rate of various soil enzymes will also be affected as soil temperature increase [37]. According to Stone et al. [38] the activity of the microbial extracellular enzyme (α-glucosidase, β-glucosidase, β-xylosidase, cellobiohydrolase, and *N*-acetyl-glucosaminidase) is more sensitive to nitrogen application and change of soil temperature. They also found that nitrogen-decaying enzymes have lesser sensitive to temperature than carbon-decaying enzymes. This finding can be attributed to the soil warm led to an increased nitrogen limitation, resulting in an increase in the production of nitrogen-decaying enzymes by soil microorganisms when compared to the production of carbon-decaying enzymes [39]. On investigating such areas, Gao et al. [12] showed that increased microbial biomass C and significantly promoted the activity of enzymatic hydrolysis of lignin and acid phosphatase as a result of short-term simulated soil warming and the addition of nitrogen fertilizer in the subtropical province, China.

Several previous studies have demonstrated the direct positive effect of warming climate on the colonization, community and structure of arbuscular mycorrhizal (AM) fungi [34, 40]. Additionally, soil temperature affects carbon allocation within arbuscular mycorrhizal networks and carbon transport from plant to fungus. For example, Hawkes et al. [41] evaluate the influence of soil temperature on the community of different species from AM fungi associated with *Plantago lanceolata* and their structure. They postulated that increased soil temperature from 14 to 26 °C significantly changed the structure and allocation of the AM fungi hyphal networks and induce a switch from more vesicles (responsible for storage) in cooler soils to more extensive extraradical hyphal networks (indicating growth) in warmer soils. Other researchers, for instance, Gavito et al. [42] and Heinemeyer and Fitter [43] also observed that higher temperature (24–30 °C) leads to promote the root colonization by AM fungi, more extensive extraradical mycelium, and more glucose uptake, nevertheless the optimum temperature was varied based on species of AM fungus. According to these authors, these effects were due to faster carbon allocation to the rhizosphere and increased respiration of the extraradical mycelium as increase soil temperature. These effects in the biomass of AM fungi are very important as AM fungi play an essential role in the plant, plant nutrition, community and ecosystem responses to world change.

Similarly, with AM fungi, soil temperature has a negative or positive effect on the performance of plant-beneficial bacterial, indicating bacterial genotype-specific favorite's temperature conditions [44]. These authors observed the shoot and root of winter wheat significantly increased when inoculated with *Mycobacterium* sp., *Pseudomonas fluorescens* and *Pantoea agglomerans* strains at 16 °C when compared to 26 °C in the loamy sand. On the other hand, *Mycobacterium phlei* and *Mycoplana bullata* performed well under both temperatures. Under Egyptian arid conditions, high soil temperature has a positive or negative effect on the performance of free-living bacteria and the life of rhizobia. The optimum temperature for growth in the medium is 28–30 °C, and many are unable to grow at 38 °C. Nevertheless, some strains of rhizobia from the woody legumes are grown well at 40 or 44 °C. Additionally, increased soil temperature leads to form ineffective nodules, but *R. leguminosarum* bv. *Phaseoli* exhibit higher heat tolerance and can form effective nodule with their host legumes [45]. In Egyptian soils, the optimum activity of nitrogen-fixing *Azotobacter* and *Clostridium* occurred when soil temperature raised to around 30 °C, while increased soil temperature over this degree is suitable for *Clostridium* than *Azotobacter* [46]. Most of the studies reported that soil microorganisms have a different response to elevated temperature or global changes. For instance, the populations of bacteria (Gram-positive or -negative bacteria) increased with increased temperature, may be due to declining the availability of substrates. However, fungi and actinobacteria biomarkers decreased at higher temperatures [47]. It has been indicated that the nitrification activities in the subtropical Egyptian clay loam soil increased with temperature and showed the maximum rate at 30 °C. The optimum temperature for the potential activation and population size of ammonium oxidizing or nitrite oxidizing bacteria, nitrate reducing bacteria and most probable numbers is 25 or 30 °C and exhibited apparent activation energies between 61 and 202 kJ mol^{-1}. However, a few nitrifiers and denitrifiers were also able to grow at 8 or 50 °C [48].

3.2 Impact of Carbon Dioxide

The Egyptian Central Agency for Public Mobilization and Statistics mentioned that the amount of carbon dioxide (CO_2) emissions from the consumption of oil products and gas in Egypt is 197.1 million tons in the year 2014 and increased to 299 million tons in the year 2015 [49]. The CO_2 is the most abundant and important well-mixed greenhouse gases contribute to human-induced climate change and responsible for 74% of global warming over the past decade [50]. Increasing atmospheric CO_2 concentration and simultaneous rises in temperature are influencing the global climate, henceforth affecting growth, development and functioning of plants [51]. The response of soil microbial activity to changes in atmospheric CO_2 concentrations is mainly indirectly and can be positive or negative, and consistent overall trends between sites and studies have not been observed. Moreover, the primary effects of the increased concentration of atmospheric CO_2 include increased belowground C allocation (root biomass and root respiration) to the root zone, changed the composition of root exudates, altered C/N ratio and impact on soil nutrients status [52]. For example, Soussana and Hartwig [53] found that increased the concentration of atmospheric CO_2 led to an increase the C:N ratio of plant residues and exudates, resulting in temporary N-immobilization and reduce the availability of soil nitrogen. Soils exposed to elevated CO_2 had higher relative abundances of fungi and higher activities of a soil carbon degrading enzyme that led to more rapid rates of soil organic matter degradation than soils exposed to ambient CO_2. The isotopic composition of microbial fatty acids confirmed that elevated CO_2 increased microbial utilization of soil organic matter. These results illustration how elevated CO_2 through altering soil microbial communities, can cause a potential carbon sink to become a carbon source [54]. Soil enzymes activity, community structure and microbial total biomass also have shown highly varied responses to elevated CO_2. For instance, Niklaus [55] investigates in situ effects of elevated CO_2 on soil microbial biomass and activity in a nutrient-poor calcareous soil exposed to long-term CO_2 enrichment. He found that soil basal respiration was significantly increased by 14%, while the metabolic quotient for CO_2 (qCO_2) (basal respiration/microbial C) was not statistically significant differences by elevated CO_2 treatments. Microbial biomass C and soil organic C and N were not significantly affected by elevated CO_2, and microbial biomass N increased by 18%. According to Dijkstra et al. [56], elevated CO_2 increased labile C and microbial biomass, but had no effect on net N mineralization, respiration of more recalcitrant C, or total soil C and N.

The previous literature indicate that in many systems exposed to elevated CO_2, mycorrhizal fungi can be sequestering the large amount from C in living, dead, and residual hyphal biomass in soil [57]. Increasing CO_2 had a positive effect on the populations of endo/or ectomycorrhizal fungi, while the effect of elevated CO_2 on plant growth promoting bacteria and endophytic fungi were more variable (Positive, neutral or negative) based on the study conditions. In general, plant-associated microorganisms had a positive or beneficial effect on the plant under elevated CO_2 [34]. Extraradical hyphae and percentage root length colonized in AM fungi in a manner proportional to the plant response under elevated CO_2 increased [58]. Increased atmospheric CO_2 concentration is much more affected by the rhizosphere zone than the bulk soil. The production CO_2 in the rhizosphere by soil microorganisms and plant roots is promoted by elevated CO_2 concentration and may be higher than the increase in root biomass [59]. Previous studies indicated that the abundance of arbuscular and ectomycorrhizal fungi in the rhizosphere had a positive response to elevated CO_2, whereas plant growth-promoting bacteria and endophytic fungi have so susceptible response to elevated CO_2. Additionally, beneficial effect on plant growth due to use plant-associated microorganisms will increase under elevated CO_2 [34]. The growth of mycorrhizae on plants have been positively or negatively response to elevated CO_2 concentration, and these responses are varied based on the differences in plant and fungal species under investigation.

Increased CO_2 attributed to the increase in soil respiration can have a potentially negative impact on soil C sequestration, but increased atmospheric CO_2 led to an increasingly global ecosystem C storage [59]. Previous studies investigated the potential of C storage of reclaimed soils under organic management. For instance, Luske and van der Kamp [60] investigated the potential of C storage of reclaimed desert soils in Egypt under two organic farms owned by Sekem (1–30 years in use). They showed that the reclaimed desert soils sequester C very quickly in the first few years after land reclamation. However, the rate of C sequestration is decreased after several years. Soil C storage increased from 3.9 to 28.8–31.8 tons C ha^{-1} in 30 years of organic agriculture. The effect of short (2030), medium (2050) and long-term (2100) climate change on soil organic carbon stock was studied by Muñoz-Rojas et al. [61] at different soil depths in a Mediterranean arid area (El-Fayoum, Northern Egypt) for different land use type. The result demonstrated that the evaluation of soil organic C contents and dynamics lengthways the soil profile and the potential for soil C sequestration, especially in the subsoil.

3.3 Impact of Soil Moisture Content

Water is a critical factor in global change. Egypt has been suffering from severe water scarcity in recent years, and according to the United Nations classification, Egypt falls into the category of high-water stress countries. The status of soil water can be described as soil water content and potential. Soil water content indicates how much water is present in soil (water balance), whereas soil water potential relates to the energy level by which the water is held in the soil (water movement). Thus, soil matric potential may be considered an important measure of water availability, whereas soil moisture content may appear to be a more useful parameter when water availability is not limiting for microbial activity [62]. Soil moisture content is changed based on time, space, a form of wet-dry cycles, flooding and drought [52, 14]. The general interaction between soil moisture and oxygen (O_2) concentrations is intuitively clear for predominate aerobic or anaerobic conditions due to regulating O_2 diffusion. Therefore, the relationship between soil moisture and soil microbial activity is more greatly variable and complicated than that of temperature. For instance, high soil moisture content leads to decrease the diffusion of O_2 in water in which the activity of aerobic microbes is declined, but these conditions led to an increase the activity of anaerobic microbes [14]. The maximum of aerobic microbial activity occurred at moisture levels between 50 and 70% of water-holding capacity. At high moisture content (matric potential > −0.01 MPa) decreases rates of organic matter decomposition, due to a low oxygen supply, while low soil moisture decreases microbial activity by reducing diffusion of soluble substrates, microbial mobility and intracellular water potential [19].

Soil microbial activity is sensitive to changes in soil moisture, but the mechanisms suppressing microbial activity under dry conditions continue to be debated. Soil microbes are substrate limited because solute diffusivity is halted due to the breaking of water film continuity under dry conditions. Therefore, the rates of soil microbial respiration were declined with decreasing soil moisture content to the point that the effective soil microbial activity is halt. This "water stress" moisture threshold corresponds to water potential −15 MPa in disturbed soil samples, regardless of the climate of origin of the soil and its texture [13]. It has been recently indicated that the precipitation affecting microbial communities and vary with moisture content. Precipitation led to a decrease in soil bacterial diversity in the Mediterranean region compared with arid sites. Using phospholipid fatty acids, real-time quantitative PCR and terminal restriction fragment length polymorphism fingerprint analyses, Bachar et al. [63] indicated that although soil bacterial abundance decreases with precipitation, bacterial diversity is independent of precipitation gradient. Moreover, the composition of the bacterial community was found to be distinct to each site along the rainfall gradient. Specifically, in Egyptian agricultural soil, the effect of soil water content on the degradation of cellulose has not been well investigated, and few studies have been conducted on the cellulolytic activity of soil fungi. However, Badran and Abdel-Rahiem [64] found that the highest population of cellulose-decomposing fungi was observed in soil amended with sugar cane straw at 60% water content, but the lowest was in soil mixed with broad bean straw at 100% water content.

There are several mechanisms or physical processes affecting soil microbial activity that vary with moisture content. Drought is one of the major constraints on agricultural productivity Egyptian and worldwide and the most common environmental stress that soil microbes experience. The drought had negatively or positively effect on mycorrhizae development and their root colonization. In general, drought reduced the colonization by mycorrhizae. However, this kind of response is varied based on the type of strain as reported by Davies Jr. et al. [65], who demonstrated that drought enhanced the formation of arbuscular and hyphal development of *Glomus* sp. Strain ZAC-19, while reducing the root colonization by *Glomus fasciculatum*. On the contrary, the results from some studies obtained specific adaptations of certain mycorrhizae strains to drought conditions [66]. On the other hand, drought has an adverse effect on soil enzyme activities, which direct influence on the potential of photosynthetic production or indirectly effect on water use efficiency and nutrient availability in Mediterranean regions. For example, Sardans and Peñuelas [16] stated that the activity of soil enzymes like urease, protease, β-glucosidase and acid phosphatase was decreased under drought conditions, but the activity of alkaline phosphatase not significantly affect. Moreover, Elwan and Mahmoud [67] investigated the response of the bacterial population to the edaphic drought of the Egyptian desert in summer. They found that spore-forming bacteria such as *Bacillus subtilis*, *Bacillus licheniformis* and *Bacillus megaterium* have predominated in the soil and can survive drought conditions. Water stress was also affecting nitrification activity through decreased the substrate availability and triggering of physiological changes (downregulation of basic metabolism and upregulation of stress-related gene expression) to tolerate the induced osmotic changes. However, little information occurs about the differences between ammonia oxidizing bacteria (AOB) and archaea (AOA) during the initial stage of moisture stress. In this regard, Vasileiadis et al. [68] suggested that the reduction of soil moisture content from 87 to 50% of the water holding capacity leads a ~99% reduction of AOB but not of AOA *amoA* transcripts that did not change significantly.

3.4 Impact of Soil Salinity

Soil salinity is one on the main environmental factors responsible for the decreasing productivity of a wide variety of crops in an arid and semi-arid area of the world that would challenge with their food production in the 21st century with climate change [69]. Unfortunately, saline, saline-sodic, and sodic soils have a strong presence in the Nile delta soils and represent an average of 37% of the total Egyptian cultivated soils.

Numerous studies indicated that saline stress in agricultural soils due to the irrigation practices and the application of chemical fertilizers. Therefore, several researches have been conducted to evaluate the detrimental effect of salinity on the soil microbial counts and their activities on natural or synthetic saline soils. In recent decades, increased soil salinity has been a critical problem in the agricultural soils of Egypt due to increasing the salinization of soils and ground waters [70]. Increasing soil salinity led to an increase in soil osmotic potential in water and nutrient uptake by plants, poor aeration and a decrease in water permeability. These conditions may affect soil microbial activities such as soil enzyme activities, reduce microbial community structure and biochemical functioning in soil [15]. Also, increased soil salinity has a negative impact on the sustainability of beneficial soil microorganisms associated with plant rhizosphere. These negative impact can be attributed to adverse effects of toxic salts and lose soil aggregates and soil structure that leading unsuitable habitat or niches for soil microbial growth [71].

Soil enzymes play an essential role in soil biochemical transformation and nutrients cycles that related to soil fertility and quality. They play a fundamental role in the decomposition of organic residues, the formation of organic matter and in intercellular metabolic reactions in the soil. Amongst the several soil enzymes, β-glucosidase, urease, invertase, phosphatases, arylsulfatase and dehydrogenase are necessary for nutrient transformations for different plants. Under controlled conditions, soil salinity had a negative effect on soil enzyme activities, but the degree of inhibition varied according to soil microbial community, the enzyme analyzed, and the nature and the amount of soil added [72]. The inhibition of enzyme activity in saline soils can be due to the osmotic hydration of the microbial cell that releases intracellular enzymes that become vulnerable to the attack by soil proteases. The salting-out effect modifies the ionic conformation of the protein-enzyme active site, and specific ionic toxicity causes a nutritional imbalance for microbial growth and subsequent enzyme synthesis [72, 15]. Some researchers in Egypt studied the effect of salinity on soil enzyme activities using clayey soil under laboratory conditions. The authors showed that the activity of urease and invertase significantly decreased with an increasing concentration of NaCl during the incubation period. Additionally, they also observed that the effect of salinity on the activity of nitrate reductase was mainly inhibitory in the majority of the treatments [15].

Other researchers, for instance Wong et al. [73] also indicated that the effect of soil salinity on soil microbial biomass and soil respiration. For example, soil microbial biomass was highest (459–565 mg kg^{-1} soil) in the high salinity treatments (30 dS m^{-1}) and lowest (158–172 mg kg^{-1} soil) in the low salinity treatments (0.5 dS m^{-1}). On the other hand, soil respiration rate was highest (56–80 mg CO_2-C kg^{-1} soil) in the low salinity treatments and lowest (1–5 mg CO_2-C kg^{-1} soil) in the mid salinity treatments (0.5 dS m^{-1}) [73].

Several investigators reported that fungi are more sensitive to osmotic stress than bacteria or actinobacteria [72, 74]. There is a significant decrease in the total count of fungi and most general fungi and species in some Egyptian soils that salinized with different concentration of sodium chloride (5–20% NaCl). Similarly, the total count of bacteria and actinobacteria was remarkably decreased with an increase in soil salinity level to above 5% NaCl [15]. In fungi, increased soil salinity decreases spore germination and the growth of hyphae and changes the morphology, resulting in the formation of spores with thick walls [72].

Sodic, saline and saline-sodic soils are the three main types of salt-affected soils dominant in the Nile delta especially in the middle and north delta soils which suffer from its location close to the Mediterranean coast and saline waterlogging [75]. These saline conditions led to an increase in the diversity and community of indigenous microorganisms especially bacteria that can be modified in their structure and physiological activity under these conditions. These bacterial groups can maintain osmotic equilibrium between the cytoplasm and the surrounding medium excluding sodium ions from inside the cell. Additionally, they can synthesis specific compatible organic osmolytes like proline, glycine and glutamine that accumulate the inorganic solutes (Na^+, K^+ and Mg^{2+}) [76]. Soil salinity in Egypt is one the common outstanding serious problems affecting crop production in cultivated and non-cultivated soils. Therefore, the influence of soil salinity on soil microbial populations and enzyme activities, especially in saline regions of Egypt has received more attention. For example, soil samples (0–15 and 15–30 cm) were collected from salt-affected soils at three different regions, the Suez Canal, Sinai peninsula and Al-Salhia regions, by the researcher Ragab [77]. He found that the total counts of bacteria, nitrifying bacteria and cellulose decomposers are negatively correlated with total soluble salts and positively correlated with organic carbon in surface layers, but nitrogen-fixing bacteria (*Azotobacter* and *Clostridium*) are not correlated with soluble salts or organic carbon. On the other hand, halophilic bacteria were positively correlated with total soluble salts and organic carbon in the soil layers. That is, the Egyptian microbiologist should be increasing the researches about behavior and physiology salt tolerant bacteria in saline soils to pursue the application of these bacteria as biofertilizers in saline environments.

4 Conclusions

Climate change has an impact on Egyptian soil microbial community and activities both directly or indirectly. The fluctuation of temperature and elevated CO_2 are the most climatic changes in Egypt that positively, negatively and neutrally affect soil microbial activity, community and enzyme activities. Additionally, predominant of saline soils in some regions in Egypt is a vital factor to the effect on soil microorganism populations and activities. However, most of researchers or studies focused on climate change impacts as a single factor or by separate the different climate change components (altered temperature and moisture or elevated CO_2) and manipulating their responses on soil microbial activity independently. Indeed, the future studies should be focused on multiple, and simultaneously interacting climate change factors such as temperature will interact with changes in moisture that both impact on the overall and relative rate of enzyme production by soil microorganisms. The behavior and response of plant growth promoting microorganisms to climate change are very variedly and complicated, especially in saline habitats. Therefore, further studies should be working on the activity of aerobic and anaerobic soil microbes and enzyme activities in Egyptian saline agricultural soils.

The structure and functions of soil microorganisms are tremendously complex and can be impacted by climate changes in several ways that prevent the ability to draw reliable conclusions. Therefore, the investigators must be adding the estimation of microbial biomass and compositions in their studies to fully understand and proceed with the influences of climate change on soil microbial activity, and design studies over the longer-term. Long-term experiments are critical to more illustrate the interactions between biotic and abiotic factors and better predict soil microbial responses to future climate changes. On the other hand, use of modern molecular techniques such as metagenomics will help and gave more details about soil microbial communities that would otherwise be undetectable.

5 Recommendations

The future impact of the above negative environmental consequences of climate change on Egyptian soil microbial activity and dynamics are the main points of concern. For instance, many newly reclaimed soils in the Suez Canal region, Egypt are already under excessive applications of chemicals and pesticides, and signs of severe environmental and health impacts have already been detected. Beneficial plant growth promoting microorganisms could be an important biotechnological tool for sustainable agriculture by their positive effect on soil fertility and crop production and quality and could reduce the costs for chemical fertilization and pest control. Staying abreast of technological innovations is vital to the future efficiency and competence of the agriculture system in the region. Therefore, the future work should be determining the best approaches to observe and quantify the response of plant-microbe interactions or microbe-microbe interactions to climate change to predict future ecosystem function. Additionally, temporal and seasonal variation and plant root exudates play a vital role in the response of soil microbial activity to climate change, subsequently should be incorporated into future climate change experimental manipulations.

Although, soil microorganisms play an essential role in nutrients cycling in agricultural soils; many microbiological processes in these soils are changed due to climate change, and these changes are likely to result in changed plant production and atmospherically active gases. However, the impacts of multiple climate changes on microbial activity, abundance, and the composition of the microbial community in agricultural soils are poorly understood because of the complexity of the microbial community in the soil. Therefore, the future work or researches should be concerning about soil microorganisms that high resistance to multiple climate changes such as climate warming, adapt to soil salinity and drought as well as resistance to elevated CO_2, especially in Egyptian climate conditions.

References

1. Zahran MA (2010) Climate–vegetation relationships: perspectives. In: Gilbert F (ed) Climate–vegetation: Afro-Asian Mediterranean and Red Sea coastal lands. Springer Netherlands, Dordrecht, pp 219–248. https://doi.org/10.1007/978-90-481-8595-5_3
2. El Kenawy A, López-Moreno JI, Vicente-Serrano SM, Morsi F (2010) Climatological modeling of monthly air temperature and precipitation in Egypt through GIS techniques. Clim Res 42(2):161–176
3. Agrawala S, Moehner A, El Raey M, Conway D, Van Aalst M, Hagenstad M, Smith J (2004) Development and climate change in Egypt: focus on coastal resources and the Nile. Working Party on Global and Structural Policies, Organization for Economic Cooperation and Development (OECD)
4. El Massah S, Omran G (2015) Would climate change affect the imports of cereals? The case of Egypt. In: Leal Filho W (ed) Handbook of climate change adaptation. Springer Berlin Heidelberg, Berlin, Heidelberg, pp 657–683. https://doi.org/10.1007/978-3-642-38670-1_61
5. Agwa HE, Al-Sodany YM (2003) Arbuscular-mycorrhizal fungi (Glomales) in Egypt. III: distribution and ecology in some plants in El-Omayed Biosphere Reserve. Egypt J Biol 5:19–26
6. Zahran H, Moharram A, Mohammad H (1992) Some ecological and physiological studies on bacteria isolated from salt-affected soils of Egypt. J Basic Microbiol 32(6):405–413
7. Abd El-Azeem SA, Mehana TA, Shabayek AA (2007) Some plant growth promoting traits of rhizobacteria isolated from Suez Canal region, Egypt. In: African crop science conference proceedings, pp 1517–1525
8. Hanna AL, Youssef HH, Amer WM, Monib M, Fayez M, Hegazi NA (2013) Diversity of bacteria nesting the plant cover of north Sinai deserts, Egypt. J Adv Res 4(1):13–26. https://doi.org/10.1016/j.jare.2011.11.003
9. Abd El-Azeem SA, Elwan MW, Sung J-K, Ok YS (2012) Alleviation of salt stress in eggplant (*Solanum melongena* L.) by plant-growth-promoting rhizobacteria. Commun Soil Sci Plant Anal 43(9):1303–1315
10. Bruce K, Jones T, Bezemer T, Thompson L, Ritchie D (2000) The effect of elevated atmospheric carbon dioxide levels on soil bacterial communities. Glob Change Biol 6(4):427–434
11. Runion G, Curl E, Rogers H, Backman P, Rodriguez-Kabana R, Helms B (1994) Effects of free-air CO_2 enrichment on microbial populations in the rhizosphere and phyllosphere of cotton. Agric For Meteorol 70(1–4):117–130
12. Gao J, Wang E, Ren W, Liu X, Chen Y, Shi Y, Yang Y (2017) Effects of simulated climate change on soil microbial biomass and enzyme activities in young Chinese fir (*Cunninghamia lanceolata*) in subtropical China. Acta Ecol Sin 37(4):272–278. https://doi.org/10.1016/j.chnaes.2017.02.007
13. Manzoni S, Katul G (2014) Invariant soil water potential at zero microbial respiration explained by hydrological discontinuity in dry soils. Geophys Res Lett 41(20):7151–7158
14. Yan N, Marschner P, Cao W, Zuo C, Qin W (2015) Influence of salinity and water content on soil microorganisms. Int Soil Water Conserv Res 3(4):316–323. https://doi.org/10.1016/j.iswcr.2015.11.003
15. Omar SA, Abdel-Sater MA, Khallil AM, Abd-Alla MH (1994) Growth and enzyme activities of fungi and bacteria in soil salinized with sodium chloride. Folia Microbiol 39(1):23–28. https://doi.org/10.1007/bf02814524
16. Sardans J, Peñuelas J (2005) Drought decreases soil enzyme activity in a Mediterranean *Quercus ilex* L. forest. Soil Biol Biochem 37(3):455–461. https://doi.org/10.1016/j.soilbio.2004.08.004
17. Fierer N, Schimel JP (2002) Effects of drying-rewetting frequency on soil carbon and nitrogen transformations. Soil Biol Biochem 34(6):777–787
18. Schjønning P, Thomsen IK, Moldrup P, Christensen BT (2003) Linking soil microbial activity to water- and air-phase contents and diffusivities. Soil Sci Soc Am J 67(1):156–165. https://doi.org/10.2136/sssaj2003.1560

19. Stres B, Danevčič T, Pal L, Fuka MM, Resman L, Leskovec S, Hacin J, Stopar D, Mahne I, Mandic-Mulec I (2008) Influence of temperature and soil water content on bacterial, archaeal and denitrifying microbial communities in drained fen grassland soil microcosms. FEMS Microbiol Ecol 66(1):110–122
20. Abdelrahman Abdalla M, Alavi Panah SK, Bationo A, Chude V, Comerma J, Gerasimova M, Hempel J, Indraratne S, Krasilnikov P, McKenzie N, Mendonça Santos M, Norbu C, Ogunkunle A, Pennock D, Reinsch T, Robinson D, Smith P, Taboada M, Yagi K, Erdogan E (2015) Status of the world's soil resources—main report
21. Alshaal T, El-Ramady H, Al-Saeedi AH, Shalaby T, Elsakhawy T, Omara AE-D, Gad A, Hamad E, El-Ghamry A, Mosa A (2017) The rhizosphere and plant nutrition under climate change. In: Essential plant nutrients. Springer, pp 275–308
22. Burns RG, Dick RP (2002) Enzymes in the environment: activity, ecology, and applications. Marcel Dekker, Inc
23. Tabatabai M (1982) Soil enzymes. In: Methods of soil analysis. Part 2. Chemical and microbiological properties (methods of soil analysis), pp 903–947
24. Yang L, Zhang Y, Li F (2012) Soil enzyme activities and soil fertility dynamics. In: Advances in citrus nutrition. Springer, pp 143–156
25. Acosta-Martinez V, Tabatabai M (2000) Enzyme activities in a limed agricultural soil. Biol Fertil Soils 31(1):85–91
26. Gianfreda L, Ruggiero P (2006) Enzyme activities in soil. In: Nannipieri P, Smalla K (eds) Nucleic acids and proteins in soil. Springer Berlin Heidelberg, Berlin, Heidelberg, pp 257–311. https://doi.org/10.1007/3-540-29449-x_12
27. Skujins J (1967) Enzymes in soil. Soil Biochem 1:371–414
28. El-Shinnawi M (1978) Urease activity in different soils of Egypt. Zentralbl Bakteriol B 133(6):483–492
29. Elsoury H, Shouman A, Abdelrazek S, Elkony H (2015) Soil enzymes and microbial activity as influenced by tillage and fertilization in wheat production. Egypt J Soil Sci 55(1):53–65
30. Hasan H (2000) Ureolytic microorganisms and soil fertility: a review. Commun Soil Sci Plant Anal 31(15–16):2565–2589
31. Abd-Alla MH, El-Enany A-WE, Nafady NA, Khalaf DM, Morsy FM (2014) Synergistic interaction of *Rhizobium leguminosarum* bv. viciae and arbuscular mycorrhizal fungi as a plant growth promoting biofertilizers for faba bean (*Vicia faba* L.) in alkaline soil. Microbiol Res 169(1):49–58
32. El-Tarabily KA, Nassar AH, Sivasithamparam K (2008) Promotion of growth of bean (*Phaseolus vulgaris* L.) in a calcareous soil by a phosphate-solubilizing, rhizosphere-competent isolate of *Micromonospora endolithica*. Appl Soil Ecol 39(2):161–171. https://doi.org/10.1016/j.apsoil.2007.12.005
33. Yasser MM, Mousa AS, Massoud ON, Nasr SH (2014) Solubilization of inorganic phosphate by phosphate solubilizing fungi isolated from Egyptian soils. J Biol Earth Sci 4(1):83–90
34. Compant S, Van Der Heijden MG, Sessitsch A (2010) Climate change effects on beneficial plant-microorganism interactions. FEMS Microbiol Ecol 73(2):197–214
35. Change IPOC (2007) Climate change 2007: The physical science basis. Agenda 6(07):333
36. DeAngelis KM, Pold G, Topçuoğlu BD, van Diepen LT, Varney RM, Blanchard JL, Melillo J, Frey SD (2015) Long-term forest soil warming alters microbial communities in temperate forest soils. Front Microbiol 104:1–13. https://doi.org/10.3389/fmicb.2015.00104 Article 6
37. Steinweg JM, Dukes JS, Paul EA, Wallenstein MD (2013) Microbial responses to multi-factor climate change: effects on soil enzymes. Front Microbiol 4
38. Stone MM, Weiss MS, Goodale CL, Adams MB, Fernandez IJ, German DP, Allison SD (2012) Temperature sensitivity of soil enzyme kinetics under N-fertilization in two temperate forests. Glob Change Biol 18(3):1173–1184
39. Wallenstein MD, McMahon SK, Schimel JP (2009) Seasonal variation in enzyme activities and temperature sensitivities in Arctic tundra soils. Glob Change Biol 15(7):1631–1639
40. Heinemeyer A, Ineson P, Ostle N, Fitter A (2006) Respiration of the external mycelium in the arbuscular mycorrhizal symbiosis shows strong dependence on recent photosynthates and acclimation to temperature. New Phytol 171(1):159–170

41. Hawkes CV, Hartley IP, Ineson P, Fitter AH (2008) Soil temperature affects carbon allocation within arbuscular mycorrhizal networks and carbon transport from plant to fungus. Glob Change Biol 14(5):1181–1190
42. Gavito ME, Olsson PA, Rouhier H, Medina-Peñafiel A, Jakobsen I, Bago A, Azcón-Aguilar C (2005) Temperature constraints on the growth and functioning of root organ cultures with arbuscular mycorrhizal fungi. New Phytol 168(1):179–188
43. Heinemeyer A, Fitter A (2004) Impact of temperature on the arbuscular mycorrhizal (AM) symbiosis: growth responses of the host plant and its AM fungal partner. J Exp Bot 55(396):525–534
44. Egamberdiyeva D, Höflich G (2003) Influence of growth-promoting bacteria on the growth of wheat in different soils and temperatures. Soil Biol Biochem 35(7):973–978
45. Zahran HH (2001) Rhizobia from wild legumes: diversity, taxonomy, ecology, nitrogen fixation and biotechnology. J Biotechnol 91(2–3):143–153
46. Abd-El-Malek Y (1971) Free-living nitrogen-fixing bacteria in Egyptian soils and their possible contribution to soil fertility. Plant Soil 35(1):423–442
47. Zogg GP, Zak DR, Ringelberg DB, White DC, MacDonald NW, Pregitzer KS (1997) Compositional and functional shifts in microbial communities due to soil warming. Soil Sci Soc Am J 61(2):475–481
48. Saad OALO, Conrad R (1993) Temperature dependence of nitrification, denitrification, and turnover of nitric oxide in different soils. Biol Fertil Soils 15(1):21–27. https://doi.org/10.1007/bf00336283
49. Fawaz MM, Soliman SA (2016) The potential scenarios of the impacts of climate change on Egyptian resources and agricultural plant production. Open J Appl Sci 6(4):270–286
50. Canadell JG, Schulze ED (2014) Global potential of biospheric carbon management for climate mitigation. Nat Commun 5:5282
51. Pathak H, Aggarwal PK, Singh S (2012) Climate change impact, adaptation and mitigation in agriculture: methodology for assessment and applications. Indian Agricultural Research Institute
52. Balser TC, Gutknecht JL, Liang C (2010) How will climate change impact soil microbial communities? In: Soil microbiology and sustainable crop production. Springer, pp 373–397
53. Soussana J, Hartwig U (1995) The effects of elevated CO_2 on symbiotic N_2 fixation: a link between the carbon and nitrogen cycles in grassland ecosystems. Plant Soil 187(2):321–332
54. Carney KM, Hungate BA, Drake BG, Megonigal JP (2007) Altered soil microbial community at elevated CO_2 leads to loss of soil carbon. Proc Natl Acad Sci 104(12):4990–4995
55. Niklaus PA (1998) Effects of elevated atmospheric CO_2 on soil microbiota in calcareous grassland. Glob Change Biol 4(4):451–458
56. Dijkstra FA, Hobbie SE, Reich PB, Knops JM (2005) Divergent effects of elevated CO_2, N fertilization, and plant diversity on soil C and N dynamics in a grassland field experiment. Plant Soil 272(1–2):41–52
57. Treseder KK, Allen MF (2000) Mycorrhizal fungi have a potential role in soil carbon storage under elevated CO_2 and nitrogen deposition. New Phytol 147(1):189–200
58. Alberton O, Kuyper TW, Gorissen A (2005) Taking mycocentrism seriously: mycorrhizal fungal and plant responses to elevated CO_2. New Phytol 167(3):859–868
59. Nannipieri P (2011) Potential impacts of climate change on microbial function in soil. In: Sustaining soil productivity in response to global climate change: science, policy, and ethics, pp 201–211
60. Luske B, van der Kamp J (2009) Carbon sequestration potential of reclaimed desert soils in Egypt. Louis Bolk Instituut & Soil and More International. http://orgprints.org/16438/1/2192.pdf. Accessed 4 Dec 2009
61. Muñoz-Rojas M, Abd-Elmabod SK, Zavala LM, De la Rosa D, Jordán A (2017) Climate change impacts on soil organic carbon stocks of Mediterranean agricultural areas: a case study in Northern Egypt. Agric Ecosyst Environ 238(Supplement C):142–152. https://doi.org/10.1016/j.agee.2016.09.001

62. Skopp J, Jawson M, Doran J (1990) Steady-state aerobic microbial activity as a function of soil water content. Soil Sci Soc Am J 54(6):1619–1625
63. Bachar A, Al-Ashhab A, Soares MIM, Sklarz MY, Angel R, Ungar ED, Gillor O (2010) Soil microbial abundance and diversity along a low precipitation gradient. Microb Ecol 60(2):453–461. https://doi.org/10.1007/s00248-010-9727-1
64. Badran R, Abdel-Rahiem A (1996) The effect of soil moisture content on the cellulolytic mycoflora of soil amended with cellulosic remains. Microbiol Res 151(3):301–308
65. Davies Jr F, Olalde-Portugal V, Aguilera-Gomez L, Alvarado M, Ferrera-Cerrato R, Boutton T (2002) Alleviation of drought stress of Chile ancho pepper (*Capsicum annuum* L. cv. San Luis) with arbuscular mycorrhiza indigenous to Mexico. Sci Horticult 92(3–4):347–359
66. Augé RM (2001) Water relations, drought and vesicular-arbuscular mycorrhizal symbiosis. Mycorrhiza 11(1):3–42
67. Elwan S, Mahmoud S (1960) Note on the bacterial flora of the Egyptian desert in summer. Arch Mikrobiol 36(4):360–364
68. Vasileiadis S, Coppolecchia D, Puglisi E, Balloi A, Mapelli F, Hamon RE, Daffonchio D, Trevisan M (2012) Response of ammonia oxidizing bacteria and archaea to acute zinc stress and different moisture regimes in soil. Microb Ecol 64(4):1028–1037. https://doi.org/10.1007/s00248-012-0081-3
69. Pitman MG, Läuchli A (2002) Global impact of salinity and agricultural ecosystems. In: Salinity: environment-plants-molecules, pp 3–20
70. Mohamed ES, Morgun EG, Goma Bothina SM (2011) Assessment of soil salinity in the Eastern Nile Delta (Egypt) using geoinformation techniques. Mosc Univ Soil Sci Bull 66(1):11–14. https://doi.org/10.3103/s0147687411010030
71. Lu H, Lashari MS, Liu X, Ji H, Li L, Zheng J, Kibue GW, Joseph S, Pan G (2015) Changes in soil microbial community structure and enzyme activity with amendment of biochar-manure compost and pyroligneous solution in a saline soil from Central China. Eur J Soil Biol 70(Supplement C):67–76. https://doi.org/10.1016/j.ejsobi.2015.07.005
72. de Souza Silva CMM, Fay EF (2012) Effect of salinity on soil microorganisms. In: Hernandez-Soriano MC (ed) Soil health and land use management. Intech Open. https://doi.org/10.5772/28613. Available from: https://www.intechopen.com/books/soil-health-and-land-use-management/effect-of-salinity-on-soil-microorganisms
73. Wong VNL, Dalal RC, Greene RSB (2008) Salinity and sodicity effects on respiration and microbial biomass of soil. Biol Fertil Soils 44(7):943–953. https://doi.org/10.1007/s00374-008-0279-1
74. Wichern J, Wichern F, Joergensen RG (2006) Impact of salinity on soil microbial communities and the decomposition of maize in acidic soils. Geoderma 137(1–2):100–108. https://doi.org/10.1016/j.geoderma.2006.08.001
75. Mohamed NN (2016) Management of salt-affected soils in the Nile Delta. In: The Nile Delta. Springer, pp 265–295
76. Zahran H (1997) Diversity, adaptation and activity of the bacterial flora in saline environments. Biol Fertil Soils 25(3):211–223
77. Ragab M (1993) Distribution pattern of soil microbial population in salt-affected soils. In: Lieth H, Al Masoom AA (eds) Towards the rational use of high salinity tolerant plants: vol. 1 deliberations about high salinity tolerant plants and ecosystems. Springer Netherlands, Dordrecht, pp 467–472. https://doi.org/10.1007/978-94-011-1858-3_48

Soils as Driver and Victim of Climate Change in Egypt

Reda Ragab Shahin

Abstract Egypt is a country that foresees to face severe effects owing to climate change. Soil may consider an important source of greenhouse gas emissions (i.e. carbon dioxide, methane and nitrous oxides). The drivers of soil GHG emissions are soil type and composition (i.e. soil texture, pH, soil organic matter (SOM), etc.), soil temperature, moisture, fertilization, soil miss-management (Tillage), rice cultivation and burning of Crop residues. Soil also considered as a victim of climate change. Global warming may induce, depletion of soil organic matter that causes the decline of soil fertility, poor soil water regime, shifting of soil microbiome and soil compaction (i.e. Increase soil compaction, surface sealing and crust formation). Global warming induces also sea level rise (SLR) on soils of Egypt which increase the area of submerged lands in northern Nile Delta and consequently soil salinization. With climate change, more frequent extreme precipitation and drought events are projected which may exacerbate the rate and soil susceptibility to accelerated erosion, salinization and other degradation processes, leading to further carbon losses. In conclusion, this chapter summarizes geographical nature of climate change impacts and the history of flooding rainstorms in Egypt.

Keywords Global warming · Soil GHG emissions · Soil miss-management · SOM depletion · Submerged lands · Nile delta · Rainstorms in Egypt

1 Introduction

Climate change and consequently global warming are of great concern for humanity in the present, near and far future. Enormous amounts of greenhouse gases (GHG) emitted every day and accumulating in the troposphere. The world temperature recorded warmest July 2016 at +0.88 °C (+1.58 °F) changes in the temperature of air relative to previous Century (1901–2000) global average [1]. This global heat and warming are thought to be a directly correlated to the emissions of potential warming gases into troposphere which estimated to reach 37 Bt CO_2eq in 2014 which was

R. R. Shahin (✉)
Department of Soil Science, Faculty of Agriculture, Cairo University, Giza, Egypt
e-mail: dredashahin@agr.cu.edu.eg; dredashahin@gmail.com

© Springer Nature Switzerland AG 2020

E.-S. E. Omran and A. M. Negm (eds.), *Climate Change Impacts on Agriculture and Food Security in Egypt*, Springer Water, https://doi.org/10.1007/978-3-030-41629-4_7

60% above emissions in 1990 [2]. For the decade from 2005 to 2014, about 44% of CO_2 emissions remain in the troposphere and 26% in the ocean (https://www.co2.earth/global-co2-emissions). The land is the natural sink of 30% of the world CO_2 emissions [3]. Methane (CH_4), nitrous oxide (N_2O) and carbon dioxide (CO_2) are the main GHG emissions resulted from the agricultural production. Greenhouse gas emissions are generally reported as an equivalent of carbon dioxide (CO_2-eq). On a global scale, agricultural activities accounted to be 6.1–5.1 Gt CO_2-eq/year of emissions in 2005 (10–12% of the entire anthropogenic GHG emissions) and a nearly 17% increase in CH_4 and N_2O emissions from 1990 to 2005 [4].

2 Agricultural GHG Invention of Egypt

The arable land area of Egypt is about 3.8 Mh, and the annual cropping area around 6.2 Mh, which represents 163% of the actual cultivated land area [5]. Egyptian agricultural land can be grouped into: alluvial old land that is the lands of the Nile Valley and the Nile Delta which have been irrigated and intensively cultivated since ancient times, and which represent about 80% of the arable area; and sandy newly cultivated land representing about 20% of the arable area (see Fig. 1).

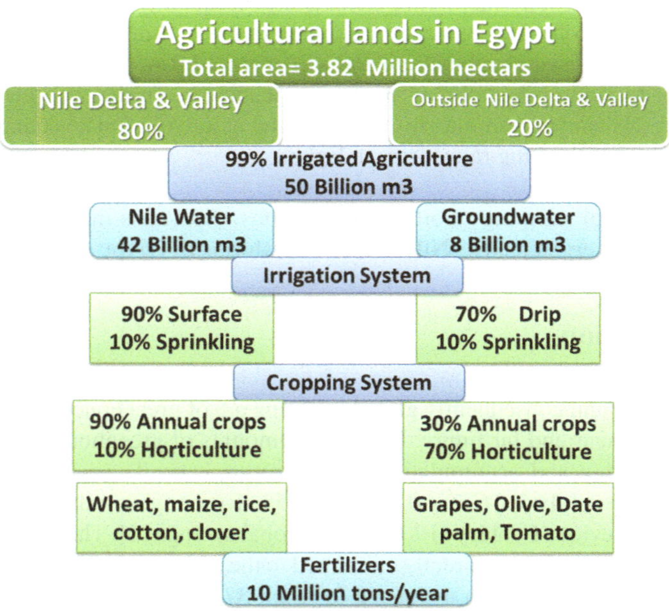

Fig. 1 Profile of agricultural lands, irrigation, cropping systems and consumed fertilizers in Egypt. Data extracted from CAMPAS [5]

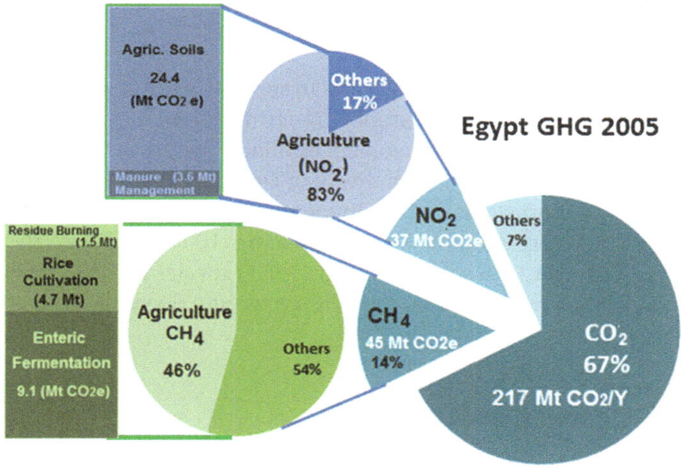

Fig. 2 Profile of agricultural GHG fluxes for Egypt in 2005. Data extracted from UNFCCC [7]

Egypt is a country that foresees to face severe effects owing to climate change, although its share in the global greenhouse gas (GHG) emissions is only 0.57% [6], and despite the fact that it is one of the countries that is not required to manage or reduce its emissions according to Kyoto protocol [7]. Nevertheless, Egypt has considered a national plan that includes mitigation measures to manage and reduce GHG fluxes from the major sectors contributing to the projected climate change which are agriculture, industry and energy wastes [7]. Figure 2 shows the trend in GHG emissions in Egypt in the different sectors as calculated by Climate Change Central Department of the Environmental Affairs Agency of Egypt (EEAA) from the year 2000 until the year 2010. Such calculations were based on emission increase rate of 5.1% and population increase rate of 1.7–2.3%.

Figure 2 concluded the profile of agricultural GHG fluxes for Egypt in 2005 as extracted from the data reported in UNFCCC [8] and [9]. Methane emissions from agriculture represented 46% of its total NH_4-budget for Egypt (45 Mt CO_2eq year^{-1}) which mainly produced from enteric fermentation (9.1 Mt CO_2eq year^{-1}) and rice cultivation (4.7 Mt CO_2eq year^{-1}) produced by 500,000 ha rice fields. The quantity of different forms of nitrogen oxides emissions from agriculture represented 83% of the total NO_2-budget (37 Mt CO_2eq year^{-1}), which mainly resulted from N-fertilized agricultural soils (24.4 Mt CO_2eq year^{-1}) besides NO_2-emissions from manure management (3.6 Mt CO_2eq year^{-1}). Methane (1.5 Mt CO_2eq) and NO_2 (0.3 Mt CO_2eq) are the main products of burning the agricultural wastes.

3 Drivers of Soil GHG Emissions

The main drivers of greenhouse gases fluxes from soils in Egypt are soil texture, soil moisture, and temperature, soil respiration, N-fertilization, rice cultivation, burning of crop wastes and type of soil management or tillage operations (Fig. 3). Moreover, soil carbon dioxide flux is the main pathway of carbon emission from soil to troposphere in terrestrial ecosystems and an important origin of atmospheric carbon [10, 11]. Annual global soil respiration is estimated at 68–100 billion tons (Bt) year^{-1} of C accounting for almost 10% of the atmospheric CO_2 cycles [12, 13], and more than 11 folds of the carbon released from the consumption of fossil fuel [14].

3.1 Soil Type and Composition

The emissions of Soil greenhouse gases could be controlled by soil type concerning physical, chemical and biological properties.

3.1.1 Soil Texture

Soil texture considered a key property that dominates and control the other soil properties. Some studies have concluded that CO_2 evolved from coarse-textured soils were higher than in clay textured ones [15, 16]. Also, higher fluxes of CH_4 and N_2O were [17] reported in sand-textured than in clay-textured soils, whereas, different findings were reported by others [16]. A recent study [18] using different textures

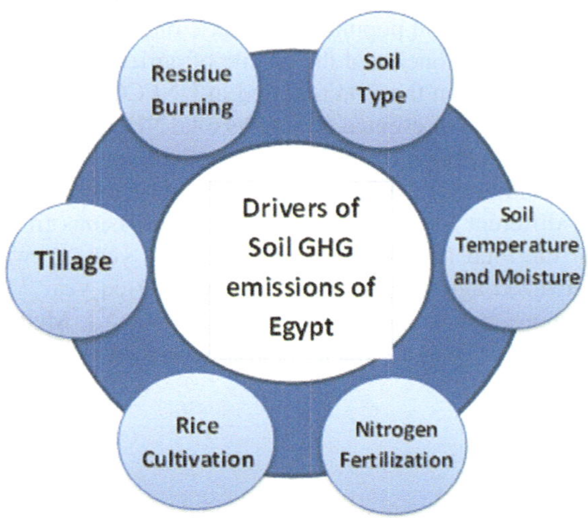

Fig. 3 Key drivers of soil greenhouse gases emissions of Egypt

of soil from sandy loam to fine clay, showed that the effects of biochar on the three GHG emissions were affected by other soil properties, especially soil C and N forms, soil reaction (pH), bulk density, and porosity (aeration), all of which were mainly also affected by soil texture. Furthermore, it was stated that in the coarse-textured, low-buffer Ultisol, cumulative CO_2 and CH_4 emissions increased with increasing volatilized carbon (VC) content of biochars; however, CO_2 emission significantly decreased at 83 mg VC g^{-1} soil [19]. By increasing VC values in the fine-textured, high-buffer Oxisol, cumulative CO_2 emission increased.

On the other hand, clayey soils tend to show greater N_2O emissions than sandy soils, and N management may increase the emission of N_2O, particularly in soils of fine texture before showing N_2O emissions induced by the applied soil management and by rain which were greater by four times in a fine textured soil than in the course-textured one [20, 21].

In the clayey soil, the small amount of large pores would increase anaerobic conditions which enhancing N_2O emissions. It was reported that emissions in coarse-textured soils occur with soil moisture higher than that necessary for the same emissions in a clayey soil [22]. Another study was conducted on GHG fluxes from corn-soybean cropping systems across 35 Wisconsin soil series different in their textures [23]. Their results suggested that soil texture is an important variable in controlling a range of N_2O emission characteristics and is critical for projecting future N_2O emissions from agricultural fields. Silt loam soils showed the largest N_2O emissions among the other soil types, exhibiting higher mean emissions (80–158%) and 100–282% greater emissions when compared to loamy sand and sandy loam soils, respectively. They suggested that denitrification may explain the most variation (74–98%) in total N_2O emissions in the different soil textures and locations.

Also, a study on heavy clay and gravelly loam soils in eastern Canada [24] concluded that the annual fluxes of N_2O were exceptionally high in the heavy clay soil, varying from 12 to 45 kg N_2O–N ha^{-1} during the 3 years of the study. This high fluxes may be not associated with N-fertilizer inputs but rather with denitrification maintained by the decomposition of relatively large contents of soil organic matter (192 Mg C ha^{-1} in the top 0.5 m). It was found that the optimal soil moisture for NO emission differed significantly between the various soil textures, and ranged between 15% water-filled porosity (WFPS) in sandy Italian floodplain soil and 65% in loamy Austrian beech forest soils (Fig. 4) [25]. Thus, each soil had its own optimum condition probably owing to differences in other soil characteristics [26].

3.1.2 Soil pH

Soil pH is a variable that supervised soil productivity and plays a significant role in controlling the chemical and biological activity in soil and consequently soils GHG emissions. Previous work has shown that liming acid soils could enhance [27] or depress N_2O emission [28]. Nitrous gas N_2O emissions may increase when soil pH was close to being at neutral value but decreased when pH exceeded the neutral value [29]. It was found that N_2O flux declines with increasing pH within the range 5–7

Fig. 4 The relationship between soil GHG emissions and water-filled pores from different soil textures. Based on data in Schindlbacher et al. [25] and adapted from Pilegaard [26]

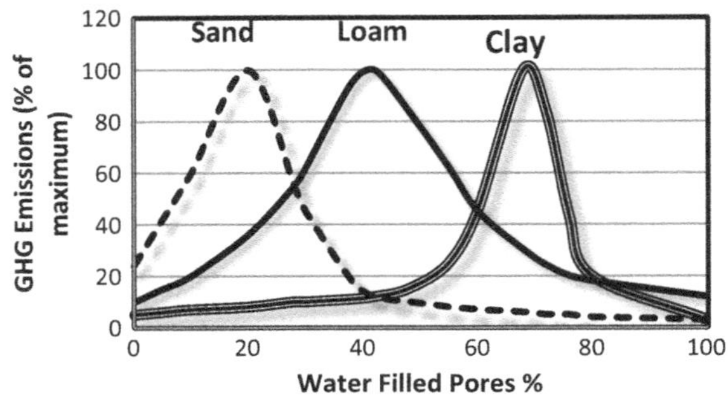

[30]. Thus, by managing the soil pH at a near neutral level appears to be a feasible way to reduce N_2O and the other GHG soil emissions. Regarding the alleviation of soil acidity by liming, a study showed the effect of lime applications on greenhouse gas (GHG) fluxes in livestock farms [31]. They stated that liming to pH 6 had reduced N_2O fluxes (and nitrate leaching) but, in CO_2–C equivalents, the GHG emissions from liming as it neutralizes acidity were about four times those saved by reducing N_2O emissions. Another study conducted an experiment to show the effects of various soil pH on the CO_2 flux from acidic black soils [32]. They concluded that CO_2 emission increased significantly due to N fertilizer addition. CO_2 emission increased positively significant with soil pH ($R^2 = 0.98$). The lowest CO_2 emission (30.2 mg CO_2–C kg^{-1}) was found in pH-3.65 soils of zero N-fertilization. For ammonium fertilization, the highest cumulative N_2O emissions showed up in the pH-8.55 soils was 199 mg CO_2–C kg^{-1}. For nitrate fertilization, the highest cumulative N_2O emissions were found in pH-6.90 soils was 184 mg CO_2–C kg^{-1}. Some researchers had suggested that soil pH controls the reduction reaction of N_2O to N_2, and examined the effect of pH and the N_2O emission in thirteen temperate pasture soils [33]. They found a strong relationship between pH and N_2O emission ratio. Soil pH was negatively correlated with N_2O emission ratio. Furthermore, Luo et al. [34] stated that because different NO producing processes are favored by different pH values, there is often no direct effect of pH on the amount of NO emitted. They found that high pH generally favors nitrification and thus NO emission. Decreasing pH values resulted to increase NO production by both biological and chemical denitrification, (NO production from NO_2^-) increases at a pH less than 4.

3.1.3 Soil Organic Matter

In general, summative CO_2 emission was strongly related to organic carbon (SOC) content in soil because the quantity of substrates for soil microbes can be greatly increased by SOC and microbial activity of soil can be further altered [35, 36]. It was found that organic farming or animal manure application can potentially sequestrate more C to the soils and thus convert the soils to a net CO_2 sink [37–39]. However, some previous studies presented different or opposite effect of SOM on soil CO_2 flux under different conditions [40].

In a recent study [41], data from long-term experiments were conducted to study the effect of crop waste (CW) incorporation on SOM and GHG emissions in Europe [41]. They concluded that the CW incorporation had increased SOC by 7%. In contrast, in some instances, CO_2 emissions were 6-times and N_2O emissions 12-times higher after CW incorporation. As the experiment proceeded, response ratio (RR) for SOC concentration increased. For N_2O emissions, RR was significantly higher in experiments with a duration period of <5 years compared with 11–20 years.

3.2 Soil Temperature and Moisture

The Intergovernmental Panel on Climate Change [4] has found that the extrapolated warming trend over the 50 years from 1956 to 2005 (0.13 °C [0.10–0.16 °C] per decade) is nearly double that for the 100 years from 1906 to 2005. This temperature rise is widespread all over the world and is higher over land and at higher latitudes. Soil temperature is considered as sensitive climate indicator and driver at the same time. Scientists regarded soil temperature data in the research on climate change [42]. Close relationships were recorded between soil temperatures and global warming.

The changes in the temperatures of soils over the last 100 years for three locations in Ireland was studied [43]. They showed an increasing attitude of 0.04–0.25 °C per decade. Higher attitude (0.43–0.66 °C/decade) in the mean temperatures of surface soil (0–40 cm) in Tibet, China, during 1961–2005 was found, especially in spring [44]. They also concluded that such positive trend with soil temperatures was stronger than that for air temperature. In a study on loamy soils of Faisalabad region, India [45], an increasing difference (ΔT) was found between soil and air temperatures as it reached +9.8 °C in summer and +6.9 °C in winter. Also, It was concluded from data of 30 climate stations across Canada during 1958–2008 that stronger trend with time in soil temperature was found at about 66% of the investigated sites at all depths below 5 cm [46]. A warming attitude of 0.26–0.30 °C per decade was consistently recorded in spring (March–April–May) at all the investigated depths between 1958 and 2008. Air temperatures (T_{min} or T_{max}) had a consistent effect on soil temperatures, showing a strong positive correlation which was less in winter when compared to other seasons. Furthermore, it was stated that the average air, minimum or maximum air temperature and average soil temperature, respectively, had significant exponential regression relationship during the whole measuring period of CO_2 gas emission [47].

On the other side, both soil temperature and moisture have strong interactive effects on soil GHG emissions. Laboratory incubations, field observations, and meta-analyses have documented changing GHG fluxes with rising temperature [48, 49]. Rates of CO_2 emission increased exponentially with increasing temperature from −1 to +16 °C during the wet season of forest soils in both Norway and Germany [50]. It is apparent that soil temperature was the dominant factor for soil carbon dioxide emissions when matric potential was higher than −20 kPa (mostly during the wet season). During the drought season, CO_2 emission rates were below 50 mg C m^{-2} h^{-1} [51].

The changes of soil moisture (SM) and temperature (T) in different seasons can affect the trace gases carbon dioxide (CO_2), nitrous oxide (N_2O), and methane (CH_4) that can be exchanged between the ecosystem and atmosphere [52]. This chapter reports the impact of year-round SM status on GHG fluxes in three semiarid vegetation zones, having different SOM contents, in southeastern Arizona. Carbon dioxide and N_2O emissions were highly affected by available soil moisture and temperature. During the rainy season (238 mm total rainfall), large differences in soil C content did not correlate with variations in CO_2 emissions. The limited rains (95 mm total rainfall) reduced CO_2 emissions by 19–40% as compared to the heavy rain season. Laboratory incubation experiments showed potentials for CH_4 oxidation from 0 to 45 cm soil layer. This suggested that when the soil surface becomes dry, CH_4 oxidation activity shifted depthwise in the sandy soils. In semiarid riparian soils of southwestern Arizona, the predicted climate change shifts in annual precipitation from arid to the wet season could reduce soil CO_2 and N_2O emissions while enhancing CH_4 oxidation rates of, potentially acting as negative feedback for future global warming.

As long as other factors such as soil organic carbon (SOC) availability and soil moisture are not limiting, NO emission increases with soil temperature due to the positive effect of temperature on enzymatic activities and microbial turnover rates. Thus, the temperature is not the main factor that controlling the amount of soil flux, but rather a regulating factor of short-term changes [53]. Generally, it has been concluded that NO emission inclines exponentially with soil temperature [25]. Based on a 5 year dataset of continuous measurements of soil-atmosphere exchange of N_2O, NO, and CO_2 at the temperate, nitrogen-saturated Norway spruce forest site Höglwald, nitrous gas NO emissions had significantly positive correlations with water-filled pores % (WFPS) up to a soil temperature of 15 °C, but at soil temperatures above 15 °C, highest NO fluxes were found at lowest WFPS values [54]. The correlation of GHG gas fluxes with soil temperature was stronger than that with soil moisture. However, soil moisture could become the crucial regulator of N_2O emission. An increased NO emission was noticed following soil re-wetting after long drought periods. The NO:N_2O ratio was controlled by WFPS rather than by soil temperature. A significant positive correlation between soil temperature and NO:N_2O ratio was observed only when WFPS was below 45%. The highest NO:N_2O ratio was found under conditions favorable to nitrification (soil temperature around 15 °C; WFPS less than 40%). A recent analysis of soil emission measurements over the period 1994–2010 at the Höglwald site Samad et al. [33] confirmed these findings, specifying that NO fluxes were highest when the temperature was high (more than 15 °C), but the soil moisture was ranged between 24 and 30% (Fig. 5).

A study on the effect of soil temperature (at 5 and 10 cm depths) on soil GHG fluxes, stated that soil temperature significantly influences CO_2 emissions by inducing the accelerated decomposition of soil organic carbon, root respiration, and microbe respiration. Insignificant correlations were detected between soil temperature and both NH_4 and NO emissions [55].

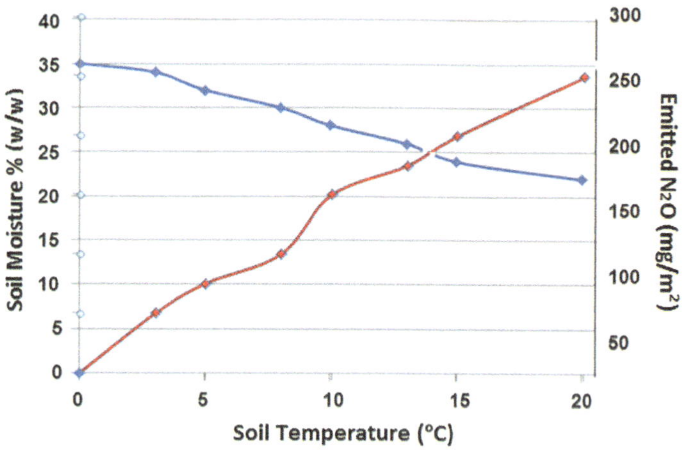

Fig. 5 The combined effect of soil temperature (5 cm depth) and moisture (10 cm depth) on monthly means of soil NO fluxes from German forest. Data from Samad et al. [33]

3.3 Fertilization as a Source of NO_2 Emission

Soil nitrogen is increased by N-fertilization, which usually leads to increase in N_2O emissions. The Intergovernmental Panel on Climate Change [56] suggests a linear relationship between N fertilization and N_2O fluxes and assigns an emission factor (EF) of 1% (1 kg of every 100 kg of applied N fertilizer is lost as N_2O–N). With a global warming potential 310 folds of carbon dioxide (CO_2), nitrous oxide (N_2O) has the highest warming potential of the three major agricultural greenhouse gases (CO_2, methane CH_4, and N_2O). Nitrous oxide gas (N_2O) emission is the major contributor to ozone depletion among the other anthropogenic sources [57], with agriculture accounting for more than 60% of world N_2O emissions [58].

It has been well known that N_2O emission proportionally increases with increasing N-fertilizer rates. However, a nonlinear relationship has often been observed, and emissions increase most rapidly when N rate exceeds crop demand [59, 60]. In addition, a nonlinear model better described the association between N_2O fluxes and fertilization rate than a linear model [61, 62]. In soil systems, N_2O is produced, especially after N-fertilization, through two microbial pathways: nitrification, which converts ammonium (NH_4^+) to NO_3^-, and denitrification, which converts NO_3^- to N_2 [63]. Both processes produce N_2O as a byproduct and can simultaneously produce in the soil. However, nitrification is an aerobic process that requires oxygen, while denitrification is an anaerobic process that is inhibited at high oxygen concentrations. In soil, the oxygen content is largely affected by soil moisture; when soil moisture is high, the oxygen content is low and enhances nitrate reduction (Fig. 6).

Fertilizer form and placement also influence emissions. Fertilizers which increased soil pH and/or highly concentrate N application, such as drip versus micro-jet irrigation or knife injection versus banding of ammonium forms or urea, have

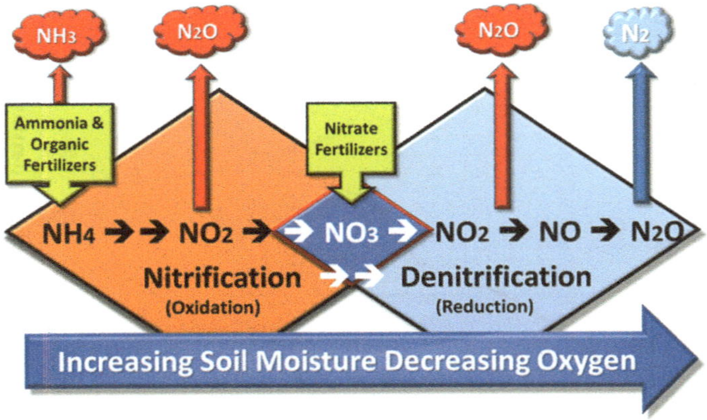

Fig. 6 Nitrogen Fertilizer's induced processes of NO emissions from soils

been found to increase emissions. The injection of anhydrous ammonium caused an escalated seasonal N_2O emissions by 44% as compared to the band-application of ammonium sulfate [64]. Both organic and mineral N fertilization could result in N_2O emission pulses. During fertigation, emissions pulses were immediate but disappeared within a short period (i.e. 1 or 2 days). The impact of fertilization on GHG emissions may occur shortly after the application of N-fertilizer, also application timing may affect net GHG fluxes in arid land cropping systems [65].

Other studies determined and compared growing season emissions of N_2O, CH_4, and CO_2 resulting from applications of different N-fertilizer types, urea (U), polymer-coated urea (PCU) urea-ammonium nitrate (UAN), ammonium nitrate (AN) and poultry litter commercially available [66]. They showed an enhanced-efficiency of N-fertilizers as follows: PCU, U, UAN, AN, poultry litter in no-till corn (*Zea mays* L.) production system. Greenhouse gas emissions were determined during two growing seasons using static, ventilated chambers. The PCU delayed the N_2O emission peak by 3–4 weeks compared with other N sources.

It was found that the application of N-fertilizer may increase N_2O fluxes by 5–6 times at 200 kg N/ha and by 10–14 times at 270 kg N/ha (Fig. 7) [55]. The optimum rate of the applied N fertilizer was 200 kg N/ha which resulted in the lowest global warming potential. Regarding organic fertilizers, a study found that CO_2 fluxes during the period of constant soil moisture of 60% water-holding capacity, were 1.2–2.0 times higher where farmyard manure was applied for a long-term periods than those with mineral N or nil fertilization [67]. Recently, in a study on rice fields of Thailand, concluded that significant positive correlations were found between N_2O emissions and N fertilizer application, with r values of 0.925 ($p < 0.01$) [68].

However, in Egypt, Agricultural fluxes of nitrous oxide during the period (2000–2014) varied from a minimum value of 12.21 Mt in 2000, to a maximum value of 20.49 Mt in 2014. More than 80% of nitrous emissions coming from N-fertilization [69]. Agricultural emissions of nitrous in Egypt have been significantly increased

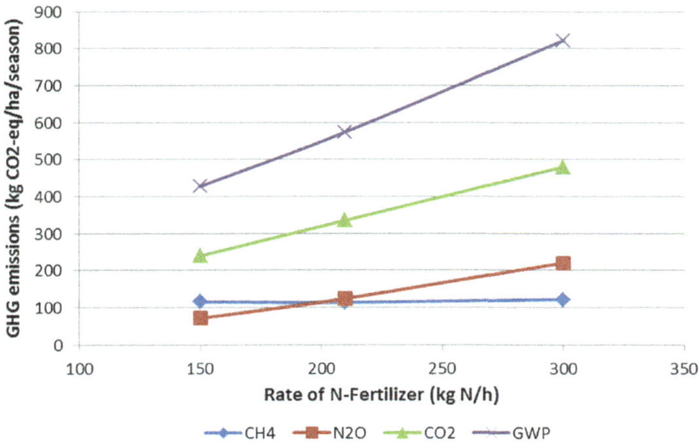

Fig. 7 Cumulative of GHG emissions from different fertilizer applications during rice growing season, GWP = Global warming potential. *Source* data Zhang et al. [55]

with an annual average growth of 2.66%. The annual average of agricultural emissions of nitrous in Egypt totaled of 16.22 Mt during the same period. The annual maximum growth rate of agricultural NO_2 fluxes in Egypt from 2000 to 2014 recorded in 2009: 20.01%, while the annual maximum fall minimum growth rate was in 2008: −10.12% [9, 8].

3.4 *Soil Miss-Management (Heavy Tillage)*

Soil management (tillage) has a very important impact on soil CO_2 emissions [70]. Heavy tillage is the main cause in losing of soil organic carbon content by about 50% due to the stimulation of aerobic decomposition processes and microbial respiration [71]. Studies had shown contrasting results where CO_2 emissions had been both decreased and increased by zero tillage compared with traditional tillage, but many researchers had determined higher soil CO_2 fluxes under heavy or conventional tillage. Implementation of heavy tillage showed changes in the soil properties and enhances favorable conditions for the organic matter decomposition and mineralization processes, i.e., microbial degradation of plant and animal wastes [72, 73]. The intensity of tillage should be reduced to reduce the soil carbon loss. Studies had stated that soil tillage under conventional practices (CT) for sugarcane cropping system "would generate a loss equivalent to 80% of the C that could potentially be accumulated in this soil layer during one year of mechanical harvesting over a period of only 44 days" [74, 75].

The magnitude of soil CO_2 flux had been affected by tillage intensity; intensive tillage, such as a deep plough, showed higher fluxes [76, 77]. Several researchers had suggested that soil management systems involving plowing and harrowing generate a greater impact on CO_2 fluxes than systems using scarification and/or sub-soiling as the sole method of soil tillage [78, 72, 79].

3.5 Rice Cultivation

Rice farms in paddy soils have been a concern of scientists all over world because they produce the most potent greenhouse gases (GHGs), carbon dioxide (CO_2), methane (CH_4), and nitrous oxide (N_2O) [80], because of their positive increases in radiative forcing and their contribution to global warming [4].

Paddy rice fields emit CH_4 due to a methanogenesis process that occurs in anaerobic conditions, during which soil organic matter (SOM) undergoes decomposition [81]. Methanogenesis in soil microbes is a form of anaerobic respiration consuming organic carbon with low molecular weights. The best two pathways which described this process involved the use of acetic acid or inorganic carbon dioxide as terminal electron acceptors [82]:

$$CO_2 + 4H_2 \rightarrow CH_4 + 2H_2O$$
$$CH_3COOH \rightarrow CH_4 + CO_2$$

At between 50 and 100 Mt of methane a year, rice agriculture is a big source of atmospheric methane, possibly the biggest of anthropogenic methane sources and are responsible for 10% of human-induced CH_4 emissions, or 20% of total agricultural CH_4 emissions [83]. The global warming potential of GHG fluxes from rice field is approximately four times higher than either wheat or maize [84]. Another study suggested that rising CO_2 and warming will approximately double the GHG flux magnitude of rice production by the year 2100, which enforcing the need for management practices that optimize rice production while reducing its GHG intensity [85].

In Egypt, the agricultural activities are the second largest source of GHG emissions. Methane emissions are mainly produced from rice cultivation and animal wastes. Rice is one of the main crops in Egypt. The regions with higher emissions are located as a rice belt in the Northern Nile Delta, where methane emission is representing about 53.25% of the agricultural GHG emissions [86].

Total CH_4 emission from rice cultivation is decreased from 235.87 Gg in 1999 to 220.78 Gg in 2005 submission, by a reduction of 6.4%. Meanwhile, it steady decreased to be 223.30 Gg for the year 2012 (5.2% decrease than 1999). This reduction is mainly may be due to the rapid switching of long-duration traditional cultivars to early-maturing short-duration rice cultivars. The conventional cultivars stay about four months under flooding, while the early maturing ones stay only about three months under flooding [87].

3.6 Burning of Crop Residues

The burning of crop residues emit greenhouse gases (GHGs) in the atmosphere, where they affect the inter-annual variability and increase the rate of CO_2, CH_4 and N_2O and other trace gases [88]. Anthropogenic emissions of non-CO_2 gases from global forest fires were 300 Mt CO_2eq year^{-1} on average over the period 2001–2010, with the prescribed burning of savanna adding 200 Mt CO_2eq year^{-1} [89]. Slash and burn practices associated with deforestation were furthermore responsible for over 1.5 Bt CO_2eq year^{-1} as direct CO_2 gas [90].

Specific fire emission categories are significant in terms of their overall regional impacts. For instance, in Africa, the recorded burning of savanna represents 25% of the entire agricultural emissions [89]. Around 730 Mt of biomass is burned every year in Asia by both human and natural activities. The burning of crop residue in open field was found to be 250 Mt of the biomass burning [91]. Rice straw is one of the major crop residues produced in Asia, and its management varies widely. Open burning of straw is ordinary practice in rice straw management in many countries. Greenhouse gas GHG emissions rates were compiled from straw burning (1460–1688 g CO_2eq/kg^{-1} dry straw) [92, 93].

In Egypt, the main sources of biomass waste are an agricultural waste (crop residues), followed by municipal solid waste, animal waste, and sewage waste [94]. Wheat, sugarcane, rice, and maize crop residues are the major sources of the agricultural wastes in Egypt. In particular, rice straw represents the largest part of unrecycled crop residue. FAO [95] stated that Egypt is the largest producer of rice in Africa, with supplying 5.9 Mt of rice in 2013 (more than 22% of rice production in Africa) [95]. It is estimated that about 3.1 Mt/year of rice straw are directly burned in open field. The mean amount of crop residues in 2013 Egypt was around 26.0 Mt, and 25% of this amount was wasted through open field burning [96]. In Egypt, more than 25 Mt of crop residues are burned annually [7]. Two separated years of annual production of the major field crops (wheat, broad been, maize, cotton, rice and sugar cane) depends on burning to disposal its residues, were obtained from FAO Statistics Division [97]. Total emissions of CH_4 from field burning of agricultural residues increased from 64.67 Gg in the year 1999 to 69.44 Gg in the year 2005. Meanwhile, it increased in 2012 to be 70.31 Gg. Moreover, N_2O emissions from the same source increased from 0.94 Gg in 1999 to 1.01 Gg in 2005 and 1.03 Gg in 2012. Therefore, rice and sugar cane were the main sources of emissions from field burning emissions. However, the total GHG emissions arising from open field burning of crop residues in Egypt amounted to 1.8 Mt CO_2eq/year [7]. The reduction of burning rice straw by 20%, the overall emission from rice burning will be reduced by 642.2 Gg (33.3% of total CO_2 less from burning rice straw) [87].

4 Impact of Climate Changes on Soil Functions

The impact of climate change on soils is a slow complex process as because soils not only be strongly affected by climate change directly (for example effect of temperature on soil organic matter decomposition and indirectly, for example, changes in soil moisture via changes in plant-related evapotranspiration) but also can act as a source of greenhouse gases and thus contribute to the gases responsible for climate change. Also changes in the functions and uses soils may be driven more by socio-economic factors than environmental ones. However, the interaction of the various soil forming processes, particularly biological ones, makes difficult to quantify the changes. However, the possible impacts of climate change on the soil functions could be figured out (Fig. 8).

4.1 Depletion of Soil Organic Matter

The effects of global warming on soil C stocks were variable, with positive, negative and neutral impacts. Specifically, the effects of global warming were negligible in areas with small initial C stocks (i.e. Aridisols) [98]. It was suggested that global soil carbon stocks in the upper soil horizons would fall by 30 Bt of carbon under one degree of warming, depending on the rate at which the effects of warming are realized. Under the conservative assumption that the response of soil carbon to warming occurs within a year, a business-as-usual climate scenario would drive the loss of 55 Bt of carbon from the upper soil horizons by 2050. This value is around 12–17% of the expected anthropogenic emissions over this period [99, 100].

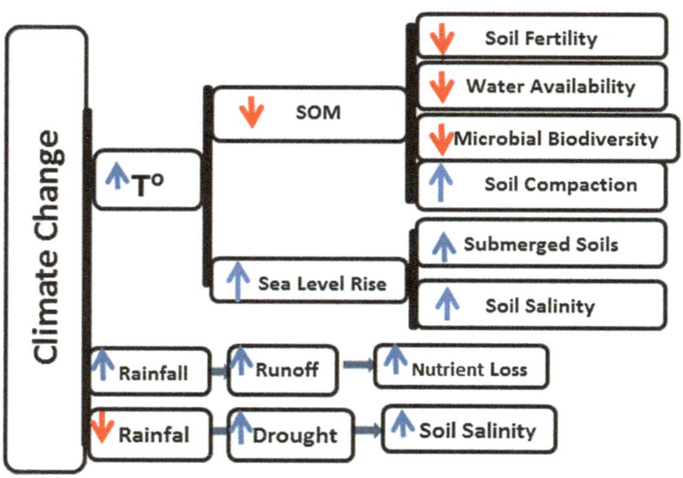

Fig. 8 The possible impacts of climate change on the soil

The predicted climate warming is proposed to enhance SOC decomposition, which may further increase soil N availability, leading to higher soil CO_2 efflux [101]. Each Tone of carbon lost from the soil releases 3.7 Tones of CO_2 in the atmosphere [102]. A recent study made detailed predictions of the future spatial evolution of topsoil SOC driven by climate change and land use change for France up to the year 2100 [103]. The climate change will have a much bigger influence on future SOC losses in mid-latitude mineral soils than land use change dynamics. High-resolution simulation of land use and climate change impacts on SOC stocks indicated that France would lose between 774 and 1221 Mt of SOC by 2100 (i.e. 20–30% of 1990 stock). The future climate change will contribute approximately 10 times more to this total SOC decrease than land use change [103].

Another study was conducted on SOC change at 0–30 cm depth compared with 1980s under B2(+2 °C) scenario in the future of China [104]. They found that loss of SOM could reach 17.7% in north China (N-China) at the year 2080 while the lowest loses (8.8%) was expected in south China (S-China) due to global warming and climate change (Fig. 9).

The potential change in major pools of organic C stocks in upland soils in response to global warming by 2100 was studied [105]. The study showed that the highest potential loss of SOC by 2100 from the upper 1 m layer of the upland was from the labile-C pool (@15%) followed by humus (@6%) and the lowest was for the most resistant recalcitrant pool (0.1%). Therefore, the turnover time of the resistant recalcitrant pool may reach 1000 years while the labile pool may disappear from the soil within 6 years only (Fig. 10).

The utilization efficiency of a more recalcitrant organic carbon "increased at higher temperatures in soils exposed to almost two decades of warming 5 °C above ambient" [106]. This work suggests that climate warming could alter the decay dynamics

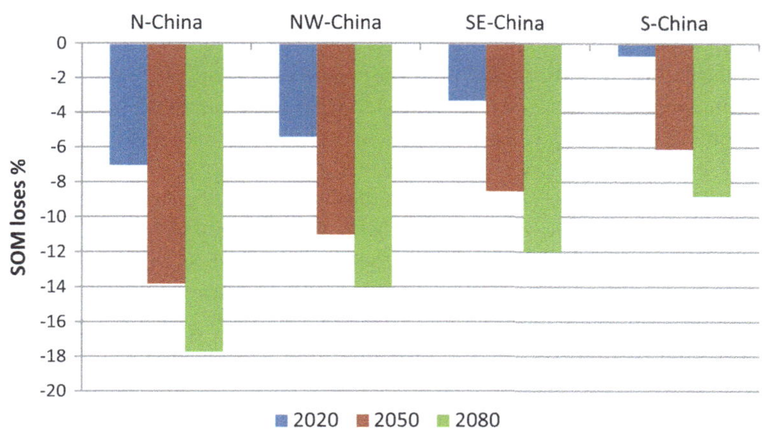

Fig. 9 SOC change at 0–30 cm depth compared with 1980s under B2 scenario in the different regions of China until 2080. Data extracted from Wan et al. [104]

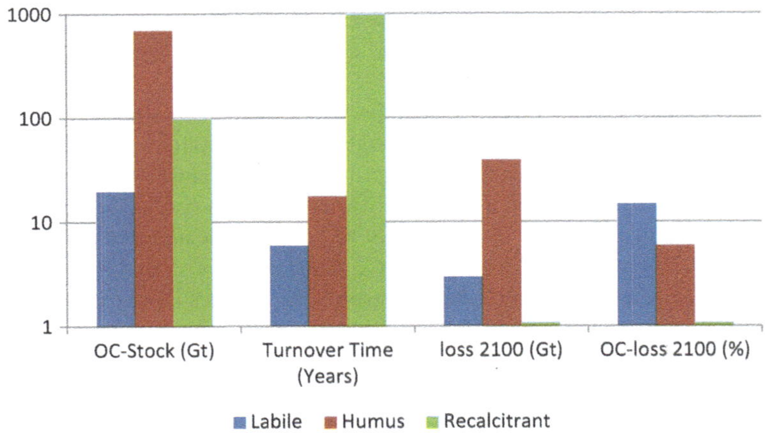

Fig. 10 The amounts of soil organic carbon pools, turnover time and the potential OC-loss at the year 2100. *Source* data Davidson and Janssens [105]

of more stable organic matter compounds, thereby having positive feedback to climate that is attenuated by a shift towards a more efficient microbial community in the longer term (http://adsabs.harvard.edu/abs/2013NatCC…3..395F). Some other studies suggested that recalcitrant carbon not sensitive to global warming variations [107, 108].

For semi-arid Mediterranean ecosystem in Spain, strong correlations between climate change and soil erosion [109] and negative impacts on aggregate stability, bulk density, water holding capacity, pH, organic matter content, total N, and soluble P in the soil, all properties important for good crop growth [110]. Therefore, it can be stated that if climate change increases soil erosion, it will also damage soil properties that are important in the production of food and fiber resources needed by humans.

The distribution of soil organic carbon in vegetated and un-vegetated sites in a Mediterranean lagoon (i.e. Lake Borullus) in Egypt was studied [111]; soil organic carbon content decreased from 22.0 g C kg^{-1} at a depth of 0–10 cm reaching a minimum of 10.2 g C kg^{-1} at a depth of 20–30 cm. Soil organic carbon content was significantly higher in the vegetated sites than in the un-vegetated sites, especially in the surface horizon. The soil organic carbon stock ranged between 760.6 Mt C in the vegetated sites and 2420.2 Mt C in the un-vegetated sites, with total soil organic carbon storage of 3180.8 Mt C. The average carbon sequestration rate of the vegetated sites was higher than the un-vegetated sites (14.9 and 8.6 g C m^{-2} $year^{-1}$). Based on the area and carbon sequestration rate, the total carbon sequestration potential of Lake Borullus was 4.04 Mt C $year^{-1}$. The present study concluded that Lake Borullus could be instrumental in formulating efficient strategies related to carbon sequestration and reduction of GHG in Mediterranean wetlands (Fig. 11).

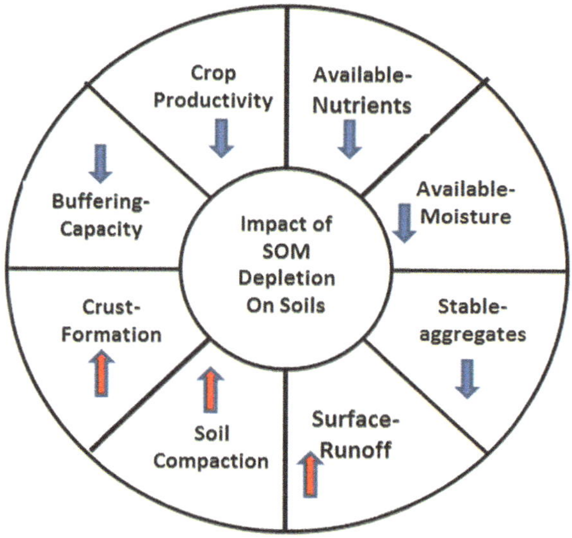

Fig. 11 Impact of the depletion of soil organic matter resulted from global warming on soil properties

4.1.1 Decline of Soil Fertility

Organic matter plays a key role in maintaining soil fertility. It holds nitrogen and sulfur in organic forms and other essential nutrients such as potassium and calcium. The decline in SOM due to global warming may induce disturbance in soil fertility balance. The causes of the decline in soil fertility could be:

- The decrease in organic matter content and soil bioactivity: The decreased levels of organic matter causes a strong reduction in soil fertility as it plays several roles in the soil. Decreased organic matter levels result in poor physical, chemical, and biological properties. The microbial activities in soil decrease due to reduced levels of organic matter. These microbial activities play a significant role in nutrient availability and recycling
- Nutrient mining
- Loss of nutrients through various routes, (leaching, volatilization, runoff)
- Soil acidification, salinization, and alkalinization
- Soil pollution.

Warmer soil temperature will accelerate soil processes, rapid decomposition of organic matter, increased microbiological activity, quicker nutrients release, increase nitrification rate and generally accentuate chemical weathering of minerals. However, soil temperatures will also be affected by the type of vegetation occurring at its surface, which may change itself as a result of climate change or adaptation management [112]. "Organic matter is particularly important as the prime habitat for immense numbers and variety of soil fauna and microflora, which play a critical role in the nutrient storage and turnover, health, and productivity of soils" [113].

Soil warming can increase nutrient uptake from 100 to 300% by enlarging the root surface area and increasing rates of nutrient diffusion and water influx [114]. Temperature increases in the rhizosphere can also stimulate nutrient acquisition by increasing nutrient uptake via faster ion diffusion rates and increased root metabolism [115]. However, any positive effects of warmer temperature on nutrient capture are dependent on adequate soil moisture. If under dry conditions higher temperatures result in extreme vapor pressure deficits then nutrient acquisition driven by mass flow will decrease [116]. In barley and wheat, the soil available K and Ca and their uptake in grain were increased with soil warming (+2.5 °C), while the concentration of Mg was slightly reduced [117, 118].

4.1.2 Poor Soil Water Regime

As climate changes, soil moisture levels will be influenced by direct climatic effects (precipitation, temperature effects on evaporation), climate-induced changes in vegetation, different plant growth rates and different cycles, different rates of soil water extraction and the effect of enhanced CO levels on plant transpiration. Changes in soil water fluxes may also feedback to the climate itself and even contribute to drought conditions by decreasing available moisture, altering circulation patterns and increasing air temperatures [119]. "Increasing temperatures will also lead to greater evapotranspiration and hence loss of water from the soil. Much will depend on land use also, which itself will change, together with its water needs" [112].

The integral influence of climate-hydrology-vegetation-land use changes is reflected by the field water balance and soil moisture regime. As for example, the rise in temperature increases the potential evaporation and transpiration. Whereas the decrease in atmospheric precipitation will result in a decrease in water infiltration (I) and water storage (S) in the soil and plants water supply; surface runoff (R), consequently water erosion hazard (but increasing the risk of wind erosion for dry surfaces), filtration losses and groundwater recharge (G) and will increase evaporation losses; the rate of transpiration (if the vegetation or crop canopy has not deteriorated due to water deficiency), drought sensitivity with its physiological, ecological and environmental consequences [120].

4.1.3 Shifting of Soil Microbiome

The rising temperatures would be expected to speed up the decomposition of labile (easily degraded) soil organic compounds such as the carbohydrates from leaf litter, whereas more biochemically resistant carbon-containing structures—such as the lignin from woody tissues and lipids from leaf cuticles—would be expected to remain stable over decades, possibly even centuries. Global warming may change present-day decomposition patterns by altering the soil microbial communities and activities, thus changing the overall flow of carbon into and out of the soil and affecting soil fertility as well [112, 121].

Global warming may not have a direct effect on the ecological composition because soil fauna and flora have a relatively broad range of optimum temperature. However, changes in ecosystems and migration of vegetation zones are likely in some areas as a result of increased temperature and changes in rainfall. Some soil flora and fauna may be seriously affected by such changes because their migration rates are likely to be too small [122]. Warming by 5 °C in a temperate forest, for example, altered the relative abundance of soil bacteria and increased the bacterial to the fungal ratio of the community [123]. Further, drought amplifies the differential temperature sensitivity of fungal and bacterial groups [124]. Even with small changes in soil moisture availability (<30% reduction in water holding capacity), soil fungal communities may shift from one dominant member to another while bacterial communities remain constant. The activity of some microorganisms increased in the warmer soil, resulting in faster degradation of carbohydrates and other labile components. However, the soil fungi numbers and activity may increase in the warmer soil, with a corresponding rise in abundance of lignin-derived compounds (reflecting decomposition by the fungi) [125]. SOM loss from 5 to 2% over a 60-year period at Rothamsted resulted in a 90% decrease in microbial biomass, but no significant effect on microbial diversity [126] or substrate utilization [127]. Smaller fungal biomass [128] and fungal-to-bacterial biomass ratios [129] have been found in soils of low SOM content compared with undisturbed and botanically rich grassland soils in the UK. A further significant impact of climate change on soil fauna and flora is through enhanced CO_2 levels in the atmosphere leading to enhanced plant growth and in turn, increased allocation of carbon below ground. The microbial population and its activity under this regime would lead to higher rates of nitrogen fixation, nitrogen immobilisation and denitrification, increased mycorrhizal associations, increased soil aggregation and increased weathering of minerals [122].

4.1.4 Soil Compaction and Crusting

Soil structure is an important property which indicates how the soil particles combine. Soil structure is responsible for the movement of gases, water, pollutants/contaminants, seepage, nutrients, maintenance of water quality, building foundations, soil fauna and the emergence of crops. The nature and quality of the structure are strongly influenced by the amount and quality of organic matter present. A decline in soil organic matter levels lead to a decrease in soil aggregate stability, infiltration rates and increase in susceptibility to compaction, run-off furthermore susceptibility to erosion (similar to the observations of [113]). In some areas, there could be an increase in flash flooding as a result of increased cracking and change in structure.

(a) Increase Soil Compaction

Soil organic matter is undoubtedly the most important soil component as it improves soil quality through the influences in soil structure, water holding capacity, soil stability, and oxygen-holding capacity. Soil organic matter is highly susceptible to changes in land use and management, soil temperature and moisture. In the last decade's changes in land use and management have already led to a significant decline in organic matter levels in many soils which increases the susceptibility to soil erosion [113]. The water stability was significantly correlated with a carbon content of the aggregates. Arable soils of low SOM content are susceptible to slumping when wetted due to aggregate instability, leading to less infiltration and greater surface runoff [130]. Specific decreases in the aggregate stability of 10–40% with a 1% decrease in SOM content have been recorded in the UK and similar locations [131], as well as a decrease in the resilience of soil to physical stresses, such as compaction [132].

Soil compaction is the compression of soil due to outside pressure. The effect of soil compaction is an increase in the density of soil and a corresponding reduction in the amount of air present in the spaces between soil particles. This process can be caused mainly by the depletion of SOM and may be harmful or beneficial depending on the circumstances:

1. Causes nutrient deficiencies
2. Restricts root development
3. Reduces soil aeration
4. Decreases soil available water
5. Reduces infiltration rate
6. Increases bulk density
7. Increases sediment and nutrient losses
8. Increases surface runoff
9. Damages soil structure, through reroute the traffic
10. Break up the hard layers
11. Reduces crop productivity.

(b) Surface Sealing and Crusting

Soil crusting refers to the formation of a compact, thin layer having high bulk density and high penetration resistance at the surface of the soil when dries after the dispersion of fine particles clogging of pores due to the beating action of raindrop and slaking of irrigation water. Soil crust is formed in two phases. (1) Surface sealing: The initial or wetting phase. (2) Surface crusting: The hardening of the surface seal in the subsequent dry phase [133]. When a seal layer exists, the chance of runoff occurrence can dramatically increase. On a vegetated surface, the seal layer can enhance the effects of heterogeneous soil properties and surface features and lead to a highly heterogeneous pattern of runoff and soil moisture distribution [134, 135]. Therefore, higher organic matter levels tend to reduce the risk of soil capping particularly on fine-textured soils through an improved soil structure.

4.2 *Impact of Sea Level Rise (SLR) on Soils of Egypt*

The potential impacts of climate change on coastal regions include progressive inundation from sea-level rise, loss of wetlands, and increased salinity from saltwater intrusion. Worldwide, currently, about 600 million people, which may increase to about one billion by 2050, inhabit low-elevation coastal zones that will be affected by progressive salinization [136, 137]. Previous research suggests significant agricultural productivity losses from rising soil salinity [138].

Results of SLR values confirmed that the values are varied from location to another west to east along northern Nile Delta, so the values of SLR at Port Said are greater than values at Alexandria. Also, the predicted values of SLR using pessimistic (4.0 °C) scenario are the double of the values of optimistic (1.8 °C) scenario. Recently, El Shinnawy [139] stated that the expected SLR till 2100 by the projected increase in air temperature (1.8 °C) is 28 cm ASLR at Alexandria 35.0 at Al-Burullus and 72.5 at Port Said (Fig. 12).

4.2.1 Submerged Lands and Soil Waterlogging

Impact of climate changes on coastal zones was investigated by Coastal Research Institute (CoRI), Netherlands (1989–1992). The study at that time has estimated the sea level rise impact on all the entire coastal zones of Egypt (3500 km) in terms of quality and quantity. The study focused on the Nile Delta coast as it has been considered the most vulnerable area in the coastal zones of Egypt [140]. El Shinnawy [139] had estimated the impacts of sea level rise on the area of submerged lands if water has raisin by 1, 2, and 3 m (Fig. 13).

Worst scenarios of SLR effects along the Mediterranean coast of Egypt was studied [141]. The computed maximum run-up heights in the front of the Delta are about 9.4

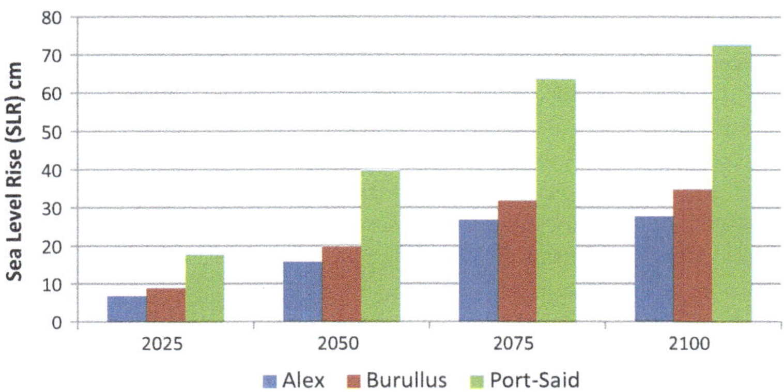

Fig. 12 The projected trend of sea level rise (ASLR) for (1.8 °C) scenario along the Nile Delta Coast until 2100. Data extracted from El Shinnawy et al. [139]

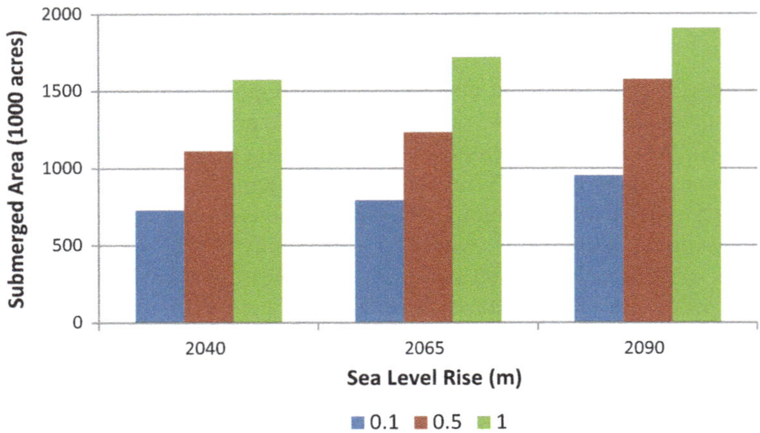

Fig. 13 Projected submerged area with different SLR scenarios until 2090. *Data source* El Shinnawy et al. [139]

and 6.1 m at Alexandria and Damietta, respectively. The maximum estimated run-up height at the western Egyptian coast (Matrouh) is about 2.2, while at the eastern side of the Egyptian coast at El-Arish it is 1.9 m. A minimal relative sea-level rise of ~100 cm is predicted between now and the year 2100 at the Nile delta's coast, where laterally variable but continued ~6.7 to ~11.4 mm/year rates of submergence had been measured [142].

4.2.2 Increase Soil Salinization

Increased salinization and alkalization would occur in areas where evaporation increased or rainfall decreased [143]. Transient salinity increases as capillary rise dominates, bringing salts into the root zone on sodic soils. Leaching during episodic rainfall events may be limited due to surface sealing. Increased subsoil drying increases the concentration of salts in the soil solution. Conversely, the severity of saline scalds due to secondary salinization may abate as groundwater levels fall in line with reduced rainfall; this development could have significant impacts on large areas semi-arid zones. In areas where salinity is a result of recharge processes, salinization would increase if the upstream recharging rainfall increased [144]. Increasing atmospheric CO_2 concentration can reduce the impact of salinity on plant growth [145].

There is an increase in the minimum temperatures by +0.79 and +0.94 °C in the last decades in Ghaza district [146], and there is an increase in the maximum temperatures by +0.29 and +0.26 °C in the same period. Normally the evapotranspiration was increased with the increased temperatures; increasing of +1 °C or + 2 °C cased an increase of the annual average evapotranspiration by 45 and 91 mm.

Therefore the irrigation requirements were increased with an average of 3.28 and 6.68% for the total cultivated area of the studied orchards. In the worse scenario of increasing temperatures +2 °C, and decreasing precipitation by 10%, the irrigation requirements will be increased by 8.69% [146].

In coastal Bangladesh, climate change poses a major soil salinization risk. Across 41 monitoring stations, the annual median projected change in soil salinity is 39% by 2050. Above the median, 25% of all stations have projected changes of 51% or higher [147]. In Egypt, It is widely assumed that SLR will increase saltwater intrusion processes in coastal aquifers. The simulations show that the effect of the global sea level rise in the period from the year 1990 up to 2100 appears clearly through the coastal topography with relatively low-level elevation adjacent to the coast [148]. A shift of the interface is expected up to the year 2100 by a maximum of 43 km covering an area of 1980 km^2 according to scenario RCP2.6 (2.6 °C), and by a maximum of 57 km covering an area of 2870 km^2 according to scenario RCP8.5 (8.5 °C). Overexploitation of groundwater leads to an increase in salinity concentration up to 5000 mg/l and covers about 10% of the Nile Delta area.

4.3 Impact of Rainfall Scenarios on Soils

With climate change, more frequent extreme precipitation and drought events are projected [149] which may have greater impacts on ecosystem dynamics than the singular or combined effects of rising CO_2 and temperature [150]. This increase in the frequency of extreme events may exacerbate the rate and susceptibility to accelerated erosion, salinization, and other degradation processes, leading to further carbon losses.

The potential impact of climate changes on soil erosion could be enhanced by higher precipitation, especially intensive rainfalls and thunderstorms, may result in an increased rate of erosion and higher surface runoff. Increased rainfall could increase atmospheric N deposition to soils, may promote soil disturbances, flooding, and subsidence which changes in wetland and waterlogged habitats and also enhance soil erosion, potentially leading to the pollution of surface waters. Significant increases in rainfall will lead to increases in leaching, loss of nutrients and increasing acidification, depending on the buffering pools existing in soils. The direction of change towards increased leaching or increased evaporation will depend on the extent to which rainfall and temperature change and consequent changes to land use and its management. In either case, the situation could lead to important changes in soils [120].

On the other hand, low precipitation generally reduces the rate of water erosion, but it may intensify wind erosion. Increased droughts will increase the likelihood of shrink-swell in clay soils, and disturbance to building foundations and increased soil temperature may also exacerbate chemical attack to foundations to engineered structures based on clay caps (e.g., in contaminated landfills), with the likelihood of increased leachate generation and release of landfill gases [112].

Evidence for rapid climate change was recorded in the Sinai Peninsula, Egypt. Analyses of data for temperature and rainfall from 1970 to 2014 show a clear tendency towards decreasing rainfall and increasing average temperatures. This trend caused severe droughts for many years that were suddenly interrupted by high and unpredictable rainfall that fluctuated heavily in space and time. If this tendency continues, the population dynamics of many plant and animal species will be negatively affected, with many of them being important for local inhabitants [151].

4.4 *Geographical Nature of Climate Changes Impacts in Egypt*

Climate changes may induce different impacts on Egypt's land which could dominate in certain geographical sites (Fig. 14) and can be summarized in the following:

- The dominance of global warming and increased evaporation from water bodies (Upper Egypt)
- Sea level Rise and seawater intrusions (North of Delta)
- Rainfall fluctuation (Northern and Eastern coastline)
- Flooding rainstorms (South-Eastern side—North Sinai)
- Unstable River Nile income.

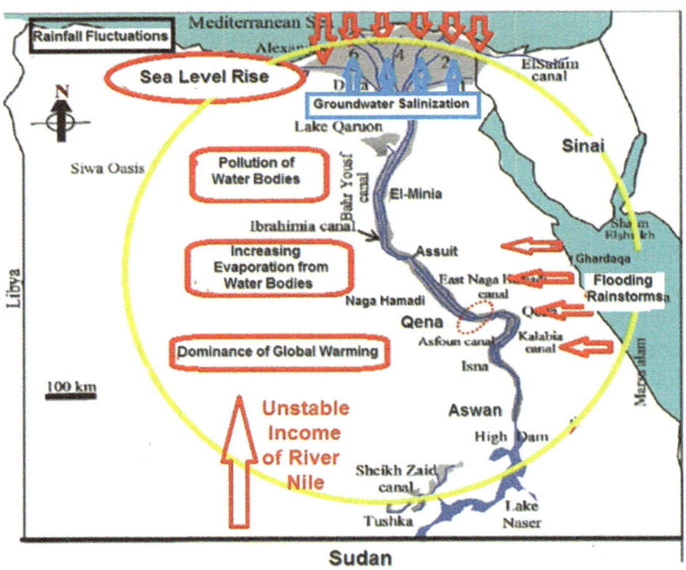

Fig. 14 Geographical distribution of different impacts of climate change on Egypt

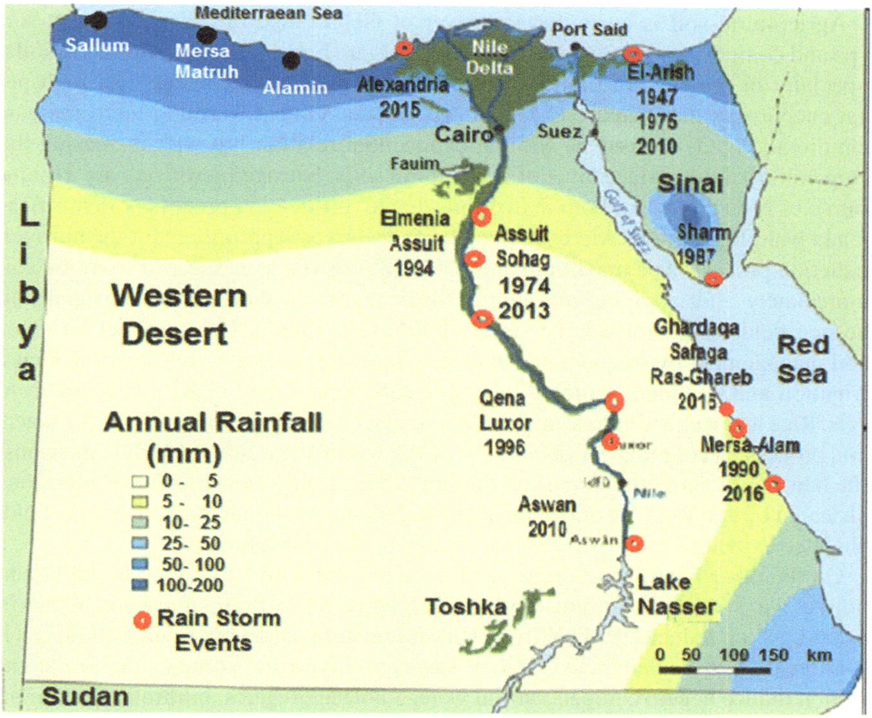

Fig. 15 Distribution of the most deadly rainstorms in Egypt since 1947

4.5 *History of Flooding Rainstorms in Egypt*

Egypt normally receives relatively little rain, and the heaviest rainfall usually occurs along the Mediterranean coast. Heavy rainfall from intense storms led to flooding in parts of Egypt including the Sohag and Bani Suef in Upper Egypt, the Southern Sinai and the Red Sea. Egypt has been experienced up to 120 deadly rainstorms since 1947 left behind about 4000 dead (unpublished data). The rainstorms events were more frequently observed through the last decade especially in Sinai and eastern desert (Fig. 15).

5 Conclusions

For a long time ago, it was believed that soils are innocent victim of the harsh climate events which cause their deterioration and degradation. Recently, the growing knowledge of climate change dilemma showed a close feedback circle with soil environment.

Agricultural soil is an important driver of GHG emissions, depending on soil type and composition. Emissions from clay soils are higher compared to sandy soils, especially in presence of high moisture and organic matter contents and high pH that encourages the increase CO_2 and N_2O fluxes. Global warming increases carbon dioxide (CO_2) emissions with an exponential relationship with increasing the temperature of the surface layer of agricultural soils. Nitrogen fertilizers are a major source of N_2O emissions both in dry soils due to nitrification process or denitrification in waterlogged soils where anaerobic conditions are appropriate for the nitrogen reduction process. The maximum amount of nitrous oxide is released from the soil immediately after applying nitrogen fertilizer and for a short period, especially in the rice fields. The emitted-N_2O is 14 times at the rate of 210 kg N per ha. Poor land management by frequent and extensive plowing, excessive fertilization, heavy irrigation and complete scraping of crop residues promoted GHG emissions from soils. Rice and wetland farms are the main source of methane emissions due to anaerobic process, accounting for about 20% of the world's agricultural GHG emissions. The burning of agricultural waste represents 25% of the total agricultural emissions Africa. In Egypt, @25 Mt of rice husk and sugarcane wastes are burned in open field, which contributes strongly to climate change in the near future.

On the other hand, agricultural land is a victim of climate change in direct and indirect ways. Global warming works to accelerate the burning of organic matter in the soil, which reduces the fertility of soil, water storage and microbial activity, as it destroys the texture and increases their susceptibility to soil compaction. Soil compaction induce negative impact on soil water stable aggregates, infiltration, aeration, root development which increase soil loss through surface sealing, crusting, runoff and erosion, which leads to deterioration of soil productivity for food. The surface layer of soil is expected to lose approximately 55 Mt of carbon by the year 2050, representing about 17% of total expected emissions. Carbon flux is more frequently higher from the surface layer of soil cultivated than from uncultivated soils. Carbon loss is low and even stored in wetlands, such as in the surrounding areas of the Manzala and Borolus lakes in northern Delta.

The changes in rainfall and temperature my lead to a considerable shifting in vegetation pattern which may seriously affect sensitive types of soil flora and fauna and increased the bacterial to the fungal ratio of the community especially in soils enriched with organic matter. Global warming generates changes in the pattern of rainfall. When it decreased, drought dominates the entire area of Egypt and increased desertification. This is offset by sudden increases in rainfall resulting in heavy floods that increase surface runoff and erosion. In both cases, large areas of land will lose their capacity to produce crops.

Global warming induces rising sea level (SLR), which dominates the lowlands. In Egypt, It is widely assumed that SLR will increase saltwater intrusion processes in coastal aquifers. The northern delta is one of the most flood-prone areas in the world. Climate change models showed up to the year 2100 that areas ranging from 1980 to 2870 Km^2 will be submerged by sea water.

Egypt's agricultural land is experiencing increased drought throughout the country. Northern coasts of the Delta are intruded by rising SLR and salt-affected soils cover @10% of the Delta area. The north-west coast is exposed to extreme fluctuation in rainfall events. Waves of heavy rainfall events and flooding frequently sweep the eastern side of Nile basin and northern Sinai. Egypt has suffered @120 of flood incidents since the year 1947, which left about 4000 dead and its frequency increased in Sinai in the last decade.

6 Recommendation

The following recommendations could be drawn up from the discussion in this chapter:

- Reduce conventional land management and adopt sustainable soil and water management by reducing deep tillage and increase organic and bio-fertilization on account of chemical fertilizers which encourage soil carbon sequestration.
- Expanding the cultivation of dry rice varieties and switching of long-duration traditional cultivars to early-maturing short-duration rice cultivars without reducing the area of this strategic rice crop.
- Stop the burning of agricultural waste to reduce emissions that increase global warming and adopt recycling strategies.
- Protection and development of wetlands which could be instrumental in formulating efficient strategies related to carbon sequestration and reduction of GHG in coast of Mediterranean north Egypt.
- Encourage and stimulate projects to convert crop residues into organic fertilizer, compost or biochar and return it to the soil as a sustainable conservation stratigy.
- Building farmers' awareness of modern methods and measures to manage soil and water in the presence of extreme climatic changes.
- Adopting Climate smart farming systems and provide farmers with appropriate guidance and on time forecasts to deal with severe weather waves. Expanding the use of smart systems in land and water management and digital farming systems that help protect the soil and provide water, fertilizers and pesticides to mitigate climate change.
- Reduce the use of heavy agricultural machinery and replace it with small light machines suitable for the small and fragmented agricultural holdings, which reduces soil degradation on the one hand and reduce carbon emissions from high fossil fuels consumption.
- Adopting future projects to protect the Nile Delta from sea level rise guided by the Dutch expertise in this regard and to protect about 1.5 million acres of submersion and salinity and the consequent serious decline in the food basket in Egypt.
- Expansion of water harvesting projects from rainfall and floods by setting up the necessary engineering constructions in the most vulnerable areas in Sinai and eastern Nile Valley to minimize the economic and human losses.

References

1. NOAA-NCEI (2017) National oceanic and atmospheric administration—national centers for environmental information. Global Climate Report, Sept 2017
2. CO_2-Earth (2017) https://www.co2.earth/global-co2-emissions
3. Olivier JGJ, Janssens-Maenhout G, Muntean M, Peters JAHW (2016) Trends in global CO_2 emissions: PBL Netherlands Environmental Assessment Agency 2016 Report publication number: 2315. http://edgar.jrc.ec.europa.eu/news_docs/jrc-2016-trends-in-global-co2-emissions-2016-report-103425.pdf
4. IPCC (Intergovernmental Panel on Climate Change) (2007) Fourth assessment report on climate change: climate change: impacts, adaptation and vulnerability. Cambridge University Press, Cambridge
5. CAMPAS (Central Agency for Public Mobilization And Statistics) (2017) Statistical yearbook. Issue no. 108. http://www.capmas.gov.eg/Pages/StaticPages.aspx?page_id=5034
6. Selim TH (2009) Egypt's carbon emissions and the kyoto protocol. Egypt, energy and the environment. Adonis & Abbey Ltd
7. UNFCCC, United Nations Framework Convention on Climate Change (2016) Egypt THIRD national communication
8. UNFCCC (2017) United Nations Climate Change Conference ("COP23") Bonn (Germany) from 6–17 November 2017. https://www.carbonbrief.org/cop23-key-outcomes-agreed-un-climate-talks-bonn
9. EEAA (Egyptian Environmental Affairs Agency) (2016) Egypt third national communication, under the United Nations Framework convention on climate change. EEAA Publications
10. Koponen HT, Duran CE, Mjanen M, Hytnoen J, Martikainen PJ (2006) Temperature responses of NO and N_2O emissions from boreal organic soil. Soil Biol Biochem 38:1779–1787
11. Nosalewicz M, Stepniewska Z, Nosalewicz A (2013) Effect of soil moisture and temperature on N_2O and CO_2 concentrations in soil irrigated with purified wastewater. Int Agrophys 27:299–304
12. Raich JW, Potter CS (1995) Global patterns of carbon dioxide emissions from soils. Global Biogeochem Cycles 9:23–36
13. Raich JW, Potter CS, Bhagavatti D (2002) Interannual variability in global soil respiration, 1980–94. Global Change Biol. 8:800–812
14. Marland G, Boden TA, Andres RJ (2000) Global, regional and national fossil fuel CO, emissions. In: Trends: a compendium of date on global change. Oak Ridge National Laboratory, U.S. Oak Ridge, Tennessee, USA
15. Fang Y, Singh B, Singh BP, Krull E (2014) Biochar carbon stability in four contrasting soils. Eur J Soil Sci 65:60–71. https://doi.org/10.1111/ejss.12094
16. Sun L, Li L, Chen Z, Wang J, Xiong Z (2014) Combined effects of nitrogen deposition and biochar application on emissions of N_2O, CO_2 and NH_3 from agricultural and forest soils. Soil Sci Plant Nutr 60:254–265. https://doi.org/10.1080/00380768.2014.885386
17. Xie Z, Xu Y, Liu G, Liu Q, Zhu J, Tu C, Amonette JE, Cadisch G, Yong JWH, Hu S (2013) Impact of biochar application on nitrogen nutrition of rice, greenhouse-gas emissions and soil organic carbon dynamics in two paddy soils of China. Plant Soil 370:527–540. https://doi.org/10.1007/s11104-013-1636-x
18. Stewart CE, Zheng J, Botte J, Cotrufo MF (2013) Co-generated fast pyrolysis biochar mitigates green-house gas emissions and increases carbon sequestration in temperate soils. GCB Bioenergy 5:153–164. https://doi.org/10.1111/gcbb.12001
19. Butnan S, Deenik JL, Toomsan B, Antal MJ, Vityakon P (2016) Biochar properties influencing greenhouse gas emissions in tropical soils differing in texture and mineralogy. J Environ Qual 45:1509–1519. https://doi.org/10.2134/jeq2015.10.0532
20. Chen ST, Huang Y, Zou JW (2008) Relationship between nitrous oxide emission and winter wheat production. Biol Fertil Soil (Berlin) 44(7):985–989

21. Tan IYS, van Esa HM, Duxburya JM, Melkoniana JJ, Schindelbecka RR, Larry D, Dean-Hivelya GW, Moebiusa BN (2009) Single-event nitrous oxide losses under maize production as affected by soil type, tillage, rotation, and fertilization. Soil Till Res (Amsterdam) 102(1):19–26. https://doi.org/10.1016/j.still.2008.06.005
22. Neill C, Steudler PA, Garcia-Montiel DC, Melillo JM, Feigl BJ, Piccolo MC, Cerri CC (2005) Rates and controls of nitrous oxide and nitric oxide emissions following conversion of forest to pasture in Rondônia. Nutr Cycl Agroecosys 71:1–15
23. Gaillard R, Duval BD, Osterholz WR, Kucharik CJ (2016) Simulated effects of soil texture on nitrous oxide emission factors from corn and soybean agroecosystems in Wisconsin. J Environ Qual 45:1540–1548. https://doi.org/10.2134/jeq2016.03.0112
24. Rochette P, Angers DA, Chantigny MH, Bertrand N (2008) Nitrous oxide emissions respond differently to no-till in a loam and a heavy clay soil. Soil Sci Soc Am J 72:1363–1369. https://doi.org/10.2136/sssaj2007.0371
25. Schindlbacher A, Zechmeister-Boltenstern S, Butterbach-Bahl K (2004) Effects of soil moisture and temperature on NO, NO_2, and N_2O emissions from European forest soils. J Geophys Res 109:d17302. https://doi.org/10.1029/2004JD004590
26. Pilegaard K (2013) Processes regulating nitric oxide emissions from soils. Phil Trans R Soc B 368:20130126. https://doi.org/10.1098/rstb.2013.0126
27. Čuhe J, Šimek M, Laughlin RJ, Bru D, Chèneby D, Watson CJ, Philippot L (2010) Insights into the effect of soil pH on N_2O and N_2 emissions and denitrifier community size and activity. Appl Environ Microbiol 76:1870–1878. https://doi.org/10.1128/AEM.02484-09
28. Barton L, Gleeson DB, Maccarone LD, Zúñiga LP, Murphy DV (2013) Is liming soil a strategy for mitigating nitrous oxide emissions from semi-arid soils? Soil Biol Biochem 62:28–35. https://doi.org/10.1016/j.soilbio.2013.02.014
29. Mørkved PT, Dörsch P, Bakken LR (2007) The N_2O product ratio of nitrification and its dependence on long-term changes in soil pH. Soil Biol Biochem 39:2048–2057. https://doi.org/10.1016/j.soilbio.2007.03.006
30. Nadeem S, Lars Bakken L, Köster JK, Mørkved PT, Simon N, Dörsch P (2015) Soil pH management without lime, a strategy to reduce greenhouse gas emissions from cultivated soils. Geophys Res Abstr 17, EGU2015-9790, EGU General Assembly 2015
31. Gibbons JM, Williamson JC, Williams AP, Withers PJA, Hockley N, Harris IM, Hughes JW, Taylor RL, Jones DL, Healey JR (2014) Sustainable nutrient management at field, farm and regional level: soil testing, nutrient budgets and the trade-off between lime application and greenhouse gas emissions. Agr Ecosyst Environ 188:48–56
32. Wang L, Han Z, Zhang X (2010) Effects of soil pH on CO_2 emission from long-term fertilized black soils in Northeastern China. In: Proceedings of conference on environmental pollution and public health 2010 (CEPPH 2010). Wuhan, China, Sept 2010. ISBN: 9781617828089
33. Samad MS, Biswas A, Bakken LR, Clough TJ, de Klein CAM, Richards KG, Lanigan G, Morales SE (2016) Phylogenetic and functional potential links pH and N_2O emissions in pasture soils. Sci Rep 6. Article number: 35990. http://dx.doi.org/10.1038/srep35990, https://www.nature.com/articles/srep35990#supplementary-information
34. Luo GJ, Brüggemann N, Wolf B, Gasche R, Grote R, Butterbach-Bahl K (2012) Decadal variability of soil CO_2, NO, N_2O, and CH_4 fluxes at the Höglwald forest, Germany. Biogeosciences 9:1741–1763. https://doi.org/10.5194/bg-9-1741-2012
35. Kirchmann H, Haberhauer G, Kandeler E, Sessitsch A, Gerzabek MH (2004) Effects of level and quality of organic matter input on carbon storage and biological activity in soil: synthesis of a long-term experiment. J Glob Biogeochem Cycles 18:GB4011. https://doi.org/10.1029/2003GB002204
36. Li LJ, You MY, Shi HA, Ding XL, Qiao YF (2013) Soil CO_2 emissions from a cultivated Mollisol: Effects of organic amendments, soil temperature, and moisture. Eur J Soil Biol 55:83–90. https://doi.org/10.1016/j.ejsobi.2012.12.009
37. Gattinger A1, Muller A, Haeni M, Skinner C, Fliessbach A, Buchmann N, Mäder P, Stolze M, Smith P, Scialabba Nel-H, Niggli U (2012) Enhanced top soil carbon stocks under organic farming. Proc Natl Acad Sci USA 109:18226–18231

38. Gregorich EG, Rochette P, Vanden Bygaart AJ, Angers DA (2005) Greenhouse gas contributions of agricultural soils and potential mitigation practices in Eastern Canada. Soil Till Res 83:53–72. https://doi.org/10.1016/j.still.2005.02.009
39. Lynch DH, MacRae R, Martin RC (2011) The carbon and global warming potential Impacts of organic farming: does it have a significant role in an energy constrained world? Sustainability 3:322–362
40. Ren R, Zhang X, Liu J, Sun N, Wu L, Li Z, Xu M (2017) A synthetic analysis of greenhouse gas emissions from manure amended agricultural soils in China. Sci Rep 7. Article number: 8123
41. Lehtinen T, Schlatter N, Baumgarten A, Bechini L, Ruger JK, Grignani CL, Zavattaro L, Costamagna C, Spiegel H (2014) Effect of crop residue incorporation on soil organic carbon and greenhouse gas emissions in European agricultural soils. Soil Use Manag 30:524–538. https://doi.org/10.1111/sum.12151
42. Sharratt BS (1992) Growing season trends in the alaskan climate record. Arctic: 45:124–127. Year Book, 2005–2006, Ministry of Food, Agriculture and Live Stock, Pakistan
43. García-Suárez AM, Butler CJ (2006) Soil temperatures at Armagh Observatory, northern Ireland, from 1904 to 2002. Int J Climatol 26:1075–1089. https://doi.org/10.1002/joc.1294
44. Du J, Li C, Liao J, Pa L, Lu H (2007) Responses of climatic change on soil temperature at shallow layers in Lhasa from 1961 to 2005. Meteorol Mon 33(10):61–67
45. Ahmad MF, Rasul G (2008) Prediction of soil temperature by air temperature; a case study for faisalabad. Pak J Meteorol 5(9):19–27
46. Qian B, Gregorich EG, Gameda S, Hopkins DW, Wang XL (2011) Observed soil temperature trends associated with climate change in Canada. J Geophys 116. https://doi.org/10.1029/2010JD015012
47. Sadek II, Youssef MA (2014) Effect of different types of fertilization and some climatic factors on soil carbon dioxide (CO_2) emission. N Y Sci J 7(7):1–12. ISSN: 1554-0200. http://www.scienceoub.net/newvork
48. Hashimoto S, Carvalhais N, Ito A, Migliavacca M, Nishina K, Reichstein M (2015) Global spatiotemporal distribution of soil respiration modeled using a global database. Biogeosciences 12:4121–4132. https://doi.org/10.5194/bg-12-4121-2015
49. Treat CC, Natali SM, Ernakovich J, Iversen CM, Lupascu M, McGuire AD, Norby RJ, Chowdhury TR, Richter A, Šantrůčková H, Schädel C, Schuur EAG, Sloan VL, Turetsky MR, Waldrop MP (2015) A pan-Arctic synthesis of CH_4 and CO_2 production from anoxic soil incubations. Glob Change Biol 21:2787–2803. https://doi.org/10.1111/gcb.12875
50. Borken W, Xu YJ, Brumme R, Lamersdorf N (1999) A climate change scenario for carbon dioxide and dissolved organic carbon fluxes from a temperate forest soil drought and rewetting effects. Soil Sci Soc Am J 63:1848–1855. https://doi.org/10.2136/sssaj1999.6361848x
51. Bond-Lamberty B, Smith AP, Bailey V (2016) Temperature and moisture effects on greenhouse gas emissions from deep active-layer boreal soils. Biogeosciences 13:6669–6681. https://doi.org/10.5194/bg-13-6669-2016
52. McLain JET, Martens DA (2006) Moisture controls on trace gas fluxes in semiarid riparian soils. Soil Sci Soc Am J 70:367–377. https://doi.org/10.2136/sssaj2005.0105
53. Ludwig J, Meixner FX, Vogel B, Förstner J (2001) Soil–air exchange of nitric oxide: an overview of processes, environmental factors, and modeling studies. Biogeochemistry 52:225–257. https://doi.org/10.1023/A:1006424330555
54. Wu X, Brüggemann N, Gasche R, Shen Z, Wolf B, Butterbach-Bahl K (2010) Environmental controls over soil-atmosphere exchange of N_2O, NO, and CO_2 in a temperate Norway spruce forest. Glob Biogeochem Cycles 24. https://doi.org/10.1029/2009GB003616
55. Zhang X, Yina S, Li Y, Zhuang H, Li C, Liua C (2014) Comparison of greenhouse gas emissions from rice paddy fields under different nitrogen fertilization loads in Chongming Island, Eastern China. Sci Total Environ 472(2014):381–388. https://doi.org/10.1016/j.scitotenv.2013.11.014
56. IPCC (2013) Supplement to the 2006 IPCC guidelines for National greenhouse gas inventories: wetlands. In: Hiraishi T, Krug T, Tanabe K, Srivastava N, Baasansuren J, Fukuda M, Troxler TG (eds) IPCC/TFI

57. Ravishankara AR, Daniel JS, Portmann RW (2009) Nitrous oxide (N_2O): the dominant ozone-depleting substance emitted in the 21st century. Science 326:123–125. https://doi.org/10.1126/science.1176985
58. Mosier A, Kroeze C, Nevison C, Oenema O, Seitzinger S, van Cleemput O (1998) Closing the global N_2O budget: nitrous oxide emissions through the agricultural nitrogen cycle. Nutr Cycl Agroecosys 52:225–248. https://doi.org/10.1023/A:1009740530221
59. McSwiney CP, Robertson GP (2005) Nonlinear response of N_2O flux to incremental fertilizer addition in a continuous maize (*Zea mays* L.) cropping system. Glob Change Biol 11:1712–1719. https://doi.org/10.1111/j.1365-2486.2005.01040
60. Van Groenigen JW, Velthof GL, Oenema O (2010) Towards an agronomic assessment of N_2O emissions: a case study for arable crops. Eur J Soil Sci 61:903–913. https://doi.org/10.1111/j.1365-2389.2009.01217.x
61. Meurer KHE, Franko U, Stange CF, Dalla Rosa J, Madari BE, Jungkunst HF (2016) Direct nitrous oxide (N_2O) fluxes from soils under different land use in Brazil—a critical review. Environ Res Lett 11:023001. https://doi.org/10.1088/1748-9326/11/2/023001
62. Shcherbak I, Millar N, Robertson GP (2014) Global meta-analysis of the nonlinear response of soil nitrous oxide (N_2O) emissions to fertilizer nitrogen. Proc Natl Acad Sci USA 111:9199–9204
63. Verhoeven E, Pereira E, Decock C, Garland G, Kennedy T, Suddick E, Horwath Wand Six J (2017) N_2O emissions from California farmlands: a review. Calif Agr 71(3):148–159. https://doi.org/10.3733/ca.2017a0026x
64. Zhu-Barker X, Horwath WR, Burger M (2015) Knife-injected anhydrous ammonia increases yield-scaled N_2O emissions compared to broadcast or band-applied ammonium sulfate in wheat. Agr Ecosyst Environ 212:148–157. https://doi.org/10.1016/j.agee.2015.06.025
65. Phillips RL, Tanaka DL, Archer DW, Hanson JD (2009) Fertilizer application timing influences greenhouse gas fluxes over a growing season. J Environ Qual 38:1569–1579. https://doi.org/10.2134/jeq2008.0483
66. Sistani KR, Jn-Baptiste M, Lovanh N, Cook KL (2011) Atmospheric emissions of nitrous oxide, methane, and carbon dioxide from different nitrogen fertilizers. J Environ Qual 40:1797–1805. https://doi.org/10.2134/jeq2011.0197
67. Jäger N, Stange CF, Ludwig B, Flessa H (2011) Emission rates of N_2O and CO_2 from soils with different organic matter content from three long-term fertilization experiments: a laboratory study. Biol Fertil Soils 47:483–494. https://doi.org/10.1007/s00374-011-0553-5
68. Arunrat N, Pumijumnong N (2017) Practices for reducing greenhouse gas emissions from rice production in Northeast Thailand. Agriculture 7(1):4. https://doi.org/10.3390/agriculture7010004
69. Farag AA, Abd-Elrahman SH (2016) Greenhouse gas emission from cauliflower grown under different nitrogen rates and mulches. Int J Plant Soil Sci 9(1):1–10. Article no. IJPSS.19880. ISSN: 2320-7035
70. Jarecki MK, Lal R (2006) Compost and mulch effects on gaseous flux from an alfisol in Ohio. Soil Sci 171:249–260
71. Birkás M (2008) Environmentally-sound adaptable tillage–solutions from Hungary. AkademiaiKiado, Budapest, pp 191–194
72. Jabro JD, Sainju U, Stevens WB, Evans RG (2008) Carbon dioxide flux as affected by tillage and irrigation in soil converted from perennial forages to annual crops. J Environ Manag 88:1478–1484
73. Kessavalou A, Mosier AR, Doran JW, Drijber RA, Lyon DJ, Heinemeyer O (1998) Fluxes of carbon dioxide, nitrous oxide, and methane in grass sod and winter wheat-fallow tillage management. J Environ Qual 27:1094–1104
74. Galdos M, Cerri C, Cerri C (2009) Soil carbon stocks under burned and unburned sugarcane in Brazil. Geoderma 153:347–352
75. Silva-Olaya AM, Cerri CEP, La Scala N Jr, Dias CTS, Cerri CC (2013) Carbon dioxide emissions under different soil tillage systems in mechanically harvested sugarcane. Env Res Lett 8(1). IOP Publishing Ltd

76. Bilandžija D, Zgorelec Z, Kisi´ I (2016) Influence of tillage practices and crop type on soil CO_2 emissions. Sustainability 8:90. https://doi.org/10.3390/su8010090
77. Bilen S, Celik A, Altikat S (2010) Effects of strip and full-width tillage on soil carbon IV oxide-carbon (CO_2–C) fluxes and on bacterial and fungal populations in sunflower. Afr J Biotechnol 9:6312–6319
78. Al-Kaisi MM, Yin XH (2005) Tillage and crop residue effects on soil carbon and carbon dioxide emission in corn-soybean rotations. J Environ Qual 34:437–445
79. La Scala N, Bolonhezi D, Pereira GT (2006) Short-term soil CO_2 emission after conventional and reduced tillage of a no-till sugar cane area in southern Brazil. Soil Tillage Res 91:244–248
80. Zheng X, Han S, Huang Y, Wang Y, Wang M (2004) Re-quantifying the emission factors based on field measurements and estimating the direct N_2O emission from Chinese croplands. Glob Biogeochem Cycle 18, GB2018
81. Jain N, Pathak H, Mitra S, Bhatia A (2004) Emission of methane from rice fields—a review. J Sci Ind Res 63:101–115
82. Corbett JE, Tfaily MM, Burdige DJ, Glaser PH, Chanton JP (2015) The relative importance of methanogenesis in the decomposition of organic matter in northern peatlands. J Geophys Res Biogeosci 120:280–293. https://doi.org/10.1002/2014JG002797
83. EPA (United States Environmental Protection Agency) (2006) Global anthropogenic non-CO_2 greenhouse gas emissions: 1990–2020, EPA 430-R-06-003
84. Linquist B, van Groenigen KJ, Adviento-Borbe MA, Pittelkow C, van Kessel C (2012) An agronomic assessment of greenhouse gas emissions from major cereal crops. Glob Change Biol 18:194–209
85. van Groenigen KJ, van Kessel C, Hungate BA (2013) Increased greenhouse-gas intensity of rice production under future atmospheric conditions. Nat Clim Change 3:288–291. https://doi.org/10.1038/nclimate1712
86. Farag AA, Radwan HA, Abdrabbo MAA, Heggi MAM (2013a) Inventory of the greenhouse gas Emissions from rice in the Nile Delta by using emission models, Egypt. J Agric Res 91(2b):917–937
87. Farag AA, Radwan HA, Abdrabbo MAA, Heggi MAM, McCarl BA (2013b) Carbon footprint for paddy rice production in Egypt Nat Sci 11(12):36–45. ISSN: 1545-0740. http://www.sciencepub.net/nature
88. Smith JB, McCarl BA, Kirshen P, Jones R, Deck L, Abdrabo MA, Borhan M, El-Ganzori A, El-Shamy M, Hassan M, El-Shinnawy I, Abrabou M, Hassanein MK, El-Agizy M, Bayoumi M, Hynninen R (2014) Egypt's economic vulnerability to climate change. Int Res Clim Res 62:59–70. https://doi.org/10.3354/cr01257
89. Tubiello FN, Salvatore M, Condor Golec RD, Ferrara AF, Rossi S, Biancalani R, Federici S, Jacobs H, Flammini A (2014) Agriculture, forestry and other land use emissions by sources and removals by Sinks. ESS working paper no. 2. Food and Agriculture Organization of the United Nations, Rome
90. van der Werf G, Randerson J, Giglio L (2010) Global fire emissions and the contribution of deforestation, svanna, forest, agricultural, and peat fires (1997–2009). Atmos Chem Phys 10:11707–11735
91. Streets DG, Yarber KF, Woo J-H, Carmichael GR (2003) Biomass burning in Asia: annual and seasonal estimates and atmospheric emissions. Global Biogeochem Cycles 17:1099–1118. https://doi.org/10.1029/2003GB002040
92. Arai H, Hosen Y, Hong VP, Th NT, Huu CN, Inubushi K (2015) Greenhouse gas emissions from rice straw burning and straw-mushroom cultivation in a triple rice cropping system in the Mekong Delta. Soil Sci Plant Nutr 61(4):719–735. https://doi.org/10.1080/00380768.2015.1041862
93. Gadde B, Bonnet S, Menke C, Garivait S (2009) Air pollutant emissions from rice straw open field burning in India, Thailand and the Philippines. Environ Pollut 157:1554–1558. https://doi.org/10.1016/j.envpol.2009.01.004
94. Said N, EL-Shatoury SA, Diaz LF, Zamorano M (2013) Quantitative appraisal of biomass resources and their energy potential in Egypt. Renew Sustain Energy Rev 24:84–91

95. FAO (2013) Rice market monitor, Nov 2013
96. Abdelhady S, Borellob D, Shabanb A, Rispolib F (2014) Viability study of biomass power plant fired with rice straw in Egypt. Energy Proc 61:211–215. https://doi.org/10.1016/j.egypro.2014.11.1072
97. FAOSTAT (2012) Economic growth is necessary but not sufficient to accelerate reduction of hunger and malnutrition. The state of food insecurity in the world, FAO, Rome, p 65
98. Crowther TW, Todd-Brown KEO, Rowe CW, Wieder WR, Carey JC, Machmuller MB, Snoek BL, Fang S, Zhou G, Allison SD, Blair JM, Bridgham SD, Burton AJ, Carrillo Y, Reich PB, Clark JS, Classen AT, Dijkstra FA, Elberling B, Emmett BA, Estiarte M, Frey SD, Guo J, Harte J, Jiang L, Johnson BR, Kröel-Dulay G, Larsen KS, Laudon H, Lavallee JM, Luo Y, Lupascu M, Ma LN, Marhan S, Michelsen A, Mohan J, Niu S, Pendall E, Peñuelas J, Pfeifer-Meister L, Poll C, Reinsch S, Reynolds LL, Schmidt IK, Sistla S, Soko NW, Templer PH, Treseder KK, Welker JM, Bradford MA (2016) Quantifying global soil carbon losses in response to warming. Nature 540:104–110
99. Moebius BN, van Es HM, Schindelbeck RR, Idowu OJ, Clune DJ, Thies JE (2007) Evaluation of laboratory-measured soil properties as indicators of soil physical quality. Soil Sci 172:895–912
100. Reynolds WD, Drury CF, Tan CS, Fox CA, Yang XM (2009) Use of indicators and pore volume-function characteristics to quantify soil physical quality. Geoderma 152:252–263
101. Dang Y, Ren W, Tao B, Chen G, Lu C, Yang J (2014) Climate and land use controls on soil organic carbon in the Loess Plateau Region of China. PLoS ONE 9(5):e95548. https://doi.org/10.1371/journal.pone.0095548
102. Garnett T (2012) Climate change and agriculture. International Institute for Environment and Development, UK. ISBN: 978-1-84369-853-1. http://pubs.iied.org/pdfs/16512IIED.pdf
103. Meersmans J, Arrouays D, Anton JJ, Rompaey V, Pagé C, De Baets S, Quine TA (2016) Future C loss in mid-latitude mineral soils: climate change exceeds land use mitigation potential in France. Sci Rep 6:35798. https://doi.org/10.1038/srep35798
104. Wan Y, Linab E, Xiong W, Li Yu'e, Guo L (2011) Modeling the impact of climate change on soil organic carbon stock in upland soils in the 21st century in China. Agr Ecosyst Environ 141(2):23–31. https://doi.org/10.1016/j.agee.2011.02.004
105. Davidson EA, Janssens IA (2006) Temperature sensitivity of soil carbon decomposition and feedback to climate change. Nature 440:165–173. https://doi.org/10.1038/nature04514
106. Frey SD, Lee J, Melillo JM, Six J (2013) The temperature response of soil microbial efficiency and its feedback to climate. Nat Clim Change 3:395–398. https://doi.org/10.1038/nclimate1796
107. Ågren GI, Wetterstedt JAM (2007) What determines the temperature response of soil organic matter decomposition? Soil Biol Biochem 39(7):1794–1798. https://pub.epsilon.slu.se/4560/1/agren_g_et_al_100304.pdf
108. Giardina CP, Ryan MG (2000) Evidence that decomposition rates of organic carbon in mineral soil do not vary with temperature. Nature 404(6780):858–861. https://doi.org/10.1038/35009076
109. García-Fayos P, Bochet E (2009) Indication of antagonistic interaction between climate change and erosion on plant species richness and soil properties in semiarid Mediterranean ecosystems. Glob Change Biol 15:306–318
110. Brevik EC (2009) Soil health and productivity. In: Verheye W (ed) Soils, plant growth and crop production. Encyclopedia of life support systems (EOLSS). Developed Under the Auspices of the UNESCO. EOLSS Publishers, Oxford. Available online: http://www.eolss.net
111. Eid EM, Shaltout KH (2013) Evaluation of carbon sequestration potentiality of Lake Burullus, Egypt to mitigate climate change. Egypt J Aquat Res 39(1):31–38. https://www.sciencedirect.com/science/article/pii/S1687428513000666
112. DEFRA (Department for Environment, Food and Rural Affairs) (2005) Impacts of climate change on soil functions. DEFRA Research and Development, UK, Final Project Report [CSG 15]. http://randd.defra.gov.uk/Document.aspx?Document=SP1601_9494_FRP.pdf

113. Bot A, Benites J (2005) The importance of soil organic matter key to drought-resistant soil and sustained food and production. In: FAO soils bulletin, vol 80. Food and Agriculture Organization of the United Nations, Rome, pp 78
114. Mackay AD, Barber SA (1984) Soil temperature effects on root growth and phosphorus uptake by corn. Soil Sci Soc Am J 48:818–823
115. Bassirirad H (2000) Kinetics of nutrient uptake by roots: responses to global change. New Phytol 147:155–169
116. Cramer MD, Hawkins HJ, Verboom GA (2009) The importance of nutritional regulation of plant water flux. Oecologia 161:15–24. https://www.ncbi.nlm.nih.gov/pubmed/19449035
117. Högy P, Poll C, Marhan S, Kandeler E, Fangmeier A (2013) Impacts of temperature increase and change in precipitation pattern on crop yield and yield quality of barley. Food Chem 136:1470–1477. https://doi.org/10.1016/j.foodchem.2012.09.056
118. Li X, Jiang D, Liu F (2016) Soil warming enhances the hidden shift of elemental stoichiometry by elevated CO_2 in wheat. Scientific Reports, 6, Article number: 23313. https://doi.org/10.1038/srep23313
119. Karmakar R, Das I, Dutta D, Rakshit A (2016) Potential effects of climate change on soil properties: a review. Sci Int 4:51–73. https://doi.org/10.17311/sciintl.2016.51.73
120. Varallyay G (2010) The impact of climate change on soils and on their water management. Agron Res 8:385–396. http://agronomy.emu.ee/vol08Spec2/p08s214.pdf
121. Jones DL, Nguyen C, Finlay RD (2009) Carbon flow in the rhizosphere: carbon trading at the soil-root interface. Plant Soil 321:5–33. https://doi.org/10.1007/s11104-009-9925-0
122. Bardgett RD, Freeman C, Ostle NJ (2008) Microbial contributions to climate change through carbon cycle feedbacks. ISME J 2:805–814. https://doi.org/10.1038/ismej.2008.58
123. DeAngelis KM, Pold G, Topcuoglu BD, van Diepen L TA, Varney RM, Blanchard J, Melillo J, Frey SD (2015) Long-term forest soil warming alters microbial communities in temperate forest soils. Front Microbiol 6. https://www.frontiersin.org/articles/10.3389/fmicb.2015.00104/full
124. Briones MJI, McNamara NP, Poskitt J, Crow SE, Ostle NJ (2014) Interactive biotic and abiotic regulators of soil carbon cycling: evidence from controlled climate experiments on peatland and boreal soils. Glob Change Biol 20:2971–2982. https://doi.org/10.1111/gcb.12585
125. Feng X, Simpson AJ, Wilson KP, Williams DD, Simpson MJ (2008) Increased cuticular carbon sequestration and lignin oxidation in response to soil warming. Nat Geosci 1(12):836–839. https://doi.org/10.1038/ngeo361
126. Hirsch PR, Gilliam LM, Sohi SP, Williams JK, Clark IM, Murray PJ (2009) Starving the soil of plant inputs for 50 years reduces abundance but not diversity of soil bacterial communities. Soil Biol Biochem 41:2021–2024. https://doi.org/10.1016/j.soilbio.2009.07.011
127. Wu Y, Kemmitt S, White RP, Xu J, Brookes PC (2012) Carbon dynamics in a 60 year fallowed loamy-sand soil compared to that in a 60 year permanent arable or permanent grassland UK soil. Plant Soil 352:51–63. https://doi.org/10.1007/s11104-011-0979-4
128. Gregory AS, Watts CW, Whalley WR, Kuan HL, Griffiths BS, Hallett PD, Whitmore AP (2007) Physical resilience of soil to field compaction and the interactions with plant growth and microbial community structure. Eur J Soil Sci 58:1221–1232. https://doi.org/10.1111/j.1365-2389.2007.00956.x
129. Bardgett RD, Hobbs PJ, Frostegard A (1996) Changes in soil fungal: bacterial biomass ratios following reductions in the intensity of management of an upland grassland. Biol Fertil Soils 22:261–264. https://doi.org/10.1007/bf00382522
130. Whitmore AP, Rickson RJ, Watts CW, Clarke MA (2004) Critical levels of soil organic carbon in surface soils in relation to soil stability, function and infiltration. Project SP0519 final report. Department for Environment, Food and Rural Affairs, London, UK. http://randd.defra.gov.uk/Document.aspx?Document=SP1601_9494_FRP.pdf
131. Riley H, Pommeresche R, Eltun R, Hansen S, Korsaeth A (2008) Soil structure, organic matter and earthworm activity in a comparison of cropping systems with contrasting tillage, rotations, fertilizer levels and manure use. Agr Ecosyst Environ 124:275–284. https://doi.org/10.1016/j.agee.2007.11.002

132. Gregory AS, Watts CW, Griffiths BS, Hallett PD, Kuan HL, Whitmore AP (2009) The effect of long-term soil management on the physical and biological resilience of a range of arable and grassland soils in England. Geoderma 153:172–185. https://doi.org/10.1016/j.geoderma.2009.08.002
133. Gakab G, Németh T, Csepinszky B, Madarász B, Szalai Z, Kertész A (2013) The influence of short term soil sealing and crusting on hydrology and erosion at balaton uplands, hungary. Carpath J Earth Env Sci 8(1):147–155. http://real.mtak.hu/4000/1/Jakab_et_al_crust.pdf
134. Chen L, Sela S, Svoray T, Assouline S (2012) Impact of soil surface sealing on Rainfall-Runoff Processes in semi-arid areas. EGU general assembly 2012, Vienna, Austria, 22–27 Apr 2012, p 7441. http://meetingorganizer.copernicus.org/EGU2012/EGU2012-7441.pdf
135. Nciizah AD, Wakindiki IC (2015) Soil sealing and crusting effects on infiltration rate: a critical review of shortfalls in prediction models and solutions. Arch Agron Soil Sci 61(9):1211–1230. https://doi.org/10.1080/03650340.2014.998203
136. Brecht H, Dasgupta S, Laplante B, Murray S, Wheeler D (2012) Sea-level rise and storm surges: High stakes for a small number of developing countries. J Env Dev 21:120–138. https://doi.org/10.1177/1070496511433601
137. Wheeler D (2011) Quantifying vulnerability to climate change: implications for adaptation assistance. Center for Global Development Working Paper No. 240, Washington
138. Vermeer M, Rahmstorf S (2009) Global sea level linked to global temperature. Proc Natl Acad Sci 106(51):21527–21532
139. El Shinnawy IA, Abo Zed AI, Ali MA, Deabes EA, Abdel-Gawad S (2010) Vulnerability to climate changes and adaptation assessment for coastal zones of Egypt. In: Proceedings of the first international journal on coastal zone management of River Deltas and Low Land Coastlines, Alexandria, Egypt, pp 145–160. ISSN 1110-4929
140. Smith P, Bustamante M, Ahammad H, Clark H, Dong EA, Elsiddig H, Haberl R, Harper J, House J, Jafari M, Masera O (2014) Agriculture, forestry and other land use (AFOLU). In: Edenhofer OR, Pichs-Madruga Y, Sokona E et al (eds) Climate change: mitigation of climate change. Contribution of working group III to the fifth assessment report of the intergovernmental panel on climate change. Cambridge University Press, Cambridge
141. Hamouda AZ (2010) Worst scenarios of tsunami effects along the Mediterranean coast of Egypt. Mar Geophys Res 31(3):197–214. https://doi.org/10.1007/s11001-010-9099-4
142. Jean-Daniel SJD, Clemente PL (2017) Increased land subsidence and sea-level rise are submerging Egypt's Nile Delta coastal margin. GSA Today 27(5):4–11. http://www.geosociety.org/gsatoday/archive/27/5/pdf/GSATG312A.1.pdf
143. Varallyay G (1994) Climate change, soil salinity and alkalinity. In: Rounsevell MDA, Loveland PJ (eds) Soil responses to climate change. Springer, Heidelberg, pp 39–54. ISBN: 978-3-642-79220-5
144. Peck AJ, Allison GB (1988) Groundwater and salinity response to climate change. In: Pearman GI (ed) Greenhouse: planning for climate change. Brill Archive, Melbourne, pp 238–251. ISBN-13: 9780643048638
145. Nicolas ME, Munns R, Samarakoon AB, Gifford RM (1993) Elevated CO_2 improves the growth of wheat under salinity. Aust J Plant Physiol 20:349–360. http://agris.fao.org/agris-search/search.do?recordID=AU19950115537
146. Ashour EK, Al-Najar H (2012) The impact of climate change and soil salinity in irrigation water demand in the Gaza Strip. J Earth Sci Clim Change 3:120. https://doi.org/10.4172/2157-7617.1000120
147. Hossain M, Huq M, Wheeler D (2015) Climate change and soil salinity: the case of coastal Bangladesh Susmita Dasgupta. Ambio 44(8):815–826. https://doi.org/10.1007/s13280-015-0681-5, http://www.plantstress.com/Articles/min_deficiency_i/nutrient%20uptake.pdf, https://doi.org/10.1073/pnas.0907765106, https://doi.org/10.1073/pnas.1209429109, https://www.ncdc.noaa.gov/sotc/global/201709
148. Wassef R, Schüttrump H (2016) Impact of sea-level rise on groundwater salinity at the development area western delta, Egypt. Ground Water Sustain Dev 2(3):85–103. https://doi.org/10.1016/j.gsd.2016.06.001

149. FAO (2017) Soil organic carbon: the hidden potential. Food and Agriculture Organization of the United Nations Rome, Italy. http://www.fao.org/3/a-i6937e.pdf
150. IPCC (2014) Climate change: synthesis, report contribution of working groups I, II and III to the fifth assessment report of the intergovernmental panel on climate change. IPCC, Geneva, http://www.ipcc.ch/report/ar5/syr/
151. Dadamouny MA, Schnittler M (2016) Trends of climate with rapid change in Sinai, Egypt. J Water Clim Change 7(2):393–414. https://doi.org/10.2166/wcc.2015.215

Soil Carbon Sequestration for Climate Change Mitigation: Some Implications to Egypt

Heba Elbasiouny and Fathy Elbehiry

Abstract The soil is the largest terrestrial carbon (C) stock, and those factors that affect C retention and release also influence on atmospheric CO_2 levels. Soil C sequestration represents about 90% of the total mitigation practices of climate change and about 10% of emission reduction. There is a great concern of soil carbon (C) sequestration and its role in absorbing atmospheric CO_2 not only because of its impacts on climate change mitigation but also because of its positive impacts on the sustainability of crop productivity, soil fertility and soil quality. Cultivation has resulted in considerable loss of soil C due to chemical and biological decomposition of soil organic carbon (SOC), as well as erosion by wind and water. However; in carefully managed croplands, soil C sequestration can be substantial and represents a potentially constructive portion for mitigating the increased levels of atmospheric CO_2. There is a general agreement that many agricultural ecosystems have a huge potential to sequester carbon in the soil, which could decrease CO_2 concentrations in the air and mitigate its global emissions. Egyptian soils are low in their C content. Thus its potential to sequester C is high. Therefore, good management practices should be considered for enhancing soil C sequestration in Egyptian soils especially in degraded and desert soil.

Keywords Soil carbon and nitrogen stocks · Soil carbon sequestration · Climate change mitigation · Egypt

H. Elbasiouny (✉)
Environmental and Biological Sciences Department, Home Economics Faculty, Al-Azhar University, Tanta 31732, Egypt
e-mail: Hebaelbasiouny@azhar.edu.eg; Hebayehia79@hotmail.com

F. Elbehiry
Central Laboratory of Environmental Studies, Kafr El-Sheikh University, Al-Geish Street, Kafr El-Sheikh 33516, Egypt
e-mail: Fathyelbehiry@gmail.com

© Springer Nature Switzerland AG 2020

E.-S. E. Omran and A. M. Negm (eds.), *Climate Change Impacts on Agriculture and Food Security in Egypt*, Springer Water, https://doi.org/10.1007/978-3-030-41629-4_8

1 Introduction: Climate Change and Carbon Cycle

Climate change, tropical deforestation, biodiversity loss and desertification are global environmental issues getting serious attention all interested of scientists, citizens, and policymakers. These issues are related to land-use systems. Climate change is a global environmental concern [1]. An increasing trend of the atmospheric temperature and global changes in the weather conditions are indicated to as climate change [2]. Climate change has the potential to impact natural systems such as forests and socioeconomic systems than their services such as food production and coastal settlements all over the world [1]. This is a major environmental and socio-economical problem, and in the lack of potential mitigation and adaptation processes, climate change can negatively affect a lot of parts worldwide, including environmental resources and ecosystem services [2].

There are strong scientific indications that anthropogenic greenhouse gases concentration is the primarily responsible for most observed increases in global average temperatures since the mid-20th century [3]. There are two anthropogenic processes that contribute greenhouse gases to the atmosphere; fossil fuels burning and changes in land use [1, 2]. The ongoing climatic changes by anthropogenic activities include emissions from fossil fuels burning (292 Pg C) and land-use change (136 ± 30 Pg C), deforestation, and soil cultivation [4]. However, it is mentioned that in the period 1750–2011, the contribution of Land Use and Land Use Change (LULUC) to carbon emission are estimated to equal to 180 ± 80 Pg C. Therefore, understanding the biological mechanism regulating the carbon exchanges (between land, oceans, and atmosphere) and their exchange responses to climate changes through climate-ecosystem feedbacks is very important [2, 4]. Scenario modeling indicates that if this issue left unchecked, the rise in average temperature globally could be 2 °C or greater by the end 21st century. This increase in global temperatures can result in significant changes in climate then ecosystems such as increased duration and intensity of extreme events, affecting food, fiber, energy security, and natural resources. To address the climate change issues, new accessible and reliable information, either nationally or locally, about land and water resources will be needed to assist mangers of these resources to make a suitable decision [3].

The soils of the world play a critical role in the global C cycle [5]. Following fossil fuels, SOC is the largest terrestrial C stock [2]. This crucial role is given, not only because of the size of the soil C reservoir (1576 Pg) (Pg = 1015), but also because of the dynamic character of some SOC fractions [6]. Climate change can significantly affect soil carbon (C), since changes in temperature, rainfall patterns and CO_2 concentrations influence C inputs to soil, and soil C decomposition. Recently, the impact climate change on global SOC stocks has been widely recognized [2].

Because of an immediate and long-term threat to the entire biosphere and human systems on the earth posed by global warming poses as reducing the emissions of GHGs and sequestering atmospheric CO_2 are the fundamental solutions. There are many different available options, terrestrial carbon sequestration, particularly in the soil is count as a ‘win–win’ strategy because generally management practices that lead to increasing carbon in the soil also enhance soil fertility [7]. Different management systems can modify SOC stocks, mitigating or worsening climate change through acting as a sink for storing C or as a source for C emission. For reducing and managing greenhouse emissions, SOC sequestration through agricultural and forestry management is important. Therefore, during the last few years, prediction of SOC stocks in the different ecosystem has become a crucial issue, because of the probable effects of C on upcoming climate change. On the other hand, understanding the effect of land cover/land use and expected climate change on SOC stocks provides valuable information to improve approaches to land planning. However, uncertainties in climate change studies are high so investigating several scenarios is essential [2].

2 Emissions of CO_2: Reasons and Mitigation

Terrestrial ecosystems release greenhouse gases (GHGs) (i.e. carbon dioxide (CO_2), methane, and nitrous oxide) and act as a sink when storing carbon (C) in vegetation and soil [4]. As levels of CO_2 and other GHGs in the atmosphere have been increasing, rising concern over the impact of these gases on world climate systems has been elevated which led to increasing the awareness towards the global carbon cycle. Therefore, international negotiations are held for reducing GHGs level in the atmosphere [8]. The actions for mitigating GHGs impacts increased with the adoption of United Nations Framework Convention on Climate Change (UNFCCC) in 1992 and the Kyoto Protocol in 1997. The prime objective of this convention is to stabilize the level of greenhouse gases at the level that would not cause further damage; especially anthropologically; to the climate system [9].

An estimated emission of GHGs from agriculture, forestry and land-use is approximately 24% (concentrated mainly in mainly CH_4 and N_2O of total global atmospheric emissions of GHGs [10]. Global cultivated soils are estimated to release about 50 Pg C into the atmosphere via mineralization of SOC [11]. Although soil is the main terrestrial organic carbon (OC) stock, representing 75% of the Earth's terrestrial OC, it can act as a sink or a source of carbon and atmospheric CO_2 with direct influence on the greenhouse effect [12].

Available data indicate that, managed soils have historically been a net source of atmospheric CO_2 contributing to more than 20% of the annual increase of CO_2; 3.2 Pg C in the atmosphere, however, soils can be a net sink for C through enhancement of long-term C storage or sequestration in the soils and thus reduction in CO_2 accumulation rate in the atmosphere [6]. Therefore, measurement of CO_2 flux has been considered as a reliable and sensitive indicator of C cycling in response to soil management practices [13]. There are many factors affecting the rate of CO_2 flux from agricultural soils and these factors vary temporally and regionally. These factors are representing in management practices (such as the addition of exogenous organic inputs, quantity and quality of litter inputs), environmental conditions (i.e. temperature and soil moisture) and the soil microbial community. It is reported soil respiration as a main source of CO_2 released from the soil; it changes with different climatic conditions, ecosystem type and SOC concentration [14]. Moreover, it is reported that a substantial amount of total global CO_2 emission comes from the soil through mineralization and OM decomposition, and roots and organism respiration in the soil. Therefore, even a small change in the flux of soil CO_2 can greatly influence atmospheric CO_2 levels [13]. In addition, it is stated that in agriculture, CO_2 is released by (1) decomposition of crop residues and SOC, which is enhanced by intense tillage practices, (2) production and application of crop inputs, (3) using fossil fuels directly in farm machinery, and (4) burning or oxidation of biomass. The only process of CO_2 uptake is photosynthesis, and the only stock that can sequester C is the soil. Subsequently, net carbon emission indicates loss of CO_2 from biomass and soil to the atmosphere through decomposition or combustion. Net carbon removal or sequestration indicates net CO_2 uptake and storage in biomass and soil [1, 15]. Hence, soil can serve as a source or sink for atmospheric CO_2 depending on the agricultural management practices employed [13].

3 Soil Carbon Sequestration for Climate Change Mitigation and Greenhouse Gases Reduction

The effects of land use changes on SOC stocks are a matter of concern in international policy agendas on greenhouse emissions mitigation [12]. There are two potential measures to balance global C cycle; either cut the emission rate of CO_2 and/or to develop natural C sink that can absorb the increasing level of CO_2. Increasing forest areas can be a sustainable way to mitigate elevated atmospheric CO_2 concentration [14]. Sequestration of SOC, as a key tool of the strategy of climate change mitigation, is ideal in natural ecosystems that have high plant primary productivity and low SOC decomposition [16]. Also, since the United Nations Framework Convention on Climate Change, there has been increasing interest in afforestation to sequester CO_2 from the atmosphere. More attention is paid to environmental restoration of land use change by afforestation as an appropriate prescription and one of the most currently effective and applicable carbon sequestration strategies. As well, the role of soils in the sequestration of SOC is so crucial that the Kyoto Protocol (article 3.3) and the Paris agreement involved it as an important portion for managing [12].

The dynamics of SOC stocks and the role that the soil may play in the long-term storage and sequestration of atmospheric CO_2 are of great concern not only because of their impacts on climate change mitigation, but also the sustainability of crop productivity, and soil fertility [17]. Although cultivation has resulted in considerable loss of soil C due to chemical and biological decomposition of SOC, in carefully managed croplands, soil C sequestration can be substantial and represents a potentially constructive measure for mitigating the increased levels of atmospheric CO_2 [18]. There is a general agreement that many agricultural ecosystems have the potential to sequester more carbon in their soil [15]. CO_2 sequestered in the soils is slowly decomposed and transferred into the atmosphere by the diffusion process between the surface soil and atmosphere [14].

Adoption of best management practices may make agricultural soils an effective C sink because the C content of the soils is below the C saturation point, which means that an equilibrium C content at which no extra additional C can be sequestered overtime under a steady C input. For example, fine-textured paddy soils are more efficient for C sequestration when compared with dry soils; because of slow decomposition rate of OC under submerged or anaerobic conditions over the cultivation season. Minimum or no tillage, cover cropping, the addition of organic amendments, balanced fertilization, and rotational cropping are the most common used best management practices to increase C sequestration in soils [19]. It is stated that in cropland, a high level of SOC could be accomplished through appropriate methods such as crop rotations, proper application rates for inorganic fertilizers and organic manures, conservation tillage methods, and integrated soil fertility management. Such enrichment of the SOC stock will help in preserving good soil health for sustainable crop production as well as in managing global climate change [17]. However, soil C sequestration relying on the traditional best management practices is not always effectual because of loss of applied OC and native soil C loss because of soil disturbance by cultivation. It is emphasized on that application of organic amendments in agricultural soils not only improvs physical, chemical, and biological soil properties and decrease nutrients leaching but also can mitigate GHGs emissions. They referred to biochar application for this purpose. As well, applying balanced fertilization with synthetic fertilizer for rice production, for example, may produce more rice plant residues, but residue incorporation may not necessarily enhance soil C sequestration because of such residues are easily decomposed during cultivation. Also, green manure application among organic nutrient sources, may not directly contribute to soil C sequestration because of its faster decomposition rate. Although, application of livestock manure compost which has recalcitrant C can enhance soil C sequestration; compost application may decline rice crop due to its low nutrient availability. In this context, the application of mineral amendments that have a high specific area, such as coal fly ash, along with organic fertilizers as a source of nutrients can be effective to enhance soil C sequestration via soil C stabilization [19–21].

Carbon sequestration in soils is typically expressed as the change in the total carbon stock through the time. However, this method does not separate the fractions of carbon that can be easily lost back to the atmosphere. For this reason, it is argued that carbon sequestration should consider only the fraction that stays in the soil for a long term. In this respect, soil C has been categorized into three stocks; active (residence time 1–5 years), slow (20–40 years) and passive (400–2000 years) based on their residence times [7]. Therefore, valuation the relative percentages of the different C stocks along with C tracing using stable C isotopes has great promise for better understanding of soil response to the changes in management or in the climate. In this respect, it is emphasized that soil microbes act as a mediator in soil C sequestration process because they are involved in most of the processes in C storage and decomposition [22].

Briefly, carbon sequestration is defined as the storage of carbon in long-term terrestrial or aquatic C stocks, resulting in reducing the atmospheric CO_2 concentration [15]. It is reported [23] that stabilization of soil organic matter (SOM), thus, soil C sequestration can occur through important biochemical and physical mechanisms. Biochemical processes include interactions of recalcitrant litter that slow or inhibit microbial degradation, and chemical conversions through humification and other oxidative reactions that lead to the synthesis of new more resistant compounds to decay than their precursors. Conceptual models of SOC define three protection mechanisms: (i) silt + clay protected SOC (i.e. mineral particles <53 μm), (ii) microaggregate protected SOC (i.e. 53–250 μm), and (iii) biochemically protected SOM because of its chemical composition and through chemical complexing processes. Also, a fourth SOC stock is defined as the unprotected carbon (C), which is comprised in part of particulate organic matter (POM) [24]. Thus; soil sequesters C through some processes such as humification of OM, the formation of organo-mineral aggregates, incorporation of OM in a subsurface layer beneath the plow zone, the addition of deep root residues and precipitation of carbonates as SIC [18]. It is reported that in photosynthesis, plants utilize atmospheric CO_2, and store organic C above and below ground biomass. Therefore, using growing plants is an important mean not only to remove CO_2 from the atmosphere but also to convert atmospheric CO_2 to usable food, fiber, and excess biomass, and improve the conversion of the excess biomass to SOC. Although field studies have long documented the ability of plants to transport C from the atmosphere to the soil, national inventories that quantify the amount of C being stored are relatively new but necessary for any type of full C accounting system [8].

4 Soil C Stock

4.1 Soil Organic C Stock

The SOC plays a fundamental role in improving soil quality, sustaining, and improving crop productivity and providing other ecosystem services, such as enhanced soil structure and water retention. As well it plays a very important role in mitigating global warming through reducing the net atmospheric CO_2 [25–28]. The SOC is the largest terrestrial C stock with a global estimate of 2344 Gt, stored in the top 3-m soil layer and 1115–2200 Pg in a meter soil profile [11, 16, 29, 30], it contains approximately twice as much as carbon in the atmosphere [31]. The global SOC stock in the top 1 m is approximately 1500 Pg (1 Pg = 1 Gt = 10^{15} g) C. Agro-ecosystems, representing 10% of the total terrestrial area, are amongst the most vulnerable ecosystems to the global climate change impacts because of their large carbon stock. Small changes in SOC stock may have considerable effects on the carbon dioxide concentration in the atmospheric [29, 30]. Moreover, it has been reported that a 1%

increment SOC content of the surface soil (0–20 cm) could increase cereal yield by 430 kg ha^{-1} and decline yield variability by 3.5% [27].

Many studies in the last decades have reinforced the role of soil as a C sink, the importance of SOC in the global C cycle and the potential feedback to climate change. Hence, scientists worldwide have paid their attention to the potential factors that may affect the storage of C in the soil. For example, land use and management systems, the application of crop residues or agro-food industry wastes, temperature and precipitation, soil properties, landscape position, slope … etc. [26, 27]. Thus, there is always an urgent need for the adoption of restorative management practices in agricultural soils for improving the soil fertility and environment.

As well, the shifts in climatic factors such as temperature and precipitation notably affect the change of SOC because C sequestration in the agriculture soil is a function of both primary production and decomposition of organic matter (OM), which effect by these factors [29]. As well, the agricultural SOC stock can be strongly affected by human activities, such as land use changes, which influence the input of OM and soil carbon source and alter the soil structure and the equilibrium of OC by changing the cultivation practice, and ultimately change soil carbon sequestration capacity. Therefore, changes in such practices as tillage and fertilization along with land use can directly affect soil physicochemical and microbiological properties and affect the ability of soil aggregates and OC content [31]. According to different studies worldwide, there are many management practices can be used for enhancing SOC such as zero or reduced tillage, mulching and residue management or composting, application of organic fertilizers and manure, improved rotations, use of improved crop varieties, and water management [32]. However, others have argued that claims about the possible benefits of increasing C inputs to the soil must be made carefully because of the uncertainties regarding the quantity that can be sequestered under different climates and soil types. Therefore, it is an urgent need to highlight the relationships between C inputs, SOC sequestration and crop productivity [27].

As well, early detection and prediction of changes in a long-term monitoring of SOC contents are crucial to achieve an effective management of SOC not only to minimize SOC loss and reductions in soil fertility, but also to focus on strategies for soil C sequestration should be focused in the subsoil together with the use of appropriate recommended management practices [32]. However, short- and medium-term changes in soil total organic C (TOC) are hard to detect due to the high background C and temporal and spatial variabilities of recalcitrant C [11, 25].

Since SOM is a complex and heterogeneous entity comprising of a continuum of materials, which have different degrees of stabilization and turnover times, the response of SOC to land use can be better understood by isolating different fractions of SOC. The SOC stock can be chemically divided into labile organic C, semi-labile organic C and recalcitrant organic C. Compared with recalcitrant OC, labile and semi-labile C stocks have smaller sizes while higher bioavailability with short or medium-term storage, thus they are more sensitive to environmental changes. Since the changes in SOC fractions may affect both nutrient supply and soil C sequestration, it is required to fractionate and quantify those stocks for a better understanding of the impact of land-use on SOC kinetic [11, 30]. Labile organic C fractions include microbial biomass C (MBC), dissolved organic C (DOC), and easily oxidizable C (EOC). As those fractions respond relatively rapidly to land use and soil management practices, it is suggested to be used as early and sensitive indicators of SOC changes [11, 25].

4.2 Soil Inorganic Carbon (SIC) Stock

The SIC is a large stock too; nevertheless, the national or global studies on SIC storage and content have been more tentative and have only focused on local or regional assessments than estimates of SOC stocks [33]. The SIC stock consists of elemental C and carbonate minerals (either primary carbonate which formed because of weathering; or secondary carbonate which formed because of CO_2 dissolution in soil water) [34, 35]. Most of the SIC exists as carbonates and is supposed to occur in soils of arid and semiarid regions [33]. Although the role of inorganic carbon (IC) in the C sequestration process is less important than OC, SIC stock determination is very important for assessing the role of soil as a sink for CO_2 [36]. Therefore; carbon stock, either organic or inorganic, is not only important to soil productivity and soil functions, but also to C sequestration and global C cycle [37].

5 Soil Nitrogen (N) Stock and Its Relation to Soil Organic C

The SOM is a primary indicator of soil quality because its influence on soil physical, chemical, and biological properties and functions, subsequently soil productivity. Therefore, quantitative assessment of SOM is important for better understanding the dynamics of SOM and providing valuable information for decision making to determine management practices for maintaining or increasing productivity. The measurements of SOM quality and quantity such as soil total N, SOC and soil C/N ratio are the most widely recommended indicators of SOM quality. The labile fraction of SOM such as microbial biomass, water extractable organic N, and light fraction and particulate organic matter N are more sensitive to management-induced changes and therefore can serve as early indicators of possible changes in SOM quality [38].

The global N stock in SOM is estimated by 1.33×10^5 Tg N (1 Tg = teragram = 10^{12} g = 1 million tons) compared with 3.9×10^{12} Tg N in atmosphere and 3.5×10^3 Tg N in terrestrial biomass. Nitrogen in soil is mainly stored in organic form (90%) and the rest 10% is stored as inorganic N (NH^{4+} and NO^{3-}). Soils vary in their capacity to fix N, for example; the sandy soil has a little capacity to fix N than clay soil [39, 40]. The N stocks also are strongly influenced by the land use and management practices, OM additions, topographical position and climatic condition [26].

Nitrogen is frequently a limiting factor for growth and re-establishment of vegetation in disturbed or degraded soils. Thus it is essential for increasing SOC concentrations. Therefore, inputs of N should be considered in any future restoration and sequestration efforts. Total N (TN) and SOC are largely related to SOM accumulation, and hence, always follow similar patterns [41]. Because of this joined cycling of C and N, assessment of the total nitrogen (TN) is important in all SOC stock studies because of the effective impact of soil N dynamic on SOC sequestration [42, 43].

There is a rising research interest in N deposition effects on carbon sequestration in forest ecosystems. Increased N deposition can affect forest ecosystems in several directions such as increment of foliar biomass and efficiency of photosynthesis, and thus increase forest ecosystem biomass, reducing allocation to fine roots, reducing respiration rates, accelerating N saturation in areas of high nitrogen in soils, and consequently cause leaching loss of N and base cations to aquatic systems. It was suggested that N deposition may increase the net primary production (NPP) at first, but then NPP declines with N saturation which will affect acquiring atmospheric therefore an affecting sequestered C. However, there is a suggestion to consider regional or site-specific in the investigation the effects of increasing N input on C sequestration [43]. It is stated that optimizing fertilizer N inputs, may increase SOC content through increasing crop productivity and subsequently the residues amount returned to soils. They also added that excessive N fertilization might suppress the microbial communities or stimulation of mineralization of old native organic which lead to blocking SOC sequestration [5]. In this context also, it is reported that the amount of N sequestered or lost was still closely associated with the C dynamics, showing that similar controls apply to both [44]. It is explained that the relationship between C and N sequestration by that SOM, as indicated by soil C and N levels, can directly impact crop production. The increase in crop production with increased SOM storage is a result of enhancing soil structure and improving soil water–nutrient–crop productivity relationships. Sequestration of C in the soil can also conserve N, because SOC and total N (TN) levels are highly related [45]. Crop residues type, and its properties (i.e. C/N ratios or quality of the residues) has an important role in C sequestration and soil aggregation. C/N ratios often govern the degradation of fresh crop residues, however, undergoing residues decomposition will make it more recalcitrant. Subsequently, degradation becomes controlled by lignin contents or lignin/N ratios. Thus, confirm on the ability of soils to sequester C is highly related to N [46].

6 Required Variables for Calculating Soil C and N Stocks

The most appropriate method to study the organic C and N content in the soil is on a unit area basis, for a specific interval depth. This requires more information about the spatial variability of different soils and their content of C and N and bulk density. Commonly, reference depth intervals of 0–0.3 m and 0–1 m are used in studies of SOC stocks. The first layer encompasses the depths that are most directly interacting with the atmosphere, and most sensitive to land use and environmental changes [47]. It is supposed that the changes in SOC related to land-use change occur almost entirely in the surface soil (i.e. 30 cm) because deep SOC is inert on timescales <100 years and is transported slowly downward relative to plant sources. Because of these assumptions, most ecosystem biogeochemistry models and researches, which estimate the response of vegetation and soil to climate change, land management, and other factors, consider C dynamics in only the upper 20–30 cm of soil [48].

The soil C or N stocks are determined from two key variables: (i) concentration and (ii) bulk density (BD), which are both prone to changes over time. To quantify the sequestration rate in soils based on frequent sampling designs, it is essential to assess both variables every time [49]. The BD is required for converting soil C or N measurements (e.g., mg C g^{-1} soil) to an area base using units of area or volume (e.g., g C m^{-2} soil) [50]. However, many studies in the past did not include BD measurements. The BD is a function of parent material, soil genesis in addition to soil aggregate formation and may change by tillage and compaction because of land use changes, erosion, seasonal wet/dry cycles, and anthropogenic disruption [49]. Bulk density is a critical soil property that influences many other soil properties and functions such as infiltration rates, aeration, root proliferation, and plant growth. Since BD is one of a critical component of stocks measurement and it can be altered by land use and land covers, careful and spatially-intensive measurements of BD are critical for accurate extrapolations. Although BD is a simple soil attribute and widely assessed by ecologists, engineers, and soil scientists, minor differences in methods can cause considerable impact estimates of C or nutrient stocks sizes [50].

Bulk density is negatively correlated with SOC or soil N concentration in many soils [49, 51, 52, 53]. The correlation between SOC and BD has been used widely to estimate BD from SOC concentrations [49, 52, 53]. The SOC stocks may also be underestimated because of increasing bulk density, as SOC stocks are a function of bulk density and SOC concentration [41]. This also is applied on soil N stocks. However, stocks inventories cannot be completed using only one of these two variables, even though both were weakly correlated with each other [49].

7 Factors Controlling of Carbon Stocks in the Soil

Many factors determine the maximum soil C stock and SOC sequestration such as the age of ecosystem, climate, parent material, physiography, drainage, and soil properties (i.e. clay content, clay minerals, nutrient reserves …). As well, soil drainage and moisture regime, beside soil aspect and landscape position [54], plant species composition, root biomass, period of the experiment are important factors controlling soil C stock [22, 25, 26]. For example, total SOC can be increased with increasing precipitation, while the opposite is true with increasing temperature. Under similar climate conditions, soil texture may be the controlling factor of SOC outputs through decomposition because there is a negative relationship between SOC mineralization rate and soil clay + silt content [22]. Indeed, land-use has been realized as a key factor that determines the size and kinetic of SOC stock because of its influencing on the balance between SOC inputs (e.g. plant litter) and outputs (e.g. mineralization of SOC) by altering plant community and land management practice [30]. Also, optimizing agricultural management practices are very important factors that can increase C inputs and decrease C outputs such as fertilizer application, straw-return, or conservation tillage [25]. The impact of topographic aspect on SOC is widely realized. However, relatively few studies have been conducted to examine its role on SOC contents globally [26].

7.1 Soil Properties

Soil parent material can have a major impact on the functions of ecosystem vegetation and soil, hence deeply influence SOM stock. Soil parent material has been noticed to be an important factor affecting regional SOC and soil N stocks [55]. Young and Spycher [56] found that C levels increased with decreasing particle size. Heavy textured soils had a larger C storage capacity than those with a light texture, which had already attained their estimated capacity [24]. Thus, the properties of coarse textured soils such as low clay content and a shortage of water-stable aggregation are major factors that limit the C and N storage capacity [57]. Zhang et al. [28] as well, emphasized on the favorability of SOC accumulation in soils with finer particles are favorable and attributed that to the stronger adsorption ability of soil clay and silt to SOC. The protective effect of soil clay on OM involves the interaction of OM with the surface of the clay particles (cation bridges, hydrogen bonds, electrostatic and van der Waals interactions) and the occlusion of organic material in the matrix of soil aggregates [58]. Nelson et al. [59] mentioned that clay with its high surface area protects OC from decomposition on developing stable clay-organic complexes. However, such a relationship between clays and OC does not always occur. Bot and Benites [60] mentioned that SOM tends to increase as the clay content increases. This increase relies on two mechanisms; (1) bonds between the surface of clay particles and OM retard the decomposition process and (2) soils with higher clay content increase the capacity for aggregate formation. Macroaggregates physically protect OM molecules from further mineralization caused by the microbial attack. They also mentioned that in similar climate conditions, the OM content in clayey soils ranges from 2 to 4 times its corresponding of sandy soils.

Soil aggregation as well have a great influence on the physical characteristics of the soil. Well-aggregated soils provide a larger pore space, a higher infiltration rate and better gaseous exchange between soil and atmosphere than poorly aggregated soils, resulting in enhanced microbial activity. Soil aggregation and SOM dynamics are largely linked. Each of aggregates and SOM plays an important role for the other. SOM thought to be physically protected within aggregates and at the same time, aggregates are formed by binding SOM with mineral particles to form of different sizes. The protection of SOM is desirable in the soil since SOM is widely recognized as a key component in nutrient cycling. Furthermore, the retention of SOC is becoming more important since the rise in atmospheric CO_2 and global warming are recent concerns [61].

7.2 *Climate Conditions*

Climatic shifts in temperature and precipitation significantly affect SOC because the soil C sequestration is a function of both primary production and decomposition of OM in agricultural soils [29]. Generally, SOC tends to accumulate in areas with lower temperatures and higher precipitation [28]. Both the increase in global temperature and the decrease of precipitation due to climate change may result in an overall decrease in the SOC storage especially in the critical areas which have typically low SOC contents such as Mediterranean areas. The agricultural soil in these areas are particularly vulnerable to climate change, i.e. most climate models predict higher-than-average increases in temperature and generalized decreases in precipitations in most of these areas [32].

7.3 *Land Use*

Land use is a key determining factor of soil carbon stocks. Any change in land use influences on the flow of soil C stock to the new equilibrium point is finally reached under the new land use conditions. The equilibrium C content is a key feature of each soil depending on climate, soil type, vegetation nature and the balance between C inflows and outflows under certain vegetation [62]. Land use change is the second main reason for carbon emissions after fuel consumption [11]. Conversion from the forest or natural grasslands for example to croplands can disturb soil and vegetation and ultimately lead to a dramatic depletion of soil carbon stocks and nutrients [11, 16, 32, 62]. As well, land use affects SOC stocks mainly by changing plant species and management practices [30]. Land use contributes approximately 25% of total global anthropogenic GHG emissions, 10–14% of this total are resulted from agricultural production either via emissions from soils or livestock management, while 12–17% are from land cover change. Land use changes mainly have an influence of food, fiber and fuel demands of a growing population can critically affect soil's capacity to store C [32]. Soils have lost 40–90 Pg C due to historical land-use, mainly through deforestation and reclamation [30]. Mathew et al. [62] estimated that soils over the world had lost between 25 and 75% of their C stocks due to some factors such as soil degradation, vegetation changes and tillage operations. Land use change and land mismanagement (such as the use of excessive plowing, residue removal and biomass burning, and farming practices where nutrient balance is often negative) have not only depleted SOC stocks dramatically but also leads to loss of C from soils to atmosphere. This is estimated by 3.5 C Pg annual increase in atmospheric C, which means that soil acts as a C source and atmosphere acts as a C sink [54, 62].

Forests is important ecosystems, however, over the time, forest surface area has decreased due to increasing human development. Recently, the attention is paid to forests not only because its providing of wood products to society in addition to other important ecosystem services, such as recreation, habitat functions, water regulation, water erosion control, and improved air quality, but also because of its potentiality of C sequestration in addition to biomass provision for bioenergy. This also illustrates the possible trade-offs even within the theme of climate change mitigation [26]. The overall decrease of SOM in croplands has become a threat to soil sustainability. The COP21 conference held in November 2015 in Paris resulted in the adoption of a “4p1000” initiative aiming at favoring C storage practices to mitigate greenhouse gas emissions such as recycling of organic waste treatments originating from urban, industrial, and agricultural activities. This will be a promising source of C stocking amendments that can also substitute synthetic fertilizers [63]. Therefore, restoration processes and converting agricultural soil to a more natural land use opposites some of the effects responsible for SOC depletion as a result of the conversion of natural to managed ecosystems. Applying ecological management concepts to natural resources, such as nutrient cycling, favorable C budget, enhancing soil biodiversity and soil mixing by macro invertebrates, may be an important factor to improve soil quality and SOC sequestration [54]. Retention of crop residues, no till farming and crop cover incorporation in a diversified rotation cycle (i.e. Conservation Agriculture), integrated nutrient management techniques of utilizing compost and biosolids, erosion control, water conservation, contour hedges with perennials, controlled grazing etc. are so important techniques of land use and soil management. Such techniques lead to C sequestration with an average long-term SOC sequestration rate 200–1000 kg/ha/year in humid temperate regions and 50–250 kg/ha/year for in tropical regions. Also, SIC sequestration rate is about 5–25 kg/ha/year in arid and semi-arid regions [54].

Hence, understanding the influence of SOC by future climate and land use changes can help land managers and policymakers to develop appropriate land planning strategies. However, examining multiple scenarios to understand the range of potential outcomes is crucial when assessing future environmental and anthropogenic changes [32].

7.4 Topography

Topography is one of the key factors of soil formation. The topography (i.e. slope, position, and topographical direction), associated with elevation [26]. Topography is one of the most affecting factors on soil C and N contents at the field level. However, topographical effects are probably differing in magnitude in different agricultural systems [64]. Topography plays an important role in temperature and moisture regimes because of differences in insolation period based on site aspect, which affect microclimate differences, then affects the distribution of plant communities and soil processes. Solar radiation influences the temperature regime, which affects soil biological and chemical processes then vegetation distribution. Consequently, soil temperature is a limiting factor controls the rate of biomass decomposition. Therefore, OM content in soils in warm climates are usually less than its corresponding in cooler climates due to higher mineralization rates. On the other hand, when the average annual of rainfall increases, the SOC content increases. Therefore, aspect, through its influence on soil temperature and water content, can affect microorganism activity which leads to inhibiting or accelerating SOC decomposition, hence on SOC distribution [26]. Moreover, topography affects soil C through erosion and distribution of both of fine soil particles and OM across the landscape, and water is leading to variance in leaching, infiltration, and runoff potentials [64]. Land use along with topographic position (aspect) are two factors affecting SOC accumulation. Therefore, considering the combined effects of these factors on SOC accumulation is very important in restoring the ecosystem and sequestering C [26].

8 The Impact of Soil Carbon Sequestration on the Sustainability of Agroecosystem and Crop Productivity

Cropland soil has a huge potential to sequester C (0.4–0.8 Pg year^{-1}), which could decrease atmospheric CO_2 and mitigate global emissions. Also, because SOC is crucial to soil physical, chemical, and biological properties, loss of SOC, the decline in soil structure and overall degradation of the soil resources are standard features of nonsustainable land use. Therefore, more SOC sequestration could help into reserve these features and in sustaining soil fertility and agronomic productivity [65, 66]. Thus, soil C sequestration along with improving agricultural productivity is a win–win strategy [67]. Hence, Lal et al. [67] stated that some reports indicate inducing improvement in soil quality can lead to more C sequestration, climate change mitigation and improvement in agricultural productivity. However, they considered some concerns of the significance of SOC sequestration to climate change mitigation and on agronomic yield [67]. Since SOC is critical for the security of food production because it provides substrate and energy, moreover it protects the biodiversity that helps to preserve soil quality and the functioning of the ecosystem. Thus, the preservation and sequestration of SOC become great challenges that humanity must face to highlight the double crisis of global change and food insecurity [30]. Although, the major goal of any agricultural management strategy is enhancing crop productivity; environmental sustainability is a major issue for the continuing stability of the agroecosystems. Therefore, maintaining SOC is essential for the long-term sustainability especially in dryland agroecosystems [13].

9 Management Practices and Soil Carbon Sequestration

The dynamics of the SOC stocks are depending on the balance between C inputs and outputs through different pathways and is strongly influenced by soil management practices. It has been shown from different studies that balanced fertilization with NPK in combination with manures increased SOC concentrations and maintained high crop yields [17]. However, soils have lost 50–75% of the original SOC stock in the managed ecosystems. The main reason of SOC stock depletion when converting natural to managed ecosystems is lower C input into the agricultural ecosystems, in addition to higher losses in managed ecosystems than natural ecosystem due to erosion, mineralization and leaching. Tillage and other soil disturbances may cause a decline in SOC stock through disrupting soil aggregates, then considerably accelerate the decomposition of SOC because of exposure of organic molecules to biological attack or air. Some studies indicate the possibility of SOC to redistribution and release into the atmosphere to the water through soil erosion [30]. The maximum soil C sink capacity approximately equals the historic C loss, is the amount of C that can potentially be stored in the soil [54]. Improved management practices, such as fertilization, applications of animal manure, the incorporation of straw in the soil and residue management practices, can increase SOC stocks of agro-ecosystems [27, 54]. However, improving and maintaining SOC stock particularly in dryland soils are major challenges due to high temperature and frequent drought stress. Therefore, application of plant nutrients and organic amendments, including the legumes incorporation in the crop rotation, enhance soil fertility and sustainability [54], crop rotation [38]. Thus, optimizing agricultural management through the efficient use of fertilizers, irrigation, and tillage operations can often enhance C sequestration while simultaneously reducing the C emissions associated with agricultural inputs such as fertilizers and on-farm fuels [15]. Wright and Hons [46] mentioned that management strategies, such as no-till and rotations with multiple crops high-intensity cropping sequences, play significant roles in SOC and SON sequestration.

10 Soil Carbon Sequestration in Arid and Mediterranean Sea Region

Agricultural soils have lost about up to 50% of C from the plow layer compared with undisturbed systems and for this reason, provide an opportunity for C sequestration. Utilizing agricultural management practices such as growing cover crops and adopting minimum tillage and no-till practices instead of conventional tillage has been shown to lead to enhanced soil C sequestration. The application of such practices regarding increase sequestration of C in the soil has received increasing attention under the aim of mitigating greenhouse gas emissions, but information is limited with respect to arid and semiarid climates [68]. Generally, soils in the Mediterranean region are characterized by low OM content (around 1%) due to the great majority of them being agricultural soils, meaning these soils have low C additions from plant residues, low-density tree cover, and they are subjected to management practices that lead to C losses. In Mediterranean forest areas, the effect is the opposite, as many of these forest soils are rich in OM; as a result, these soils store a large amount of C affecting the global C cycle. This means that these soils are characterized by high SOC [26].

In addition to C sequestration, several land-use options that increase OM and tighten the soil nitrogen (N) cycle can yield powerful synergies, such as enhanced fertility and productivity, increased soil biodiversity, reduced erosion, and runoff and water pollution. Although strategies involving land use conversions from crop production to pasture or forestry have huge potential to increase SOC stocks [62].

11 Implication of Climate Change and Soil Carbon Sequestration on Egypt

Egypt is one of the most affected countries in the world by climate change, therefore, soil carbon sequestration should be considered in the climate change mitigation in Egypt.

11.1 Soil C and N in Nile Delta, Egypt in Current State and the Context of Climate Change

Nile Delta, Egypt is one of the most fertile area in the world, however it subjected to many challenges, such as agricultural soil contamination, shrinking agricultural area due to urban sprawl. Climate change is one of the most expected challenges that will effect on Egypt and Nile Delta specially. Due to the Kyoto Protocol, national estimations and inventories of soil C and N stocks are needed for identifying and implementing strategies of stabilizing the atmospheric concentration of GHGs concentration [69–72].

11.1.1 The Current State of Soil C and N in Nile Delta, Egypt

Agricultural soils generally have a low content of OM compared to natural lands due to reduced C input because of annual harvest and removal of crop residue, etc., high organic C decomposition because of frequent tillage, increased soil erosion in addition to other factors [73]. As indicated from many references based on the information of OM, N, and other nutrients; soil C and N are low in Egyptian soils and in Delta's soils although the Nile Delta is one the most fertile area in Egypt. In this context; Abd El Hadi [74] mentioned that the cultivated area in Egypt covering three different production zones. Those zones are (1) the old irrigated lands (in the Nile Valley and Delta and most is fertile soils); (2) the newly reclaimed lands (the soil is poor in OM and macro-and micronutrients), and (3) the rainfed area (sandy soil sited in the Northwest Coast and North Sinai). However, Abd El Hadi [74] mentioned that numerous studies had established the poverty of most Egyptian soils in OM and N content. Additionally; Hegazi et al. [75] reported that extensive and frequent cropping, under unsustainable irrigation water management and inappropriate agricultural practices, in the Nile Valley and Delta has led to depletion and deficiency in many nutrient elements. This situation has been aggravated afterward the High Dam construction, which sharply declined the annual supplements of the fertile sediments to the soils. Consequently, all Egyptian soils are poor in their content of OM, total N, and other nutritive elements. As well; FAO [76] confirmed this fact where soils are poor in OM, with less SOM due to soil warming. Higher air temperatures are likely to speed the natural decomposition of OM and to increase the rates of other soil processes that affect fertility. Under drier soil conditions, root growth and OM decomposition are significantly inhibited, and as soil cover diminishes, vulnerability to wind erosion becomes higher, especially in case of intensive winds. However, it is mentioned in IFOAM [77] that with the right type of agriculture, emissions causing climate change can be reduced and nature capacity of the ecosystem to mitigate climate change can be employed to sequester high quantities of atmospheric carbon dioxide—especially in the soil. In this context Altieri and Koohafkan [78] reported that over the world, management practices that used by small farmers such as crop rotation, composting, green manures and cover crops, agroforestry, in addition to all practices that increase biomass production, will consequently build active OM.

11.2 Influence of Changing Climate on Soil C and N Stocks and Sequestration

Thornley and Cannell [79] mentioned that the response of soils to increasing temperature is an important issue. Re-examining the global patterns of soil carbon, giving a clear trend of decreasing carbon with mean annual temperature for very dry and very wet biomes. Zhang et al. [80] reported that climatic conditions, especially temperature and precipitation, may be responsible for the spatial variations in soil carbon sequestration. Elbasiouny [81] studied the correlation between soil C and N and temperature in North Nile Delta, Egypt (mean temperature data was obtained from Central Laboratory for Agricultural Climate (CLAC), Agricultural Research Center, Egypt). Precipitation is not investigated in this study because its amount in July (sampling month) is zero in the study area. As presented in Table 1, she found that the correlation between temperature and soil C and N stocks is negative. The most affected stock is total soil N stock followed by total soil C stock and SOC stock equally. The SIC stock was the least affected one. As well the correlation between temperature and sequestration rate of these stocks is negative, but total soil

Table 1 Correlation between soil C and N stock, sequestration rate and temperature in the studied locations [81]

Temp. and SOCS	Temp. and SOCSseq.	Temp. and SICS	Temp. and SICSseq.	Temp. and TSCS	Temp. and TSCSseq.	Temp. and TSNS	Temp. and TSNSseq.
−0.68	−0.60	−0.41	−0.35	−0.68	−0.63	−0.70	−0.62

Temp. temperature; *SOCS* SOC stock; *SICS* SIC stock; *TSCS* total soil C stock; *TSNS* total soil N stock; *Seq.* sequestration

C stock sequestration is the most affected stock as shown in the table. Of course, this negative correlation means that if any of the studied factors is increased, the other will be decreased. The temperature has already increased by 0.6 ± 0.2 °C, and is projected to increase by 2–4 °C towards the end of the 21st century increasing because of increasing greenhouse gases in the atmosphere [82]. The meteorological data of temperature and precipitation data in Egypt last few years revealed that the temperature in the study area is increasing and precipitation is decreasing. Thus, the soil C and N stocks and sequestration are supposed to be decreased because of the above-mentioned correlation. Therefore; this result is supported by the statements of many authors [79, 80, 83] the higher temperature will affect negatively on soil C.

Although the effect of precipitation is not examined in this study, it is supposed to have a negative effect on soil C and N. This is because precipitation level is decreasing in the study area as shown in Fig. 2b. Many researchers found that increasing precipitation will increase soil C or N such as [83–85]. To explain the negative impact of higher temperature, Ahmad et al. [84] mentioned that effects of climate change on soil carbon storage and its distribution differ between different regions, and temperature and rainfall levels are among the main influencing factors causing the differences. Temperature and humidity are important factors affecting the rate of organic material decomposition. Decomposition doubles with every 10 °C increase in temperature, while the increase in soil moisture increases the amount of OM. Also, He et al. [85] stated that soil C sequestration depends on the balance between inputs from biomass and outputs through the decomposition of residues, all of which could be affected by climate change. Higher temperature or precipitation influences the C cycle by altering NPP and soil respiration. In the semi-arid region, soil moisture has especially effective influence on the rates of plant growth and SOM decomposition, thus affecting SOM accumulation from litter and plant root inputs. Lal [86] reported that strategies for adaptation to climate change and reducing risks to agriculture on the farm level are representing in improving soil quality and adjusting management operations to buffer against the adverse effects of climatic disruption. Improvements in soil quality can be achieved by recommended management practices of soil management such as mulch farming, conservation agriculture, integrated nutrient management, water harvesting and recycling through drip sub-irrigation and judicious landscape management. Also, Lal [86] emphasized the increasing use efficiency of N fertilizer, which is essential to increasing crop yields per unit consumption of energy-based inputs. As will he have mentioned that replacement of flooded rice by aerobic rice is important for saving water and sustaining crop yields. Application of biosolids as soil amendments (e.g., compost, manure) is extremely important to improving productivity, and creating a positive C budget and enhancing the ecosystem C stock.

12 Potential Soil Carbon Sequestration in Egypt and Nile Delta

The potential of the world's degraded soils and ecosystems to sequester SOC is high; estimated at 1216 Mha, and agricultural soils estimated at 4961 Mha. As these soils have lost a significant part of their original SOC stock, they can sequester C by converting to restorative land use and adopting recommended management practices. The potential of SOC sequestration is in the following order: degraded soils and desertified ecosystems > cropland > grazing lands > forest lands and permanent crops. A considerable part of the historic C loss (estimated at 66–90 Pg) can be sequestered over 25–50 years. The rates of SOC sequestration on cropland range from 0.02 to 0.76 Mg C/ha/year when improved systems of crop management are adopted, and 0.25–0.5 Mg C/ha/year for rice land management [65]. As well, Ahirwal et al. [14] emphasized on the large potentiality of atmospheric C sequestration on the reclaimed sites, moreover its variance with the climatic conditions and plant species that could be used for reclamation. They mentioned that with an increase in the reclamation age, the development of vegetation cover and soil properties on these sites could sequester more atmospheric CO_2. However, the emission of CO_2 from the soil surface into the atmosphere can result from the root and microbial respiration. Thus, a reclaimed site in Egypt also has a huge potentiality to sequester more atmospheric CO_2. Ahirwal et al. [14] also reported that accumulation of nutrient stocks in the soil in the initial reclamation stages plays an important role in the development of soil horizons and plant growth. In this context, enhancing SOM is major importance because of bounding SOC and N, which may be mineralized and enhance the soil fertility.

Although north Nile Delta, Egypt produce tiny percent (less than 1%) of total global GHGs emissions, it is one of the most vulnerable areas to the expected impacts and risks of climate change [87]. Also, there are largely degraded and salt-affected areas in North Nile Delta [69, 70]. Coastal soils are salt-affected because of special hydrologic and geographical conditions. The salinity in coastal areas is a limiting factor for crop productivity as in the degraded coastal area in North Nile Delta, Egypt. The northern coastal zone of the Nile Delta is generally low lands as well. Hence it is vulnerable to direct and indirect impacts of climate change impacts especially sea level rise. Therefore, the Nile Delta is expected to suffer emundation, severe soil salinization and gradual merging under the groundwater logging and seawater transgression. Recently conversion of coastal lands to agricultural usages is a vital process in the deserts and coastal areas nearby Nile Delta mainly under the limitation of soil and water resources in these drylands [70]. Therefore, rising demands for water and food production in this area have led to the extension of reclamation projects in the surrounds degraded and desert areas [70, 88]. Zhang et al. [28] stated that coastal wetlands are valuable resources play a significant role in carbon sequestration. Therefore, coastal wetlands reclamation has become a common practice for the acquisition of new agricultural soil in many countries over the world. The reclamation process in such areas may have a significant effect on SOC due to altering soil condition and inputs and outputs of organic C. Such as other areas climate conditions, soil properties, topography, and anthropogenic activities are general affecting factors on SOC dynamics. However, SOC dynamics in such coastal reclaimed soils is also affected by certain specific factors such as salinity, pH and reclamation duration [28]. As the SOC is a key indicator of soil quality and productivity. Thus, restoring the quality of degraded soils require increasing soil C sequestration through increasing SOC density, improving depth distribution of SOC and stabilizing and protecting SOC in stable micro-aggregates. Therefore, adopting recommended management practices in agroecosystems, in addition to practices that can create a positive soil C budget, is a crucial strategy for SOC sequestration in the terrestrial ecosystems [54]. In a study of [70] to compare between C and N stocks in such degraded and uncultivated soils and cultivated restored (reclaimed 20 years ago) sites in North Nile Delta, and to investigate the impact of cultivation on soil C and N sequestration in this area. Soil C and N stocks increased significantly in upper soil from 2.99 and 0.43 Mg h^{-1} in the degraded soil to 19.26 and 1.66 Mg h^{-1} in restored ones. As well, salinity was reduced, and NPP was increased because of leaching and reclamation. Therefore, such findings support the suggestion that restored area have a high capacity to sequester N.

Moreover, IFOAM [77] in a case study on Egypt stated that Composting in the Egyptian desert in Sekem organic farm had been demonstrated to be high sequestration by sequestering over 3 metric tons of carbon dioxide per hectare per year. It is conducted from this trial that compost trials in the Egyptian desert show that organic farming practices—recycling of organic waste to compost and adding to the soil in dryland regions have the potential to mitigate climate change by avoiding emissions and sequestrating high levels of carbon. Furthermore, these systems improve soil fertility and increase food security in the area.

13 Sequestration Rate of Soil C and N Stocks in North Nile Delta Egypt

As the largest terrestrial reservoir of C, the world's soil may strongly act as a sink or a source of atmospheric carbon dioxide (CO_2) [23, 89, 90]. Qiu et al. [5] reported that there could be a large capacity for soil C and N sequestration when protective management practices were implemented. In a study on north Nile delta Egypt [81], the sequestration rate of soil C and N data was calculated based on actual data of soil C and N (and estimated data of past soil C and N. The actual data are field data analyzed after sampling from this area [Fig. 1 in 2007, while the estimated data are calculated based on regression equation of actual values of soil C and N and the values of Normalized Difference Vegetation Index (NDVI)]. The NDVI was calculated based on different landscape images in different time intervals (i.e. 1987, 1990 and 1998). The rate of C and N sequestration according to the equation of sequestration rate as described by [51, 91]. It is noticed from this study that there were a declining in C and N contents in some sampling locations, however, there are an increment in other sampling locations in north Nile delta which means that those sites represented sink of atmospheric C or N emissions (i.e. other profiles consider the source of CO_2 emission to the atmosphere). These data are presented in Table 1.

It is observed from Elbasiouny [81] that some profiles especially 10 and 11 in all cases (i.e. SOCS, SICS, TSCS and TSNS) are positive which means that those profiles represented a sink for C or N emissions in the atmosphere (i.e. other profiles consider the source of CO_2 emission to the atmosphere). Those profiles were cultivated by rice and may be this the reason for positivity of sequestration rate of soil C and N in them. However, there are some observations, where in SOCS; profiles 6 and 12 are sinks not sources of CO_2. Hou et al. [66] reported that since SOC is crucial affecting soil physical, chemical, and biological properties, more SOC sequestration in the soil can assist in sustaining soil fertility and agronomic productivity. Jacinthe et al. [6] showed that taking the forest soil as a reference for SOM data, the data revealed SOM depletion in the cultivated soils and SOM enrichment in the depositional areas. The SOC inventory (0–30 cm) in the drained and undrained cultivated fields was 5.6 and 6.0 kg C m^{-2}, respectively representing 72–77% of the original SOC stock (forest: 7.9 kg C m^{-2}). These data can be translated into a total SOC depletion of 2.2 and 1.8 kg C m^{-2} in the drained and undrained croplands, respectively. However; Alvarez et al. [92] found a negative C balance in their studied soil of the Rolling Pampa. Consequently, a net loss of C from the soil is apparently in progress. Chung et al. [93] stated that evidence has confirmed approximately 40–70% of OC in the active soil layer (1.0 m topsoil layer) was lost in nearly 50 years. Most of this loss recorded in the first 25 years, as carbon emission showed an almost exponential decline with time year. Therefore, the 25-year timescale is generally used to calculate the current carbon emission produced by land-use change.

On the other hand; the lowest negative sequestration rate in SICS was (−19.33 Mg ha^{-1} $year^{-1}$), while the highest is recorded as (20.34 Mg ha^{-1} $year^{-1}$) as presented in Table 1. It is also observed that SIC represent a sink of C in some profiles as recorded in the previous table. Mei et al. [94] mentioned that although the role of SIC in C sequestration is less well understood, depending on the site-specific conditions, SIC may act as a sink or source or have no effect upon C sequestration. Lal [95] reported the additional potential for C sequestration beside SOC through SIC sequestration. Moreover, the emphasis on existing a strong correlation between net NPP and formation of secondary carbonates, and between SOC and secondary carbonates. The rate of formation of secondary carbonates depends on the rainfall and quality of irrigation water. For SIC sequestration rate of 20–400 kg $ha^{-1}year^{-1}$ over 344 Mha of arid and semiarid soils, total potential of SIC sequestration is 7–138 Tg C $year^{-1}$.

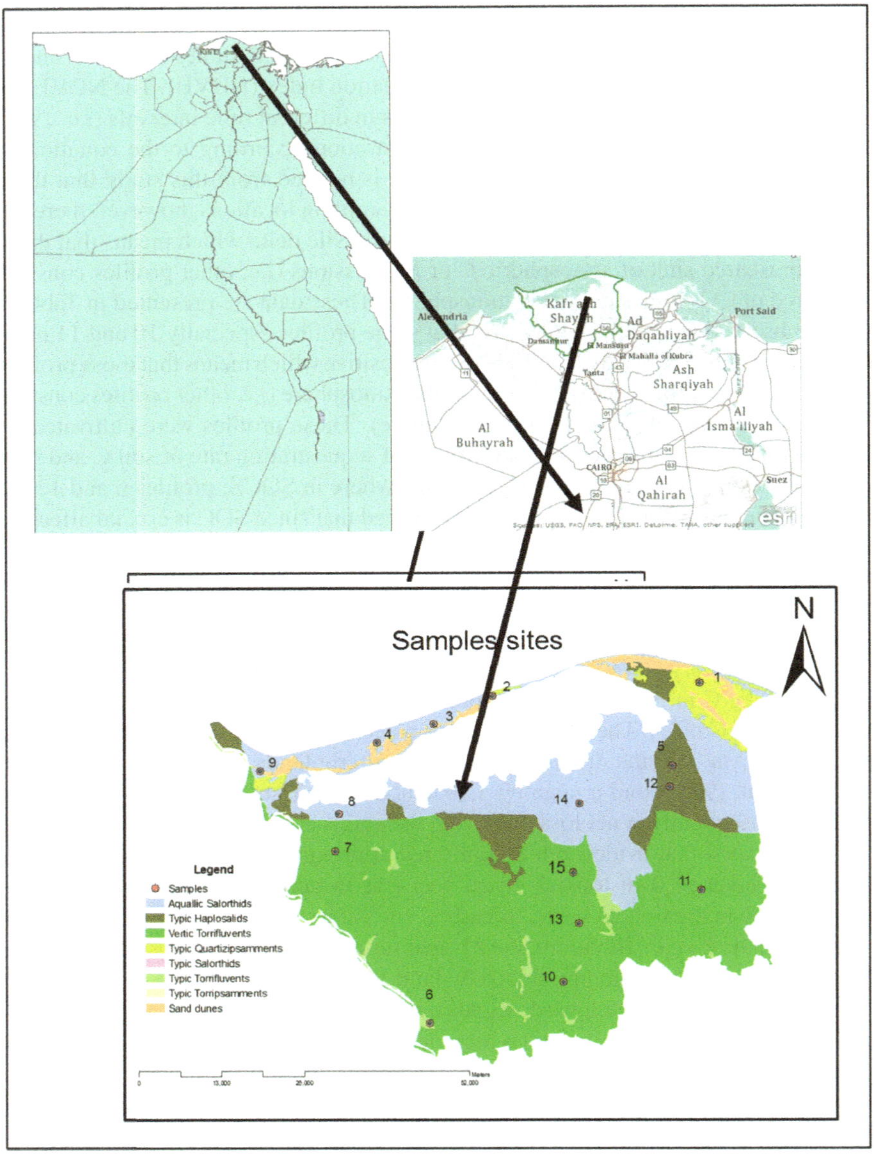

Fig. 1 Elbasiouny's [81], Elbasiouny et al. [69] study area: location of the study area on Egypt map (top left) and the administration boundaries of Kafr El_sheikh Gov. and sampling location (bottom). *Source* Elbasiouny et al. [69] and Elbasiouny [81]

Table 2 Sequestration rate of C and N stocks as calculated from 1987 and 2007 C and N stock data (Mg ha^{-1} year^{-1}) [81]

Profile no.	SOCSseq.	SICSseq.	TSCSSeq.	TSNSseq.
1	−21.30	−11.55	−40.50	−1.37
2	−7.19	−2.07	−7.84	1.26
3	−11.75	−5.26	−17.08	0.59
4	−7.80	−2.96	−10.82	1.23
5	−35.79	−18.93	−57.89	−2.51
6	1.73	−8.93	13.13	3.47
7	−16.18	−8.35	−26.47	−0.32
8	−20.85	20.34	−10.23	−1.15
9	−35.48	−19.33	−69.79	−3.35
10	7.33	9.18	17.30	3.22
11	13.11	3.64	18.19	3.25
12	0.98	−7.87	−14.42	1.66
13	−0.04	−13.68	−24.77	0.58
14	−10.38	2.15	0.46	3.11
15	−5.05	−8.26	−33.10	0.09
Average Mg/ha/year	−9.91	−4.79	−17.59	0.65

SOCS soil organic C stock; *SICS* soil inorganic C stock; *TSCS* total soil C stock; *TSNS* total soil N stock; *seq.* sequestration

Regarding TSCS, there were 4 of 15 profiles represent sink of C emissions. Those profiles are cultivated. Therefore this result is in accordance with [66] where they motioned that cropland soil has a huge potential as a C sink (0.4–0.8 Pg year^{-1}), although this rate does not exist here. However, they emphasized that this potentially could decrease CO_2 concentrations in the air and mitigate global emissions.

It is noticed that the sequestration rate of TSNS was positive as indicated in Table 2. Types of crop residues have critical roles in C sequestration and soil aggregation because of their C/N ratios or quality. C/N ratios often govern the degradation of fresh crop residues, but as residues undergo decomposition, they become more recalcitrant, and degradation then becomes controlled by lignin contents or lignin/N ratios. Thus, the ability of soils to sequester C is closely related to N [46] (Fig. 2).

Generally, the average sequestration rate (based on the average of all profiles) is negative in SOCS and SICS. Oppositely TSNS is positively sequestered in the studied locations with the rate of 0.65 Mg/ha/year. Although the TSN content in most salt-affected profiles in the study area is very low, it is a surprise to be positively sequestered in TSNS. This confirms the last information about enhancing or restoring the salt-affected soils will provide a potential for C and N sequestration [35]. Of course, this was the status in these profiles although there was no enhancing for OM inputs in there, except through growing some tolerant salinity plants without any human interface, let us imagine if they have been improved. Li et al. [96] reported that

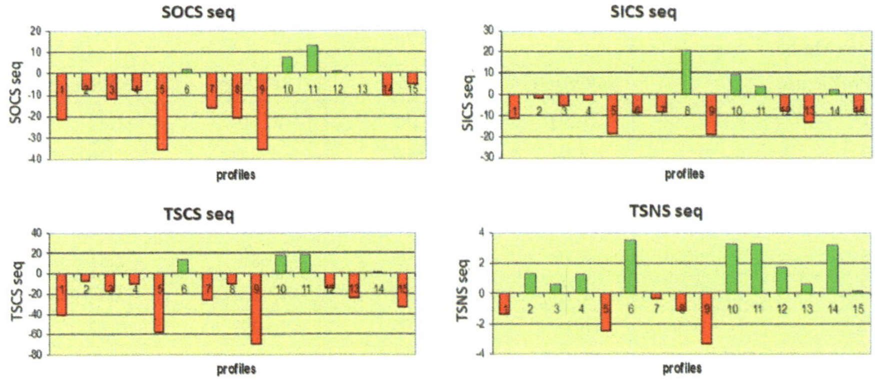

Fig. 2 Negative and positive sequestration rates of soil C and N stocks as drawn from (Table 2). *Source* Elbasiouny [81]. SOCS: soil organic C stock; SICS: soil inorganic C stock; TSCS: total soil C stock; TSNS: total soil N stock; seq.: sequestration

although increasing salinity and/or alkalinity will likely decrease soil OM content due to suppressed plant productivity, with large areas of land are salt-affected worldwide; it may be assumed that considerable reserves of terrestrial C exist in those soils. The C and nutrients contents (especially N) and their cycling in soils are controlling response factors of an ecosystem to environmental stresses, assigning special importance for the need of understanding C cycling in naturally salt-affected ecosystems.

The sequestration rate in some other is negative may be due to excessive N fertilizing that may restrain SOC sequestration through microbial population suppression or stimulation of mineralization in old native organic C [5]. Addition reasons for this negative sequestration rate may be their salinity and hence low productivity which decline OM inputs, or may be cropping intensity and cultivation which deplete OM and C in most cases as a result of non-offsetting of the inputs. However, cropping intensity would be supposed to provide higher quantities of crop residues to the soil, which should increase soil OM in the long term [97]. In the continued context, Hou et al. [66] reported that generally, no-tillage with residue left in place has the potential for sequestering more SOC than conventional tillage in the upper soil depths because the tillage destroys the protection provided by crop residue on the surface as well it increases the oxidization of SOM which could be avoided by no-tillage treatment. The finding by Puget et al. [98] should be carried out in the study area; namely SOC and N stocks are affected strongly by land use changes and management practices. Alterations in the soil C or N budget due to anthropogenic activities have resulted in changes in their net fluxes, hence affecting the atmospheric concentration of CO_2 and N_2O and the global climate change. Therefore, adoption of sustainable agricultural practices such as conservation tillage, residue returning to the soil, and using animal or green manure can not only increase SOM concentration but also enhance soil quality. Lal [99] and Hou et al. [66] stated that addition of crop residue plays an important role in SOC sequestration, improving soil structure and soil water-holding capacity, and soil erosion prevention. Crop residue is important to soil nutrient cycling and soil fertility. Crop residue removal will cause the depletion of soil nutrient (e.g. NPK) which could decline the agronomic productivity and increase soil degradation. Thus, there are some factors may be considered as limiting factors in the sequestration process. For example, the removal of crop residue either by burning or by moving from another place, as well lack the attention of utilizing the organic fertilizers and manure, the dry climatic conditions which encourage the decomposition of OM, as well the irrigation does as mentioned above.

Finally; the adoption of good management practices can enhance C and N sequestration. In this context, Wright and Hons [46] motioned that management strategies, such as no-till and rotations with multiple crops or high-intensity cropping sequences, play significant roles in SOC and SON sequestration. In this context, they interpreted lower soil organic nitrogen (SON) storage for continuous soybean systems than for multiple cropping and double cropping systems which provide greater residue inputs, by lower crop residue inputs. This may lead to N may not be immobilized to the extent as would occur under high-intensity cropping, with SON possibly undergoing mineralization and loss from the soil. As well; [46] concluded the enhancement of SOC and SON sequestration under no-till compared with conventional-tillage

It is worth noticing after presenting the results from the sequestration concept point of view that the Nile Delta region is one of the most fertile areas of the world. However; there were some negative changes in this region such as what explained by [100]. They reported that elevated soil salinity level is one of the principal causes of reduced agricultural productivity in the Delta. Agricultural management practices have been changed after the construction of the Aswan High Dam in 1964, also, perennial, furrow irrigation replaced basin flooding, and insensitive cropping replaced the single crop per year leading to shorter fallow periods. Soil nutrients declined as a result of intensive cultivation along with a lack of regular nutrient replacement and a loss of the alluvium deposits. Also, in some areas, overirrigation combined with insufficient drainage led to increased soil salinity and elevated water tables. Although these changes, there is a potential opportunity to enhance and sustain agricultural soil in Egypt especially in Nile Delta and enhancing C and N sequestration through land use and management practices. In this context Wu et al. [18] mentioned that soil C increases linearly with increased additions of crop residues and root OM, however, suggesting that irrigated agriculture will increase soil C with time. It is typical that SOC commonly declines at the early stages, approaches a steady state in 25–50 year, and shows an evident increase in about 50 year after native soil conversion to agricultural ones. This pattern suggests that time is an important factor for C sequestration in irrigated cropland [18]. Thus, understanding the effects of agricultural management on SC stocks and dynamics under specific soil-climatic conditions is warranted by the necessity of sequestering atmospheric C in agricultural soils to mitigate the climate change. Adoption of sustainable agricultural practices (e.g., conservation tillage, residue retention, and using animal or green manure) can increase SOC concentration and enhance soil quality. Reduced tillage practices such as no-till and chisel till may increase both concentration and amount of SOC in topsoil. However, such increases may not occur in fine-textured soils, under cold and poorly drained conditions, or where the antecedent SOC concentration is high [98].

14 Summary

Soil C sequestration is a very important tool for C storage, atmospheric CO_2 reduction and climate change mitigation. As well, it is critical for enhancing soil fertility, agricultural productivity, and sustainability. Adoption of good management practices may make agricultural soils an effective C sink because of their content of C content is below the C saturation point, which means that an additional C can be sequestered with time. Soil C and N are low in Egyptian soils and in Delta's soils although the Nile Delta is one the most fertile area in Egypt. Although North Nile Delta, Egypt produces a very small percentage of global GHGs emissions, it is one of the most vulnerable areas to the expected impacts and risks of climate change. In addition, although, Nile Delta in Egypt is one of the most fertile areas in the world, it contains a considerable area of degraded and un-reclaimed soils. Therefore, the reclamation process in such areas is very important not only for agricultural extension in Egypt but also it may have a significant effect on SOC by alternation of soil condition and inputs and outputs of organic C. Thus, such soil has huge potential to sequester more C and N, subsequently reducing atmospheric CO_2 if improved management practices are adopted.

References

1. Ravindranath NH, Ostwald M (2008) Carbon inventory methods: handbook for greenhouse gas inventory, carbon mitigation and roundwood production projects. In: Advances in global change research, vol 29. Springer Science + Business Media BV
2. Soleimani A, Hosseini SM, Bavani ARM, Jafari M, Francaviglia R (2017) Simulating soil organic carbon stock as affected by land cover change and climate change, Hyrcanian forests (northern Iran). Sci Total Environ 599–600:1646–1657
3. Sofi JA, Lone AH, Ganie MA, Dar NA, Bhat SA, Mukhtar M, Dar MA, Ramzan S (2016) Soil microbiological activity and carbon dynamics in the current climate change scenarios: a review. Pedosphere 26(5):577–591. https://doi.org/10.1016/s1002-0160(15)60068-6
4. Waltman SW, Olson C, West L, Moore A, Thompson J (2010) Preparing a soil organic carbon inventory for the United States using soil surveys and site measurements: why carbon stocks at depth are important. In: 19th world congress of soil science, soil solutions for a changing world
5. Qiu SJ, Ju XT, Ingwersen J, Qin ZC, Li L, Streck T, Christie P, Zhang FS (2010) Changes in soil carbon and nitrogen pools after shifting from conventional cereal to greenhouse vegetable production. Soil Till Res 107:80–87
6. Jacinthe PA, Lal R, Kimble JM (2001) Organic carbon storage and dynamics in croplands and terrestrial deposits as influenced by subsurface tile drainage. Soil Sci 166(5):322–335
7. Pandey D, Agrawal M, Bohra JS, Adhya TK, Bhattacharyya P (2014) Recalcitrant and labile carbon pools in a sub-humid tropical soil under different. Tillage combinations: a case study of rice–wheat system. Soil Till Res 143:116–122. https://doi.org/10.1016/j.still.2014.06.001

8. Eve MD, Sperow M, Paustian K, Follett RF (2002) National-scale estimation of changes in soil carbon stocks on agricultural lands. Environ Pollut 116:431–438
9. Lokupitiya E, Paustian K (2006) Agricultural soil greenhouse gas emissions: a review of national inventory methods. J Environ Qual 35:1413–1427
10. Victor DG, Zhou D, Ahmed EHM, Dadhich PK, Olivier JGJ, Rogner H-H, Sheikho K, Yamaguchi M (2014) Introductory chapter. In: Edenhofer O, Pichs-Madruga R, Sokona Y, Farahani E, Kadner S, Seyboth K, Adler A, Baum I, Brunner S, Eickemeier P, Kriemann B, Savolainen J, Schlömer S, von Stechow C, Zwickel T, Minx JC (eds) Climate change 2014: mitigation of climate change. Contribution of working group III to the fifth assessment report of the intergovernmental panel on climate change. Cambridge University Press, Cambridge
11. Geraei DS, Hojati S, Landi A, Cano AF (2016) Total and labile forms of soil organic carbon as affected by land use change in southwestern Iran. Geoderma Reg 7:29–37. https://doi.org/10.1016/j.geodrs.2016.01.001
12. Nadal-Romero E, Cammeraat E, Pérez-Cardiel E, Lasanta T (2016) How do soil organic carbon stocks change after cropland abandonment in Mediterranean humid mountain areas? Sci Total Environ 566–567:741–752
13. Singh KP, Ghoshal N, Singh S (2009) Soil carbon dioxide flux, carbon sequestration and crop productivity in a tropical dryland agroecosystem: influence of organic inputs of varying resource quality. Appl Soil Ecol 42:243–253
14. Ahirwal J, Maiti SK, Singh AK (2017) Changes in ecosystem carbon pool and soil CO_2 flux following post-mine reclamation in dry tropical environment, India. Sci Total Environ 583:153–162. https://doi.org/10.1016/j.scitotenv.2017.01.043
15. West TO, Marland G (2002) Net carbon flux from agricultural ecosystems: methodology for full carbon cycle analyses. Environ Pollut 116:439–444
16. Rolando JL, Dubeux JC, Perez W, Ramirez DA, Turin C, Ruiz-Moreno M, Comerford NB, Mares V, Garcia S, Quiroz R (2017) Soil organic carbon stocks and fractionation under different land uses in the Peruvian high-Andean Puna. Geoderma 307:65–72. https://doi.org/10.1016/j.geoderma.2017.07.037
17. Fan J, Ding W, Xiang J, Qin S, Zhang J, Ziadi N (2014) Carbon sequestration in an intensively cultivated sandy loam soil in the North China Plain as affected by compost and inorganic fertilizer application. Geoderma 230–231:22–28. https://doi.org/10.1016/j.geoderma.2014.03.027
18. Wu L, Wood Y, Jiang P, Pan G, Lu J, Chang AC, Enloe HA (2008) Carbon sequestration and dynamics of two irrigated agricultural soils in California. Soil Sci Soc Am J 72(3):808–814
19. Lim S, Choi W, Chang SX, Arshad MA, Yoon K, Kim H (2017) Soil carbon changes in paddy fields amended with fly ash. Agr Ecosyst Environ 245:11–21. https://doi.org/10.1016/j.agee.2017.03.027
20. Fernández JM, Nieto MA, López-de-Sá EG, Gascó G, Méndez A, Plaza C (2014) Carbon dioxide emissions from semi-arid soils amended with biochar alone or combined with mineral and organic fertilizers. Sci Total Environ 482–483:1–7
21. Mandal S, Sarkar B, Bolan N, Novak J, Ok YS, Singh L Van, Swieten L, Singh BP, Kirkham MB, Choppala G, Spokas K, Naidu R (2016) Designing advanced biochar products for maximizing greenhouse gas mitigation potential. Crit Rev Env Sci Technol. https://doi.org/10.1080/10643389.2016.1239975
22. Li C, Fultz LM, Moore-Kucera J, Acosta-Martínez V, Horita J, Strauss R, Zak J, Calderón F, Weindorf D (2017) Soil carbon sequestration potential in semi-arid grasslands in the conservation reserve program. Geoderma 294:80–90. https://doi.org/10.1016/j.geoderma.2017.01.032
23. Marín-Spiotta E, Swanston CW, Torn MS, Silver WL, Burton SD (2008) Chemical and mineral control of soil carbon turnover in abandoned tropical pastures. Geoderma 143:49–62
24. Oca PCM (2009) Effect of land use, climate and soil structure on soil organic carbon in Costa Rican Ecoregions. Presented in partial fulfillment of the requirements for the Degree Master of Science in the Graduate School of the Ohio State University

25. Zhu L, Hu N, Zhang Z, Xu J, Tao B, Meng Y (2015) Short-term responses of soil organic carbon and carbon pool management index to different annual straw return rates in a rice–wheat cropping system. CATENA 135:283–289. https://doi.org/10.1016/j.catena.2015.08.008
26. Lozano-García B, Parras-Alcántara L, Brevik EC (2016) Impact of topographic aspect and vegetation (native and reforested areas) on soil organic carbon and nitrogen budgets in Mediterranean natural areas. Sci Total Environ 544:963–970
27. Zhang X, Sun N, Wu L, Xu M, Bingham IJ, Li Z (2016) Effects of enhancing soil organic carbon sequestration in the topsoil by fertilization on crop productivity and stability: evidence from long-term experiments with wheat-maize cropping systems in China. Sci Total Environ 562:247–259
28. Zhang H, Wu P, Yin A, Yang X, Zhang M, Gao C (2017) Prediction of soil organic carbon in an intensively managed reclamation zone of eastern China: a comparison of multiple linear regressions and the random forest model. Sci Total Environ 592:704–713. https://doi.org/10.1016/j.scitotenv.2017.02.146
29. Zhang L, Zhuang Q, He Y, Liu Y, Yu D, Zhao Q, Shi X, Xing S, Wang G (2016) Toward optimal soil organic carbon sequestration with effects of agricultural management practices and climate change in Tai-Lake paddy soils of China. Geoderma 275:28–39. https://doi.org/10.1016/j.geoderma.2016.04.001
30. Liu X, Li L, Qi Z, Han J, Zhu Y (2017) Land-use impacts on profile distribution of labile and recalcitrant carbon in the Ili River Valley, northwest China. Sci Total Environ 586:1038–1045
31. Wang H, Guan D, Zhang R, Chen Y, Hu Y, Xiao L (2014) Soil aggregates and organic carbon affected by the land use change from rice paddy to vegetable field. Ecol Eng 70:206–211. https://doi.org/10.1016/j.ecoleng.2014.05.027
32. Lozano-García B, Muñoz-Rojas M, Parras-Alcántara L (2017) Climate and land use changes effects on soil organic carbon stocks in a Mediterranean semi-natural área. Sci Total Environ 579:1249–1259
33. Guo Y, Amundson R, Gong P, Yu Q (2006) Quantity and spatial variability of soil carbon in the conterminous united states. Soil Sci Soc Am J 70:590–600
34. Cheng HH, Kimble JM (2001) Characterization of soil organic carbon pools. In: Lal R, Kimble JM, Follett RF, Stewart BA (2001) Assessment methods for soil carbon. Lewis Publishers, 676pp
35. Lal R (2009a) Carbon sequestration in saline soils. J Soil Salin Water Qual 1(1–2):30–40
36. Diaz-Hernandez JL (2010) Is soil carbon storage underestimated? Chemosphere 80:346–349
37. Qadir M, Noble AD, Schubert S, Thomas RT, Arslan A (2006) Sodicity-induced land degradation and its sustainable management: problems and prospects. Land Degrad Dev 17:661–676
38. Luce MS, Ziadi N, Zebarth BJ, Grant CA, Tremblay GF, Gregorich EG (2014) Rapid determination of soil organic matter quality indicators using visible near infrared reflectance spectroscopy. Geoderma 232–234:449–458. https://doi.org/10.1016/j.geoderma.2014.05.023
39. Johnson RS, Uriu K (1989) Mineral nutrition. In: Larue JH, Johnson RS (1989) Peaches, plums, and nectarines growing and handling for fresh market. Cooperative extension. University of California. Division of agriculture and natural resources. Publication 3331
40. Nieder R, Benbi DK (2008) Carbon and nitrogen in the terrestrial environment. Springer Science + Business Media BV, Springer, Netherlands, p 432
41. Wong VNL, Murphy BW, Koen TB, Greene RSB, Dalal RC (2008) Soil organic carbon stocks in saline and sodic landscapes. Aust J Soil Res 46:378–389
42. Jelinski NA, Kucharik CJ (2009) Land-use effects on soil carbon and nitrogen on a U.S. midwestern floodplain. Soil Sci Soc Am J 73(1):217–225
43. Wei X, Blanco JA, Jiang H, Kimmins JPH (2012) Effects of nitrogen deposition on carbon sequestration in Chinese fir forest ecosystems. Sci Total Environ 416:351–361
44. Paul EA, Morris SJ, Six J, Paustian K, Gregorich EG (2003) Interpretation of soil carbon and nitrogen dynamics in agricultural and afforested soils. Soil Sci Soc Am J 67:1620–1628
45. Sainju UM, Senwo ZN, Nyakatawa ZN, Tazisong IA, Reddy KC (2008) Soil carbon and nitrogen sequestration as affected by long-term tillage, cropping systems, and nitrogen fertilizer sources. Agr Ecosyst Environ 127:234–240

46. Wright AL, Hons FM (2004) Soil aggregation and carbon and nitrogen storage under soybean cropping sequences. Soil Sci Soc Am J 68:507–513
47. Batjes NH, Dijkshoorn JA (1999) Carbon and nitrogen stocks in the soils of the Amazon region. Geoderma 89:273–286
48. Baisden WT, Parfitt RL (2007) Bomb 14C enrichment indicates decadal C pool in deep soil? Biogeochemistry 85:59–68
49. Don A, Schumacher J, Scherer-Lorenzen M, Scholten T, Schulze E (2007) Spatial and vertical variation of soil carbon at two grassland sites—implications for measuring soil carbon stocks. Geoderma 141:272–282
50. Throop HL, Archer SR, Monger HC, Waltman S (2012) When bulk density methods matter: implications for estimating soil organic carbon pools in rocky soils. J Arid Environ 77:66–71
51. Jarecki MK, Lal R (2005) Soil organic carbon sequestration rates in two long-term no-till experiments in Ohio. Soil Sci 170(4):280–291
52. Howard PJA, Loveland PJ, Bradley RI, Dry FT, Howard DM, Howard DC (1995) The carbon content of soil and its geographical distribution in great-Britain. Soil Use Manag 11(1):9–15
53. Crowe AM, McClean CJ, Cresser MS (2006) Application of genetic algorithms to the robust estimation of soil organic and mineral fraction densities. Environ Model Softw 21(10):1503–1507
54. Srinivasarao C, Lal R, Kundu S, Babu MBBP, Venkateswarlu B, Singh A (2014) Soil carbon sequestration in rainfed production systems in the semiarid tropics of India. Sci Total Environ 487:587–603
55. Barré P, Durand H, Chenu C, Meunier P, Montagne D, Castel G, Billiou D, Soucémarianadin L, Cécillon L (2017) Geological control of soil organic carbon and nitrogen stocks at the landscape scale. Geoderma 285:50–56. https://doi.org/10.1016/j.geoderma.2016.09.029
56. Young JL, Spycher G (1979) Water-clispersible soil organic-mineral particles: I. Carbon and nitrogen distribution. Soil Sci Soc Am J 43:324–328
57. Su WZ, Liu WJ, Yang R, Chang XX (2009) Changes in soil aggregate, carbon, and nitrogen storages following the conversion of cropland to Alfalfa Forage Land in the Marginal Oasis of Northwest China. Environ Manage 43:1061–1070
58. Gama-Rodrigues EF, Nair PKR, Nair VD, Gama-Rodrigues AC, Baligar VC, Machado RCR (2010) Carbon storage in soil size fractions under two cacao agroforestry systems in Bahia, Brazil. Env Manag 45:274–283
59. Nelson PN, Ladd JN, Ades's JM (1996) Decomposition of 14C-labelled plant material. In: A salt-affected soil. Soil Biol Biochem 28(4/S):433–444
60. Bot A, Benites J (2005) The importance of soil organic matter-key to drought-resistant soil and sustained food production. FAO Soils Bulletin 80. FAO Land and Plant Nutrition Management Service. Food and Agriculture Organization of the United Nations, Rome
61. Bossuyt H, Six J, Hendrix PF (2005) Protection of soil carbon by microaggregates within Earthworm Casts. Soil Biol Biochem 37:251–258
62. Mathew I, Shimelis H, Mutema M, Chaplot V (2017) What crop type for atmospheric carbon sequestration: results from a global data analysis. Agr Ecosyst Environ 243:34–46. https://doi.org/10.1016/j.agee.2017.04.008
63. Zaouche M, Bela L, Vaudour E (2017) Geostatistical mapping of topsoil organic carbon and uncertainty assessment in Western Paris croplands (France). Geoderma Reg 10:126–137. https://doi.org/10.1016/j.geodrs.2017.07.002
64. Senthilkumar S, Kravchenko AN, Robertson GP (2009) Topography influences management system effects on total soil carbon and nitrogen. Soil Sci Soc Am J 73:2059–2067. https://doi.org/10.2136/sssaj2008.0392
65. Lal R (2004) Soil carbon sequestration to mitigate climate change. Geoderma 123:1–22
66. Hou R, Ouyang Z, Li Y, Tyler DD, Li F, Wilson GV (2012) Effects of tillage and residue management on soil organic carbon and total nitrogen in the North China Plai. Soil Sci Soc Am J 76(1):1–11
67. Lal R, Griffin M, Apt J, Lave L, Morgan MG (2004) Manag Soil Carbon. Science 304(5669):393

68. Steenwerth KL, Pierce DL, Carlisle EA, Spencer RGM, Smart DR (2010) A vineyard agroecosystem: disturbance and precipitation affect soil respiration under Mediterranean conditions. Soil Sci Soc Am J 74:231–239
69. Elbasiouny H, Abowaly M, Abu_Alkheir A, Gad A (2014) Spatial variation of soil carbon and nitrogen pools by using ordinary Kriging method in an area of north Nile Delta, Egypt. CATENA 01(113):70–78
70. Elbasiouny H, Abowaly M, Abu_Alkheir A, Gad A, Elbehiry F (2017) Restoration and sequestration of carbon and nitrogen in degraded northern coastal area in Nile Delta, Egypt for climate change mitigation. J Coast Conserv 21:105–114. https://doi.org/10.1007/s11852-016-0475-3
71. Elbehiry F, Elbasiouny H, Elhenawy A (2017) Boron: spatial distribution in an area of North Nile Delta, Egypt. Commun Soil Sci Plant Anal 48(3):294–306. https://doi.org/10.1080/00103624.2016.1269795
72. Elbehiry F, Mahmoud MA, Negm A (2018) Land use in Egypt's coastal lakes: opportunities and challenges. In: The handbook of environmental chemistry. Springer, Berlin. https://doi.org/10.1007/698_2018_250
73. Yan H, Cao M, Liu J, Tao B (2007) Potential and sustainability for carbon sequestration with improved soil management in agricultural soils of China. Agr Ecosyst Environ 121:325–335
74. Abd El Hadi AH (2004) Country report on Egyptian agriculture. IPI regional workshop on Potassium and Fertigation development in West Asia and North Africa. Rabat, Morocco, 24–28 Nov 2004
75. Hegazi AM, Afifi MY, EL Shorbagy MA, Elwan AA, El-Demerdashe S (2005) Egyptian national action program to combat desertification. Arab Republic of Egypt, Ministry of Agriculture and Land Reclamation Desert Research Center (DRC)
76. FAO (2005) Fertilizer use by crop in Egypt. Land and plant nutrition management service land and water development division. Food and Agriculture Organization of the United Nations, Rome
77. IFOAM (2009) The contribution of organic agriculture to climate change mitigation. With support from European Community, DG Environment
78. Altieri MA, Koohafkan P (2008) Enduring farms: climate change, smallholders and traditional farming communities. TWN (Third World Network), Penang, Malaysia
79. Thornley JHM, Cannell MGR (2001) Soil carbon storage response to temperature: An hypothesis. Ann Bot 87:591–598
80. Zhang WJ, Wang XJ, Xu MG, Huang SM, Liu H, Peng C (2010) Soil organic carbon dynamics under long-term fertilizations in Arable Land of Northern China. Biogeosciences 7:409–425
81. Elbasiouny H (2012) Soil carbon and nitrogen pools inventory in North Delta—Egypt for climate change mitigation. Thesis submitted in partial fulfillment of the requirements for the degree of Ph.D. Faculty of Agriculture, Kafrelsheikh University
82. Lal R (2009b) Soil carbon sequestration for climate change mitigation and food security. Platinum jubilee celebrations of the Indian society of soil science. Souvenir, 39–46
83. Jobbagy EG, Jackson RB (2000) The vertical distribution of soil organic carbon and its relation to climate and vegetation. Ecol Appl 10(2):423–436
84. Ahmad H, Athar F, Manochehr G (2010) Carbon sequestration under different physiographic and climatic conditions in North Karaj River Basin. In: 19th world congress of soil science, soil solutions for a changing world
85. He N, Chen Q, Han X, Yu G, Li L (2012) Warming and increased precipitation individually influence soil carbon sequestration of inner Mongolian Grasslands, China. Agr Ecosyst Environ 158:184–191
86. Lal R (2009c) Soil quality impacts of residue removal for bioethanol production. Soil Till Res 102:233–241
87. Fahim MA, Hassanein MK, Khalil AA, Abou Hadid AF (2013) Climate change adaptation needs for food security in Egypt. Nat Sci 11(12):68–74
88. Arnous MO, El-Rayes AE, Green DR (2015) Hydrosalinity and environmental land degradation assessment of the East Nile Delta region, Egypt. J Coast Conserv 19:491–513. https://doi.org/10.1007/s11852-015-0402-z

89. Eshel G, Fine P, Singer MJ (2007) Total soil carbon and water quality: an implication for carbon sequestration. Soil Sci Soc Am J 71:397–405
90. Oelbermann M, Voroney RP (2007) Carbon and nitrogen in a temperate agroforestry system: using stable isotopes as a tool to understand soil dynamics. Ecol Eng 29:342–349
91. Lal R, Kimble JM, Follet RF, Cole CV (1998) The potential of US cropland to sequester carbon and mitigate the greenhouse effect. Ann Arbor Press, Chelsea, MI
92. Alvarez R, Russo ME, Prystupa P, Scheiner JD, Blotta L (1998) Soil carbon pools under conventional and no-tillage systems in the Argentine Rolling Pampa. Agron J 90:138–143
93. Chung H, Grove JH, Six J (2008) Indications for soil carbon saturation in a temperate agroecosystem. Soil Sci Soc Am J 72(4):1132–1139
94. Mei Z, Wang GZ, Guangmeng G, Lui H, Linhe W (2007) Estimate soil organic carbon and nitrogen distribution in Huolin Wetland with MODIS data. Remote sensing and modeling of ecosystems for sustainability IV. In: Gao W, Ustin SL (eds) Proceedings of the SPIE, vol 6679, 66790p
95. Lal R (2002) Soil carbon sequestration in China through agricultural intensification, and restoration of degraded and desertified ecosystems. Land Degrad Develop 13:469–478
96. Li X, Rengel Z, Mapfumo E, Bhupinderpal-Singh. (2007) Increase in pH stimulates mineralization of 'native' organic carbon and nitrogen in naturally salt-affected sandy soils. Plant Soil 290:269–282
97. Franzluebber AJ (2002) Soil organic matter stratification ratio as an indicator of soil quality. Soil Till Res 66:95–106
98. Puget P, Lal R, Izaurralde C, Post M, Owens L (2005) Stock and distribution of total and corn-derived soil organic carbon in aggregate and primary particle fractions for different land use and soil management practices. Soil Sci 170(4):256–279
99. Lal R (2009d) Soil carbon sequestration for climate change mitigation and food security. Platinum jubilee celebrations of the Indian society of soil science. Souvenir, 39–46
100. Lenney MP, Woodcock CE, Collins JB, Hamdi H (1996) The status of agricultural lands in Egypt: the use of multitemporal NDVI features derived from landsat TM. Remote Sens Environ 56(1):8–20

Mapping Egypt Vulnerability to Sea Level Rise Scenarios

Aymen Solyman and Tarek Abdel Monem

Abstract The human induced increase in global average temperature over the past two centuries has led to an increase in the global mean sea level. The Nile Delta is one of the three most vulnerable areas to the sea level rise threat in the world. The saltwater intrusion caused by sea level rise is expected to negatively affect agricultural activities in the Nile Delta through its detrimental effects on groundwater and soil. Given its significant contribution to Egypt's agricultural production, the Nile Delta's adaptation to the anticipated rise in sea level is necessary for social and economic reasons. In order to investigate and analyze such environmental problems and their impacts, the EGSLR model has been developed. The EGSLR covers the area from Alexandria to Port Said which extends to about 285 km along the Mediterranean coast. The development of the EGSLR is using the assessment of the IPCC definition of vulnerability. National and sub-national geographic and socio-economic data have been gathered for the area of study from different sources including Shuttle Radar Topography Mission (SRTM), land cover, urban areas, industrial zones, protected areas, power plants and population data. The scenarios of sea level rise are the preliminary basis for ministries, sectors and provinces to assess possible impacts on socio-economics sectors to develop and implement their respective plans for responding to and reducing potential impacts of future changes.

Keywords Climate change · Emissions · Sea level rise · Vulnerability · Mapping · Land · Agriculture · GIS · Nile Delta · Egypt

A. Solyman
Bibliotheca Alexandrina, Alexandria, Egypt
e-mail: aymensolyman@yahoo.com

Present Address:
T. Abdel Monem (✉)
International Fund for Agricultural Development (IFAD), Giza, Egypt
e-mail: monem.tarek@yahoo.com

© Springer Nature Switzerland AG 2020

E.-S. E. Omran and A. M. Negm (eds.), *Climate Change Impacts on Agriculture and Food Security in Egypt*, Springer Water, https://doi.org/10.1007/978-3-030-41629-4_9

1 Introduction

There is no doubt that human-induced climate change has become one of the biggest challenges facing the world nowadays. According to IPCC [1], the global average of land and ocean surface temperature has increased by 0.85 °C since the end of the nineteenth century. Such a phenomenon is caused by anthropogenic activities that intensified with the start of the industrial revolution and is mainly linked with the greenhouse gas emissions released by burning fossil fuels. Climate change threatens almost every economic sector either directly or indirectly with some of its effects already being felt. The agriculture sector is expected to suffer due to climate variability, change in rainfall patterns and sea level rise (SLR). Although some higher latitude locations will benefit from higher crop yields due to higher temperatures, major crops in lower latitude regions are expected to be negatively affected by a 2 °C rise in temperature [2].

Stern [2] reported that around 2 million km^2 are only 1 m above sea level and are expected to be highly impacted by the anticipated rise in sea levels. Such impacts include coastal flooding, loss of wetlands and displacement of millions of people [2]. However, climate change impacts are not equally distributed, with developing countries expected to have the bigger share [2]. In particular, some areas are more vulnerable to SLR than others.

In the Middle East and North Africa (MENA) region, although the land area affected is not the biggest globally, the impacts are much more severe [3]. For a 1 m rise in sea level, 3.2% of the population and 1.49% of the GDP of the MENA region will be negatively impacted, with these figures being higher than the global impact figures [3]. Among the MENA region, Egypt's population is the most exposed to a 1 m SLR with 10% of it expected to be affected where most of the impact is upon the Nile Delta area [3]. More than 1 million people will be directly affected by 2050 in three megadeltas: the Ganges–Brahmaputra delta in Bangladesh, the Mekong delta in Vietnam and the Nile Delta in Egypt [4]. Hence, the Intergovernmental Panel for Climate Change (IPCC) declared the Nile Delta in Egypt as one of three hotspots that are most vulnerable to SLR [4]. The Nile Delta is also predicted to experience the highest share of impact in the MENA region with respect to the agricultural sector with 12.5% of the agricultural extent expected to be negatively affected by 1 m SLR [3]. The 2011 Egyptian adaptation strategy for climate change has recommended a set of policies designed for the agricultural sector to cope with the future climate challenges. The policies mainly focused on flexibility, the use of technological innovation, systems diversity and developing new systems for risk management [5].

In this chapter, we present a methodology to build an integrated web map solution (SLRS) using Geographic Information System to access the effect of the SLR scenarios on the northern coast of Egypt. SLRS covers the area from Alexandria to Port Said which extends to about 285 km along the Mediterranean coast. The overall objective of the SLRS is to develop a web vulnerability mapping solution to study the different sea level rise scenarios and to identify and map the areas that are most vulnerable to SLR on the Nile Delta coast. Four scenarios are implemented to assess the impact of SLR, for 25, 50, 75 and 100 cm.

2 Study Area: Nile Delta

The Nile Delta is a lotus flower shaped area that starts from Cairo where the Nile divides into two main branches ending in Damietta and Rosetta governorates and then empties into the Mediterranean Sea as shown in Fig. 1. According to the IDRC [6], the Nile Delta contributes to 40% of Egypt's agricultural production with almost half of the country's population living in that area. The population density in the Nile Delta is around 1,000 inhabitants/km^2 with a population growth rate of 2% [7]. The significance of the study area comes from its high contribution to the agricultural sector and its high vulnerability to SLR.

Agriculture is the pillar upon which the ancient Egyptian civilisation was based thousands of years ago and represents an important historical value that constitutes part of the Egyptian culture today. There is also a significant economic value for agriculture as its added value is 14.5% of total Egyptian GDP in 2013 with 29% of the labour working in the sector in 2011 [8]. El-Nahrawy [9] reported that agriculture provides livelihood for around 55% of Egyptians and contributes to about 20% of all foreign exchange earnings. Agriculture has the highest share of water use among all sectors with 81.1% of the total water use [9].

The economic, social and cultural values of the Nile Delta suggest that there is an urgent need for effective climate change adaptation policies for the various sectors that are active in the area with a particular focus on the agriculture sector. Currently, Egypt has started implementing the "Enhancing Climate Change Adaptation in the

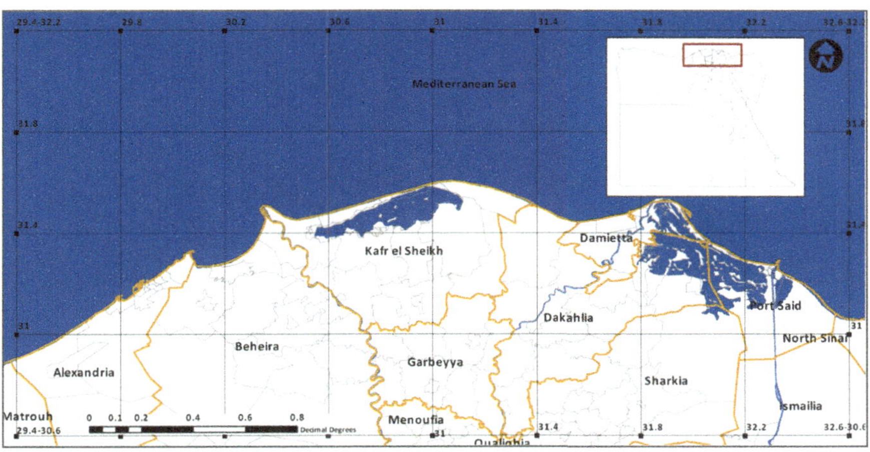

Fig. 1 Study area

North Coast and Nile Delta Regions in Egypt" project funded by the Green Climate Fund (GCF) and will be implemented by UNDP [10]. In addition, the Shore Protection Agency has allocated around EGP 1.7 billion (approximately US$ 97 million) for adaptation measures to protect the coastline of the Mediterranean Sea and Red Sea [10].

3 Climate Change Observations and Potential Impacts

3.1 Greenhouse Gas Emissions and Anthropogenic Climate Change

There is enough evidence that the global climate is changing and the debate around the causes and drivers of this phenomenon is settled with only some differences around regional and local impacts. The most authoritative assessments of climate change among policymakers are the IPCC reports. According to IPCC [1], there has been a warming of 0.85 °C in globally averaged land and ocean surface temperature over the period 1880–2012. The main driver of this warming is the positive total radiative forcing caused mainly by the greenhouse gases emissions where the greatest contribution is made by the significant CO_2 increase since 1750 [1]. The atmospheric CO_2 concentration has become 40% higher in 2011 than it was in the pre-industrial era [1]. Figure 2 shows the trend of increase since 1958.

Other gases contributed to that rise in temperature. According to IPCC [1], the total anthropogenic radiative forcing recorded a 43% increase in 2011 compared to 2005 caused by changes in the concentrations of carbon dioxide, methane, halocarbons, nitrogen monoxide, short-lived gases and aerosols in addition to the albedo change due to land use. Figure 3 shows the relative contributions of the different gases and drivers (including the natural driver).

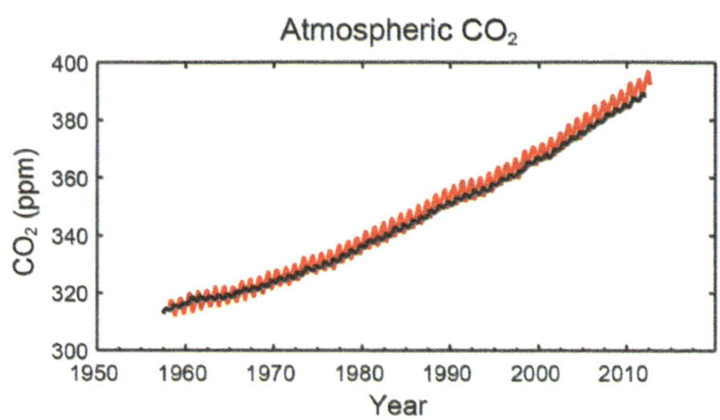

Fig. 2 Atmospheric concentrations of CO2 since 1958 [1, p. 12]

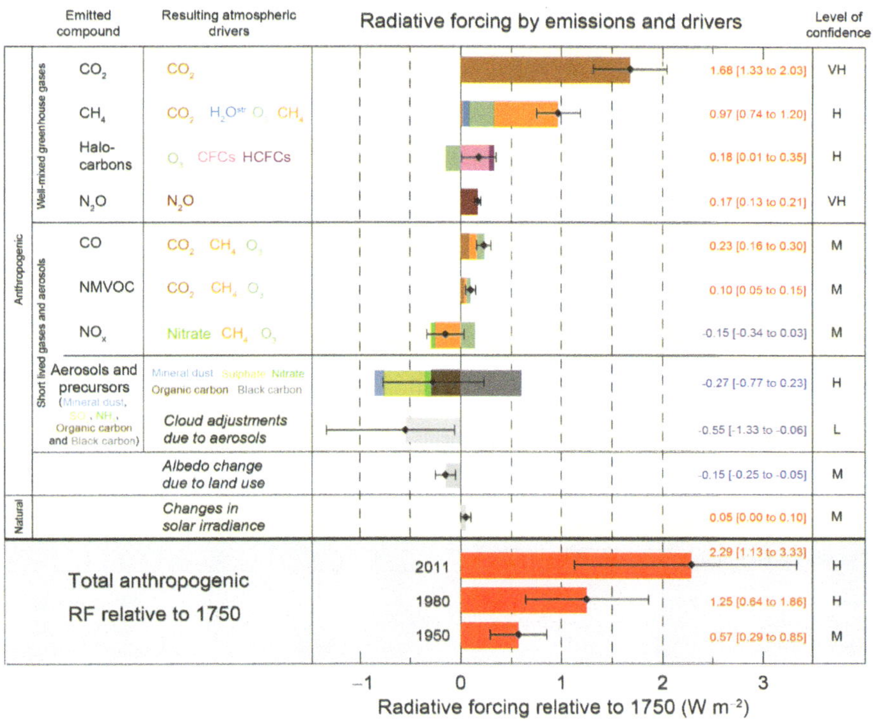

Fig. 3 Radiative forcing estimates in 2011 relative to 1750 for the main drivers of climate change [1, p. 14]

3.2 SLR and Agriculture in the Nile Delta

As for sea level, there was a mean rise of 0.19 m over the period 1901–2010 [1]. Figure 4 summarises the processes responsible for controlling sea level and Fig. 5 quantifies the contributions of each of them in the global mean SLR. The report identifies the loss of glacier mass and ocean thermal expansion as the main reasons for 75% of this rise since the 1970s. The IPCC [1] predicts a likely rise in sea level in the range of 0.52–0.98 m by the end of this century under the highest emissions scenario (RCP 8.5).

According to Stern [2], rising sea levels will have significant impacts globally with 2 million km^2 of land and a $1 trillion worth of assets lying only 1 m above sea level. There will be an increase in coastal flooding, loss of wetlands, saltwater intrusion with surface and groundwater and millions of people at the risk of permanent displacement [2]. Thomas et al. [12] classified the impacts of sea level change into physical and socio-economic. On the one hand, they listed hydrological changes including saltwater intrusion and flooding, morphological changes including erosion and ecological changes mainly through loss of habitats all under physical impacts.

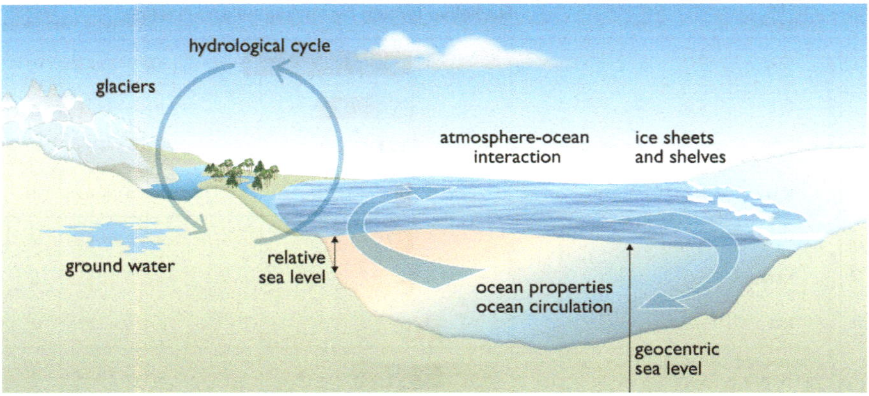

Fig. 4 Climate-sensitive processes and components that can influence global and regional sea level [11, p. 1143]

Source	1901–1990	1971–2010	1993–2010
Observed contributions to global mean sea level (GMSL) rise			
Thermal expansion	–	0.8 [0.5 to 1.1]	1.1 [0.8 to 1.4]
Glaciers except in Greenland and Antarctica[a]	0.54 [0.47 to 0.61]	0.62 [0.25 to 0.99]	0.76 [0.39 to 1.13]
Glaciers in Greenland[a]	0.15 [0.10 to 0.19]	0.06 [0.03 to 0.09]	0.10 [0.07 to 0.13][b]
Greenland ice sheet	–	–	0.33 [0.25 to 0.41]
Antarctic ice sheet	–	–	0.27 [0.16 to 0.38]
Land water storage	−0.11 [−0.16 to −0.06]	0.12 [0.03 to 0.22]	0.38 [0.26 to 0.49]
Total of contributions	–	–	2.8 [2.3 to 3.4]
Observed GMSL rise	1.5 [1.3 to 1.7]	2.0 [1.7 to 2.3]	3.2 [2.8 to 3.6]

Fig. 5 Global mean sea level budget (mmyr-1) [11, p. 1151]

On the other hand, socio-economic impacts included negative impacts on human well-being, damage to infrastructure and effects on industry and services [12]. Stern [2] also underlined the high vulnerability of developing countries- in particular- to climate change due to their geographic locations, low incomes and their overdependence on sectors that are sensitive to climate change such as agriculture. In that sense, countries like Egypt are left with high risk and low adaptive capacity to deal with climate change impacts. As for the agriculture sector, FAO [13] named three production systems that will be most affected by SLR: vegetable production that tends to be mainly located around urban areas that are threatened by saltwater intrusions, aquaculture systems sited in areas at or below sea level, and coastal fisheries based in mangrove swamps and other coastal wetlands threatened by SLR. Relevant to the research study area, Stern [2] highlighted the potential threat of SLR upon the agricultural land in the Nile Delta as well as the impacts on Egypt's coastal cities.

Dasgupta et al. [3] analysed the potential impacts of the different SLR scenarios in different regions around the world including the Middle East and North Africa region. They found that Egypt had the highest percentage of population exposed to SLR with around 10% expected to be impacted by a 1 m rise and more than 20%

exposed by a 5 m rise as illustrated by Figs. 6 and 7. They also predicted that most of these impacts would occur in the dense Nile Delta region.

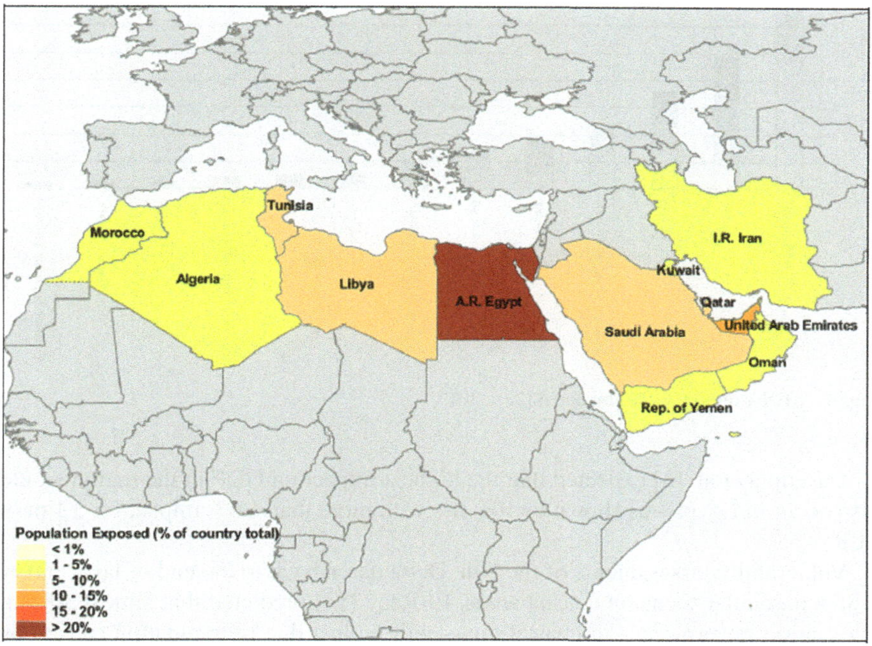

Fig. 6 MENA region: exposed population in a 5 m SLR scenario [3, p. 19]

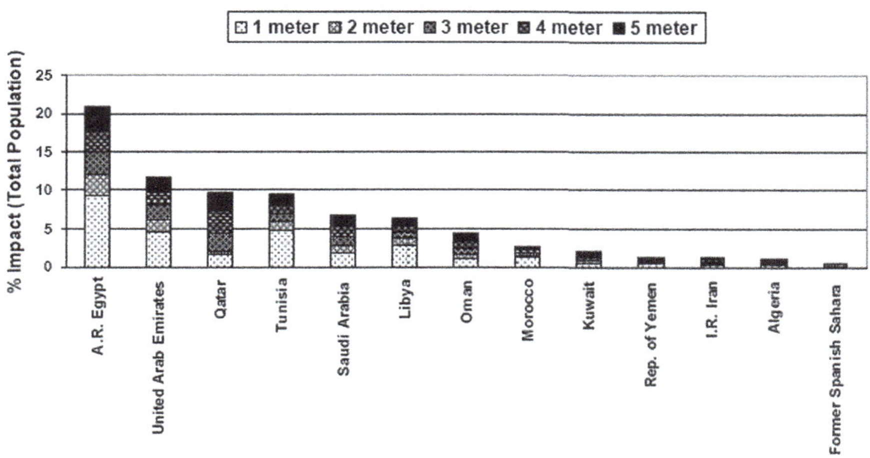

Fig. 7 MENA region: population impacted [3, p. 20]

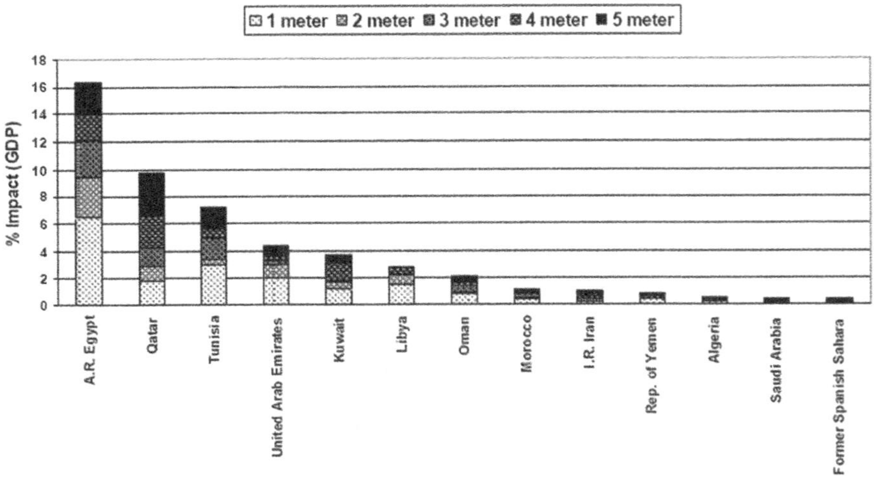

Fig. 8 MENA region: GDP impacted [3, p. 20]

Dasgupta et al. [3] expected that the highest impact on GDP in the region would also occur in Egypt—as shown by Fig. 8—with more than a 6% impact for a 1 m of SLR.

Vulnerability assessments of the Nile Delta dated back to the end of last century with a particular focus on coastal areas. El-Raey [14] predicted that almost 30% of Alexandria governorate—a coastal city—will be lost due to inundation caused by SLR with 2 million people expected to migrate. However, Frihy [15] argued that not all coastal areas are vulnerable and then classified those areas according to their degree of vulnerability to beach erosion—including SLR as one of its drivers. The vulnerable areas constituted around 30% of the coastal areas where the coastal plain is low-lying (0–1 m above mean sea level) such as in Abu Quir Bay, sand barriers are narrow such as in Burullus, areas affected by subsidence such as Manzala-Port Said area, or shores are covered in fine sand such as Ras El Bar [15]. The naturally protected areas constitute 55% of the coastal areas as follows: beaches supported by high-elevated dunes or carbonate ridges, areas experiencing land uplifting such as a part of Alexandria, and the prograding coastal zones [15]. The remaining 15% of the coastal areas are the artificially protected ones using physical structures such as lagoon and harbour jetties, walls and breakwaters [15].

Stern [2] classified the Nile Delta as a highly vulnerable area to SLR in Africa-alongside the continent's west coast. The IPCC agreed with this as it declared the Nile Delta as one of three hotspots that are most vulnerable to SLR [4]. Sefelnasr and Sherif [16] concluded that about 19 and 32% of the total area of the Nile Delta would be submerged under the 0.5 and 1 m seawater level rise scenarios, respectively.

Some literature highlighted the impacts on the agriculture sector in the Nile Delta. Frihy [15] listed saltwater intrusion caused by SLR as a major negative impact on agriculture in the Nile Delta area because of the disruption caused to freshwater

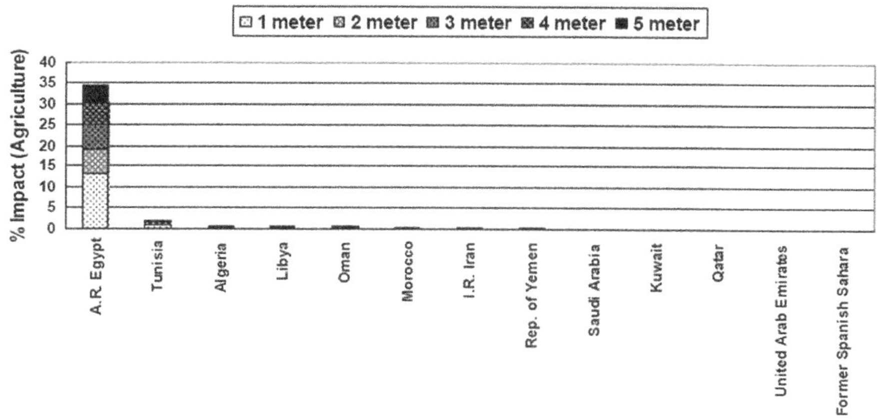

Fig. 9 MENA region: agriculture extent impacted [3, p. 21]

aquifer underlying the area. Dasgupta et al. [3] also predicted that in the Middle East and North Africa region, Egypt would be exposed to the highest impact on the agricultural sector. They predicted that the impact on the sector in Egypt would be around 12.5% for a 1 m SLR scenario as shown in Fig. 9. Given that the Delta contributes with around 40% of the agricultural production in the area [6], such potential impact is expected to have major negative consequences on the sector as a whole on the national level.

According to Abdrabo and Hassaan [17], the overall decline in crop yield values due to higher levels of groundwater table induced by SLR was projected to be as much as L.E. 6.43 billion (around $842 million) up to the year 2100 in Damietta Governorate—one of the Nile Delta's governorates. Moreover, they added that the loss in crop yield would also carry indirect costs related to loss of incomes and employment opportunities as some people's livelihoods mainly depend on the various agricultural activities taking place in the Nile Delta.

4 The Methodology of the Development of Interactive Web Enabled SLR Scenarios (SLRS)

The most important factors being included in the design and implementation of SLRS as a web mapping application for the internet are functional modules, a graphic user interface, download time and system performance. The system must be portable and extendable to accommodate future changes in hardware, software and networking. With this is mind, SLRS was developed based on two components: the server side and the web client side, which run in the browser. The framework of the technical development process of the SLRS focuses on:

1. Spatial and attribute database collection, analysis and design
2. Data conversion and import
3. Development environment and tools
4. SLRS database model
5. SLRS architecture
6. SLRS main features
7. SLRS main components
8. SLRS main functions.

The database in the system consists of two types, tabular data and spatial data. The tabular data is stored in SQL format and the spatial data is stored in Geo-database format. The structure of the geodatabase is optimized for both performance and storage for multi levels of GIS data layers as shown in Fig. 10. The SLRS database model is structured to be stored physically in different format although the relationships between the two categories of data are preserved regardless of whether the division is physical or logical (in a hybrid database model). This means that two separate databases were used in the system, one for spatial data and one for attribute data.

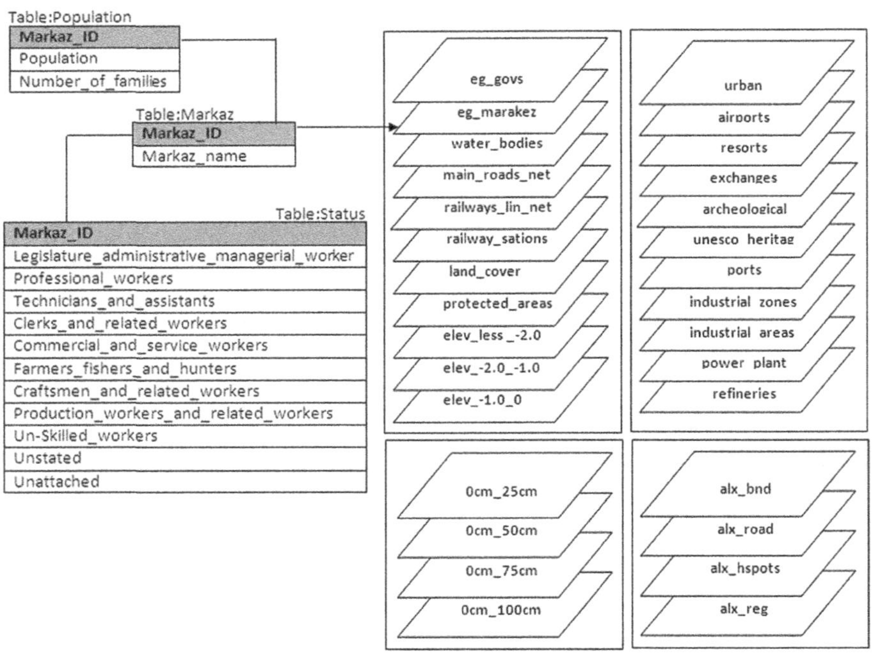

Fig. 10 SLRS database model

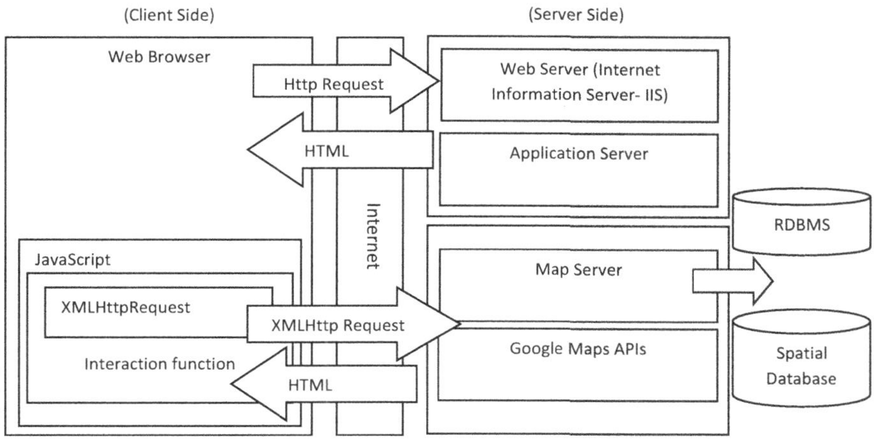

Fig. 11 SLRS architecture

5 SLRS Architecture

There are two types of architecture for developing web-based GIS applications: client-side, and server-side. In a client-side GIS web application, the client (web browser) is enhanced to support GIS functionality, while in a server-side GIS application, a web browser is used to generate server requests and display the results. SLRS is a server side application (with a thin client architecture) where users can send a request to a server (i.e. an address), and the server processes the request and sends the results back as an image embedded in an HTML page via standard HTTP (Hypertext Transfer Protocol) as shown in Fig. 11. The response is shown as a standard web page that a generic browser can view in server-side web GIS applications, all the complex and proprietary software, in addition to the spatial and tabular data, remain on the server. This architecture has several advantages because the application and data are centralized on a server.

6 SLRS Main Components

SLRS is divided into a set of functional units (modules). Each module represents a set of related tasks or functions. The modules are independent of one another, but they do communicate with each other. SLRS is developed and deployed using standard web development tools and is comprised of two elements, the website framework and the functional tools. The framework presents the SLRS supporting information to the user via a graphical user interface. "The second element is the functional tools that enable access to GIS functions such as SLR scenarios, mapping and query functions. The main modules are" (https://www.ee.co.za/article/web-mapping-solution-assess-effect-sea-level-rise.ht):

1. Display functions—Map tools.
2. Cartographic presentation.
3. Utilities.
4. Mashup module.
5. Layer manager module.
6. Sea level rise scenarios module: This module contains four levels to study:
   a. Scenario 1: sea level rise from 0 to 25 cm.
   b. Scenario 2: sea level rise from 0 to 50 cm.
   c. Scenario 3: sea level rise from 0 to 75 cm.
   d. Scenario 4: sea level rise from 0 to 100 cm.
      For each scenario, the user can:

- Select any scenario and show all affected districts by the selected scenario also the user can display more than one scenario at the same map view to compare. The results can be displayed in different format (maps, tabular and graphs).
- Select any Markaz/district in the study area and locate it geographically on the base map. For any selected Markaz, the application provides:
  - The number of population and the number of families.
  - The affected area of the selected markaz divided by total markaz area for each scenario.
- For each scenario, the user can select any affected district and retrieve the labor force information about it. The labor force is classified into eleven categories according to the Central Agency for Public Mobilization & statistics classification.
- Select any hotspot point. The hotspots data include urban areas, railway stations, archaeological sites, airports, industrial zones, exchanges, resorts and industrial spots. For any selected hotspot the application shows its location on the map and which SLR scenario(s) may affect it.
- Locate any GPS point by inputting its latitude and longitude in decimal degree format to check if it is affected by any SLR scenario or not.

7. Population distribution module.

7 Results

According to the SLRS, the impact of SLR on the coastal areas by each scenario is shown in Table 1.

1. The high-risk governorates are part of Alexandria, Behira, Dakahlia, Damietta, Ismailia, Kafr Alshikh, Port Said and Sharkia governorates as shown in Fig. 12.
2. The most vulnerable governorates are Kafr Alshikh, Behira and Port Said governorates which present about 70–75% of the total affected area by each scenario.
3. The total number of districts affected by the different scenarios is 58 districts in 8 governorates as shown in Table 2.
4. The highest vulnerable governorate is Kafr El-Shikh, and the highest impacted district is Fouh as shown in Table 3.

Figure 13 shows extracts from the SLRS. Figures 14, 15, 16 and 17 demonstrate results from Table 1 on SLRS and Figs. 18 and 19 demonstrate results from Tables 2 and 3.

8 Discussion

Nile Delta is vulnerable to the impacts of climate change especially when it comes to the agriculture sector. However, some areas in the Nile Delta are more vulnerable

Table 1 The total areas affected by each scenario

Scenario	Total areas affected in Km^2
Scenario (1)—sea level rise 0–25 cm	581
Scenario (2)—sea level rise 0–50 cm	1058
Scenario (3)—sea level rise 0–75 cm	1578
Scenario (4)—sea level rise 0–100 cm	2200

ID	Governorate	Total_Population	No_of_Families	Area_affected_by_25cm km^2	Area_affected_by_50cm km^2	Area_affected_by_75cm km^2	Area_affected_by_100cm km^2
1	Alexanderia	4123869	1071582	4.13	5.53	6.68	28.03
2	Behira	4747283	1094141	120.22	233.90	325.10	430.85
3	Dakahlia	4989997	1264276	41.62	89.70	142.70	221.20
4	Damietta	1097339	283944	38.20	75.85	116.60	169.70
5	Ismalia	953006	227713	16.67	28.70	78.00	102.96
6	Kafr Alshikh	2620208	614392	173.78	322.20	483.07	670.80
7	Port Said	570603	141982	153.72	234.80	316.90	416.10
8	Sharkia	5354041	1261502	33.31	67.65	109.60	160.70
Total affected Area in km^2				581.65	1058.33	1578.65	2200.34

Fig. 12 The total areas affected in each governorate by each scenario

Table 2 The total number of districts affected by the different scenarios

Governorate	Number of impacted districts			
	Scenario 1	Scenario 2	Scenario 3	Scenario 4
Alexanderia	9	9	10	13
Behira	11	12	13	14
Dakahlia	6	7	7	8
Damietta	6	6	6	6
Ismailia	1	1	2	2
Kafr Alshikh	6	7	9	9
Port Said	3	3	3	4
Sharkia	1	1	2	2
Total	43	46	52	58

Table 3 The highest impacted districts by each scenario

Scenario	Number of affected districts	The highest impacted district	Governorate	% Area affected
Scenario 1 (25 cm) SLR	43	Kesm-Port_Fouad	Port Said	14.12
Scenario 2 (50 cm) SLR	46	Mitubas	Kafr El-Shikh	22.5057
Scenario 3 (75 cm) SLR	52	Mitubas	Kafr El-Shikh	30.0684
Scenario 4 (100 cm) SLR	58	Fouh	Kafr El-Shikh	39.5121

than others. Geographic Information System (GIS) is capable of creating, analyzing, and displaying sea level rise scenarios enabling local officials and decision makers to address the negative effects of elevated sea levels by allowing them to identify affected communities that are at risk, assess the situation, and develop adaptation strategies. Results for the SLR impact on Nile Delta show that total area affected by 25, 50, 75 and 100 cm SLR is 581.6, 1058.3, 1578.6 and 2200.3 km^2 respectively. The highest vulnerable governorate is Kafr El-Shikh and the highest impacted district (Markaz) is Fouh. The results of the analysis shall be used to rank the priority areas for adaptation to SLR.

9 Conclusion

The Nile Delta contains high population densities, significant economic activities and ecosystem services. This area is already subject to climate change has the potential to pose increasing risks in the future. Consequently, a range of negative socio-economic impacts is expected to follow such as food insecurity, loss of income sources and unemployment of small farmers as well as the loss of livelihoods of other people who

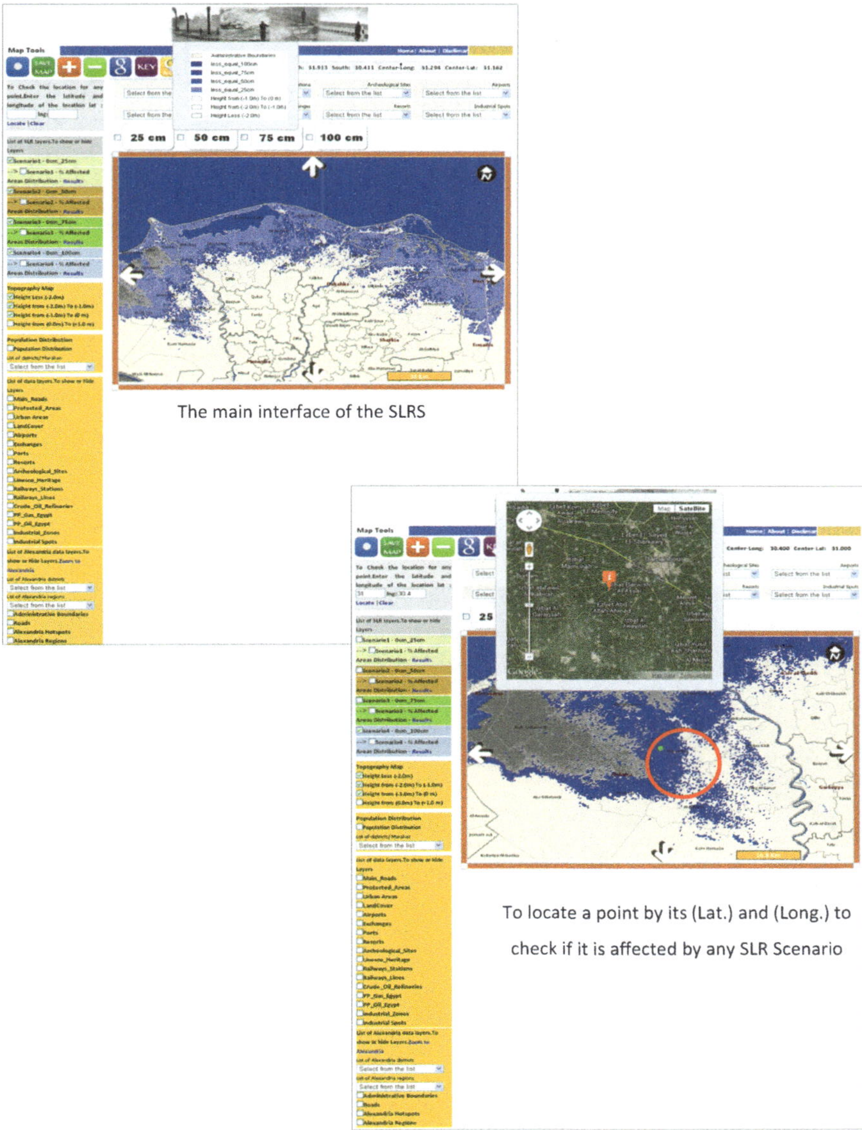

Fig. 13 Extracts from SLRS

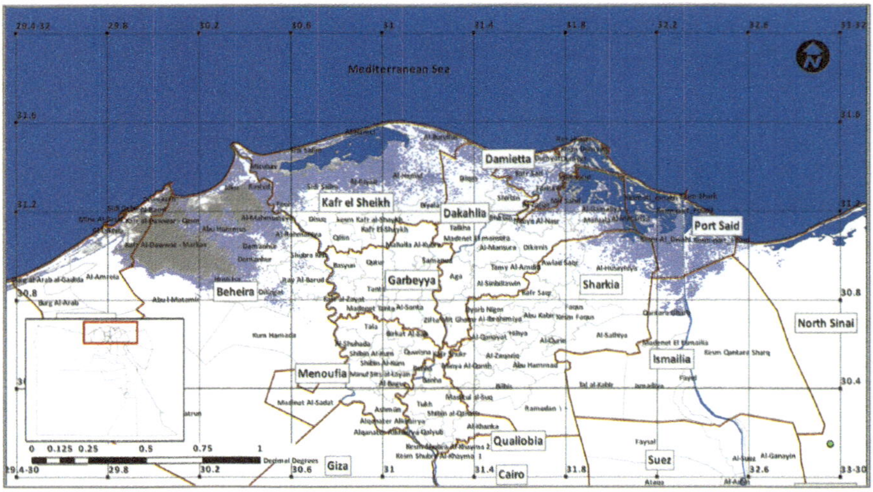

Fig. 14 The expected impact of a 25 cm SLR—Scenario (1)

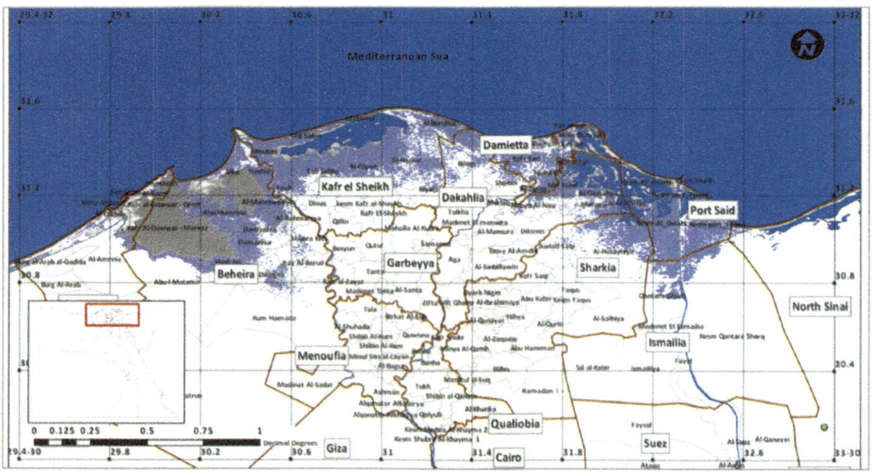

Fig. 15 The expected impact of a 50 cm SLR—Scenario (2)

are currently dependant on the agricultural activities carried out in the Nile Delta. These predictions assure the need for the agricultural sector in Egypt to adapt- on the long term- to the potential threat of SLR as part of its adaptation to climate change in general. The SLRS analysis can be used to identify priority areas for adaptation based on vulnerability of cities and districts.

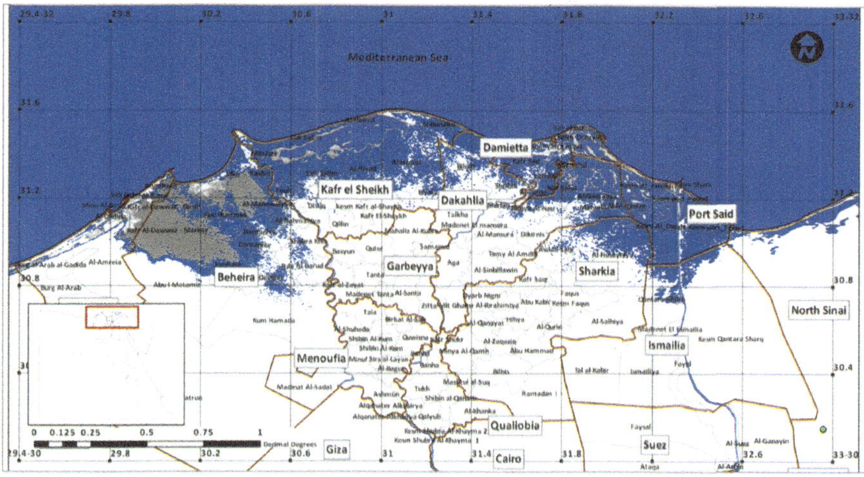

Fig. 16 The expected impact of a 75 cm SLR—Scenario (3)

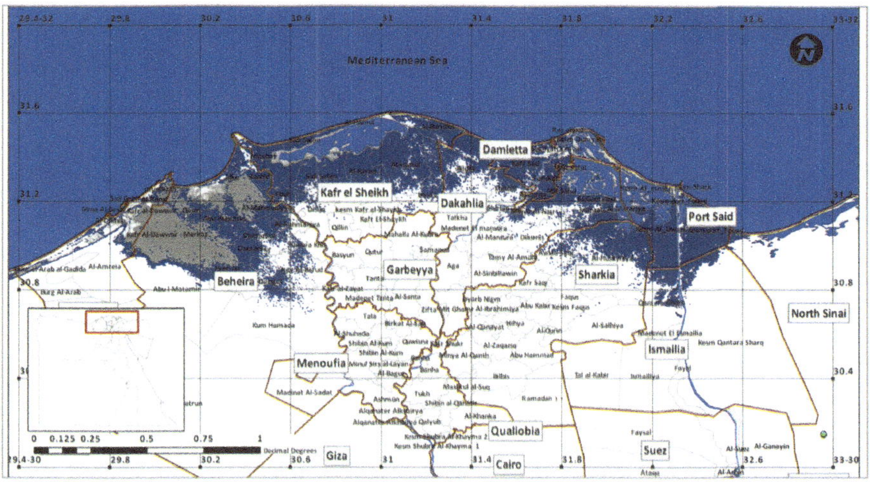

Fig. 17 The expected impact of a 100 cm SLR—Scenario (4)

10 Recommendations

Research around institutional reform regarding the distribution of power and the flow of knowledge in the policymaking chain with respect to SLR adaptation is needed. From an agricultural perspective, in depth studies are needed to identify the different physical, social, economic and institutional barriers that stand in the way

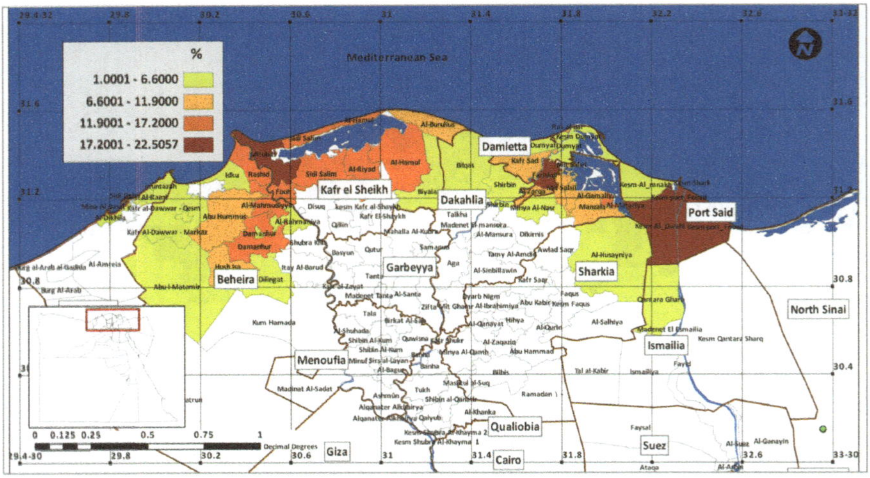

Fig. 18 Map shows the impacted districts by 50 cm SLR—Scenario (2)

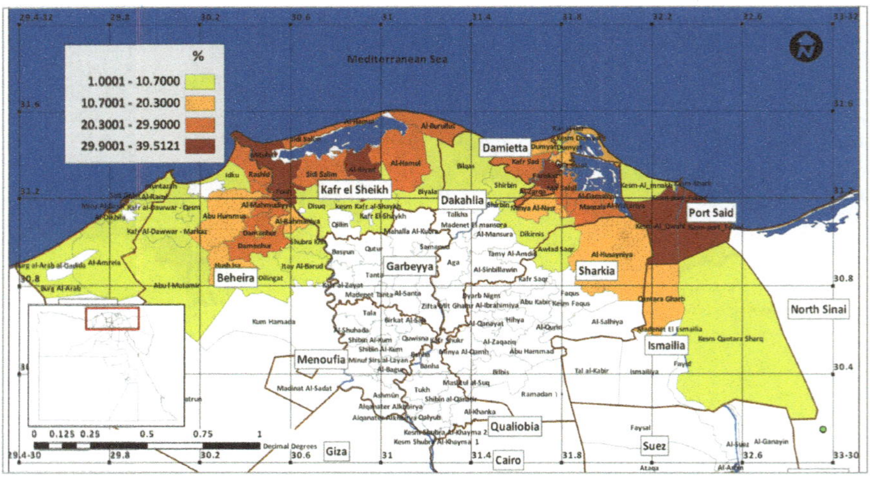

Fig. 19 Map shows the impacted districts by 100 cm SLR—Scenario (4)

of agricultural adaptation in the Nile Delta region. In addition, regarding farm level adaptation and the resilience of the Nile Delta communities facing potential SLR are also worth studying.

References

1. IPCC (2013a) Summary for policymakers. Climate change 2013: the physical science basis. Contribution of working group I to the fifth assessment report of the intergovernmental panel on climate change, Cambridge and New York
2. Stern N (2006) The economics of climate change. Cambridge
3. Dasgupta S, Laplante B, Meisner C, Wheeler D, Yan J (2007) The impact of sea level rise on developing countries: a comparative analysis. World bank policy research working paper 4136
4. IPCC (2007) Full report. Climate change 2007: impacts, adaptation and vulnerability. Contribution of working group II to the fourth assessment report of the intergovernmental panel on climate change, Cambridge
5. IDSC and UNDP (2011) Egypt's national strategy for adaptation to climate change and disaster risk reduction
6. IDRC (2011) Annual report 2010–2011. Climate change adaptation in Africa programme
7. Bucx T, Marchand M, Makaske B, van de Guchte C (2010) Synthesis report: comparative assessment of the vulnerability and resilience of 10 deltas. Delft-Wageningen, The Netherlands, Delta Alliance International
8. World Bank (2015) Indicators. https://data.worldbank.org/indicator, 24 Aug 2015
9. El-Nahrawy MA (2011) Egypt. Country pasture/forage resource profiles. FAO
10. UNDP (2018) National adaptation plans in focus: lessons from Egypt.
11. IPCC (2013b) Sea level rise. Climate change 2013: the physical science basis, Chap 13. Contribution of working group I to the fifth assessment report of the intergovernmental panel on climate change. Cambridge and New York
12. Thomas M, Pidgeon N, Whitmarsh L, Ballinger R (2015) Mental models of sea-level change: a mixed methods analysis on the Severn Estuary, UK. Glob Environ Change 33:71–82
13. FAO (2003) World agriculture: towards 2015/2030—an FAO perspective. London
14. El-Raey M, Dewidar K, El-Hattab M (1999) Adaptation to the impacts of sea level rise in Egypt. Mitig Adapt Strat Glob Change 4:343–361
15. Frihy OE (2003) The Nile delta-Alexandria coast: vulnerability to sea-level rise, consequences and adaptation. Mitig Adapt Strat Glob Change 8:115–138
16. Sefelnasr A, Sherif M (2014) Impacts of seawater rise on seawater intrusion in the Nile delta aquifer, Egypt. Groundwater 52:264–276
17. Abdrabo MA, Hassaan MA (2014) Economic valuation of sea level rise impacts on agricultural sector: Damietta governorate. Egypt. Journal of Environmental Protection 5:87–95

Smart Farming

Agricultural Production in Egypt: Assessing Vulnerability and Enhancing Adaptive Capacity and Resilience to Climate Change

Mohamed A. S. Abdel Monem and Biljana Radojevic

Abstract The international community has developed the 2030 Agenda and the Sustainable Development Goals (SDGs) that include specific elements to enhance sustainable agricultural production and food security. SDG 13 "take urgent actions to combat climate change and its impacts" calls for enhancing the adaptive capacity and build resilience to climate change as well as mainstreaming climate change measures into policies and planning processes of the countries. According to the IPCC assessment, Near East North Africa is one of the world regions that are most vulnerable to climatic change and consequently, agriculture is the most affected sector. It is anticipated that such a climate change impact would reduce crop yield up to 30%. Within the '*Regional Initiative for the Assessment of the Impact of Climate Change on Water Resources and Socio-Economic Vulnerability in the Arab Region*' (RICCAR), the impact of climate change on specific agriculture sectors was projected using the two considered climate scenarios, based on selected Representative Concentration Pathways (RCP), specifically RCP 4.5 (moderate-case scenario) and RCP 8.5 (worst-case scenario). Based on the RICCARD projections, the vulnerability of the agricultural sectors in the Arab Region to climate change have few aspects. Environmental challenges in the Arab world include water scarcity, with the lowest freshwater resource endowment in the world and very low precipitation with excessive exposure to extreme events. The Arab Region is one of the most impacted regions by climate change with half of its area's cropland systems producing wheat, maize, olives, potatoes, olives and vegetables are considered high vulnerable to climate change. Livestock production as well is expected to decline due to drought, increase in degradation of rangelands and desertification. Expected more drought cycles in addition to increasing deforestation rate add to the challenges facing the forestry sector in the region. Egypt is considered as one of the countries that are greatly impacted by the climate change. The vulnerability of Egypt to climate change is significant in all agriculture sectors: cropping, fishery and livestock. Developing strategy to

---

M. A. S. Abdel Monem (✉) · B. Radojevic
Regional Office for the Near East and North Africa, Food and Agriculture Organization of the United Nations, FAO, P.O. Box: 2223, Dokki, Cairo, Egypt
e-mail: mohamed.abdelmonem@fao.org

M. A. S. Abdel Monem
Soil and Water Department, Egyptian Atomic Energy Authority, Cairo, Egypt

© Springer Nature Switzerland AG 2020
E.-S. E. Omran and A. M. Negm (eds.), *Climate Change Impacts on Agriculture and Food Security in Egypt*, Springer Water, https://doi.org/10.1007/978-3-030-41629-4_10

enhancing adaptive capacity and resilience to climate change in Egypt requires coordination and integration between the main related national strategies such as Egypt's Vision 2030 [1], the Sustainable Agricultural Development Strategy [2], and Egypt's National Strategy for Adaptation to Climate Change and Disaster Risk Reduction [3]. Climate change adaptation is of paramount importance to agriculture, given the reliance of the sector on climate. Climate change adaptation policies should be based on science, and incorporate knowledge of indigenous peoples and traditional practices.

Keywords Resilience to climate change · Vulnerability · Adaptive capacity · Agricultural production in Egypt

Acronyms

ACSAD	The Arab Center for the Studies of Arid Zones and Dry Lands
COP	Conference of the Parties
ESCWA	The United Nations Economic and Social Commission for Western Asia
FAO	Food and Agriculture Organization of the United Nations
GIZ	The German Development Agency
IFAD	International Fund for Agricultural Development
IPCC	The Intergovernmental Panel on Climate Change
RICCAR	The Regional Initiative for the Assessment of the Impact of Climate Change on Water Resources and Socio-economic Vulnerability in the Arab Region
SDG	The Sustainable Development Goal
UNEP	The United Nations Environment Program
UNFCCC	United Nations Framework Convention on Climate Change

1 Introduction

It is expected that population of the world will increase by one third by 2050 (about 2 billion people), most of them will be in the developing countries. According to FAO, agricultural production should be increased by 60% by 2050 to provide the necessary food and feed demand, if the growth rate of income and consumption maintain the same level. To provide the growing population with its food requirement and contribute to poverty reduction, It became essential for agricultural to be transferred into more productive systems. However climate change is introducing new challenge, under the business as usual scenario, that will hinder the efforts to achieve food security due to impact of climate change on agricultural production through increasing temperature, reducing precipitation and rising the see level [4].

In order to implement the 2030 Agenda, the international community has endorsed five principles of sustainable food and agriculture providing a framework for policy dialogue and for developing appropriate policies, strategies, regulations and incentives. FAO [5] identified these principles as:

- Improving efficiency in the use of resources is crucial to sustainable agriculture;
- Sustainability requires direct action to conserve, protect and enhance natural resources;
- Agriculture that fails to protect and improve rural livelihoods, equity and social well-being is unsustainable;
- Enhanced resilience of people, communities and ecosystems is key to sustainable agriculture;
- Sustainable food and agriculture require responsible and effective governance mechanisms.

According to the definition of The United Nations Framework Convention on Climate Change (UNFCCC); climate change is "a change of climate which is attributed directly or indirectly to human activity that alters the composition of the global atmosphere and which is in addition to natural climate variability observed over comparable time periods". UNFCC differentiate between climate change due to human activities altering the atmospheric composition, and climate variability in the short term that take place due to natural causes [6]. The global agreement on climate change was adopted by 196 countries at the twenty-first Conference of the Parties (COP 21) of UNFCC held in Paris in December 2015. By June 2017, 195 country had signed the agreement, including 148 that ratified it. Global commitment to this agreement was reasserted during the COP 22 of UNFCC, held in Marrakesh, Morocco, in November 2016.

It is well recognized that climate change and its effects pose a major threat to achieve sustainable development goals. In the Arab region, climate change could undermine past successes and prospects, although the region contributes less than 5% to global greenhouse gas emissions [6]. The Sustainable Development Goal (SDG13) urges countries to take urgent action to combat climate change and its impacts, which requires new and improved statistical measures on resilience, adaptive capacity and resource mobilization.

The term 'vulnerability' may, therefore, refer to the vulnerable system itself, e.g., low-lying islands or coastal cities; the impact to this system, e.g., flooding of coastal cities and agricultural lands or forced migration; or the mechanism causing these impacts, e.g., the disintegration of the West Antarctic ice sheet. The vulnerability is understood to be the "function of a system's climate change exposure, sensitivity and adaptive capacity to cope with climate change effects" [7]. According to the IPCC [8], these terms are defined as follows:

- Exposure refers to changes in climate parameters that might affect socio-ecological systems. Such parameters are for example temperature, precipitation and wind speed, which climate change alters with regard to their quantity and quality as well as their spatial and temporal distribution [6].

- Sensitivity tells us about the status quo of the physical and natural environment of the affected systems that makes them particularly susceptible to climate change. For example, a sensitivity factor could be topography, land use land cover, distribution and density of population, built environment, proximity to the coast, etc.
- Potential Impact is determined by combining exposure and sensitivity to climate change on a system.
- Adaptive capacity "the ability of a system to adjust to climate change (including climate variability and extremes), to moderate potential damages, to take advantage of opportunities, or to cope with the consequences" [8].

The focus of the global process lead by the UNFCCCis on identifying the most vulnerable countries to climate change. At this level, assessment is focused on the degree of exposure to climate variability and change; however, some of the proposed approaches also consider economic factors, such as poverty data. The poorest, often are most vulnerable people to climate change, in particular, those who face social exclusion, marginalization in their communities and countries, and inequality in terms of rights, power and access to resources and services. Evaluation of capacities and priorities of the most vulnerable groups are the main focus of vulnerability assessments in any region.

2 Vulnerability of the Near East and North Africa Region to Climate Change

Between 2006 and 2010 the Arab Region was exposed to one of its most severe drought cycles in the past century, leading to a loss of livelihoods, high food prices, and decreased purchasing power for the average citizen. For example in Syria, the drought devastated the livelihoods of about 20% of the rural population, displacing up to 1 million people and fueled mass internal displacement and social tensions in advance of broader conflicts that have more recently emerged. In Somalia in 2011 with approximately 4 million people were displaced and about 900,000 people displaced solely due to the drought between November 2016 and August 2017 [9]. More than 40% of the people in the Arab region have already been exposed to drought and other climatic disasters [10].

Food and water shortage is a critical issue in the Near East North Africa Region (NENA region) with strong implications for the national security. However, among major trends expected to impact the Region, climate change is perhaps the most critical, acting as the "ultimate threat multiplier" [11]. According to the IPCC [12] assessment, Near East North Africa is one of the world regions that are most vulnerable to climatic change. It is anticipated that such a climate change impact would reduce crop yield up to 30%. The livelihoods of a large segment of the region's population, particularly those depending on small scale rainfed agriculture, will be at great risk.

The environment will be impacted as well due to the anticipated increase of desertification, land degradation and biodiversity losses. Ultimately, people migration will be accelerated.

Recently the 'Regional Initiative for the Assessment of the Impact of Climate Change on Water Resources and Socio-Economic Vulnerability in the Arab Region' (RICCAR) was established through a project involving all the countries of the region in partnership between FAO, GIZ, ACSAD and ESCWA. The outcome of the RICCAR that was launched in November 2017 focused on the change in maximum and minimum temperatures, change of precipitation, changes in evapotranspiration (ET). The future climate change data were generated through downscaling projections from three global circulation models, EC-EARTH, CNRM-CM5, and GFDL-ESM, exercised on the Arab region using the regional climate model RCA4 to provide higher-resolution information (50 km).

The effect of climate change on agriculture in the region is expected to vary by according to the existing farming system Table 1. In some areas, farmers can respond to changes through adaptation measures. In other regions, agriculture may

Table 1 Impact of climate change on different farming systems in the Aran region

Farming system	Exposure Expected climate change-related events	Sensitivity Likely impact on farming systems
Irrigated	Increased temperatures Reduced supply of surface irrigation water Dwindling of groundwater recharge	More water stresses Increased demand for irrigation and water transfer Reduced yields when temperatures are too high Salinization due to reduced leaching Reduction in cropping intensity
Highland mixed	Increase in aridity Greater risk of drought Possible lengthening of the growing period Reduced supply of irrigation water	Reduction in yields Reduction in cropping intensity Increased demand for irrigation
Rain-fed mixed	Increase in aridity Greater risk of drought Reduced supply of irrigation water	Reduction in yields Reduction in cropping intensity Increased demand for irrigation
Dryland mixed	Increase in aridity Greater risk of drought Reduced supply of irrigation water	A system very vulnerable to declining rainfall Some lands may revert to rangeland Increased demand for irrigation
Pastoral	Increase in aridity Greater risk of drought Reduced water for livestock and fodder	A very vulnerable system, where desertification may reduce carrying capacity significantly Non-farm activities exit from farming, migration

face serious challenges, and ruler inhabitants might need to change their livelihood to off- farm employment or relocate [13].

The two considered climate scenarios are based on selected Representative Concentration Pathways (RCP), specifically RCP 4.5 (moderate-case scenario) and RCP 8.5 (worst-case scenario). The projected climatic and hydrological data are presented as the change between the twenty-year periods of 2046–2065 (mid-century) and 2081–2100 (end-century) concerning the baseline period of 1986–2005. Results [14] illustrate that change in maximum temperature to the middle of the century shows a varied increase between 0.3 and 2.4 °C in RCP 4.5 and from 1.1 to 3.4 °C under RCP 8.5. Towards the end of the century, the change "shows an increase between 1 and 3 °C in RCP 4.5 and from 2 to 5 °C with RCP 8.5".

The general change in annual average minimum temperatures to the middle of the century shows a varied increase under RCP 4.5 that goes from less than 1 °C at the Mediterranean coasts to more than 2 °C in the Sahara desert, the Atlantic mountains and Euphrates-Tigris basin. Under RCP 8.5 the rise exceeds 1 °C almost everywhere in the Arab region and reaches 3–3.5 °C in some areas of the Sahara desert. The change of minimum temperatures is more evident towards the end of the century increasing between 1 and 3 °C in RCP 4.5 and from 2 to 5 °C and more with RCP 8.5. The increasing change in temperature becomes much more evident throughout the region with scenario RCP 8.5, when the increased temperature by more than 3 °C would cover most of the Arab region.

Decreasing trends in precipitation will occur in most of the Arab region towards the midcentury, with a reduction of about 90 mm in average annual precipitation for the Atlas Mountains when worse-case scenario is considered (RCP 8.5). For the end of the century, both scenarios show a reduction of the average annual precipitation reaching 90–120 mm/year in the coastal areas and mainly around the Atlas Mountains in the west and upper Euphrates and Tigris river basins in the East. The projected annual change in precipitations is compliant with projected changes in evapotranspiration (ET). The most affected areas are spread out in the coastal zones of the Northern Africa (especially in the West: Morocco, Algeria and Tunisia), but also in the Sahel, in the north of Horn of Africa as well as in the coasts of Arabian Peninsula (mostly in the south-west).

The overall RICCAR assessment indicates a consistent warming trend; increased minimum and maximum temperatures coupled with reduced precipitations, reduced runoff and increased evapotranspiration in most of the region, though with important spatial variations. More specifically, the vulnerability of the agricultural sectors in the Arab Region to climate change could be explained in the context of impact of climate change on water sector. Water scarcity as the one of the main environmental challenges in the Arab Region, with the lowest freshwater resource Per-capita in the world; very low and variable precipitation. Impact of climate change on the Arab Region will add more stress to the already limited resources. It is expected to have more weather extreme events while the temperature will increase in several parts of the Region. Combination of less rainfall and increase in temperature added to the other factors such as growing population will adversatively contribute to the vulnerability of the region to climate change.

Farmers in the region face a series of challenges, to which climate change will be risk multiplier. They include limited and degraded natural resource that are poorly managed (especially of water and land), limited land tenure security, small farm sizes, low technological access, low market access and limited investment [15]. Of the various agricultural communities, it is small-scale farmers who will be disproportionately impacted by climate change. This is partly due to their direct dependence on natural resources and detachment from the extension services and social protection systems that could enable them to build their capacity and resilience. Rain-fed areas (farming systems that depend mainly on precipitation) are likely to be the most dramatically impacted along with pastoralists, farm laborers and landless. Nevertheless, the impacts will be locally specific and difficult to predict [15]. In addition, part of the challenge in estimating the impacts of climate change on smallscale farmers is that there is no unanimous definition of this group. Nevertheless, analysis of the impacts of climate change at the farming system level regarding crop and livestock production and productivity, as well as the implications that this has for the livelihoods of those dependent on them, will elucidate some of the common trends and challenges. Most of the farmers live in the areas of mixed cereals-livestock farming systems keep goats, which are the most vulnerable to climate change after cattle. Their vulnerability comes from the fact that while the potential impact of climate change on livestock is less than that on crops, there is limited adaptive capacity for any of them [5].

3 The Farming System in Egypt

The farming system along the banks of the Nile River in Egypt is irrigated, and 94.8% of all small-scale farmers in Egypt depend on the Nile water for irrigation. Cereals, legumes, sugar crops, oilseeds and forages are cultivated by small-scale farmers for their own consumption with the surplus being sold at the nearest market. Rice, wheat and maize are the major cereal crops cultivated, occupying most of the irrigated land. Unlike rice and wheat, maize is used largely for livestock and poultry feed. Sugar cane and sugar beet are grown along the Nile River in Upper (southern) Egypt as well as the Nile Delta. Broad beans and soybeans are also cultivated [16].

Where these field crops are not grown, horticultural crops (fruits and vegetables) are grown instead and used as cash crops. These include Egyptian clover, berseem, oranges, grapes, stone fruits and pome fruits. Clover and berseem are commonly cultivated in the Nile Delta. Regarding fruit, oranges are the most important, representing 85% of total citrus production. For vegetables, tomatoes are the most important even though they represent only around three percent of Egypt's total planted area. Potatoes are the second most important vegetable after tomatoes. Medium and large farms focus on the cultivation of horticultural crops instead of field crops, whereas the small-scale farmers focus more on field crops.

Furthermore, traditional farming systems such as those employed by small-scale farmers are typically mixed and integrated crop-livestock systems. In 2010, 68.5%

of small-scale farmers had large ruminants (i.e. cows and/or buffaloes), and 47.3 had small ruminants (i.e. goats and/or sheep) [16]. Most of these small-scale farmers supplement their livelihoods with off-farm income as the produce from their crops and livestock is insufficient to sustain them. They diversify into a variety of primarily rural activities with a smaller portion obtaining part of their livelihood in urban areas [16].

4 Vulnerability of Egypt to Climate Change

Although Egypt is one of the least contributors to the global CO_2 emission, with only 0.63%, it is considered as one of the countries that are greatly impacted by the climate change. The vulnerability of Egypt to climate change is aggravated by the significant dependence on climate-sensitive sectors, the concentration of populations in certain areas, and the economic activity in flood-prone urban coastal zones. It has been recognized that the sectors of water resources, agricultural resources and food security and coastal resources are all highly vulnerable with serious socioeconomic implications [17]. In addition to the impact of climate change on increasing the temperature and decreasing precipitation in the Near East North Africa Region including Egypt, the IPCC [8], has singled out the Nile Delta in Egypt as one of the low-lying river deltas that are considered as one of the most vulnerable systems to climate change and sea level rise.

The Nile Delta in Egypt accounts for more than 50% of Egypt's economic activity through agriculture, industry and fisheries. It contributes about 20% of the Egypt's GDP and account for the largest source of employment, around 30% of the labor force. According to UNDP [18], most of Egypt's population and activities including industry, agriculture, and tourism infrastructure and development is located along the northern coastal area, and the reliance on the Nile Delta for prime agricultural land is critically important to the country's economy. Studies on the vulnerability of the City of Alexandria, showed that sea level rise of 0.3 m would lead to infrastructure damage worth billions of dollars, displacement of over half a million inhabitants, and a loss of about 70,000 jobs. The lagoons of the Delta's coastal area are considered among the most productive natural systems in Egypt and they are internationally renowned for their abundant bird life. It has been estimated that approximately 60% of Egypt's annual fish catch are from three main Delta lagoons, Idku, Burullus and Manzalla, separated from the Mediterranean by 0.5–3 km sand belt and dune system. Coastal flooding and/or permanent inundation of these areas would lead to a decline in water quality in coastal freshwater lagoons and corresponding adverse impacts on fisheries and biodiversity.

4.1 Vulnerability of Water Resources to Climate Change

The vulnerability of Egypt's water resources to climate change entails those affecting Nile flows including sensitivity to precipitation on the Ethiopian highlands; sensitivity to a temperature increase in equatorial lakes and Bahr El Ghazal, and uncertainty due to significant differences in the Global Circulation Models output of water flow into the Nile. The sensitivity of the Nile River, as the main source of water in Egypt, to climate change, has been well documented. The expected effects of climate change on Egyptian water resources at the southern part of Aswan High Dam Reservoir, Lake Nubia, were investigated by Elshemy [19].

The United Nations Environment Program [20] report indicted the impact of climate change on the Nile flow, as illustrated by Fig. 1. The study shows the annual average stream levels on the Nile at the Atbara station which was established early in the twentieth century. The hydrograph covers a 90 years period (1907–1997) of monitoring. The Study shows mainly three distinct periods of water level rise and decline can be identified: from 1907 to 1961 water level was slightly high, with sharp decline from 1962 to 1984, and from 1987 to 1997 a water level rise which is an indicator of the high sensitivity of the natural flow to rainfall on the Ethiopian highlands.

Given the current hydro-political transboundary configurations in the basin, institutional and cooperative structures play a fundamental role of either alleviating or exacerbating the vulnerabilities due to a changing climate in their consideration and implementations of plans, policies, and projects addressing climate change issues.

It is expected that climate change will increase saline intrusion associated with Sea Level Rise directly impacting Egypt's entire economy. The Nile Delta's coastal lakes, are important key ecosystems that act as a protective zone for inland economic activities. Lake Manzala, Burullus, Idku and Maryut, however, are only separated from the Mediterranean by a 0.5–3 km stretch of land that is being impacted by erosion and sand dune destabilization [18].

Also climate change is expected to have its critical impact on the local population of the lakes, which will be exacerbated by existing stresses including population

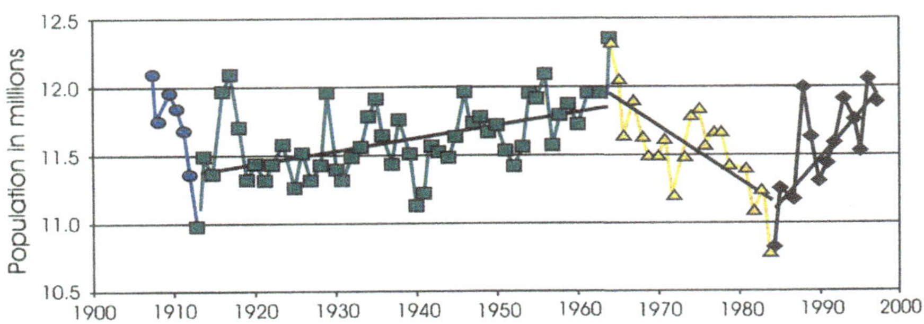

Fig. 1 Annual average stream levels on the Nile at Atbara [20]

growth, poverty and poor nutrition, accumulating levels of air, land and water pollution and ever-growing gender and class inequalities. Sea Level Rise would destroy parts of the protective offshore sand belt, already weakened by reduced sediment flows. In the future decade and in the absence of adaptive action, 3.3% of total land area of the Nile Delta will be lost to the sea, including the submersion of approximately 16 km^2 of fertile currently cultivated land [18].

To combat coastal erosion, the Egyptian Shore Protection Authority has been focusing on the construction of 'hard' coastal protection structures including jetties, groins, seawalls and breakwaters. The total cost of the required activities was estimated at US$200 million. Only a small fraction of these infrastructural solutions were implemented. Even if fully implemented, these solutions would lead to more negative impacts such as sediment transport and blockage of waterways due to an improper understanding of longer-term coastal dynamics associated with climate change [18].

4.2 Vulnerability of the Crop Production System

The 'Regional Initiative for the Assessment of the Impact of Climate Change on Water Resources and Socio-Economic Vulnerability in the Arab Region' [14] indicted that, the highest vulnerability of the cropland systems in the Arab Region is assessed to three regions including the Nile valley. The vulnerability increases from the moderate-case to the worse-case scenario and from mid-century to end-century.

Climate change studies Abou-Hadid [21] and Hassanein and Medany [22], predicted reduction in the productivity of two major crops in Egypt—wheat and maize—by 15% and 19% respectively by 2050. Losses in crop productivity are mainly attributed to the projected temperature increase, crop-water stress, pests and disease, as well as the inundation and salinization.

Because of its exceptional importance to the Egyptian diet, wheat received great attention with regards to the future impact of climate change on its production; Khalil et al. [23] studied effect of climate change on the yield of three wheat varieties (Sids1, Sakha 93 and Giza 168) in Egypt using CropSyst model with two climate change scenarios. These scenarios were A2 (temperature increase by 3.1 °C and CO_2 concentration is 834 ppm) and B2 (temperature increase by 2.2 °C and CO_2 concentration is 601 ppm). Results indicated that A2 scenario predicted greater reduction in wheat yield, compared with B2 scenario in the year of 2038 and also indicated that varieties showed different vulnerability to climate change scenarios.

However, results of the projected impact of climate change on crops should be carefully examined taking into considerations the physiological effects of the elevated CO_2 enrichment on crop production. Degener [24] highlighted two mechanisms that responsible for the impact of elevated, CO_2 on crop yield. One mechanism is when photosynthesis rate raises, resulting in more energy and thus a quicker development of the plant in what is referred to as "the fertilization effect of CO_2". The second mechanism through which an increase in CO_2 reduces the amount of water needed to produce an equivalent amount of biomass. This improvement in water use efficiency (WUE) is due to a closing of the stomata to regulate the flux of CO_2 molecules and affects both C3- and C4-plants alike [24].

Wang et al. [25] analyzed 59 peer reviewed papers and found that, elevated CO_2 (between 450 and 800 ppm) significantly increased the yield of wheat grain by 24%. This yield increase was found to be 44% less in free-air CO_2 enrichment than in the enclosed experimental environment. They concluded that key targets for future plant breeding programs are to select new wheat genotypes which have higher sink capacity for photosynthetic products and are capable of increasing nitrogen uptake under elevated CO_2.

Ruiz-Vera et al. [26] investigated the effects of rising CO_2 and warmer temperature, independently and in combination, on Maize grown in the field throughout a full growing season. Growing Maize, as C4 crop, in mimicking conditions predicted by the second half of this century, in which both CO_2 and temperature were elevated, they found that elevated CO_2 did not stimulate the photosynthesis, the biomass, or the yield of Maize. On the contrary, they found that warming had caused a reduction in photosynthesis of Maize and it shifted above ground carbon allocation away from the reproduction processes of the crop, resulting in reduced yield. The study concluded that future warming would reduce yield in maize, and this will not be mitigated by higher atmospheric CO_2 unless appropriate adaptation traits can be introduced into future cultivars.

Ruiz-Vera et al. [26] reviewed more than twenty years of work on open-air enrichment with CO_2 of C3 crops of cotton, wheat, ryegrass, clover, potato, grape, rice, barley, sugar beet, soybean, cassava and mustard, and C4 crops of sorghum and maize. They concluded that: elevated CO_2 (550 ppm from an ambient concentration of about 353 ppm in 1990) biomass and yield were increased by in all C3 species, but not in C4 species except when water was limiting and that the yields of C3 grain crops were increased on average about 19%.

4.3 Vulnerability of the Livestock Production System

The livestock production contributes significantly to the national economy of the countries of the Near East and North Africa region. It has been more than doubled during the last 40 years from 186 to 412 million head [4]. It has been estimated that 38% of the rural population lives in agro-pastoral livestock production areas, while 59% of the rural population lives in mixed livestock production systems.

The likely impacts of climate change on food production are not limited to crops and agricultural production. Climate change will have great consequences for dairy, meat and wool production, mainly arising from its impact on grassland and rangeland productivity. In areas threatened by global warming such as the Near East and North Africa, higher incidences of transboundary diseases can illustrate a major challenge for the animal production system.

Climate change effects livestock production through its impact on (i) the drinking water resources (ii) quality and quantity of feed (iii) biodiversity and the genetic breeding and (iv) the health of animals [27]. Climate change impact on the different components of the livestock environment in dry areas was summarized by Hassanein and Medany [22] as:

- High ambient temperatures compromise the reproductive efficiency of farm animals (both sexes) and hence affect milk and meat production.
- Humidity also affects animal farm productivity. The combination of temperature and humidity is referred to as the "Temperature Humidity Index (THI)", and it measures the discomfort experienced by animals. At THI above 72, production of cow's milk starts to decline for a cow with no access to shade, while at over 78, milk production starts to decline for cows that have access to both shade and sprinkler system.
- Heat stress hits lactating cows at the rectal temperature of above 39.4 °C. To cope with the stress, the cow allocates more energy to maintain the body temperature at the expense of the amount of energy needed for milk production.
- High temperatures impact the productivity of livestock as well as their growth performance. They impact the uterine environment, reduce the total number of embryo cells resulting in smaller **lamb** sizes, which are susceptible to dehydration during their early stage of life.

SADS [2] estimated that livestock in Egypt contribute significantly to the agricultural gross domestic production with a value of about USD 6.1 billion in 2007. The agricultural development strategy 2030 aims to increase dairy production to 9.5 million tons, red meat to 1 million tons, poultry to 1.4 million tons, and fish to 1.4 million tons per year. This requires—among other activities—the protection of such resources from farm and common diseases [3].

Small scale farmers are the main of livestock in Egypt, while the government sector contribute with less than 2% for the number of animals. Animal production activity is very well integrated within the cropping system, since Egypt has limited natural pastures and depends mainly on planting forages, mainly clover (Barseem) for animal feed in addition to straw and hay. Animal production is highly dependent on cattle and buffaloes as milk-producing animals, as well as male animals and un-reproductive females, are fattened for meat.

Eid [28] explained that livestock/crop production in Egypt is an excellent example of an integrated production system where fodder crops and agricultural residues provide the feed for animals. Majority of small farmers (about 90% of farmers) practice this system. Animal manure with its addition of organic matter that will increase the soil fertility. It is estimated that more than 50 million m^3 of animal manure are produced annually.

However, the livestock production system in Egypt will face even more challenges with the predicted future climate change causing high mortality of young animals and low daily gain well below the genetic potential particularly the high frequency and severity of drought. Resource degradation, in general, contributes to the decrease of livestock production, including the decline of aquifers and consequentially available water for animals; soils salinization due to sea level rise in the North of Delta and over exploitation of ground water; in addition to degradation of range land due to overgrazing, and cultivation of marginal lands.

In addition to the harmful heat stress impacts on animals' productivity, new animal diseases appeared in Egypt causing serious negative impacts on livestock production, such as the blue tongue disease and rift valley fever. With regards to the feed production, availability of fodder is subject to decrease due to climate change impacts on crops productivity, and the increasing competition over the land and water resources between fodder and cereal crops. In this respect Habeeb et al. [29] reported that exposure of animals to heat stress causes a series of drastic changes in the biological functions, including decrease in feed intake, feed efficiency and utilization, disturbances in water, protein, energy and mineral balances, enzymatic activities, hormonal secretions and blood metabolites ending to impairment the productive and reproductive performance.

4.4 The Vulnerability of the Fishery Production System

According to FAO [30], climate change impacts the fisheries and aquaculture sectors either direct physical effects of changing sea levels, flooding and storms patterns, or through biological and ecological responses to physical changes (e.g. productivity, species abundance, ecosystem stability, stock locations, pathogen levels and impacts). These impacts are expected to affect fishers, fish farmers and coastal inhabitants through their impact on livelihoods, changes in the availability and quality of fish for food, and through rising risks to human health, safety and homes.

FAO report which summaries the findings of the fifth assessment of the intergovernmental panel on climate change [16], suggested that:

- Ecosystems of the Arabian Gulf, the Red Sea and the Mediterranean Sea are linked to continuous stratification, rising temperatures, changes in pH values, reduced O_2 concentration, and subsequent effects on corals and the primary production and larger commercially valuable fish stocks.
- Increasing temperatures, algal bloom changes in the diversity of zooplankton and the "tropicalization" of the fauna have been observed in the Mediterranean.
- Several northern African countries that border the Mediterranean and Red Seas including Egypt are especially vulnerable to physical and chemical climate drivers and are heating up much faster than the global mean ocean temperature due to their shallow physical structure and their land-locked location.
- The catch potential in 28 locations around the Meditation and the Red Sea (including Egypt) is expected to reduce by as high as 50%.

Egypt's National Strategy for Adaptation to Climate Change and Disaster Risk Reduction [3], reported that, there is a chance that the water of the northern lakes in Egypt may change, due to the rise in temperature and intrusion of the sea water from the Mediterranean Sea as a result of its level rise. This may cause destruction and transgression of the narrow sand barrier separating the sea from these lakes, which will then lead to its gradual sink in.

This may cause a change in the biodiversity of marine life and in fisheries located south of the lakes. Salinity will increase, and will gradually reach that of the open sea. The fish ecosystem will change as the rate of nutritional assimilation rises. This will lead to a higher need for nutrition, and increased competition between different species, thus raising the Biological Oxygen Demand (BOD). In terms of the hydrological aspect, the Mediterranean Sea level rise—whether it is 50 cm or 1 m—will have a significant impact on the rates and locations of egg hatching. The increased salinity will limit the spread of fresh water fish in the northern areas of the Delta, and alter specific mix of the fish.

4.5 Enhancing Adaptive Capacity and Resilience to Climate Change in Egypt

Developing strategy for enhancing adaptive capacity and resilience to climate change in Egypt requires coordination and integration between the main related national strategies. These strategies include Egypt's Vision 2030 (2016), the Sustainable Agricultural Development Strategy [2], and Egypt's National Strategy for Adaptation to Climate Change and Disaster Risk Reduction [3] are as well as activities that cut across a number of stakeholders and national processes.

The objective of Egypt's National Strategy for Adaptation to Climate Change and Disaster Risk Reduction [3] is to increase the resilience of the Egyptian communities to climate change impacts at various sectors and activities. The Egyptian Strategy has been developed based on the expectation of the IPCC reports that average temperature increase will not exceed two degrees Celsius until the year 2100 and under two scenarios of sea level rise by 2100 that are, 0.5 and 1 m.

The Strategy presents an assessment of the current situation in all sectors, including water resources and irrigation, agriculture, health, urban areas, housing, roads and tourism. The strategy focuses on the importance of coastal zones as impacted by sea level rise. Also, the Strategy provides suggestions for adaptation measures based on extensive research efforts of the Egyptian scientific institutions to face the climate change challenges.

In the water sector, the strategy identifies population increase, fragmented agricultural holdings, unstable cropping system, lack of financial resources, and weak legislations as main challenges. Adaptation measures have been identified to cope with the impact of climate change on water.

The most important specialized adaptation programs proposed in the National Strategy for Adaptation to Climate Change and the Sustainable Agricultural Development Strategy that can be reflected in the Climate Smart Agriculture strategy include the following:

- Introduction of breeding programs for the economic field and horticultural crops to introduce new varieties that are capable of adapting to the expected changes in climate including gradual temperature increase, high soil salinity, less available water, resistance to insects and plant disease.
- Development of new short growing crop varieties to reduce their water requirements.
- Expansion of cultivation of crops of economic importance, which can adapt to climate change (e.g. rice), while taking into account the use of new farming systems to adapt to changing expectations in water supply and the gradual replacement of sugar cane by sugar beet.

The research institutions in Egypt contribute to enhancing the adaptive capacity of the country to cope with impacts of climate change. Colleges of Agriculture in the Egyptian Universities, The Agricultural Research Center and the Desert Research Centers are the main pillars of the agricultural research system in Egypt that provide the scientific bases for impact of climate change on agricultural sector. In addition to the National Research Center, The The National Water Research Center, National Institute of Oceanography and Fisheries and The Egyptian Atomic Energy Authority and institutes of Environmental studies.

5 Climate Change Adaptation Efforts

Although impact of climate change on agricultural and water sectors in Egypt has been well documented in the regional and global reports, the national scientific research institutions did not give the issue the attention it deserves. Changing sowing dates and management practices are among the important measures for climate change adaptation to cope with its adverse impact. Changing crop varieties to those tolerant to drought, heat, salinity and pests, and changing crop pattern and rotation are the most promising adaptation measures at the national level. Using low efficient irrigation practices and improved surface irrigation systems with higher efficiencies and applying deficit irrigation are considered as important adaptation measures that contribute to cope the impact of climate change.

Efforts have been done to improve the current low productivity of livestock in Egypt under the drought and heat stress as well as considering their feeding programs. However, there is no clear adaption measures have been identified for the fisheries. More studies should be conducted as well as tools and guidelines should be developed to address assessment and impact of climate change as well as proposing adaptation measures.

In addition research and innovation; capacity development and awareness rising are important components of any program or project that will address climate change in Egypt.

6 Conclusions

Climate change adaptation is of paramount importance to agriculture, given the reliance of the sector on climate. Climate change adaptation policies should be based on science, and incorporate knowledge of indigenous peoples and traditional practices. Adaptation considerations should be mainstreamed into sectoral and cross-sectoral policymaking, and promote good adaptation practices to confront the heterogeneity and uncertainty of climate change impacts. Ultimately, adaptation efforts should contribute towards sustainable food production and food security for all [5].

7 Recommendations

- Strengthening the climate change institutionally at the country level to include environmental institutions in addition to those of agriculture and water.
- Mainstream issues related to impact of climate change in the national policies and strategies of the agricultural and water sector.
- Enhance the scientific research sector to address climate change concerns.
- Improve irrigation water productivity for sustainable agricultural production and food security.

References

1. Egypt's Vision 2030 (2016) Ministry of planning, monitoring and administrative reform, Cairo, Egypt
2. SADS (Sustainable Agricultural Development Strategy) (2009) Sustainable agricultural development strategy towards 2030. Agricultural Research & Development Council, Arab Republic of Egypt, Ministry of Agriculture & Land Reclamation, Oct 2009
3. UNDP (2011) Egypt's national strategy for adaptation to climate change and disaster risk reduction. UNDP and Egyptian Cabinet for Information, Decision Support Center, Cairo, Egypt
4. FAO (2011) "The state of the world's land and water resources for food and agriculture-SOLAW"—managing systems at risk. Food and Agriculture Organization of the United Nations, Rome, Italy
5. FAO (2017) The state of food security and nutrition in the world. Food and Agriculture Organization of the United Nations, Rome, Italy
6. ESCWA (2017) Climate change-related statistics in the Arab region: a proposed set of indicators. Special issue of the compendium of environment statistics in the Arab region 2017. The United Nations Economic and Social Commission for Western Asia (ESCWA), United Nations House, Beirut, Lebanon
7. Kimball BA (2016) Crop responses to elevated CO_2 and interactions with H_2O, N, and temperature. Curr Opin Plant Biol 31:36–43
8. IPCC (2007). Intergovernmental Panel on Climate Change report: impacts, adaptation and vulnerability. Contribution of Working Group II to the fourth assessment report of the Intergovernmental Panel on Climate Change. Cambridge University Press, Cambridge, United Kingdom and New York, NY, USA

9. UN (2017) UN Office for the Coordination of Humanitarian Affairs. Humanitarian Bulletin Somalia, 26 Sep 2017. https://reliefweb.int/report/somalia/humanitarian-bulletin-somalia-september-2017-issued-26-september-2017
10. Koday K (2017) The International Policy Centre for inclusive growth. Social protection after the Arab Spring, vol 14, no 3, December
11. Adelphi (2015) A new climate for peace—taking action on climate and fragility risks. Executive summary. The Wilson Center, EUISS
12. IPCC: Climate Change (2014) In: Edenhofer O, Pichs-Madruga R, Sokona Y, Farahani E, Kadner S, Seyboth K, Adler A, Baum I, Brunner S, Eickemeier P, Kriemann B, Savolainen J, Schlömer S, von Stechow C, Zwickel T, Minx JC (eds) Mitigation of climate change, contribution of Working Group III to the fifth assessment report of the Intergovernmental Panel on Climate Change. Cambridge University Press, Cambridge
13. OECD-FAO (OECD/Food and Agriculture Organization of the United Nations) (2018) The Middle East and North Africa: prospects and challenges. In: OECD-FAO agricultural outlook 2018–2027. OECD Publishing, Paris/Food and Agriculture Organization of the United Nations, Rome
14. RICCAR (2017) Regional initiative for the assessment of the impact of climate change on water resources and socio-economic vulnerability in the Arab region. The Economic and Social Commission for Western Asia (ESCWA), Beirut, Lebanon
15. Morton J (2007) The impact of climate change on smallholder and subsistence agriculture. Proc Natl Acad Sci U S A 104(50):19680–19685
16. FAO (2018) Impacts of climate change on farming systems and livelihoods in the Near East and North Africa—with a special focus on small-scale family farming. FAO, Cairo, 92 p
17. INCCC (Initial National Communication on Climate Change) (1999) EEAA report to UNFCCC. Cairo, Egypt
18. UNDP (2018) Climate change adaptation in the Arab states. United Nations Development Program, New York, USA
19. Elshemy M (2013) Climate change impacts on water resources in semiarid regions: case study of Aswan high dam reservoir. In: Younos T, Grady C (eds) Climate change and water resources. The handbook of environmental chemistry, vol 25. Springer, Berlin
20. UNEP (2005) Facing the facts: assessing the vulnerability of Africa's water resources to environmental change. Report of the United Nations Environment Program, Nairobi, Kenya
21. Abou-Hadid AF (2006) Assessment of impacts, adaptation and vulnerability to climate change in North Africa: food production and water resources. Final report. Assessments of Impacts and Adaptations to Climate Change (AIACC), Project No. AF 90
22. Hassanein MK, Medany MA (2007) The impact of climate change on production of maize (Zea mays L.). In: International conference on climate change and their impacts on coastal zones and river deltas: vulnerability, mitigation and adaptation, 23–22 Apr 2007, Alexandria, Egypt, pp 268–281
23. Khalil FA, Farag H, El Afandi G, Ouda SA (2009) Vulnerability and adaptation of wheat to climate change in Middle Egypt. In: Thirteenth international water technology conference, IWTC 13 2009, Hurghada, Egypt
24. Degenr JF (2015) Atmospheric CO_2 fertilization effects on biomass yields of 10 crops in northern Germany. Environ Sci 3:48
25. Wang L, Fengb Z, Schjoerring J (2013) Effects of elevated atmospheric CO_2 on physiology and yield of wheat (*Triticum aestivum* L.): a meta-analytic test of current hypotheses. Agric Ecosyst Environ 178
26. Ruiz-Vera UM, Siebers MH, Drag DW, Ort DR, Bernacchi CJ (2015) Canopy warming caused photosynthetic acclimation and reduced seed yield in maize grown at ambient and elevated CO_2. Glob Change Biol 21
27. IFAD (International Fund for Agricultural Development) (2011) Rural poverty report. Final report. IFAD, Rome
28. Eid YZ (2014) Climate change and livestock vulnerability and adaptation. Alexandria Research Center for Adaptation to Climate Change (ARCA), Alexandria

29. Habeeb AA, Gad AE, EL-Tarabany AA, Atta MA (2018) Negative effects of heat stress on growth and milk production of farm animals. J Anim Husbandry Dairy Sci 2(1):1–12
30. FAO (2016) Climate change and food security: risks and responses. Food and Agriculture Organization of the United Nations, Italy, Rome

Vulnerability of Crop Pollination Ecosystem Services to Climate Change

Mohamed A. M. Osman and Mohamed A. Shebl

Abstract Crop pollination is one of the most valuable ecosystem services. It plays an important part in human food security. Pollination services are mainly provided by wild pollinator species as solitary bees and by commercially managed honeybees. Pollinators also have a key role in maintaining other ecosystem services including ensuring biodiversity. Declines in bee diversity over the last two decade have been recorded in many countries worldwide. Some drivers generate many stressors for pollinators such as loss of habitat, nest fragmentation, urbanization, reduced floral resource supply, increasing of pests and diseases, extensive use of pesticides and climate change. Climate change is potentially the most serious threat to pollinator biodiversity, reducing crop productivity and negatively impacting global food security. This chapter focuses on explaining the evidence of the biotic responses to the slight climate changes that occurred recently. In Egypt, many studies have proven that many bee species including honeybees are under huge threats affecting the agriculture production and plant vegetation. To mitigate this great problem, several attempts were conducted in order to conserve and propagate the most vulnerable solitary bee species through re-nesting them in artificial nests to be used for crop pollination.

Keywords Pollination · Bees · Decline · Climate change · Conservation

1 Introduction

The pollination of flowering plants is one of the most important ecological services in natural and agricultural ecosystems. The majority of angiosperm plants rely on animals for pollination [1]. Not all insects are pollinators, nor all pollinators are insects. Most insect pollinators are Hymenopteran (Bees), Dipteran (flies), Lepidopteran (butterflies and moths) and Coleopteran (beetles and weevils). Bees and sphecid wasps belong to the superfamily Apoidea are considered the major plant

M. A. M. Osman (✉) · M. A. Shebl
Plant Protection Department, Faculty of Agriculture, Suez Canal University, Ismailia 41522, Egypt
e-mail: naeim70@hotmail.com

M. A. Shebl
e-mail: mohamedshebl2002@hotmail.com

© Springer Nature Switzerland AG 2020

E.-S. E. Omran and A. M. Negm (eds.), *Climate Change Impacts on Agriculture and Food Security in Egypt*, Springer Water, https://doi.org/10.1007/978-3-030-41629-4_11

pollinators. The bees (called Apiformes) can be distinguished from the wasps by the presence of erect, plumose hairs on their faces [2]. Bees and their life-sustaining relationships with flowering plants occupy keystone in both natural and agricultural ecosystems [3–5]. Given their ecological services, native bee faunas are an important natural resource that can complement honey bees and other managed pollinators in their ecosystem services [6, 7]. Most bees are dependent on products produced by angiosperm-flowers which include pollen, nectar and oils, which used for their progeny. The co-evolution of bees and flowers has resulted in special morphological adaptations for both insects and plants, and the needs of some plants for pollination by bees are absolute [8]. Bees comprise over 20,000 species worldwide, and most of them are critical to ecosystem health through their pollination activities, except for cuckoo bees. Native bees can enhance the services provided by managed honey bees via behavior interactions [9]. Thus the complementary foraging activities of honey bees plus wild bees can result in enhancing pollination [10]. The decline of some species is well documented globally due to several reasons such as climate change, lack of natural foraging resources, epidemic of some pests and diseases and elimination of nesting habitats; these declines have been reported worldwide [11–14].

2 Pollination as an Essential Ecosystem Service

Ecosystem services are the set of processes through which the natural ecosystems contribute to human well-being [15]. These include provisioning services such as food and water; regulating services for instance regulation of climate, floods, drought, land degradation, pests, diseases and pollination; supporting services such as soil formation, photosynthesis and nutrient cycling; and cultural services such as recreational, spiritual enrichment and other non-material benefits [5, 16].

The natural pollination of crops is one of the most recognized ecosystem services. It is a pivotal function in all terrestrial ecosystems, supporting ecological stability and food security worldwide. Pollination service, supplying production inputs to agriculture from wild biodiversity, provides one of the strongest cases for valuing and managing natural habitats and resources for the services they provide to livelihoods.

Pollination is defined as a process by which plants sexually reproduce. The process is defined as the transfer of pollen grains from the anther of a flower to the stigma of the same or another flower of the same plant species. So, pollination can be accomplished by cross-pollination or by self-pollination. This process is one of the oldest complicated basic biology processes of the evolution history of at least 100 million years [17]. Pollination service is one of the key ecosystem services associated with agro-ecosystems, as one-third of crop species worldwide require insect pollinators [18, 19, 20, 13, 21, 22, 23].

Pollination in flowering plants provided by animals is considered as a regulating as well as a supporting ecosystem service. As a regulating service, pollination is essential for the reproduction of pollinator-dependent plants that supply humans with foods, forage, fiber, timber, firewood, biofuels and medicine. As a supporting service, it is also necessary for maintaining populations of pollinator-dependent wild plants [24] that subsequently provide additional ecosystem services, such as water filtration, carbon storage, erosion control, and habitat for biodiversity [25].

Pollination services may be directly or indirectly contribute to human well-being. It contributes significantly to the agricultural production of a broad range of crops, in particular seeds, fruits, vegetables, nuts, fiber and forage crops (e.g., clover, orange, apple, strawberry, cucumber, cotton, melon, squash, bean, soy and sunflower). Indirect contributions of pollination include, for example, the reproduction of much valued timber tree species (e.g., mahogany, [26]), and plant-extracted medicines, conservation of plant genetic diversity resources, and ecosystem resilience [27].

3 Role of Pollinators in Sustaining Agricultural Productivity

There are 250,000–400,000 species of angiosperms (flowering plants) known worldwide, including both wild and cultivated plants. These plants play an important role in sustaining our atmosphere, soils, and other elements of global functioning, as well as supporting an enormous number of animals by providing food, shelter, and defense. On the other hand, most flower visitors, in particular insect pollinators, play a key role to fertilize these plants through cross-pollination. Thus, the loss of one partner affects the other, which seriously affects the entire ecosystem. Unfortunately, pollination is not considered as one of the major aspects of world food security [20]. Without pollination services, especially insect pollination services, probably our valuable crops would require another way of pollination. Otherwise, without insect pollination services, crop productivity will be reduced, which means not providing adequate sources of food [28]. Likewise, the scarcity of pollinator populations reduces the productivity of food, timber and fiber plants, which lead to hunger and poverty [29]. The decline in agricultural production can result from insufficient pollination rather than using agrochemicals or pest infestation or other insufficient agriculture input. Pollinators are part of very specific relationships with their plants; lack of these pollinators can have a great impact across the natural and agriculture ecosystems. Several literatures discussed the contribution of pollinators to our ecosystem services in terms of monetary value [30–34]. Up to now, there are no any assessments for the contribution of pollinators and bees in particular to the agriculture production in Egypt. Still, there is no public awareness of the importance of the pollinators and their contribution for agricultural production. The whole concern of producers is honeybee and honey production regardless of their main importance for crop and plant pollination. The bees are considered one of the major insect pollinators for several crops worldwide such as almond, berries, citrus, clover, vegetables and others.

4 Climate Change and Crop Pollination

Climate is a key environmental factor that has a significant impact on various ecosystems. Climate change is any substantial change in Earth's climate that refers to seasonal changes over a long period with regard to the rising accumulation of greenhouse gases in the atmosphere. It is one of the major challenges of our time, adding considerable stress to our communities and the environment.

Changing climate could lead to substantially alter or eliminate certain ecosystem services in the future. It affects ecosystems in different ways, for instance, warming may force species to migrate to higher latitudes or higher altitudes where temperatures are more favorable to their survival [35]. Climate change is already affecting our climate, agriculture, native ecosystems, infrastructure, health and biosecurity. It has broad impacts on plant phenology and growth, carbon and nutrient cycling, as well as biodiversity and extinction risk. It is affecting the ability of landscapes to provide these environmental benefits. Climate change also has the potential to severely impact ecosystem services such as pollination process.

Crop pollination is a vital ecosystem service and essential for sustainment agricultural and horticultural productivity. Pollination is a keystone process in natural and managed ecosystems. The reproductive success of plants, and consequently their survival and evolution depend on it. Pollination interventions have been shown to enhance seed production, quality of seeds, and fruits in various crops. Over 75% of the major world crops and 80% of all flowering plant species rely on some amount of animal pollinators [36, 37]. The total economic value of animal pollination to agricultural production has been estimated at €153 billion (~$200 billion) annually [38], which represented about 9.5% of the total value (€1618 trillion) of agricultural production used for human food worldwide in 2005.

Animal pollinators include many insect species (for example, bees, flies, butterflies, moths, wasps, beetles and thrips), as well as several species of birds, bats and other mammals [37]. Bees are considered the most important group of insect pollinators, which help pollinate about 84% of crop species, increasing the yield and quality of approximately 35% of crops worldwide. The leading pollinator-dependent crops are vegetables and fruits, representing about €50 billion each, followed by edible oil crops, stimulants, nuts and spices. The most pollinator-dependent crop categories ranked in descending order according to the economic value of insect pollination were vegetables, fruits, and edible oil crops (Table 1).

Climate change may pose a serious threat to pollination services [39–41]. It is an important driver affecting pollinators and plant-pollinator interactions [42–44]. Climate change is considered as one of the multiple environmental pressures acting in concert [41]. The Intergovernmental Panel on Climate Change (IPCC) [45] has developed a list of many observed changes of the global climate, including increased temperatures, reduced snow and ice cover, and variable rainfall frequency and intensity of precipitation [46]. Increasing global temperatures, droughts and more frequent extreme events are all suggested to affect plant-pollinator interactions [47]. In this context, the most important impact of climate change is the increasing trend in the

Table 1 Economic impacts of insect pollination of the world agricultural production used directly for human food and listed by the man categories ranked by their rate of vulnerability to pollinator loss [38]

Crop category	Average value of a production unit	Total production economic value (EV)	Insect pollination economic value (IPEV)	Rate of vulnerability (IPEV/EV)
	€ per metric ton	109€	109€	%
Stimulant crops	1225	19	7.0	39.0
Nuts	1269	13	4.2	31.0
Fruits	452	219	50.6	23.1
Edible oil crops	385	240	39.0	16.3
Vegetables	468	418	50.9	12.2
Pulse	515	24	1.0	4.3
Spices	1003	7	0.2	2.7
Cereals	139	312	0.0	0.0
Sugar crops	177	268	0.0	0.0
Roots and tubers	137	98	0.0	0.0
All categories pooled together		1618	152.9	9.5

global temperatures that may considerably impacts on pollinator interactions and thus the whole pollination process.

Due to the experimental studies that focused on the impacts of climate change on the interactions of wild plant pollinators and those in crop pollination are rare, the negative effects of climate change on pollinators and pollination services had been remained fully unclear for several decades.

However, over recent decades, several authors [48, 49] have argued that including species interactions when analyzing the ecological effects of climate change is of utmost importance. The future scenarios on climate change reported by the IPCC suggest that the seasonal activity, abundance, disrupting life cycles, diversity of wild pollinator species, and interactions between species, as well as community composition may change differentially correlated with observed changes in terrestrial climate [50].

4.1 Impact of Climate Change on Crop Diversity

Crop diversity is the variance in genetic and phenotypic characteristics of plants used in agriculture. It comprises the genetic diversity within each crop (within-crop diversity), and the number of species commonly grown (between-crop diversity). Over the past 50 years, there has been a great decline in both components of crop diversity. The loss of biodiversity is considered one of today's most serious environmental concerns by the Food and Agricultural Organization (FAO) [51]. This diversity is therefore important for all scales of crops and food security. If a particular crop variety fails due to any environmental stresses entailing shifts in temperature, precipitation, drought, flooding or disease, another variety can survive to avoid food shortages.

Changes in long-term environmental conditions that known as climate change had enormous impacts on current plant diversity patterns; further impacts are expected in the future. It is recognized that climate change will remain one of the major drivers of biodiversity patterns in the future. Climate change plays a key role in plant diversity, as well as its productivity.

Climate change having substantial, often negative, impacts on farming systems worldwide. It is affecting a diversity of species in a variety of ways [52–54]. In general, it is predicted that the tropics will suffer the most from the effects of increased and sporadic temperature spikes and decreased and sporadic precipitation [55, 54]. There is already evidence that plant species are shifting their ranges in altitude and latitude as a response to changing regional climates. Climate change has already led to shifts in the species distribution in many parts of the world. In the future, it is expected that there will be increasing impacts; however few studies have aimed for a general understanding of the regional basis for species vulnerability.

Africa is one of the most vulnerable continents to climate variability, a situation exacerbated by multiple stresses that interact in various levels and the low adaptive capacity [56]. In Africa, the distribution of crops will vary in the future according to the expected impact of climate change as follows [57]:

- In the event of a change to a very hot climate, more fruits and vegetables will be cultivated.
- In the event of a change to a mild and humid climate, more millet will be grown, except in low dry savannas and semi-arid lowlands.
- In all possible scenarios, maize selection will be significantly reduced in all regions.
- Wheat cultivation will be decreased across Africa as the climate warms.

In many areas, the crop varieties and species currently grown by farmers cannot tolerate these stresses in climate change, with resultant losses in productivity, and potentially negative consequences for food security. According to IPCC and other studies, temperature increases of 1–2 °C will increase production of some of the world's major staples, while temperature increases of more than 2 °C will have increasingly negative impacts [58].

Researchers predict that 20–30% of the species in the Arab world will become extinct if the temperature rises only 1 degree. Yemen has the highest number of threatened species, with 159 species, Somalia has 17 threatened species, and Jordan, Egypt, Saudi Arabia and others combined more than 80 endangered species, including the cedar forests in Lebanon and Syria. Amongst the many threatened species are wild relatives of our crops that possess valuable traits for crop breeding, such as pest and disease resistance. Almost 6% of relative wild species of cereal crops such as wheat, maize, rice, and sorghum, about 18% of legumes as beans, peas and lentils, as well as 13% of family Solanaceae that includes potato, tomato, eggplant, and peppers are at risk of extinction.

Crop diversity is an aspect of biodiversity provides a number of benefits within agricultural production systems. It contributes directly to global food security, ecosystem function and human well-being. International organizations are working across the world to preserve crop diversity, using means such as seed banks. More than 70% of all crop wild relative species worldwide are in urgent need of further attention through more collection and conservation to improve their representation in genebanks.

4.2 Effect of Climate Change on Crop Flowering Time

Climate change not only affects the natural species diversity and distribution but also extends its influences to other phenological events of crops [59, 60]. The timing of phenological events such as flowering is often related to environmental variables. Changing climate is therefore expected to affect the timing of flowering in many plant species [61, 62].

Flowering time refers to the time after vegetation growth in which a plant switches to reproductive growth through the appearance of the first floral bud. It is a critical phenological stage for subsequent reproductive phase. The timing of flowering is regulated by environmental factors and by genetic factors. It is a key parameter for both crop adaptation to environmental conditions and yield potential. The optimum time of flowering is the most important for agricultural production, especially for crops yield and seeds quality.

The flowering time of cultivated crops within a season is largely determined by responses to temperature, photoperiod, atmospheric CO_2 [63, 64], water availability and other ecological cues; all represent climate changes typically associated maturation with global warming [62, 65]. In recent decades, earlier flowering has been observed and documented in crop plants [66–70], as well as in natural communities [61]. Almost, 78% of a total of 542 plant species in 21 European countries showed earlier flowering, with an average of 2.5 d per decade [68]. In Germany, the early phenological events of agricultural and horticultural crops were recorded between 1951 and 2004 with an average of 1.1–1.3 d per decade [66].

Earlier flowering and the relative duration of the pre and post-flowering phases (vegetative and reproductive phases, respectively) are critical determinant of crop yield and seeds quality [71]. In general, warmer temperatures shorten the developmental stages of crops and, eventually lead to lower yields of a given plant species. In this respect, the expected changing in crop growing cycle reduction (number of days) were reported in different Mediterranean countries in response to an increase of 2 °C (2031–2060) respect to present scenario (1960–1990) (Table 2) [72].

In general, plant flowering tends to occur earlier than it did a decade ago, with the greatest phenological changes observed in early spring [61, 68]. The flowering times (FT) in 461 plant species showed a trend of earlier flowering with climate warming

Table 2 Crop growing cycle reduction (number of days) in response to an increase of 2 °C (2031–2060) concerning present scenario (1960–1990)

Country	C4 summer crops	Legumes	C3 summer crops	Tuber crops	Cereals
Algeria	−12.3	−8.0	−9.9	−7.6	−3.3
Egypt	−9.5	−6.0	−7.5	−7.4	−4.2
France	−13.6	−10.6	−16.2	−19.6	−8.3
Jordan	−8.3	−4.9	−6.2	−4.9	−10.3
Greece	−4.5	−2.6	−3.6	−1.9	−7.1
Italy	−4.1	−2.3	−5.0	−5.4	−10.1
Libya	−7.9	−5.0	−6.5	−6.7	−5.7
Morocco	−14.0	−10.1	−11.2	−9.0	−8.8
Portugal	−12.1	−8.3	−10.0	−9.5	−8.9
Serbia	−2.9	−1.7	−1.6	−10.2	−12.2
Spain	−19.4	−7.1	−15.7	−15.4	−7.3
Tunis	−9.2	−5.4	−7.7	−12.1	−1.2
Turkey	−18.1	−11.6	−18.4	−17.6	−8.9

Legend C4 crops = maize; Legumes = bean, lentil; C3 summer crops = sunflower; Tuber crops = potato; Cereals = barley, wheat [73]

[74]. These shifts are largely attributed to rising temperatures, but anticipated changes in precipitation may also affect phenology, especially in arid regions [75]. With regard to warmer temperatures, crop yield can be affected at any time from sowing to grain maturity, but it is the time around flowering and during the grain-filling stage, high temperatures have the most impact on the final harvestable crop, as found in cereals [76, 77]. These changes in phenology can also lead to asynchrony between plants and their insect pollinators (as a plant may no longer flower at the same time that wild bees emergence are occurring, for example), with potentially drastic effects on yield reproduction. The interactions among flowering plants and pollinators are ecologically important and economically valuable. However there are no adequate studies of these important consequences for plant-pollinator interactions.

5 Vulnerability of Pollinating Insects to Climate Change

5.1 Decline of Insect Pollinators in Egypt

There are no enough field expeditions conducted across Egypt for clarifying the exact number of bee fauna in the country. Several field works were conducted during the 1970s for listing several bee taxa and species in some areas [78]. By then some other works were conducted during the last two decade but on a small scale compared to that one in the 1970s. In previous studies, some authors mentioned that Egypt is one of the richest areas with solitary bees [79–91]. The agricultural cultivated areas all over Egypt have been changed dramatically over the last 20 years. Some agricultural areas have been changed to be urban and suburban communities. Since then, a lot of impacts have been affected by several issues in the agriculture ecosystem with special concern to flora and fauna including wild and domestic bees. On the other hand, the newly reclaimed areas are shortened due to the limited water resources and low rainfall which mainly affected by climate changes. This has been proved with the bee of the genus *Anthophora*, which Egypt was so rich with over 80 species, but right now the species richness of this genus was fewer than the past [90]. Also, the increase of temperature is well recognized in both seasons; winter season becoming short and the summer season becoming longer, which will effect on the species thermal tolerance levels, distributions and the whole lifecycle. In additions, some species may disappear, and some other species will have more superiority for dominating in the ecosystem [92, 39]. There are a lot of studies in this regard worldwide, but unfortunately there is no any information or data about the real or expected scenarios for insect pollinators in Egypt. Recently, *Apis florea*, the dwarf honeybees were recorded in Egypt, which considered as a biological invasion, and it is expected to compete with domestic honeybees *Apis mellifera* on their foraging resources [93]. So, this species in the coming two-decades probably will invade the whole of North Africa.

5.2 *Solitary Bees and Nest Fragmentation*

Several bee species, which are "cavity and ground-nesting bees" are threatened by nesting fragmentation, habitat elimination, degradation and urbanization [94, 78]. The populations of some bee species from the family: Megachilidae was less than their populations in the past, such species as *Chalicodoma siculum* (Rossi, 1792), *Chalicodoma nigripes* (Spinola, 1838), *Chalicodoma flavipes* (Spinola, 1838), *Megachile minutissima* (Radoszkowski, 1876), *Osmia latreillei* (Spinola, 1806) and *Osmia submicans* (Morawitz, 1870) [89, 78]. The same case was found with another bee species, *Eucera nigrilabris* (Lepeletier, 1841). The whole nesting area of this species was eliminated for to national project of covered drainage [95]. Therefore, a lot of attempts were carried out in the Bee Research Center of Suez Canal University for conserving several cavity nesting bee species. This was done successfully with three major species *Chalicodoma siculum* (Rossi, 1792), *Megachile minutissima* (Radoszkowski, 1876) and *Osmia latreillei* (Spinola, 1806).

5.3 *Bees and Shortage of Foraging Resources*

One of the major threats for bees is the threat of plant pollinator interactions. This could be recognized in several studies worldwide. The climate change is affecting the synchronization of the host plant blooming and their bee pollinators and misses matching and changes the plant-pollinator phenology [96]. So, the bees could miss the blooming season of their major floral plants, which will affect the bee survival and disappearance, this might be happening and was proved in rare cases [96, 97]. Also, with climate change, it is expected to have some more changes on the bee generations and their inter correlations with their foraging resources. It is known that bee's voltinism (the number of generations per year) divided into univoltine, bivoltine, multivoltine and an uncountable number of generations with social species. So, within the time when the whole ecosystem will change, some species will have more dominant superiority and others will completely disappear and decline. It is not easy to prove this assumption, especially with insufficient studies in this regard in Egypt. However, there are a lot of agricultural areas across the whole country completely changed from a cultivated area to urban and suburban cities. Beekeeping activities have been stopped in these areas, and many species of solitary bees that were recorded in the past are not present at this time. This is clearly showed in our research works on major pollinators of broad bean; Anthophorids, which considered as one of the major bee pollinators of this economic plant [98, 99, 90]. Egypt was one of the richest Mediterranean countries with this group of solitary bee, and the total recorded number was over 80 species [86] but the total species number is currently less than the number recorded in the past [89]. With a clean environment and more diverse vegetation and floral resources, we will get more number and a diverse number of bee species richness and abundance. So, it is recommended to encourage farmers, local agriculture producers, and natural protectorate conservative areas and national parks mangers for cultivated bee-friendly plants. We introduced *Phacelia tanacetifolia* to our conservation area of bees at the Suez Canal University campus in 2015. Fortunately, the plant attracts several plant pollinators of butterflies, honeybees and solitary bees (Fig. 1). Several bee species were recorded visiting this plant such as *Megachile cinnamomea*, *Megachile mintissiuma*, *Chalicodoma siculum*, *Stelis murina* (Megachilidae); *Apis mellifera*, *Amegilla* sp., *Certaina trasata*, *Xylocopa pubscences* (Apidae); *Nomidiodes* sp. (Halictidae), *Colletes lacunatus*, *Hylaeus* (Colletidae) and *Andrena flavipes* (Andrenidae) (Shebl, unpublished data).

5.4 Honeybee Pests and Diseases

Many species of bee pollinators are in decline; these declines have been reported worldwide [11]. The impacts of diseases, parasites, and pathogens on pollinator populations are of growing concern globally [12–14]. Managed honey bees and bumble bees are affected by parasitic mites, many diseases and pathogens, and there is evidence that these can spread from commercial colonies to wild populations and species [100–103]. There are some pests have a great impact on the number of honeybee colonies across Egypt especially the red date palm wasp, *Vespa orientalis*. This was attacking a lot of colonies, especially in the late summer season from July and August in addition to some other minor pests such as ants and wax moths [104].

Fig. 1 *Phacelia tanacetifolia* bee-friendly plant introduced to Egypt as a forage resource during the winter season. Photographed by M. Shebl

A newly recorded pest, *Senotainia tricuspis*, was recorded for the first time in Egypt in 2015 [105] (Fig. 2). A survey was conducted for the latter new recorded pest, and the result is shown in Table 3 (Shebl, unpublished data).

One of the most severe problems of honeybee colonies is the widespread of *Varroa* mite, but *Acarapis woodi* has less impact on honeybee health. Other parasites also such as *Nosema ceranae*, *N. apis*, fungal, viral, and bacterial diseases have been found to be highly damaging to honeybee colonies [104]. One of the major problems of honeybee colonies is the high infection of the Deformed Wind Virus which transmitted by the *Varroa* mite, and the infection is wised spread in Egypt [106]. In the Arab world and the Middle East, Colony Collapse Disorder is recognized since 2006 [107] but there is clear published data for CCD in Egypt. There are some losses on the total number of honeybee colonies all over the country. Also, a lot of beekeepers have been quite their jobs due to their hives loss. Climate change and human interferences such as miss management of honey bee races have a great influence on the spreading and weakness of honeybees. With the less adequate number of honeybee colonies, our food will be less, and a lot of hunger problems will be created worldwide. Some

Fig. 2 Some pests of honeybee colonies in Egypt, **a–b** *Vespa orientalis*; **c** *Cataglyphus bicolor*; **d** *Senotainia tricuspis*. Photographed by M. Shebl

Table 3 Infestation rate of *Senotainia tricuspis* at different localities of N, E, W, and S of Egypt

Sample no.	Total bees no.	City	Coordinates	Coordinates	Infection percentage (%)	+
			N (North)	E (East)		–
1	35 (32 Healthy and 3 infected)	Alexandria	31°16′ N	29°88′ E	8.60	+
2	15 Healthy	Alexandria	31°16′ N	29°88′ E	0	–
3	97 Healthy	Alexandria	31°17′ N	30°13′ E	0	–
4	98 Healthy	Alexandria	31°17′ N	30°13′ E	0	–
5	25 Healthy	Ismailia	30°26′ N	32°16′ E	0	–
6	21 Healthy	Alexandria	31°16′ N	29°88′ E	0	–
7	76 Healthy	N Sinai	31°07′ N	33°46′ E	0	–
8	105 Healthy	N Sinai	31°07′ N	33°46′ E	0	–
9	125 Healthy	Alexandria	31°16′ N	29°88′ E	0	–
10	114 Healthy	Alexandria	31°16′ N	29°88′ E	0	–
11	97 Healthy	Qena	25°42′ N	32°38′ E	0	–
12	92 Healthy	Qena	25°42′ N	32°38′ E	0	–

Unpublished data by M. Shebl

Egyptian beekeepers have reported some symptoms of CCD [108] so it is a great challenge for bee scientists for figuring out some solutions for protecting honeybees and solitary bees. By this, we will assure better pollination services and more plant and crop production.

5.5 Honeybee and Pesticides

Several studies have highlighted the great impact of the application of chemical pesticides on honey bees and other pollinators. Pesticides are a key factor behind colony collapse disorder (CCD), directly and in tandem with two leading co-factors, pathogens and shortage of natural resources. Most pesticides are intended to serve as plant protection products (also known as crop protection products), which in general, protect plants from weeds, fungi, or insects. Prominent insecticide families include organochlorines, organophosphates, and carbamates. Potential exposure of bees to pesticides can vary greatly depending on the type of pesticide, formulation, application method, label restrictions, and other factors. Also, the exposure to the pesticides might have a sublethal effect on honeybees and solitary bees as well, and with long-term of exposure. This leads to the decline of bees and less pollination services [109–119].

The intensive use of different pesticides and, in particular, a new class of neuro-active insecticides, called neonicotinoid insecticides (NIs) play a significant role in honey bee colony losses [120]. Neuro-active insecticides are extremely dangerous to honeybees because they disturb the organism's neurobehavioral and immune system [121]. The mode of action of these compounds is to bind to the nicotine acetyl-choline receptors (nAChRs), acting agonistically, causing paralysis and death. NIs bind more strongly to nAChRs of invertebrates, making them more toxic at lower doses than other compounds. The sub-lethal toxicity of NIs is of particular concern because the most common field exposure scenarios are likely at the sublethal rather than acute level. It includes behavioral disruptions such as disorientation, reduced foraging, impaired memory and learning, and shifts in communication behaviors. Other important sub-lethal effects might include compromised immunity, delayed development and a host of indirect, potentially cascading effects that impact the hive's ability to sustain itself, leading to large-scale population extinctions in bee populations.

Egypt, particularly the Nile Delta, is the most important honey and honey-producing region in the Middle East and North Africa region with an estimated 1.3 million hives [122]. There have been reports in recent decades of great losses in honey bee colonies for unspecified reasons [114]. However, there is increasing evidence that the intensive use of pesticides, especially NIs is one of the most important factors that have led to such recent losses. According to the statistics of the Agricultural Pesticides Committee, Egypt consumed about 10,600 metric ton of pesticides (active ingredient) during 2016 which represent 0.2% of the global consumption (5 million metric tons of a value of 52 billion dollars). A number of registered pesticides based on trade name recorded 1159 from about 264 common names at 2016 [123]. However, there is limited information upon the use of NIs in Egypt, concentrations and frequency of detection indicates that use is similar to that of the US and European farms [124].

6 Newly Approaches for Bee and Other Pollinators Conservation

There are a lot of efforts from several scientists, conservationists, environmentalists and others for protecting and conserving plant pollinators with a particular interest with honey and solitary bees. Many current projects, programs and organizations are working to conserve pollinating fauna worldwide. Some closely related projects include the CBD (Convention on Biological Diversity), the IUCN (The World Conservation Union), the WWF (World Wildlife Fund), the FAO (Food and Agriculture Organization), the NAPPC (The North American Pollinator Protection Campaign), Status and Trends of European Pollinators (STEP), Integrated Crop Pollination (ICP, USDA), USA and the CANPOLIN (Canadian Pollination Initiative).

Here we are summarizing some of our efforts for protecting the habitats of some solitary bees present in Egypt. In the last decade, we noticed that several cavity-nesting bees are in decline due to the natural habitat fragmentation, degradation and destruction. Several artificial nesting materials (e.g. wood, paper straws and polystyrene foam) were evaluated for nesting several bee species in the Canal region as conservation efforts of wild bee fauna of Egypt to ensure better pollination services of some cultivated crops. These nests were used as traps in the natural nesting areas and established at the Bee Research Center of the Suez Canal University (Fig. 3). Three bee species have been nested successfully in the artificial nests (cavity-nesting bees). All of them belong to the family Megachilidae including *Chalicodoma siculum* (Rossi), *Megachile minutissima* (Radoszkowski, 1876) and *Osmia latreillei* (Spinola). The three species were used for pollination of three major crops in Egypt broad bean, alfalfa and sunflower [125, 126, 127, 128, 129, 130, 131, 91].

Last but not least, the global decline of bees is a real problem and will have a great influence on our food security. Therefore, a political welling must be taken into considerations for several governments and in particular in Egypt. Several efforts must be taken into account such as enhancing the health of pollinators "honeybees and solitary bees", avoid extensive using of chemical pesticides which harm to pollinators, encourage local farmers for cultivating bee-friendly plants and improving the nutrition of bees and protecting the natural nesting habitats of solitary bees. Fundraising should be subjected for solving several problems of honeybee health, a decline of bee pollinators and the importance of the pollination services for our ecosystem.

Fig. 3 **a–d** Artificial made nesting structures for Megachilid bees in Egypt; **e**, **f** *Osmia latreillei* Spinola nesting in wooden nests and pollinate sunflower. Photographed by M. Shebl

7 Potential Mitigation Measures of Climate Change Impacts on Pollinators

Many studies in the recent years have addressed the adverse impact of climate change on crop pollination ecosystem services, especially the pollination process. However, little progress has been made in identifying and testing mitigation measures for maintaining the native pollinator communities, protecting against declines and enhancing their role in pollinating crops. Risk mitigation usually focused on honey bees, while efforts for non-managed bees were negligible. Actions that maintain or restore the diversity of pollinators are therefore critically important. One of the most important measures to be taken into consideration to reduce the impact of climate change on pollinators is to follow the strategy of ecological intensification, or the improvement of crop yield through enhancement of biodiversity, which may be a sustainable pathway toward greater food supplies. Ecological intensification of agriculture represents a strategic alternative to improve these drivers of pollinator decline while supporting sustainable food production, by promoting biodiversity beneficial to agricultural production through management practices such as intercropping, crop rotations, farm-level diversification and reduced agrochemical use. Ecological intensification in arable fields has been demonstrated to enhance within-field species richness and abundance of plants considered weeds, by decreasing the use of inorganic fertilizers and herbicides [132, 133]. Such measures can build highly diverse pollinator community that is more resilient to climate change.

Therefore, it is important to enact policies and implement actions that provide different floral resources, including agricultural, urban, pastoral, and plantation landscapes. Such plant diversity can support high diversity of pollinators under proper management (e.g., [134–138]). Maintaining the diversity and abundance of floral resources is a key factor in maintaining the diversity of pollinators [139]. For added resilience to climate, specific attention should be paid to ensure that a phenological sequence of plants by switching between crop species and varieties is established that provide bloom along the activity season of insect pollinators [140, 141].

In this context, longer-term monitoring of multiple pressures of climate change on pollinators and pollinator-dependent plants will help elucidate the mechanisms underpinning observed shifts, and assessment of mitigation and adaptation options for these impacts.

8 Conclusion

Pollinators play a key role in food security due to their substantial intrinsic value by providing pollination services for agriculture ecosystems. Bees are considered the most important pollinators. Managed honeybees and other wild bee species have an enormous, unchallenged impact as plant pollinators, both for wild plants and crops. The value of annual global food production relies on direct contributions by pollinators is substantial, with estimates between 235 billion and 577 billion dollars. Recently, there is growing concern about the decline in insect pollinators across the world. Pollinator decline has become a worldwide issue as its potential consequences to global food production. The decline in bees is caused by multiple stresses, including habitat fragmentation and loss, exposure to pathogens, parasites and pesticides, shortage of food sources, and the impacts of climate change. Climate change considered the most relevant factor responsible for the decline of insect pollinators. It is likely to affect the activity of insect pollinators, thereby influencing its role in crop pollination. Climate change also is likely to alter the pollinator lifecycle, distribution and their interaction with the prevalence of pests and diseases, affect the bee-plant interaction, the mismatch between the flowering time and seasonal activity of bees, lack of floral resources, biological invasion, habitat loss, degradation and fragmentation. Thus, the impact of climate change can often lead to increased risk of extinction for bees and some native insect pollinators.

It is worth bearing in mind that the insect pollinators are closely linked to human well-being through the maintenance of ecosystem, as well as immediate pollination of wild plants, crop production and assure food security. Therefore, it is strongly recommended to mitigate the potential adverse impacts of climate change on the diversity and efficiency of insect pollinators especially, managed honeybees and other solitary bee species. So that, it is necessary to urge conservationists, ecologists, environmentalists, researchers and policymakers to seriously figure out some appropriate solutions for protecting plant pollinators and in particular bees.

9 Recommendations

In this context, the current study suggests several recommendations and new approaches that can be taken out for protecting our native bees from the potential risks of climate change that will be listed in the following points:

- Raising awareness on the adverse impacts of climate change on ecosystem services and food security;
- Enhance the public appreciation of the importance of natural pollination by bees as one of the most substantial ecosystem services;
- Increasing awareness of bee pollinator diversity and what need to survive and richness;
- Fill the knowledge gaps in pollinator ecology, declines and intact management;
- Encourage public engagement in the protection and conservation of honeybees and other wild pollinators;
- Conserving and restoring natural nesting habitats of solitary bees and enhancing nesting materials and nesting sites by creating bee hotels (artificial nests) for the propagation of the most promising solitary bees;
- Build on the latest scientific researches in increasing the tolerance of bees against pests and diseases with special concern to the cutting edge approach to protecting honeybees against *Varroa* mite;
- Encourage farm owners for cultivation bee-friendly plants and other crops with a continuous succession of flowering;
- Promote the ecological intensification strategy such as intercropping, crop rotations, farm-level diversification and reduced agrochemical use for promoting biodiversity beneficial to agricultural production.

References

1. Waser NM (2005) Specialization and generalization in plant-pollinator interactions. Chicago University Press, pp 1–441
2. Michener CD (2007) The world of the bees, 2nd edn. Johns Hopkins University Press, Baltimore, xvi+953 pp
3. Buchmann SL, Nabhan GP (1996) The forgotten pollinators. Island Press, Washington, D.C.
4. Dafni A, O'Toole C (1994) Pollination syndromes in the Mediterranean: generalizations and peculiarities. In: Arianoutsou M, Groves R (eds) Plant-animal interactions in Mediterranean-type ecosystems. Tasks for vegetation science, vol 34. Dordrecht, Boston & London, pp 125–135
5. Daily GC (1997) Nature's services. Societal dependence on natural ecosystems. Island Press, Washington, D.C., 392 pp
6. Freitas BM, Paxton RJ (1998) A comparison of two pollinators: the introduced honey bee *Apis mellifera* and an indigenous bee *Centris tarsata* on cashew *Anacardium occidentale* in its native range of NE Brazil. J Appl Ecol 35(1):109–121
7. Garibaldi LA, Dewenter IS, Winfree R, Aizen MA, Bommarco R, Cunningham SA, Kremen C, Carvalheiro LG, Harder LD, Afik O, Bartomeus I, Benjamin F, Boreux V, Cariveau D, Chacoff NP, Dudenhöffer JH, Freitas BM, Ghazoul J, Greenleaf S, Hipólito J, Holzschuh A, Howlett B, Isaacs R, Javorek SK, Kennedy CM, Krewenka KM, Krishnan S, Mandelik Y, Mayfield MM, Motzke I, Munyuli T, Nault BA, Otieno M, Petersen J, Pisanty G, Potts SG, Rader R, Ricketts TH, Rundlöf M, Seymour CL, Schüepp C, Szentgyörgyi H, Taki H, Tscharntke T, Vergara CH, Viana BF, Wanger TC, Westphal C, Williams N, Klein AM (2013) Wild pollinators enhance fruit set of crops regardless of honey bee abundance. Science 339:1608–1611
8. James RR, Pitts-Singer TL (2008) Bee pollination in agricultural ecosystems. Oxford University Press, New York, 232 pp
9. Greenleaf SS, Kremen C (2006) Wild bees enhance honey bees' pollination of hybrid sunflower. Proc Natl Acad Sci USA 103:13890–13895
10. Chagnon M, Gingras J, Deoliveira D (1993) Complementary aspects of strawberry pollination by honey and indigenous bees (Hymenoptera). J Econ Entomol 86:416–420
11. Brown, MJF, Paxton RJ (2009) The conservation of bees: a global perspective. Apidologie 40:410–416
12. Dicks LV, Abrahams A, Atkinson J, Biesmeijer J, Bourn N, Brown C, Brown JFM, Carvell C, Connolly C, Cresswell EJ, Croft P, Darvill B, De Zylva P, Effingham P, Fountain M, Goggin A, Harding D, Harding T, Hartfield C, Heard SM, Heathcote R, Heaver D, Holland J, Howe M, Hughes B, Huxley T, Kunin W, Little J, Mason C, Memmott J, Osborne J, Pankhurst T, Paxton JR, Pocock JOM, Potts GS, Power FE, Raine EN, Ranelagh E, Roberts S, Saunders R, Smith K, Smith MR, Sutton P, Tilley ANL, Tinsley A, Tonhasca A, Vanbergen JA, Webster S, Wilson A, Sutherland JW (2013) Identifying key knowledge needs for evidence-based conservation of wild insect pollinators: a collaborative cross-sectoral exercise. Insect Conserv Divers 6:339–353
13. Potts SG, Biesmeijer JC, Kremen C, Neumann P, Schweiger O, Kunin WE (2010) Global pollinator declines: trends, impacts and drivers. Trends Ecol Evol 25(6):345–353. https://doi.org/10.1016/j.tree.2010.01.007
14. Vanbergen AJ, Baude M, Biesmeijer JC, Britton NF, Brown MJF, Bryden J, Budge GE, Bull JC, Carvell C, Challinor AJ, Connolly CN, Evans DJ, Feil EJ, Garratt MP, Greco MK, Heard MS, Jansen VAA, Keeling MJ, Kunin WE, Marris GC, Memmott J, Murray JT, Nicolson SW, Osborne JL, Paxton RJ, Pirk CWW, Polce C, Potts SG, Priest NK, Raine NE, Roberts S, Ryabov EV, Shafir S, Shirley MDF, Simpson SJ, Stevenson PC, Stone GN, Termansen M, Wright GA (2013) Threats to an ecosystem service: pressures on pollinators. Front Ecol Environ 11:251–259
15. TEEB (2010) The economics of ecosystems and biodiversity: ecological and economic foundation. Earthscan, Cambridge

16. MEA (Millennium Ecosystem Assessment) (2005) Ecosystems and human well-being: synthesis. Island Press, Washington, DC, 2005
17. Pudasaini R, Chalise M, Poudel PR, Pudasaini K, Aryal P (2015) Effect of climate change on insect pollinator: a review. N Y Sci J 8(3):39–42
18. Albrechdt M, Schmid B, Hautier Y, Müller CB (2012) Diverse pollinator communities enhance plant reproductive success. Proc R Soc Lond B Biol Sci 279(1748):4845–4852. https://doi.org/10.1098/rspb.2012.1621
19. Breeze TD, Vaissière BE, Bommarco R, Petanidou T, Seraphides N, Kozák L, Scheper J, Biesmeijer JC, Kleijn D, Gyldenkærne S, Moretti M, Holzschuh A, Steffan-Dewenter I, Stout JC, Pärtel M, Zobel M, Potts SG (2014) Agricultural policies exacerbate honeybee pollination service supply-demand mismatches across Europe. PLoS One 9(1):e2099. https://doi.org/10.1371/journal.pone.0082996
20. Klein A-M, Vaissière BE, Cane JH, Steffan-Dewenter I, Cunningham SA, Kremen C, Tscharntke T (2007) Importance of pollinators in changing landscapes for world crops. Proc R Soc Lond B Biol Sci 274(1608):303–313. https://doi.org/10.1098/rspb.2006.3721
21. Prescott-Allen R, Prescott-Allen C (1990) How many plants feed the world? Conserv Biol 4(4):365–374. https://doi.org/10.1111/j.1523-1739.1990.tb00310.x
22. Roubik DW (1995) Pollination of cultivated plants in the tropics. FAO, Rome
23. Winfree R, Bartomeus I, Cariveau DP (2011) Native pollinators in anthropogenic habitats. Annu Rev Ecol Evol Syst 42(1):1–22. https://doi.org/10.1146/annurev-ecolsys-102710-145042
24. Aguilar R, Ashworth L, Galetto L, Aizen M (2006) Plant reproductive susceptibility to habitat fragmentation: review and synthesis through a meta-analysis. Ecol Lett 9:968–980
25. Kremen C, Williams NM, Aizen MA, Gemmill-Herren B, LeBuhn G, Minckley R, Packer L, Potts SG, Roulston T, Steffan-Dewenter I, Vázquez DP, Winfree R, Adams L, Crone EE, Greenleaf SS, Keitt TH, Klein AM, Regetz J, Ricketts TH (2007) Pollination and other ecosystem services produced by mobile organisms: a conceptual framework for the effects of land-use change. Ecol Lett 10:299–314
26. White GM, Boshier DH, Powell W (2002) Increased pollen flow counteracts fragmentation in a tropical dry forest: an example from *Swietenia humilis* Zuccarini. Proc Natl Acad Sci USA 99:2038–2042
27. Bauer DM, Wing IS (2010) Economic consequences of pollinator declines: a synthesis. Agric Resour Econ Rev 39(3):368–383
28. Tirado R, Simon G, Johnston P (2013) Bees in decline: a review of factors that put pollinators and agriculture in Europe at risk. Greenpeace Research Laboratories report (review) 1
29. LeBuhn G, Droege S, Connor E, Gemmill-Herren B, Azzu N (2016) Protocol to detect and monitor pollinator communities: guidance for practitioner. Food and Agriculture Organization of the United Nations, pp 1–64
30. Johannsmeier MF (2001) Beekeeping in South Africa, 3rd edn. Plant Protection Research Institute, Pretoria
31. Kevan P, Phillips T (2001) The economic impacts of pollinator declines: an approach to assessing the consequences. Conserv Ecol 5(1):8
32. Morse RA, Calderone NW (2000) The value of honey bees as pollinators of U.S. crops in 2000. Bee Cult 128:15 pp inserts
33. Robinson WS, Nowogrodzki R, Morse RA (1989) The value of honey bees as pollinators of USA crops. Part II of a two-part series. Am Bee J 129:477–487
34. Southwick EE, Southwick LJR (1992) Estimating the economic value of honey bees (Hymenoptera: Apidae) as agricultural pollinators in the United States. J Econ Entomol 85(3):621–633
35. EPA (2017) Climate impacts on ecosystems. United States Environmental Protection Agency. https://19january2017snapshot.epa.gov/climate-impacts/climate-impacts-ecosystems_.html
36. Kevan PG, Imperatriz-Fonseca VL (2002) Pollinating bees: the conservation link between agriculture and nature. Ministry of Environment, Brasilia-DF, Brazil

37. Nabhan GP, Buchmann SL (1997) Services provided by pollinators. In: Daily GC (ed) Nature's services: societal dependence on natural ecosystems. CoVelo: Island Press, Washington, D.C.
38. Gallai N, Salles J-M, Settele J, Vaissière BE (2009) Economic valuation of the vulnerability of world agriculture confronted with pollinator decline. Ecol Econ 68:810–821
39. Hegland SJ, Nielsen A, Lázaro A, Bjerknes AL, Totland Ø (2009) How does climate warming affect plant-pollinator interactions? Ecol Lett 12:184–195
40. Memmott J, Craze PG, Waser NM, Price MV (2007) Global warming and the disruption of plant-pollinator interactions. Ecol Lett 10:710–717
41. Schweiger O, Biesmeijer JC, Bommarco R, Hickler T, Hulme P, Klotz S, Kühn I, Moora M, Nielsen A, Ohlemuller R, Petanidou T, Potts SG, Pysek P, Stout JC, Sykes M, Tscheulin T, Vilà M, Wather GR, Westphal C (2010) Multiple stressors on biotic interactions: how climate change and alien species interact to affect pollination. Biol Rev 85:777–795
42. Bartomeus I, Park MG, Gibbs J, Danforth BN, Lakso AN, Winfree R (2013) Biodiversity ensures plant-pollinator phenological synchrony against climate change. Ecol Lett 16:1331–1338
43. Papanikolaou AD, Kühn I, Frenzel M, Schweiger O (2016) Semi-natural habitats mitigate the effects of temperature rise on wild bees. J Appl Ecol 54(2):527–536
44. Thomson DM (2016) Local bumble bee decline linked to recovery of honey bees, drought effects on floral resources. Ecol Lett 19:1247–1255
45. IPCC (2014) Climate change 2014: mitigation of climate change. In: Edenhofer O, Pichs-Madruga R, Sokona Y, Farahani E, Kadner S, Seyboth K, Adler A, Baum I, Brunner S, Eickemeier P, Kriemann B, Savolainen J, Schlömer S, von Stechow C, Zwickel T, Minx JC (eds) Contribution of working group III to the fifth assessment report of the intergovernmental panel on climate change. Cambridge University Press, Cambridge
46. IPCC (2007a) Climate change 2007: synthesis report. In: Core Writing Team, Pachauri RK, Reisinger A (eds) Contribution of working groups I, II and III to the fourth assessment report of the intergovernmental panel on climate change. IPCC, Geneva, Switzerland, 104 pp
47. Kjøhl M, Nielsen A, Stenseth NC (2011) Potential effects of climate change on crop pollination. Food and Agriculture Organization of the United Nations (FAO), Rome
48. Sutherst RW, Maywald GF, Bourne AS (2007) Including species interactions in risk assessments for global change. Glob Change Biol 13:1843–1859
49. van der Putten WH, de Ruiter PC, Bezemer TM, Harvey JA, Wassen M, Wolters V (2004) Trophic interactions in a changing world. Basic Appl Ecol 5:487–494
50. Potts SG, Imperatriz-Fonseca V, Ngo HT, Aizen MA, Biesmeijer JC, Breeze TD, Dicks LV, Garibaldi LA, Hill R, Settele J, Vanbergen AJ (2016) Safeguarding pollinators and their values to human well-being. Nature 540(7632):220–229. https://doi.org/10.1038/nature20588
51. Cardinale BJ, Duffy JE, Gonzalez A, Hooper DU, Perrings C, Venail P, Narwani A, Mace GM, Tilman D (2012) Biodiversity loss and its impact on humanity. Nature 486(7401):59–67. https://doi.org/10.1038/nature11148. PMID 22678280
52. Hughes L (2000) Biological consequences of global warming: is the signal already apparent? Trends Ecol Evol 15:56–61
53. Parmesan C (2006) Ecological and evolutionary responses to recent climate change. Annu Rev Ecol Evol Syst 37:637–669
54. Walther G-R, Post E, Convey P, Menzel A, Parmesan C, Beebee TJ, Fromentin JM, Hoegh-Guldberg O, Bairlein F (2002) Ecological responses to recent climate change. Nature 416:389–395
55. Gornall J, Betts R, Burke E, Clark R, Camp J, Willett K, Wiltshire A (2010) Implications of climate change for agricultural productivity in the early twenty-first century. Philos Trans R Soc B Biol Sci 365(1554):2973–2989. https://doi.org/10.1098/rstb.2010.0158
56. IPCC (2007b) Climate change 2007: impacts, adaptation and vulnerability. In: Parry ML, Canziani OF, Palutikof JP, van der Linden PJ, Hanson CE (eds) Contribution of working group II to the fourth assessment report of the intergovernmental panel on climate change. Cambridge University Press, Cambridge, UK, 976 pp

57. World Bank (2016) Climate change: likely impacts on African crops, livestock & farm types. http://go.worldbank.org/NKNKPX8FL0
58. Andy J, Upadhyaya H, Gowda CLL, Aggarwal PK, Fujisaka S, Anderson B (2008) Climate change and its effect on conservation and use of plant genetic resources for food and agriculture and associated biodiversity for food security. Monograph. Food and Agriculture Organization of the United Nations, UK
59. Dale VH, Joyce LA, McNulty S, Neilson RP, Ayres MP, Flannigan MD, Hanson PJ, Irland LC, Lugo AE, Peterson CJ, Simberloff D, Swanson FJ, Stocks BJ, Wotton BM (2001) Climate change and forest disturbances. Bioscience 51:723–734
60. Doney SC, Ruckelshaus M, Duffy JE, Barry JP, Chan F, English CA, Galindo HM, Grebmeier JM, Hollowed AB, Knowlton N, Polovina J, Rabalais NN, Sydeman WJ, Talley LD (2012) Climate change impacts on marine ecosystems. Annu Rev Mar Sci 4:11–37. https://doi.org/10.1146/annurev-marine-041911-111611
61. Fitter AH, Fitter RSR (2002) Rapid changes in flowering time in British plants. Science 296(5573):1689–1691. https://doi.org/10.1126/science.1071617. PMID: 12040195
62. Inouye DW (2008) Effects of climate change on phenology, frost damage, and floral abundance of montane wildflowers. Ecology 89(2):353–362. https://doi.org/10.1890/06-2128.1
63. Roberts EH, Hadley P, Summerfield RJ (1985) Effects of temperature and photoperiod on flowering in chickpeas (*Cicer arietinum* L.). Ann Bot 55:881–892
64. Wallace DH, Yan W (1998) Plant breeding and whole-system physiology. Improving adaptation, maturity and yield. CAB International, Wallingford, UK, 390 pp
65. Rosenzweig C, Karoly D, Vicarelli M, Neofotis P, Wu Q, Casassa G, Menzel A, Root TL, Estrella N, Seguin B, Tryjanowski P, Liu C, Rawlins S, Imeson A (2008) Attributing physical and biological impacts to anthropogenic climate change. Nature 453:353–357. https://doi.org/10.1038/nature06937
66. Estrella N, Sparks T, Menzel A (2007) Trends and temperature response in the phenology of crops in Germany. Glob Change Biol 13:1737–1747
67. Hu Q, Weiss A, Feng S, Baenziger P (2005) Earlier winter wheat heading dates and warmer spring in the U.S. Great Plains. Agric For Meteorol 135:284–290
68. Menzel A, Sparks T, Estrella N, Koch E, Aasa A, Ahas R, Alm-Kübler K, Bissolli P, Braslavská O, Briede A, Chmielewski FM, Crepinsek Z, Curnel Y, Dahl Å, Defila C, Donnelly A, Filella Y, Jatczak K, Måge F, Mestre A, Nordli Ø, Peñuelas J, Pirinen P, Remišová V, Scheifinger H, Striz M, Susnik A, Van Vliet AJH, Wielgolaski F-E, Zach S, Zust A (2006) European phenological response to climate change matches the warming pattern. Glob Change Biol 12:1969–1976
69. Tao F, Yokozawa M, Xu Y, Hayashi Y, Zhang Z (2006) Climate changes and trends in phenology and yields of field crops in China, 1981–2000. Agric For Meteorol 138:82–92
70. Williams T, Abberton M (2004) Earlier flowering between 1962 and 2002 in agricultural varieties of white clover. Oecologia 138:122–126
71. Sadras VO, Connor DJ (1991) Physiological basis of the response of harvest index to the fraction of water transpired after anthesis: a simple model to estimate harvest index for determinate species. Field Crops Res 26:227–239
72. Moriondo M, Bindi M (2007) Impact of climate change on the phenology of typical Mediterranean crops. Ital J Agrometeorol 3:5–12
73. Bindi M, Moriondo M (2005) Impact of a 2°C global temperature rise on the Mediterranean region: agriculture analysis assessment. In: Climate change impacts in the Mediterranean resulting from a 2°C global temperature rise. A WWF report. http://assets.panda.org/downloads/medreportfinal8july05.pdf
74. Franks SJ, Sim S, Weis AE (2007) Rapid evolution of flowering time by an annual plant in response to a climate fluctuation. Natl Acad Sci USA. PNAS 104(4):1278–1282. https://doi.org/10.1073/pnas.0608379104
75. McCarthy JJ, Canziani OF, Leary NA, Dokken DJ, White KS (2001) Climate change 2001: impacts, adaptation & vulnerability, contribution of working group II to the third assessment report of the intergovernmental panel on climate change (IPCC). Cambridge University Press, 1000 pp

76. Barnabás B, Järgen K, Fehér A (2008) The effect of drought and heat stress on reproductive processes in cereals. Plant Cell Environ 31:11–38
77. Morison JIL, Lawlor DW (1999) Interactions between increasing CO_2 concentration and temperature on plant growth. Plant Cell Environ 22:659–682
78. Shebl M, Kamel S, Mahfouz H (2013) Bee fauna (Apoidea: Hymenoptera) of the Suez Canal region, Egypt. J Apic Sci 57(1):33–44
79. Blüthgen P (1933) Ein beitrag zur kenntnis der bienenfauna Agyptens (Hymenoptera: Apidae-Halictidae-Halictinae). Bull Soc R Entomol Egypte 17:1–27
80. Blüthgen P (1934) Zweiter beitrag zur kenntnis der Halictinenfauna Agyptens (Hymenoptera: Apidae-Halictidae-Halictinae). Bull Soc R Entomol Egypte 18:188–200
81. El Akkad MK, Kamel SM (2002) Revision of the genus *Sphecodes latreille* of Egypt (Hymenoptera: Halictidae) Egypt. Agric Res J 80(4):1579–1595
82. Ibrahim MM (1973) Breeding and propagation of some efficient insect pollinators in newly reclaimed land in Egypt. Final technical report, pp 1–75
83. Ibrahim MM (1979) Breeding and propagation of some efficient insect pollinators in newly reclaimed land in Egypt. Project report no. 4, pp 1–67
84. Moustafa MA, El Berry AA (1976) Morphological studies on *Osmia submicans*. Mitt Zool Mus Berlin 52:209–225
85. Moustafa MA, El-Hefny AM, Abd El-Salam AL, Salem MM (1979) Taxonomical studies on some common *Andrena* spp. In: Egypt. 185-bee symposium. (Affiliated to 3rd Arab pesticide conference), Tanta University, pp 185–238
86. Priesner H (1957) A review of the *Anthophora* species of Egypt. Bull Soc Entomol Egypte, XLI, pp 1–115
87. Rashad SE (1979) Utilization of non-*Apis* bees as crop pollinators. Annual report no. II, pp 1–40
88. Rashad SE (1980) Utilization of non-*Apis* bees as a crop pollinator. Annual report no. III, pp 1–34
89. Shebl AM, Patiny S, Michez D (2015) Supplementary note on the solitary bee fauna from the Suez Canal region of Egypt (Hymenoptera: Apoidea). J Melittol 47:1–5
90. Shebl MA, Mahmoud MF (2015) The bee diversity (Hymenoptera: Apoidea) visiting broad bean (*Vicia faba* L.) flowers in Egypt. Zool Middle East 61(3):256–263. https://doi.org/10.1080/09397140.2015.1069245
91. Shoukry AM, El Akkad MK, Kamel SM, El Akkad S (2004) Revision of the subfamily Nomiinae of Egypt (Hymenoptera: Halictidae). Agric Res J Suez Canal Univ 3(2):145–158
92. Deutsch CA, Tewksbury JJ, Huey RB, Sheldon KS, Ghalambor CK, Haak DC, Martin PR (2008) Impacts of climate warming on terrestrial ectotherms across latitude. Proc Natl Acad Sci USA 105:6668–6672
93. Shebl MA (2017) Discovery of *Apis florea* colonies in northeastern Egypt. Afr Entomol 25(1):248–249
94. Kamel SM, Abu Hashesh TA, Osman MA, Shebl MA (2007) A new model of polystyrene foam for renesting leafcutting bees (*Megachile* spp., Megachilidae, Hymenoptera). Agric Res J Suez Canal Univ 7(2):97–101
95. Shebl MA, Al Aser RM, Ibrahim A (2016) Nest architecture and nest soil characteristics of long-horned bee *Eucera nigrilabris* (Hymenoptera, Apidae). Sociobiology 63(4):1031–1037
96. Forrest JRK (2015) Plant—pollinator interactions and phenological change: what can we learn about climate impacts from experiments and observations? Oikos 124:4–13. https://doi.org/10.1111/oik.01386
97. Gordo O, Sanz JJ (2005) Phenology and climate change: a long-term study in a Mediterranean locality. Oecologia 146:484–495
98. Kamel SM (1981) Studies on insect pollinators at Ismailia Governorate with special reference to the biology and ecology of *Anthophora atriceps* (Hymenoptera: Anthophoridae). M.Sc. thesis, Faculty of Agriculture, Cairo University, Egypt
99. Osman MAM (1998) Ecological studies on some insect pollinators of broad beans *Vicia faba* with special reference to *Chalicodoma siculum* Rossi. M.Sc. thesis, Faculty of Agriculture, Suez Canal University, Ismailia, Egypt

100. Graystock P, Yates K, Darvill B, Goulson D, Hughes WOH (2013a) Emerging dangers: deadly effects of an emergent parasite in a new pollinator host. J Invertebr Pathol
101. Graystock P, Yates K, Evison SEF, Darvill B, Goulson D, Hughes WOH (2013b) The Trojan hives: pollinator pathogens, imported and distributed in bumblebee colonies. J Appl Ecol
102. Murray TE, Coffey MF, Kehoe E, Horgan FG (2013) Pathogen prevalence in commercially reared bumble bees and evidence of spillover in conspecific populations. Biol Conserv 159:269–276
103. Singh R, Levitt AL, Rajotte EG, Holmes EC, Ostiguy N, van Engelsdorp D, Lipkin WL, dePamphilis CW, Toth AL, Cox-Foster DL (2010) RNA viruses in Hymenopteran pollinators: evidence of inter-taxa virus transmission via pollen and potential impact on non-*Apis* Hymenopteran species. PLoS One 5:e14357
104. Pirk CW, Strauss U, Yusuf A, Démares F, Human H (2015) Honeybee health in Africa—a review. Apidologie. https://doi.org/10.1007/s13592-015-0406-6
105. Haddad N, Adjlane N, Loucif-Ayad W, Shebl MA, Saba M, Albaba I, El-Obeid D, Sabah M, Giusti M, Felicioli A (2015) Presence and infestation rate of *Senotainia tricuspis* (Meigen) (Diptera, Sarcophagidae) on honey bees in the Mediterranean region. J Apic Res 54(2):121–122
106. Haddad N, Noureddine A, Al-Shagour B, Loucif-Ayad W, El-Niweiri MA, Anaswah E, Hammour WA, El-Obeid D, Imad A, Shebl MA, Almaleky AS, Nasher A, Walid N, Bergigui MF, Yañez O, de Miranda JR (2017) Distribution and variability of deformed wing virus of honeybees (*Apis mellifera*) in the Middle East and North Africa. Insect Sci 24:103–113. https://doi.org/10.1111/1744-7917.12277
107. Haddad N (2011) Honey bee viruses, disease and hive management in the Middle East and their relation to the colony collapse disorder and bee losses. Uludag Bee J 11(1):17–24
108. Hassan AR (2009) Proceedings of the 4th COLOSS conference
109. Alaux C, Ducloz F, Crauser D, Le Conte Y (2010) Diet effects on honeybee immunocompetence. Biol Let 6(4):562–565
110. Blacquiere T, Smagghe G, van Gestel CAM, Mommaerts V (2012) Neonicotinoids in bees: a review on concentrations, side-effects and risk assessment. Ecotoxicology 21:973–992
111. Colin ME, Bonmatin JM, Moineau I, Gaimon C, Brun S, Vermandere JP (2004) A method to quantify and analyze the foraging activity of honey bees: relevance to the sublethal effects induced by systemic insecticides. Arch Environ Contam Toxicol 47:387–395
112. Desneux N, Decourtye A, Delpuech JM (2007) The sublethal effects of pesticides on beneficial arthropods. Annu Rev Entomol 52:81–106
113. Heylen K, Gobin B, Arckens L, Huybrechts R, Billen J (2010) The effects of four crop protection products on the morphology and ultrastructure of the hypopharyngeal gland of the European honeybee, *Apis mellifera*. Apidologie. https://doi.org/10.1051/apido/2010043
114. Roat TC, Carvalho SM, Nocelli RCF, Silva-Zacarin ECM, Palma MS, Malaspina O (2013) Effects of sublethal dose of fipronil on neuron metabolic activity of Africanized honeybees. Arch Environ Contam Toxicol 64:456–466
115. Rossi CA, Roat TC, Tavares DA, Cintra-Socolowski P, Malaspina O (2013) Brain morphophysiology of Africanized bee *Apis mellifera* exposed to sublethal doses of imidacloprid. Arch Environ Contam Toxicol 65:234–243
116. Schmuck R (2004) Effects of a chronic dietary exposure of the honeybee *Apis mellifera* (Hymenoptera: Apidae) to imidacloprid. Arch Environ Contam Toxicol 47:471–478
117. Williamson SM, Wright GA (2013) Exposure to multiple cholinergic pesticides impairs olfactory learning and memory in honeybees. J Exp Biol 216(10):1799–1807
118. Williamson SM, Baker DD, Wright GA (2013) Acute exposure to a sublethal dose of imidacloprid and coumaphos enhances olfactory learning and memory in the honeybee *Apis mellifera*. Invert Neurosci 13:63–70
119. Yang EC, Chuang YC, Chen L, Chang LH (2008) Abnormal foraging behavior induced by sublethal dosage of imidacloprid in the honey bee (Hymenoptera: Apidae). J Econ Entomol 101:1743–1748

120. Smith KM, Loh EH, Rostal MK, Zambrana-Torrelio CM, Mendiola L, Daszak P (2013) Pathogens, pests, and economics: drivers of honey bee colony declines and losses. EcoHealth 10:434–445. https://doi.org/10.1007/s10393-013-0870-2
121. Schacker M (2008) A spring without bees: how colony collapse disorder has endangered our food supply. The Lyons Press, Guilford, Connecticut
122. Al Naggar Y, Codling G, Vogt A, Naiem E, Mohamed M, Seif A, Giesy JP (2015) Organophosphorus insecticides in honey, pollen and bees (*Apis mellifera* L.) and their potential hazard to bee colonies in Egypt. Ecotoxicol Environ Saf 114:1–8
123. APC (2016) The Agricultural Pesticides Committee. Ministry of Agriculture and Land Reclamation, Egypt
124. Codling G, Al Naggar Y, Giesy JP, Robertson AJ (2016) Concentrations of neonicotinoid insecticides in honey, pollen and honey bees (*Apis mellifera* L.) in central Saskatchewan, Canada. Chemosphere 144:2321–2328
125. Hassan AH, Kamel SM, Osman MAM, Mahmoud MF, Bedeir EH, Shebl MA (2016a) Conservation of the mason bees *Osmia latrellei* Spinola (Hymenoptera Megachilidae) in Egypt. Egypt J Appl Sci 31(6):73–84
126. Hassan HH, Kamel SM, Osman MAM, Mahmoud MF, Bedeir EH, Shebl MA (2016b) Ecological notes on the nest construction and cell provisioning of *Osmia latrellei* Spinola (Hymenoptera: Megachilidae) in Egypt. Egypt J Appl Sci 31(6):85–95
127. Shebl MA, Kamel SM, Abu Hashesh TA, Osman MA (2008a) Seasonal abundance of leafcutting bees (*Megachile minutissima*, Megachilidae, Hymenoptera). World J Agric Sci 4(2):280–287
128. Shebl MA, Kamel SM, Abu Hashesh TA, Osman MA (2008b) The impact of leafcutting bees (*Megachile minutissima*, Megachilidae, Hymenoptera) (Radoszkowski, 1876) artificial nest sites on seed production of alfalfa, Ismailia, Egypt. Agricultura 5:33–35
129. Shebl MA, Kamel SM, Abu Hashesh TA, Osman MA (2009a) The most common insect species in Alfalfa field in Egypt. Acad J Entomol 1(2):27–31
130. Shebl MA, Kamel SM, Abu Hashesh TA, Osman MAM (2009b) The impact of using leafcutting bees (Megachilidae: Hymenoptera) with different fertilization treatments on alfalfa seed production. J Soil Plant Nutr 9(2):134–141
131. Shoukry AA, Kamel SM, Abu-Hashish TA, Osman MA (1997) Ecological studies of mason bee *Chalicaodoma siculum* Rossi at Ismailia Governorate. In: Proceeding of 7th national conference of pests & diseases of vegetables & fruits, vol 2, Ismailia, Egypt, pp 362–379
132. Kleijn D, Kohler F, Báldi A, Batáry P, Concepción ED, Clough Y, Díaz M, Gabriel D, Holzschuh A, Knop E, Kovács A, Marshall EJP, Tscharntke T, Verhulst J (2009) On the relationship between farmland biodiversity and land-use intensity in Europe. Proc R Soc B 276:903–909
133. Kovács-Hostyánszki A, Kőrösi A, Orci KM, Batáry P, Báldi A (2011) Set-aside promotes insect and plant diversity in a Central European country. Agric Ecosyst Environ 141:296–301
134. Batary P, Andras B, Kleijn D, Tscharntke T (2011) Landscape-moderated biodiversity effects of agri-environmental management: a meta-analysis. Proc R Soc B Biol Sci 278:1894–1902
135. Cane JH, Minckley RL, Kervin LJ, Roulston TH, Williams NM (2006) Complex responses within a desert bee guild (Hymenoptera: Apiformes) to urban habitat fragmentation. Ecol Appl 16:632–644
136. Holzschuh A, Steffan-Dewenter I, Tscharntke T (2008) Agricultural landscapes with organic crops support higher pollinator diversity. Oikos 117:354–361
137. Kremen C, Williams NM, Thorp RW (2002) Crop pollination from native bees at risk from agricultural intensification. Proc Natl Acad Sci 99:16812–16816
138. Winfree R, Griswold T, Kremen C (2007) Effect of human disturbance on bee communities in a forested ecosystem. Conserv Biol 21:213–223
139. Potts SG, Vulliamy B, Dafni A, Ne'eman G, Willmer PG (2003) Linking bees and flowers: how do floral communities structure pollinator communities? Ecology 84:2628–2642
140. Memmott J, Carvell C, Pywell RF, Craze PG (2010) The potential impact of global warming on the efficacy of field margins sown for the conservation of bumble-bees. Philos Trans R Soc Lond B Biol Sci 365:2071–2079

141. Menz MHM, Phillips RD, Winfree R, Kremen C, Aizen MA, Johnson SD, Dixon KW (2011) Reconnecting plants and pollinators: challenges in the restoration of pollination mutualisms. Trends Plant Sci 16:4–12

Exploring Changes in the Agricultural Calendar as a Response to Climate Variability in Egypt

El-Sayed Ewis Omran

Abstract Adaptation to climate change should involve changes in agricultural management practices in response to changes in climate conditions. This chapter reviews agricultural adaptation strategies in Egypt in cushioning the effects of climate change. Widely known agricultural adaptation techniques utilized by farmers included the use of drought-resistant crop varieties, crop diversification, changes in crop pattern and planting schedule, soil moisture conservation via suitable tillage practices, improved irrigation efficiency, and forestry and agro-forestry. Crop calendars include information on the timing of periods of crop sowing, growing, and harvesting. A huge number of literature take into consideration the techniques established by farmers to deal with climate variability. However, there is no literature on how exactly the agricultural calendar is presently moving. The current chapter is exploring observed changes in the agricultural calendar as a response to climate variability in Egypt. Land surface phenology (LSP) metrics were used as a proxy for crop calendars and criteria like those of season start and end (SOS and EOS respectively) were applied to classify the pixel-level growth period of active agricultural vegetation. Indeed, this information is not crop-specific, and therefore it is still applicable to use crop calendars from reliable sources which provide crop-specific phenological timing like sowing, growing, and harvesting. It was observed that seasons have shifted and shortened. For instance, in Egypt, the short dry season, which started in August, has shifted to July. In conclusions, there is a shift in cropping areas in Egypt due to gradual climate changes. The chapter has demonstrated the need for farmers to adjust their agricultural calendar and switch to agricultural practices that make better use of natural resources.

Keywords Agricultural calendar · Climate variability · Egypt · Climate change · Adaptation

---

E.-S. E. Omran (✉)
Soil and Water Department, Faculty of Agriculture, Suez Canal University, Ismailia 41522, Egypt
e-mail: ee.omran@gmail.com

Department of Natural Resources, Institute of African Research and Studies, Aswan University, Aswan, Egypt

© Springer Nature Switzerland AG 2020
E.-S. E. Omran and A. M. Negm (eds.), *Climate Change Impacts on Agriculture and Food Security in Egypt*, Springer Water, https://doi.org/10.1007/978-3-030-41629-4_12

1 Introduction

Climate change has now become a reality, and this is not possible to overestimate its negative effect on agricultural production as well as other social and economic endeavors, which cannot be under estimated. Warming triggered by emissions can affect climate variability, and climate change would be among the most complicated issues for future development, particularly in Africa [1]. Agriculture in Africa will be negatively affected by climate change [2]. Climate variability has attracted great attention, not only because of the unparalleled persistence of abnormally low precipitation but also because of the low power of social and economic systems to deal with the hazards related to climate change [3]. Due to this low capacity, extreme climate variability, like those of drought, is sometimes followed by an environmental decline the decimation of livestock herds, widespread food scarcity, mass migration and huge losses of human life [4].

On the one hand, agriculture is the cornerstone on which the ancient Egyptian civilization was established thousands of years ago and represents a significant historical value, which is part of today's Egyptian culture. There is also a significant economic value for agriculture as its added value is 14.5% of total Egyptian GDP in 2013 with 29% of the labor working in the sector in 2011 [5]. El-Nahrawy [6] reported that agriculture offers a livelihood for around 55% of Egyptians and contributes to about 20% of all foreign exchange earnings. Agriculture has the highest share of water use among all sectors with 81.1% of the total water use [6]. On the other hand, agriculture is the most vulnerable sector to climate change, which affects the two most essential direct agricultural production input; precipitation and temperature [7]. It is predicted that long-term changes in rainfall patterns and temperature shifts would have harmful effects on agriculture [8].

Climate change often impacts agricultural production indirectly by affecting the development and distribution of crop pests and animal diseases, increasing the rate and dissemination of harmful diseases, weather conditions, reducing water supplies and irrigation; and enhancing the severity of soil erosion [9, 10]. Such climatic hazards became the driving forces, which challenge many farmers' livelihoods. The rural population, whose primary source of food, direct and/or indirect employment and income is agriculture, will be most influenced because of the susceptibility of agriculture to climate change.

Adaptation to climate change refers to modification in natural or human systems in response to actual or predictable climatic stimuli [9, 11]. It enables farmers to produce their food, income, and livelihood protection goals in light of climate and socio-economic change [12]. Common methods of adaptation in agriculture involve new crop varieties and animal species that are suitable to drier situations, irrigation, divarication of crops, implementation of mixed crop and livestock farming systems, and changes in the dates of agricultural activity [13, 14]. Some of these methods (e.g., changes in agricultural activity dates) undertaken in response to short-term climate variability are classified as coping responses. Several factors and critical information are considered amongst them crop calendars [15]. The chronological sequence of the occurrence of different phenological stages of a crop in its growth cycle defines the so-called crop calendar [16]. Crop calendars include information on the timing of periods of crop sowing, growing, and harvesting. Different information sources on crop calendars (i.e. [17–19]) were employed. These calendars normally list the timing of land-level planting and harvesting of the major crop types.

The great influence of decision-making and innovations on crop production presents a major challenge in the historical past to detect, attribute and understand the impacts of climate change on crop patterns. For example, the potential contribution of climate change over the past decades to increase the crop area has long been observed and proposed by several research exploring agro-climatic indices (e.g., growing-season graduation days with a certain base temperature), but very few studies have confirmed this hypothesis. A huge number of studies take into consideration the techniques established by farmers to deal with climate variability [13, 14] have reported changes in sowing dates. However, there is no literature on how exactly the agricultural calendar is presently moving. While it is recognized that seasons and even agro-ecological zones are changing because of climate change, there is no research focusing on how farmers are adapting their whole agricultural calendar in the face of climate change. Thus, this chapter aims at exploring observed changes in the agricultural calendar as a response to climate variability in Egypt. It also attempts to highlight the factors affecting the farmers' decision to adjust their agricultural calendar as a coping mechanism against climate variability.

2 Climate Change and Sustainable Agriculture

Agriculture is a part of the problem and a part of the solution to climate change as well. On the one hand, climate change has created challenges for the agricultural sector—and will continue to do so.

Climate change induced temperature increases, variations in rainfall, and the frequency and intensity of extreme weather events add to pressures on global agricultural and food systems. In most regions, climate change is expected to have a negative impact on both crop and livestock production systems, although some countries may actually benefit from changing conditions. The changing climate also adds to resource issues such as water scarcity, pollution, and degradation of soil. Also, the effects of agriculture on the environment may change as the climate changes. In the future, climatic extremes such as droughts, heat waves, and excess moisture are expected to increase in frequency.

On the other hand, agriculture not only contributes to climate change and is affected by it; it also forms part of the solution. Climate change can be mitigated by sustainable agriculture. However, sustainability is a concept that is relatively easy to understand but difficult to define in practice. We need to start transforming the way we adapt to such changes in order to manage intensifying climate impacts. Measures for adaptation could include the choice and change of species and varieties, the adaptation of the field works to the calendar (more flexibility), the adaptation of plant production practices (i.e. fertilization, plant protection, irrigation, etc.) or the adoption of plant production practices that increase the soil organic matter content or the soil coverage by plants, manure management and agroforestry practices.

The international community has approved five principles of sustainable food and agriculture in order to achieve the 2030 agenda, which provide a framework for policy discussion and the development of adequate policies, strategies, regulations, and opportunities. FAO [20] identified these principles as:

> Improving efficiency in the use of resources is crucial to sustainable agriculture; Sustainability requires direct action to conserve, protect and enhance natural resources; Agriculture that fails to protect and improve rural livelihoods, equity, and social well-being is unsustainable; Enhanced resilience of people, communities, and ecosystems is key to sustainable agriculture; and Sustainable food and agriculture require responsible and effective governance mechanisms.

Climate change and its impacts are well identified as posing a significant threat to achieving sustainable development goals. Climate change in the Arab region could undermine past success and prospects, although less than 5% of the region contributes to global greenhouse gas emissions [21]. The Sustainable Development Goal (SDG13) advises countries to accept effective action to combat climate change and its effects, enabling new and improved statistical measures on resilience, adaptability, and mobilization of resources.

3 The Climate (Temperature and Precipitation) Variability of Egypt, Is There Any Change?

Egypt is one of the developing countries, which are vulnerable to climate impacts over different vital sectors such as agriculture, water resources and health. Egypt (Lat. 22° N, 33° N, Lon. 24° E 37° E), bounded by Med. Sea at the north, Red Sea to the east, Sudan and Libya to south and west respectively.

Egypt has been divided into four climate-based affective management zones in combination with physiography, natural resources, agriculture and other socio-economic factors (Fig. 1).

GRiSP [18] investigates the distinction between the two periods (1951–2016) and (1901–1950) to assess if the temperature (TN and TX) or precipitation changes take place. It is evident that there is a positive rise in TX across all Egypt, especially in Lower Egypt about 0.6 °C and 0.4 °C respectively in winter and summer, which is due to the growth in urbanization in northern Egypt than in southern Egypt. For TN, the winter difference in the Greater Cairo and its surroundings ranges from 0.85 °C to about 0.5 in southern Egypt, but in the summer, the significant difference in southern Egypt was about 0.7 °C and 0.45 °C in the north. For rainfall, there is no significant difference except a relatively increase at the northern coast of Egypt.

There has been a general warming over Egypt for the current climate since 1951 with higher warming in winter than in summer, and this seems to be more TN than TX

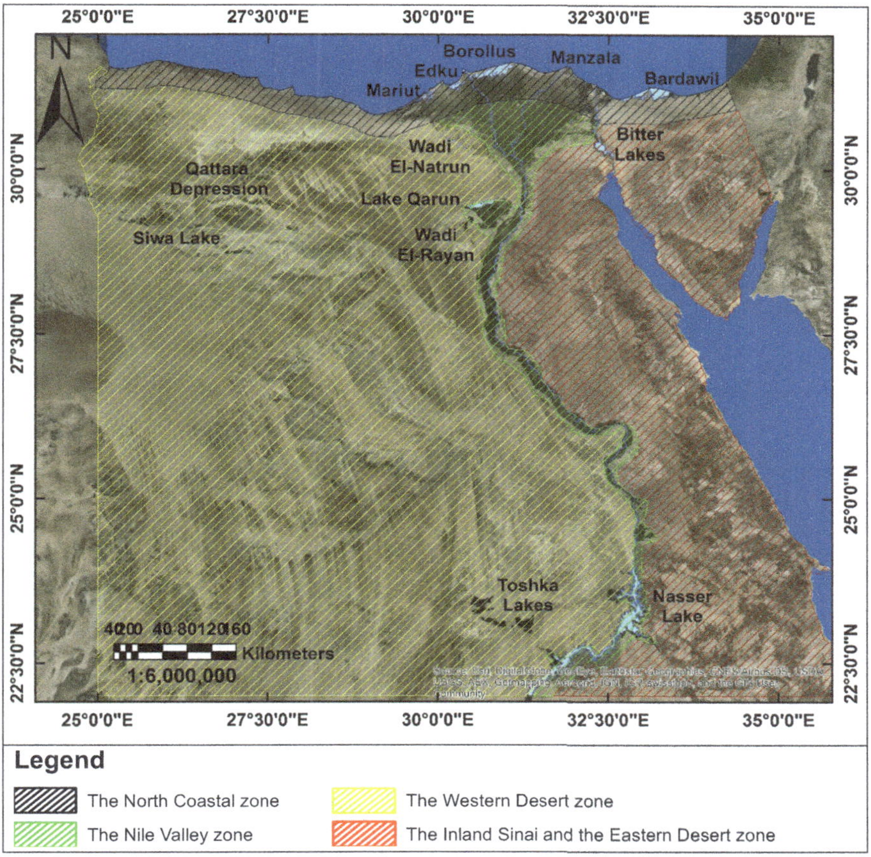

Fig. 1 Adaptive management zones in Egypt [22]

and is distinct from northern parts of Egypt to the south. All models have projected a TX and TN increase to 3 °C with a higher increase in summer than winter.

There is a significant decline over Egypt for the percentage of cold days and nights under the set of predicted models in the near future, while the percentage of warm days and nights for the same period is significantly increased.

It was observed that seasons have shifted and shortened. For instance, in Egypt, the short dry season, which started in August, has shifted to July. Egypt is expected to experience precipitation declines primarily, with a strong agreement between the ensembles. Smaller southeast changes are projected.

Changes in temperature and precipitation in Egypt were estimated using General Circulation Model (GCM). This model is a coupled gas-cycle/climate model that drives a spatial climate change scenario generator. GCM estimates of temperature and precipitation changes in Egypt (Table 1).

Table 1 General Circulation Model estimates of temperature and precipitation changes in Egypt [23]

Year	Temperature change (°C) mean (standard deviation)			Precipitation change (%) mean (standard deviation)		
	Annual	DJF[a]	JJA[b]	Annual	DJF	JJA
2030	1.0 (0.15)	0.8 (0.21)	1.1 (0.18)	−5.2 (7.93)	−8.9 (3.01)	10.7 (26.35)
2050	1.4 (0.22)	1.2 (0.30)	1.7 (0.26)	−7.6 (11.46)	−12.8 (4.35)	15.4 (38.07)
2100	2.4 (0.38)	2.1 (0.52)	2.9 (0.45)	−13.2 (19.95)	−22.3 (7.58)	26.9 (66.28)

[a]DJF is December, January, and February
[b]JJA is June, July, and August

4 Crop Weather Calendar and Crop Activity Calendar

In a particular moment, the demand for human activity in agriculture relies first and primarily on the growing season, the type of crop and its stage of development. In other contexts, the timing of the growing season relies on the temperature and/or availability of water. Even if the distribution of temperature and soil moisture permits those crops to grow, situations may not be suitable for all crops or varieties of crops. If an existing day length and temperature regime suit one cultivar of a crop species, not all cultivars of that specific crop may mature within the available growing season. Therefore, just a small number of crops or varieties of crops are appropriate for a specific environment in many other cases. The demand for human activity often depends on the time preparing the land, an activity, which is closely connected to the soil's physical condition.

A Crop Weather Calendar (CWC) contains the typical life history of the crop, from sowing through vegetative growth, flowering, and grain growth to a period of maturity. These CWCs include information on crop growth stages, normal crop growth weather, alerts to be provided depending on predominant weather conditions, crop water requirements throughout their distinct phytophases, favorable meteorological situations for crop pest and disease development. These calendars are beneficial for crop planning, irrigation management and plant protection measures that are essential for appropriate crop planning and for optimizing and strengthening the country's food production. In a broader context, the descriptive information presented in these calendars provides wide evidence of the direction of development that can be helpful to planners, agricultural managers, plant breeders and farmers in constructing political issues attributed to plant breeding, crop adaptation, drought testing, supplemental irrigation, maximizing the yield, etc.

The farmer must complete a couple of field operations throughout the growing period of a crop in order to build or preserve the most optimal conditions for growing crops. Among these is the reduction of competition by unwanted plants (by destroying them), optimizing the availability of nutrients and water (by fertilizer application, drainage, and irrigation), and protection of the crop (by pest and disease control). Harvesting is needed at the maturity of the crop. There may also be some on-farm processing, and the products will be placed or marketed. Most of these operations should be practiced in a specific order and a short period each of them. Therefore a crop activity calendar can be constructed, indicating the most favorable timing of the various activities and the type of operations required.

The rate of progress of planting influences the degree of the area planted in a particular time interval and therefore describes the variations in the cropped area from year to year in part. While attempts to demonstrate the global planting date by climate have been made [24, 25], the large errors between the estimated and reported planting dates reconfirm the fact that climate is an important, but not the sole determinant of planting date [26]. The timing and number of plantings (and thus harvesting) in a year relies significantly on both economic and climatic conditions (i.e., availability of labor and fresh water and seasonal rainfall patterns) [27].

The delay of some operations (seedbed preparation, planting, harvesting, etc.) could lead to cropping failure and decreased the number of harvests [28]. The timing of operations, including planting and harvesting, is affected by that of the previous operation in a work calendar [27] and needs to fulfill the field workability when heavy machines are used.

5 Data, Geographical Domain, and Satellite-Derived Phenology

The crop calendars for setting the dates for planting and harvesting specific crops were acquired from three sources: (1) the United Nations Food and Agriculture Organization [17]; (2) the United States Department of Agriculture [19]; and, (3) the International Rice Research Institute [18]. The FAO crop calendar includes several countries all over the world, with an emphasis on developing nations, especially in Africa [26]. This database comprises the seeding and harvesting range for each of the FAO, IRRI and USDA crop calendars. In the last column, named "Land Surface Phenology" are presented the crop seasons derived from LSP and matched with a crop type from the calendars. The empty space in this column indicates no match with the corresponding crop type.

The use of satellite data with high temporal resolution has emerged as an effective tool for examining crop phenology [29]. Remote sensors include a frequent, reliable and consistent vegetation response measurement at different crop growth stages [16]. Land Surface Phenology (LSP) refers to the spatiotemporal development of the vegetated land surface as confirmed by satellite sensors. Usually, LSP metrics describe [30]: (i) time of onset of greening, (ii) time of onset of senescence, (iii) timing of the maximum development during the growing season, and (iv) growing season length.

Since the ASAP immediate warning classification system provides warnings linked to all crop types which are in the growing phase at the time of analysis in a given spatial unit, it is essential to include analysts with more detailed information about the specific crop types that are likely to occur. Using moderate to coarse time-series vegetation index such as MODIS NDVI, it is possible to predict phenological timing like those of start of season (SOS) and end of season (EOS) per pixel and for all vegetation areas [31]. Whitcraft et al. [32] utilizing MODIS surface reflectance imagery, the timing of the agricultural season for all major crops within a given geographical area has been derived, delivering the earliest crop cycle SOS and the latest crop cycle EOS.

Land surface phenology used in the Anomaly hot Spots of Agricultural Production (ASAP) system is defined by the satellite-derived phenology computed on the long-term average of 10-day MODIS NDVI data produced by land surface phenology process is determined by satellite-derived phenology calculated on the long-term average of 10-day MODIS NDVI data generated by BOKU University [33] starting from to MOD13A2 and MYD13A2 V006 16-day Global data at 1 km resolution. Phenology was extracted and applied to the historical average of the smoothed NDVI.

From the phenological analysis, the following key parameters are retrieved for each land pixel [34]: number of growing season per year (i.e. one or two); start of season (SOS, occurring at the time at which NDVI grows above the 25% the ascending amplitude of the seasonal profile); time of maximum NDVI (TOM); start of senescence period (SEN, when NDVI drops below 75% of the descending amplitude); and end of the season (EOS, when NDVI drops below 35%). Phenology of the surface (Table 2) used in the system is defined by the satellite-derived phenology computed on the long-term average of 10-day MODIS NDVI data [33]. The data set of crop calendars was consistent with terrestrial surface phenology originating from remote sensing.

The following data (expressed in dekads) were obtained after the processing of the initial data, namely digitization for crop calendars and cluster extraction from LSP: (i) the range for sowing and harvesting period for major food crops from the calendars, and (ii) the range for SOS and EOS for each identified cluster.

Utilizing moderate to coarse time-series vegetation index like MODIS NDVI, it is possible to estimate phenological timing including start of season (SOS) and end of season (EOS) per pixel and for all vegetation areas.

We have information on standard seeding and harvesting periods in crop calendars. Five crops in the FAO crop calendar were acknowledged that could suit the measured cluster (Table 2). Three of them have the same seasonal characteristics (i.e., millet, maize, and rice), meaning sowing between 11 and 12 dekad and harvesting between 23 and 29 dekad.

Agricultural statistics were particularly vital at this stage of the analysis, as the knowledge of the area harvested, as well as the quantity produced for each crop, was used to verify the presence of a crop type. FAOSTAT data, confirm that these crop types are present.

Two main growing seasons for Egypt are identified in LSP with this information at hand (Table 2). The growing seasons in LSP have been compared with the FAO calendar and assessed in line with the timing of the growing season of five FAO crops (Table 2).

Table 2 Digitized data from FAO crop calendar for Egypt

Crop	FAO				IRRI				USDA				Land surface phenology			
	SOW		HARV		SOW		HARV		SOW		HARV		SOS		EOS	
	min	max	min	max	min	max	min	max	min	max	min	max	min	max	min	max
Barley	32	35	11	14	–	–	–	–	–	–	–	–	29	35	9	14
Maize	12	14	29	32	–	–	–	–	–	–	–	–		–	–	–
Millet	11	17	23	29	–	–	–	–	–	–	–	–	15	19	23	29
Rice	12	19	29	32	13	15	28	30	–	–	–	–	–	–	–	–
Wheat	31	33	13	16	–	–	–	–	25	33	10	24	25	35	9	15

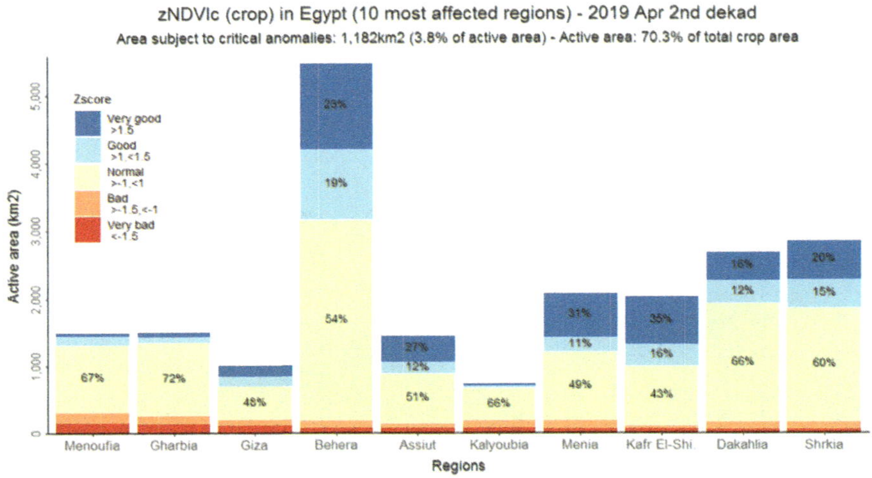

Fig. 2 zNDVI, and zNDVIc for crop in 10 most affected regions in Egypt at April 2019 [34]

Figures 2 and 3 shows the zNDVI, and zNDVIc, zWSI, SPI3, and SPI3 for the crop in 10 most affected regions in Egypt in April 2019.

In the crop weather calendar for the wheat crop, during the anthesis and grain filling stage of wheat (February–March months), the maximum temperature should be within 17–25 °C and minimum temperature within 4–9 °C. However, if the temperature rises by 3–4 °C from the maximum temperature limit for more than 4–5 days in a week, then if no remedial measures are taken, the productivity of the wheat crop is adversely affected. Therefore, the agrometeorologist may immediately give an advisory to farmers for applying additional irrigation to offset the negative effect of a heat wave in wheat crop.

Figure 4 shows the start and end of the first and second season in Egypt.

Finally, the following ASAP—adapted calendar (Fig. 5) is provided, introducing crops described in the FAO crop calendar which fits with LSP. In the legend of the calendar, the growth stages addressed, planting and harvesting, are correlated with SOS and EOS, respectively. Similar to the FAO crop calendars, the residual interval period was labeled as "growth." It is stated that the real growth period may extend to any sowing dekad from any planting.

Early warning system (Anomaly hot Spot of Agricultural Production, ASAP) offers timely warnings of agricultural production deficits in Egypt at high risk of food insecurity.

Since the automatic warning classification system produces warnings for all crop types, which are in the growing phase in a provided spatial unit at the time of assessment, more specific information is useful (Fig. 6).

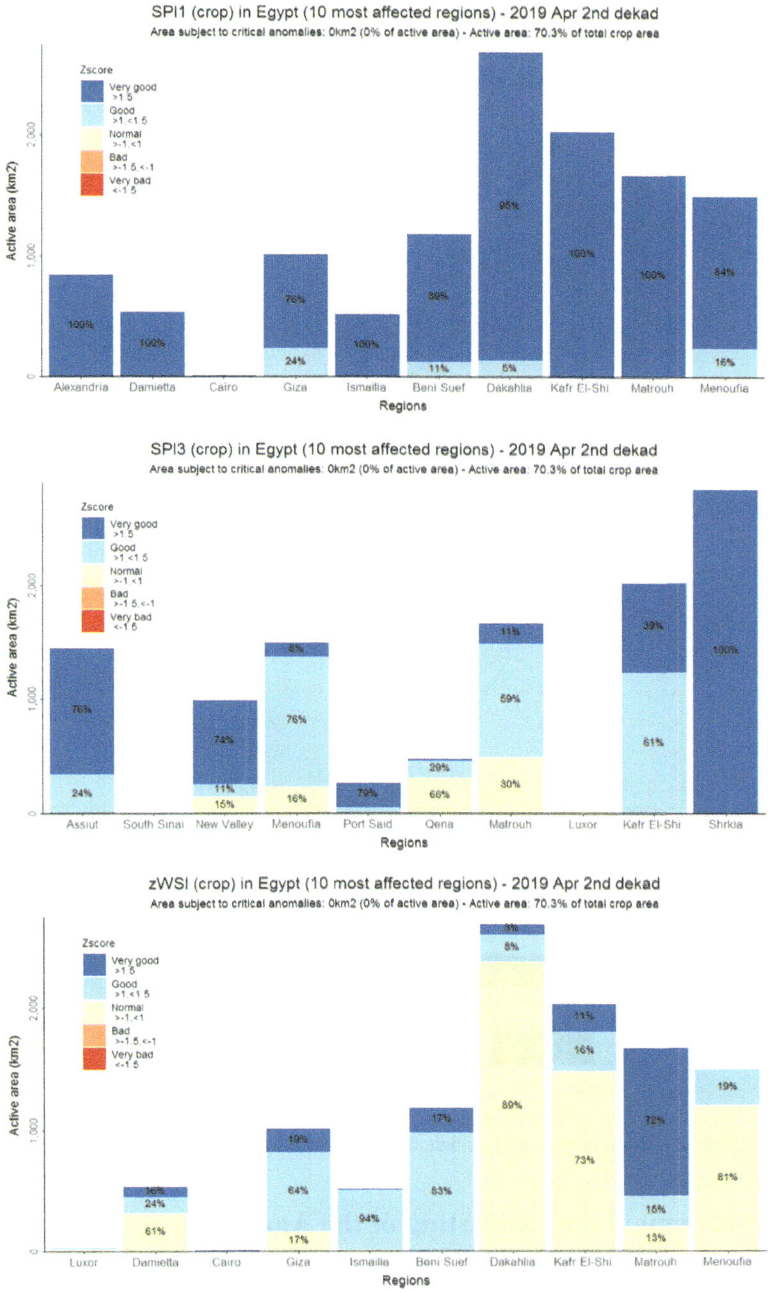

Fig. 3 zWSI, SPI3, and SPI3 for crop in 10 most affected regions in Egypt at April 2019 [34]

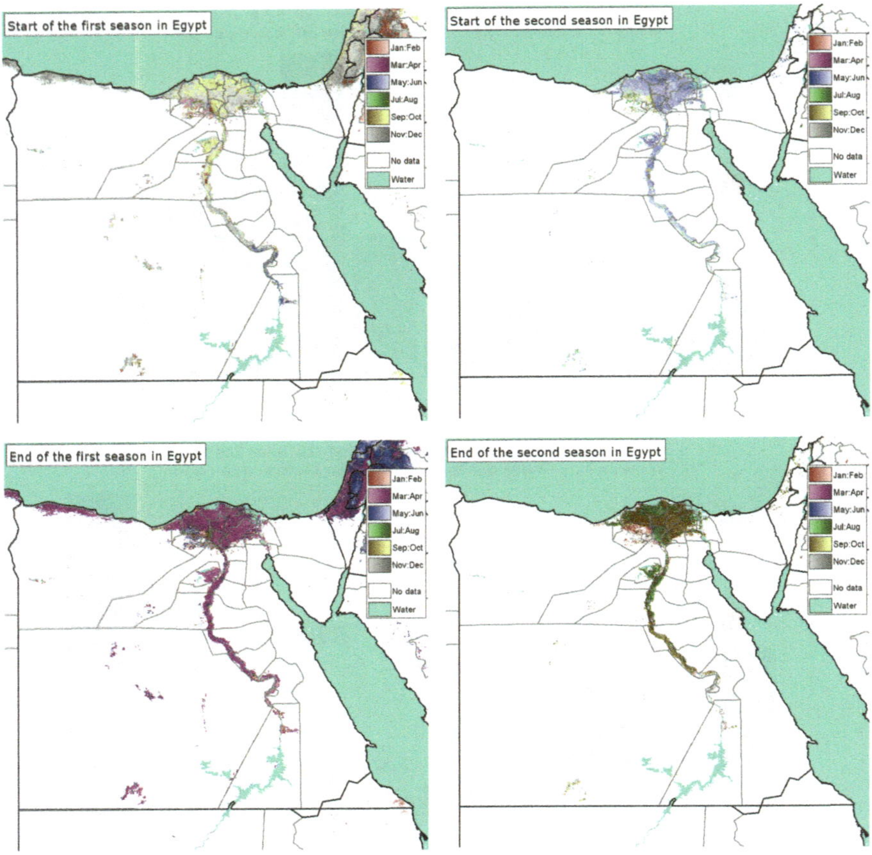

Fig. 4 Start and end of the first and second season in Egypt [34]

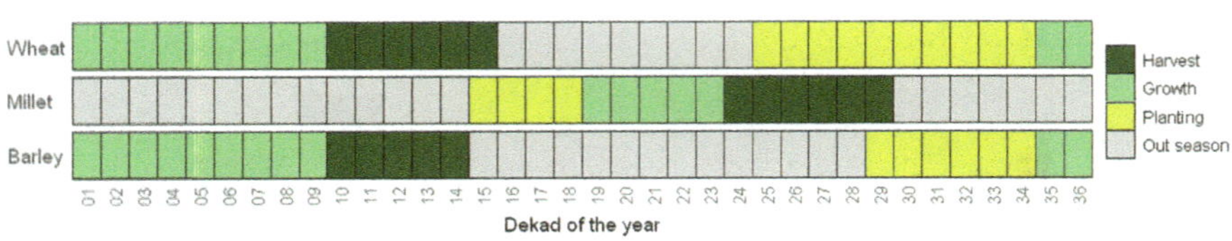

Fig. 5 Selection of FAO crop calendars matching with ASAP phenology in Egypt [34]

6 Management and Adjusting of the Agricultural Calendar as a Coping Mechanism to Climate Variability

Climate change is the goal number 13 in the 2030 agenda for sustainable development adopted by the United Nations, which addresses the target action to combat climate

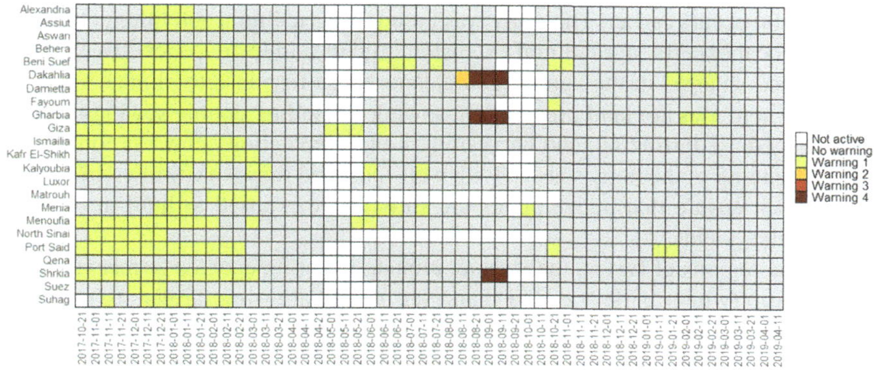

Fig. 6 Warnings for the crop in Egypt in April 2019 [35]

change and its impact to improve the lives of people around the globe (www.un.org/sustainabledevelopment/climate-change-2/).

In order to deal with the effects of climate change on agricultural fields, new technologies and innovations have been examined and confirmed via agricultural research projects that contribute to efforts to mitigate climate change and allow farmers and policymakers to establish their strategies and actions to tackle climate change effects.

To eliminate the adverse effects of climate change, farmers should be able to adopt different strategies to climate change. Farmers are suggesting particular adaptation options. Even though the majority of them have started to adapt themselves by cultivating short duration heat tolerant crops like pulses maize millets, vegetables, and floriculture crops especially Jasmine, as they understood from their farming experience that these crops could withstand heat and water stress as an adaptation response to the climate variability.

Farmers understood climate change via changes in their farming practices. Agricultural calendar interruption is evidence of climate variability and change. Farmers stated that existing rain starting and stopping times do not interfere with the time in their prior agricultural calendar. They are therefore confused regarding regularizing their crop calendar and evaluating the most suitable crops to grow. Some specific months and dates over the years are regarded to be key climate reference indicators. As soon as a change takes place in these times, it is combined by farmers as evidence of climate change. The intensity of the sun (sunshine) and the power of winds have changed. Before, the winds were less intense, but now, strong wind, which precedes the rain, chases out the stormy clouds.

Farmers change land preparation, and sowing dates primarily in terms of agricultural calendar adjustment. Because of all the existing operations rely on the sowing date, the dates of other activities were also shifted forward or backward with regard to the prior calendar. Because of uncertainties linked to climate variability, farmers have to start land preparation activities earlier to be ready in case the onset of the rainy season occurs earlier. The study was based on agricultural activities conducted out with local experts (agricultural extension officers and local leaders) to examine the detected changes in the agricultural calendar as a response to climate variability. These activities are land preparation, sowing (including double sowing in some cases), weeding, fertilizing, and harvesting. Land preparation generally took place at the beginning of the rainy season, with well-loosened ridges or furrows 15–30 cm deep, which ensure that maize plants are sufficiently exposed to light in order to obtain higher yields.

In some regions of Egypt, the warmer climate in spring enables farmers to plant the first crop (winter barley) earlier than ever before, allowing farmers to grow the second crop for the remaining portion of the growing season in order to increase income.

Adjustment in crop sequence and planting schedule: climate change negatively affects crop production across long-term rainfall changes resulting in changes in the cropping pattern and operating schedule. The first steps against a global assessment would involve the development of detailed work calendar information for various crops grown on a farm, allowing an improved understanding of the pattern of activities and weather-related workability in specific areas.

7 Innovations and Technologies for Climate Change Adaptation Measures

In addition to traditional and native knowledge, utilizing emerging technologies and innovations evaluated and validated by agricultural research projects will contribute to attempts to deal with climate change and allow both farmers and policymakers to face climate change problems.

Management of water resources: Water is one of climate change's most influenced natural resources. Choices under water shortages to enhance water management in Egypt involve; efforts to increase irrigation efficiency; enhancing saving water techniques and water harvesting—among others, to conserve the already limited water resources under different climate change scenarios. Water harvesting can also be enhanced through the application of satellite remote sensing.

Crop simulation models: Models of crop simulation can estimate many key crop characteristics across a wide variety of climatic conditions, like flowering timing and physiological maturity, by correctly defining phonological responses to temperature and daytime. Also, yield accumulation should be anticipated by predicting appropriately the development and loss of the leaf area and therefore the ability of a crop to intercept radiation, accumulate biomass and split it into harvestable parts including grain. Crop water use is also needed to be precisely predicted by correctly predicting evapotranspiration and the extraction of soil water by plant root [36].

8 Challenges Still Facing Climate Change Governance in Egypt

The influence and mitigation of climate change is not only a political decision but also involves academic research to examine the phenomena, assist in drawing a clearer picture and proposing creative scientific solutions. As climate change becoming a serious man-made an environmental challenge with its severe impact on livelihood, the role of science and innovations and technologies became even more critical than ever. Egypt vision 2030 stated, "Knowledge, innovation, and scientific research contribute in achieving the goals of sustainable development in general, whether national or international, where the increase in the contribution of a large number of economic sectors in economic activity can be achieved by connecting scientific research outputs and innovation by the needs of these sectors".

A cross-cutting issue like climate change can not be controlled by a single department of government or ministry. Assessing the effects of climate change and incorporating adaptation and remediation measures efficiently must include practically every sector of government with its administrative system that holds climate change in its strategies and development policies and plans. All Egypt's economic sectors are linked in one way or another with climate change because they emit greenhouse gas (GHG) emissions or are themselves impacted by climate change. The most important restriction for useful environmental policy-making and development in Egypt is the lack of reliable and timely information which suggests how specific sectors of society affect the environment and whether or not advancement is becoming more sustainable. Various constraints related to the processes of environmental information collection, production and dissemination are evident in Egypt. These include uncoordinated institutional set-ups for monitoring activities, the lack of an important information system for controlling organizations to feed data and findings, the lack of comprehensive systematic monitoring methodologies, the lack of valuation and/or undervaluation of several natural resources, and the lack of financial resources to preserve monitoring processes.

9 Satellite Images, GIS, Mathematical Models in Studying Climate Change in 2030

Satellite images, GIS, mathematical models, and scenarios have been used in studying the impacts of climate change on agriculture and natural resources in Egypt. Egypt's National Strategy used simulation models in a study for Adaptation to Climate Change and Disaster Risk Reduction (2011) on different agricultural areas over 25–40 years to predict the statues of the major crops in Egypt and their water use under different climate change scenarios.

Under climate change in 2030 and because of the rise in air temperature, it is expected that planting date will be earlier by 5–7 days. Morsy [37] visualized the impact of early planting under climate change for wheat and maize in 2030 and figured that early planting of both crops resulted in yield losses being reduced. Also, seasonal length of the cultivated crops is supposed to be decreased caused by climate change. Khalil et al. [38] reported that wheat season length was reduced by 5 days in 2030, as a result of acceleration in its growing season. Similar results were obtained by Ouda et al. [39] for maize under climate change in 2030. Table 3 presented the projected planting and harvest dates as well as its season length for some field, fruit, and vegetable crops planted in Egypt. It should be noted that it was believed that the planting date for the fruit crops observed was 5 days earlier. An assumption was made that sowing date for these crops in 2030 was 5 days earlier than its sowing date in 2016. Furthermore, the length of the growing season for the selected vegetable crops in 2030 is expected to be 3–5 days shorter.

Climate change studies [40, 41], predicted a reduction in the productivity of two major crops in Egypt—wheat and maize—by 15% and 19% respectively by 2050. Losses in crop productivity are mainly attributed to the projected temperature increase, crop-water stress, pests, and disease, as well as the inundation and salinization.

The results concluded that the productivity of wheat would fall by 9% if the temperature rises by 2 °C. This crop's water consumption will increase by 6.2% compared to the current weather conditions. A 4 °C rise will increase the deficit to 18%. Crop productivity of Maize will fall by 19% by the middle of this century if temperature increases by 3.5 °C, in comparison with the situation under the current weather conditions. Accordingly, water consumption will rise by 8%. Productivity will fall by 11% in comparison with the situation under the current weather conditions. The consumption of water will increase by 16%. If the temperature increases by 2 °C, the productivity of tomatoes will fall by 14%. Water consumption of the crop will increase by 4.2–5.7% in comparison with the situation under the current weather conditions. The drop in productivity will reach 51% if the temperature rises by 3.5 °C. The outcomes of climate change studies indicate a 24.5% decreases in productivity and a 2.3% increases in water consumption in the production of sugar cane. The crop return will drop by 25.6% per unit of water. However, climate change will have a positive impact on the productivity of cotton. Production will rise by 17% if the temperature rises by 2 °C. On the one hand, the water consumption of this crop will increase by 4.1–5.2% in comparison with the situation under the current weather conditions. If the temperature rises by 4 °C, production will increase by 31%. On the other hand, under the current weather conditions, water consumption will increase by 10% compared to the situation.

Table 3 The projected planting and harvest dates for the selected field, fruits, and vegetable crops in 2030 [37]

Crop	Planting date	Harvest date	Season length (days)
Fruits crops			
Apple	10-Feb	9-Feb	365
Citrus	10-Feb	9-Feb	365
Date	10-Feb	9-Feb	365
Grape	10-Feb	9-Feb	365
Olive	10-Feb	9-Feb	365
Stone fruit	10-Feb	9-Feb	365
Field crops			
Barley	10-Nov	23-Mar	134
Bean (faba)	20-Oct	18-Apr	150
Clover	10-Oct	23-Mar	165
Cotton	10-Mar	5-Aug	149
Flax	10-Nov	5-Apr	147
Lentil	20-Oct	18-Mar	150
Maize	10-Mar	20-Aug	103
Rice	10-May	7-Sep	121
Sorghum	10-May	20-Aug	103
Soybean	10-May	16-Aug	99
Sugarbeet	10-Oct	5-Apr	178
Sugarcane	15-Feb	14-Feb	365
Sunflower	10-May	7-Aug	90
Wheat	10-Nov	9-Apr	151
Vegetable crops			
Cucumber	10-Mar	7-Jun	90
Eggplant	25-Mar	29-Jul	127
Onion	10-Nov	7-Apr	149
Peas	28-Aug	23-Nov	88
Pepper	28-Mar	3-Aug	129
Potato (winter)	28-Oct	26-Jan	91
Potato (summer)	28-Jul	23-Nov	119
Squash	10-Mar	8-Jun	91
Strawberry	28-Aug	5-May	251
Sweet potato	28-Sep	25-Feb	151
Tomato (winter)	28-Apr	23-Aug	118
Tomato (summer)	28-Sep	23-Feb	149
Watermelon	10-Feb	23-May	103

10 Impact of Climate Change on Field Crops' Production and Net Value

The increase in temperature is predicted to influence the agricultural land area in the governorates of the Nile Delta. Due to their insufficient access to fresh water, remote and marginal land areas could be spoiled [42]. Nevertheless, the potential area to be lost due to the temperature increase is not quantified in this chapter.

Temperature increase would affect field crops' production in the Nile Delta governorates. Table 4 shows that the temperature increase would decrease the total field crops' production in Egypt by 4.4% compared to 2015. Depending on their cropping pattern, however, the impact on the governorates is expected to differ. For example, pulses should be most affected by a 9% decrease in total production in Egypt. The production of pulses is concentrated in Alexandria, Damietta and Kafr El-Sheikh where it covers 2.1%, 1.9%, and 1.2% of their total field crops' area respectively, followed by vegetables' production that would suffer from a reduction of 7.2%. The impact on vegetable-growing governorates is predicted to be higher, however, as vegetables cover larger agricultural areas in Alexandria, Ismailia, and Beheira, amounting about 43%, 35% and 22% respectively. In contrast, the production of cotton is to increase by about 9% of total cotton production in Egypt (based on 2015 data). This increase is to mainly benefit Kafr El-Sheikh, and Beheira governorates as cotton cover 7.1% and 3.5% of their respective agricultural land areas.

Also, the production of strategic field crops will suffer as the production of cereals will decrease by 5%, the production of sugar crops by 0.3% and the production of oil crops by around 1.1% of their total production in Egypt based on data from 2015. Wheat, sugar crops and oil crops are labeled strategic crops for the Egyptian government. These crops are essential for the diet and clover of the Egyptian populations

Table 4 Impact of temperature increase on agriculture in the Nile Delta governorates in 2030

Field crops' group	P, %	N, %	W, %
Cereals	5.00	4.29	2.02
Sugar crops	0.33	0.30	0.56
Oil crops	1.08	1.07	0.89
Pulses	9.01	3.95	1.57
Vegetables	7.19	6.01	1.42
Clover (berseem)	5.13	6.93	2.53
Medicinal crops	0.21	4.66	0.29
Cotton	(+8.75)	(+10.59)	2.32
% Reduction of total Egypt	4.39	4.95	1.75

P = % reduction/increase of field crops production of total field crops production in Egypt in 2015. N = % reduction/increase of field crops net value of total field crops net value in Egypt in 2015. W = % increase of field crops water requirement of total water requirement in Egypt in 2015 for livestock breeding (meat production) [43]. This is to affect almost all the Nile Delta governorates; however, Daqahlia, Sharkia, and Qalyubia governorates are to be affected the most as about 70% of their agricultural land is covered with strategic crops.

This decrease in production percentage relies on the accessibility of the necessary water requirements, as the increase in temperature would increase water requirements per field crop per feddan. Water scarcity would, therefore, exacerbate the yield deficiency of the field crops. The cultivation of field crops' in the Nile Delta governorates in 2030, would increase the field crops' water requirements by about 2% of the total field crops water requirements in Egypt based on 2015 data that are equivalent to about 0.7 billion m^3.

The net value of total field crops in the governorates of the Nile Delta is projected by about 5% of the net value of total field crops in Egypt based on information from 2015. The net value of clover is to be influenced with such a decrease of roughly 7%, followed by vegetables with a net decrease of 6%, and medicinal crops with a decrease of approximately 5%, while the net value of cotton is to increase by approximately 11% of the total net value of cotton in Egypt based on 2015 data. The governorates to be affected the most due to these reductions are: Damietta and Menufia as about 50% and 40% of their field crops' net value respectively comes from clover. Alexandria and Ismailia as about 70% and 65% of their total field crops' net value comes from vegetables. Daqahlia and Sharkia as their net value from cereals is about 50% and 40% respectively. Based on the average yield per field crop per feddan per governorate the farm level income would differ from one governorate to the other, which would consequently affect the livelihood of farmers.

11 Overall Discussion

Agricultural history discards a series of adaptations by farmers to a wide range of climatic, social and agronomic factors. They have been able to establish their livelihood techniques in a manner that allows them to continuously deal with climate variability, dangerous pest attacks and changing local, national and global policies [44].

The dates of adjustments in agricultural operations, in general, and especially the seeding date, are reported widely in the literature on adaptation to climate change [13, 14]. The entire agricultural calendar is adapted when the sowing date has been changed. As a response to climate variability, the detected changes in the agricultural calendar display that the cropping season is getting longer because farmers have to start the land preparation activities earlier while the onset of the rainy season might occur later than usual.

In Sub-Saharan West Africa in general, farmers indeed tend to sow maize after the first main rainfall (>20 mm) occurring at the start of the rainy season [45]. Farmers tend to face dry spells after the first sowing due to the high climate variability.

They performa second sowing to cope with this lack of rain. As a result, the cropping season's length is longer than normal, posing the question of time-inefficiency. At present, adjustments to the agricultural calendar do not seem consistent across locations.

12 Conclusions

Climate change is the goal number 13 in the 2030 agenda for sustainable development adopted by the United Nations, which addresses the target action to combat climate. Climate and weather have different influences on crop area, intensity, and yield. These influences are modified by farmer decision-making and innovations. Farmers are developing coping strategies, such as adjusting some of their farming practices, given climate variability. Enhancing our understanding of climate impact and management contributions to crop area and intensity as well as yield is crucial for lowering uncertainty regarding long term effects of climate change on agricultural production and creating more targeted responses to climate adaptation.

Crop calendars are a core component of monitoring systems for agricultural production as they help experts focus on the seasons when different crop types grow in the field. Land surface phenology (LSP) metrics have been used as a proxy for crop calendars and apply criteria such as the start and end of the season (SOS and EOS respectively) to identify the period of pixel-level growth of active agricultural vegetation. This information, indeed, is not crop-specific and so it remains relevant to use crop calendars from independent sources which provide crop-specific phenological timing like those of sowing, growing, and harvesting. Under climate change in 2030 and because of the rise in air temperature, it is expected that planting date will be earlier by 5–7 days. Climate change studies predicted a reduction in the productivity of two major crops in Egypt—wheat and maize—by 15% and 19% respectively by 2050.

13 Recommendations

In future research, interactions between technology, decision-making, and climatic influences are necessary to help climate adaptation and enhance food security. The goal is to enhance our understanding of technology interactions, farmer decision-making and climatic impacts on crop area, intensity, and yield. While the capacity for climate adaptation has been evaluated using crop models, there is limited technology considered in these evaluations. These analyses tell us about the upper limit of the potential for adaptation without limiting access to technology. The real relationships between technology, farmer decision-making and climate impacts are, however, complex.

We know little, as this chapter shows, about how weather and climate, mediated by farmer decision-making and available technology, influence crop area, and intensity, especially on a global level.

More research needs to be done to enhance our understanding of: (1) the impacts of relatively less studied extreme weather events on crop area, intensity, and production; (2) climate influences on crop production by changing the work schedule and the workability of the field; and (3) responses of farmers to climate shocks under different economic conditions and different technology access. To develop more climate-resilient crop production systems, a global analysis of these issues is crucial.

We are now taking into consideration certain specific gaps in knowledge, which need to be addressed in order to deepen our understanding of climate impacts on crop production.

In view of global climate change, food trade globalization and the growing importance of food imports to maintain national food balance in many countries, we claim that addressing these knowledge gaps globally is crucial.

References

1. Adger N, Agrawala S, Mirza MMQ, Conde CO, Brien K, Pulhin J, Pulwarty R, Smit B, Takahashi T (2007) Assessment of adaptation practices, options, constraints and capacity. In: Parry ML, Canziani OF, Palutikof JP, van der Linden PJ, Hanson CE (eds) Climate change 2007: impacts, adaptation and vulnerability contribution of working group II to the fourth assessment report of the intergovernmental panel on climate change. Cambridge University Press, Cambridge, UK, pp 717–743
2. Müller C, Cramer W, Hare WL, Lotze-Campen H (2011) Climate change risks for African agriculture. Proc Natl Acad Sci USA 108:4313–4315
3. Mengistu D (2011) Farmers' perception and knowledge of climate change and their coping strategies to the related hazards: case study from Adiha, Central Tigray, Ethiopia. Agric Sci 2(2):138–145. https://doi.org/10.4236/as.2011.22020
4. Tarhule A, Lamb P (2003) Climate research and seasonal forecasting for West Africans: perceptions, disseminations, and use? Bull Am Meteor Soc 84:1741–1759. https://doi.org/10.1175/BAMS-84-12-1741
5. World Bank (2015) Indicators. http://data.worldbank.org/indicator. 24 Aug 2015
6. El-Nahrawy MA (2011) Egypt. Country pasture/forage resource profiles. FAO
7. Deschenes O, Greenstone M (2006) The economic impacts of climate change: evidence from agricultural output and random fluctuations in weather. Am Econ Rev 97(1):354–385
8. Charles N, Rashid H (2007) Micro-level analysis of farmers' adaptation to climate change in Southern Africa. IFPRI discussion paper 00714 Washington DC, USA
9. Intergovernmental Panel on Climate Change (2001) Climate change 2001: impacts, adaptation, and vulnerability. In: McCarthy JJ et al (eds) Contribution of working group II to the third assessment report of the intergovernmental panel on climate change. Cambridge University Press, Cambridge, UK
10. Watson R, Zinyoera M, Moss R (1998) The regional impacts of climate change: an assessment of vulnerability. Cambridge University Press, Cambridge
11. IPCC (2014) Climate change. Impacts, adaptation, and vulnerability. In: Field CB, Barros VR, Dokken DJ, Mach KJ, Mastrandrea MD (eds) Part A: Global and sectoral aspects, contribution of working group II to the fifth assessment report of the intergovernmental panel on climate change

12. Kandlinkar M, Risbey J (2000) Agricultural impacts of climate change: if adaptation is the answer, what is the question? Clim Change 45:529–539
13. Nhemachena C, Hassan R (2007) Micro-level analysis of farmers' adaptation to climate change in Southern Africa. IFPRI discussion paper no 00714, International Food Policy Research Institute, Washington, DC
14. Yegbemey RN, Yabi JA, Tovignan DS, Gantoli G, Kokoye SEH (2013) Farmers' decisions to adapt to climate change under various property rights: a case study of maize farming in Northern Benin (West Africa). Land Use Policy 34:168–175. https://doi.org/10.1016/j.landusepol.2013.03.001
15. Meroni M, Rembold F, Urbano F, Csak G, Lemoine G, Kerdiles H, Perez-Hoyoz A (2018) The warning classification scheme of ASAP—anomaly hot spots of agricultural production. 2.1. https://doi.org/10.2760/798528
16. Patel JH, Oza MP (2014) Deriving crop calendar using NDVI time-series. https://doi.org/10.5194/isprsarchives-xl-8-869-2014
17. FAO (2005) FAO global information and early warning system (GIEWS). Available online: http://www.fao.org/giews/countrybrief/index.jsp
18. GRiSP GRSP (2013) GRiSP: partnership in motion. International Rice Research Institute, Los BaZos (Philippines)
19. United States Department of Agriculture (USDA) (1994) Major world crop areas and climatic profiles. In: USDA agricultural handbook. JAWF, USDA, Washington, DC
20. FAO (2017) The state of food security and nutrition in the world. Food and Agriculture Organization of the United Nations, Rome, Italy
21. ESCWA UNESCWA (2017) Arab climate change assessment report—main report. Beirut, E/ESCWA/SDPD/2017/RICCAR/Report. www.unescwa.org/publications/riccar-arab-climate-change-assessment-report
22. Omran E, Negm A (2018) Adaptive management zones of Egyptian coastal lakes. In: The handbook of environmental chemistry. Springer, Berlin, Heidelberg
23. Agrawala S, Moehner A, El Raey M, Conway D, Van Aalst M, Hagenstad M, Smith J (2004) Development and climate change in Egypt: focus on coastal resources and the Nile. Organisation for Economic Co-Operation and Development
24. Deryng D, Sacks WJ, Barford CC, Ramankutty N (2011) Simulating the effects of climate and agricultural management practices on global crop yield. Glob Biogeochem Cycles 25, GB2006. https://doi.org/10.1029/2009GB003765
25. Waha K, van Bussel LGJ, Müller C, Bondeau A (2012) Climate-driven simulation of global crop sowing dates. Global Ecol Biogeogr 21:247–259
26. Sacks WJ, Deryng D, Foley JA, Ramankutty N (2010) Crop planting dates: an analysis of global patterns. Glob Ecol Biogeogr 19:607–620
27. Kotera A, Nguyen KD, Sakamoto T, Iizumi T, Yokozawa M (2014) A modeling approach for assessing rice cropping cycle affected by flooding, salinity intrusion, and monsoon rains in the Mekong Delta, Vietnam. Paddy Water Environ 12:343–354
28. Sawano S, Hasegawa T, Goto S, Konghakote P, Polthanee A, Ishigooka Y, Kuwagata T, Toritani H (2008) Modeling the dependence of the crop calendar for rain-fed rice on precipitation in Northeast Thailand. Paddy Water Environ 6:83–90
29. You X, Meng J, Zhang M, Dong T (2013) Remote sensing based detection of crop phenology for agricultural zones in China using a new threshold method. Remote Sens 5(7):3190–3211
30. de Beurs KM, Henebry GM (2010) Spatio-temporal statistical methods for modelling land surface phenology. In: Keatley M, Hudson I (eds) Phenological research. Springer, Dordrecht
31. Meroni M, Verstraete MM, Rembold F, Urbano F, Kayitakire F (2014) A phenology based method to derive biomass production anomalies for food security monitoring in the Horn of Africa. Int J Remote Sens 35(7):2472–2492. https://doi.org/10.1080/01431161.2014.883090
32. Whitcraft AK, Becker-Reshef I, Justice CO (2015) Cloud cover throughout the agricultural growing season: impacts on passive optical earth observations. Remote Sens Environ 156:438–447

33. Klisch A, Atzberger C (2016) Operational drought monitoring in Kenya using MODIS NDVI time series. Remote Sens 8. https://doi.org/10.3390/rs8040267
34. Rembold F, Meroni M, Urbano F, Csak G, Kerdiles H, Perez-Hoyos A, Lemoine G, Leo O, Negre T (2018) ASAP: a new global early warning system to detect anomaly hot spots of agricultural production for food security analysis. Agric Syst
35. Rembold F (2018) Ten-daily automatic warnings about agricultural vegetation anomalies. European Commission, Joint Research Centre (JRC). http://data.europa.eu/89h/jrc-10112-10001
36. Reichter GM, Semenov MA (2005) Modeling impacts of climate change on wheat yields in England and Wales: assessing drought risks. Agric Syst 84(1):77–97
37. Morsy M (2015) Use of regional climate and crop simulation models to predict wheat and maize productivity and their adaptation under climate change. PhD thesis, Faculty of Science Al-Azhar University
38. Khalil F, Farag H, El Afandi G, Ouda S (2009) Vulnerability and adaptation of wheat to climate change in Middle Egypt. In: Proceeding of the 13th international conference on water technology, Egypt, 12–15 Mar
39. Ouda S, Khalil F, Yousef H (2009) Using adaptation strategies to increase water use efficiency for maize under climate change conditions. In: Proceeding of 13th international conference on water technology, Egypt
40. Abou-Hadid AF (2006) Assessment of impacts, adaptation and vulnerability to climate change in North Africa: food production and water resources. Final report, assessments of impacts and adaptations to climate change (AIACC) project no. AF 90
41. Hassanein MK, Medany MA (2007) The impact of climate change on production of maize (Zea mays L.). In: International conference on climate change and their impacts on coastal zones and river deltas: vulnerability, mitigation and adaptation, Alexandria, Egypt, 23–25 Apr 2007, pp 268–281
42. El-Raey M (1999) Impact of climate change on Egypt. Environmental software and services. http://www.ess.co.at/GAIA/CASES/EGY/impact.html. Accessed July 2018
43. SADS 2030 (2009) Sustainable agricultural development strategy towards 2030. Ministry of Agriculture and Land Reclamation (MALR), Cairo, Egypt
44. Mertz O, Mbow C, Reenberg A, Diouf A (2009) Farmers' perceptions of climate change and agricultural adaptation strategies in rural Sahel. Environ Manage 43:804–816
45. Agbossou EK, Toukon C, Akponikpè PBI, Afouda A (2012) Climate variability and implications for maize production in Benin: a stochastic rainfall analysis. Afr Crop Sci J 20(s2):493–503

Projected Crop Coefficients Under Climate Change in Egypt

Samiha Ouda

Abstract The objective of this chapter was to quantify how climate change will affect the value of Kc for several important crops in Egypt. One way to do so is to develop a procedure to accurately estimate Kc values for 14 field crops, 7 fruit crops and 13 vegetable crops in the five agro-climatic zones of Egypt in 2030. Monthly values of evapotranspiration (ETo) in 2016 were calculated using Penman-Monteith equation (P-M) and Hargreaves-Samani equation (H-S). Then, the monthly ETo(H-S) values were regressed on monthly ETo(P-M) values, and prediction equations were developed for each agro-climatic zone of Egypt. These equations were used to project ETo values under climate change in 2030 using RCP6.0 climate change scenario resulted from MIROC5 climate change model. These values of ETo were used to run BISm model and to calculate Kc values for the studied crops, the date of each Kc growth stage and its water consumptive use in 2030. Comparison between Kc values in 2016 and 2030 for field and vegetable crops revealed that the values of Kc_{ini} were higher in 2016, compared to its counterpart values in 2030. The values of Kc_{mid} and Kc_{end} were similar or lower in 2016, compared to its counterpart values in 2030. Whereas, there was no change in the values of Kc for fruit crops between 2016 and 2030.

Keywords Penman-Monteith equation · Hargreaves-Samani equation (H-S) · MIROC5 climate change model · RCP6.0 climate change scenario

1 Introduction

The relationships between climate, crop, water, and soil are complex with many biological processes involved [1]. An increase in temperatures is expected under climate change [2] and could lead to a net deficit in atmospheric water content. Thus excessive evaporation from soil, water, and plant surfaces would occur [3]. Climate change is expected to have many adverse impacts on various sectors, yet

S. Ouda (✉)
Water Requirements and Field Irrigation Research Department, Soils, Water and Environment Research Institute, Agricultural Research Center, 9 El-Gamah Street, Giza 11211, Egypt
e-mail: samihaouda@yahoo.com

© Springer Nature Switzerland AG 2020

E.-S. E. Omran and A. M. Negm (eds.), *Climate Change Impacts on Agriculture and Food Security in Egypt*, Springer Water, https://doi.org/10.1007/978-3-030-41629-4_13

agriculture is considered to be the most tangible affected sector, as any alteration in the prevailing temperature or precipitation patterns will disturb the agricultural sector as a whole, including crop yields, crops water requirements, and soil fertility [4]. Climate change has already caused significant impacts on water resources and food security [5]. Land ecosystems would require more water to match increased water demand, and consequently to prevent drought [6]. Furthermore, it is expected that competition between agriculture and other economic sectors will increase in the future, which can reduce the allocation of water to agriculture. Reference and crop evapotranspiration rates are likely to increase, as well as irrigation requirements for the cultivated crops under climate change conditions [7] due to higher temperature, solar radiation and wind speed [8]. It was reported that climate change is expected to negatively affect crops productivity [9] and cause increases in water requirements for crops in Egypt [10]. Thus, it is essential to improve water management in agriculture and reduce unnecessary losses in the present time, adapt this behavior and carry it with us for our future generations.

Despite technological advances in crops production, such as improved crop cultivars and irrigation systems, weather and climate are still key factors in agriculture productivity. Evapotranspiration (ETo) is the key factor in determining crop water requirements in various development stages. ETo is the combination of soil evaporation and crop transpiration [11]. Efficient water management of crops requires accurate irrigation scheduling which, in turn, requires accurate measurement of crop water requirement. Because climatic parameters are the only factors affecting ETo, it is the most important hydrological and meteorological variable to reflect climate change [12]. Previous research in Egypt on the effect of climate change on ETo values revealed that temperature rise by 1 °C might increase ETo rate by about 4–5%, whereas a rise by 3 °C may increase ETo rate by about 15% [13]. Attaher et al. [14] and Khalil [15] concluded that the future climate change in 2100 would increase potential irrigation demands, due to the increase in ETo. Ouda et al. [16] stated that the value of ETo would increase by an average of 9% in 2030 and by 13% in 2040 in Egypt.

Crop evapotranspiration (ETc) and crop coefficient (Kc) are two important factors for crop production. Crop type, stage of growth, soil moisture, health of plants, cultural practices are affecting both ETc and Kc [17]. Several studies were done in Egypt to project the expected increase in water requirements for several crops under climate change conditions [18–20, 16, 21–23]. Their results proved that water requirements for crops would increase with different percentages depending on crop type, growing season and geographic location.

On the other hand, Kc is a key component in ETc determination on the field level. The variation and magnitude of Kc are important for the accurate determination of ETc. Thus, the most known and used technique to estimate ETc is the one based on the Kc approach [24] where the ETc is calculated using standard agro-meteorological variables and a crop-specific coefficient, which should take into account the relationship between atmosphere, crop physiology and agricultural practices [25]. Although several studies discussed the role of Kc in crops water requirements calculations [26, 27, 25, 28], few international studies dealt with the projected effect of climate change on Kc values of different crops. Furthermore, there is no local studies were done to tackle this subject.

Thus, the objective of this chapter was to quantify how climate change will affect the value of Kc for several important crops in Egypt. One way to do so is to develop a procedure to accurately estimate Kc values for 14 field crops, 7 fruit crops and 13 vegetable crops in the five agro-climatic zones of Egypt in 2030.

2 Methodology

2.1 Climate Change Scenario

Climate change scenario RCP6.0 resulted from MIROC5 model in 2030 was used in this analysis. It is available from the following website: http://www.ccafs.cgiar.org/marksimgcm#.Ujh1gj-GfMY. The MIROC5 model is one on the CMIP5 General Circulation Models developed by Atmosphere and Ocean Research Institute (The University of Tokyo), National Institute for Environmental Studies, and Japan Agency for Marine-Earth Science and Technology. The model has a horizontal resolution equal to $1.40° \times 1.40°$.

MIROC5 model produces four RCPs climate change scenarios. RCP6.0 climate change scenario is one of them, where it represents a larger set of mitigation scenarios and radiative forcing targets in 2100. In this scenario, the total radiative forcing is stabilized shortly after 2100, without overshoot by the application of a range of technologies and strategies for reducing greenhouse gas emissions. Thus, it is considered as a stabilization scenario [29, 30].

2.2 The Projected Value of ETo Using Hargreaves-Samani Equation

The RCP6.0 climate change scenario contained maximum and minimum temperature, as well as solar radiation data, which are suitable to calculate ETo using Hargreaves-Samani (H-S) equation [31]. However, the accuracy of Hargreaves-Samani (H-S) equation is lower than its counterpart of Penman-Monteith (P-M) equation [16], which requires using solar radiation, maximum, minimum and dew point temperature, as well as wind speed. Furthermore, Penman-Monteith (P-M) equation is widely recommended because of its detailed theoretical base and its accommodation of small time periods [32]. To solve this problem, a linear regression equation can be established with ETo values resulted from P-M plotted as the dependent variable and ETo values from H-S equation to be plotted as the independent variable. The intercept (a) and calibration slope (b) of the best-fit regression line can be used as regional calibration coefficients. This methodology was developed by

Shahidian et al. [32] as follows:

$$\mathrm{ET(P\text{-}M)} = a + b * \mathrm{ET(H\text{-}S)} \tag{1}$$

Thus, we used BISm model [17] to calculate ET(P-M) and ET(H-S) using 2016 weather data. The agro-climatic zones developed by Ouda and Noreldin [33] were used to ease calculations. Figure 1 illustrated these five agro-climatic zones.

An equation for each agro-climatic zone was developed, where different (a) and (b) values were estimated. The quality of the fit between ET(P-M) and ET(H-S) was presented in terms of the coefficient of determination (R^2), which is the ratio of the explained variance to the total variance and through calculation of root mean square error per observation (RMSE/obs), which gives the standard deviation of the model prediction error per observation [34].

Thus, the above-developed equations were used to project ETo value similar to ET(P-M) using the data of RCP6.0 scenario and apply it to BISm model [17] in 2030. The projected value of ET(P-M) was then used in BISm model [17] to calculate Kc values for the selected crops in each of the five agro-climatic zones of Egypt.

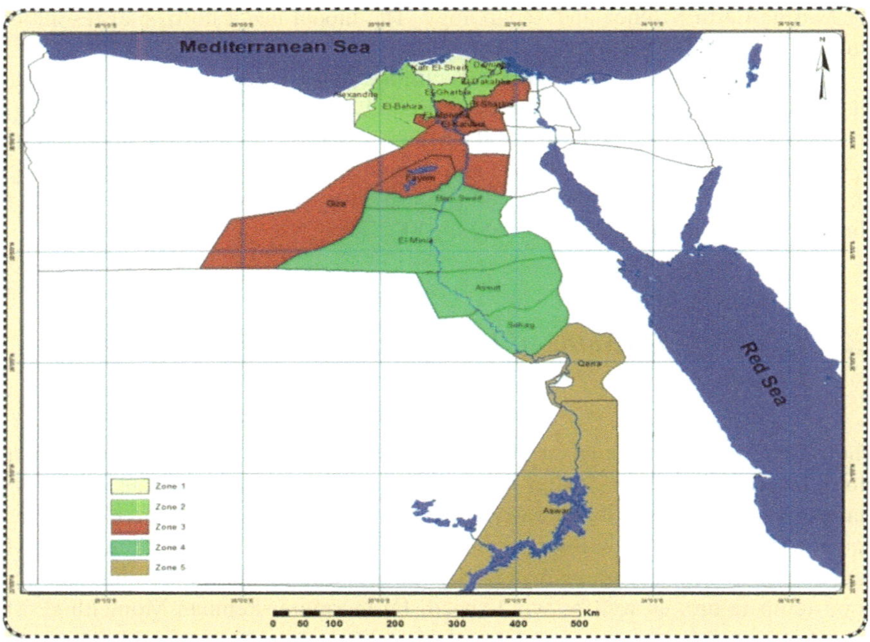

Fig. 1 Map of agro-climatic zones of Egypt. *Source* Ouda and Noreldin [33]

Table 1 Prediction equations, the coefficient of determination (R^2) and root mean square error per observation (RMSE/obs) for ET(P-M) values in the agro-climatic zones of Egypt

Zone	Prediction equation	R^2	RMSE/obs
Zone 1	ET(P-M) = 0.95 + 1.21 * ET(H-S)	0.97	0.35
Zone 2	ET(P-M) = 0.98 + 1.16 * ET(H-S)	0.97	0.31
Zone 3	ET(P-M) = 0.58 + 1.13 * ET(H-S)	0.99	0.22
Zone 4	ET(P-M) = 0.30 + 1.40 * ET(H-S)	0.99	0.35
Zone 5	ET(P-M) = 0.04 + 1.42 * ET(H-S)	0.99	0.34

2.3 The Predicted Values of ETo Using Penman-Monteith Equation in 2030

The Developed prediction equation for ET(P-M) for each agro-climatic zone are presented in Table 1. The results in the table showed that R^2 values between ET(P-M) and ETo(H-S) was between 0.97 and 0.99 in the five agro-climatic zones. Furthermore, RMSE/obs values between them were between 0.22 and 0.35 mm/day.

Figure 2 presented the monthly ET(P-M), ET(H-S) and predicted ET(P-M) values in each agro-climatic zone of Egypt. Figure showed that the predicted ET(P-M) values were close to the calculated values in most of the months in the five agro-climatic zones.

Similar results were obtained by Ouda et al. [16] when they used the above procedure to predict ET(P-M) for 17 governorates in Egypt, where R^2 values were between 0.95 and 0.99 and RMSE/obs values were between 0.03 and 0.06 mm/day.

3 Kc Values for Field Crops in the Agro-Climatic Zones in 2030

3.1 The Projected Planting and Harvest Dates for the Studied Field Crops in 2030

Under climate change in 2030 and as a result of the rise in air temperature, it is expected that planting date will be earlier by 5–7 days. Morsy [35] simulated the effect of early planting for wheat and maize under climate change in 2030 and found that early planting of both crops resulted in a reduction of yield losses. Furthermore, it is also expected that season length of the cultivated crops will be reduced under climate change. Khalil et al. [36] reported that wheat season length was reduced by 5 days in 2030, as a result of acceleration in its growing season. Similar results were obtained by Ouda et al. [37] for maize under climate change in 2030.

Table 2 presented the projected planting and harvest dates for the studied field crops planted in the five agro-climatic zones of Egypt.

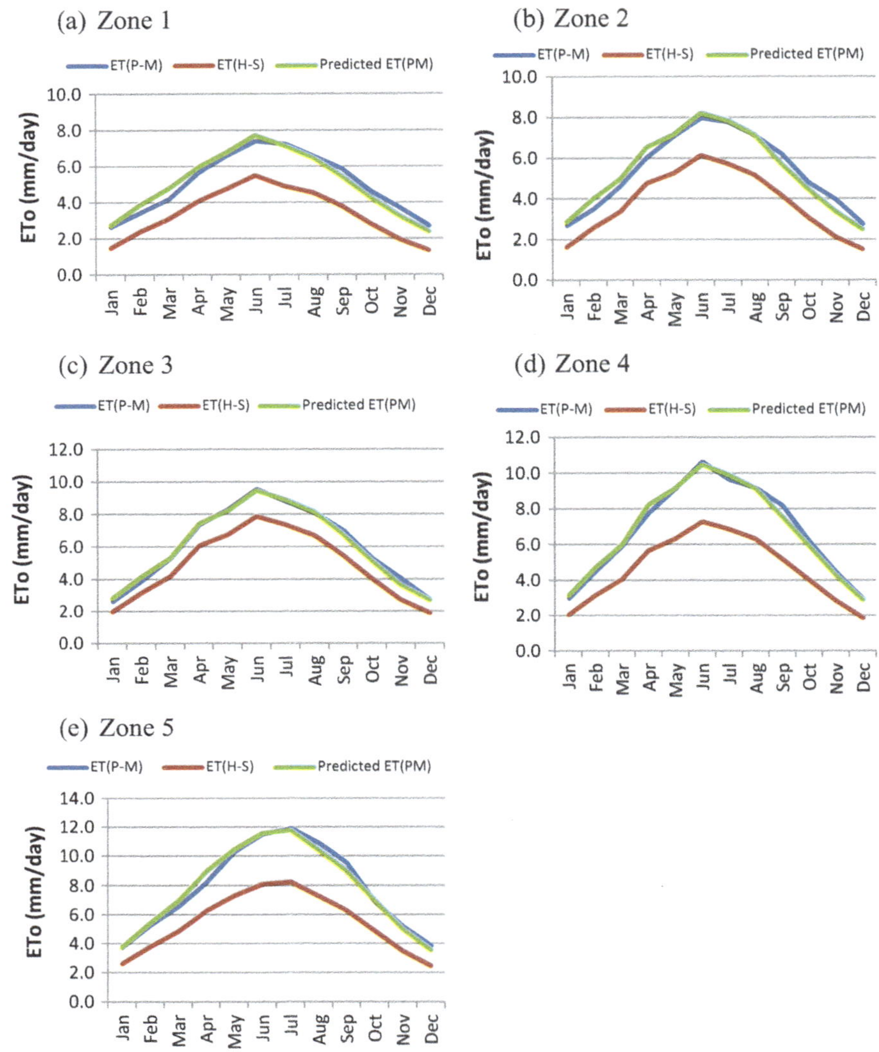

Fig. 2 Monthly ET(P-M), ET(H-S) and predicted ET(P-M) values in the agro-climatic zones of Egypt in 2016

3.2 The Projected Date of Kc Growth Stages and Its Values for the Studied Field Crops in 2030

The results of the analysis indicated that the projected date of each Kc growth stage of the selected field crops was similar in the first, second and the third agro-climatic zone (Table 3).

Table 2 The projected planting and harvest dates for the selected field crops in 2030

Crop	Planting date	Harvest date	Season length (days)
Barley	10-Nov	23-Mar	134
Bean (faba)	20-Oct	18-Apr	150
Clover	10-Oct	23-Mar	165
Cotton	10-Mar	5-Aug	149
Flax	10-Nov	5-Apr	147
Lentil	20-Oct	18-Mar	150
Maize	10-May	25-Aug	107
Rice	10-May	7-Sep	121
Sorghum	10-May	20-Aug	103
Soybean	10-May	16-Aug	99
Sugar beet	10-Oct	5-Apr	178
Sugarcane	15-Feb	14-Feb	365
Sunflower	10-May	7-Aug	90
Wheat	10-Nov	9-Apr	151

Table 3 The projected date of Kc growth stages for the studied field crops in the first, second and third agro-climatic zones in 2030

Crop	Data of growth stages		
	Kc_{ini}	Kc_{mid}	Kc_{end}
Barley	6-Dec	9-Jan	23-Mar
Bean (faba)	25-Nov	19-Dec	18-Mar
Clover	21-Oct	28-Nov	23-Mar
Cotton	1-Apr	16-Apr	5-Aug
Flax	4-Dec	15-Jan	5-Apr
Lentil	12-Nov	19-Dec	18-Mar
Maize	30-May	26-Jun	20-Aug
Rice	8-Jun	23-Jun	7-Sep
Sorghum	26-May	22-Jun	20-Aug
Soybean	29-May	23-Jun	16-Aug
Sugar beet	5-Nov	29-Dec	5-Apr
Sugarcane	13-Apr	16-Oct	09-Feb
Sunflower	28-May	19-Jun	7-Aug
Wheat	10-Dec	16-Jan	9-Apr

The analysis also showed that the projected dates of Kc growth stages were similar in the fourth and fifth agro-climatic zones. Table 3 indicated that the projected dates of Kc growth stages were different from its counterpart in the first, second and third agro-climatic zones (Table 4).

Table 4 The projected date of Kc growth stages for the studied field crops in the fourth and fifth agro-climatic zones in 2030

Crop	Data of growth stages		
	Kc_{ini}	Kc_{mid}	Kc_{end}
Barley	5-Dec	8-Jan	23-Mar
Bean (faba)	24-Nov	18-Dec	18-Mar
Clover	20-Oct	27-Nov	23-Mar
Cotton	31-Mar	15-Apr	5-Aug
Flax	3-Dec	14-Jan	5-Apr
Lentil	5-Nov	18-Dec	18-Mar
Maize	29-May	24-Jun	20-Aug
Sorghum	25-May	21-Jun	20-Aug
Soybean	28-May	22-Jun	16-Aug
Sugar beet	4-Nov	28-Dec	5-Apr
Sugarcane	12-Apr	14-Oct	09-Feb
Sunflower	26-May	17-Jun	7-Aug
Wheat	9-Dec	15-Jan	9-Apr

The results in Tables 5 and 6 indicated that the projected values of Kc_{ini} and Kc_{end} were different in the three agro-climate zone, where both were decreasing in the third agro-climatic zone, compared to the first zone. Furthermore, the value of Kc_{mid} was similar in these three agro-climatic zones.

The value of Kc_{ini} is affected by soil evaporation, which is determined by the ETo rate and irrigation frequency [24]. Thus, ETo rate is increasing from the first agro-climatic zone to the third agro-climatic zone. Furthermore, the interval between irrigation is long due to the characteristics of the prevailing clay soil in the agro-climatic zones, which results in a reduction in the value of Kc_{ini}. Regarding to Kc_{mid} and Kc_{end}, both depend on the difference in daily net radiation, soil heat flux density, crop morphology effects on turbulence and physiological differences between the crop and reference crop [17].

3.3 Comparison Between Kc Values for Field Crops in 2016 and 2030

The values of Kc for the studied field crops in 2016 were obtained from Ouda [38] to conduct a comparison between Kc values in 2016 and the projected Kc values in 2030. Figure 3 showed that in the first, second and third agro-climatic zones, the values of Kc_{ini} were lower in 2030, compared to its counterpart values in 2016, except for faba bean. The values of Kc_{mid} were higher in 2030, compared to its counterpart values in 2016 and the values of Kc_{end} were higher or similar in 2030, compared to its counterpart values in 2016.

Table 5 The projected values of Kc of the studied field crops in the first, second and third agro-climatic zones in 2030

Crop	Kc in the first agro-climatic zone			Kc in the second agro-climatic zone			Kc in the third agro-climatic zone		
	Kc_{ini}	Kc_{mid}	Kc_{end}	Kc_{ini}	Kc_{mid}	Kc_{end}	Kc_{ini}	Kc_{mid}	Kc_{end}
Barley	0.29	1.11	0.20	0.28	1.11	0.19	0.27	1.11	0.18
Bean (faba)	0.29	1.00	0.21	0.25	1.00	0.20	0.25	1.00	0.19
Clover	0.24	1.15	0.39	0.23	1.15	0.38	0.22	1.15	0.37
Cotton	0.28	0.95	0.49	0.27	0.95	0.49	0.25	0.95	0.49
Flax	0.29	1.12	0.25	0.28	1.12	0.25	0.27	1.12	0.24
Lentil	0.22	1.00	0.21	0.21	1.00	0.20	0.20	1.00	0.19
Maize	0.23	1.06	0.60	0.22	1.06	0.60	0.20	1.06	0.60
Rice	0.35	1.00	0.77	0.33	1.00	0.77	0.30	1.00	0.77
Sorghum	0.19	1.06	0.50	0.18	1.06	0.50	0.17	1.06	0.50
Soybean	0.23	1.11	0.40	0.22	1.11	0.40	0.20	1.11	0.40
Sugar beet	0.25	1.16	0.96	0.24	1.16	0.96	0.23	1.16	0.96
Sunflower	0.23	1.08	0.38	0.22	1.08	0.36	0.20	1.08	0.36
Wheat	0.29	1.08	0.19	0.28	1.08	0.18	0.28	1.08	0.17

Table 6 The projected values of Kc of the studied field crops in the fourth and fifth agro-climatic zones in 2030

Crop	Kc in the fourth agro-climatic zone			Kc in the fifth agro-climatic zone		
	Kc_{ini}	Kc_{mid}	Kc_{end}	Kc_{ini}	Kc_{mid}	Kc_{end}
Barley	0.26	1.11	0.17	0.24	1.11	0.16
Bean (faba)	0.23	1.00	0.18	0.22	1.00	0.17
Clover	0.20	1.15	0.36	0.19	1.15	0.35
Cotton	0.23	0.95	0.49	Not cultivated		
Flax	0.26	1.12	0.25	0.24	1.12	0.25
Lentil	0.19	1.00	0.18	0.18	1.00	0.17
Maize	0.19	1.06	0.60	0.17	1.06	0.60
Sorghum	0.16	1.06	0.50	0.15	1.06	0.50
Soybean	0.19	1.11	0.40	0.17	1.11	0.40
Sugar beet	0.21	1.16	0.96	Not cultivated		
Sugarcane	0.40	1.25	0.75	0.40	1.25	0.75
Sunflower	0.19	1.08	0.38	0.17	1.08	0.36
Wheat	0.26	1.08	0.16	0.24	1.08	0.16

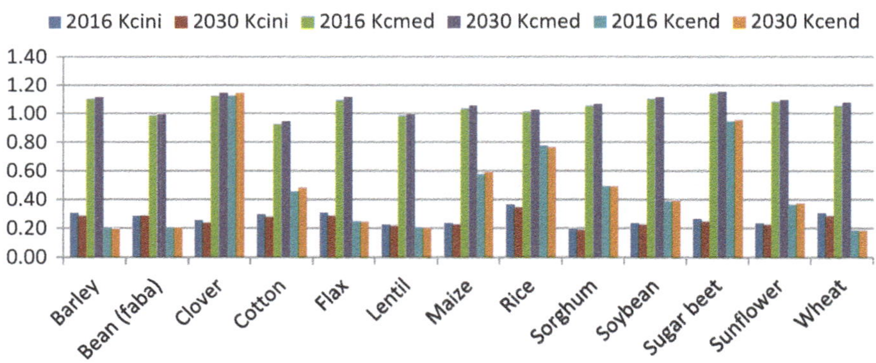

Fig. 3 Comparison between Kc values in 2016 and 2030 for studied field crops in the first, second and third agro-climatic zones

A similar trend was found in the fourth and fifth agro-climatic zones, where the values of Kc_{ini} were lower in 2030, compared to its counterpart values in 2016, except for maize, sorghum and soybean. The values of Kc_{mid} were higher in 2030, compared to its counterpart values in 2016 and the values of Kc_{end} were higher or similar in 2030, compared to its counterpart values in 2016. This increase in the Kc values, especially in the middle of the growing season, where maximum growth existed, could result in increases in the required irrigation amounts to satisfy the needs of these crops (Fig. 4).

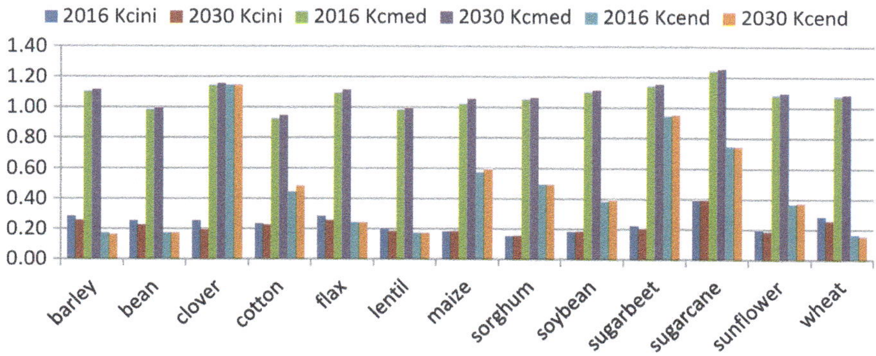

Fig. 4 Comparison between Kc values in 2016 and 2030 for studied field crops in the fourth and fifth agro-climatic zones

3.4 *The Projected Values of ETc for the Studied Field Crops in 2030*

The results in Table 7 indicated that the projected values of ETc in 2030 for the studied crops followed an increasing trend, where it was the lowest in the first agro-climatic zone and were the highest in the fifth agro-climatic zone. This is attributed to the rise in the value of climatic elements in the fifth agro-climatic zone, compared to the first agro-climatic zone. It can also be noticed from the table that cotton and sugar beet are not cultivated in the fifth agro-climatic, as a result of the unsuitability

Table 7 The values of ETc (mm) for the studied field crops in the five agro-climatic zones in 2030

	Zone 1	Zone 2	Zone 3	Zone 4	Zone 5
Barley	333	346	365	423	512
Bean (faba)	376	391	410	451	539
Clover	594	623	657	728	862
Cotton	890	952	1098	1192	Not cultivated
Flax	416	434	465	518	607
Lentil	367	382	401	442	530
Maize	637	676	760	834	915
Rice	778	827	922	Not cultivated	
Sorghum	641	682	769	845	928
Soybean	634	672	757	830	909
Sugar beet	581	610	653	721	Not cultivated
Sugarcane	Not cultivated			3127	3514
Sunflower	563	596	674	736	802
Wheat	393	410	441	488	571

Table 8 The projected planting and harvest dates for the selected fruits crops in 2030

Crop	Planting date	Harvest date	Season length (days)
Apple	10-Feb	9-Feb	365
Citrus	10-Feb	9-Feb	365
Date	10-Feb	9-Feb	365
Grape	10-Feb	9-Feb	365
Olive	10-Feb	9-Feb	365
Stone fruit	10-Feb	9-Feb	365

of the weather conditions. Rice is not cultivated in the fourth and fifth agro-climatic zones as it is prohibited by law. In addition, sugarcane is only cultivated in the fourth and fifth agro-climatic zones because weather conditions are suitable for its growth.

4 The Projected Kc Values for the Studied Fruit Crops in the Agro-Climatic Zones in 2030

4.1 The Projected Planting and Harvest Dates for the Studied Fruit Crops in 2030

Table 8 presented the projected planting and harvest dates for fruit crops, as well as its season length. It worth noting that planting date for the studied fruit crops was assumed to be 5 days earlier.

4.2 The Projected Date of Kc Growth Stages and Its Values for the Studied Fruit Crops in 2030

Table 9 presented the projected dates of Kc growth stages and its values for the studied fruit crops. The analysis indicated that the dates and the values were similar in the five agro-climatic zones of Egypt. This could be attributed to that fact that fruit trees established ground cover all year long, could make it less responsive to the weather of winter and summer seasons.

Table 9 The projected date of Kc growth stages and its values for the studied fruit crops in the agro-climatic zone in Egypt in 2030

Crop	Data of growth stages			Kc value		
	Kc_{ini}	Kc_{mid}	Kc_{end}	Kc_{ini}	Kc_{mid}	Kc_{end}
Apple	10-Feb	10-Aug	9-Feb	0.55	1.05	0.80
Citrus	10-Feb	10-Jun	19-Feb	1.00	1.00	1.00
Date	10-Feb	10-Jun	10-Feb	0.95	0.95	0.95
Grape	10-Feb	28-Apr	9-Feb	0.45	0.80	0.35
Olive	10-Feb	10-Jun	9-Feb	0.80	0.80	0.80
Stone fruit	10-Feb	10-Aug	9-Feb	0.55	1.05	0.65

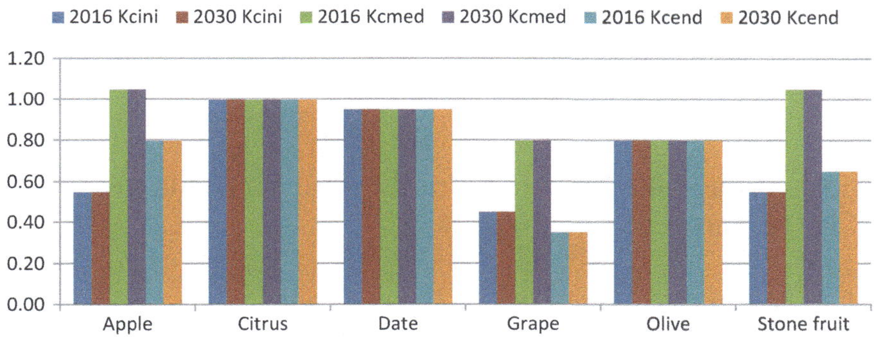

Fig. 5 Comparison between Kc values in 2016 and 2030 for studied fruit crops in the agro-climatic zones of Egypt

4.3 Comparison Between Kc Values for the Studied Fruit Crops in 2016 and 2030

The values of Kc for the studied fruit crops in 2016 were obtained from Ouda [38] to compare it with the obtained values in 2030. Figure 5 showed that there were no differences in the value of Kc in its three growth stages in 2030, compared to its counterpart values in 2016 in the five agro-climatic zones of Egypt.

4.4 The Projected Value of ETc for the Studied Fruit Crops in 2030

Table 10 showed the values of ETc for the studied fruit crops in the five agro-climatic zones of Egypt in 2030. Although the projected value of Kc for fruit crops will have similar values to its counterpart in 2016, the table showed that the projected values

Table 10 The projected values of ETc (mm) for the studied fruit crops in the five agro-climatic zones in 2030

	Zone 1	Zone 2	Zone 3	Zone 4	Zone 5
Apple	1903	2027	2259	2509	2834
Citrus	2114	2252	2520	2793	3149
Date	2008	2140	2394	2654	2992
Grape	1368	1463	1661	1841	2048
Olive	1691	1802	2016	2235	2519
Stone fruit	1914	2038	2270	2521	2849

of ETc would follow an increasing trend from the first agro-climatic zone to the fifth agro-climatic zone. This increase can be attributed to the expected rise in weather elements that could lead to a net deficit in atmospheric water content, consequently excessive evaporation from soil, water, and plant surfaces [4].

5 The Projected Kc Values for the Studied Vegetable Crops in the Agro-Climatic Zones in 2030

5.1 The Projected Planting and Harvest Dates for the Selected Vegetable Crops in 2030

The recommended planting and harvest dates for selected vegetable crops, as well as its season length are presented in Table 11. An assumption was made that sowing date for these crops in 2030 was 5 days earlier than its sowing date in 2016. Furthermore, the length of the growing season for the selected vegetable crops in 2030 is expected to be 3–5 days shorter.

5.2 The Projected Dates of Kc Growth Stages and Its Values for the Selected Vegetable Crops in 2030

Similar to the studied field crops, the results of the analysis indicated that the projected dates of each Kc growth stage of the selected vegetable crops were similar in the first, second and the third agro-climatic zone (Table 12).

Furthermore, Table 13 indicated that the date of Kc growth stages was similar in the fourth and fifth agro-climatic zones and it was different from its counterpart in the first, second and third agro-climatic zones.

The results in Table 14 indicated that the projected values of Kc_{ini} would have

Table 11 The projected planting and harvest dates of selected vegetable crops in 2030

Crop	Planting date	Harvest date	Season length (days)
Cucumber	10-Mar	7-Jun	90
Eggplant	25-Mar	29-Jul	127
Onion	10-Nov	7-Apr	149
Peas	28-Aug	23-Nov	88
Pepper	28-Mar	3-Aug	129
Potato (winter)	28-Oct	26-Jan	91
Potato (summer)	28-Jul	23-Nov	119
Squash	10-Mar	8-Jun	91
Strawberry	28-Aug	5-May	251
Sweet potato	28-Sep	25-Feb	151
Tomato (winter)	28-Apr	23-Aug	118
Tomato (summer)	28-Sep	23-Feb	149
Watermelon	10-Feb	23-May	103

Table 12 The date of Kc growth stages for the studied vegetable crops in the first, second and third agro-climatic zones in 2030

Crop	Data of growth stages		
	Kc_{ini}	Kc_{mid}	Kc_{end}
Cucumber	27-Mar	21-Apr	7-Jun
Eggplant	23-Apr	1-Jun	29-Jul
Onion	24-Nov	18-Dec	7-Apr
Peas	14-Sep	8-Oct	23-Nov
Pepper	22-Apr	25-May	3-Aug
Potato (winter)	15-Nov	7-Dec	26-Jan
Potato (summer)	20-Aug	19-Sep	23-Nov
Squash	28-Mar	24-Apr	8-Jun
Strawberry	5-Oct	18-Dec	5-May
Sweet potato	28-Oct	4-Dec	25-Feb
Tomato (winter)	27-Apr	26-Jun	23-Aug
Tomato (summer)	4-Nov	12-Dec	23-Feb
Watermelon	2-Mar	2-Apr	23-May

Table 13 The date of Kc growth stages for the studied vegetable crops in the fourth and fifth agro-climatic zones in 2030

Crop	Data of growth stages		
	Kc_{ini}	Kc_{mid}	Kc_{end}
Cucumber	26-Mar	20-Apr	7-Jun
Eggplant	22-Apr	1-Jun	29-Jul
Onion	23-Nov	19-Dec	7-Apr
Peas	13-Sep	7-Oct	23-Nov
Pepper	21-Apr	24-May	3-Aug
Potato (winter)	14-Nov	6-Dec	26-Jan
Potato (summer)	19-Aug	18-Sep	23-Nov
Squash	27-Mar	23-Apr	8-Jun
Strawberry	4-Oct	17-Dec	5-May
Sweet potato	27-Oct	3-Dec	25-Feb
Tomato (winter)	26-Apr	25-Jun	23-Aug
Tomato (summer)	3-Nov	11-Dec	23-Feb
Watermelon	1-Mar	1-Apr	23-May

different values and it will be decreasing in the three agro-climate zone. Regarding Kc_{mid} and Kc_{end}, both were similar in the first, second and third agro-climatic zones.

Moreover, in Table 15, Kc_{ini} was decreasing in the fifth agro-climatic zone, compared to the fourth zone. On the other hand, Kc_{mid} and Kc_{end} values were the same or lower, compared to its counterpart values in Table 14.

5.3 Comparison Between Kc Values for the Selected Vegetable Crops in 2016 and 2030

The values of Kc for the studied vegetable crops in 2016 were obtained from Ouda [38], and it was compared to its counterpart values in 2030. Figure 6 showed that in the first, second and third agro-climatic zone the values of Kc_{ini} were lower in 2030, compared to its counterpart values in 2016. The values of Kc_{mid} were higher in 2030 and Kc_{end} values were lower in 2030, compared to its counterpart values in 2016, except for potato, squash, and sweet potato. The projected increase in the Kc values, especially in the middle of the growing season, where maximum growth existed, will cause increases in irrigation amounts required to satisfy the needs of these crops.

The values of Kc_{ini} in the fourth and fifth agro-climatic zones were equal or lower in 2030, compared to 2016. Furthermore, the values of Kc_{mid} were higher and the values of Kc_{end} for the studied vegetable crops were lower in 2030, compared to its counterpart values in 2016, except potato, tomato and watermelon (Fig. 7).

Table 14 The projected values of Kc of the studied vegetable crops in the first, second and third agro-climatic zone in 2030

Crop	Kc in the first agro-climatic zone			Kc in the second agro-climatic zone			Kc in the third agro-climatic zone		
	Kc_{ini}	Kc_{mid}	Kc_{end}	Kc_{ini}	Kc_{mid}	Kc_{end}	Kc_{ini}	Kc_{mid}	Kc_{end}
Cucumber	0.25	0.85	0.85	0.24	0.85	0.85	0.22	0.85	0.85
Eggplant	0.22	0.89	0.84	0.21	0.89	0.84	0.19	0.89	0.84
Onion	0.28	1.21	0.54	0.27	1.21	0.54	0.26	1.21	0.54
Peas	0.20	1.01	1.01	0.19	1.01	1.01	0.18	1.01	1.01
Pepper	0.22	1.02	0.86	0.21	1.02	0.86	0.19	1.02	0.86
Potato (winter)	0.26	1.07	0.65	0.25	1.07	0.65	0.25	1.07	0.65
Potato (summer)	0.22	1.11	0.69	0.21	1.11	0.69	0.20	1.11	0.69
Squash	0.25	0.89	0.68	0.24	0.89	0.68	0.22	0.89	0.68
Strawberry	0.21	0.70	0.70	0.20	0.70	0.70	0.18	0.70	0.70
Sweet potato	0.23	1.08	0.67	0.23	1.08	0.67	0.22	1.08	0.67
Tomato (winter)	0.24	1.11	0.66	0.22	1.11	0.66	0.20	1.12	0.66
Tomato (summer)	0.24	1.09	0.64	0.23	1.09	0.64	0.22	1.09	0.64
Watermelon	0.28	1.00	0.75	0.27	1.00	0.75	0.26	1.00	0.75

Table 15 The projected values of Kc of the studied vegetable crops in the fourth and fifth agro-climatic zone in 2030

Crop	Kc in the fourth agro-climatic zone			Kc in the fifth agro-climatic zone		
	Kc_{ini}	Kc_{mid}	Kc_{end}	Kc_{ini}	Kc_{mid}	Kc_{end}
Cucumber	0.21	0.84	0.84	0.19	0.85	0.84
Eggplant	0.18	0.89	0.84	0.18	0.90	0.84
Onion	0.25	1.21	0.54	0.23	1.22	0.50
Pea	0.16	1.01	1.01	0.15	1.02	1.00
Pepper	0.18	1.02	0.86	0.18	1.00	0.85
Potato (winter)	0.23	1.07	0.65	0.22	1.00	0.70
Potato (summer)	0.18	1.11	0.69	0.16	1.10	0.70
Squash	0.20	0.89	0.68	0.19	0.90	0.70
Strawberry	0.17	0.70	0.70	0.15	0.70	0.70
Sweet potato	0.20	1.08	0.67	0.18	1.10	0.70
Tomato (winter)	0.19	1.12	0.66	0.18	1.01	0.65
Tomato (summer)	0.20	1.09	0.64	0.19	1.01	0.65
Watermelon	0.24	1.00	0.75	0.22	1.00	0.75

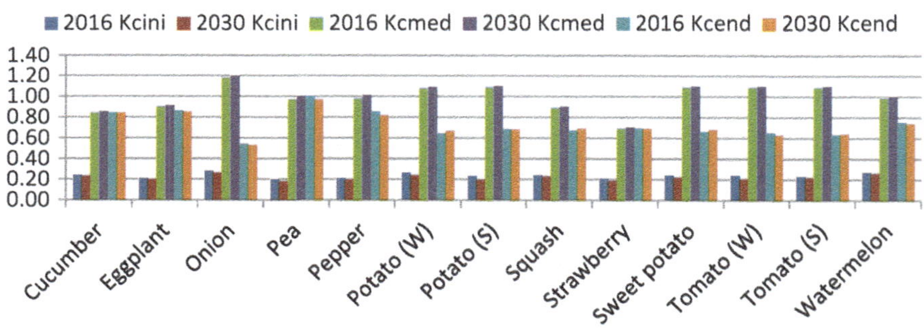

Fig. 6 Comparison between Kc values in 2016 and 2030 for studied vegetable crops in the first, second and third agro-climatic zones of Egypt

5.4 *The Value of ETc for the Selected Vegetable Crops*

Similar to field and fruit crops, Table 16 showed that there would be an increasing trend in the values of ETc from the first agro-climatic zone to the fifth agro-climatic zone of Egypt in 2030.

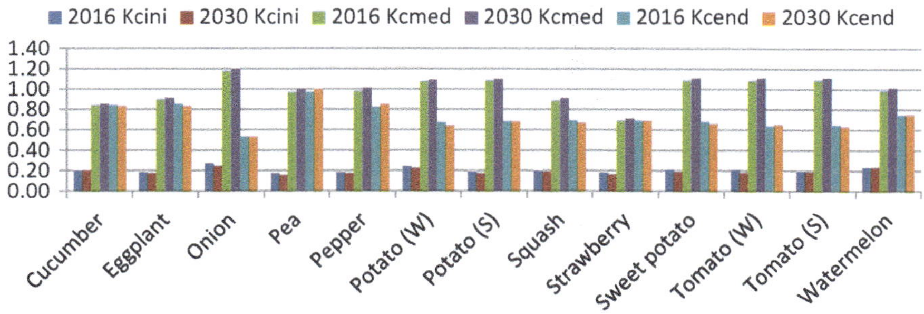

Fig. 7 Comparison between Kc values in 2016 and 2030 for studied vegetable crops in the fourth and fifth agro-climatic zones of Egypt

Table 16 The values of ETc (mm) for the studied vegetable crops in the five agro-climatic zones in 2030

	Zone 1	Zone 2	Zone 3	Zone 4	Zone 5
Cucumber	397	427	498	534	613
Eggplant	645	687	784	852	929
Onion	519	544	582	645	736
Pea	335	352	377	424	474
Pepper	779	832	951	1035	1133
Potato (W)	219	229	227	247	305
Potato (S)	532	561	609	686	771
Squash	400	431	502	539	617
Strawberry	610	643	701	765	868
Sweet potato	398	415	424	464	559
Tomato (W)	371	386	394	430	519
Tomato (S)	727	773	868	953	1047
Watermelon	641	700	816	881	958

6 Conclusions

Quantification of the impact of climate change on Kc values for several crops is very important for policymakers when developing their future water management plans. This requires an accurate equation to calculate ETo values. Because only monthly air temperature and solar radiation are available in the RCP6.0 climate change scenario, it is impossible to use P-M equation. Instead, monthly ETo can be calculated using H-S equation, and the developed prediction equations for ET(P-M) can be used to calculate the values of monthly ETo using the developed calibration coefficients for each agro-climatic zone. Our results showed that this method was accurate and the predicted ETo values were close to the calculated ETo values by the P-M equation.

Thus, it is recommended to use this procedure in case of unavailability of wind speed and dew point temperature values.

The BISm model was used to project Kc values in 2030. The results indicated that Kc_{ini} values for field and vegetable crops were lower in 2030, compared to its counterpart values in 2016. The values of Kc_{mid} were higher in 2030, and the value of Kc_{end} was similar or higher in 2030, compared to its counterpart values in 2016. This increase in the Kc values, especially in the middle of the growing season, where maximum growth existed, results in increases in the required irrigation amounts to satisfy the needs of these crops. Whereas, there was no change in the values of Kc for fruit crops between 2016 and 2030. This could be attributed to that fact that fruit trees established ground cover all year long, which makes it less responsive to weather variation between growing seasons. However, the projected values of ETc will increase. This increase can be attributed to the expected rise in weather elements that could lead to a net deficit in atmospheric water content, consequently excessive evaporation from soil, water, and plant surfaces.

7 Recommendations

1. The policymakers should take into account the expected consequences of climate change in their future plans, regulate the amount of available water for agriculture, and distribute it on the basis of crops needs in each agro-climatic zone.
2. The policymakers should make efforts to improve irrigation water transport, distribution, and application efficiency to reduce water losses through evaporation and deep percolation.
3. The prevailing irrigation system in Egypt is surface irrigation, which endures high water losses; therefore policymakers should change this type of system to more efficient one.

References

1. Rao KPC, Ndegwa WG, Kizito K, Oyoo A (2011) Climate variability and change: farmer perceptions and understanding of intra-seasonal variability in rainfall and associated risk in semi-arid Kenya. Exp Agric 47:267–291
2. IPCC: Intergovernmental Panel on Climate Change (2007) Intergovernmental panel on climate change fourth assessment report: climate change 2007. Synthesis report. World Meteorological Organization, Geneva, Switzerland
3. Kimball BA, Kobayashi K, Bindi M (2002) Responses of agricultural crops to free-air CO_2 enrichment. Adv Agron 77:293–368
4. El-Massah S, Omran G (2014) Would climate change affect the imports of cereals? The case of Egypt. In: Handbook of climate change adaptation. Springer-Verlag Berlin Heidelberg, Berlin

5. IPCC: Intergovernmental Panel on Climate Change (2013) Summary for policymakers. In: Stocker TF, Qin D, Plattner GK, Tignor M, Allen SK, Boschung J, Nauels A, Xia Y, Bex V, Midgley PM (eds) Climate change. The physical science basis. Contribution of working group I to the fifth assessment report of the intergovernmental panel on climate change. Cambridge University Press, Cambridge, United Kingdom and New York, NY, USA
6. Karmakar R, Das I, Dutta D, Rakshit A (2016) Potential effects of climate change on soil properties: a review. Sci Int 4(2):51–73
7. Onol B, Semazzi FHM, Unal YS, Dalfes HN (2006) Regional climatic impacts of global warming over the eastern Mediterranean. In: International conference on past, present and future climate change and the Middle East, pp 20–23. Istanbul Technical University, Turkey
8. Abtew W, Melesse A (2013) Climate change and evapotranspiration. In: Evaporation and evapotranspiration: measurements and estimations. Springer Science Business Media, Dordrecht. https://doi.org/10.1007/978-94-007-4737-113
9. Ouda S, Khalil F, Noreldin T (2013) Simulation of adaptation to climate change on wheat and maize in Egypt. LAMBERT Academic Publishing. ISBN: 978-3-659-44279-7
10. Ouda S, Noreldin T, Abd El-Latif K (2015) Water requirements for wheat and maize under climate change in North Nile Delta. Span J Agric Res 13(1):1–7
11. Gardner FP, Pearce RB, Mitchell RL (1985) Physiology of crop plants. Iowa State University Press, Ames, USA
12. Rahman A, Karim NN, Kadir N, Naher T (2015) Impacts of climate change on crop coefficient and reference crop evapotranspiration of boro rice in north-east hydrological region of Bangladesh. In: 5th international conference on water and flood management, pp 345–350
13. Eid H (2001) Climate change studies on Egyptian agriculture. Soils, Water and Environment Research Institute (SWERI) ARC, Ministry of Agriculture, Giza, Egypt
14. Attaher S, Medany M, Abdel Aziz AA, El-Gindy A (2006) Irrigation-water demands under current and future climate conditions in Egypt. In: The 14th annual conference of the MISR society of agricultural engineering, pp 1051–1063
15. Khalil AA (2013) Effect of climate change on evapotranspiration in Egypt. Researcher 51:7–12
16. Ouda S, Noreldin T, Hosny M (2016) Evapotranspiration under changing climate. In: Major crops and water scarcity in Egypt. Springer Publishing House, pp 1–22. ISBN: 978-3-319-21770-3
17. Snyder RL, Orang M, Bali K, Eching S (2004) Basic irrigation scheduling (BIS). http://www.waterplan.water.ca.gov/landwateruse/wateruse/Ag/CUP/Californi/Climate_Data_010804.xls
18. Mahmoud A, Ouda S, Abd El-Hafez S (2016) High water consuming crops under control: I. Case of rice crop. In: Major crops and water scarcity in Egypt. Springer Publishing House, pp 69–82. ISBN: 978-3-319-21770-3
19. Noreldin T, Ouda S, Taha A (2016) Combating adverse consequences of climate change on maize crop. In: Major crops and water scarcity in Egypt. Springer Publishing House, pp 53–67. ISBN: 978-3-319-21770-3
20. Ouda S, Zohry AA (2016) Significance of reduction of applied irrigation water to wheat crop. In: Major crops and water scarcity in Egypt. Springer Publishing House, pp 33–50. ISBN: 978-3-319-21770-3
21. Taha A, Ouda S, Zohry AA (2016) High water consuming crops under control: II. Case of sugarcane crop. In: Major crops and water scarcity in Egypt. Springer Publishing House, pp 85–96. ISBN: 978-3-319-21770-3
22. Zohry AA, Ouda S (2016a) Crops intensification to face climate induced water scarcity in Nile Delta region. In: Management of climate induced drought and water scarcity in Egypt: unconventional solutions. Springer Publishing House. ISBN: 978-3-319-33659-6
23. Zohry AA, Ouda S (2016b) Upper Egypt: management of high water consumption crops by intensification. In: Management of climate induced drought and water scarcity in Egypt: unconventional solutions. Springer Publishing House. ISBN: 978-3-319-33659-6
24. Allen RG, Pereira LS, Raes D, Smith M (1998) Crop evapotranspiration—guidelines for computing crop water requirements. FAO irrigation and drainage paper 56. FAO, Rome, Italy, p D05109

25. Reddy KC, Arunajyothy S, Mallikarjuna P (2015) Crop coefficients of some selected crops of Andhra Pradesh. J Inst Eng (India) Ser A 96:123–130
26. Annandale JG, Stockle CO (1994) Fluctuation of crop evapotranspiration coefficients with weather. A sensitivity analysis. Irrig Sci 15:1–7
27. Ko J, Piccinni G, Marek T, Howell T (2009) Determination of growth-stage-specific crop coefficients (Kc) of cotton and wheat. Agric Water Manag 96:1691–1697
28. Wright JL (1982) New evapotranspiration crop coefficients. ASCE J Irrig Drain Div 108:57–74
29. Fujino J, Nair R, Kainuma M, Masui T, Matsuoka Y (2006) Multi-gas mitigation analysis on stabilization scenarios using AIM global model. Multi-gas mitigation and climate policy. Energy J 27:343–353
30. Hijioka Y, Matsuoka Y, Nishimoto H, Masui M, Kainuma M (2008) Global GHG emissions scenarios under GHG concentration stabilization targets. J Glob Environ Eng 13:97–108
31. Hargreaves GH, Samani ZA (1985) Reference crop evapotranspiration from temperature. Appl Eng Agric 1(2):96–99
32. Shahidian S, Serralheiro R, Serrano J, Teixeira J, Haie N, Francisco S (2012) Hargreaves and other reduced-set methods for calculating evapotranspiration. In: Irmak A (ed) Evapotranspiration—remote sensing and modeling. In Tech. ISBN: 978-953-307-808-3. Available from: http://www.intechopen.com/books/evapotranspiration-remote-sensing-and-modeling/hargreaves-and-otherreduced-set-methods-for-calculating-evapotranspiration
33. Ouda S, Noreldin T (2017) Evapotranspiration data to determine agro-climatic zones in Egypt. J Water Land Dev 32(I–III):79–86
34. Jamieson PD, Porter JR, Goudriaan J, Ritchie JT, van Keulen H, Stol W (1998) A comparison of the models AFRCWHEAT2, CERES-Wheat, Sirius, SUCROS2 and SWHEAT with measurements from wheat grown under drought. Field Crops Res 55:23–44
35. Morsy M (2015) Use of regional climate and crop simulation models to predict wheat and maize productivity and their adaptation under climate change. PhD thesis, Faculty of Science, Al-Azhar University
36. Khalil FA, Farag H, El Afandi G, Ouda SA (2009) Vulnerability and adaptation of wheat to climate change in Middle Egypt. In: Proceeding of the 13th international conference on water technology, Egypt, 12–15 Mar
37. Ouda SA, Khalil FA, Yousef H (2009) Using adaptation strategies to increase water use efficiency for maize under climate change conditions. In: Proceeding of 13th international conference on water technology, Egypt
38. Ouda S (2019) Accurate estimation of crop coefficients for better irrigation water management in Egypt. In: Technological and modern irrigation environment in Egypt: best management practices & evaluation (in press)

Rice Production in Egypt: The Challenges of Climate Change and Water Deficiency

Heba Elbasiouny and Fathy Elbehiry

Abstract Climate change has become one of the major global environmental problems of the 21st century. Rice is the main cereal crop for over 50% of the world's population. Rice cultivation is known as an important emitter of greenhouse gases emission especially methane due to rice management practices and burning of rice straw after harvesting. However, many studies confirmed that rice soils accumulate carbon higher than other crops such as wheat and corn. The cultivated area of rice in Egypt is approximately 650,000 ha from the whole cultivated area in Egypt; approximately 3.3 million ha; i.e. around 20% of the cultivated area in Egypt. Egypt relies on the Nile for 97% of its water requirements. The expected scenario of water deficiency in Nasser lake due to the Grand Ethiopian Renaissance Dam construction, with pulling of deficiency from Dam Lake; is emphasizing on wasting approximately 1.7 million ha of Egypt's cultivated area. As well, the expected high scenario of a relative sea level rise in Egypt; especially Nile Delta increases the amount of land that lying under risk from inundation in the north Nile Delta by 300 km^2, which estimated by one-fifth of the total agricultural land in the northeast Nile Delta only. Also, all crops are projected to have a decrease in yields and an increase in irrigation needs. Thus; all these challenges will increase the stresses on rice production and decrease soil C storage in Egypt as a result of climate change and water shortage due to establishing GERD. Therefore, the changing in rice management practice; such as decreasing ploughing, creating another alternative to rice straw burning and balanced fertilizer application; will lead to mitigating of greenhouse gases emission from rice cultivation and improving soil organic matter (SOM) stocks, subsequently soil quality and productivity.

H. Elbasiouny (✉)
Environmental and Biological Sciences Department, Home Economics Faculty, Al-Azhar University, Tanta 31732, Egypt
e-mail: Hebaelbasiouny@azhar.edu.eg; Hebayehia79@hotmail.com

F. Elbehiry
Central Laboratory of Environmental Studies, Kafr El-Sheikh University, Al-Geish Street, Kafr El-Sheikh 33516, Egypt
e-mail: Fathyelbehiry@gmail.com

© Springer Nature Switzerland AG 2020

E.-S. E. Omran and A. M. Negm (eds.), *Climate Change Impacts on Agriculture and Food Security in Egypt*, Springer Water, https://doi.org/10.1007/978-3-030-41629-4_14

Keywords Rice production · Climate change · Water deficiency · Soil carbon sequestration · Egypt

1 Introduction

Rice (*Oryza sativa*) is one of the main field crops and a major staple food in most parts of the world. More than 160 million ha of arable land dedicated to its production, and more than 730 million tons produced in 2012. The top producers in the world are China, India, and Indonesia [1]. Flooding/Basin irrigation is the most common pattern for irrigating rice; it is always called paddies or paddy basins. Basin irrigation is useful if leaching is required to remove salts from the soil [2]. Rice crop is stunningly diverse and unique because of its ability to grow in wet ecosystems, while other crops cannot survive. Rice paddies are estimated by a large portion of the wetland ecosystem; mainly in Asian countries [3]. It is the most broadly grown under irrigation [2]. Paddy soils are featured by high organic material inputs with rather a low decomposition rate under anaerobic conditions, which favors organic matter accumulation [4]. Also, paddy rice is best grown on clayey soils which are almost impermeable to decline losses by percolation. Rice could also be cultivated on sandy soils, but losses by percolation are high without maintaining a shallow water table. Such conditions sometimes take place in valley bottoms. Loamy soils are preferred with basin irrigation to avoid waterlogging (permanent saturation of the soil) [2].

Climate change has been one of the main topics in environmental policy. It is highlighted as a major security issue. Climate change impacts include increased frequency of flash flood events, droughts, or periods of water deficiencies and rising temperatures. As well, sea level rise is also an important issue. Future observed trends and projections indicate a strong susceptibility to changes in hydrological regimes, a rising general deficiency of water resources and thus threats to water availability and management. The volume of water consumed in Egypt is about 68 km^3 of which 86% is dedicated to agriculture. Ninety-five percent of the water consumed in Egypt derives from the Nile River. In the context of the construction of the Grand Ethiopian Renaissance Dam under establishment on the upper Nile, geopolitical stability requires better knowledge of water threats over the region [5].

Mitigation atmospheric GHGs (CO_2, CH_4 and N_2O) which contributes considerably to global warming is generating a major challenge in recent day agriculture. Soil organic carbon (SOC) content is one of the most important indicators of soil fertility or quality. Soil carbon, as included globally in an important biogeochemical C cycle, is of great significance to climate change. Therefore, enhanced agricultural management practices have been studied to reduce the loss of SOC and crop residues, sequestering SOC, reducing atmospheric GHGs, especially there is need to reduce GHG emissions from agriculture and rice production while increasing or maintaining rice production levels which consider a major challenge [6, 7].

2 Rice Production in the World

Rice is one of the fundamental agricultural commodities and food sources, feeding approximately 50% of the world's population (more than 3.5 billion people) as a staple food, which directly nourishes people more than any other crop. It is covering about 160 Mha of the world's land surface [1–4, 7–11]. Moreover, rice is a key source of employment and income for the rural people most of whom live in developing countries. The crop occupies one-third of the world's total area cultivated by cereals [2]. Approximately 75% rice is produced worldwide in irrigated lowlands [10]. Staple foods are well known as reasonable, regularly available, energy-rich and essential for daily life. Mainly staple foods are resulted from cereals, such as rice, maize, wheat, and barley. Rice and maize either single and double-crop rice, spring, and summer maize) are the most staple food crops [4]. Rice is a healthy, nutritious, and versatile food. Its content of complicated carbohydrates is transformed to glycogen by digestive processes in the human body. Glycogen is stored in muscle tissues and released as a required energy for the activities. As the rice being the staple food of most Egyptians, its local consumption rate is 35–40 kg/capita/annum [12].

Over 90% of rice is produced and consumed in Asian, and the African countries, where mostly all rice is produced by smallholders on farms ranges from 0.5 to 3 ha. In many Asian counties, rice yield/ha has doubled up through 20 years of the Green Revolution. There are many reasons contributing in this yield such as the adoption of high yielding rice varieties and increasing the cropping intensity and farm inputs. Therefore, during the 1970s and 1980s, global rice production increased at a rate of 2.3–2.5% per year. Consequently, however, the growth rate declined to 1.5% during 1990s as well during the first decade of 21st century, primarily due to land degradation problems. The population in major rice consuming countries are increased at an annual rate of 1.5%, and thus, the present growth rate of rice production is not enough to feed them; therefore, a gradual annual increase of about two million tons in production is required to fulfill the rising demand [13]. Rice is cultivated in 16 countries in the Near East region; Afghanistan, Algeria, Azerbaijan, Egypt, Iran, Iraq, Kazakhstan, Kyrgyzstan, Mauritania, Morocco, Pakistan, Somalia, Sudan, Tajikistan, Turkey, and Turkmenistan. Approximately, 92% of this region rice yield is produced mainly in Pakistan, Egypt, and Iran [2].

It is stated that to cope with the world's rising rice consumption level, the world's rice production also expanded. Rice production has increased greatly since the 1960s. During the last decades of 20th century, rice has seen steady increases in demand. Its growing importance is evident in the strategic food security planning policies. This is applied in many countries, except in few countries in the Near East Region that have accomplished self-sufficiency in rice production and imported the large quantities of rice that overtake the demand to meet demand at a huge charge in hard currency [2]. It is reported that it is estimated that a 40% increase in rice production is needed to meet the surging demand from the hastily increasing population by the end of 2030. Numerous strategies of agricultural management are now being developed for improving rice productivity as well as soil C sequestration, such as water and fertilization management, cultivation methods, developing new rice cultivars, and utilizing of interesting materials such as biochar and rice straw [14].

3 Rice Production in Egypt

Only about 4% of Egypt's total area is agricultural land. This area includes one of the heaviest population densities worldwide. The remaining 96% of Egypt's land area is arid desert [15]. Egypt is the largest producer of rice in the Near East region, as well it has the most productive rice farms worldwide with an average yield of 9×5 ton ha^{-1} based on 2012 data. As well, with a whole production of about 6 million-ton $year^{-1}$, Egypt is the highest rice producer in Africa [1]. Therefore, rice is one of the major field crops in Egypt. It engages about 0.65 million ha with about 6 million metric tons yearly of rough rice contributing about 20% to the per capita cereal consumption. Rice, in Egypt, is one of the potential export crops that can supply foreign exchange to the country. Moreover, it has an important socio-economic influence since many of the labor force is employed in the rice production sector. The inability to accomplish self-sufficiency in rice production in some countries is resulting from several major limitations; mainly water scarcity [2].

Rice is mainly grown in the Nile Delta [1, 12]. However, the country has several production zones and grades as one of the highest producers per area unit in the world [2]. Rice cultivation is intensified in the Delta mainly because of its soils that consisted of a thick clay layer. The formation of this layer was associated with the deposition of sediments that were carried along the Nile river path by historic floods. Sediment depositions in the northern part of the Delta (coastal area) led to forming a compact muddy clay layer that is almost impermeable. Thus, soil drainage is hindered by heavy clays that extended over large areas of this region, endangering the cultivation of non-flooded crops. It worthy noticed that the northern strip of the Delta is also distinguished by highly saline groundwater because of subsurface intrusion of seawater and/or marine entrance which attributed to continuous submergence of this Delta part underneath sea water in historical periods. Intensive irrigation of the low-permeable soils in the northern Delta would result in long flooding period, thus creating an appropriate environment for high water-consuming plants, such as rice and berseem [12]. It is stated that in Egypt and most other developing countries that produce rice, it is commonly cultivated under continuous flooding with about 5 cm depth of standing water during the growing season [1]. Rice cultivation in such method would also lead to leaching salts and/or pushing away salty groundwater from the root zone. Subsequently, poorly drained parts of the northern Delta may be considered as areas for rice and berseem cultivation only, being inappropriate for rotation alternatives. The other area of the Delta (the southern part) varies in its suitability for crop rotations. Therefore, cultivated rice in the southern part of the Delta may be considered as useful for a cash crop as well as for its role in leaching salts and/or improving soil conditions [12]. A comparison of the net profits among crops shows that cultivating rice has higher benefit to farmers; subsequently, many farmers in the Nile Delta have converted to intensive paddy rice despite the implications of some of commonly used practices on soil quality, environment and natural resources which have been mostly ignored in most of the developing countries [1].

According to the Egyptian agriculture calendar, rice is a summer crop. Rice areas in Egypt have steadily increased after the construction of the nationwide irrigation network in the 19th century. Rice cultivations is usually rotated with cotton and maize cultivation in the two types of crop rotation commonly carried out in the Nile Delta (i.e. two-year and three-year rotation systems) [12]. Those rotation systems present two patterns of crop rotation in Egypt, one of them include rice with the rotation (i.e. berseem, wheat, rice, cotton, maize, and beans) and other without rice (i.e. berseem, cotton, wheat, maize, beans, and vegetables) [1]. There is limited potential for an additional increase of the rice area in Egypt because of rice is high water consumption crop thus all area has to be irrigated therefore the supply of irrigation water is the most important limiting factor. Moreover, many other factors, including soil type, climate, also controlling the choice of suitable areas for rice cultivation. On the other hand, there are many economic factors should be considered by the farmers like yield, cost, farm-gate price, and net return to take a decision regarding cultivate or not cultivate rice [2].

4 The Potential of Paddy Soils for Carbon Sequestration

The soil is an important part of the global carbon (C) cycle and has the double potential to store C than the atmosphere. The SOC plays a vital role in enhancing soil fertility as well as sustaining soil productivity because of its influences on soil physical, chemical, and biological properties. Furthermore, climate change feedback and crop productivity in agricultural soils essentially depend on SOC dynamics and C storage [16–19]. The SOC sequestration in cropland could reduce agricultural GHGs by approximately 90% by improved management practices, such as minimum or no tillage, fertilization, perennial or extended cropping systems, manure application, crop residue recycling, and irrigation practices … etc. [20, 21]. As well, changes in SOC are affected by many management practices, such as fertilizer application, straw return, and tillage. However, SOC is always not sensitive to short-term changes in agricultural management practices because of large background levels of SOC [21].

The concentration of atmospheric CO_2 in has increased from 280 μmol mol^{-1} before the industrial revolution to 391 μmol mol^{-1} in 2011. Much attention has been paid to carbon (C) sequestration for reducing the CO_2 concentration to mitigate global climate change [14, 22, 20, 23]. Soil acts both as source for greenhouse gases (GHGs) (by releasing CO_2 and CH_4 to the atmosphere through soil respiration and anaerobic decomposition) and sink of GHGs by sequestering SOC [18, 23] depending on soil use and management [22]. Concerns regarding rising atmospheric CO_2 levels have driven considerable interest recently concerning the potential of SOC as a sink for atmospheric CO_2. Because of the important role of SOC in terrestrial ecosystems and its large stock, minor changes in SOC due to disturbances, such as changes in land use or climate, may influence not only long-term ecosystem functions but also the global atmospheric carbon budget and. Cropland soils contain slightly more than 10% (about 170 Pg C) of the total SOC pool. Therefore, a great attention is paid to carbon sequestration in agricultural soils [24] because of its potential impact on global climate change [22]. Subsequently, boosting C sequestration and minimizing GHG emissions has become one of the essential tasks worldwide to the effective combat of upcoming climate change. In considering the magnitude of soil C stock, even small changes in this reservoir can exert a substantial influence on gaseous C emissions and concentrations of atmospheric CH_4 and CO_2, thus affecting global climate change. Under active human interference and cultivation, the soils in paddy fields usually have a larger potentiality of C sequestration than natural wetland soils [14].

Respecting soil C stocks, submerged rice ecology has also emerged as a potential C sink. Very few studies have demonstrated the unique soil C chemistry in rice soils. Slow decomposition of organic substances is common in rice soils under extended waterlogging, anaerobic conditions due to depletion of O_2 levels and the absence of iron oxides and hydroxides as electron acceptors. This leads to higher accumulation of stable fractions of C or in the other meaning, SOC sequestration in rice systems [18]. However, under such conditions of submergence and increasing the quantity of SOM, the degradation of soil quality because of the breakdown of stable aggregates and deterioration of soil organic matter occurs. Crop rotations are known to favor the enforcement of SOC and improving soil nutrients comparing to monocultures. Continuous monoculture will not be active in sequestering C [25]. It is stated that in the past decades, the SOC declined in high-yielding cropping systems, especially in rice production systems, due to using chemical fertilizers and pesticides instead cover crops and organic mature to retain crop growth and to increase grain yields [17]. However, SOC accumulation was attributed to the increased application of chemical fertilizers that stimulate greater rice yields, higher biomass production, and higher returning of crop residues to the soil, over the last several decades [23].

Owing to its high accumulation rate of SOC, rice cultivation may play a substantial role in mitigation CO_2 in the atmosphere [23]. The dynamics of soil carbon (C) and nitrogen (N) in submerged rice soils are different from those of aerobic, because of maintaining submerged rice soils at lower redox potentials. Recently, stagnation or decline in yields has been observed worldwide under the intensive rice-based cultivation systems; this is attributed to the loss of quality and quantity of SOC which influenced nutrient supply, specifically N [3]. It is reported that SOC in the surface layer (0–20 cm) of paddy field is higher than its corresponding in the upland croplands. They explained that the strong aggregate stability of paddy soil boosts the SOC conservation and the enrichment of SOC in macro-aggregates, resulting in a greater carbon sequestration potential in this soil. They also reported a declining tendency of SOC after paddy conversion into the vegetable field [26].

On the other hand, rice agriculture contributes meaningfully to global straw production. These agricultural residues are spread in the field, removed from the field, burned in situ, piled, incorporated to the soil, or mulched on the next crop. In the past, straw was regularly removed from the field and used as fuels or construction materials in many countries. Agricultural lands worldwide suffer from the increment of pest attacks and hardening of soil agglomerates because of the excessive fertilizer application. This has led to enlarge the practice of applying straw return to the field after harvesting, which has led to improve soil fertility, upgrade soil physical and chemical properties, enhance crop yield, boost soil C sequestration, and mitigate GHG emissions [14]. However, the farmers, to save time and labor, were used to directly burn straw in fields, causing serious atmospheric problems [27]. It is stated that utilizing of plant residues as mulch, instead of burning, has valuable effects for refilling SOC, and returning to the soil of 1 Mg ha^{-1} of rice, wheat, and maize straw each year, thus sequestering about 130 kg C ha^{-1} year^{-1} [24]. It is reported that the rates of rice residue inputs in rice production systems managed by farmers are variable, so the change of SOC in field experiments does not sufficiently represent the actuality of C sequestration in paddy ecosystems. Therefore, it is important to know more information about SOC responding to variation in C inputs due to changes in paddy rice production [23].

5 Rice Fields as a Source of Greenhouse Gas (GHG) Emissions

Climate change has been one of the main global environmental issues of the 21st century, with increasing anthropogenic GHG emissions being the principal reason. Nitrous oxide (N_2O) and methane (CH_4) are the two significant GHGs, due to their positive increases for radiative forcing and the permanence in the atmosphere, responsible for global warming, which contribute to shape the overall earth's climate system [14, 27, 28]. It is mentioned that agriculture is responsible for about 13% of annual GHG emissions that are related to all human activities [29]. While, it is stated that agriculture only contributes to approximately 20% of the current atmospheric GHG concentrations, with methane (CH_4) and carbon dioxide (CO_2) as the two most effective carbon-containing GHGs released from agricultural activities [14]. However, it is reported that approximately 50.1 $GtCO_2e$ GHG emissions were released from anthropogenic sources in 2010 worldwide. Agriculture only accounts for approximately 5.0–5.8 $GtCO_2e$ (i.e. 10–12%) of these emissions. They also pointed to these percentages only are the direct sources. If indirect sources are also taken into consideration, agriculture probably accounts for an additional 3–6% of the global emissions. Modern intensive farming, which heavily depends on irrigation and chemical fertilizer application, is the largest source of CH_4 and nitrous oxide (N_2O) emissions [13, 30]. It is counted approximately 50% or more than 50% of the global anthropogenic emissions of N_2O and CH_4 released from agricultural soils [27, 29]. While it is reported that flooded-rice ecosystems accounted for approximately for 20% of the total global CH_4 budget [31]. It is considered that edaphic condition and meteorological factor, in addition to agricultural management such as tillage, irrigation, fertilization, and organic amendment application could significantly affect N_2O and CH_4 flux [27].

Rice paddy fields play an important role in the global budget of GHGs, such as CO_2 and CH_4 [10]. Rice farming systems tend to consume higher energy and have a higher carbon footprint than many further comparable cropping systems. As well, production, packaging, transportation, and the utilizing of extra farm inputs need more energy. Subsequently it is likely to emit more GHGs. For example, since 1990–2005, the global agriculture emissions increased by approximately 14%, an average rate of 49 $MtCO_2e$ year^{-1} [13]. Rice is an important emitter of CH_4 which contribute to the warming as 19–25 times higher than that of CO_2 per unit of weight based on 100-year global warming potentials [32]. Rice production does not only play a major role in sustaining global food security, creates wealth and jobs in the cultivating areas, but also results in significant environmental impacts such as atmospheric GHG emissions [7, 31]. It is mentioned that CH_4 is produced by methanogens in flooded soil and released to the atmosphere through the rice growing season. N_2O is produced by nitrification and denitrification processes primarily from agricultural soil management activities, such as OM application, fertilization, and irrigation. They added that the global paddy rice cultivation in 2000–2010 emitted 22–25 Mt CH_4 year^{-1} (i.e. 472–518 Mt CO_2eq year^{-1}). The annual total non-CO_2 GHG emissions from agriculture in 2000–2010 was reported to range from 4.6 to 5.1 Gt CO_2eq year^{-1}, representing 57% from N_2O emission and 43% from CH_4 emission. Approximately, 75% of worldwide rice production is performed in continuously flooded paddies. Farmers believe that this practice has many advantages such as retaining soil moisture and temperature, increasing soil C, and suppressing the soil-borne disease and weeds. However, flooding causes anaerobic conditions and therefore promotes methanogenesis and methane emissions [8, 10, 13]. It is also emphasized on paddy rise and pond aquaculture as major sources of atmospheric CH_4 and N_2O, mainly due to the periodic dry/wet alteration episodes and intensive inputs of organic material and nitrogen fertilizers [28]. Methane emissions from flooded-rice cultivation have been revealed to be affected by various soil and plant properties, especially soil texture, soil management practices, former crop, and selected cultivars. They also mentioned that up to 90% of the produced CH_4 in flooded-rice cultivation is emitted into the atmosphere. The remaining 10% of CH_4 in the soil is often re-oxidized into CO_2 and released into the atmosphere. Methane emissions are differentially regulated by rice growth stage and vary extensively among rice cultivars such as hybrids, inbred lines, and conventional ones. This variation is a result of physiological differences among cultivars in the production of CH_4 and methanotrophic activity in the rhizosphere [31].

Furthermore, rice stubble is left on the ground to decay or burnt to ashes, which is likely to produce both CH_4 and N_2O emissions [13]. The Intergovernmental Panel on Climate Change [7, 32, 33] estimated the annual global emission rate from paddy fields averages 60 Tg year^{-1}, with a range of 20–100 Tg year^{-1} which counted about 5–20% of the total CH_4 emissions from anthropogenic sources. This figure is primarily based on field measurements from different paddy fields in different countries such as the United States, Spain, Italy, China, India, Australia, Japan, and

Thailand. It is referred to that observed seasonal rice CH_4 emissions worldwide demonstrate large ranges, which reflects the effects of local and regional variations in agricultural, biological, and climatic factors [29]. It is also reported that although the rice production, among the crops, is being the world's second most produced staple crop, it is one of the largest anthropogenic sources of CH_4 emission. Rice cover 11% of the global arable land area and it is responsible for 10.1% of total agricultural GHG emissions and about 1.3–1.8% of the global human-source GHG emissions [13].

In Egypt, rice is important in Egyptian agriculture sector, as Egypt is the largest rice producer in the Near East region. The total area used for rice cultivation in Egypt is about 600 thousand ha or approximately 22% of all cultivated area in Egypt during the summer. The mean yield is 8.2 tons ha^{-1} with an estimated straw production of approximately 5–7 tons ha^{-1} [29]. The carbon footprint of crop production depends on various factors, such as soil types, crop types, cultivation practices, management factors, types and amounts of farm inputs, irrigation conditions, etc. For a specific crop type, these factors also differ among different countries and even within a country, for example in Australia, irrigate barley, chickpea, wheat, and rice cropping produce about 2.5 tCO_2e ha^{-1}, 2.6 tCO_2e ha^{-1}, 2.8 tCO_2e ha^{-1} and 1.7 tCO_2e $ha^{-1}$of GHGs, respectively. The sources of CO_2 simply represented in: on-farm fuel and electricity consumption; production, packaging, storage, and transportation of agrochemicals (fertilizers, herbicides, insecticides, fungicides, plant regulators etc.); N_2O emissions resulted from soils associated with application of synthetic nitrogen fertilizers; and farm machinery usage [13].

Therefore, a concern of GHGs-C emission and anxiety about global warming has resulted in grown attention on soil C storage, which is a function of climate, soil type, cropping systems, management practices such as tillage and fertilizers application. Particularly, the net C emissions from the paddy soil (i.e. CO_2 or CH_4) are governed by several factors including soil types, crop biomass, growing condition, type of cultivars, fertilizer practices, amendments use, water management, air transport mechanisms, and cultural practices [16]. Recently, more studies are greatly required to focus also on GHG emissions from aquaculture wetlands, mainly because of intensive input of organic feeding materials and frequent loading of chemical nutrients. The available budgets of global CH_4 and N_2O emissions from aquaculture were obtained from modeling approaches data based on surface water dissolved CH_4 and N_2O concentrations. However, there is still a lack of direct field estimations of CH_4 and N_2O fluxes to get a perception of regional or global estimations of CH_4 and N_2O source strengths from aquaculture wetlands. Particularly, it is not well known if the current shift in agricultural land use from rice paddies to inland aquaculture would point to what extent of shape the direction and rate of CH_4 and N_2O fluxes [28]. Therefore, with the accumulating evidence on climate change, there has been concerning about investigating the GHGs contribution of production practices and products to identify intensive emitting options that could be the target of GHG mitigation actions [29]. Since the global warming potential of CH_4 and N_2O is 25 and 298 times higher than CO_2 respectively, it is well realized that attention on reducing CH_4 or N_2O emissions may be an effective climate change mitigation strategy [30].

In many places, rice straw is not commercially valuable and is disposed of in different ways [9]. It is burned in fields which makes rice residue as another emission source yielding CO_2, CH_4, and N_2O, in addition to, pollutants such as carbon monoxide (CO), particulate matter (PM), and toxic polycyclic aromatic hydrocarbons (PAHs). Thus, the burning of these residues causes many problems such as air pollution, damaging to human health, and loss of soil C and nutrients [7, 29]. Rice straw residue also is left in the situ or removed from the fields for animal feed. Therefore, the incorporation of rice straw residues into the fields has been widely encouraged to preserve SOM and nutrients in the soil. Nonetheless, the incorporation of crop residues into paddy rice fields significantly increases CH_4 emissions because CH_4 production in flooded fields mainly relies on the availability of readily decomposable OM [7]. In 2008, about 620 million tons of rice straw was produced in Asia alone, with increasing quantity every year. Rice straw consists of about 0.6% (w) N, 0.1% P, 1.5% K, 5% Si, 0.1% S and 40% C, being a convenient source of plant nutrients because of its on-the-spot availability in amounts varying from 2 to 10 t ha^{-1}. Although residue retention is crucial for the sustainable soil management of mixed and non-rice cropping, incorporation of rice straw directly into soil commonly causes CH_4 emissions due to its anaerobic breakdown. Therefore, the promoting a proper rice straw treatment method before its incorporation into the soil is urgently needed [9]. As well, with a target to reduce GHG emissions from agriculture and rural development sectors, there is a critical requiring to mitigate rice CH_4 emissions while increasing or maintaining rice production levels which presents a major scientific and societal challenge [7].

6 The Impact of Climate Change and Sea Level Rice on Land Loss and Rice Production in Egypt

Climate change characterized by global warming has already had noticeable impacts on the ecological systems and human societies. The historical records indicate increasing the global mean surface temperature by 0.89 °C from 1901 to 2012. This warming trend expected to continue in the upcoming decades and would lead to more significant impacts on ecosystems and human societies. The main reason for current global warming is the human-induced GHG emissions [30]. The energy sector emissions are generated from fuel combustion in the different sectors such as industry, transportation, electricity generation, agriculture, residential and commercial, petroleum production and industries. Industrial emissions are related to heavy industries such as cement, iron, and steel production. Agricultural emissions are mainly released from rice cultivation, enteric fermentation of livestock, manure management, agricultural soils and field burning of agricultural residues [34].

Crop cultivation has moved northward since the 1940s. Rapid urbanization, policy changes in agriculture and land use, mechanization, and changes in farm management have driven this movement. As well, climate change has also played an important part

[4]. Climatic factors are key factors that are controlling crop production processes. Solar radiation, rainfall and temperature fluctuations lead to water deficit, changing in soil moisture content, flood, pest, and diseases outbreak that limit crop growth and can account for 15–80% of the variation of inter-annual yield resources [35]. It has already been a global consent that climate factors and their changes play an important role in agriculture, particularly in grain production which is the main essential sector for feeding a population over the world. Therefore, agricultural production is sensitive to climate change, not only due to the direct impacts of changes in temperature and precipitation but also due to changes in hydrological processes such as availability of water resources and the demand for irrigation water. Agricultural irrigation has an important role in the utilization of water resources as well as the distribution of farmland [4]. Meanwhile, extreme climate events such as drought, flooding, or drought immediately followed by flooding in recent years have led to severe negative impacts on crop production [23].

It is well known from the rich literature of climate change and its impacts on grain production, that there is growing evidence of crop yield response to current global warming. Various approaches were used in such studies as using crop model to calculate the contribution of weather to grain yield, simulation, and empirical statistical approach such as linear regression analysis that has been mostly used for calculating the impact of climate factors on grain yield. It was reported that had the increased temperature by 1–2 °C during the paddy earring stage, could have decreased the paddy rice production by 10–20% [36]. It is reported that climate change implies further pressure on the world's food supply system. It affects food production directly through changes in agro-ecological conditions which affects the overall food supply. Extensive research shows that high temperatures, variable rainfall, floods, droughts, and cyclones may cause a large decrease in global food production, especially in developing countries. For example, the increase in temperature reduces the phenological phases of crops (i.e. planting, flowering, and harvesting) and affects plant growth and development [37].

Several researchers have studied the probable impact of climate change on rice production and conducted that it is pivotal to determine how climate change affects rice production and water use efficiency to prepare plans and policies for adapting the agricultural system versus the changing climate [37]. Chen et al. [23] indicated a decreased rice yields associated with global warming in Philippines. They also referred that in other Asian countries, long-term field experiments have also conducted declining or stagnating yields in rice-based cropping systems. However, the impacts of climate change on the future rice production is still under debate in the three highest producing countries; China, India, and Indonesia. This is attributed to the counterbalance between the beneficial effects of increasing atmospheric CO_2 concentration and the exacerbation of abiotic stresses on crop growth, such as heat, drought, and salinity. The available estimations are discordant and vary between an increase of 10–15% in 2020 to 7–10% yield losses per every 1 °C increase in air temperature. With the prominent role of rice as a staple food for humans and the necessity of doubling crop production by 2050 to meet the proposed demand of the global population, the efforts are still required to better understand the current knowledge. Currently, rice cultivators and stakeholders of the rice sector are already applying adaptation strategies, varying from individual autonomous local reactions to planned policy interventions, to mitigate the negative impacts of climate change in the main producing environments. Nevertheless, less research has been dedicated to analyzing the impacts of climate change on rice production in Europe, where rice is the 6th most produced cereal. Even though not being a staple food crop in Europe, rice plays a focal sociocultural and ecological role in many Mediterranean countries, where the human consumption is steadily increasing. With elevated temperatures because of climate change, heat stress could be more common on temperate rice crop. Recent findings on rice physiology reveal that heat stress causing sterility in the exerted part of the panicle may occur even in Mediterranean countries, especially in warm and humid years [4].

Egypt is one of the countries assuming to face damaging impacts of climate change [34, 38], although its contribution to the global greenhouse gas (GHG) emissions is 0.57% and despite the fact of not considering it in a non-annexe I country not requiring any certain emission reduction targets under the Kyoto protocol. However, Egypt has included in its National plans mitigation actions to reduce greenhouse gas emissions from its main sectors contributing to climate change; energy, industry, agriculture, and waste [34]. Agriculture in Egypt is one of the significant sectors that is expected to be affected by climate change. Rice in Egypt is a strategic crop, and it occupies the second class of the cultivated crops in Egypt [39]. He employed the CERES-RICE model (a mechanistic, process-oriented model from grain cereals that includes crop development) to understand physiological processes and yield of rice production. He used actual measurements for rice production characteristics in the comparison between present and future. He extracted meteorological data representing in temperature, solar radiation, and precipitation for the years 1998 and 1999 from outputs of RegcM3. The climate data is used from HADCM3 for SRES—A2 Scenario for 2040 and switched in RegcM3. The results of his work indicate changes in flowering dates, physiological maturity, and production. He expected over the country with the studied climate change scenario, the rice maturity period is projected to diminish by 20% and yield to decrease by 25.8% on average. As well, it is conducted that rice production in Egypt will be declined under expected climate change. Subsequently, a gap between rice production and consumption will be created. Furthermore, when the increased population is considered, this gap will widen. Therefore, in the face of climate change, better understanding of climatic impacts on rice production is vital to identifying solutions that will enhance current food production and increase the adaptability of these systems in the future [35, 40].

The acceleration in global Sea Level Rise since the last decades of the 20th century is one of the most evident consequences of the higher temperature as the main impact of climate change. This phenomenon has been considered in all the Assessment Reports published by the Intergovernmental Panel on Climate Change (IPCC). IPCC provided different scenarios of the sea level rice in 2065 and 2100, with the worst one estimated as 0.98 m in 2100 [41]. Sea level rise as a result of both natural and human-induced climate change, it is one of the main reasons of salinity intrusion into soil and groundwater. Considering other forms of lands, deltas are easily vulnerable to sea level rise. For instance, several studies found that Nile Delta and some other deltas are facing constant inundation and saline intrusion because of exposure to the sea [42]. He added that the predicted sea level rise due to climate change would submerge a lot of low-lying lands worldwide by 2050, and salinity intrusion will be more severe. It is also emphasized on that the worldwide low-lying areas; the main alluvial coastal plain; is accounted as one of the most vulnerable areas by sea level rise. They highlighted the Nile Delta, Egypt, in the Mediterranean area and mentioned that currently, about 10% of the global population already live in low-lying coastal regions, between 0 and 10 m above sea level. They also reported that the sea level rise associated with global warming is also intensified by local factors, such as the subsidence, due to both natural and anthropogenic causes [41]. It is found in the study of [43] on the impact of sea level rice on Nile Delta, Egypt that that mean sea level is projected (based on IPCC scenarios which generate by MAGICC/SCENGEN program) to rise by 14.0 to 18.9 cm between 1990 and 2100, for the full range of all scenarios. The range between the best case A1 (low) and the worst case B1(low) is 4.8 cm. This will affect the developmental activities in these areas due to heavy population and high productivity of the best fertile areas in Egypt. The results of all scenarios show that A1+ land subsidence was the highest case, ranged between (48.2 and 48.9 cm), while B1+ land subsidence was the lowest case (44.0–44.1 cm) till the year 2100. Therefore, in this context, the impact of climate change on agriculture productivity, water shortage and shifting growing seasons can't be neglected in the Nile Delta. This is because the area is vulnerable to negative impacts of climate change such as loosing, salinization of considerable areas of cultivated lands (because of rising sea level) and rapid decomposition of OM as result of increasing temperature. It is expected also in the study of [44] that relative sea level rise in the high scenario increases the amount of land risk from inundation in the northern Nile Delta by 300 km^2, or more than one-fifth of the total agricultural land in the northeast Nile Delta only. As well they added; all crops are projected to have a decrease in yields and an increase in irrigation needs. All these will lead to a decline in C and N in agricultural soil. Thus; under such previous stresses on C and N in North Nile Delta Egypt as a result of climate change and water shortage due to establishing GERD, the actions should be taking towards either mitigation or adaptation of this stresses.

7 The Impact of Water Deficiency on Rice Production in Egypt

Egypt covers an area of about 1.0 million km^2 [45]. The Nile valley plus Delta region cover about 4% of this area [46]. Agriculture is practiced over an area of approximately 3.5 million ha, involving recently reclaimed lands. As a highly populated country with a population of approximately up to 90 million, Egypt is an agriculturally based country. Agriculture remains a major sector and a very lifelike component of the economy. Even though its performance stayed relatively modest in the last few years, it has successfully attracted considerable investments. Agriculture employs about 31% of the labor force, and about 14% of the GDP is produced by agricultural production. Because of the favorable agro-climatic, perennial water supplies and rich fertile soils, Egypt produces a diversity of crops, vegetables, and fruits for feeding its population and earning foreign exchange through exports. The country's main crops include cotton, rice, wheat, sugarcane, beet, clover, fodders, vegetables, sesame, peanut, sunflower, beans, lentils, and onion, in addition to fruits such as citrus and dates [45].

Water resources in Egypt are limited to the Nile River, rainfall, and deep groundwater [47]. The water of the Nile River has a great value for Egypt as it provides approximately 95% of its water needs. Egypt's portion of the water of the Nile River is almost 55.5 billion m^3. Agriculture uses most of it; about 85.6% of the used water. This verifies the saying "Egypt is the Nile's Gift" as Egypt does not have a large portion of the rain, so we find that life in Egypt is concentrated on the sides of the Nile and any shortage in the of supplied water in the river inevitably causes a disaster. The Nile River stems from two springs in the upper lands in Ethiopia and the lakes that cover parts of Uganda, Kenya, Tanzania, and Congo. Then the Nile flows till North Cairo and reaches to its two main branches, Damietta, and Rashid (about 6680 km from its springs till the Mediterranean Sea) that end in the Mediterranean Sea [48, 49]. The river revenue is featured by two main periods, flood period during August, September, and October in which the river increases, and Althariq period in the other months of the year in which the river revenue is declining a lot. The annual revenue of the three major tributaries of the Nile is around 84 billion m^3, where approximately 48.7 billion m^3 form the Blue Nile, 24.4 billion m^3 of form the White Nile and 10.9 billion m^3 for the branch of Atbara [49]. With the fast growth in the population and increasing water consumption in different fields, such as agriculture, industry, domestic use etc., it is supposed that Egypt will relay somewhat on the groundwater to develop some new projects such as East Eweinat. There are two main ways that can reduce concise demand with supply by reducing demand or by increasing supply. Despite the increasing shortage of water, there are almost no indications of efforts to reduce water demand in the three-main water-consuming sectors [48].

In Egypt, water planning established in 1933 for using extra storage capacity that was available after the second elevating of both dams of old Aswan and the Gabal El-Awlia in Sudan. This plan generated programs for many purposes such as land reclamation, conversion of some basin irrigation to perennial irrigation, and increases the rice cultivation area. This strategy was first revised in 1974 and again in 1975 when a new plan was prepared to hold the further volumes of water resulting from the construction of the Aswan High Dam. The multi-year regulatory storage capacity posed by the Aswan High Dam was a reason for stability to Egypt water resources by delivering a reserve storage capacity during years more than the requirements and providing additional resources during loan year. Currently, adopting more efforts are running for water management and application of different available water resources [48]. Although there are several uses for water in Egypt, agriculture engages about

85.6% of the used water; estimated by about 69.30 billion m^3. A large portion of this water is groundwater used directly, or after mixing with Nile fresh water, and treating to be appropriate for using [49]. Rains are not sufficient and effective enough to depend upon for production. Rain-fed agriculture is practiced in only 2% of the total area. The per capita rate of Nile water is nearly 850 m^3 $year^{-1}$—under the water scarcity—whereas the minimum per capita needs should be up to one thousand m^3 $year^{-1}$. Egypt is among 35 water deficit countries in the world. In Egypt, 87.7% of the total water is being consumed by agriculture, 5.4% by industry while the total human consumption touches the figure of 6.8% of the total water. For the irrigation of the new land, each drop of water has become the focus of the government in Egypt. Efforts are being increased to focus in the future on the enhancement of the irrigation systems, the introduction of real irrigation technologies that could be effectively utilized to irrigate the newly reclaimed agricultural areas. Modern irrigation methods such as drip irrigation and sprinkler need to be used for overcoming the water shortage and scarcity [45].

Rice is the second major staple crop and is considered the most profitable export crop of the summer season. Egypt reported a 2.3% growth in production to a total of 6.9 million tons in 2008 due to a 5.6% extension in productive land. The increased rice cultivation that has been resulted in a hike on international markets has added pressure on water resources since rice cultivation is outstandingly water intensive. Farmers practice fish-farming in rice fields to enhance a rice farmer's income. In 2008, because of water management policy, the restrictions were imposed on the area that will be brought under rice cultivation, but policy proved ineffectual. Therefore, the government placed an export forbid in for reducing the domestic price of rice. The raising of the ban has allowed exports to increase again, and to benefit from the high international prices of the commodity. Rice Cultivation also helped farmers to realize higher profits as compared to other traditional summer crops [45].

Rice is easy cultivation crop with assured results. Local rice productivity is estimated at around 3.5 tons/acre. The present practice of rice cultivation is requiring fewer farming efforts, can satisfy farmers' daily nutritional requirements as well as can wash salty lands. A tendency to grow large patterns of rice throughout the Delta is also created among farmers due to the free delivery of irrigation water, according to which rice becomes of better profitability compared to other crops. The deformation in making cropping decisions promotes the extension of rice areas even with deficiencies concerning water requirements for the country reclamation strategy [12]. Rice water consumption is approximately 2–3 times higher than is required for producing other cereals, such as maize or wheat [10, 37]. In addition to the land and energy issues, with most of the rice being grown under irrigated conditions, water is another scarce resource that is crucial for rice production. Producing one kg of rice requires around 2672 L of water-about 2.5 times the amount of water needed to grow a kg of wheat or maize [13]. Despite the importance of rice production, it also adversely affects the environment. For instance, besides the highwater extraction required for rice production, heavy pesticide usage is another burning concern. Also, when rice is flooded, it undergoes anaerobic processes, resulting in the formation and release of large amounts of methane into the atmosphere [37].

In Egypt, the Per capita fresh water availability was dropped from 1893 cubic meters in 1959 to 900–950 m^3 in 2000, to about 670 m^3 in the year 2017 and the author expected to decline further to 536 by 2025. The main reason behind this rapid fall is the mixed water resources and the rising pressure from population growth. Water resources in Egypt are becoming rare. Surface water resources originating from the Nile are currently fully exploited, while groundwater sources are being brought into full production. Egypt is facing growing water needs, demanded by rapidly increasing population, increased urbanization, higher standards of living and by an agricultural policy which emphasizes intensifying production to feed the growing population. The population is presently increasing by more than one million persons a year. With a population of Egypt is expected to increase to about 100 million by 2025. The most critical limitation facing Egypt is the growing deficiency of water resources associated with degradation of water Quality [47]. As previously shown, rice is usually grown in the Delta under continuous flooding. During most of the growing season, expanding from May to October, rice fields are flooded understanding water layers of variable depths. The outlined irrigation process reflects the intense need for rice for water diversions. Exposed to temperatures ranging from 30 to 40 °C, rice fields are subject to excessive evaporation in addition to percolation, causing significant rates of water loss. Almost 50% of water amount diverted to rice fields is consumed by evapotranspiration, and the rest is lost via percolation [12]. Despite all the prior efforts for planning water resources management in Egypt, we still have to make more attempts to achieve the required balance between the available water resources and the high growth population and water needs subsequently [48].

The economic evaluation of crop production requires realizing the value of water used in irrigation. Rural water pricing is not applied in Egypt based on a rule implying free delivery of irrigation water to farmers. Nevertheless, assumed the underlying natural resource limitations facing the country, it is necessary to guarantee efficient use of water across introducing a value that reflects the vital concern for water as a most important restrictive factor of Egypt's agricultural production [12]. Upcoming stresses on the Nile's water promotes Egyptian demand to Egypt's "historic rights" of the Nile. Egypt relies on the Nile for 97% of its water needs. In line with current aspects of water misusing, population growth and the possible redistribution of the Nile's resources to other riparian countries, Egypt confronts the challenge of coping with severe future water scarcity. Water deficiency and limited arable land mean that Egypt already depends heavily on food imports to satisfy its population demands for food (Egypt imports 60% of its total food needs). Egypt's agricultural sector currently uses 80% of the country's water resources. As the population growths, water requirements will increase because of household and industrial use as well to ensure the country's food security by producing the food. Egypt's dependency on food imports makes it vulnerable to global food price rise and supply scarcities. To mitigate the security risk, Egypt has to continue in land reclamation plans in desert areas, which require huge water quantities and will place additional restricts on the portions of other agricultural, industrial, and municipal water consumers. As the population grows, the country will need more water than its current available share; however, shifting geostrategic alliances among upstream countries mean that its allocation is likely to decline. Unless it goes on a large-scale modification of its inefficient water networks, Egypt could go through upcoming major water crises that could cause conflicts with its neighbors [50].

Currently, major challenge confronting Egypt is the crucial need for better development and management of the available restricted resources of water, land, and energy to face the needs of population growth [48]. The decreasing availability and increasing costs of water menace the traditional way of cultivated rice under irrigated conditions. Hence, due to the increasing deficiency of water for agriculture and competition from non-agriculture sectors, there is an urgent calling for better understanding water use efficiency in irrigated agricultural ecosystems [10]. In addition to environmental and demand pressures and potential conflict, the Nile is threatened by many environmental stresses, such as climate change, pollution, and degradation. Climate change will put sever challenges for the Nile, including decreased river flow, land degradation, the increased droughts and floods probability, and rising rates of disease. Dam establishment on the Nile is responsible for watershed land degradation. Population growth models in Egypt and upstream Nile countries, such as Uganda and Ethiopia, will undoubtedly cause future environmental issues such as raising in municipal, industrial, and agricultural wastes. Egypt has depleted the Nile's water resources by overdrawing its share, through projects such as the desert reclamation in the Toshka Depression and the Sinai Desert by the Al-Salam Canal system [50].

7.1 Water Deficiency Due to Climate Change

Agricultural sector is one of the main water consuming sectors. The water used for agriculture has reached about 59.30 BCM in 2009/2010. Under conditions of climate change, required water to irrigate various crops are expected to increase to about 61.8 BCM in 2024/2025 as a direct impact of high temperature. This will be associated with high efficiency of using water by some crops because of the increasing concentration of CO_2 [49]. Hence, the issue of water deficiency in Egypt, associated with its probable exacerbating factors such as economic, population and food demand growth, climate change and the current debate over the share of the Nile's water among its ten basin countries) is widely documented in the recent literature [51]. Climate change is impacting the Mediterranean region in myriad and distinct ways such as sea level rise, increased frequency of flash flood events, droughts, or periods of water shortage and rising temperatures. Observed trends and projections for the future indicate a strong vulnerability to changes in hydrological regimes, an increasing shortage of water resources and following threats to water availability and management [5].

Population growth and economic development are motivating significant raise in agricultural and industrial water demand. Water use in agriculture accounts for more than two-thirds globally, including as much as 90% in developing countries. Much of the demand is resulting from expected increases in the world population from 6.6 billion to about 8 billion by 2030 and more than 9 billion by 2050. Climate change will likely increase water demand for agriculture, mainly for irrigation, due to long dry periods and severe drought, especially it is estimated at over 40% increase in irrigated land by 2080. As well, it will likely increase water demand for billions of farm animals due to higher atmospheric temperatures thus hydration needs. Also climate change will likely increase water quantities that needed for industrial cooling because of increased atmospheric and water temperature [52].

In the future, water may become more expensive, less available and allocations will be less secure, especially for water-intensive activities such as rice. Rice industry stake-holders and growers have already adopted a risk-averse approach. At some place in the world such as California and Papua New Guinea, there is a flexible and effective global supply network for warranting and processing to ensure continuous supply through critical periods. At the farm level, during years of low water availability, rice growers trade water as a tactical response and move to low water-intensive or dryland farming. This, however, has caused highly variable rice supplies and has filled some production gaps [53]. It is added that water has many competing usages, but in the currently, climate change is further aggravating the water scarcity issues by decreasing its availability for irrigation purposes. Approximately 15–20 million ha of irrigated rice may suffer from water scarcity by the year 2025. Furthermore, the upcoming increased concentration of atmospheric CO_2 may increase the GHG intensity of rice production. Therefore, improving rice varieties and better management practices that need less water, land, and energy, and those that improve the rice productivity is urgent in the 21st century. Also, a farmer seeking to cultivate a given crop under increasingly the stress of water resources will utilize in the available adaptation options to improve the efficiency of water usage and will amplify the adaptive effort because access to water resources becomes more restricted. At some point, adaptation efforts under the existing regime will become inconsistent to the benefits and a new adaptation action such as a novel irrigation system or altering a crop will be needed to retain a farming livelihood [13, 54].

7.2 Water Deficiency Due to Grand Ethiopian Renaissance Dam Construction

The Nile River is one of the most important rivers in the world. Eleven countries are depending on the Nile River water; Burundi, Eritrea, Tanzania, Uganda, Ethiopia, Rwanda, the Democratic Republic of Congo, Kenya, Sudan, and Egypt. Egypt share from the Nile River fresh water is limited by a covenant signed in 1959 between Sudan and Egypt where Egypt share is 55.5 Billion Cubic Meters (BCM) year^{-1} and Sudan share is 18.5 BCM year^{-1}. Most of the Nile River water comes mainly from the Ethiopian plateau cross the Blue Nile and Atbara in the flood period of the flood that starting from August to December. Ethiopia's tributaries provide approximately 86% of the Nile River water. Ethiopia began constructing the GERD. In 2011, on Guba that located on the Blue Nile approximately 60 km from Sudan and 750 km northwest of Addis Ababa. The GERD reservoir will extend for an area of 1874 km^2 with a full supply of 640 m above mean sea level, with total and effective storage volumes of 74 BCM. The main concern about the construction of Grand Ethiopian Renaissance Dam (GERD) is the filling period of its reservoir which will decline Egypt's share from Nile River water and as a result affecting water security of Egypt [55]. Thinking about the construction of the dam for power generation on the Blue Nile started in the sixties in the 20th century. The U.S. Office of Land Reclamation in 1964, has studied the establishment of around 11 dams, most importantly the four large dams on the Blue Nile with a total capacity of 80 BCM and these dams have been involved in the Nile Basin Initiative. However, the establishment of the Renaissance Dam was announced with a total capacity of 74 BCM [i.e. five times the size and capacity of the total of the four old dams in the American Studies (14.5 BCM)] [49].

Construction of GERD will have some negative implication on many sectors in Egypt including agriculture and water deficiency. Lack of supplied water to Nasser Lake due to water storage in front of the Renaissance Dam (by approximately 25–33 BCM $year^{-1}$ unless there is pulling of shortage from Dam Lake) means wasting approximately 3–5 million feddan of Egypt's cultivated area. Each feddan requires about 5 thousand cubic meters (CM) of water according to the estimation of the Irrigation Ministry in Egypt. Therefore, the minimum of wasting cultivated lands will lead to a lack of cultivated area of approximately 46.9%, while the maximum wasting limit of cultivated land will be approximately 67.6% of cultivated lands. As well, deficiency of irrigation water will lead to increasing the use of agricultural drainage water for irrigation up to about 7 BCM $year^{-1}$, which duplicate the salinity 3 times following irrigation and thus reaching the water to the banks again with higher concentrations in each irrigation, which lead to increasing in water salinity of agricultural lands in the Delta. Additionally; lack of water flow means stopping all land reclamation projects, agricultural expansion, and ending of some great agriculture projects in Egypt such as Toshka project, Al-Salam Canal, and El Hamam Canal in the Northwestern coast of Egypt. Anyhow, there will be an urgent need for establishing many wastewater treatment plants, for treating contaminated water to become suitable for irrigation, and also treating industrial wastewater and sewage, which costs Egyptian country millions of dollars [49].

Based on the previous mention data, there will be a decline in rice cultivated area which store C and N more than other cultivated areas, especially in the surface layer. Already it is noticed obviously in Egypt this year that farmers are suffering from drought and inability to cultivate their preferred crop; rice; as a result of water shortage. Thus, loss of 50% of cultivated area in Egypt based on the previous scenario of [49] will lead to a loss of approximately 50% of cultivated rice area also, the loss of 50% of C and N storage in those soils (equal approximately third of C and N of Egypt's cultivated soils) because of converting rice to dry crops. Also; some soils in Egypt, especially in North Nile Delta (more 95% or rice cultivation in Egypt is concentrated in this area), is salt-affected soil; thus, rice cultivation is one of the most important practices that lead to salinity leaching. Saline soil requires a huge amount of water irrigation for salt leaching to be suitable for agricultural activities. In such case, rice cultivating could be the best solution, while, other areas with non-saline soil could be cultivated with other crop types [56]. Furthermore, the negative impacts on salinity on soil properties and productivity, salinity is one of the major sources of declining net primary production the OM, C, and N in soil [57, 58].

8 The Effect of Agriculture Management Practice on Enhanced Rice Production for Confronting the Challenges of Climate Change and Expected Water Deficiency

Enhancement agricultural productivity in the last five decades has mainly been related to energy-intensive systems for growing crops such as rice and wheat. A high input conventional tillage and intensive weed management systems consist of primary and secondary tillage implements. Furthermore, transplanted paddy cultivation demands a huge amount of energy regarding labor for land preparation, puddling and transplanting. Transplanted rice paddy requires 4000–5000 L kg^{-1} of water for rice production. Moreover, a huge amount of energy is required for application of water, as well caused significant CH_4 emission. On the progress of second-generation farm apparatus and global concern about energy savings and GHGs emissions issues, zero tillage, residue retention, green manuring, using of small farm machinery, real-time N management through leaf color chart offer a platform to highlight these issues. Among these techniques, zero-till transplanting without puddling provides a new opportunity for energy and C saving [16]. As rice cultivation requires a great quantity of water, the cultivation of rice in rotation with other crops that need low water consumption in the dry season is an interesting option. Crop rotation in rice fields can enhance the utilization of agricultural land [8].

As previously reported, the OM plays a fundamental role in a various biological, chemical, and physical processes in the soil ecosystem; the SOC stock is one of the essential indicators of soil health and quality. The SOC represents a large C stock in the global C cycle, acting as a dynamic balance between C inputs (through photosynthesis and deposition) and losses (via respiration, erosion, and leaching). The SOC accumulation can convert atmospheric CO_2 into stable organic C stocks in the soil and sequester atmospheric CO_2, to mitigate climate change. Subsequently, increasing and sustaining the SOC stock is a critical issue for reaching desired soil functions and sequestering atmospheric CO_2. Therefore, in agroecosystems, the SOC balance is affected by management practices as OM additions, fertilization, tillage intensity, irrigation, and crop rotation. Thus, the utilization of organic amendments such as farmyard manure, green manure, and crop residues has been recognized as the most practical method for increasing the SOC stock [59]. The potential carbon sequestration by improving soil C stocks via sustainable land management has now been realized for world agriculture. Management practices such as crop rotations, soil tillage, fallow periods and water management could either reduce or increase soil C sequestration. Paddy fields are documented to have higher SOC storage, and sequestration compared to drier croplands. Organic C accumulation in paddy ecosystems was faster and more distinct than other arable ecosystems because OM decomposition is reduced in lowland rice fields apparently due to extremely reduced conditions. As well, the deficiency of oxygen for microbial activity under submerged conditions results in a decline in the decomposition rate. Incomplete decomposition of organic materials has been reported in addition to decreased humification of OM under submerged conditions, causing net accumulation of OM in paddy soils. In the long term, soil management governs the weathering and mineral formation as well as accumulation of organic nitrogen in paddy soils. The suggested mechanisms for the accumulation of SOM in paddy soils are known as an occlusion in aggregates, the formation of organo-mineral associations, the addition of pyrogenic OM [25].

Over recent years, the scarcity and competition for water have been growing globally, and opportunities for developing new water resources for irrigation became increasingly limited. Rice as a dominant irrigated crop, representing about 30% of the total irrigated area. More than 50% of the rice area in the world is irrigated. With water becoming scarcer, the future of rice production will, therefore, rely heavily on developing and adopting efficient water use strategies and practices in irrigation schemes. This situation applies in many parts of the world; however, it is particularly critical for the Near East region because water is among the most limiting factors for general development and particularly for agriculture [2].

Rice cultivation is known as an important emitter of greenhouse gases emission especially methane (as a result of rice management practices, rice straw burning after harvesting and machinery activities [29]. However, some studies such as [2, 29, 60] recommended that changing management practices (such as decreasing ploughin, converting to agriculture conservation, replace burning of rice straw to some other uses and balanced fertilizer application) will lead to mitigating of greenhouse gases emission from rice cultivation. As well, such management practices improve build-up of soil organic matter stocks and enhance the environmental impacts of current management practices.

9 Conclusions

Agricultural soils in North Nile Delta Egypt is vulnerable to water shortage and climate change. Rice system in this area has higher C and N pools. However the area of rice production is expected to decline because of many reasons such as Establishing GERD and climate change. Management practices should be taken into consideration to avoid such decline in those soils and consequently decline in SOC which considers one of the most important methods to mitigate climate change if sequestered in the soil. Sequestering SOC decline CO_2 in the atmosphere where CO_2 is one of the most important GHG that cause climate change. Therefore, the attention should be paid for soil C sequestration in the North Nile Delta Egypt for balancing the excepted decline in C and N, due to reducing rice cultivation area because of many challenges such as climate change, water deficiency and sea level rice, as well for climate change mitigation.

10 Recommendations

1. Enhancing soil C stock should be considered not only for climate change mitigation but also for improving soil quality and agricultural productivity.
2. Rice field is one of the most important fields maintaining soil organic matter and soil C, therefore the quality of these fields should be enhanced or conserved.
3. Under the expected deficiency in water, rice cultivation should be directed in saving water and maintaining soil C through finding new cultivars consume less water and new methods reduce irrigation water.

References

1. El-Shahway AS, Mahmoud MMA, Udeigwe TK (2016) Alterations in soil chemical properties induced by continuous rice cultivation: a study on the arid Nile delta soils of Egypt. Land Degrad Dev 27:231–238
2. FAO (2003) Rice irrigation in the near East: current situation and prospects for improvement. FAO Regional Office for the Near East, Cairo, Egypt
3. Shahid M, Nayak AK, Puree C, Tripathi R, Lal B, Gautam P, Bhattacharyya P, Mohanty S, Kumar A, Panda BB, Kumar U, Shukla AK (2017) Carbon and nitrogen fractions and stocks under 41 years of chemical and organic fertilization in a sub-humid tropical rice soil. Soil Tillage Res 170:136–146. https://doi.org/10.1016/j.still.2017.03.008
4. Zhang Y, Wang Y, Niu H (2017) Spatio-temporal variations in the areas suitable for the cultivation of rice and maize in China under future climate scenarios. Sci Total Environ 601–602:518–531
5. Sušnik J, Vamvakeridou-Lyroudia LS, Baumert N, Kloos J, Renaud FG, Jeunesse I, Mabrouk B, Savić DA, Kapelan Z, Ludwig R, Fischer G, Roson R, Zografos C (2015) Interdisciplinary assessment of sea-level rise and climate change impacts on the lower Nile delta, Egypt. Sci Total Environ 503–504:279–288
6. Li S, Li J, Li C, Huang S, Li X, Li S, Ma Y (2016) Testing the RothC and DNDC models against long-term dynamics of soil organic carbon stock observed at cropping field soils in North China. Soil Tillage Res 163(2016):290–297. https://doi.org/10.1016/j.still.2016.07.001
7. Mohammadi A, Cowie A, Mai TLA, Rosa RA, Kristiansen P, Brandao M, Joseph S (2016) Biochar use for climate-change mitigation in rice cropping systems. J Clean Prod 116:61–70. http://dx.doi.org/10.1016/j.jclepro.2015.12.083
8. Cha-un N, Chidthaisong A, Yagi K, Sudo S, Towprayoon S (2017) Greenhouse gas emissions, soil carbon sequestration and crop yields in a rain-fed rice field with crop rotation management. Agr Ecosyst Environ 237:109–120. https://doi.org/10.1016/j.agee.2016.12.025
9. Liu Y, Yao S, Wang Y, Lu H, Brar SK, Yang S (2017) Bio- and hydrochars from rice straw and pig manure: inter-comparison. Biores Technol 235:332–337. https://doi.org/10.1016/j.biortech.2017.03.103
10. Wang Y, Zhou L, Jia Q, Yu W (2017) Water use efficiency of a rice paddy field in Liaohe Delta, Northeast China. Agric Water Manag 187:222–231. https://doi.org/10.1016/j.agwat.2017.03.029

11. Elbasiouny H, Elbehiry F (2019a) Potential soil carbon and nitrogen sequestration in future land use under stress of climate change and water efficiency in northern Nile Delta, Egypt. Agricultura 10:3–4. http://dx.doi.org/10.15835/agrisp.v110i3-4.13399
12. Ahmed TA (1998) Worth of rice cultivation in the Nile delta. In: 24th WEDC conference, Sanitation and water for all. Islamabad, Pakistan
13. Maraseni TN, Deo RC, Qu J, Gentle P, Neupane PR (2017) An international comparison of rice consumption behaviours and greenhouse gas emissions from rice production. J Cleaner Prod xxx:1–13. https://doi.org/10.1016/j.jclepro.2017.11.182
14. Wang W, Lai DYF, Wang C, Pan T, Zeng C (2015) Effects of rice straw incorporation on active soil organic carbon pools in a subtropical paddy field. Soil Tillage Res 152:8–16. https://doi.org/10.1016/j.still.2015.03.011
15. Abd El-Kawy OR, Rød JK, Ismail HA, Suliman AS (2011) Land use and land cover change detection in the western Nile delta of Egypt using remote sensing data. Appl Geogr 31(2):483–494
16. Dash PK, Bhattacharyya P, Shahid M, Roy KS, Swain CK, Tripathi R, Nayak AK (2017) Low carbon resource conservation techniques for energy savings, carbon gain and lowering GHGs emission in lowland transplanted rice. Soil Tillage Res 174:45–57. https://doi.org/10.1016/j.still.2017.06.001
17. Chen S, Xu C, Yan J, Zhang X, Zhang X, Wang D (2016b) The influence of the type of crop residue on soil organic carbon fractions: an 11-year field study of rice-based cropping systems in southeast China. Agric Ecosyst Environ 223:261–269. http://dx.doi.org/10.1016/j.agee.2016.03.009
18. Deb S, Chakraborty S, Weindorf DC, Murmu A, Banik P, Debnath MK, Choudhury A (2016) Dynamics of organic carbon in deep soils under rice and non-rice cropping systems. Geoderma Reg 7:388–394. https://doi.org/10.1016/j.geodrs.2016.11.004
19. Elbasiouny H, Elbehiry F (2019) Soil carbon and nitrogen stocks and fractions for improving soil quality and mitigating climate change: review. Egypt J Soil Sci 59(2):131–144. https://doi.org/10.21608/ejss.2019.9984.1251
20. Zhu L, Li J, Tao B, Hu N (2015) Effect of different fertilization modes on soil organic carbon sequestration in paddy fields in South China: a meta-analysis. Ecol Ind 53:144–153. https://doi.org/10.1016/j.ecolind.2015.01.038
21. Chen Z, Wang H, Liu X, Zhao X, Lu D, Zhou J, Li C (2017) Changes in soil microbial community and organic carbon fractions under short-term straw return in a rice–wheat cropping system. Soil Tillage Res 165:121–127. https://doi.org/10.1016/j.still.2016.07.018
22. Kim GW, Jeong ST, Kim PJ, Gwon HS (2017) Influence of nitrogen fertilization on the net ecosystem carbon budget in a temperate mono-rice paddy. Geoderma 306:58–66. https://doi.org/10.1016/j.geoderma.2017.07.008
23. Chen A, Xie X, Dorodnikov M, Wang W, Ge t, Shibistova O, Wei W, Guggenberger G (2016a) Response of paddy soil organic carbon accumulation to changes in long-term yield-driven carbon inputs in subtropical China. Agric, Ecosyst Environ 232:302–311. http://dx.doi.org/10.1016/j.agee.2016.08.018
24. Dossou-Yovo ER, Brüggemann N, Ampofo E, Igue AM, Jesse N, Huat J, Agbossou EK (2016) Combining no-tillage, rice straw mulch and nitrogen fertilizer application to increase the soil carbon balance of upland rice field in northern Benin. Soil Tillage Res 163:152–159. http://dx.doi.org/10.1016/j.still.2016.05.019
25. Ratnayake RR, Perera BMACA, Rajapaksha RPSK, Ekanayake EMHGS, Kumara RKGK, Gunaratne HMAC (2017) Soil carbon sequestration and nutrient status of tropical rice based cropping systems: rice-rice, rice-soya, rice-onion and rice-tobacco in Sri Lanka. Catena 150:17–23. https://doi.org/10.1016/j.catena.2016.11.006
26. Wang H, Guan D, Zhang R, Chen Y, Hu Y, Xiao, L (2014) Soil aggregates and organic carbon affected by the land use change from rice paddy to vegetable field. Ecol Eng 70:206–211. http://dx.doi.org/10.1016/j.ecoleng.2014.05.027
27. Jiang CM, Yu WT, Ma Q, Xu YG, Zou H (2017) Alleviating global warming potential by soil carbon sequestration: a multi-level straw incorporation experiment from a maize cropping

system in Northeast China. Soil Tillage Res 170:77–84. https://doi.org/10.1016/j.still.2017.03.003
28. Wu S, Hu Z, Hu T, Chen J, Yu K, Zou J, Liu S (2018) Annual methane and nitrous oxide emissions from rice paddies and inland fish aquaculture wetlands in southeast China. Atmos Environ. https://doi.org/10.1016/j.atmosenv.2017.12.008
29. Farag AA, Radwan HA, Abdrabbo MAA, Heggi MAM, McCarl BA (2013) Carbon footprint for paddy rice production in Egypt. Nat Sci 11(12):36–45
30. Tian Z, Niu Y, Fan D, Sun L, Ficsher G, Zhong H, Deng J, Tubiello FN (2017) Maintaining rice production while mitigating methane and nitrous oxide emissions from paddy fields in China: evaluating tradeoffs by using coupled agricultural systems models. http://dx.doi.org/10.1016/j.agsy.2017.04.006
31. Brye KR, Nalley LL, Tack JB, Dixon BL, Barkley AP, Rogers CW, Smartt AD, Norman RJ, Jagadish KSV (2016) Factors affecting methane emissions from rice production in the Lower Mississippi river valley, USA. Geoderma Reg 7:223–229. https://doi.org/10.1016/j.geodrs.2016.04.005
32. IPCC (2007) Climate change 2007: mitigation. In: Metz B, Davidson OR, Bosch PR, Dave R, Meyer LA (eds) Contribution of working group III to the fourth assessment report of the intergovernmental panel on climate change. Cambridge University Press, Cambridge, United Kingdom and New York, NY, USA
33. IPCC: Climate Change (2014) Mitigation of climate change, contribution of working group III to the Fifth Assessment Report of the Intergovernmental Panel on Climate Change. In: Edenhofer O, Pichs-Madruga R, Sokona Y, Farahani E, Kadner S, Seyboth K, Adler A, Baum I, Brunner S, Eickemeier P, Kriemann B, Savolainen J, Schlömer S, vonStechow C, Zwickel T, Minx JC (eds) Cambridge University Press, Cambridge
34. Nakhla DA, Hassan MG, El Haggar S (2013) Impact of biomass in Egypt on climate change. Nat Sci 5(6):678–684. https://doi.org/10.4236/ns.2013.56083
35. Chung NT, Jintrawet A, Promburom P (2015) Impacts of seasonal climate variability on rice production in the central highlands of Vietnam. In: 1st international conference on asian highland natural resources management, AsiaHiLand 2015. Agric Agric Sci Procedia 5:83–88
36. Wen-juan L, Hua-jun T, Zhi-hao Q, Fei Y, Xiu-fen W, Chang-li C, Jianhua J, Xiu-mei L (2014) Climate change impact and its contribution share to paddy rice production in Jiangxi, China. J Integr Agric 13(7):1565–1574. https://doi.org/10.1016/S2095-3119(14)60811-X
37. Shrestha S, Chapagain R, Babel MS (2017) Quantifying the impact of climate change on crop yield and water footprint of rice in the Nam Oon Irrigation Project, Thailand. Sci Total Environ 599–600:689–699. https://doi.org/10.1016/j.scitotenv.2017.05.028
38. Elbehiry F, Mahmoud MA, Negm A (2018) Land use in Egypt's coastal lakes: opportunities and challenges. In: Negm AM et al (eds) Egyptian coastal lakes and wetlands: Part I—characteristics and hydrodynamics. Hdb Env Chem https://doi.org/10.1007/698_2018_250. © Springer International Publishing AG 2018
39. Wahab M (2006) Impact of climate change on rice production over Egypt. Int J Meteorol 31(314):351–359, +362–366
40. Ouda S (2015) Major crops and water scarcity in Egypt: irrigation water management under changing climate, 126 p. Springer. (3319217712, 9783319217710)
41. Aucelli PPC, Paola GD, Incontri P, Rizzo A, Vilardo G, Benassai G, Buonocore B, Pappone G (2017) Coastal inundation risk assessment due to subsidence and sea level rise in a Mediterranean alluvial plain (Volturno coastal plain e southern Italy). Estuar Coast Shelf Sci 198:597–609. https://doi.org/10.1016/j.ecss.2016.06.017
42. Khanom T (2016) Effect of salinity on food security in the context of interior coast of Bangladesh. Ocean Coast Manag 130:205–212. https://doi.org/10.1016/j.ocecoaman.2016.06.013
43. Medany M, Elbasiouny H, El-Hefnawy NN (2007) The impact of climate change on sea level rise at North of Nile delta. In: Conference: climate change and their impacts on coastal zones and river deltas: vulnerability, mitigation and adaptation, Alexandria, Egypt

44. Smith J, Deck L, McCarl B, Kirshen K, Malley J, Abdrabo M (2013) Potential impacts of climate change on the Egyptian economy. United Nations Development Programme Cairo, Egypt
45. Shalaby MY, Al-Zahrani KH, Baig MB, Straquadine GS, Aldosari F (2011) Threats and challenges to sustainable agriculture and rural development in Egypt: implications for agricultural extension. J Anim Plant Sci 21(3):581–588
46. Elbaiouny H, Elbehiry F (2019c) The geology of Egypt. In: Elramady H (ed) The soil of Egypt. Springer Nature Switzerland AG
47. Abdin AE, Gaafar I (2009) Rational water use in Egypt. In: El Moujabber M, Mandi L, Trisorio-Liuzzi G, Martín I, Rabi A, Rodríguez R (eds) Technological perspectives for rational use of water resources in the Mediterranean region. CIHEAM, Bari, pp 11–27 (Options Méditerranéen nes: Série A. Sémin aires Méditerran éen s; n. 88)
48. Ashour MA, El Attar ST, Rafaat YM, Mohamed MN (2009) Water resources management in Egypt. J Eng Sci 37(2):269–279
49. El Agroudy N, Shafiq FA, Mokhtar S (2014) The impact of establishing the Ethiopian Dam renaissance on Egypt. J Basic Appl Sci Res 4(4):1–5
50. Nunzio JD (2013) Conflict on the Nile: the future of transboundary water disputes over the world's longest river. Strategic analysis paper, Future Directions International Pty Ltd.
51. Osman R, Ferrari E, McDonald S (2015) Water scarcity and irrigation efficiency in Egypt. In: International conference of agriculture economists, Italy
52. Morrison J, Morikawa M, Murphy M, Schulte P (2009) Water scarcity & climate change: growing risks for businesses & investors. Pacific Institute, 654 13th Street, Preservation Park, Oakland, CA 94612, www.pacinst.org
53. Mushtaq S (2016) Economic and policy implications of relocation of agricultural production systems under changing climate: example of Australian rice industry. Land Use Policy 52(2016):277–286
54. Mushtaq S (2017) Managing climate risks through transformational adaptation: economic and policy implications for key production regions in Australia. Climate Risk Management
55. El-Nashar WY, Elyamany AH (2017) Managing risks of the Grand ethiopian renaissance dam on Egypt. Ain Shams Eng J (in press)
56. Arafat S, Afify A, Aboelghar M, Belal A (2010) Rice crop monitoring in Egyptian Nile delta using Egyptsat-1 data. Joint U.S.-Egypt workshop for space technology & geo-information for sustainable development, NARSS-Cairo 2010
57. Elbasiouny H, Abowaly M, Abu_Alkheir A, Gad A (2014) Spatial variation of soil carbon and nitrogen pools by using ordinary Kriging method in an area of north Nile Delta, Egypt. Catena 01(113):70–78
58. Elbasiouny H, Abowaly M, Abu_Alkheir A, Gad A, Elbehiry F (2017) Restoration and sequestration of carbon and nitrogen in degraded northern coastal area in Nile Delta, Egypt for climate change mitigation. J Coast Conserv 21:105–114. 10.1007/s11852-016-0475-3
59. Hwang HY, Kim G, Kim SY, Haque MM, Khan MI, Kim PJ (2017) Effect of cover cropping on the net global warming potential of rice paddy soil. Geoderma 292:49–58. https://doi.org/10.1016/j.geoderma.2017.01.001
60. Shibu ME, Leffelaar PA, Van Keulen H, Aggarwal PK (2006) Quantitative description of soil organic matter dynamics—a review of approaches with reference to rice-based cropping systems. Geoderma 137:1–18

Nano-technology for Real-Time Control of the Red Palm Weevil Under Climate Change

El-Sayed Ewis Omran

Abstract The Red Palm Weevil (RPW) is a devastating insect on the date palm in the Arabic countries. It causes considerable damage to date palm plantations. All the stages of insect are hidden inside the palm tree and larvae feed on the interior tissue of trunk. For this reason, the detection of infection in the early stages is very difficult or impossible and when the action of the pest is discovered, normally it is too late for recovering the tree. The current measures used to control the insect are not effective enough to succeed in eliminating the insect because of the great difficulty in early detection of infection and reaching all life stages inside the trunk. The objective of this study is to detect the presence of living stages of RPW, which are hidden in the palm tree. Sensors and nanoparticles have useful applications in detection and treatment of the date palm. The study revealed that using Nano-techniques such as an acoustic sensor, thermal sensors as fingerprints, the infection of RPW can be discovered in early stages. The results of the audio analysis would be reported wirelessly to a control station for subsequent processing to send warning messages to be accessible via the Internet. Nano-thermal sensors are efficient to detect the existence of the RPW through its thermal properties. Also, using Nano-minerals as a treatment for the already Red Date Palm Weevil is important. The nanoparticle minerals (Rutile, Anatase or Brookite) were used and prepared in *Moringa oleifera* leaves extract to control RPW. The plant-mediated biosynthesis of nanoparticles is advantageous over chemical and physical methods because it is a cost-effective and environmentally friendly method, where it is not necessary to use high pressure, energy, temperature, and toxic chemicals. Also, using ultra-sonic sensors prevent the RPW to attack the date palm trees. Unlike traditional methods, which are time-consuming, and labor intensive, these types of methods offer the advantages of keeping the palm trees intact; reduce costs, as well as saving time and money in the process of pest infestation detection. Nano-minerals extract (NME) drastically decreased developmental stages of *R. ferrugineus*. The compound is a 100% natural solution. It is derived from natural

E.-S. E. Omran (✉)
Soil and Water Department, Faculty of Agriculture, Suez Canal University, 41522 Ismailia, Egypt
e-mail: ee.omran@gmail.com

Department of Natural Resources, Institute of African Research and Studies, Aswan University, Aswan, Egypt

© Springer Nature Switzerland AG 2020
E.-S. E. Omran and A. M. Negm (eds.), *Climate Change Impacts on Agriculture and Food Security in Egypt*, Springer Water, https://doi.org/10.1007/978-3-030-41629-4_15

plants and minerals that can significantly aid in controlling RPW. Also, it is safe to use on all plants and crops including natives.

Keywords Pest detection · Acoustic signal · Wireless sensor networks · Red palm weevil · Arab countries

1 Introduction

Date palm, *Phoenix dactylifera* L. (Arecaceae), is one of the earliest plants that had been cultivated for its fruit for at least 5000 years BC [1] and most important fruit trees cultivated in the Middle East and North Africa including Egypt [2]. One of the major threats to dactylifera over the world is the red palm weevil (RPW), *Rhynchophorus ferrugineus* (Olivier), (Coleoptera: Curculionidae) [3]. RPW is an economically important invasive tissue borer that has a broad host range restricted to palm trees, mostly young trees less than 20 years old [4, 5]. Climate change will profoundly affect the distribution and abundance of all species, including insects, and promote ecology-driven changes in phenology [6]. Climate change may result in a potential expansion of insect populations to polar regions and higher elevations [7]. Increases in global temperatures will expand or reduce the range of insects by converting climatically unsuitable habitats into suitable ones or vice versa [8]. Red palm weevil is 'global threat' after spreading to 60 countries. Therefore, identifying the potential distribution of *R. ferrugineus* under different global warming scenarios is crucial if we are to develop effective prevention and control measures. The integrated pest management strategy has been applied successfully to deal with the RPW problem [9, 4, 10]. The duration of all life cycle parameters varies significantly [5], which takes about 45–298 days. Several overlapping generations may also be seen inside the infested palm [4]. RPWs are considered to be the most serious blight, which threatens millions of palm trees in the Gulf region and the whole world.

Early detection and control timing of the RPW is an important topic to avoid the extensive damage to palm trees, as well as the emergence and migration of adult weevils [4]. The early prediction of insect pests helps the farmers to avoid heavy sprays of pesticides and take the necessary actions to restrict dangerous infestations [11]. Early warning of this pest may represent the only way to set up efficient actions to fight the coleopteron in trees where it takes over, thus limiting it's spreading in contiguous palms. Different technologies have been applied to detect the initial stages of RPW pest infestation [12]. Smell detection of infestation by Dogs (Golden Retriever) could be taught to recognize the chemical signatures produced by plant feeding stages [13]. These dogs detect the smell of the ooze that exudes from RPW infested date palm trees. Also, the gas sensor is one of this approach to detect the smell of some volatiles, generated during the fermentation processes in the infested palms [13]. "Anyway, gas sensors are not very selective and their response is influenced by many other volatiles" [14].

On the one hand, the most common method of detection is the visual inspection [15]. Visual detection and control of the RPW is difficult due to: the concealed nature of the pest life cycle [9] and the possibility of re-infestation of the treated palms with migrating adults from neighboring trees. Since direct visual detection of the RPW is quite difficult, the alternative detection method is needed. In nature, different sounds and vibrations have different origins and different functions. Bioacoustics technology and X-rays enable the detection of the early phase of infestation, however the usage of X-rays is expensive, but the acoustic technology [16, 17] has the potential for reducing the expense and dangers involved in the tree inspection. The acoustic recordings from insects in trees often reveal signals with spectral and temporal features that make them distinctive and easily detectable [18]. Sensitive microphones and dedicated amplifiers enable detection of the movement and feeding sounds of RPW larvae in palm trees [19]. Due to its high reproduction rate, the RPW prefers to live with no other insects in one trunk, which gives a good base for acoustic detection [20].

Seasonal activity of RPW catches varied significantly among months being highest in the warmer seasons in March, April, and May in KSA, and Egypt [21, 22]. The weevil activity was low during the monsoon between June and July. It became high after the monsoon between October and November in Egypt while in KSA the pest was more active during May with a second peak during November, and was low during August and February [9]. Abe et al. [23] stated the difficulty for RPW to overwinter in Japan, where winter temperatures is below 0 °C. When they measured the temperature of damaged palm tissues, it was between 30 and 40 °C, even in winter. The temperature inside a palm is between 15 and 35 °C [24]. This suggested that the temperature plays an important role in the detection of the RPW. The RPW feeding causes intensive plant tissue fermentation, which increases the local temperature (30 °C and above 45 °C) inside the crown/trunk [23, 25]. The temperature rise in the center of the crown of heavily infested palms could be detected only when viewed from above. The natural insulation of the palm tissue prevents detection in lateral view and also the solar radiation interferes with the thermal imaging.

On the other hand, thermal imaging is a viable alternative to point measurements. The tunneling insects destroy the vascular system of the palm and create local conditions of water stress. This "crop water status" could be sensed through inspection of the thermal portion of the spectrum of the reflected irradiation. Advances in remote thermal images offer the potential to acquire spatial information on surface temperature [17] and thus facilitate the mapping of canopy temperature variability over large areas. High-resolution thermal imaging systems have been used to evaluate the water status of cotton [26, 27], wheat [28], vineyards [29, 30], olives [31] and peanut [32, 33]. Aerial thermal images are a promising tool to map the water status of date palm trees [34].

Currently, there is no reliable instrument to detect infestation in the field [13]. To the best of my knowledge, until now, rapid and comprehensive early detection and control of RPW are not available in practice. There is a need to develop a system that can replace traditional techniques. Therefore, the first objective of this study was to propose an early warning system (portable and automatically) to detect the presence of RPW along with its larva by sensing its (sound and heat) activity in offshoots. The second objective was to propose Nano-natural (plants and minerals) extract for RPW treatment and cure. The last objective was to propose a Nano-ultrasonic system to prevent the date palm field from infected RPW.

2 Life Cycle and Geographical Distribution of RPW in Arabian Countries

Female of RPW bore into a palm tree to form a hole into which they lay an average of 250 eggs, which take about three days to hatch. Larvae emerge and tunnel toward the interior of the palm, inhibiting the palms ability to transport water and nutrients upward to the crown. RPW larvae bore deep into palm crowns, trunks and offshoots, generally concealed from visual inspection until the palms are nearly dead. Adults live for 2–3 months, during which time they feed on palms, mate multiple times and lay eggs. The adult is a strong flier that can move > 900 m in a single flight, and as much as 7 km in 3–5 day [35]. All RPW stages (egg, larva, pupa, and adult) as described in Fig. 1 are spent inside the palm itself and the life cycle (about 4 months) cannot be completed elsewhere. Different laboratories suggest that a single RPW female lays 58–531 eggs (28–30 °C and 70–75% relative humidity). The larvae live for 25–105 days. The pupal period ranges from 11 to 45 days. The duration of the life cycle from egg to adult stage was reported to vary from 45 to 139 days while adult weevils were reported to live for 50–90 days [36].

Damage because of RPW may result in one of the subsequent symptoms (Fig. 1) contingent on the phase of infestation: occurrence of channels on the trunk and base of leaf petiole, gnawing sound due to feeding by grubs, oozing out of thick brown fluid from the tunnels, presence of chewed plant tissues in and around opening of channels with a representative fermented odor, fallen empty pupal cases and dead adults around a heavily infested palm, breaking of the trunk or toppling of the crown in case of severe and prolonged infestation and drying of offshoots in date palm [9]. Screening for infection with the RPW is very difficult because infected palm does not quickly symptoms. This pest is especially destructive. This is because of the visible symptoms only appear when the infestation is severe. By then, it is too late to save the palm tree; therefore, only preventive actions are effective. Among these actions, early detection systems are crucial to fighting against RPW pest. They can quickly detect it in the early infestation stages and trigger the actuation protocol to save the rest of the plantation.

The RPW is reported to attack 19 palm species worldwide. “This expansion has been due to the movement of infested planting material from contaminated to uninfected areas” [37]. RPW has been identified as a pest of date palm in the Middle-East by FAO [4]. The RPW has detected in the Gulf area in the mid-1980s [38] then it was able to cross the Red Sea as it was found in Egypt (Ismailia and Sharkyia governorates) in 1992 [39]. Table 1 summarized the comprehensive information on the different host species of the RPW and the different Arabian countries of record.

Fig. 1 Life cycle stages and damage symptoms of the red palm weevil

3 Materials and Methods

3.1 Quantifying of Individual Palm Trees

The image analysis is valuable for the counting of palm trees. Quantifying palm trees manually are very time-consuming and complex task, because many factors involve as physical and environmental conditions. High-resolution IKONOS images have expanded, which can have the accuracy to view the plantation from many angles.

The system that merges both spatial imaging and spectroscopy is hyperspectral remote sensing. Images produced from hyperspectral sensors contain much more data than images from multispectral sensors and have a greater potential to detect differences between land and water features. For example, multispectral imagery was used to map forested areas. While hyperspectral imagery can be used to map tree species within the forest. Jusoff [46] differentiated between the three activities of remote sensing, i.e. detection, discrimination and identification. Detection involves a

Table 1 Distribution of RPW and number of productive date palms in the Arab countries

Country of records[a]	Year of first detection	Governorate or Province of the first detection	Number of productive trees (×1000) [40]	References
Iraq	1918	–	8024.00	[41]
United Arab Emirates (UAE)	1985	Rass El Khaima	16,342.19	[42]
Sultanate of Oman	1985	Northern Oman, particularly in Buraimi, Mahdah and Musandam areas	2457.00	[43]
The Kingdom of Bahrain	1985	–	379.98	[42]
Qatar Doha	1985	–	335.30	[42]
Kingdom of Saudi Arabia (KSA)	1987	In a Palm Nursery in El Qatif district in the Eastern Province	12,000.00	[37, 42]
Egypt	1992	Al Sharqiyah (Al Salihyah), Ismailia (Al Qassasin)	12,039.42	[39]
Kuwait	1993	–	312.21	[42]
Jordan	1999	–	104.38	[44]
Palestinian Authority Territories	1999	–	301.20	[44]
Syria	2005	Lattakia city, at Bouka Agriculture school	72.60	www.redpalmweevil.com
Morocco	2008	Tangier	5760.00	[45]
Libya	2009	Tobruk in the North East of Libya	2100.00	www.redpalmweevil.com
Lebanon	2010	Sidon	–	www.redpalmweevil.com

[a]Host plant for all Countries is *Phoenix dactylifera*, except Morocco, which is *Phoenix canariensis*

spectral signal that is consequential above the noise level. Discrimination is based on the spectral signal to be detectable and difference from adjacent material. However, identification is both discrimination and a spectral band. The abundance of spectral information and power in distinguishing objects are the advantages of hyperspectral data. However, it does not mean that the more bands are used, the better because of the following points. There is an evident correlation between bands, which will destroy the normal distribution of spectra and influence the classification accuracy. If all the bands are used in classification without selection, the classification precision will decrease instead of increase. Second, the more bands are selected, the more training samples are needed to classify correctly. It is very difficult for hyperspectral data to find sufficient and correct training samples that can meet the demand of the classifier if too many bands are chosen in classification. Third, an increase of band number in the classification will inevitably result in the increase in processing time and cost that will reduce the processing speed and benefit.

3.2 *Overall Nano System for RPW*

Figure 2 shows the overall Nano system technology for RPW detection and control. The starting point to achieve the main goal of detection and control RPW is the building nano scale system. The presented Fig. 2 describes the overall design of the

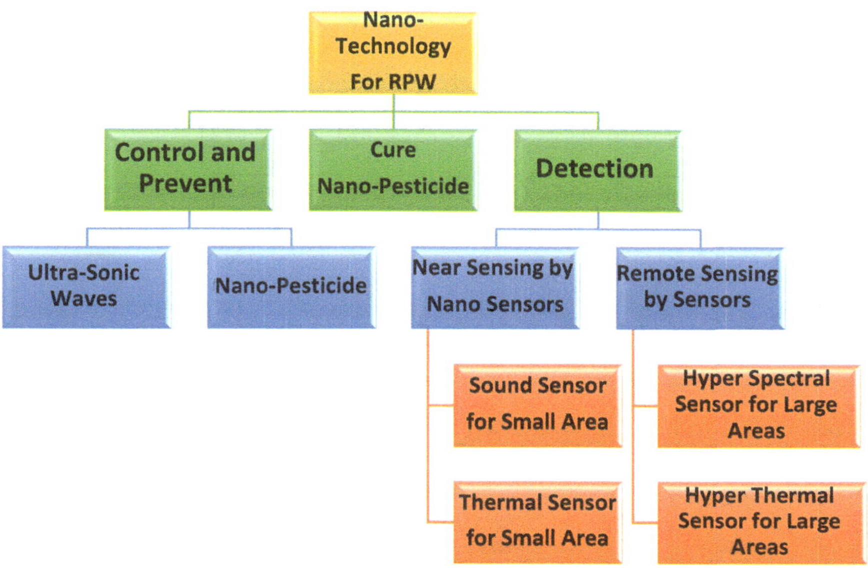

Fig. 2 Overall nano system for RPW detection and control

nano system for RPW detection and control. Three main approaches in the proposed nano system are detection, cure, and control.

3.2.1 Proposed Audio Sensor Architecture

Three different kinds of sounds come from RPW larvae: "eating", "squealing" and "moving" have been identified [12]. "The first one corresponds to the characteristic crunch sound produced when the RPW larvae chew internal palm fibers. The second one "squealing" is also a characteristic RPW sound, but its cause is not clear" (http://www.mdpi.com/1424-8220/13/2/1706/htm). Finally, the last one corresponds to the larvae movement through galleries inside the palm trunk. For the current study, the first one "eating" was chosen because (1) it is the loudest sound, (2) its frequency is clearly superior to the other identified sounds, and (3) it is representative of the RPW larvae feeding actions in the first stages of their evolution. Consequently, this one is the proper target sound to analyze in our early detection sensor. The main components of the proposed sensor architecture are as follows (Fig. 3):

- An audio probe, in charge of acquisition of sounds from the RPW, conditioning and properly amplifying the captured audio signal, making it suitable to be processed by the detection sensor.
- A low-power processor and supply, that will be able to run and "process the sound captured by the audio probe and determine the RPW presence" (http://www.mdpi.com/1424-8220/13/2/1706/htm).

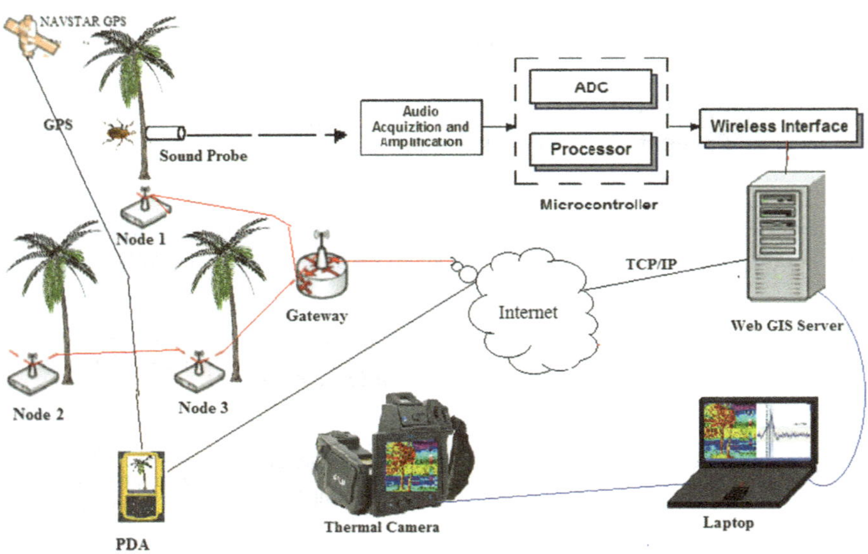

Fig. 3 A prototype block diagram for the proposed RPW detection system

- "A wireless communication interface, able to deliver data messages reporting the results of RPW activity" (http://www.mdpi.com/1424-8220/13/2/1706/htm).

The audio probe provides the signal captured from the palm tree with the highest possible quality. The weight and size of the sensor must be small enough to be installed on the palm tree. Concerning the audio acquisition task, there are two possible ways: First option uses four-wire audio interface. The use of an external audio Coder/Decoder (CODEC) is mandatory. Some CODECs include an amplification stage, which may be an advantage. The second option uses the Analog-to-Digital Converter (ADC). The audio implementation should be done using the on-chip ADC of the microcontroller to digitalize the audio. The advantage of this option is that the parameters and power consumption of the amplification stage may be fine-grain designed, so we have decided to use this approach in our sensor. Regarding the power supply, the sensor measured consumption rises up to 200 mA when performing the detection process. Four stages for RPW detection are identified: sound acquisition, digitalization, audio analysis, and transmission of results. Finally, the radio interface is up to create a reliable point-to-point outdoor communication over distances of 30 m. The 30 m maximum reliable range has been proposed after performing some experimental field tests with two sensor prototypes.

The audio probe is composed of three elements: the microphone, the probe and the signal conditioning stage. A wireless microphone contains a radio transmitter. It transmits the audio as a radio rather than via a cable. It sends its signal using a small FM radio transmitter to a nearby receiver connected to the sound system. To assign the sound sensor to the palm tree, two distinctive probe designs were considered. The first comprises of appending the sensor to the palm tree surface. The fundamental benefit of this approach is the low or even invalid effect on the palm tree structure. A 1 cm diameter circle has been sanded down to accommodate the sensor, which is held with a strap around the palm tree. Experiments have demonstrated that it is very difficult to settle the sensor close to the stem. Likewise, this appending strategy leaves the sensor visible to environmental noise, which in urban zones and parks may be higher than sounds originating from the palm tree trunk. The second option was to utilize a nail to embed the sensing or detecting device inside the palm tree. This alternative gives two focal points: The sensor is nearer to the sound source and it is confined from outside noise. The fundamental disservice is the damage caused to the palm tree, leaving a 1 cm diameter hole. It requires being amazingly cautious when removing the probe, filling the influenced area with proper putty material and/or painting it with a fungicide. An aluminium probe was used because it is resistant to corrosion and easy to work, with a 10 mm of interior diameter, and around 10 cm of length to get the maximum sound transmission and to avoid an excessive injury to the palm tree. Finally, the microphone has been situated on the outer side of the probe to get a better sound transmission, leaving the other end opened.

Once the audio sensor and the probe have defined (Fig. 4), some signal conditioning was performed to deliver the microphone output signal to the A/D converter with the highest possible quality (amplification stage). A printed circuit board (PCB) has been planned, where two of the operational enhancers are utilized to amplify the

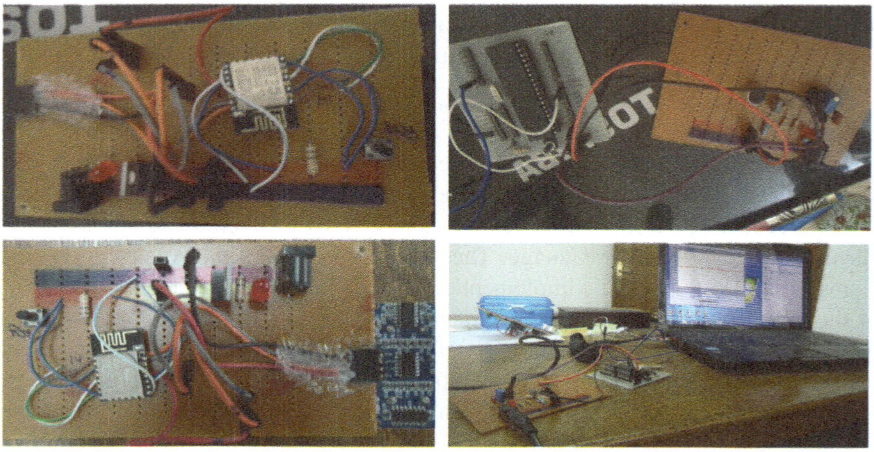

Fig. 4 Prototype of sensor installation

microphone output. Two extra speakers are utilized to create an active low-pass filter at 10 kHz, to avoid aliases in the digital part of the acquisition chain. This board is inserted in a plastic box, with an RCA connector for the microphone, a potentiometer to adjust the desired gain, and two jack outputs, one for the headphones and the other for the connection to an analog input of a microcontroller (a prototype is shown in Fig. 4). The objective of this plan is the measured quality and adaptability of the device: distinctive microphones might be associated with the amplifier, it tends to be utilized by a human operator listening through headphones, or it can be connected to the microcontroller for the self-ruling recognition.

During the sensor installation, the operator fixes the sensor (audio probe plus sensor board) to the palm tree at the desired monitoring location. Then, the sensor is configured using the Sensor Deployment Software (SDS) running on a portable device (smartphone, tablet, laptop, etc.) with GPS support and equipped with the control station radio interface (i.e., a USB dongle) which allows direct communication with the sensor. This device acts as a control station only during the sensor node initialization/configuration stage, without requiring the synchronization with other network nodes.

3.2.2 Thermal (Heat) Sensing for Detection of RPW Infected Trees

One of this study goal was to examine the ability to detect infected trees using thermal images. The hypothesis was that the tunneling of RPW destroys the vascular system of the palm and create local conditions of water stress. Mozib and El-Shafie [47] show that the average temperature of the infested date palm was significantly higher (32.60 °C) than the average temperature recorded at the same time both inside the healthy trees (29.53 °C) and in the ambient atmosphere (29.35 °C). The average temperature of infested date palm increased slightly with an increase in the infestation level with the highest being 32.80 °C. It is evident that temperature differences between infested date palm on one side and healthy date palm and ambient atmosphere on the other side could well provide a platform for identifying infested trees.

Real-time analysis of a thermal image (TI) reveals if the tree is healthy if it needs cutting or other remedial actions or has to be felled. A uniform surface temperature distribution is present and the TI shows a uniform coloring if the tree is being healthy, but if the color is not uniform, then decay/cavities may be present. 4–5 thermal images are normally enough to understand the conditions of a whole tree. A quick assessment normally requires 2–3 min, while an in-depth analysis (with detailed visual shootings, a record of the tree characteristics and details, etc.) which is less than 10 min. The used technique requires less time than any other technologically advanced investigation system. The system has been used in all weather conditions, by night and day, in the summer and winter, with temperatures ranging from +2 to +35 °C.

3.2.3 Nano-minerals Technology for Controlling RPW Under Climate Change

There have been several setbacks in controlling RPW using synthetic insecticides. The conventional pesticides are costly and may result in: high mammalian toxicity, high level of persistence in the environment, health hazards, toxic residues on food, adverse effects on non-target organisms and pest resistance as well as a toxic effect for the users [48]. Therefore, the feasibility of new and more safety biopesticides like natural oils, pathogenic bacteria, pheromones, and nematode is needed. Many investigations have been conducted on the antifeedant effects, growth inhibition and abnormal development in RPW caused by using natural insecticides with little or no negative impacts on the environment and not toxic to mammals [49, 50]. "One solution would be to replace synthetic chemicals with compounds that occur naturally in plants" [51]. Khalid et al. [1] used the powders of Argel (*Solenostemma argel* Del. Hyne.), usher (*Calotropis procera* Ait.), and two desert plants to control the insect pests. They found that there was an increase in the mortality of the insect as well as the yield of date palm. The antifeedant activity of three essential oils extracted from two plants of the Asteraceae family namely, Crofton weed, Eupatorium adenophorum Spreng and Indian wormwood, *Artemisia nilagirica* was evaluated by Paraj et al. [52]. The essential oils were found to be effective against the weevil. Several workers [53, 54] had also reported the efficacy of neem extracts against the weevil.

Materials at the nanoscale behave differently than the same material at the bulk scale. Nanoparticles are used due to their special properties stemming from their surface area, shape, electronic, and optical behavior. "Their size is typically between 1 and 100 nm for at least two of their dimensions" [55]. Titanium dioxide or Titanium (IV) oxide (TiO_2, CAS no. 13463-67-7) exists in nature in one of three crystalline forms, which are anatase (CAS no. 1317-70-0), rutile (CAS no. 1317-80-2),

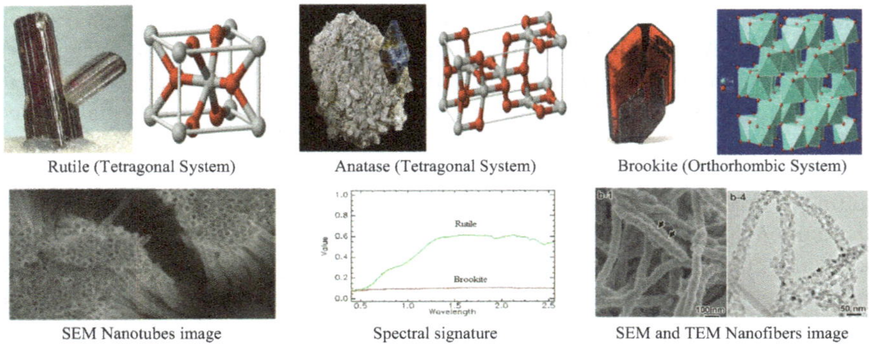

Fig. 5 Different minerals and their structures used as a source for Titanium dioxide. Transmission electron micrograph (TEM) and scanning electron micrograph (SEM) images. Adopted from Vijayalakshmi and Rajendran [56] and Wang [57]

and brookite (12188-41-9). These three crystalline forms of nanoparticles minerals, which potential for RPW controlling have 95% TiO_2 (Fig. 5). However, other titanium minerals such as ilmenite, $FeTiO_3$ (52.6% TiO_2), perovskite, $CaTiO_3$ (58% TiO_2), sphene, $CaTiSiO_5$ (35–40% TiO_2) have low TiO_2.

The nanoparticle minerals (rutile, anatase or brookite) were used and prepared in *Moringa oleifera* leaves extract to control RPW. Figure 5 shows the spectral signature of titanium minerals and the procedure used to control RPW. The plant-mediated biosynthesis of nanoparticles is advantageous over chemical and physical methods because it is a cost-effective and environmentally friendly method.

In the current study, five doses (1, 2, 3, 4, and 5 ml L^{-1}) of Nano-Minerals extract (NME) were applied for studying its effect on the insect duration.

"Rutile is the thermodynamically stable form of titanium dioxide; anatase rapidly transforms to rutile above 700 °C" (https://echa.europa.eu/registration-dossier/-/registered-dossier/15560/4/1). Rutile melts between 1830 and 1850 °C [58]. TiO_2 was obtained by a one-step procedure. The hydrothermal synthesis process was performed under acidic conditions in a 0.59 M H_2SO_4 solution at 160 °C for 24 h or at 200 °C for 6 h using a titanium metal complex with glycolic acid. Titanium dissolves rapidly in hydrofluoric acid. The dissolution of titanium in hydrofluoric acid is probably an electrochemical process, based on the activity of local elements. Metal salts added to the acid during the dissolution of titanium increased the rate in the following order: Fe^{2+}, Ni^{2+}, Ag^{+}, Cu^{2+}, Au^{3+}, Pt^{4+}, that is, nearly the order of decreasing over voltage of the corresponding metals. Mg^{2+} proved to be inactive while Pb^{2+} hampered the dissolution reaction. A visible grayish blue film was formed on titanium in concentrations below 0.5 *N*.

4 Results and Discussion

4.1 Early Detection of the RPW Using Heat, Sound, and Spectral Sensors (HSSS)

4.1.1 Sound Fingerprint for RPW Detection

Figure 6 shows an example of "eating" sound that was extracted from the system. We can see the "eating" sound in both temporal, and frequency domains. Energy distribution across wavelet subbands may be considered a spectral fingerprint of RPW feeding activity, is an important feature to be included in our RPW "eating" sound model. The fundamental frequency was found to be in the range of 200–300 Hz, which is taken as the pass band of the Band pass filter. The detected signals are clear and easy to identify. Every sound that is heard is caused by activities in the trunk. Several different sounds of the RPW could be isolated. These sounds represent different steps of the weevil development. The recorded sound stream is imported to the Band Pass filter and we listen to the output of the Band Pass filter, which consists of only the clicks through a speaker. If the clicks are heard, the tree is said to be infested, else the tree is not infested. The following typical sounds caused by the RPW was recorded: eating sounds from larvae, moving sounds from larvae, larvae spinning a kokon, moving of a pupa, and sounds of digestion from larvae.

The "eating" sample in the temporal domain, determining its main duration in terms of audio samples was analyzed. We need to know the beginning and ending positions of "eating" sound inside a captured audio window using three parameters: audio signal level, signal variance, and SNR level. These parameters are empirically established to properly identify potential RPW sounds.

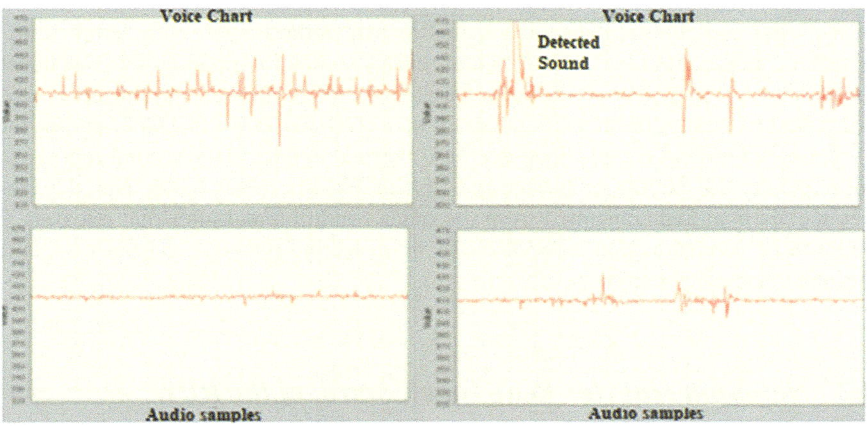

Fig. 6 A sample window containing the RPW "eating" sound captured at temporal, and frequency domain

Notice that not all of the "eating" samples have been recorded with the same quality as the one shown in Fig. 6. There are "eating" samples with (1) different noise levels, mainly due to the recording conditions, like sound probe placement and environmental noise present at recording time, (2) different sound levels due to the distance between the source RPW and the audio probe resulting in strong, average, and weak "eating" samples, and (3) also different temporal extensions of "eating" samples.

4.1.2 Thermal Map of RPW Defected Date Palm

By measurements, imaging and analyzing of infested and uninfested trees, results partially showed that the RPW creates water stress and affects canopy temperature. Analysis of thermal images of date palm plantation successfully detected infected trees. Figure 7 shows the thermal image (TI), which can be considered as a real "thermal map" of the tree under study. "It only shows the surface temperature distribution, which is generated by the different thermal properties (especially conductivity) of the underlying areas of healthy or decaying tissue" (http://www.treethermography.it/about_treethermography.htm). The thermal conductivity in wood is mainly linked to tissue humidity and decreases as the liquid content decreases: that is why the areas with a cavity (no tissue) or decay (less heat-conveying liquids) have a lower surface temperature than healthy areas. The more the decay, the lower the surface temperature in the area covering it. The thermal camera detects this deficit in liquids at the collar level because this decreases conductivity and consequently surface-bound heat transmission.

Figure 7 shows that limited areas having a warmer temperature for no apparent reason were sometimes spotted: when no rejuvenating tissue and vital tissue formations were found. The hypothesis that heat was emitted because of the initial activity of a fungal or bacterial attack was put forward. The thermal camera does not "see inside", like the clinical thermometer that measures the human body temperature. A surface temperature variation in a tree reveals something inside it that can cause this variation. The area covering decay/cavities is colder than surrounding areas covering healthy tissue, because the absence of woody tissue (in the hole) and/or the decrease in fluids in rotted tissue decrease thermal conductivity around there, and subsequently the amount of surface-bound heat. Heat is somewhat delivered by the tree metabolism and consumed from the outer condition and after that released. It ought to be viewed as that wood is a decent protecting material, yet has a specific thermal conductivity anyway.

4.2 Nano-minerals for Real-Time Control of the RPW

The present study also investigates the effects of *Moringa oleifera* on the activities of *Rhynchophorus ferrugineus* larvae. Results showed that mortality increased in the

Fig. 7 Thermal image used as a real "thermal map" of the palm tree under study

larvae stage by using this prepared solution of natural extract with titanium dioxide nanoparticles at concentration 100 mg L^{-1} and this was the best concentration used. This innovative natural and non-toxic to humans does not affect the rights and health of human, animal and bird. It is used throughout the year, even during the fruiting period. So they are used in agriculture safety. It kills the larva through the digestive system when it enters the digestive tract of an insect. It also kills by contacting and so when the insect touches the treated surface upside insect. In this case, the material penetrates the cover of the outer insect and affects the nervous system and therefore die. It also kills through the respiratory tract or other natural openings in the body where it affects them and succumbs. The effect of the killing depends on one property of the three properties. If the three properties exist at the same time, it becomes very toxic to insect and eliminates decay in 180 s.

4.2.1 Injection Apparatus

The low-pressure injection unit was locally adapted as in Fig. (8) by using 1.0 L plastic bottles, which were connected to canola apparatus by a rubber tube. The aluminium probe was inserted into the hole and the flow rate was adjusted to prevent overflow out of the injection pore. The bottles were fixed on the trunk 1 m above the injection pore. The injection pores drilled within the dimensions of 25 × 0.9 cm with a slope of 45° situated at 1.5 m above the ground. One pore per date palm drilled, where the canola fitted in the pore by a stopper. According to our experience,

Fig. 8 Real-time control of the RPW and low-pressure injection apparatus

it seems that there is no need to drill more than one hole per palm trunk to introduce the pesticides inside the trunk as long as the pesticides moved out of injection sites and distributed fairly in date palm tissues. The rapid pesticides distribution on date palm trunk depends on upon the distribution of xylem and phloem bundles in the trunk.

4.2.2 Effect of Proposed Nano-minerals Extract on the Different Developmental Stages of RPW

Data in Table 2 show that Nano-minerals extract (NME) was high powerful against larvae. The higher mortality (100%) occurred when larvae of *R. ferrugenius* was treated with NME at a dose of 5 ml L^{-1} and the lower toxicity (35%) was obtained after the larva was treated with the lower dose (1 ml L^{-1}). The Pupal stage was less affected by NME. The Pupal mortalities were 92% after treatment with the higher dose (5 ml L^{-1}) of NME. The lower adult mortality (18%) produced at a dose of 1 ml L^{-1}. Similar results were obtained against the red palm weevil, *R. ferrugineus* reported by many authors. Abdullah and Nassar [59] reported that *B. chinensis* oil, Abamectin, *N. oleander* and *A. viridis* extracts were effective and reduced the eggs fertility of *Spodoptera littoralis* larvae. Bream et al. [60] suggested that the lethal effect of *B. chinensis* oil was increased by increasing the age of pupae. Nassar and Abdullah [61], reported that Azadirachtin was very toxic against the different developmental stages of *R. ferrugineus*.

From the results recorded in Table 2, it shows that, duration was significantly ($p < 0.05$) shortened by the effect of NME on larvae and adults of *Rhynchophorus ferrugineus*. Also, duration was dose dependent term, it decreased as increasing doses of NME. The pupal duration decreased, 30 min as affected by the high dose (5 ml L^{-1}) of NME, while the lower dose (1 ml L^{-1}) caused increasing (160 min) of the pupal duration (Table 2). Some authors obtained similar results by using precoceneII or some plant extracts. Shabana et al. [62] use an aqueous extract of the whole plant, *Ageratum conyzoides* contained precoceneII, verified reduction of larvae emergence

Table 2 Effect of nano-minerals extract on the different developmental stages and adult durations of red palm weevil

Doses ml L^{-1}	% of larval mortality	% of pupal mortality	% of adult mortality	Larval duration in minuets	Pupal duration in minuets	Adult duration in minuets
1	35	29	18	105	160	78
2	44	41	34	65	110	48
3	60	58	46	2.5	69	6.5
4	80	69	58	1.9	58	2.6
5	100	92	87	1.3	30	1.5
Control	1	0.25	0.21	–	–	–

of *Meloidogyne incognita*. Plant extract of *A. conyzoides* in *Citrus orchards* sheltered predators of the spider *Panonychus citri*, suggested that its development in orchards is beneficial [63].

Treatment of insects, or their food with precoceneII causes growth inhibition and increasing doses of precoceneII in larval instars result in different forms of effect, one of them is extending the life period of larvae, which remain as 'over-aged" larvae of a wide variety of insects. Duration of the different developmental stages of the red palm weevil, *R. ferrugineus* was dose-dependent shortened after application of Azadirachtin [60, 61].

4.2.3 Spread and Direction of Solution Movement Inside the Trunk

To know the spread of the proposed aqueous solution, methylene blue dye was injected into the trunk at a depth of 20 cm with a 45° angle and the volume of 1 L. The anatomy of the Palm after 24 h of the treatment chain saw to where to take them cross sections (Fig. 9) and longitudinal above and below the injection zone.

Fig. 9 Spread and direction of solution movement inside the trunk

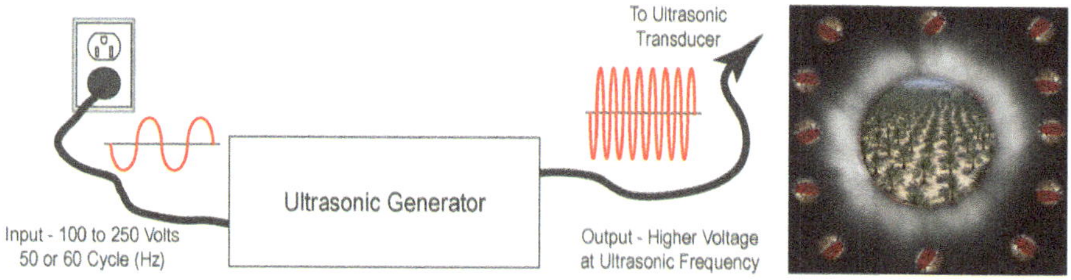

Fig. 10 Procedure used to control RPW

4.3 Echolocation and Ultrasound as an Acoustically Wall to Protect Palm Trees from the RPW

Each type of behavior results in a certain oscillation of the cells in the insect. Exposing the insect to the acoustic vibrations frequently leads to a change in the way the vibrating cells. In other words, a change in the cellular vibrational frequencies. There are frequencies make the brain cells vibrate dynamically active and positive and increase the positive energy of the cells and vice versa is true. Therefore, the correct frequencies may lead to the death or escape of the insect pest. The system (Fig. 10) emits high-frequency sound waves (ultrasound) between 28 and 80 kHz, which are much higher than the earshot of human beings and common pets (cats, dogs, birds, fish, etc.). However, both rodents and other pests hear these frequencies that are unpleasant to them. This is because the ultrasound uncomfortable for pests, and are looking for a respite from these new disturbances in the surrounding areas.

5 Conclusions

The Red Palm Weevil (RPW) is a highly destructive pest of date palms in several countries. Control of the RPW is difficult due to the concealed nature of the life cycle of the pest, since palm trees do not show visual evidence of infection until it is too late for them to recover. In this chapter, Nano-sensor system was developed for significantly recognition of RPW in an earlier phase of the infestation and applying the appropriate treatment. The early detection system proposed is based on acoustic monitoring, as the activity of RPW larvae inside the palm trunk is audible for human operators under acceptable environmental noise levels. The sensor can be installed in palm tree and is able to analyze the caught audio signal during large periods of time. The results of the sound analysis would be accounted wirelessly to a control station, to be subsequently processed and conveniently stored. That control station is to be open by means of the Internet. It is modified to send warning messages when predefined alarm thresholds are achieved, thereby enabling managers to check online the status and evolution of the palm trees. The results imply the efficiency of the developed system to detect the existence of the RPW through its feeding sound. Nano-sensors for RPW thermal analysis is efficient to detect the existence of the RPW through its thermal properties. Nano-minerals were used and prepared in natural leaves extract to control RPW. Nanoparticles of TiO_2 could be used along with *Moringa oleifera* extract against *Rhynchophorus ferrugineus* larvae. The present study revealed that NME extract drastically decreased developmental stages of *R. ferrugineus*. The nanoparticle minerals are advantageous over chemical and physical methods because it is a cost-effective and environmentally friendly method, where it is not necessary to use high pressure, energy, temperature, and toxic chemicals. Ultrasonic sensors were used to prevent the RPW to attack the date palm trees. Unlike traditional methods, which are time consuming, and labor intensive, this system offers the advantages of keeping the palm trees intact; reduce costs, as well as saving time and money in the process of pest's infestation detection. The compound is a 100% natural solution. It is derived from naturally plants and minerals that can significantly aid in controlling RPW. Also, it is safe to use on all plants and crops including natives.

6 Recommendations

Current method recommended for the management of Red Palm Weevil, which have focused on integrated different techniques involving Nano-sensor system, Ultrasonic sensors, and Nano-minerals extract. Raising awareness, applying incentive and punishment system, re-evaluation of the current recommended management practices, effective early detection technique, support to set up surveillance program etc. are some of the major recommendations proposed to overcome the challenges of RPW management.

The following is the proposed improvements/recommendations:

1. Raising awareness on RPW measures among all the stakeholders.
2. Strengthen the coordination and engagement of all stakeholders (farmers/farmers' cooperatives, NGOs, other law enforcement agencies etc.).
3. Urgent need to develop a fast and reliable, cost effective, and easy to handle early detection device for RPW.
4. Implement a GIS aided monitoring system for efficient mapping, data collection, analysis and management.
5. Use of natural/organic products to control RPW.
6. Develop a user friendly mobile application for reporting, data collection and transmission.
7. Develop a GIS and spatial database to be used operationally by countries.
8. Use a remote sensing imagery to geo-reference palm trees in countries to be used as primary base map in the GIS.
9. Establish a training programme for different categories of the users of the tools (mobile apps, GIS, software).
10. Use of social media to expedite transmission of information is essential.

Acknowledgements Special thanks go to Prof. Dr. N. Mandour for his comments on a draft of this chapter.

References

1. Khalid OAM, Eldoush AKT, Tag EM, Omar A, Sidahmed A, FakhrEldeen A, Hatim GM (2011) Application of plant based extracts for the control the green pit scale insect (*Asterolicanium phoenicis* Rao.) with yield enhancement on date palm. Emir J Food Agric 23(5):404–412
2. Sawaya WN (2000) Proposal for the establishment of a regional network for date-palm in the Near East and North Africa. A draft for discussion, FAO/RNE
3. Dembilio Ó, Jacas JA, Llácer E (2009) Are the palms *Washingtonia filifera* and *Chamaerops humilis* suitable hosts for the red palm weevil, *Rhynchophorus ferrugineus* (Col. Curculionidae). J Appl Entomol 33:565–567
4. Faleiro JR (2006) A review on the issues and management of red palm weevil *Rhynchophorus ferrugineus* (Coleoptera: Rhynchophoridae) in coconut and date palm during the last one hundred years. Int J Trop Insect Sci 26:135–154
5. Kaakeh W (2005) Longevity, fecundity, and fertility of the red palm weevil, *Rynchophorus ferrugineus* Olivier (Coleoptera: Curculionidae) on natural and artificial diets. Emir J Agric Sci 17:23–33
6. Nooten SS, Andrew NR, Hughes L (2014) Potential impacts of climate change on insect communities: a transplant experiment. PLoS ONE 9(1):e85987. https://doi.org/10.1371/journal.pone.0085987
7. Chen Y, Ma CS (2010) Effect of global warming on insect: a literature review. Acta Ecol Sin 30(8):2159–2172
8. Sambaraju KR, Carroll AL, Zhu J, Stahl K, Moore RD, Aukema BH (2012) Climate change could alter the distribution of mountain pine beetle outbreaks in western Canada. Ecography 35(3):211–223
9. Faleiro JR (2005) Insight into the management of red palm weevil *Rhynchophorus ferrugineus* Olivier: based on experiences on coconut in India and date palm in Saudi Arabia, Fundación Agroalimed. Jornada Internacional sobre el Picudo Rojo de las Palmeras 27–29:35–57
10. Justin CGL, Leelamathi M, Thangaselvabai T, Johnson SBN (2008) Bioecology and management of the red palm weevil, *Rhynchophorus ferrugineus* Oliv. on coconut: a review. Agric Rev 29:117–124
11. Yones MS, Arafat SM, Abou Hadid AF, Abd Elrahman HA, Dahi HF (2012) Determination of the best timing for control application against cotton leaf worm using remote sensing and geographical information techniques. Egypt J Remote Sens Space Sci 15:151–160
12. Martínez RM, Migallón GH, López GO, Perez MM, Martí CA, José SM (2013) On the design of a bioacoustic sensor for the early detection of the red palm weevil. Sensors (Basel) 13(2):1706–1729
13. Nakash J, Osam Y, Kehat M (2000) A suggestion to use dogs for detecting red palm weevil (*Rhynchophorus ferrugineus*) infestation in date palm in Israel. Phytoparasitica 28:153–154
14. Mielle P, Marquis F (1999) An alternative way to improve the sensitivity of electronic olfactometers. Sens Actuators B Chem 58:526–535
15. Mahmud AI, Farminhao J, Viez RA (2015) Red palm weevil (*Rhynchophorus ferrugineus* Olivier, 1790): threat of palms. J Biol Sci 15:56–67
16. Omran ESE (2012) Detection of land-use and surface temperature change at different resolutions. J Geogr Inf Syst 4(3):189–203. https://doi.org/10.4236/jgis.2012.43024
17. Omran ESE (2012) On-the-go digital soil mapping for precision agriculture. Int J Remote Sens Appl 2(3):20–38

18. Mankin R, Osbrink WL et al (1995) Acoustic detection of termite infestations in urban trees. J Econ Entomol 981–988
19. Hetzroni A, Mizrach A et al (2004) Developing spectral model to monitor activity of red palm weevil. Alon Hanotea 58:466–469
20. Hussein WB, Mohamed AH, Thomas B (2009) Application of the signal processing technology in the detection of red palm weevil. In: 17th European signal processing conference (EUSIPCO 2009) Glasgow, Scotland, 24–28 Aug
21. El-Garhy ME (1996) Field evaluation of the aggregation pheromone of the red palm weevil, *Rhynchophorus ferrugineus*. In: Proceeding of international conference on Egypt, brighten crop protection conference: pest and diseases, Brighton, UK, 18–21, pp 1059–1064
22. Soroker V, Blumberg D, Haberman A, Hamburger-Rishard M, Reneh S, Talebaev S, Harari AR (2005) Current status of red palm weevil infestation in date palm plantations in Israel. Phytoparasitica 33:97–106
23. Abe F, Ohkusu M, Kubo T, Kawamoto S, Sone K, Hata K (2010) Isolation of yeasts from palm tissues damaged by the red palm weevil and their possible effect on the weevil overwintering. Mycoscience 51:215–223
24. El-Mergawy RAAM, Al-Ajlan AM (2011) Red palm weevil, *Rhynchophorus ferrugineus* (Olivier): economic importance, biology, biogeography and integrated pest management. J Agric Sci Technol A 1:1–23
25. Suma P, Longo S (2009) Applicazioni di termografia, endoscopiaedanalisiindiretta per la diagnosiprecocedegliattacchi di punteruolorosso. In: RegioneSiciliana - AssessoratoAgricoltura e Foreste. La ricercascientificasulPunteruolorosso e glialtrifitofagidellepalme in Sicilia., 1(SBN Pal0217180.), pp 103–106
26. Alchanatis V, Cohen Y, Cohen S, Möller M, Sprinstin M, Meron M, Sela E (2010) Evaluation of different approaches for estimating and mapping crop water status in cotton with thermal imaging. Precision Agric 11:27–41
27. Meron M, Tsipris J, Orlov V, Alchanatis V, Cohen Y (2010) Crop water stress mapping for site-specific irrigation by thermal imagery and artificial reference surfaces. Precision Agric 11:148–162
28. Tilling AK, O'Leary GJ, Ferwerda JG, Jones SD, Fitzgeral GJ, Rodriguez D, Belford R (2007) Remote sensing of nitrogen and water stress in wheat. Field Crops Res 104:77–85
29. Grant OM, Tronina L, Jones HG, Chaves MM (2007) Exploring thermal imaging variables for the detection of stress responses in grapevine under different irrigation regimes. J Exp Bot 58:815–825
30. Möller M, Alchanatis V, Cohen Y, Meron M, Tsipris J, Naor A, Cohen S (2007) Use of thermal and visible imagery for estimating crop water status of irrigated grapevine. Int J Exp Bot 58:827–838
31. Ben-Gal A, Koolb D, Agam N, vanHalsema GE, Yermiyahu U, Yafe A, Dag A (2010) Whole-tree water balance and indicators for short-term drought stress in non-bearing 'Barnea' olives. Agric Water Manag 98:124–133
32. Omran ESE (2016) The Egyptian journal of remote sensing and space sciences
33. Omran ESE (2016) Early sensing of peanut leaf spot using spectroscopy and thermal imaging. Arch Agron Soil Sci 1–14. https://doi.org/10.1080/03650340.2016.1247952
34. Cohen Y, Alchanatis V, Prigojin A, Levi A, Soroker V (2012) Use of aerial thermal imaging to estimate water status of palm trees. Precision Agric 13:123–140
35. Abbas MST, Hanounik SB, Shahdad AS, Aibagham SA (2006) Aggregation pheromone traps, a major component of IPM strategy for the red palm weevil, *Rhynchophorus ferrugineus* in date palms (Coleoptera: Curculionidae). J Pest Sci 79:69–73
36. Abraham VA, Faleiro JR, Nair CPR, Nair SS (2002) Present management technologies for red palm weevil *Rhynchophorus ferrugineus* Olivier (Coleoptera: Curculionidae) in palms and future thrusts. Pest Manag Hortic Ecosyst 8:69–82
37. Ferry M, Gomez S (2002) The red palm weevil in the Mediterranean area. Palms 46:172–178
38. Abozuhairah RA, Vidyasagar PSPV, Abraham VA (1996) Integrated management of red palm weevil, *Rhynchophorus ferrugineus*, in date palm plantations of KSA. In: Proceedings of international congress of entomology, Firenze, Italy, 25–31 Aug, pp 541

39. Cox ML (1993) Red palm weevil, *Rhynchophorus ferrugineus* in Egypt. FAO Plant Prot Bull 41:30–31
40. AOAD, Arab Organization for Agricultural Development (2008) Arab agricultural statistics yearbook, 28, Year-2008 PART III. Plant Production, Statistics Division
41. Buxton PA (1920) Insect pests of dates and the date palm in Mesopotamia and elsewhere. Bull Entomol Res 11:287–303
42. Zaid A, de Wet PF, Djerbi M, Oihabi A (2002) Diseases and pests of date palm. In: Zaid A, Arias-Jiménez EJ (eds) Date palm cultivation, FAO plant production and protection. Review 1, Rome, Italy, Paper 156 110
43. Kaakeh W, Khamis AA, Aboul-Nour MM (2001) The red palm weevil: the most dangerous agricultural pest. UAE University Printing Press, p 165
44. Kehat M (1999) Threat to date palms in Israel, Jordan and the Palestinian Authority by the red palm weevil, *Rhynchophorus ferrugineus*. Phytoparasitica 27:107–108
45. EPPO, European and Mediterranean Plant Protection Organization (2009) Reporting service, No. 1. 2009-01-01/2009/01-First record of *Rhynchophorus ferrugineus* in Morocco and 2009/02-First record of *Rhynchophorus ferrugineus* in Curaçao, Netherlands Antilles No. 3. 2009-03-01/2009/051—*Rhynchophorus ferrugineus* found on *Howea forsteriana* in Sicilia (IT) and No 11. 2009-11-01/2009/207, First report of *Rhynchophorus ferrugineus* in Albania and 2009/208, First report of *Rhynchophorus ferrugineus* in Ceuta (ES) and 2009/209, *Rhynchophorus ferrugineus* found again in Liguria, Italy and 2009/2010, Distribution of *Rhynchophorus ferrugineus* in China and 2009/2011, Control strategy against *Rhynchophorus ferrugineus*
46. Jusoff K (2009) Mapping of individual oil palm trees using airborne hyperspectral sensing: an overview. Appl Phys Res 1(1):15–30
47. Mozib ME, El-Shafie HA (2013) Effect of red palm weevil, *Rhynchophorus ferrugineus* (Olivier) infestation on temperature profiles of date palm tree. J Entomol Nematol 5(6):77–83
48. Ileke KD, Oni MO (2011) Toxicity of some plant powders to maize weevil, *Sitophilus zeamais* (Coleoptera: Curculionidae) on stored wheat grains. Afr J Agric Res 6(13):3043–3048
49. Ahmed IA, Umma M, Kutama AS, Hassan KY (2015) Insect pests of date palm (*Phoenix dactylifera* L.) and potentials of botanical insecticides for their control in the tropics: a review. Glob Adv Res J Agric Sci 4(7):275–279. ISSN: 2315-5094
50. Ileke KD (2008) Insecticidal activity of five edible plants powders against lesser grain borer, *Rhyzopertha dominica* on stored sorghum grains. Sci Res Ann 5(1):72–80
51. Ileke KD, Odeyemi OO, Ashamo MO (2012) Insecticidal activity of *Alstonia boonei* De Wild powder against cowpea bruchid, *Callosobruchus maculatus* (Fab.) [Coleoptera: Chrysomelidae] in stored cowpea seeds. Int J Biol 4(2):125–131
52. Paraj S, Polana SP, Saleh AA, Mahmoud A (2012) Antifeedant activity of three essential oils against the red palm weevil (*Rhynchophorus ferrugineus*). Bull Insectol 65(1):71–76
53. Abuhussein MO (2008) The effect of neem seed extract (Azadiractin) on red palm weevil, *Rhnchophorus ferrugineus* (Olivier) (Rhchophoridae: Coleoptera). J Sci Res 17:89–94
54. Faleiro JR (2006) A review of the issues and management of red palm weevil, *Rhynchophorus ferrugineus* (Curculionidae: Coleoptera) in coconut and date palm during the last one hundred years. Int J Trop Sci 26(3):135–154
55. Min Y, Mustafa A, Kai K, Yuval G, Jacob I (2008) The role of interparticle and external forces in nanoparticle assembly. Nat Mater 7:527–538
56. Vijayalakshmi R, Rajendran V (2012) Synthesis and characterization of nano-TiO_2 via different methods. Arch Appl Sci Res 4(2):1183–1190
57. Wang C, Shaohua L, Yingying D, Zhehao H, Shunai C (2015) Hard-templating of chiral TiO_2 nanofibres with electron transition-based optical activity. Sci Technol Adv Mater 16(5):1–7
58. Kirk O (2006) Kirk othmer encyclopedia of chemical technology, 5th edn, vol 19. Wiley, New York, pp 387–393
59. Abdullah MA, Nassar MI (2001) Effect of some natural products, (*Boxus chinesis*, *Nerium oleander* and *Amaranthus viridis* extracts) against the cotton leafworm *Spodoptera littoralis* (Boised) (Lepidoptera-Noctruidae). J Egypt Ger Soc Zool Inverteb Zool Parasitol 36:177–189

60. Bream AS, Ghoneim KS, Tanani MA, Nassar MM (2001) Respiratory metabolic responsiveness, during the pupal stage, of the red palm weevil, *Rhynchophorus ferrugineus* (Coleoptera: Curculionidae) to certain plant extracts. Med Fac Landbouww Univ Gent 66(2a):491–502
61. Nassar MM, Abdullah MA (2001) Evaluation of Azadirachtin for control of the red palm weevil *Rhynchophorus ferrugineus* (Coleoptera: Curculionidae). J Egypt Ger Soc Zool Entomol 36:163–173
62. Shabana N, Husain SI, Nisar S (1990) Allelopathic effects of some plants on the larval emergence of Meloidogyne incoming. J Indian Appl Pure Bid 5:129–130
63. Ming LCh (1990) *Ageratum conzoides*: a tropical source of medicinal and agricultural products. In: Janik J (ed) ASHS Press, Alexandria, VA, pp 469–473

Facing Climate Change: Urban Gardening and Sustainable Agriculture

Eid M. Koriesh and Islam H. Abo-Soud

Abstract Measures to combat climate change encompass two primary titles: removal of the maximum greenhouse effective gases and decreasing causes of greenhouse gas emissions. The direct greenhouse gases are carbon dioxide, nitrous oxide, and methane. Nitrous oxide is approximately 300 and methane around 30 times than carbon dioxide at trapping heat in the atmosphere. Climate change affects and is affected by all communities, but its treatment must begin with the actions of individuals. Trees in the urban area strongly reduce pedestrian level heat stress by absorbing and reflecting solar irradiance. Vegetation in gardens is one of the most important components affecting climate change. Urban gardening, mainly consist of trees resource, is a valuable asset. Trees' benefits to the human being were most pronounced in their contribution to environmental benefits. Thus, plants in gardens were found to provide a particularly important function in mitigating climate change and maintaining environmental quality of communities. Gardeners can help lessen the global warming pollutants associated with waste disposal by turning leaves, grass, woody garden clippings, and dead garden waste into mulch or compost, then using it in the garden. Recycling these wastes will not only reduce *methane* emissions from landfills but also improve garden's soil and help it store carbon. One of the innovative methods that reduces greenhouse gases emissions is to make and use biochar. Because nitrous oxides is an important greenhouse gas, better management of nitrogen fertilizers can reduce its emissions. The four main management factors that help reduce nitrous oxide emissions from applied nitrogen fertilizer are commonly known as the 4R's: right application rate; right formulation (fertilizer type); right timing of application; right placement at the plant's root zone as possible. So, selecting right plants for urban gardens have a potential to influence Earth's climate by altering regional and global circulation patterns and changing the amount of CO_2 in the atmosphere. Also, it is important to deal with the soil to minimize the harmful impact it could cause to the environment. Planting appropriate tree species near industrial complexes is critical for aesthetic value and gases mitigation.

E. M. Koriesh (✉) · I. H. Abo-Soud
Department of Horticulture, Suez Canal University, Ismailia 41522, Egypt
e-mail: quriesheid@yahoo.com

I. H. Abo-Soud
e-mail: islamhassan@hotmail.com

© Springer Nature Switzerland AG 2020

E.-S. E. Omran and A. M. Negm (eds.), *Climate Change Impacts on Agriculture and Food Security in Egypt*, Springer Water, https://doi.org/10.1007/978-3-030-41629-4_16

Keywords Climate change · Gardening · Urban gardening · Trees · Shrubs · Biochar · Global warming

1 Introduction

Egypt is one of the potential susceptible countries to the effect of global warming [1] and regarded as the fifteenth most populated country in the world with affected human-induced of environment that would worsen the common problems [2]. This due to the large and tightly packed population, and if climate change makes Egypt's climate drier or warmer pressure on agriculture would aggravate. Also, competition among the limited water resources states for water could expand (even without climate change) in addition to increased warming, droughts and evaporation, reduced flow in the Nile would further aggravate. Egypt's problems and the country could face an unstable situation.

We can summarize the impact of climate change on the Egyptian Agriculture Challenges for Egypt in the 21st century as follows [3].

1. Sea level rise (as a result to temperature increasing)
2. Food Security (Land-, water-, and nutrients availability)
3. Poverty (income of farmers in rural areas)
4. Adverse impacts on human health
5. harm Egypt's tourism sector.

Sea level is additional influence of climate change. Nile Delta is at present dropping at a rate of 3–5 mm per year. A rise of 100 cm would flood one-fourth of the Nile Delta, forcing about 10.5% of Egypt's population from their family units (homes, farms, animals [1] Fig. 1). The effects would be more overwhelming if Egypt's population, as predictable, doubles to about 160 million by the middle of the century, with the current population density in Delta of around 4000 person per mile^2. As a consequence of sea level rise and flooding a great portion of Nile Delta (the most cultivated area of Egypt's land), food production and sources would not be

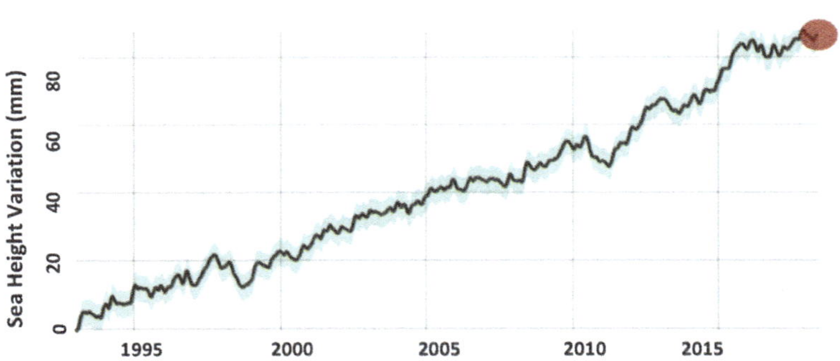

Fig. 1 Average global changes in sea level from 1995 to 2015 [4]

achieved Nearly half of Egypt crops, including wheat, rice, oranges and bananas, are cultivated in the Delta. Contrariwise the rest of Nile Delta areas (not under water) would also be affected, with salt water from the Mediterranean Sea which will pollute the fresh underground water used for irrigation. Also, agriculture activities and the self-sufficiency of food will expose to an additional effect as a result of temperature increase.

The decline in agriculture activities due to temperature increases is expected to range from 10 to 60% (production of main crops will be 18% for wheat, 11% for rice and 19% for maize). Such losses, if it is not planned and financed, will increase the risk of malnutrition and hunger among the population [5]. The increase in temperature will be associated with some changes in the ecological system and increase in air pollution as a result of emissions, soil erosion, and wind speed.

The new climate situation would increase the amount of Saharan dust carried across the country causing health troubles and economic problems. In all these dimensions' agriculture plays a crucial role.

Agriculture is threatened by climate change, responsible for food supply and employs about 30% of Egypt's labor force. Thus, a new criterion, "sustainable agriculture", is needed which addresses all these issues simultaneously! "Sustainable Agriculture" and climate change adaptation potential adaptation by adopting cropping patterns:

(a) Salt resistant crops (where sea level rise in agricultural areas)—jojoba, quinoa, sugar beets, row barley, safflower, sunflower, winter wheat, spring wheat, canola, and corn
(b) Less water-intensive crops. Farming practices such as using seed varieties that require less water and mature in less time should be widely promoted in water-deficient regions of India to help farmers reduce their vulnerability to climate change. vegetables such as cluster beans, watermelon, gourds and cucurbits
(c) Heat resistant crops (millet and sorghum, multipurpose grain legumes (cowpea, chickpea), barley, maize)
(d) Adaptation by adopting crop characteristics
(e) It is expected by 2050 that climate change will raise water demand by an average of 5% [6].
(f) Adapting soils, reducing the sources of greenhouse gases

Climate change is probable to have adverse effects on human wellbeing in Egypt, which is able be aggravated by high population densities. These may incorporate increments within the predominance and seriousness of asthma, and infectious diseases, vector-borne diseases, eye cataracts, heat strokes and skin cancer. Additional deaths from respiratory and cardiovascular diseases, dysenteric infection and diarrhea are expected. Children mortality frequency and malnutrition cases are expected to be further frequent. Climate change will harm Egypt's tourism sector through sea level rises. The Nile Delta is home of many of Egypt's tourism, and for cities like Alexandria and Matruh City, the threat of a rising sea level will reduce both their capability to sustain tourism as well as the desire of tourists to visit them. 49% of Alexandria's tourism industry would be underwater if sea level rose 0.5 m [7].

2 Gases that Contribute to the Greenhouse Effect

Most climate scientists agree that the main cause of the current global warming trend is the human expansion of the "greenhouse effect" warming that results when the atmosphere traps heat radiating from Earth toward space. Certain gases in the atmosphere block heat from escaping. Long-lived gases that remain semi-permanently in the atmosphere and do not respond physically or chemically to changes in temperature are described as "forcing" climate change. Gases that contribute to the greenhouse effect include [4, 8, 9]:

1. Carbon dioxide (CO_2)
2. Methane
3. Nitrous oxides
4. Sulfur oxides
5. Chlorofluorocarbons (CFCs)
6. Water vapor

Carbon dioxide (CO_2), a minor but very important component of the atmosphere, carbon dioxide is released through natural processes such as respiration and human activities such as deforestation, land use changes and burning fossil fuels. Humans have increased atmospheric CO_2 concentration by more than a third since the Industrial Revolution began. This is the most important long-lived "forcing" of climate change.

The gardener may cause extra carbon dioxide to be added to the atmosphere in several ways:

1. Using potting compost or peat;
2. Digging soil and leaving it bare so that the carbon in soil organic matter is oxidized;
3. Using power tools which working with burning fossil fuel;
4. Heating greenhouses and swimming pools; and
5. Burning garden pruning.

Climate change is an issue requiring urgent action by everyone who is a part of the global community. Reducing atmospheric carbon dioxide (CO_2) concentration through enhanced terrestrial carbon storage may help slow or reverse the rate of global climate change. Since desertification is projected to increase in the future, management strategies that increase carbon sequestration or decrease carbon loss are especially important [10].

2.1 Methane

A hydrocarbon gas produced both through natural sources and human activities, including the corruption of wastes in landfills, agriculture especially paddy cultivation, as well as ruminant digestion and manure administration associated with domestic livestock. On a molecule-for-molecule basis, methane is a more extra active greenhouse gas than CO_2, but also one which is much less abundant in the atmosphere.

Methane is one of a group of volatile organic compounds (VOCs) which include, also, chlorofluorocarbon (CFC), formaldehyde, benzene and others [11, 4]. It is produced both naturally and from man's activities. Although methane has adverse effects on global warming, there are little scientific studies concerning the importance of plant species that can absorb methane from the atmosphere. Studies have proposed aerobic methane (CH_4) emissions from plants, which increase the imbalance in the global CH_4 budget [9].

Plant management practices in forestry can alter soil conditions, affecting the consumption and manufacturing techniques that manage soil methane (CH_4) exchange. Forestry control practices encompass quick-term outcomes of thinning, clean-reducing and stump harvesting of coniferous forests usually consisting of pines, spruces, and larches [12]. Also, studies of Sundqvist et al. [13] showed a net uptake of (CH_4) by way of birch (*Betula pubescens*), spruce (*Picea abies*) and pine (*Pinus sylvestris*) which might be of significance for the methane resources.

There's an indirect study to reduce methane emissions related to decreasing the portions produced by way of cows, the biggest source of methane in the world. The reduction of methane produced or deamination within the rumen are facilitated with the aid of oregano oils [14]. Studies have observed that in a few instances, the complement of dairy cow diets with *Origanum vulgare* leaves resulted in better milk production, better feed performance and lowered the methane gas produced in the rumen [15, 14]. Again, there are a few studies related to different sorts of unstable organic compounds to which methane belongs and may absorb by ornamental plants. Gases studied were formaldehyde [16, 17] and benzene [18].

Agricultural practices inclusive of irrigation structures can affect methane creation through gardens. Drip irrigation combined with break up utility of fertilizer nitrogen injected inside the water (fertigation) is generally taken into consideration high-quality techniques for water and nutrient performance. As an outgrowth, its use is turning into common. A number of the primary factors (water-filled pore space, NH^{4+}, and NO^{3-} regulating the emissions of GHG (N_2O, CO_2 and CH_4) and no from agroecosystems can without difficulty be manipulated with the aid of drip fertigation without yield penalties. Fertigation with urea instead of nitrate of calcium accelerated N_2O and NO emissions through a component of 2.4 and 2.9, respectively. Day by day irrigation reduced No emissions via 42% however elevated CO_2 emissions with the aid of 21% in contrast with weekly irrigation. Generally, weekly fertigation with a NO_3 ion based fertilizer is the satisfactory option to integrate agronomic productiveness with environmental sustainability. No relation among irrigation frequency and N_2O emissions. Observe shows that suitable management of drip fertigation, at the same time as contributing to the realization of water and food security, might also offer the hope for climate change mitigation [19].

2.2 Nitrous Oxide (N_2O)

Nitrous oxide (N_2O) is an important GHG that contributes to climate change. It has a long atmospheric lifetime (100 years) and is about 300 times better at trapping heat than is CO_2, even small quantities of N_2O affect the climate. Nitrous oxide is produced by microbes in soils. N_2O is emitted mainly from fertilized soils and animal wastes. Of the three major GHG emitted naturally—CO_2, N_2O and CH_4; N_2O is the most important. Better management of N fertilizer can reduce N_2O emissions [20].

Gardeners can help lessen the global warming pollution associated with waste disposal by turning leaves, grass, woody garden clippings, dead garden plants, and kitchen waste into mulch or compost, then using it in the garden. "Recycling" these wastes will not only reduce *methane* emissions from landfills but also improve your garden's soil and help it store carbon [21]. Compost, which can be any mixture of decaying organic materials (for example, plant leaves, animal manure, food scraps), is created organisms that break down wastes into a nutrient-rich soil amendment. While composting does produce global warming gases, studies indicate that the best practices for creating and using compost have a smaller climate impact than landfills [22]. Climate-friendly gardeners may also choose to reduce their own personal greenhouse gas emissions by growing and using carminative plants such as fennel and garlic (can reduce intestinal gases such as *methane* [23]. Also, reduction in livestock product consumption on arable land use (which is a critical component of the link with deforestation) will depend on how consumers compensate for lower intakes of meat, eggs and dairy products. A switch from beef and milk to highly refined livestock product analogues such as tofu actually increase the quantity of arable land needed. A broad-based switch to plant-based products through simply increasing the intake of cereals and vegetables is more sustainable. We estimate that a 50% reduction in livestock production consumption would release about 1.6 Mha of arable land [24].

2.3 Nitrous Oxide (N_2O) and It's Importance

Nitrous oxide (N_2O) is an important greenhouse gas that contributes to climate change [25]. Because it has a long atmospheric lifetime (over 100 years) and is about 300 times better at trapping heat than carbon dioxide, even small emissions of N_2O affect the climate. Nitrous oxide is produced by microbes in almost all soils. In agriculture, N_2O is emitted mainly from fertilized soils and animal wastes—wherever nitrogen (N) is readily available. In the United States, agriculture accounts for approximately 8% of all greenhouse gas emissions but contributes about 75% of all N_2O emissions linked to human activity. Of the three major greenhouse gases emitted naturally—carbon dioxide, *methane*, and N_2O—N_2O is the most important. Better management of N fertilizer can reduce N_2O Emissions Natural part of the nitrogen cycle, but human activities add more [26].

Gardeners, also, may add nitrous oxide to the environment by applying nitrogenous fertilizers, especially if it is applied to young plants when plants cannot make use of the nitrogen, Burning garden waste [27]. So, Nitrogen dioxide is a major component of the complex of photochemical air pollutants in urban areas.

Plant leaves can absorb many kinds of gaseous air pollutants, including NO_2, through their stomata, and consequently, they have a capacity to remove this gas from the surrounding atmosphere [28]. NO_2 absorbed by the leaves can be converted into nitrate and nitrite, and then rapidly assimilated into organic nitrogenous compounds.

A powerful greenhouse gas produced by soil cultivation practices, especially the use of commercial and organic nitrogenous fertilizers, fossil and gas fuel combustion, nitric acid production, and biomass burning.

Soil Nitrous oxide emissions originate from three sources [29].

(a) Soil microorganism activity (55%);
(b) Organic manure applications (18%); and
(c) Nitrogen fertilizer applications (27%).

2.4 How Does Nitrogen Fertilizer Increase Nitrous Oxide Emissions?

Gardeners add new N to parks and all types of gardens as synthetic fertilizers such as urea or anhydrous ammonia, or rarely as organic fertilizers. When not taken up by plants, most fertilizer N is mobile, hard to contain in the field and susceptible for loss. Nitrogen from fertilizer can be lost as nitrate to groundwater or as the gases N_2O, dinitrogen (N_2) or ammonia. Nitrogen applied in excess is particularly susceptible for loss. Though the amounts of carbon and oxygen available in the soil also affect microbial N_2O production [30].

2.5 Management of Nitrogen Fertilizer to Decrease Nitrous Oxide Emissions

Numerous management strategies can keep soil N in check and minimize N_2O emissions. Many of these strategies also help to keep other forms of N from being lost, including nitrate and ammonia. In general, practices that reduce N_2O emissions increase N use efficiency (NUE), which keeps more of the added N in the crop. The four main management factors that help reduce N_2O emissions from applied N fertilizer are commonly known as the 4R's [31].

- Right N application rate;
- Right formulation (fertilizer type);
- Right timing of application; and,

- Right placement.

Nitrogen availability is the single best predictor of N_2O fluxes in ecosystems [32]. N_2O emissions are especially high when N fertilizer is applied at rates greater than plant need. The emission rate grows exponentially with increases in fertilizer rate, so at higher rates of fertilizer application N_2O emissions increase disproportionately. Fertilizer formulations also can alter N_2O emissions. The trend toward using more urea in may help reduce N_2O emissions. Fertilizer additives can also reduce N_2O emissions. Nitrification inhibitors such as nitrapyrin, which delay the microbial transformation of soil ammonium to nitrate, can delay the formation of nitrate until closer to the time that plants can use it [33].

Applying N fertilizer when it is most needed by plants can also help reduce N_2O emissions. Also, Placing N fertilizer close to plant roots can reduce N_2O emissions. For example, applying urea in narrow bands next to the trees and shrubs in hedges can reduce N_2O emissions [34]. An integrated approach is best suited to reduce N_2O emissions. The same principles of N fertilizer best management practices for increased nitrogen use efficiency (NUE) hold true for reducing emissions [35].

1. Apply fertilizer at the economically optimum rate;
2. Use an appropriate fertilizer formulation;
3. Apply as close to the time of plant need as possible; and,
4. Apply as close to the plant's root zone as possible.

2.5.1 Sulfur Oxides (SO_x)

Sulfur oxide (SO_2) is the component of greatest concern and is used as the indicator for the larger group of gaseous sulfur oxides (SO_x). Other gaseous SO_x (such as SO_3) are found in the atmosphere at concentrations much lower than SO_2. The sulfur oxide is the most voluminous chemically active gas emitted from some industrial processes such as chemical preparation, refining, pulp-making and solvent extraction and burning of fossil fuels such as coal, oil and natural gas. Coal-fired power stations, in particular, are major sources of sulfur dioxide, with coal burning accounts for 50% of annual emissions, as explained by the Tropospheric Emission Monitoring Internet Service (TEMIS). But trace amounts of SO_2 exert significant influence on climate. Large volumes of SO_2 erupted frequently appear to overdrive the oxidizing capacity of the atmosphere resulting in very rapid warming [36].

A study carried out by Zhang et al. [37] at China conditions (which is one of the highest countries in, SO_2 emitter in the world [38]) revealed that urban native street trees have a potential for purifying SO_2 [39]. SO_2 is a major air pollutant in developing countries. They added that many trees are seriously impaired by SO_2, while other species can mitigate air pollution by absorbing this gas. Planting appropriate tree species near industrial complexes is critical for aesthetic value and pollution mitigation. In their study, six landscape tree species were investigated for their tolerance of SO_2: *Cassia surattensis* Burm. f., *Ceiba insignis* (Kunth) P. E. Gibbs &

Semir, *Cinnamomum camphora* (L.) J. Presl., *Ilex rotunda* Thunb., *Lysidice rhodostegia* Hance and *Michelia chapensis* Dandy [37]. They measured net photosynthesis rate, leaf sulfur content and other physiological parameters under 1.31 mg m^{-3} SO_2 fumigation for eight days. Their results showed that the six landscape tree species differed in their responses under SO_2 stress. Based on these data, the most appropriate species for planting in SO_2 polluted areas (in China) was *Ilex rotunda*, since it grew normally under SO_2 stress and can able to absorb SO_2. This study confirms the importance of planting trees to absorb pollutants. Besides, the selection of trees to be cultivated in different locations according to their efficiency in the disposal of the most common polluter in this location. So, this makes us say that planting appropriate tree species nearby industrial complexes is critical for pollution mitigation besides aesthetic value.

El-sadek et al. [40] doing a test for some ornamental plants to absorb SO_2; NO_2 and other gases. Among the tested plants, *Chlorophytum comosum Variegatum* and *Spathiphyllum wallisii* displayed superior removal efficiency (Fig. 2).

Hence, SO_2 is playing a far more active role in starting and controlling global warming than known by the Intergovernmental Panel on Climate Change [41, 42]. Huge reduction of sulfur oxides should be a top priority for decreasing both global warming and acid rain. Nonetheless, man is also adding two to three orders of magnitude more CO_2 per year to the climate than one "large" volcanic explosion added in the past. Thus CO_2, a greenhouse gas, is contributing to global warming and should be reduced [43]. Urban Landscape Plants and Sulfur Dioxide (SO_2).

The total potential of purifying SO_2 by trees was studied in China. The integrated value of purifying SO2 including absorption by plant leaves in the Chinese area Shenyang. Studies indicated that the maximum daily potential of purifying SO_2

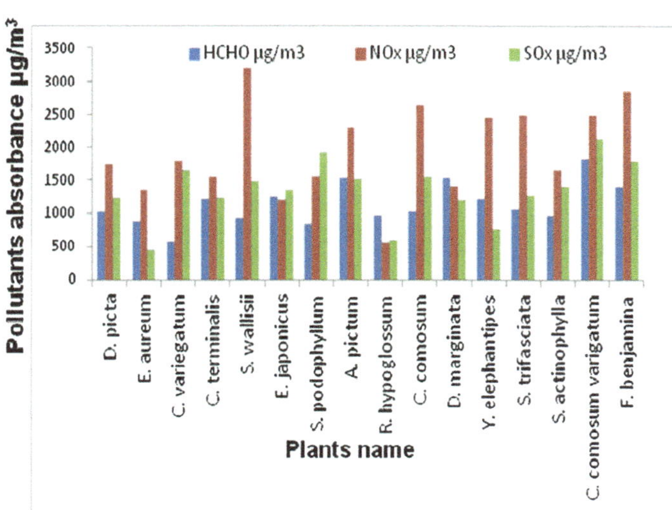

Fig. 2 Removal of formaldehyde, nitrogen and Sulphur oxides from the air inside sealed chamber

of trees might be 12.19 kg km^{-2}[39]. Supposed that the average annual absorption capacity for sulfur of trees in Shenyang area was 0.70 kg $tree^{-1}$ and the average annual absorption capacity for sulfur of broad-leaved trees was 1.35 kg $tree^{-1}$, 88.88 million broad-leaved trees would be required to purify the annual released SO_2 thoroughly. This number was equivalent to 12.9 times of existing trees. They added that on the conditions of strengthening environmental protection and decreasing 20% of annual released sulfur, if the broad-leaved trees were double of the present number, SO_2 content in the air might be below 0.06 mg m^{-3} and Shenyang would become a standardizing city with clear air. In another study, six landscape tree species belongs to subtropical area were investigated for their tolerance of SO_2. They were *Cinnamomum camphora* (L.) J. Presl., *Ilex rotunda* Thunb., *Lysidice rhodostegia* Hance, *Ceiba insignis* (Kunth) P. E. Gibbs & Semir, *Cassia surattensis* Burm. f., and *Michelia chapensis* Dandy. *They determined* net photosynthesis rate, stomatal conductance, relative water content, proline sulfur content and other parameters under 1.31 mg m^{-3} SO_2 fumigation for eight days. Their results revealed that the tree species under investigation differed in their biochemical appearances under SO_2 stress. Based on these data, the most appropriate species for planting in SO_2 polluted areas was *I. rotunda*, because it grew normally under SO_2 stress and could absorb SO_2 [37]. Trees are adapted to remove SO_2 from the atmosphere, thus donating to air quality and community health [44].

2.5.2 Chlorofluorocarbons (CFCs)

Chlorofluorocarbons (CFCs) are compounds doing as greenhouse gases. CFCs, as VOCs are unlikely to have any direct effect on the environment in the immediate vicinity of their release. They may be somewhat involved in reactions to outcome ground-level ozone, which consider cause injury to plants and other resources on a local scale. At a global level, releases of CFCs have severe environmental impacts. Their extended lifetimes in the atmosphere mean that roughly end up to the stratosphere where they can abolish the ozone layer, which protect the world from the sun's harmful UV rays. Although the amounts emitted are comparatively small, they propagate then have a high powerful warming effect (a very high "Global Warming Potential") [45].

3 Soil Management Help Mitigate Climate Change

Good soil management can help to regulate emissions of three key greenhouse gases (carbon dioxide, *methane* and nitrous oxide) from agriculture, which contribute to climate change.

Climate favorable landscaping is an innovative method of cultivation ornamental plants in urban gardens in ways which reduce emissions of different types of greenhouse gases and encourage the absorption of carbon dioxide and *methane* by plants in order to aid the reduction of global warming. To be a Climate favorable landscaper means considering both what happens in a garden and the materials brought into it and the impact they have on and climate. It can also include garden features or activities in the garden that help to reduce greenhouse gas emissions elsewhere.

The gardens of ancient Egypt included Palace gardens, Pleasure gardens, Temple gardens and Funeral gardens. These gardens also produced medicinal herbs and spices such as cumin, marjoram, anise, and coriander. These herbs can herb, indirectly, for reducing methane gas produced by cows. In this concern Nowers [14] in her thesis concluded that when oregano material was top-dressed and mixed with a portion of the total mixed ration. Also, Tekippe et al. [46] fed cattle with 500 g of common oregano (*Origanum vulgare*) leaves, for each cow per day for three weeks, and he found that the *methane* produced by the animal rumen was reduced by 40%.

The environment is the framework in which man's life. Human behavior, air quality and the emission of responsible gases affect the warming of the atmosphere and the consequent effects of changes in the global map. The rising temperature is a real and tangible phenomenon and future expectations are becoming clearly seen by residents of the northern cold areas in the melting of ice and seen by residents of the southern regions in the deadly hurricanes and the collapse of rocks. This is what the researchers found through their study of tree trunks and how to form the ring formations in tree trunks and study the evolution of the ice mountains, which confirms the continued warming of the atmosphere, which is known as global warming. Climate change can alter rainfall, influence crop yields, affect human health, cause changes to forests and other ecosystems, and even impact our energy supply. Climate-related impacts are occurring over many sectors of the economy [25]. Global warming is the result of human behavior, malpractice, industrial development and consumption of energy in all its forms, resulting in different pollutants. Also, cows, which human being man needs in his food chain, are known to let off a lot of gases including *methane*. Their belching and flatulence are estimated to contribute to almost 20% of global greenhouse gas emissions [47].

Tree foliage filter pollutants and particulates from the air, including dust, ozone, methane, carbon, sulfur, nitrous oxides and other air pollutants according to their morphological and physiological characters [48]. This mainly done through the process of photosynthesis. Trees remove CO_2 (a greenhouse gas, GHG) and release O_2 into our air (Fig. 3). Trees store the CO_2 (carbon sequestration), and according to the foliage size, leaves survival and area of the tree—can hold between 15 and 350 kg of CO_2 each year [25].

4 The Balance Between the Carbon and Nitrogen Cycles in Urban Forestation

Increasing Nitrogen deposition to urban forests can impact the balance between the carbon and Nitrogen cycles. This Nitrogen source, if taken up and used by trees can increase growth and carbon storage. Foliar uptake and utilization of inorganic

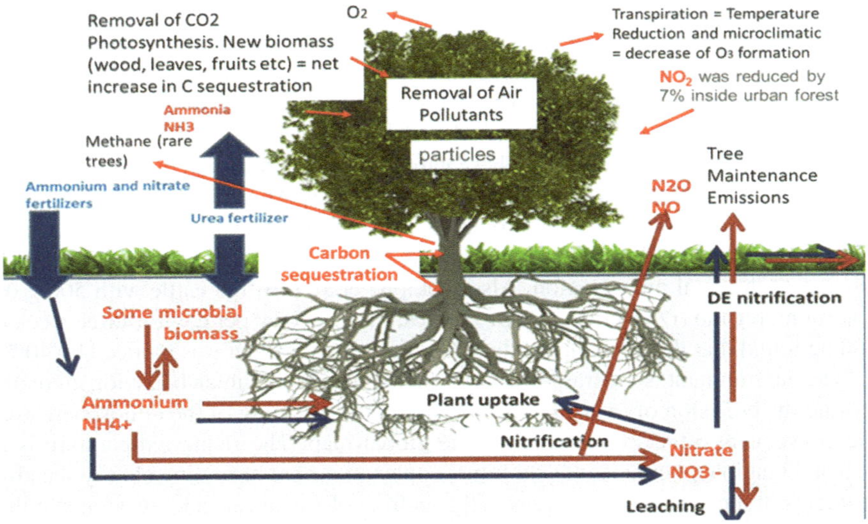

Fig. 3 Relationship between trees (vegetation) and climate change

Nitrogen, such as NO_2 and NH_3, has been shown to occur, but utilization of organic Nitrogen has not been demonstrated directly [49].

Slow-release formulations such as polymer coatings can have the same effect.

To improve the effectiveness of plant growth, more fertilizers are required, which may become an environmental hazard, unless adequate technical and socioeconomic impacts are addressed. Multifunctional slow-release nitrogen fertilizer has been developed to improve fertilizer use efficiency and reduce environmental pollution [37]. Same conclusion recorded by Ball et al. [50]. They found that NPK slow-release fertilizer also maintained grass yields and was the most effective substitute for the conventional NPK fertilizer for mitigation of N_2O fluxes.

5 Impact of Soils and Plant Nutrients on Climate Change

Soils are highly variable due to differences in local geology, topography, climate, vegetation, and management over thousands of years. Good soil management will provide a range of wider environmental benefits.

The soil has the potential to be an effective regulator of climate change by storing carbon and therefore reducing losses of carbon to the atmosphere [51].

Means to enhance storage of carbon in soil or avoid its loss from soil are, technologies that have until now received little attention. The main means by which soil carbon might be increased are first listed, these are [52]:

– Increasing the rate of input of organic matter;

- Decreasing the rate of its decomposition by biological or chemical means;
- Increasing the rate of its stabilization by physio-chemical protection within aggregates and organo-mineral complexes; and
- Increasing the depth or more correctly the total soil volume sequestering carbon at the maximum rate.

Urban forest or trees canopy can take up and use organic Nitrogen is really important. Increasing Nitrogen deposition to forests can impact the balance between the carbon and Nitrogen cycles. This Nitrogen source, if taken up and used by forests, can increase growth and carbon storage [53].

Drip irrigation combined with the fragmented application of fertilizer nitrogen (N) dissolved in the irrigation water (fertigation) is commonly considered best management practice for water and nutrient efficiency in gardens (lawns, trees, shrubs and shrubbery, grasses climbers and creepers, topiary hedges etc., some of the foremost factors (water-filled hole space, $NH^{4+,}$ and NO_3^- managed the emissions of GHG (i.e. N_2O, CO_2 and CH_4 and NO) from agroecosystems can simply be operated by fertigation without yield disadvantages. Fertigation with urea (UR) instead of calcium nitrate increased NO and N_2O emissions by a factor of 2.9 and 2.4, respectively. Generally daily irrigation reduced NO emissions by 42% but increased CO_2 emissions by 21% compared with weekly irrigation [19]. Dobbie and Smith [54] added that the variances between N fertilizers are not stable during the year. They found that at the end of spring and early summer applications, variations from ammonium nitrate (AN) were larger than variations from UR applications (5.2 ± 1.1 kg N_2O–N ha^{-1} from the AN application, compared to 1.4 ± 1.0 kg N_2O–N ha^{-1} from the UR applications). Nevertheless, through the late summer application, there was no changes between AN and UR (3.3 ± 0.7 kg N_2O–N ha^{-1} from AN and 2.9 ± 1.1 kg N_2O–N ha^{-1} from UR).

5.1 Mineral Nutrition of Garden Plants and Climate Change

Efficient use of nutrient inputs on gardens is important for the climate, plant growth as well as plant efficiency in absorbing CO_2. By improving nitrogen efficiency, less nitrous oxide will be released, reducing agriculture's contribution to climate change. Planning is necessary to balance nutrient supply to the demands of a plant type. A planning approach essentially involves [55]:

- Testing, to establish what nutrients are lacking for every plant type (evergreen or deciduous; woody or herbaceous; tree or vine; lawn or groundcover etc.)
- An assessment of the nutrient resources (not recommended to use manures after garden establishment).
- A good plan must include all nutrient-based activities: safe and secure nutrient storage, management and timely use according to recognized recommendations.

It is important to estimate, as accurately as possible, the amount of nitrogen required by the plant and to apply it in response to periods of plant demand. The amount of fertilizer required will be influenced by growth potential, but also by soil type, previous fertilizer and winter rainfall which will affect soil nutrient supply.

5.2 *Opportunities from Powerful Plant Nutrient Making Plans*

1. Ability reducing within the lawn's direct and indirect greenhouse fuel emissions of nitrous oxide and in ranges of diffuse pollution.
2. Fertilizers should be accurately incorporated to meet each plant kind needs.
3. Greater green use of manufactured fertilizer can result in monetary financial savings.

5.3 *Risks of Not Carrying Out Plant Nutrient Planning*

1. Using excess nutrients is uneconomic and might cause multiplied emissions of nitrous oxide.
2. the usage of too little nutrient risks mineral deficiency and a tailing-off of plant efficiency in decreasing pollutants [56].
3. Neglecting to measure soil nutrient substances via assessment or soil analysis can result in incorrect choices on the amount of nutrient carried out, risking additional greenhouse gas emissions and water and air pollutants.
4. If soil pH, phosphate, potash or Sulphur supplies are restricting then the nitrogen carried out cannot be utilized successfully by garden plants.
5. Uneven spreading of implemented nutrients because of inconsistent substances, or the insufficient preservation, placing and calibration of spreading device can lead to excess nitrogen in a few components of the sphere and/or nutrient deficiencies in different elements of the sphere, also main to poor nitrogen utilization.

6 Innovative Methods that Reduce GHG Emissions

6.1 *Biochar*

Is charcoal used as a soil amendment. Biochar is a stable solid, rich in carbon, and can endure in soil for thousands of years. "Biochar is under investigation as an approach to carbon sequestration" [57]. Biochar thus has the potential to help mitigate climate change via carbon sequestration [58]. The efficacy of applying plant residues to agricultural soils as a carbon (C) source for microorganisms and C sequestration is dependent on soil physiochemical properties, which can be improved by aggregation using soil conditioners. Biochar and BP accelerated the decomposition of plant residues as indicated by $14CO_2$ efflux, and resulted in reduced stabilization of residues in both soils relative to that observed in the control and polyacrylamide treatments [57].

"Biochar" is thought to be a useful soil amendment to reduce greenhouse gas (GHG) emissions [59]. Biochar also improves the physical and biological properties of soil, thereby allowing it to retain nutrients and enhance plant growth [60] and large applications enhance C stock in soil [61].

6.2 Irrigation Methods and Fertilization

A field study was conducted to compare the emissions of N_2O and methane (CH_4) as GHG under interaction between some irrigation methods (drip irrigation and furrow irrigation) and fertilization regimes. The emissions of N_2O and methane were significantly lower by fertilization through irrigation (fertigation). The key factor affecting GHG production in agricultural soil is soil moisture. Optimum irrigation can decrease GHG emissions by regulating the N and carbon turnover process in soil through manipulating soil water content [62].

Drip irrigation is one of the water saving irrigation methods generally extended in semiarid or arid regions, ever since it can reduce surface evaporation, surface runoff, and deep percolation [63].

Water and mineral N fertilizer are directly supplied to the crop root zone through a drip irrigation method to adjust to the plant requirements. Therefore, improving the water and nitrogen use efficiency (NUE). Therefore, drip irrigation may have a large influence on the nitrogen and carbon income in soil and decrease the N fertilizer induced N_2O or carbon-related greenhouse gas (e.g., CH_4) creation, in relation to ordinary furrow irrigation method [59].

7 Gardens and National Parks Soil Management Assist Mitigate Climate Change [64]

Exact soil management can assist to regulate emissions of three key greenhouse gases (carbon dioxide, methane and nitrous oxide) from agriculture, which make contributions to climate change.

- Including the addition of nitrate to soils, whether or not in fertilizer or mineralized natural be counted, will increase the chance of nitrous oxide emissions. The risks of nitrous oxide release are finest in wet, warm and clay or compacted soils, and where levels of decomposable natural organic matter are excessive.

- Decreasing compaction will improve nutrient uptake in the soil and assist lessen the discharge of nitrous oxide.
- Green nitrogen control can reduce nitrous oxide emissions, by minimizing residual nitrogen.
- Use an identified nutrient control plan and observe fertilizer recommendations.
- Reducing the intensity and frequency of disturbance will assist guard soil carbon sinks.

7.1 Role of Urban Landscaping to Reduce Greenhouse Gas Emissions and Absorb Carbon Dioxide

7.1.1 The Relationship Between Gardens and Climate

Scientists agree that global warming is properly underneath manner, the result of a buildup within the ecosystem of carbon dioxide (CO_2) and different heat-trapping gases generated using human activities consisting of the burning of fossil fuels. When an excessive amount of worldwide warming pollution is released into the air, it acts like a blanket, trapping warmness in our atmosphere and changing climate patterns around the world. This climate disruption is in all likelihood to have huge-ranging consequences no longer most effective for our health and nicely-being, but that of other living things as well.

Woody plants can uptake atmospheric organic nitrates [53], so increasing nitrogen deposition to city trees can impact the balance of the carbon and nitrogen cycles. This nitrogen source, if taken up and used by urban trees, can increase growth and carbon storage. Foliar uptake and usage of inorganic nitrogen, such as NO_2 and NH_3, has been proven to occur. However utilization of natural nitrogen has not been proven without delay. Woodland canopies can suck up organic nitrogen compounds made from pollution and turn them into beneficial amino acids.

Vegetation is already known to use their leaves to absorb inorganic airborne nitrogen molecules, which includes ammonia or nitrogen dioxide, and turn them into amino acids, and a distinctly reactive compound referred to as peroxyacetyl nitrate may be absorbed through leaves.

Many trees emit reactive volatile organic compounds (VOCs), the most commonplace of which is known as isoprene. These VOCs are so reactive that they speedy get consumed up in the atmosphere, and some react with nitrogen oxides (NO_x), emitted from combustion in engines, to form longer-lived organic nitrate compounds, extra stable than peroxyacetyl nitrate. Studies about how seedlings of the trembling aspen, a significant North American local and an isoprene emitter, reacted to an analogue of those compounds referred to as 1-nitroxy-three-methyl butane [53]. The researchers tracked two amino acids: glutamate, the primary amino acid the tree makes from absorbed nitrogen compounds, and aspartate, that is shaped further downstream in the biochemical method.

Many homeowners already see proof of global warming in their backyards. Summers are getting hotter, and generally milder winters. These adjustments could make re-planning your garden, which flowers to select, how soon to put new seedlings within the ground, and while to harvest veggies, greater of a challenge. At the same time as a longer developing season will gain a few gardeners, weather disruption is likewise projected to boom the frequency and severity of extreme climate [65]. Gardeners and landscapers can do extra than merely adapt to worldwide warming, but, wherein they could make alternatives in their gardens that don't add more problem to mankind. That is due to the fact every patch of soil (and the plant life that develop in it) takes in and gives off numerous sorts and quantities of warmth-trapping gases, depending on how it is far managed.

7.1.2 Urban Gardens Include

1. National parks (recreation areas)
2. Street gardens
3. Parks and playgrounds
4. Community gardens
5. Private or home gardens
6. Others (sanatorium gardens, college and universities gardens, governmental gardens, and so on.)

Careful interest in the world's garden soils is one piece of solving the change problem. Agriculture and forestry, which dominate managed soils worldwide, virtually have the most important roles to play. But, as cities enlarge, there is a growing possibility for urban and suburban areas to play a part. Recent researches advise that city urban green areas (lawns, gardens, parks, golf course, and trees planted alongside streets) have the capability to capture CO_2, and store that carbon through time years [66]. Practices that maximize carbon storage without producing too much global warming pollution in the process (cover crops, crop rotation, woody planting, and "low-input" lawn and garden renovation) can help sluggish the tempo of climate change. If you are a gardener or landscaper or garden designer, you should take into account the application of these results in your garden and in the gardens responsible for you.

Planting forests to soak up carbon dioxide from the atmosphere can have a range of side effects, including drying up water flow and making soil pungent. The discovery highlights the accommodation involved in tree-planting projects [67]. There are many ways in which climate-friendly gardeners may reduce their contribution to climate change and help their gardens absorb carbon dioxide from the atmosphere.

7.2 Protective and Enhancing Carbon Stores

Climate-friendly gardening includes actions which protect carbon stores beyond gardens. The largest carbon stores are:

- Soil
- woods and
- wetlands;

Construction of Woods in woody plants (trees, shrubs and some climbers) absorb more carbon dioxide per area per year than most other habitats. Climate-friendly gardeners, therefore, purpose to make certain that nothing they do will damage these habitats.

A climate-friendly gardener, therefore, does not contain large irrigated lawns, but instead includes water-butts to collect rainwater; water-thrifty plants which survive on rainwater and do not need watering after they are established; trees, shrubs and hedges to shelter gardens from the drying effects of sun and wind; and ground-cover plants and organic mulch to protect the soil and keep it moist. There are numerous methods wherein climate-friendly gardeners can also lessen their contribution to climate change and assist their gardens to soak up carbon dioxide from the environment.

7.2.1 Climate-Friendly Plants

All vegetation that soaks up carbon dioxide. However, a few plant life are more weather-pleasant than others, and it's what we do with them that counts. If we make an orchard garden with many layers, then we can boom the ground covers or lawn's productiveness and absorb greater carbon dioxide. Trees, bushes, hedges, and shrubs are lovely, supply our gardens structure, provide meals and safe haven for wild animals, and shelter our gardens from the drying consequences of wind and sun, and so assist to hold soil moist and wealthy in carbon. If we grow ordinarily perennial vegetation rather than annuals, and if we grow ground-cover vegetation, then we'll protect the carbon inside the soil and forestall it being oxidized by using cultivation. Almost any water in a lawn makes it feel extra peaceful. We want to consist of oxygenating plants inside the water to keep it oxidized and reduce the threat of methane being produced from anaerobic sediments. Climbers, like trees, shrubs, and hedges, can insulate our buildings and decrease the fossil-gasoline impact of heating or air conditioning [68].

7.2.2 Climate-Friendly Gardens

Climate friendly gardens consists of [69, 70, 25]:

1. Trees
2. Shrubs and hedges

3. Herbaceous perennials
4. Some annuals and biennials
5. Ground-cover plant life
6. Garden and meadow vegetation in small areas
7. Pond, wetland flowers
8. Climbers
9. Portable gardens

Portable gardens (Fig. 4) can help reducing suddenly polluted area with carbon dioxide Master gardener must always remember other things that make any of those plant life extra climate-pleasant:

1. Soil protection;
2. Feeding soil with carbon and nitrogen, consisting of nitrogen-fixing flowers, deep-rooted flora, and vegetation with proper root structures;
3. Long existence, which means more carbon dioxide absorbed and extra soil carbon created;
4. Resilience in extraordinary situations, especially water-thriftiness;
5. Usefulness—food plant life, craft plant life, medicinal plants;
6. Vigorous growth.

Also, he must bear in his mind other criteria which observe in all gardens:

1. Right plant in the right place; choosing plants which fit the garden's conditions.
2. Beauty, whatever form it takes in our viewer's eyes;
3. Plants as flora wildlife and for flora or fauna wildlife;
4. Multipurpose plants which benefit from pollutants and kitchen.

The master gardener or administrator must know that there is not one or more plants that can achieve all the required in place. Simply ensure it is balanced by using

Fig. 4 Portable gardens

plenty of weather-pleasant vegetation. But we want to keep away from plants with a few characteristics which would be bad for the climate so try to:

1. Avoid fire-inclined vegetation (lichen, grass leaves, deciduous leaves, conifer needles, western larch, aspen, western hemlock, western white pine, grand fir, douglas fir, Eucalyptus, Engelmann spruce);
2. Avoid thirsty flowers which cannot continue to exist without introduced tap-water;
3. Avoid vegetation which inhibits different flora.
4. Use fire-retardant plants (Yucca, Lemonade Berry, Oak, Woolly blue curl, Catalina cherry).

7.2.3 The Priorities for Climate-Friendly Gardeners Are in Ways to Reduce Emissions of GHG from Different Types of Gardens, and to

- Keep the soil's current carbon stores;
- Rise the soil's carbon stores.

To protection the soil, climate-friendly gardens:

(a) Based on plants rather than buildings and paving [71];
(b) Have soil that is kept at a fairly stable temperature by hedges from trees, shrubs and some times climbers [72];

Have soil that is constantly kept covered and therefore moist and at a moderately constant temperature by lawns and groundcover plants, fast-growing legume and herbs grown as green manures (which can be used as an intercrop in kitchen gardens of annual vegetables) and/or organic mulches [72]. Climate-friendly gardeners avoid things which may harm the soil. They do not step on the soil when it is damp, because it is then greatest weak to compressed. They dig as little is possible, and firstly when the soil is saturated rather than damp because increases the oxidation of soil organic matter and produces CO_2 [73]. To rise soil carbon sink, climate-friendly gardeners confirm that their gardens generate optimum conditions for strong, healthy growth of plants and other garden plants either above or underground, and diminish the impact of any preventive factors. In general, the extra biomass that the plants can form each year, the more carbon will be added to the soil [74].

7.2.4 Climate-Friendly Gardens Therefore Include

The changing climate will have a vast impact on the method we garden. Plants are considered the main material of gardens, by contrast with the attention of structures in cities. Climate-friendly gardens will contain group of plants that prepared for design purposes in numerous ways [75, 51, 76, 77, 78]:

1. *Annuals and Perennials*: Perennials' longer have roots allow them to clamp onto moisture and carbon, making soil better-off, even throughout droughts and floods, that many scientists have connected to climate change.
2. *Shrubs, Hedges and* Borders for living quarters: Climate change produces more shrub vegetation that can reduce $CO_{2.}$
3. *Trees*: Trees are growing further quickly due to climate change. When a tree receipts in carbon from the atmosphere, used it through its lifetime through carbon sequestration. Planting trees in areas anywhere they have not historically grown, help mitigate climate change.
4. *Ornamental Grasses:* Ornamental Grasses can Take the Heat, so reduce climate change effects.
5. *Groundcovers*: shrubs, creeping plants or lawns as groundcovers retain soil moist and at relatively constant temperatures. Generally, lawns, like other grasslands, can accumulate good levels of soil carbon, and in accordance, they will grow more strongly and store extra carbon.
6. Nitrogen-fixing plants, for example cover crops and legume plants may show an important role in mitigating the effects of climate change on agriculture.

8 Radionuclide's and Gardens

The Arctic is the fastest and severe climate change on the earth. over the following a hundred years, weather change is predicted to accelerate, contributing to important physical, ecological, social and monetary adjustments, lots of which have already begun. No matter that the contaminants of radioactive substances aren't commonplace in Egypt, however, scientific studies indicate that the global warming of the world is showing a robust impact now inside the Arctic. Scientific research had proven that some pollutants which have a tremendous impact on global warming had been located in the arctic are [79]:

1. Persistent organic pollutants (POPs),
2. Heavy metals and
3. Radionuclides.

So, it can say that gardeners must installation a design that makes use of greater absorbent plants from the ornamentals for radioactive materials. The occurrence of incidents causing the leakage of radioactive material is not the content of non-availability. Also, some radioactive materials may also move from other distant geographical areas that ultimately affect the place. Here, a brief notes about some plants that may have an ability to get rid of radioactive materials. We hope to plant them in Egyptian gardens.

8.1 Uptake of Radionuclides by Plants

Measurements of some plants showed that radioactivity in plants depended on the area of surface exposed, the developmental seasons, and on the external morphology of their above and underground parts.

Auto-radiographs on several plants showed the presence of radioactivity in certain spots (hot spots) which were distributed randomly on the surface of the plants and differed in size and activity [80]. Studies of Itoh et al. [81] showed that radioactive particles attached to outdoor plant leaves persisted for nearly 1 year.

8.2 Accumulation of Radionuclides by Plants as a Monitor System

The accumulation of radionuclides by plants acting as a monitoring system in the environment may occur by two modes; foliar absorption by the leaves and shoot of the plant, or by root uptake from the soil [82, 83]. The epidermal features of plant foliage may exert an effect upon particle retention by leaves, and subsequent uptake of radionuclides from the surface. The role of plants as monitors of radionuclides is threefold:

- as monitors of recent atmospheric releases of radionuclides; and
- as indicators of the long-term behavior of aged deposits of radionuclides in the soil [82].
- different plant species had different uptake activities for the elements [84].

Pulhani et al. [85] found that fertilizer application and soil nutrient content can affect the absorption of radionuclides by plants.

Phytoremediation of radioactive waste is a process that uses plants to remove, transfer, or immobilize radionuclides from the contaminated soil, sediment, sludge, or water, and it is a useful method for treating large-scale low-level radionuclide contamination.

There have not been established criteria which can be utilized to screen out suitable plant species that are capable of remediating the radioactivity. Important factors influencing the selection of a natural plant to remediate radioactivity, including the vegetation plant species and vegetation community composition and the criteria based on the phytoremediation factor (PF) for the selection of a natural plant to remediate radioactive waste. The vegetation plant species and vegetation community composition are important for selecting the natural plant Species [84]. An equation for Phytoremediation Factor (PF) was defined using three parameters including the target radionuclide concentration in the plant, the plant biomass, and the target radionuclide concentration in the tailings [86].

$$PF = \frac{\text{Target radionuclide concentration in the plant shoot} \times \text{biomass of the plant shoot}}{\text{Target radionuclide concentration in the tailings}}$$

Using the PF as the criteria, *Phragmites australis* (grown wild in Egypt). Through applying numerical classification for the plants recorded by Hu et al. [84] using the accumulation levels in different areas (from 1 to 9) as variables, we can arrange the most plants which can grow in Egypt for reducing nucleotides as follow:

Pteris nervosa	28
Pteris multifida	26
Phragmites australis	23
Oxalis corymbosa	21
Broussonetia papyrifera	20
Cyperus rotundus	19
Rubus hanceanus	19
Solanum nigrum	18
Artemisia capillaris	18
Clerodendrum cyrtophyllum	18
Phytolacca acinosa	16
Kyllinga brevifolia	16
Ixeris chinensis	15
Paspalum scrobiculatum	15
Cyperus iria	15
Herba gnaphaii	13
Ricinus communis	13
Vitex negundo	11
Paspalum distichum	11
Senecio scandens	10
Artemisia lavandulaefolia	10
Hibiscus syriacus	9
Melia azedarach	9
Xanthium sibiricum	9
Polygonum posumbu	8
Portulaca oleracea	8
Potamogeton pectinatus	8
Moraceae Morus alba	7
Youngia japonica	6
Solanum lyratum	6
Oxalis violacea	6
Rubus corchorifolius	6
Rosa laevigata	5
Alpinia japonica	5
Gynura crepidioides	5
Ligustrum quihoui	4
Amaranthus spinosus	4
Cyperus difformis	4
Paulownia fortunei	4
Cinnamomum camphora	4
Rubus alceaefolius	3
Herba taraxaci	3
Plantago asiatica	3
Plantago major	3
Medicago sativa	3
Salix matsudana	2

The best plants are the ones with the highest numbers

8.3 Shortwave and Longwave Radiation

Profound changes in vegetation composition in the Arctic tundra have been observed and are predicted in a warmer future climate. Shrub expansion may positively feedback to climate warming by decreasing the shortwave albedo. In Canadian studies [87], they measured an average transmission of 36% of the incoming shortwave radiation below dwarf shrub (*Betula nana*), the transmission of wet sedge (*Eriophorum angustifolium*) was 28%.

Global warming can be affected by shortwave and longwave radiation. Results of Donohoea et al. [88] suggest that, although greenhouse gas forcing predominantly acts to reduce outgoing longwave radiation, the resulting global warming is likely caused by enhanced absorbed solar radiation. They added that global conservation of energy is a powerful constraint for understanding Earth's climate and its changes. Variations in atmospheric composition that result in a net positive energy imbalance at the top of atmosphere drive global warming, with the world ocean as the primary reservoir for energy accumulation. In turn, increasing global surface temperature enhances emission of longwave (LW) radiation to space. Anthropogenic radiative forcing is dominated by LW active constituents, such as CO_2 and *methane*, and shortwave (SW) forcing agents, such as sulfate aerosols (used for a suspension of fine solid particles of a sulfate or tiny droplets of a solution of a sulfate [89]), are thought to be acting to reduce absorbed solar radiation compared with their preindustrial levels. Reduced outgoing longwave radiation, thus, seems the likely cause of the observed global energy accumulation, although the limited length of satellite TOA radiation measurements precludes determination of the relative contributions of ASR and OLR by direct observation [90].

Trees in the urban area strongly reduce pedestrian level heat stress by absorbing and reflecting solar irradiance although air temperature reduction effect is weak. Modeled results are verified by measuring shortwave radiation under trees and roof of the adjacent building [91].

Absorption profiles of incoming solar radiation determine incoming fluxes between vegetation-covered ground and the atmosphere. Model calculations demonstrate significant deviations of absorption profiles from an exponential distribution in the near-IR spectral region and high dependence of profiles on foliage orientation and the solar zenith angle. The set of input parameters of the model coincides with that of the Nilson-Kuusk canopy reflectance model. These parameters can be estimated from remote measurements [92].

9 Particulate Matter

9.1 What Is the Influence of Particulate Matter on the Climate?

Generally, it can be said there is a coherence between climate change and air quality. Some studies state elementary or carbon particulates would be the second or third most important pollutant in terms of global warming. Elementary carbon is a fraction of particulate matter and originates mainly from incomplete burning processes [93]. When it comes to predicting climate change, researchers need to take into account the effects of particles in the air. The ways in which atmospheric particles, such as mineral dust, affect climate are important.

The particle number and size fractions were compared between 14 plant species [94]. $PM_{2.5}$ particles dominated the particle distribution for all studied species. *Salix matsudana*, *Euonymus japonicus*, *Magnolia denudate*, *Salix japonica*, *Fraxinus chinensis* and *Ginkgo biloba* were efficient species in accumulating $PM_{2.5}$. PM capturing ability of shrubs and leaf undersides of plant species should be paid more attention.

Particulate matter that accumulated on leaves were comprised mainly of C, O, Si, Ca, Fe, and Pb. The densities and size fractions of the particles deposited on the adaxial and abaxial sides of leaves were significantly different, and 24% of the particles were deposited on the abaxial side of leaves. The densities of the particles in four size fractions differed significantly among the species at the two sampling sites. *Salix matsudana*, *Euonymus japonicus*, *Magnolia denudate*, *Sophora japonica*, *Amygdalus persica*, and *Salix babylonica* efficiently captured all particle sizes on their leaf surfaces. *S. matsudana*, *E. japonicus*, *M. denudate*, *S. japonica*, *Fraxinus chinensis,* and *Ginkgo biloba* efficiently captured submicron and fine particles, which can have serious effects on human health. These differences among species provide more insight into the sink capacity of green tree species, and the efficiencies of plant species for trapping particulates can be used to guide urban tree planning and decrease air pollution.

Plants not only improve air quality by adsorbing particulate matter (PM) on leaf surfaces but can also be affected by their accumulation. In a study [95, 96] about the relationship between seven leaf traits and the accumulation of three different sizes of PM (PM_{11}, $PM_{2.5}$ and $PM_{0.2}$) on leaves. The retention abilities of plant leave with respect to the three sizes of PM differed significantly at different sites and species. The average PM retention capabilities of plant leaves and specific leaf area (SLA) were significantly greater in a seriously polluted area. The findings from their study provide useful information regarding the selection of plants to reduce atmospheric pollution.

9.2 Removal of Particulate Pollutants

Suspended particles in the atmosphere are deposited on plant surfaces by three processes:

1. sedimentation under the influence of gravity,
2. impaction under the influence of eddy currents, and
3. deposition under the influence of precipitation.

Sedimentation regularly results in the deposition of particles on the different aboveground plant parts and increases onupper surface of leaves and is most important with large particles. Sedimentation velocity varies with particle mass, size [97], particle density, nutrient and organic matter content [98], shape, and trace elements [99] and other factors. Impaction happen when air flows fast on the obstacle, and the air stream divides, but particles in the air be likely to continue in a straight track due to their momentum and strike the obstacle.

The efficiency of the collection via impaction is the principal means of deposition if:

(a) particle size is of the order of tens of micrometers or greater,
(b) obstacle size is of the order of centimeters,
(c) approach velocity is of the order of meters per second or more, and
(d) the collecting surface is wet, sticky, hairy or otherwise retentive.

Ingold [100] presented data indicating that leaf petioles are considerably more efficient particulate impactors than either twigs (stems) or leaf lamina. For particles of dimension 1–5 μm, impaction is not efficient, and interception by fine hairs on vegetation is possibly the most efficient retentive mechanism [101].

The transfer of particles from the atmosphere to natural surfaces is commonly expressed via deposition velocity. For small particles, for example, condensation aerosols less than 1 μm, deposition velocities are much less than for large particles:

For example, spores and pollen 20–40 μm in diameter.

Trace metals, especially heavy metals, are most commonly associated with fine particles in contaminated atmospheres.

Trace element investigations conducted in roadside, industrial, and urban environments have dramatically demonstrated the impressive burdens of particulate heavy metals that can accumulate on vegetative surfaces.

Based on a literature survey, particulate removal efficiencies for trees have been estimated (Table 2). It is based on the following assumptions:

Particulate average deposition velocity of 1 cm s^{-1} for trees and 0.8 cm s^{-1} for grass and weeds, leaf area index of 5.1 for deciduous trees and 2.3 for conifers, and approximately 2 ha of deciduous tree surface and 1 ha of coniferous tree surface ha^{-1} of land area.

Black carbon and particulate matter are not a gas, but it acts as a greenhouse gas. It can be hanging in the air and absorb heat [102]. Gardeners may cause more carbon particulates in the atmosphere by burning garden wastes.

Particulate matter $PM_{2.5}$ alone is responsible for over 2 million deaths a year around the world [103]. PM also causes damage to the circulatory and respiratory systems and are also the second largest cause of lung cancer [104] and autism [105]. The literature on particulate matter uptake from the air by outdoor-growing plants is very extensive, most probably because of this pollutant's increasing negative impact worldwide on human health and the environment. The results of Gawrońska and Bakera [106] demonstrated that the amount of PM accumulated on aluminum plates was always significantly lower than that accumulated on the plants' leaves, showing that more than simply gravity forces are involved in PM accumulation on leaf blades.

Broad-leaved taxa with a rough surface (*Ulmus* sp.) efficiently capture airborne particulate matters [107]. The importance of plant functional traits differs by site and plant characteristics.

The literature on the atmospheric particulate matter (PM), and atmospheric aerosol, has augmented extremely over the last two decades. Fuzzi et al. [108] concluded that increasing scientific awareness in atmospheric PM or aerosol is owed to their great significance for environmental policy. Particles hanging in air interrelate with shortwave radiation from the sun and longwave radiation from the Earth, these relations greatly affect the energy budget and climate of the planet [109]. Actually, PM starts one of the most interesting problems for air quality and climate change policies. Atmospheric aerosol particles play an important role in radiation budget of the Earth's as they scatter and absorb both shortwave solar radiation and longwave terrestrial radiation [110]. Particulate matter can directly relate with cloud formation developments and may control both the amount and location of rainfall [111]. Similarly, the degree of the global effects of PM on climate remains extremely undefined. In spite of the uncertainty, there are numerous things that could be done to alleviate local and global harms of atmospheric PM.

Studies have shown that dropping black carbon (BC), one of the further main PM emissions, using known control measures, would reduce global warming and delay the period when anthropogenic effects on global temperature would exceed 2 °C. Therefore, there is ample that could be done to lessen the effects of atmospheric PM on the climate and the health of the environment and the human population.

9.3 Landscape Plants and Particulates

With the rapid progress of urbanization and industrialization, urban environmental pollution has become a public concern. This has directed to a huge studies on the urban environmental pollution issue [112]. Urban air pollution is an ambiguous type of environmental pollution. Among the pollution types, particulate pollution, notably PM_{10} and $PM_{2.5}$ directly or indirectly affect human health.

The sources of urban particulate matters are various, and the most significant ones come from the emissions of road traffic, fossil fuel combustion and construction dust [113]. Controlling and reducing urban particulate pollution has become a pressing problem. Using plants to absorb atmospheric particle matter has been confirmed as an effective method [114]. A plants' ability to remove atmospheric particulates mainly relies on its leaf function, and leaf structure, such as leaf surface texture, hair, grease, and moisture, along with other beneficial features for atmospheric particles absorption, and a huge leaf area supported by a complex stem structure can fix a lot of atmospheric particles.

Unfortunately, the atmospheric particles can crimp stomas and reduce the chlorophyll content of leaves so that gas exchange action is blocked and photosynthesis is decreased [115]. Therefore, analyzing plant stagnation differences with respect to atmospheric particulates in different contaminated areas by the use of plant leaves it is a good method.

The adsorption capacities for airborne particulates of different tree species display differences because the different tree species have different crown profiles, branches and leaves ratios and leaf surface characteristics (including waxiness, epidermis, stomata's and pubescence length, etc.).

In England, Freer-Smith et al. [116] studied the adsorption capacities for particulates of five coniferous and broadleaved species in England, and the results showed that the adsorption capacity for particulates of *Pinus tabuliformis* is the highest, that of *Sorbus aria* is the second highest and that of *Populus deltoides* is the lowest.

Urban landscape plants are an important component of the urban ecosystem, playing a significant role in the adsorption of airborne particulates and air purification. In a study carried at China Zhang et al. [117] on six common landscape plants in Beijing as research subjects, and the adsorption capacities for each different plant leaf and the effects of the leaf structures for the adsorption capacities for particulates. Preliminary results show that needle-leaved tree species adsorbed more airborne particulates than broad-leaved tree species for the same leaf area. *Pinus tabuliformis* exhibits the highest adsorption capacity, at $3.89 \pm 0.026\,\mu g\,cm^{-2}$ of PM_{10}, almost two times as much as that of *Populus tomentosa* ($2.00 \pm 0.118\,\mu g\,cm^{-2}$) as compared with same tree species leaves in the Botanical Garden. They added that there are significant adaptive changes to the leaf structures, and when compared with the slightly polluted region. In the seriously polluted region the epidermis cells of the plant leave shrinked, the surface textures of the leaves became rougher, and the stomas' frequency and the pubescence length increased.

10 Importance of Urban Landscape Plants Especially Trees for Climate Change

10.1 How Do Trees and Forests Relate to Climate Change?

Deforestation, particularly the damage of rainforests, is a vastly significant contributor to climate change. Scientists estimate that forest injury and other changes to the use of the land account for around 23% of current man-made CO_2 emissions—which compares to 17% of the 100-year warming impact of all current greenhouse-gas emissions.

As children are qualified at school, all plants absorb CO_2 from the air as they grow. Using solar energy, they use the carbon captured from the CO_2 molecules to built carbohydrates building blocks for their roots, main trunks and branches, foliage, flowers and seeds.

A mature (long lived) forest doesn't certainly absorb much more CO_2 that it releases, but when every tree dies and either rots down or is burned, for human activities, much of its stored carbon is released once again. Briefly, in the background of climate change, the most important thing that mature forests is not that they reduce the amount of CO_2 in the air but that they are big reservoirs of stored carbon. If such a forest is burned or cleared, then much of its carbon is released back into the atmosphere, totaling to atmospheric CO_2.

Obviously, the similar process also works in contrary. If new trees are planted where previously there weren't any, they will on soak up CO_2 according to the stages of growth reducing the amount of GHG in the atmosphere. Generally, it is thought that plants as "carbon sinks" currently soak up about 25% of all the CO_2 that humans add to the air each year. This figure could change as the planet warms.

Obviously, the relationship between trees and local and global temperature is very complex than the simple question of the GHGs they absorb and emit. Afforestation have a major impact on local weather structures and can affect the amount of sunlight trapped by the planet [118].

10.2 How Do Timber and Forests Relate to Climate Change?

Deforestation, and particularly the destruction of rainforests, is an enormous contributor to climate exchange. Scientists estimate that wooded area loss and other changes to using the land account for round 23% of the present-day guy-made CO_2 emissions—which equates to 17% of the warming effect of all current greenhouse-gasoline emissions.

As kids are taught at faculty, timber and different flowers absorb CO_2 from the air as they grow. The usage of electricity from the solar, they flip the carbon captured from the CO_2 molecules into building blocks for his or her trunks, branches and foliage. This is all part of the carbon cycle.

A mature forest doesn't always soak up a good deal more CO_2 that it releases, however, because while each tree dies and either rots down or is burned, tons of its stored carbon is released once more. in other phrases, in the context of weather trade, the most critical element approximately mature forests is not that they reduce the quantity of CO_2 inside the air however that they may be huge reservoirs of saved carbon. If such a wooded area is burned or cleared then a lot of that carbon is launched again into the environment, including to atmospheric CO_2 ranges.

Unsurprisingly, the connection among trees and nearby and worldwide temperature is greater complicated than the simple query of the greenhouse gases they soak up and emit. Forests have a prime impact on local weather structures and can also have an effect on the amount of daylight absorbed by means of the planet: a new vicinity of bushes in a snowy area may additionally create extra warming than cooling general via darkening the land surface and lowering the quantity of daylight reflected again to the area.

Types of Urban gardens that affect global climate change

- Street gardens
- Parks and playgrounds
- Community gardens
- Home or private gardens
- *Greenbelt*s
- Special gardens (hospital gardens, school and universities gardens, governmental gardens, etc.).

Careful attention to the world's garden soils is one piece of solving the climate problem. Agriculture and forestry, which dominate managed soils world-wide, clearly have the largest roles to play. However, as cities expand in the United States, there is a growing opportunity for urban and suburban areas to play a part.

Recent studies suggest that urban green spaces (lawns, gardens, parks, golf courses, and trees planted along streets) have the potential to capture CO_2 and "store" that carbon over time [66].

Practices that maximize carbon storage without producing moreover global warming pollution in the procedure cover crops, crop rotation, tree and shrubs planting, and groundcovers and garden maintenance) can help unhurried the jump of climate change. Garden managers and owners can lead the way in their gardens and yard.

10.3 Impression to a Climate Friendly Garden

A garden can be thought of as climate-friendly if it stores (or prevents the release of) more heat-trapping gases and generate oxygen. Researchers in landscape architecture, horticulture, climate, forestry, and urban ecology have determined some practices can change garden in facing climate change, these include:

Minimize Carbon-Emitting Inputs through [119].

- Select kinds of tools and chemicals, "inputs" which can affect the amount of heat-trapping gases your garden absorbs.
- Gasoline-powered tools are sources of CO_2.
- Synthetic nitrogen fertilizers, require a lot of energy to manufacture, which generates a significant amount of CO_2.
- Use liquid organic fertilizer after sterilization. Not recommend using non-sterilized organic nitrogen-based fertilizers (compost, animal manure), especially if the timing and amount of fertilizer are not precise.
- Understanding the Types of Lawn Fertilizer.

- Using a fertilizer for ground covers is an effective way to prevent disease and pest damage and minimize maintenance.
- Use time-release fertilizers which are beneficial at feeding the lawn over many months.
- Use humic acids
- Liquid or water-soluble fertilizers are spread using a hose and provide a quick, effective way to introduce nutrients rapidly to grassroots.
- Studies have suggested that fertilizer over-use can be a source of nitrous oxide [27].

10.3.1 Don't Leave Garden Soil Naked

Whether you grow summer vegetables, annual flowers, or perennial borders, there are periods of time perhaps as much as half the year, depending on your local climate—when plants are not actively growing. During these times, bare soil is vulnerable not only to erosion and weeds but carbon loss as well.

The use of cover crops grasses, cereal grains, or legumes that can be grown when other plants cannot is a mainstay of organic farming systems because it helps develop healthy and productive soil, reduce the need for energy-intensive chemical fertilizers and pesticides, and store large amounts of carbon. Cover crops are not meant to be harvested but to stabilize, build, and add nutrients to the soil that would otherwise remain bare.

Non-legume covers crops such as rye and winter wheat are also useful. When planted at the correct time, they can capture excess nitrogen from fertilizers or decaying plant matter; several studies showed these cover crops reduced nitrogen losses into groundwater by about 70%. By holding nitrogen in their tissues, they also prevent soil microbes from converting it into heat-trapping nitrous oxide. And non-legume cover crops typically grow larger and faster compared with legumes, absorbing more CO_2 from the atmosphere and returning more organic matter to the soil.

10.3.2 Plant Trees and Shrubs

Governments ask people, please if you have space in your home garden, plant—and keep one or more trees or large shrubs because it is an excellent way to remove more heat-trapping CO_2 from the atmosphere over a long period of time. Entirely terrestrial plants captivate CO_2 through their green organs, storing the carbon in their tissues. Trees and shrubs, because they are large and woody, and long-lived, can store larger quantities of carbon than other plants, for longer periods of time [120].

The geometric experiments [121] show that through a hypothetical scenario of worldwide sustainable forestry, the world's forests (mainly trees) could offer a large carbon sink, about one gigatonne (GT) year^{-1}, due to improvement of carbon stock in woody plants biomass. In addition, a total amount of wood, 11.5 GT of carbon year^{-1}, could be removed for reducing CO_2 emissions by replacement of wood for fossil fuels. A recent multi-city study estimated that, as a whole, the urban trees of the contiguous United States accumulate nearly 23 million tons of carbon in their tissues per year. While all trees store carbon, urban and suburban trees can also prevent CO_2 emissions because they help reduce energy use in nearby homes and buildings. Well-placed trees can shade buildings from the summer sun or buffer them from cold winter winds, reducing heating. Through evapotranspiration, urban and suburban trees can lower temperature during the summer. Thus, one study found that in addition to storing between 4 and 10 kg of carbon annually, a single shade tree in Los Angeles helps residents and businesses save enough energy to avoid the release of nearly 18 kg of carbon from power plants each year [122].

10.3.3 Expand Recycling to the Garden

According to the USEPA [123], yard trimmings and food wastes make up about one-quarter of the country's municipal solid waste (EPA [124]. When these organic wastes are disposed of in landfills, ample of the carbon is deposited underground, but the waste that breaks down releases methane gas—a heat-trapping gas 23 times strong than CO_2 [125].

Horticulturists can help decreasing the global warming pollution linked with waste disposal by turning leaves, grass, woody garden clippings, dead garden plant parts, besides kitchen waste into biochar, mulch or compost, then using it in the garden or reclamation of sandy soils. "Recycling" these wastes will not only reduce methane emissions but also improve garden's soil and help it store carbon [126, 127].

Compost, which can be any mixture of decaying organic materials (for example, plant leaves, animal manure, food scraps), is created by a and other organisms break down wastes into a nutrient-rich soil amendment. While composting does produce global warming gases, studies indicate that the best practices for creating and using compost have a smaller climate impact than landfills [128, 22].

That's because efficient composting takes place aerobically (that is, in the presence of oxygen), which minimizes the formation of *methane*. By contrast, landfills lack oxygen circulation, so organic materials are broken down primarily by bacteria that thrive in the absence of oxygen and produce *methane*. One study suggests composting is also better than incinerators for reducing heat-trapping emissions from organic waste disposal [129].

d. Think a lot about private and public Lawns

lawns can be make climate-friendly

A growing frame of research indicates that soils covered in lawn grasses can capture and store noteworthy amounts of carbon.

It appears that practices designed to maximize lawn growth and health with minimal inputs of fertilizer and water may achieve the best balance of carbon storage and nitrous oxide emissions.

Every plant from grasses absorbs CO_2 from the air. Grass have a seasonal cycle of fast growth periods beside flowering. But when grasses frequently mowed no flowering can achieved. Throughout this cycle, lawn grasses continuously cover root system and depositing carbon in the soil. After mowing grass, clippings left on the lawn to decompose, these clippings are extra source of carbon that can be kept in the soil below. Contrasting to gardens, which are often ploughed and ploughed, lawns are naturally left undisturbed, allowing the soil to store carbon for long periods of time [130]. Rates of carbon storage vary depending on the climate, soil type, and how the grasses are managed [131].

Zirkle et al. [132, 133] calculated efficiency of lawns relating to Soil organic carbon (SOC) sequestration for U.S. lawns. SOC was 46.0 to 1216.1 g C/m^2/year. They added that additional C sequestration can result from biomass gains attributable to fertilizer and irrigation management. However, nitrous oxide emissions from lawns have been related to the addition of nitrogen fertilizer and large watering [27].

10.4 The Role of Plants, Especially Trees, Solve Climate Change Problems

Climate favorable landscaping is an innovative method of cultivation ornamental plants in urban gardens in ways which reduce emissions of different types of greenhouse gases and encourage the absorption of carbon dioxide and *methane* by plants in order to aid the reduction of global warming. To be a Climate favorable landscaper means considering both what happens in a garden and the materials brought into it and the impact they have on and climate. It can also include garden features or activities in the garden that help to reduce greenhouse gas emissions elsewhere.

Plants in forests play an important role in *climate change*. Afforestation or planning a forest garden can help mitigate against *climate change* by removing CO_2 from the atmosphere. It is stored in this "biomass" until being returned back into the atmosphere, whether through natural processes or human interference [134].

Permaculture is most frequently applied in gardening. Urban gardening provides habitat for wildlife, carbon sequestering, biodiversity and others. Simple urban gardening must be contained (containing or contained) three layers: trees, shrubs, and ground plants. But for those who like to take advantage of every planting opportunity, a deluxe urban garden can contain at least six tiers of vegetation.

As the illustration below shows, a seven-layered urban garden contains (Fig. 5):

1. Big or tall trees
2. Small or low trees
3. Hedges and shrubs
4. Bedding plants, perennial or herbs
5. Lawns or ground covers or annuals

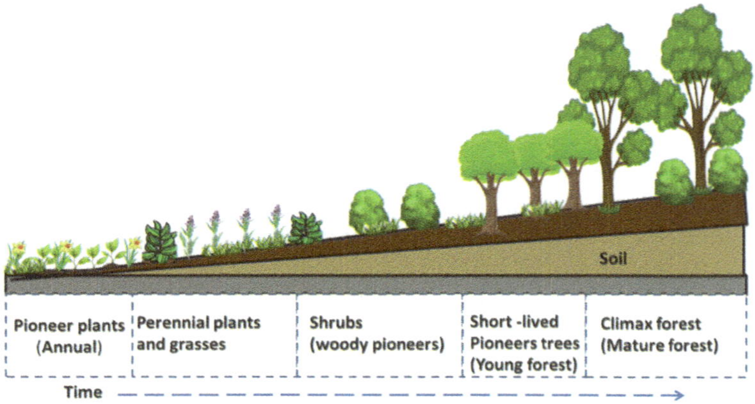

Fig. 5 Ideal layers of urban garden

11 Street Trees: Species Selection Criteria [135, 61, 56, 136]

Street trees are one of the greatest important gears of urban landscaping and they play a vital role in town pollution and climate change besides street aesthetics. When choosing trees, consider the following:

1. *Absorb pollutants*: Has a great ability to absorb specific pollutants dominant in the area.
2. *Suited to climate and soil characters*: only species which will tolerate and succeed in the specific setting are recommended.
3. *Reliable form*: A higher and taller canopy on a single trunk keeps paths clear for pedestrians and similarly keeps sightlines open for traffic safety and personal security.
4. *Scale elements*: Trees of a size which will provide a substantial degree of comport when mature are preferred. In this setting, the occurrence of overhead power lines will require the selection of trees with a tall narrow canopy, or a decision not to plant beneath wires. Longevity: Some fast-growing plants are short-lived. A long life—several decades—is required to defend the cost of establishment, and maintenance, and to supply an appropriate level of comfort.
5. *Extend or support a worthy existing theme*: native or exotic will depend on the situation. Selection should be governed by site valuation Not a weed and unlikely to become a weed. Species which yield large amounts of either fleshy fruits or light windblown seeds are discouraged. Some domestic plants are weeds, and the possible for hybridization between introduced "exotic" natives and local gene pools is regarded as a risk.
6. *Resistance to local diseases and insects*: Trees are not carriers or hosts of pests that infect plants in the area.

7. *Quality of wood*: Once the trees reach the age known to them, the municipality may resort to re-culture, and in rare cases may require digging out trees. In these cases, wood quality is required.
8. *No Street Tree is Perfect*: There is no such thing as the street tree that will fulfil perfectly all aspects of our selection criteria.
9. *Root system characteristics*: Trees species which do not have a history of problematic root behaviour are preferred.
10. *Branch drop and windthrow:* Choose trees which are not known for unexpected branch drop, or likely to blow over in strong winds.
11. *Deciduous versus evergreen*: The requirement for summer shade and winter sun should be factored into the selection process.
12. *Non-grafted cultivars*: carefully consider if a grafted species has been short-listed. When suffering hardship, grafted specimens can shoot from the under-stock, creating a maintenance burden.
13. *Fruit production*: Apart from potential weed production from fruit, fleshy or woody fruit can create a hazard when they fall on pavements.
14. *Habitat linking*: strategic location of particular indigenous tree species to link fragmented habitat or continue a wildlife corridor, can be important to some classes and populations of animals. This must be considered on a site-by-site basis.

11.1 Big or Tall Trees

Tree planting and plantation urban forestry are well established both in the private and public sectors. The data released by the UN's Food and Agriculture Organisation suggest that plantation forests comprised an estimated 7% of global forest area in 2010 [137]. Trees are also planted as part of efforts to restore natural forests as well as in agroforestry, which involves increasing tree cover on urban and agricultural lands.

- Under certain conditions plantations can grow relatively fast, thus absorbing CO_2 at higher rates. In the absence of major disturbances, newly planted or regenerating forests can continue to absorb carbon for 20–50 years or more. Tree planting could sequester (remove from the atmosphere) around 1.1–1.6 GT of CO_2 per year (that compares to total global greenhouse gas emissions equivalent to 50 GT of CO_2.
- Tree planting and afforestation were included as activities eligible for finance under the Kyoto protocol. Kyoto's rules and procedures, however, restricted the scale and scope of these activities. As a result, projects have struggled to get off the ground and the carbon sequestered has been almost negligible.
- Dense, spreading species—the classic shade trees such as *Ficus retusa*, *F. benghalensis*, *Delonix regia*, and *Cassia nodosa* don't work well in the forest garden because they cast deep shadows over a large area. Better choices are *F. religiosa*, *Khaya senegalensis*, and *Bauhinia variegatum*.

Nitrogen-fixing trees will help build soil and most bear blossoms that attract insects. In Egypt, these include *Erythrina indica, Cassia nodosa, Delonex regia, Jacaranda ovalifola* and *Albizia lebbeck*. The canopy trees will describe the chief patterns of the forest garden, thus they necessity be chosen wisely. Plant them with careful in respect to their expected final size so enough light will fall among all above-ground tree parts [138]. Most Important Street Trees According to Their Efficiency for Reducing Carbon Dioxide from Atmosphere.

1. *Albizzia lebbeck*
2. *Bauhinia variegate*
3. *Bombax malabaricum*
4. *Cassia fistula*
5. *Cassia nodosa*
6. *Chorisia speciosa*
7. *Erythrina variegate*
8. *Eucalyptus camaldulensis*
9. *Ficus retusa*
10. *Ficus* benghalensis
11. *Ficus religiosa*
12. *Ficus sycamores*
13. *Khaya senegalensis*
14. *Tipuana speciose*

Table 2 shows the role of Egyptian street trees in reducing carbon emissions or absorbing carbon dioxide from the surrounding air. The results deal with specific leaf area, i.e. calculated for every 1000 cm^3 of one leaf surface only. It is better to calculate the total leaf area of the entire tree. Because it is difficult to find different trees in the same age at the same environmental conditions. Data in Table 1 and Fig. 6 show that *Ficus religiosa* trees are best in CO_2 uptake through the first 5 min after

Table 1 Efficiency of some ornamental trees grown in Egypt for reducing CO_2/1000 cm^2 leaves area and transpiration rate, arranged ascendingly [139]

CO_2	mmol H2O m−2 min−1
1. *Ficus religiosa*	1. *Cassia fistula*
2. *Bauhinia variegate*	2. *Ficus religiosa*
3. *Cassia nodosa*	3. *Ficus benghalensis*
4. *Eucalyptus camaldulensis*	4. *Eucalyptus camaldulensis*
5. *Ficus sycamores*	5. *Ficus sycamores*
6. *Khaya senegalensis*	6. *Chorisia speciosa*
7. *Albizzia lebbeck*	7. *Albizzia lebbeck*
8. *Chorisia speciosa*	8. *Khaya senegalensis*
9. *Cassia fistula*	9. *Bauhinia variegata*
10. *Bombax malabaricum*	10. *Ficus retusa*
11. *Erythrina variegate*	11. *Tipuana speciose*
12. *Ficus retusa*	
13. *Ficus benghalensis*	
14. *Tipuana speciose*	

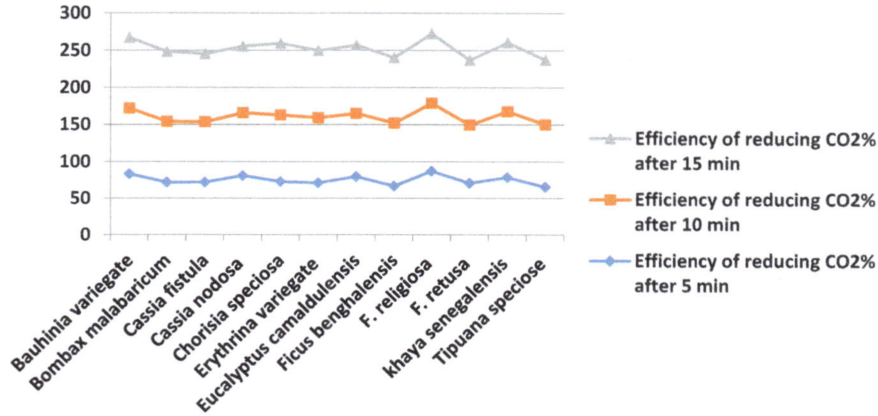

Fig. 6 Efficiency of some ornamental trees grown in Egypt for reducing $CO_2/1000\ cm^2$ leaves area [139]

the start of the carbon dioxide pollution, as well as after 10 min and after 15 min. This is followed by *Bauhinia variegate* trees as in the following table.

On the other hand, studies were conducted on the most common trees and shrubs cultivated in Egypt, from these studies on the transpiration of plants from their leaves in the plant and this property depends on various factors:

(a) Plant type, including the anatomical structure of the leaves
(b) Climatic conditions prevailing in the region such as temperature and humidity
(c) Different soil conditions.

Scientific studies show that for transpiration of 1 g of water, plant require 633 Cal, this amount of energy is withdrawn by the plant from the surrounding environment. This energy is believed to affect heat reduction in the atmosphere. However, the arrangement of the trees studied was not the same with the arrangement of trees in the disposal of carbon dioxide.

The following is an explanation of some of the common species cultivated in Egypt and have a role in the disposal of carbon dioxide that causes climate change which is also important in tempering the atmosphere, which also affects climate change (Table 3; Fig. 7).

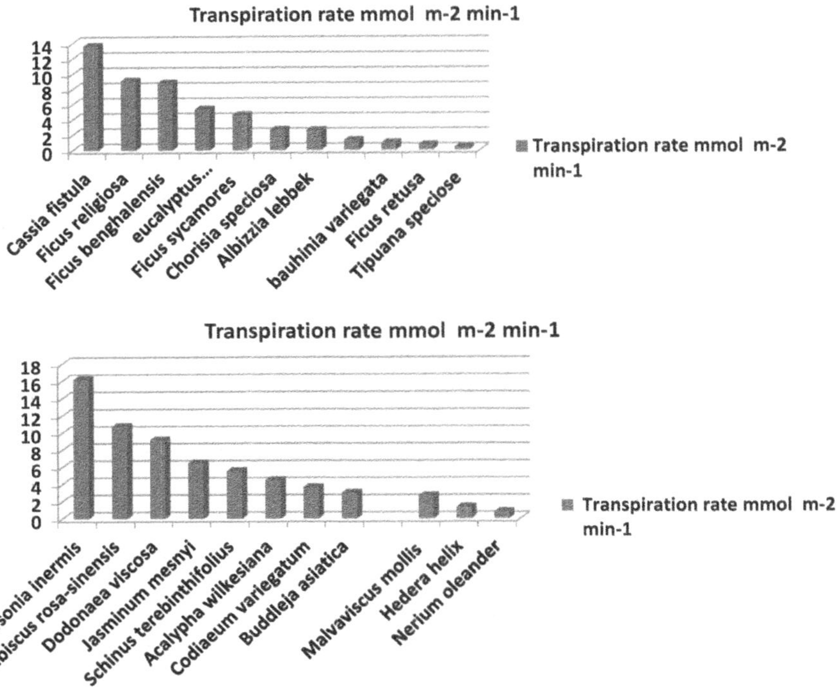

Fig. 7 Efficiency of some ornamental trees (**a**) and shrubs (**b**) grown in Egypt for transpiration rate/1000 cm^2 leaves area [17]

11.2 The Relationship Between Some Landscape Plants and Climate Change (Tables 2 and 3; Figs. 7 and 8)

11.2.1 *Ficus religiosa*

It is a large tree for dry season, semi-deciduous in Egypt, up to 30 m tall. Because the leaves have a tall petiole, it move continuously even when the air around is immobile, and no perceptible wind is blowing. This motion can be explained due to the long leaf petiole and the broad leaf blade. This movement may be the reason why this tree is one of the highest trees in the disposal of carbon dioxide in our studies [17, 139], resulting in more movement contributing to the entry of larger quantities of air gases. *Ficus religiosa* is grown as an ornamental tree, in public gardens, community and universities gardens, large private gardens and parks. They prefer full sunlight and can grow in most soil types. While it is possible for the plant to grow indoors in a pot (as a bonsai plant), it grows best outside [140]. Some scientific studies [141] have shown that the genus of *Ficus* in general, including *F. religiosa* produces moderate amounts of volatile organic compounds (VOCs) in the atmosphere, it is a pollutant.

Table 2 The efficiency of some ornamental trees and shrubs grown in Egypt for reducing percentages of carbon dioxide

	CO_2 reduction percentages after 5 min	CO_2 reduction percentages after 10 min	CO_2 reduction percentages after 15 min
Trees			
Ficus religiosa	87.65	92.11	92.64
Bauhinia variegate	83.76	89.04	94.33
Cassia nodosa	81.53	85.15	88.78
Eucalyptus camaldulensis	80.15	85.66	91.18
Ficus sycamores	79.61	90.92	92.11
Khaya senegalensis	79.04	89.35	92.27
Albizzia lebbeck	75.84	87.21	89.02
Chorisia speciose	73.29	90.36	95.92
Cassia fistula	72.54	81.82	91.10
Bombax malabaricum	72.34	82.76	93.19
Erythrina variegate	71.81	88.14	89.80
Ficus retusa	71.33	78.93	86.53
Ficus benghalensis	67.20	85.72	87.45
Tipuana speciose	65.99	84.77	86.35
Shrubs			
Bougainvillea glabra	58.68	62.28	62.47
Dodonaea viscosa	43.28	47.38	48.25
Hedera helix	25.12	25.57	25.61
Lantana camara	20.92	22.21	22.32
Schinus terebinthifolius	19.11	20.57	20.81
Malvaviscus mollis	14.88	16.62	16.83
Nerium oleander	6.32	6.96	7.04
Acalypha wilkesiana	1.73	2.06	2.11

11.2.2 *Erythrina indica*

- Has large natural distribution in the tropics
- For cultivation in gardens
- Deciduous tree
- Growing up to 27 m
- Cultivated as an ornamental tree.

Table 3 Transpiration rate of some ornamental trees and shrubs grown in Egypt as an indirect method for prediction the efficiency plants for reducing air temperature (every 1 g needs about 600 calories for evaporating from plants)

Plant	Transpiration rate as mmol m^{-2} min^{-1}
Trees	
Cassia fistula	13.68
Ficus religiosa	9.10
Ficus benghalensis	8.82
Eucalyptus camaldulensis	5.38
Ficus sycamores	4.59
Chorisia speciosa	2.63
Albizzia lebbek	2.58
Khaya senegalensis	1.27
bauhinia variegata	0.96
Ficus retusa	0.66
Tipuana speciose	0.35
Shrubs	
Lawsonia inermis	16.19
Hibiscus rosa-sinensis	10.76
Dodonaea viscosa	9.17
Jasminum mesnyi	6.45
Schinus terebinthifolius	5.54
Acalypha wilkesiana	4.47
Codiaeum variegatum	3.64
Buddleja asiatica	3.00
Malvaviscus mollis	2.65
Hedera helix	1.34
Nerium oleander	0.79

11.2.3 *Ficus benghalensis*

- The Genus are among the largest trees Egypt
- The species are among the largest trees in the world by canopy coverage
- Due to the large size of the tree's canopy, it provides useful shade in hot climates.

11.2.4 *Eucalyptus camaldulensis*

- One of around 800 species within the genus
- It is a plantation species in many parts of the world
- Can grow to 45 m tall

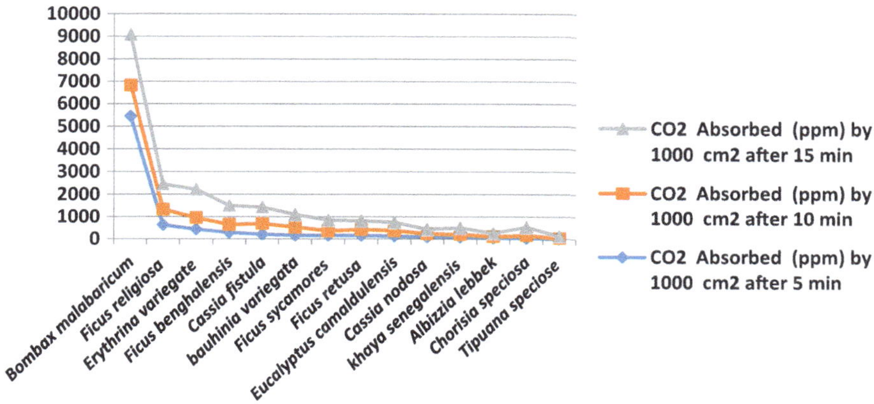

Fig. 8 Efficiency of some ornamental shrubs grown in Egypt for reducing CO_2/1000 cm^2 leaves area Koriesh [139]

- The tree has a large, dense crown of leaves
- It may contain a specific phytoncides
- Self-pruning. Saving water
- The speed of growth of the tree makes it a useful plantation.

11.2.5 *Bombax ceiba*, L.

- Has a straight, tall trunk
- Deciduous in winter
- Red flowers appear in the spring before the new foliage
- Produces a capsule which, when ripe, contains white fibers like cotton
- Grows to an average of 20–60 m
- Widely planted in parks and on roadsides there because of its beautiful red flowers which bloom in March/April.

11.2.6 *Chorisia speciosa*

- Deciduous trees
- In younger trees, the trunk is green due to its high chlorophyll content, which makes it capable of performing photosynthesis when leaves are absent; with age, it turns to grey
- The leaves are composed of five to seven long leaflets.

11.2.7 *Khaya senegalensis*

Khaya senegalensis is a species of tree native to Africa.

- The medium-sized tree which can grow up to 15–30 m
- The bark is dark grey to grey-brown
- Used for its timber
- It is now used more locally and is planted ornamentally as a roadside tree.

11.2.8 *Ficus retusa*

- Rapidly growing tree, rounded, broad-headed, evergreen
- Can reach 15 m
- Massively spreading canopy
- The tree has small, dark green leaves.

11.2.9 *Tipuana tipu*

- Semi-deciduous tree
- Height up to 10–35 m
- The flowers are abundant in number from the end of spring until the beginning of the summer
- Cultivated parks and large recreation areas
- Leaves fall briefly at the end of winter and grow again after a few weeks.

11.3 Small or Low Trees

The Low-Tree Layer

Here are many of the dwarf and semi dwarf plants. Plus, we can plant naturally small trees such as *Bauhinia variegata*, *Cassia fistula*, and *Cassia nodosa*. In a smaller forest garden, these small trees may serve as the canopy.

Both large and small nitrogen-fixing trees as *Erythrina variegata, Bauhinia variegata, Cassia fistula, Cassia nodosa and Tipuana speciose* grow quickly and can be pruned heavily to generate plenty of effective leaves.

11.3.1 *Bauhinia variegata*

- Reach a height of 6–12 m and their branches spread 3–6 m
- Flowers are 7.5–12.5 cm diameter, in shades of red, pink, purple, orange, or yellow
- Flowers are often fragrant
- Flowering in late winter and often continues to flower into early summer
- Prefer acidic soils and do not tolerate salty conditions
- Full sun exposure is preferred, but they can be grown under the partial sun.

11.3.2 *Cassia fistula*

- Popular ornamental plant and used in herbal medicine
- Medium-sized, up to 10–20 m with fast growth
- Flowers are produced in pendulous racemes with yellow petals
- Flowers with a pungent odor
- Strong and very durable wood
- Widely grown as an ornamental plant
- Trees were covered with yellow flowers, many times with almost no leaf being seen
- Grow well in dry climates
- Relatively drought-tolerant and slightly salt-tolerant.

11.3.3 *Cassia javanica* (*C. nodosa*)

- Worldwide as a garden tree owing to its beautiful crimson and pink flower bunches
- Fast growing, semi-deciduous tree which flowers in spring
- The trunk that reaches heights of 25–40 m
- Planted as a shade and ornamental tree on streets and in parks
- Used medicinally.

11.4 Hedges and Shrubs

1. *The Shrub Layer.* This tier includes flowering, fruiting, wildlife-attracting, and other useful shrubs. A small sampling: *Lawsonia inermis, hibiscus rosa-sinensis, Dodonaea viscosa, Jasminum mesnyi, Schinus terebinthifolius, Acalypha wilkesiana, Codiaeum variegatum, Buddleja asiatica, Malvaviscus mollis, Hedera helix, Nerium oleander*, and dozens of others

2. The broad palette of available shrubs allows the gardener's inclinations to surface, as shrubs can be chosen to emphasize food, crafts, ornamentals, birds, insects, native plants, exotics, or just raw biodiversity
3. Shrubs come in all sizes, from dwarf to nearly tree-sized
4. Shade-tolerant varieties can lurk beneath the trees, sun-loving types in the sunny spaces between.

11.5 Bedding Plants or Herbs

- The Herb Layer is used in the broad botanical sense to mean nonwoody vegetation: flowers, culinary herbs, and cover crops, as well as mulch producers and other soil-building plants.
- Emphasis is on perennials and annuals.
- Shade-lovers can peek out from beneath taller plants, while sun-worshiping species need the open spaces.
- It is better to use more efficient shrubs in reducing CO_2.

11.6 Lawns or Ground Covers

- It is best to use high-efficiency shrubs that absorb more carbon dioxide as *Bougainvillea glabra, Dodonaea viscosa, Hedera helix, Lantana camara, Schinus terebinthifolius* and *Malvaviscus mollis*
- They play a critical role in weed prevention, occupying ground that would otherwise succumb to invaders.

11.7 Climbers and Vines

- This layer is for climbing plants that will twine up trunks and branches, filling the unused regions (https://www.motherearthnews.com/organic-gardening/garden-planning/backyard-fores).
- It is best to use high-efficiency shrubs that absorb more carbon dioxide as *Bougainvillea glabra* and *Hedera helix.*
- Some of the perennial vines can be invasive or strangling; hence, they should be used sparingly and cautiously.

11.8 Efficiency of Trees for Reducing $CO_2/1000\ cm^2$

The following is an explanation of some of the common species cultivated in Egypt [142, 143] and has a role in the disposal of carbon dioxide that causes climate change which are also important in tempering the atmosphere, which also affects climate change.

12 Vertical Gardens to Reduce *Climate Change*

12.1 Benefits of Using the Vertical Garden System

1. The benefits of the use of the walls vary according to several criteria. Perhaps the most important of which is the location of the origin and the method and manner of cultivation of the outer envelope and the nature of the surrounding environment. In addition to the aesthetic character and the establishment of a distinct environment and an enjoyable environment of existing architecture, these benefits can be arranged under three main titles [144]. On the environment scale
2. On the scale of the building
3. Scope of users

On the environmental scope, Results of Safikhani et al. [145] showed that the living wall and green facade reduced indoor temperature up to 4.0 and 3.0 °C, respectively. The living wall and green facade also reduced cavity temperatures by 8.0 °C and 6.5 °C, respectively.

> The beauty of this system is that more than 90% of the water goes directly to the plant and the absorption rate is lower as compared to normal absorption of water by the plants (http://www.talkafrica.co.ke/vertical-gardens-technology-to-reduce-climate-change).

Vertical gardens, especially, which use solar energy have come as a savior to those living in high polluted areas, water stressed and less space environments. The beauty of this arrangement is that more than 90% of the water drives directly to the plant and the absorption amount is lesser as compared to regular absorption of water by the plants [146].

One of the major problems in landscape design in crowded cities, like Cairo, Giza, Tanta, Assiut etc. in Egypt, is looking to find free areas to establish new gardens, also old methods of street trees maintenance still used. We recommend following the new systems of external surfaces cultivated at least for the governmental buildings, using different plant species with higher efficiency in absorbing carbon dioxide (and other common pollutants in every city). Research of Koriesh [17, 139] indicated that the use of *Sciendapsis aureus, Dracaena with narrow leaves (D. marginata), Chlorophytum comosum* and *Aspidistra elatior* plants for cultivating façades and Vertical gardens showed the highest efficiency in the elimination of carbon dioxide.

Also planting *Ficus religiosa, Bauhinia variegate, Cassia nodosa, Eucalyptus camaldulensis and Khaya senegalensis* trees is better in new gardens or replacing old ones when implementing targeted maintenance programs [17, 139].

12.2 Green Walls [147, 144]

The concept of green walls is representative of the green roofs vertical or near the vertical position and is based on the elements of construction vary according to the type of the wall itself and is covered by the structures of plants. This concept is historically the presence of climbing plants next to existing architecture and then used in the coordination and beautification of sites. It is also considered the attention and spread of that system recently because it provides a quick and effective environmental solution to the scarcity of green areas and the impossibility of the development of green areas within the existing building.

Green walls are Vertical gardens on the surfaces where plants are attached to the surface. It is classified as 'green facades' or 'living wall'. On the right hand, green facades are created by directly attaching hanging pots or shrubs to the wall, or attached the plants to the wall using any supporting methods as wires, ropes and meshes either in the form of double skinned green facades or indirect green facades. On the left hand, living walls are created by attaching growing media to the vertical wall, and this relatively new technique is classified as 'continuous living walls' or 'modular living walls' [148, 149]. Green walls can improve air quality. Dunnett and Kingsbury [144] recommended green walls as one of the planting concepts to reduce particulates through deposition. Pollutant reduction along with a footpath in open areas [150]. This system is defined as a wall consisting of a set of plants supported by vertical anchors either on the inner or outer wall and sometimes stand alone. It also includes plants, irrigation and drainage system all within one system.

12.3 Greening with Self Climbing Plants

Often at walls 'self-climbers' are set like *Jasminum grandiflorum Lonicera japonica* and English Ivy, with which climbing assistance is unnecessary. The wall crest must deflect all precipitation, so that no water from above can penetrate. All joints should be closed, so that the impulses of the plants mentioned cannot penetrate there. *Climbing hydrangea, Clerodendrum inerme* climbing roses are also suitable [151, 152]

Common climbing plants such as ivy was found suitable for the green wall. The removal potential of pollutants using a green wall was shown to be influenced by plant species, methods of climbing wind, humidity and leaf area index (LAI). We suggest that air pollution tolerance index (APTI) and anticipated performance index (API) should be measured prior to selecting species for the urban gardens.

These considerations were made based on limited defined research, and further investigations research are required to produce guidance for determining the role of green walls on climate change.

12.4 Living Walls

This term refers to innovative technological methods based on prefabricated media, homogeneous soil, or a water system on a vertical surface. This concept can also be referred to as Vertical garden. The need for continuous maintenance and the need for irrigation and drainage systems and the need for fertilizers to be distributed, the cost of installation of living walls is three to five times higher than the cost of installation for the Green Facade system and the cost of replacement of unsuitable plants [153].

It is known that these walls, which occupy some forms of agriculture, are also composed of climbing plants, as *Euonymus fortune, Hedera algeriensis, Hedera colchica, Hedera helix, Hydrangea seemannii, Clematis, Lonicera henryi, Lonicera japonica, Jasminum humile* (yellow jasmine), *Jasminum grandiflorum* and *Jasminum mesnyi* whether of one type or more.

Recently, additional structural elements were used to support the development and growth of plants. These walls are characterized by their environmental benefits of biodiversity, thermal insulation, noise reduction and reducing carbon dioxide and increasing oxygen [154].

12.5 Green Façade

Are the façades that rely on climbing plants that are cultivated in the soil surrounding the building directly or are grown in containers. These plants need structural elements that are supported, cared for and monitored continuously throughout the growth periods. This system is scalable and depends on a wide range of plant species that can be used. These plants (as *Cissus antarctica, Clematis aristata, Ficus pumila, Vitis vinifera* and *Wisteria sinensis*) can adapt to the risks and seasonal changes (Considerations for Advanced Green Facade Design).

12.6 Brief History (Green Screen) [153]

In 1920, the use of cabling and cable systems began as a supporting system for climbing plants, allowing them to grow and spread and were used in gardens and houses.

A. In 1988, the use of stainless steel wire systems for the facades and the walls of the plant was started
B. In the early 1990s, the United States of America began to enter the market of green walls and cultivated walls for wider use through systematic innovation
C. In the mid-1990s Patrick Blanc, who was the founder and pioneer of Vertical gardens in the modern era, began to pay attention to this trend due to its great environmental importance.
D. After 1994, the use of green walls in the spaces was started, supported by Bio-Filtration System in Toronto, Canada.

Plants for living wall

(A) Plants for sun	(B) Plants for shade	(C) Edible plants
(i) Bacopa	(i) Ajuga	i. Lettuce
(ii) Campanula	(ii) Campanula	ii. Salad leaves
(iii) Fuchsia	(iii) Heuchera	iii. Herbs
(iv) Impatiens	(iv) *Vinca*	iv. Tomatoes
(v) Lobelia	(v) Viola	v. Strawberries
(vi) Nasturtiums	(vi) Tiarella	vi. *Anethum graveolens* = Dill
(vii) Petunia	(vii) Pansies	vii. *Coriandrum Sativum* = Coriander
(viii) Pelargonium	(viii) Wallflowers	viii. *Petroselinum crispum* = Parsley
(ix) Salvia		
(x) Scabious		

12.7 Bio-filtration System

Biofilters represents a new technology that holds great promises for GHG reduction and air pollution control [155]. This system is known as the process of drawing air through organic matter (such as algae, soil, plants, etc.) leading to the removal of carbon oxudes, sulfur oxides, nitrogen oxides, organic gases, volatile organic compounds and contaminants [11].

12.8 Reduced Urban Heat Island (UHI) Effect

Climate change and the heat island effect interact in different ways.

1. First, our warming climate will increase already higher temperatures in heat island areas.
2. Second, cooling strategies to reduce heat islands can help communities adapt to the impacts of climate change.

3. Third, lower the greenhouse gas emissions that cause climate change [25].

The average temperature of the built-up areas is higher than the surrounding rural areas and may be more than 5 °C. This is known as the Urban Heat Island (UHI) phenomenon, which causes a loss of thermal comfort for users and a significant loss of energy used for air conditioning. The use of façade and roofing works to reduce the temperature caused by solar brightness and thermal reflections, thus reducing the phenomenon of thermal islands within the existing building, leading to better temperatures within the architectural spaces [156].

12.9 Purify Air and Reduce Dust

Plants may play a major role in eliminating carbon dioxide (CO_2), where we can eliminate more than half the concentration of carbon compounds by using some plant species [139, 150]:

Acalypha wilkesiana	*Ficus religiosa*
Bauhinia variegata	*Ficus retusa*
Cassia fistula	*Hedera helix*
Chorisia speciosa	*Khaya senegalensis*
Dodonaea viscosa	*Malvaviscus mollis*
Eucalyptus camaldulensis	*Nerium oleander*
Ficus benghalensis	*Schinus terebinthifolius*
	Tipuana speciose

Green walls also can dispose of more than 200 volatile organic compounds (VOC) compounds at their intermediate concentrations and can eliminate other harmful gases such as NO_x, SO_x, formaldehyde, etc. In addition to the ability of plants to dispose of particulate matter and dust, which have many negative effects on the respiratory system and the general health of users. Different types of Vertical gardens that affect climate change through decreasing ambient temperature and carbon dioxide:

Although the walls are cultivated as a principle and method of environmental treatment and aesthetic of the facades, but have three classifications through which we can review all types of Vertical gardens and the three categories differ between them through principle and methods of installation and methods of handling and the quality of plants used and those classifications are:

1. Green Facades
2. Wall Vegetation
3. Living Walls

There is no doubt that the green area is an important element to restore balance in the environment which is in flagrant violation of its natural resources due to the wave of urbanization sweeping many countries and the negative effects. It has on living organisms and on human beings in particular.

Some modern countries have realized the dangers of vegetation cover in disrupting the balance of nature and the environment. This led them to go to the cultivation of land through the revival of the soil again. The vast and extended areas of trees and shrubs in a beautiful aesthetic harmony, as the streets roam Residential neighborhoods until you find yourself in a singing garden, stretching along the opposite street to the houses. Facades are lined with tree-lined trees, shaded with flowers, which usually begin to reveal their splendor in the spring. This green chain of plants has contributed to the city's remarkable urban landscape. Modern cities have become more oriented towards increasing the area of parks and public gardens through the cultivation and beautification of the main streets and sub-streets as well as the streets of residential neighborhoods. The main target is to break the stalemate of buildings and concrete blocks, and make this area shows some of the living aesthetic components that affect the surrounding environment. The comfort that is quickly realized in the hearts of passers-by comes from the visual enjoyment in coordination with aesthetic arrangement. This is shown by cultivation groups of trees and shrubs along the length of the streets and on both sides.

13 Cultivation of Shrubs in Cities to Combat Climate Change

Shrubs are the second largest group of plants cultivated in cities. Because they are always dense vegetative and the number of leaves on them have an important role in reducing the heat of the surrounding environment and absorbing large amounts of CO_2 and losing water by evaporation. It is known that one gram of water requires 633 cal for evaporation. So increasing evaporation water contributes to the reduction of surrounded air then atmospheric temperature. High humidity also contributes to reducing the impact of atmospheric temperature. These two factors contribute to reducing the impact of climate change. Here a present of some of these plants as mentioned from the results of the experiments carried by Koriesh [17].

The shrubs should be carefully selected by specialists so that they can play their role and function to the fullest extent in terms of preventing dust, dust and carbon dioxide.

Afforestation of the suburbs and residential neighborhoods has become an important point in the process of beautifying the cities and making the inhabitants abler to enjoy the outdoor atmosphere and to practice their sport activities without problems or inconvenience.

Taking the simple engineering design to walk at one pace in the decoration and beautification of the streets of the neighborhoods, For the formation and flowering throughout the year, with the existence of green spaces are not enough to be a station to sit on and enjoy with nature.

The use of Hedera helix climbers and Bougainvillea *shrubs which find scattered and abundant in this type of street ornamental agriculture, since they are evergreen and are up to 6 m long, they are suitable as a climber or a shrub.*

They are very sparse, and their leaves are evergreen, simple, reciprocating, oval, full-edged, dark green. Also, *Hedera helix* has a good role in formaldehyde pollutant abatement. The flowers of *Bougainvillea* are multicolored including white, red, purple, orange and sometimes double colors. *Bougainvillea* includes more than 20 varieties, and we find some varieties of hybrid that come in different sizes, and be attractive with flowers that radiate joy and beauty. It increases its coordination value and favors warm, humid climates, but it can withstand high temperatures and grow under direct sunlight and withstand moisture while resisting drought.

13.1 Dodonaea viscosa

Except for Bougainvillea bush, the *D. viscosa* shrub can be cultivated for its low water requirements and high transpiration rate (plants used about 2% for metabolism and transpiration and evaporate the rest [157], beauty and their role for reducing CO_2. It can be cultivated in the form of shrubs and hedges that appear in their natural, drooping and arched shape.

13.2 Schinus terebinthifolius

Brazilian pepper-tree is a shrub or small tree that reaches over 30 feet in height, typically with a short trunk hidden in a thicket of branches. Some trees can live over 30 years. Brazilian pepper is widely grown as an ornamental plant for its foliage and fruit.

Brazilian pepper is hard to control because it produces basal shoots if the trunk is cut. Trees also produce abundant seeds that are dispersed by birds and ants. This same hardiness makes the tree highly useful for reforestation in its native environment, but enables it to become invasive outside of its natural range.

Transpiration of high amounts of water through plants contributes to the reduction of ambient air temperatures. Its efficiency is relatively high in the absorption of CO_2, making it suitable for reducing climate change.

13.3 Lantana camara

A comparative response of *lantana camara* and *Bauhinia variegata* to elevated CO_2. Raizada et al. [158] showed that these plants are likely to emerge more competitive in the global change scenario [159].

The rise in CO_2 availability directly impacts photosynthetic processes, evoking a wide range of physiological and morphological responses in plants. These vary among species, depending on differences in photosynthetic pathways, intrinsic growth rates and other properties. Prominent responses include changes in growth rates [160], water use efficiency [161] and nutrient uptake rates [162].

Elevated CO_2 significantly promoted the growth of seedlings of different plants. At the end of experiment carried by Raizada et al. [158], maximum percent enhancement was observed in *L. camara,* meanwhile, *B. variegata* responded least to elevated CO_2. *L. camara* shrubs are therefore more efficient to remove carbon dioxide than *B. variegata* (Table 1). Also, maximum mass was allocated to roots in *L. camara* under both ambient and elevated CO_2. Increased mass allocation to stem was observed for *L. camara* under elevated CO_2.

NAR ($mg\ mg^{-1} d^{-1}$) of *Lantana camara* increased in response to elevated CO_2, but NAR of *Bauhinia variegata* was not affected, negatively or positively, by increasing the concentration of carbon dioxide in the environment around plants. These studies of Raizada et al. [158] showed that growth responses of ornamental trees and shrubs plants to elevated CO_2 varied among species, some showing greater increases in shoot length and biomass accumulation than others. However, biomass accumulation under elevated CO_2 was greater in *Lantana camara* than in the woody trees as *Bauhinia variegata*. Other studies have also shown an increase in overall biomass of *Lonicera japonica* [159]. NAR as well as RGR increased under elevated CO_2 in *Acacia catechu*, *Bauhinia variegata*, *Dalbergia latifolia*, *Tectona grandis*, *Hyptis suaveolens* and *Lantana camara*.

In conclusion, under the elevated CO_2 in future, photosynthesis and growth of plants increased [163].

Trees also used for afforestation include other types of plants, *Ficus religiosa, Bauhinia variegate, Cassia nodosa, Eucalyptus camaldulensis* and *Khaya senegalensis* (Table 2).

Afforestation and Environmental Gardens and Their Impact on Climate Change

The environment is the framework in which man's life. Human behavior, air quality and the emission of responsible gases affect the warming of the atmosphere and the consequent effects of changes in the global map.

13.4 Climate Change and Changes in the Global Map

The climate is changing as a result of global warming. Many areas will become warmer and fewer will be colder. The countries where the temperature will rise are Russia, Scandinavia, Canada, Egypt and others. The climate will be adjusted to moderate and suitable for the growth of many crops, strategic or otherwise, for positive feedback cycles.

We do not deny that many countries are warming the world in their favor, so they will not do anything seriously to reduce the accumulation of carbon dioxide, even though it is responsible for the pollution that increases the temperature of the earth which increases the melting of not less than the ice sheet on the southern continent.

14 The Impact of Industrial Development and Agricultural Development on the Environment

We have to ask ourselves why the land has been suitable for life for more than three billion years, despite its vulnerability to many disasters. Nevertheless, it maintains balance through feedback through the rotation of elements in the food chain to preserve its existence. What is now suffering from pollution is due to the industrial boom and the economic growth rate of many countries, in addition to changing the pattern of consumption of the population. The industrial boom has caused many pollutants, whether in air, water or as solid wastes. Agriculture and technologies try to cover the high requirements required by growth have also affected the hectic race by using fertilizers to increase crop rates, as well as the use of pesticides.

Tree or plant leaves filter pollutants from the air. After Wang [164] one urban park, tree cover detached 48 lb of PM, 9 lb of NO_2, 6 lb of SO_2, 0.5 lb of CO and 100 lb of carbon—daily. Beside other pollutants including ozone and methane through the most important in the World photosynthesis. Trees sucking carbon dioxide (greenhouse gas) and releases oxygen in air. Trees store or sequester the carbon dioxide, and—depending on the size of the tree—can hold between 15 and 350 kg of carbon dioxide each year [25, 165]. Many trees take decades to reach full size, and in this fast-paced world, not everyone has the patience to wait that long. The impact of climate change on many countries in the world, particularly those in North Africa, including Egypt, is rapidly increasing due to the rapid deterioration of the global environment. Trees contribute to reducing global climate degradation. So, gardeners are looking for the fastest growing and efficient trees with high efficiency in CO_2 or other GHG. It is also best to have an efficiency in improving air temperature to help reduce the effects of CO_2 on climate change.

14.1 Fast Growing Trees

Some scientific views say trees that grow quickly should be avoided, because of weak wood, numerous pests and diseases, invasive roots, a short life span, or other undesirable qualities. Meanwhile, environmental protection workers say that fast-growing trees have high absorption efficiency in their early growth and the old trees often have very few vegetative parts that can clean pollutants. Here are some of the fastest growing trees that grown in Egypt.

1. *Albizia* (Mimosa Tree)
2. *Morus* spp. (White and Red Mulberry)

3. *Pinus* spp. (Pine)
4. *Populus* (Lombardy poplar)
5. *Paulownia tomentosa* (Empress Tree)
6. *Salix matsudana*
7. White willow (*Salix alba*)

14.2 Shrubs

Cultivable shrubs vary in different areas. In landscaping, the best to help in climate change reduction factors should be chosen. All are always looking for the most CO_2 absorption plants for either the leaf area unit or the hole mature plant. We also look for the most plants that have evaporation of water that requires energy taken by the plant from the air. Taking into account the rest of the contaminants (Tables 1 and 2).

Tree crown size is a key variable in this context as it correlates with the space. A tree occupies as well as with the physiological tree functions. Crown prolongation area and crown volume, can be used as proxy variables for leaf area and leaf biomass [166].

14.3 Benefits of Planting Trees

Planting trees is massively beneficial to the world in the face of hastening climate change [167]. There is nowadays a strong international scientific agreement that human activity is beginning global warming. A considerable drop in the planet's forest cover over last and recent centuries is a major sponsor to this climatic change [137]. As trees grow they absorb carbon dioxide (CO_2), the first main 'greenhouse gas' responsible for global warming, thereby reducing the concentration of this gas in the atmosphere. Methane is the second greenhouse gas responsible for global warming, some trees can reduce the concentration of this gas in the atmosphere.

Carbon absorption rates in trees depend on the species and the location. Fast-growing species in tropical climates can sequester CO_2 many times faster than the average European forest, and plantation projects aimed at earning carbon credits are typically designed to maximize the sequestration potential.

Over a 20-year lifecycle, the right species in the right conditions can absorb over 40,000 tons of CO_2 per km^2. So, a plantation of 100 km^2 can absorb 4 million tons of CO_2 over 20 years. That is equivalent to taking 50,000 cars or more off the road during that time (based on annual emissions of 3–4 tones for the average car and its usage).

15 The Foundations and Rules of Environmental Afforestation in Cities

15.1 Introduction

1. Governments are doing a great deal to increase afforestation, parks, public parks, children's playgrounds and the multiple services needed to improve the environment.
2. Some home owners also contribute to this work.
3. Afforestation and the increase of green areas are among the most important facilities that municipalities compete to show their care for the city especially in areas that subjected to high temperatures, dust and sand storms.
4. This is due to the importance of plants in preserving the environment and adapting the local climate and its diversity.
5. It is important to choose the most efficient plant species adapted to the environment.
6. The scientific foundations of the afforestation within the cities should be followed and the plants should be distributed in the streets and parks.
7. Be careful not to plant useless plants in the streets and parks that may cause harm such as species that produce positive ions [168–170].
8. For afforestation foundations and controls must be observed in the gardens of the streets and the islands and the middle and in front of houses, parks and public parks to achieve the environmental role, climate change and health of plants and increase environmental efficiency in cities.

15.2 Some Characteristics of Local-Friendly Species

The environmental conditions of the area to be cultivated should be identified in order to select appropriate and suitable plant species to be cultivated for planting

When selecting these trees, the following characteristics shall be met:

a. Have a high capacity to withstand the local environmental conditions of the area where they are grown in terms of high temperature and pollution.
b. Its high efficiency in carbon dioxide absorption.
c. The wood has a high density, which indicates efficient storage of carbon dioxide.
d. High efficiency in lowering the temperature in the surrounding air.
e. It has high resistance to insect and pathogenic infections or worms.
f. Fast and dense growth and abundant branch.
g. It has a deep root and is not spread horizontally so as not to hinder the growth of other plants and does not affect neighboring facilities.
h. The nature of its growth, the shape of its crown and height, corresponds to the place where it is grown and the purpose of its cultivation.

i. Plant seedlings are of sufficient size and age at planting to ensure their success
j. Its condition is good in terms of vegetative and root growth and sound from fractures and pests.
k. Are locally available and require minimal care and costs during their cultivation and growth. Reproduce easily.

15.3 Plant Cultivation and Use Functions: Environmental Function

Plants have great value in the urban environment. Its lack or lack of numbers harms the environmental balance in the region. This role can be summarized in the following points:

- The plant absorbs carbon dioxide and is one of the main causes of pollution and the release of oxygen gas. This process is the most important process in the universe.
- Reduce pollution levels from most types of pollutants by absorbing them.
- Plants emit oxygen in the atmosphere through photosynthesis, thereby reducing the proportion of pollutants.
- Optimizing the atmosphere through the process of transpiration and improvement of the climate the presence of plants in a place to reduce the temperature, especially during the summer. It is known that every 1 g of water lost by the plant requires 633 kcal, which the plant withdraws from the surrounding medium, thus lowering the temperature.
- Relieve the glare of the sun and dazzle the eyes with intense light. Where the leaves of the trees intercept sunlight, absorbing part of it and reflect some of the other rays.
- Contribute to the absorption of sounds and reduce the noise, especially in crowded places in cities and succession.
- Stop sand crawling and reduce desertification.
- Protect cities from severe winds and break their limits.

The studies we conducted in Egypt showed that the shrubs cultivated in Egypt differ in their application to get rid of carbon dioxide, and some shrubs have greater efficiency (Table 1).

Bougainvillea glabra is superior to all shrubs that have been studied in the disposal of carbon dioxide after 5, 10 and 15 min of exposure to this pollutant. The efficiency value was more than twice that of *Hedera helix* and thrice as *Lantana camara* and 4 times as *Malvaviscus mollis* and 10 times as *Nerium oleander*.

From this, it is important to plant shrubs intensively in the city parks:

1. Fast growth
2. High efficiency of some varieties to get rid of carbon dioxide that causes global warming

3. Its efficiency in transpiration rate as mmol m^{-2} min^{-1} is higher than the trees (Table 2), which has a high effect, at least theoretically, in the absorption of atmospheric temperature.

Shrubs, generally, are evergreen and may reach a height of four meters at full growth. They are usually planted in small gardens instead of trees. The shrubs are used in many coordination purposes, which can be summarized as follows:

(a) Cultivation in groups close together to form a single vegetable mass used in the design of natural parks such as *Schinus* and *Lantana*.
(b) Cultivation as individuals each with its characteristic and specific and attractive to look like Queen of the night.
(c) Cultivation in front of buildings such as jasmine.
(d) Cultivation in the corners to fill spaces such as *Duranta*.
(e) Cultivation to establish gradient between trees and grasses, such as Lantana
(f) Cultivation for hytonocides as roses.
(g) Cultivation as groundcover plants such as *Dodonaea*.
(h) Cultivation to give the color of the gardens.
(i) Cultivation as Plant fencing.

16 Future Roles and Regulations in Urban Afforestation

It is best to cultivate the streets of all types of plants, taking into account the relationship with the type of coordination and nature of plant growth.

The types of plants grown in a street or garden that have been established in a specific area of a particular city vary according to the following criteria:

1. Plants planted on a street or a specific area shall be suitable for the quantity and quality of pollution in the area
2. Be resistant to diseases and insect pests
3. The distance of agriculture depends on the amount of pollution and the number of cars and street view
4. The distance between each tree depends on the nature of tree growth and the value of LAI
5. Highly polluted areas can be cultivated more than a row or the use of high-efficiency shrubs
6. Take into account the pruning of trees in a scientific manner sound where there is a difference between pruning and cutting used in many cities in Egypt
7. All trees have a lifetime to give them the highest efficiency of their functions, so must work out a plan for replacement and continuous renewal to ensure their role environment
8. Some of the older trees can be renewed or vaccinated with new, more active branches

9. All trees are not renewed at the same time to ensure the safety of the environment
10. Some falling leaf species such as *Ficus religiosa* can be used to ensure that there is sufficient vegetation
11. Scientifically known that carbon dioxide in the air is about 18–20% lower in the winter
12. The streets were planted inside the cities according to the specialists and each case separately
13. It is better not to use high trees in VOC emissions
14. The importance of tree planting in Egypt should be eliminate and store the largest amount of carbon dioxide
15. Reduce the temperature in the surrounding environment provide adequate shade in different places. The other objectives then come
16. The nature and size of plant growth must be consistent with the size and nature of the street and the conditions of the site in which it is grown and the surrounding facilities
17. The branches of trees shall not be exposed to intertwining with wires and others
18. In terms of protecting the environment, it is best to plant each street with more than one type of tree, which will vary in efficiency in combating pollution
19. In my view, the use of organic fertilizers reduces the cultivation of trees in the streets for the following reasons. Reduce the presence of a host of microbes
20. It is scientifically known that organic matter in the agricultural environment is a source of methane emissions and is more dangerous to climate change than CO_2.

16.1 Central Islands

1. Planted with high soil coverings to increase the absorption efficiency of CO_2 or dense dendritic groups
2. The central islands are an important means of increasing the number of plants on the streets
3. It is usually planted with shrubs, long grass plants, soil coverings and green areas
4. The type of plant depends on the presentation of the place
5. If it is broad, it can be planted with trees and shrubs of environmental importance
6. If they are narrow, they are planted with vegetative structures that can be formed and are environmentally beneficial.
7. If the islands are of great lengths and wide as in the roads between cities or different sectors can be planted with types of trees with water needs converged to be easy to maintain, and preferably trees flowering to beautify the streets.

Regulations to Consider When Selecting and Growing Plants in the Central Islets:

1. The intensity of plant cultivation depends on the density or amount (number) of contaminants that may be present in the area

2. It is preferable to be specially planted trees in the middle of the islands with high legs and fork high so as not to hinder the movement of pedestrians or cars
3. Do not grow large trees at the intersection of roads or at the top of the islands in the streets so as not to impede the path of sight of drivers, noting that leave at least 30–50 m around the road crossings without planting trees or large plants
4. This area can be cultivated with low-altitude grass plants and soil coverings
5. It is preferable to cultivate medium-sized islands using high-growth plants that are highly efficient in absorbing pollutants and low in water and contribute to heat reduction (Fig. 8)
6. It is enough to plant hedges or soil reservoirs with a height of not more than 50 cm for the roads that are being constructed and to re-examine the previously created roads and remove trees obstructing the vision.

16.2 Main Steps for Afforestation Within Cities

1. Determine the objective of agriculture.
2. Determine the location of agriculture Is it a street or a middle island.
3. Is the chosen location close to industrial areas?
4. Determine the appropriate plant quality according to the pollution intensity and temperature.
5. Implementation of the process of agriculture according to sound scientific basis and according to the type and nature of plant growth.

16.3 The Roles or Foundations of Afforestation for Environmental Conditions

A common mistake is to cultivate any plant species before they are studied, especially for important traits such as tolerant type of pollutants, CO_2 absorption efficiency, high heat reduction, drought tolerance and high salinity tolerance.

Existing trees should be reviewed according to recent scientific information related to the efficiency of CO_2 absorption and heat reduction. It may be necessary to change trees or other types of plants because of their lack of success and performance of their local environmental functions or contain a toxic substance or contain some undesirable qualities.

17 Types of Urban Garden Plants that Contribute to the Direct or Indirect Climate Change. Common Problems in Egypt

17.1 High-Temperature Tolerant Plants

Palm trees: *Phoenix dactylifera;* trees: *Tamarix* spp.—*Acacia* spp.—*Poinciana. Delonix regia*—*Opuntia* spp.—*Bogainvillea* spp.—*Nerium oleander*—*Dodonaea viscosa*—*Cupressus* spp.—*Schinus molle*—*Casuarina* spp. —*Albizzia Lebbek*—*Ziziphus spina-christi*—*Hyphaene thebaica*—*Azadirachta indica*—*Eucalyptus* spp. *Parkinsonia aculeata*—*Populs* spp.—*Washtunia palms*—*Creatonia siliqua*—*Melia azedarach*—*Olea europaea*—*Conocarpus erectus;* shrubs: *Lantana camara.*

17.2 Plants Reduce Methane

- *Araucaria* spp.
- Pine.

17.3 Plants for Phytoncides (Aromatic Volatile Substances)

- *Cassia nodosa*—*Bauhinia*—*Acacia farnesina.*

17.4 Plants Resistant to Smoke and Dust

- *Tamarix* spp.—*Nerium oleander*—*Melia azedarach*—*Ficus* spp.—*Hibiscus* spp.—*Eucalyptus* spp.—*Populs* spp.—*Phoenix* spp.

18 Attenuation of Gaseous Pollutants by Greenbelts

Importance of greenbelts in attenuating gaseous pollutants and presents the meteorological, physicochemical, biological, and horticultural dimensions associated with effective greenbelt design [171, 168]. They have presented the gist of a system of methodologies, developed by us, for greenbelt design. A case study demonstrating the applicability of the system has been presented.

18.1 What Is Greenbelt?

Greenbelt (Fig. 9) is a strip of trees of such species, and such a geometry, that when planted around a source, would significantly attenuate the air pollution by intercepting and assimilating the pollutants in a sustainable manner [172].

In reality greenbelts almost always include other vegetation, especially shrubs, which also play some role in capturing pollutants. However, trees are the mainstays of greenbelts, and often greenbelt plantation is simply referred as 'trees' even though other types of vegetation is also established and nurtured.

The concept of greenbelt as a source of pollution abatement was recognised initially by three nations:

1. The U.S.A.,
2. Britain and
3. Kenya [172].

Ebenezer Howard, a British social reformer, advanced the concept of greenbelt in 1898, in connection with the planning of 'new towns' located outside the periphery of London, which was then sprawling far into the countryside. Howard proposed 'garden cities' which would not only be free of pollution but would also be antidote of polluted cities, each of which would be surrounded by an agriculture 'country belt'.

It was British architect and planner Raymond Unwin, a town designer and contemporary of Howard's, who coined the term greenbelt [172].

In Britain, Howard's concept took two forms: the greenbelts surrounding the new towns in rural Britain (the first of these was Letchworth, built in 1903); and, beginning in the 1930s, the application of the idea to London itself.

The London Green Belt Act was passed by the British Parliament in 1938. A more decorative plan was created in 1944 by Patrick Abercrombie, who proposed a belt, five or more miles wide, consisting of both public open spaces and private properties that would be regulated to preclude runaway suburban development [173].

Three such towns were built with Greenbelt, Maryland, being the best known. In the American concept of greenbelt, relatively wide band of rural land or open space surrounded a town or city. The term greenbelt meant. Generally, any swath

Fig. 9 Greenbelt and central islets to reduce air pollutants (Ismailia Governorate, Egypt)

of open space separating or interrupting urban development. A land so designated is controlled through regulation or public or quasi-public ownership (such as the Nature Conservancy) to retain its natural character and provide a resemblance of rural ambience in urban or urbanizing areas [172].

The central goal of Roosevelt's greenbelt programme, an innovative and radical intervention in American city building, was not to create better urban communities, but rather to generate jobs in a declining national economy.

18.2 Air Pollutant Uptake by Vegetation

Vegetation acts as CO_2 sink, and some species can utilize air pollutants effectively. Plants also intercept tones of dust and serve as acoustic screens on busy highways and noisy factory areas. Several plants can collect the dust suspended in the atmosphere and dilute the concentration of toxic and harmful gases like SO_2, CO_2, etc. Trees have been reported to remove air pollutants like hydrogen fluoride, SO_2, and some compounds of photochemical reactions and collect heavy metals like mercury (Hg) and lead (Pb) from the air [174, 175].

After absorbing the air pollutants, trees change them to harmless metabolites through various physiological processes. Of course, every plant can not be an agent of air pollution control; only those which can tolerate pollutants can act as attenuators [176]. All-in-all, increasing vegetation in the cities, towns, and industrial establishments holds great potential to combat air pollution.

Plants remove pollutants from the air in three ways, viz.

1. Absorption by the leaves,
2. Deposition of particulate and aerosols on leaf surface, and
3. Fallout of particulate on the leeward (downwind) side of the vegetation because of the slowing of the air movement [177, 101].

18.3 Removal of Gaseous Pollutants

Amble evidence is available to support the view that plants in general and trees in particular function as sinks for gaseous pollutants (and other pollutants by roots, phytoremediation). The latter are transferred from the atmosphere to vegetation by the combined forces of diffusion and flowing air movement. Once in contact with plants, gases may be bound or dissolved on exterior surfaces or be taken up by the plants via stomata. If the surface of the plant is wet and if the gas is water soluble, the former process can be very important.

When the plant is dry or in the case of gases with relatively low water solubility, the mechanism is assumed to be the most important [178]. Plant uptake rates increase as the solubility of the pollutant in water increases. Hydrogen fluoride, sulphur dioxide, nitrogen dioxide, and ozone which are soluble and reactive are readily sorbet pollutants. Nitric oxide and carbon monoxide, which are very insoluble, are absorbed relatively slowly or not at all by vegetation.

During daylight periods when plant leaves are releasing water vapor and taking up carbon dioxide, other gases, including trace pollutant gases, in the vicinity of the leaf are also taken up through the stomata [178]. The rate of pollutant gas transfer from the atmosphere to interior leaf cells is regulated by the resistance posed by atmospheric, stomatal, and mesophilic factors, more specifically wind speed, leaf size and geometry, gas viscosity and diffusivity. Stomatal resistance is regulated by the stomatal aperture, which is influenced by water deficit, carbon dioxide concentration, and light intensity. Mesophilic resistance is regulated by gas solubility in water, gas-liquid diffusion, and leaf metabolism [179]. Because the rate of pollutant uptake is regulated by numerous forces and conditions, the rate of removal under field conditions is highly variable.

If leaf characteristics, wind speed, atmospheric moisture, temperature, and light intensity are quantified, the pollutant uptake rate can be estimated. Under growth-chamber conditions, wind velocity, canopy height, and light intensity were shown to affect the rate of pollutant removal by vegetation. As previously stressed, light plays a critical role in determining physiological activities of the leaf and stomatal opening, and as such exerts a great influence on the foliar removal of pollutants. Under conditions of adequate soil moisture, however, pollutant uptake by vegetation was judged almost constant throughout the day, as the stomata were fully open. Pollutants were absorbed most efficiently by plant foliage near the canopy surface where light-mediated metabolic and pollutant diffusivity rates were greatest.

Sulphur and nitrogen dioxides were taken up by respiring leaves in the dark, but uptake rates were greatly reduced relative to rates in the light. The Removal of sulphur dioxide is also described by tree uptake process. Because of its high solubility in water, large amounts of sulphur dioxide are absorbed into external tree surfaces when they are wet. In the dry condition, sulphur dioxide is readily absorbed by tree leaves and is rapidly oxidised to sulphate in mesophyll cells [178].

18.3.1 All Plants Absorb Carbon dioxide, but Trees are Best

While all living plant matter absorbs CO_2 as part of photosynthesis, trees process significantly more than smaller plants due to their large size and extensive root structures.

In essence, trees, as kings of the plant world, have much more "woody biomass" to store CO_2 than smaller plants, and as a result, are considered nature's most efficient "carbon sinks". It is this characteristic which makes planting trees a form of climate change mitigation. According to the U.S. Department of Energy (DOE), tree species that grow quickly and live long are ideal carbon sinks. Unfortunately, these two attributes are usually mutually exclusive. Given a choice, foresters interested in maximizing the absorption and storage of CO_2 (known as "carbon sequestration") usually favor younger trees that grow more quickly than their older cohorts. However, slower growing trees can store much more carbon over their significantly longer lives.

18.4 Fastest Growing Trees (Ideal as Carbon Sink According to [180])

1. Royal Empress, *Paulownia* (growth per year: 3–5 m);
2. Willow Hybrid, *Salix* (growth per year: 3–4 m);
3. Lombardy Poplar, *Populos* (growth per year: 2.5–3 m);
4. Eucalyptus Tree, *Eucalyptus* (growth per year: 2–3 m);
5. Red Rocket Crape Myrtle, *Lagerstroemia* (growth per year: 1–2 m);
6. Cleveland Flowering Pear, *Pyrus calleryana* (growth per year: 1–2 m) and
7. River Birch, *Betula nigra* (growth per year: 0.5–1 m).

Plant the Right Tree in the Right Location

Scientists are busy studying the carbon sequestration potential of different types of trees depending, besides the species, upon location, climate, and soils.

Dave Nowak, a researcher at the U.S. Forest Service's Northern Research Station in Syracuse, New York has studied the use of trees for carbon sequestration in urban settings across the United States.

A 2002 study he co-authored lists the Common Horse-chestnut, Black Walnut, American Sweetgum, Ponderosa Pine, Red Pine, White Pine, London Plane, Hispaniolan Pine, Douglas Fir, Scarlet Oak, Red Oak, Virginia Live Oak and Bald Cypress as examples of trees especially good at absorbing and storing CO_2. Nowak advises urban land managers to avoid trees that require a lot of maintenance, as the burning of fossil fuels to power equipment like trucks and chainsaws will only erase the carbon absorption gains otherwise made.

19 Conclusion

Trying to save our global, urban or city gardening as a sort of horticultural strategies can dealing with climate change. We want to begin thinking now not just about what used to be, but what may be, generally contributing the city regions to be sustainable socially, economically and environmentally. Climate change goes to pressure us to paintings difficult at something *Homo sapiens* has never been good. An urban gardens like our own family's is a great location to begin. Climate change is a international problem, but its remedy begins with people or communities. Measures to combat climate change encompass 2 primary titles:

1. Removal of the maximum greenhouse effective gases, specifically methane, carbon dioxide, sulfur oxides, nitrogen oxides, particulates and nucleoids
2. Decreasing causes of greenhouse gas emissions.

Urban gardens, including street trees, public gardens, recreation parks and greenbelts especially that designed around the industrial areas can help mitigation of climate change. Additionally, include non-public gardens of a wide variety in or outside residential privacy, and modern or new gardens along with one-of-a-kind sorts of vertical gardens. New plant species stand up in these situations. The soil in gardens is a reminder that things can change.

20 Recommendations

Based on the identification of the United Nations of factors affecting climate change, and the scientific information on the modern environmental roles of plants, both in the elimination of one or more of the causes of global warming in addition to absorption of different radiation and its role in reducing air temperature. Also, water professionals count on rapidly increasing water stress and water scarcity. To cope with those demanding situations, there may be a need for the creation of social corporations so one can work in partnership with governments and the international community on mitigating climate change by way of growing sustainable urban gardens. Urban agriculture additionally provides social, environmental and educational advantages to residents.

So, the main recommendations of this work is to place out the advantages of urban agriculture for reducing climate change and preserving sustainable urban environments, and a few advisees, briefly are below:

1. Municipality planning and landscape gardeners in conjunction with the Ministry of the Environmental Affairs have to review planning methods and undertake scientific foundations while choosing garden plants and verifying non-emission farming structures that may be harmful to the environment when dealing with soil during soil preparing for agriculture.
2. Look for ways or means to change the plants that are not environmentally friendly plants, with no harm to the soil.
3. Climate change affects and is affected by all countries, but its treatment must begin with the actions of individuals, which requires the provision of appropriate means of awareness in urban communities.
4. Gardeners should not lose sight of the importance of reducing water consumption, at least through selecting C_4 and CAM plants for their low water needs. Taking into consideration their environmental importance and their efficiency in eliminating greenhouse gases.
5. Reduces climate change as urban agriculture contributes to the prevention of the over-heating of city environments (Urban Heat Islands).
6. Urban agriculture offers proper land control and use for urbanized areas.

7. Social and psychological benefits, such as food security in home gardens, personal psychological effects, community cohesion and well-being; and ecological benefits, such as air quality, soil quality.
8. Urban agriculture has contributed for a long-time to keep all kinds of pollutants inside the metropolitan inside positive affordable limits.
9. Gardens are one of the maximum valuable legacies that we are able to go away for future generations.

Acknowledgements We would like to express our deepest appreciation to Prof. Dr. Fouad Hassan, Prof. Dr. M. Abd El Mohsen, Prof. Dr. Khaled Ibrahim and Dr. Mahmoud Mokhtar Hefni who provided us with the possibility to complete this chapter.

References

1. World Bank (2009) Training material case study, Egypt: the Nile delta drainage project Agriculture and Rural Development Department. The World Bank, USA Public Disclosure Authorized EGYPT, The Nile-delta drainage project. http://documents.worldbank.org/curated/en/251421468246925594/pdf/UNN1780Box338884B01official0use0only1.pdf
2. El-Shahawi M (2004) Climate change and some potential impacts on Egypt. First national symposium in climate change and its impacts on Environment in Egypt. National IGBP & SCOPE Committee and Suez Canal University, Ismailia, Egypt
3. Abdalla Z, Yunsheng L (2015) Impact of climate change on agriculture sector in Egypt and China. In: 4th International conference on agriculture and horticulture. https://doi.org/10.4172/2168-9881.s1.016
4. NASA (National Aeronautics and Space Administration) (2018) Global climate change. Vital signs of the planet. A blanket around the Earth. Available at: https://climate.nasa.gov/vital-signs/sea-level/
5. UNEP/EWWP (1998). Major environmental assessments. Global environment outlook. United Nations Environment Programme (UNEP). Oxford University Press, New York. http://www.grid.unep.ch/geo2000/
6. Eid MH (1999) Climate change studies on Egyptian agriculture. Egypt–US workshop on Global Climate Change, NARSS, EEAA and USEPA, Cairo, Egypt
7. Knickmeyer E (2008) In Egypt upper crust gets the bread. Published in the Washington Post Foreign Service. http://www.washingtonpost.com/wp-dyn/content/article/2008/04/04/AR2008040403937.html?hpid%253Dmoreheadlines&sub=AR
8. NASA (National Aeronautics and Space Administration) (2019) Global climate change. Warming seas may increase frequency of extreme storms. https://climate.nasa.gov/news/2837/warming-seas-may-increase-frequency-of-extreme-storms/
9. Wang ZP, Han SJ, Li HL, Deng FD, Zheng YH, Liu HF, Han XG (2017) Methane production explained largely by water content in the heartwood of living trees in upland forests. J Geophys Res 122:2479–2489
10. Thomey ML, Ford PL, Reeves MC, Finch DM, Litvak ME, Collins SL (2014) Climate change impacts on future carbon stores and management of warm deserts of the United States. Rangelands 36(1):16–24
11. Darlington (2004) An integrated indoor air biofiltration system for municipal infrastructure. Nedlaw Living Walls, Air Quality Solutions Ltd. University of Guelph, Canada, pp 1–50
12. Sundqvist E, Vestin P, Crill P, Persson T, Lindroth A (2014) Short-term effects of thinning, clear-cutting and stump harvesting on methane exchange in a boreal forest. Biogeosciences 11(21):6095–6105

13. Sundqvist E, Crill P, Mölder M, Vestin P, Lindroth A (2012) Atmospheric methane removal by boreal plants. Geophys Res Lett 39(21):1–6, L21806
14. Nowers N (2016) The effect of an oregano oil extract in a lactating dairy cow diet on production responses of Holstein cows. Doctoral dissertation, Stellenbosch University, Stellenbosch
15. Hristov AN, Lee C, Cassidy T, Heyler K, Tekippe JA, Varga GA, Corl B, Brandt RC (2013) Effect of *Origanum vulgare* L. leaves on rumen fermentation, production, and milk fatty acid composition in lactating dairy cows. J Dairy Sci 96(2):1189–1202.
16. Kim KJ, Kil MJ, Song JS, Yoo EH, Son KC, Kays SJ (2008) Efficiency of volatile formaldehyde removal by indoor plants: contribution of aerial plant parts versus the root zone. J Am Soc Horticult Sci 133(4):521–526
17. Koriesh EM (2018a) Studies on air pollution 3: evaluation of some ornamental trees and shrubs for transpiration and reducing atmosphere temperature. In: The 8th international Arabic conference, Ismailia, Egypt, Ismailia, Egypt
18. Ralph L, Wood R, Tarran J, Torpy F, Burchett M (2004) Removal of benzene by the indoor plant/substrate microcosm and implications for air quality. Water Air Soil Pollut 157:193–207
19. Abalos D, Sanchez-Martin L, Garcia-Torres L, Van Groenigen JW, Vallejo A (2014) Management of irrigation frequency and nitrogen fertilization to mitigate GHG and NO emissions from drip-fertigated crops. Sci Total Environ 490:880–888
20. Ussiri D, Lal R (2013). The role of fertilizer management in mitigating nitrous oxide emissions. In: Soil emission of nitrous oxide and its mitigation. Springer, Netherlands. 978-94-007-5363-1, 978-94-007-5364-8
21. Ingram DS, Vince-Prue D, Gregory PJ (2008) Science and the garden: the scientific basis for horticultural practice. Blackwell Publishing, Chichester
22. Lou XF, Nair J (2009) The impact of landfilling and composting on greenhouse gas emissions—a review. Biores Technol 100(16):3792–3798
23. Van de Kamp ME, van Dooren C, Hollander A, Geurts M, Brink EJ, van Rossum C, Biesbroeka S, Ede Valkc IB, Toxopeusand, Temme EH (2017) Healthy diets with reduced environmental impact? The greenhouse gas emissions of various diets adhering to the Dutch food based dietary guidelines. Food Res Int 104, 14–24
24. Audsley E, Brander M, Chatterton J, Murphy-Bokern D, Webster C, Williams A (2009) How low can we go? An assessment of greenhouse gas emissions from the UK food system and the scope to reduce them by 2050. WWF-UK
25. EPA (Environmental Protection Agency) (2017) What to plant. EPA Government. Available at: https://www.epa.gov/watersense/what-plant
26. Sutton M, Reis S, Skiba U, Nemitz E, Beier C, Butterbach-Bahl K, Cellier P, de Vries W, Erisman JW (2011) Nitro Europe IP—the nitrogen cycle and its influence on the European greenhouse gas balance. Final Report to the European Commission (Publishable Activity Report), pp 1–44
27. Livesley S, Dougherty B, Smith A, Navaud D, Wylie L, Arndt S (2010) Soil-atmosphere exchange of carbon dioxide, methane and nitrous oxide in urban garden systems: impact of irrigation, fertilizer and mulch. Urban Ecosyst 13:273–293
28. Okano K, Machida T, Totsuka L (1988) Absorption of atmospheric NO_2 by several herbaceous species, estimation by the ^{15}N dilution method. New Phytol 109:203–210
29. Butterbach-Bahl K, Baggs E, Dannenmann K, Zechmeister-Boltenstern S (2013) Nitrous oxide emissions from soils: how well do we understand the processes and their controls? Electronic supplementary material is available at http: http://rstb.royalsocietypublishing.org
30. Signor D, Cerri C, Conant R (2013) N_2O emissions due to nitrogen fertilizer applications in two regions of sugarcane cultivation in Brazil. Environ Res Lett Bristol 8(1):1–9
31. Follett RF, Shafer SR, Jawson MD, Franzluebbers AJ (2005) Main content area research and implementation needs to mitigate greenhouse gas emissions from agriculture in the USA. Soil Tillage Res 83(1):159–166
32. Shcherbak I, Millar N, Philip Robertson G (2014) Global metaanalysis of the nonlinear response of soil nitrous oxide (N_2O) emissions to fertilizer nitrogen. Proc Natl Acad Sci USA 111(25):9199–9204

33. Millar N, Doll JE, Robertson GP (2014) Management of nitrogen fertilizer to reduce nitrous oxide (N_2O) emissions from field crops. In: Climate change and agriculture fact sheet series, MSU Extension Bulletin E, 3152. Available at: https://lter.kbs.msu.edu/wp-content/uploads/2014/12/Nitrogen-fertilizer_climate-fact-sheet_FINAL.pdf
34. Hultgreen G, Leduc P (2003) The effect of nitrogen fertilizer placement, formulation, timing, and rate on greenhouse gas emissions and agronomic performance. Agricult Agr-Food Canada & Prairie Agricultural Machinery Institute, Swift Current
35. Cavigelli MA, Del Grosso SJ, Liebig MA, Snyder CS, Fixen PE, Venterea RT, Leytem AB, McLain JE, Watts DB (2012) US agricultural nitrous oxide emissions: context, status, and trends. Front Ecol Environ 10(10):537–546
36. EPA (Environmental Protection Agency) (2016) Sulfur dioxide (SO_2) pollution. https://www.epa.gov/so2-pollution/sulfur-dioxide-basics
37. Zhang X, Zhou P, Zhang W, Zhang W, Wang Y (2013) Selection of landscape tree species of tolerant to sulfur dioxide pollution in subtropical China. Open J For 3(4):104–108
38. Mohajan HK (2014) Mini review Chinese sulphur dioxide emissions and local environment pollution. Int J Sci Res Knowl 2(6):265–276
39. Han Y, Li Z, Liu R (2002) Potential of purifying SO_2 of main tree species and their planting quota in Shenyang area. Ying yong sheng tai xue bao = J Appl Ecol 13(5):601–604
40. El-Sadek M, Koriesh E, Fujii E, Moghazy E, Abd El-fatah Y (2012) Correlation between some components of interior plants and their efficiency to reduce formaldehyde, nitrogen and sulfur oxides from indoor air. Int Res J Plant Sci 3(10), pp 222–229. ISSN: 2141–5447
41. IPCC (Intergovernmental Panel on Climate Change) (2006) Solid waste disposal. In: IPCC guidelines for national greenhouse gas inventories, vol 5: waste. Online at http://www.Ipcc-nggip.iges.or.jp/public/2006gl/pdf/5_Volume5/V5_3_Ch3_SWDS.pdf
42. IPCC: Climate Change (2014) Mitigation of climate change, contribution of working group III to the fifth assessment report of the intergovernmental panel on climate change. In: Edenhofer EO, Pichs-Madruga R, Sokona Y, Farahani E, Kadner S, Seyboth K, Adler A, Baum I, Brunner S, Eickemeier P, Kriemann B, Savolainen, Schlömer S, von Stechow C, Zwickel T, Minx JC (eds). Cambridge University Press, Cambridge
43. Ward PL (2009) Sulfur dioxide initiates global climate change in four ways. Thin Solid Films 517(11):3188–3203
44. Smith J (2012) Urban air quality. Woodland Trust. https://www.woodlandtrust.org.uk/mediafile/100083924/Urban-air-quality-report-v4-single-pages.pdf
45. SEPA (Scotch Environment Protection Agency) (2006) Chlorofluorocarbons (CFCs) fact sheet. State of the Scottish Environment 2006. www.sepa.org.uk. http://apps.sepa.org.uk/spripa/pages/substanceinformation.aspx?pid=114
46. Tekippe JA, Hristov A, Heyler K, Cassidy T, Zheljazkov V, Ferreira J, Karnati S, Varga G (2011) Rumen fermentation and production effects of *Origanum vulgare L.* leaves in lactating dairy cattle. J Dairy Sci 94:5065–5079
47. Wilson C, Grubler A, Gallagher KS, Nemet GF (2012) Marginalization of end-use technologies in energy innovation for climate protection. Nat Clim Change 2(11):780–788
48. Zhang W, Wang B, Niu X (2017) Relationship between leaf surface characteristics and particle capturing capacities of different tree species in Beijing. Forests 8(3):92–103
49. Hyvönen R, Ågren GI, Linder S, Persson T, Cotrufo MF, Ekblad A, Freeman M, Grelle A, Janssens IA, Jarvis PG, Kellomäki S, Lindroth A, van Loustau DM, Lundmark T, Norby RJ, Oren R, Ryan MG, Sigurdsson BD, Strömgren M, Oijen MV, Wallin G (2007) The likely impact of elevated [CO_2], nitrogen deposition, increased temperature and management on carbon sequestration in temperate and boreal forest ecosystems: a literature review. New Phytol 173(3):463–480
50. Ball B, Taggart M, Scott A (2004) Mitigation of greenhouse gas emissions from soil under silage production by use of organic manures or slow-release fertilizer. Soil Use Manag **20**:287–295
51. Farming Futures (2009) Climate change: be part of the solution focus on: soil management. Fact Sheet No. 20. Available at: http://adlib.everysite.co.uk/resources/000/241/422/FS20.pdf

52. Whitmore A, Kirk G, Rawlins B (2015) Technologies for increasing carbon storage in soil to mitigate climate change. Soil Use Manag 31:62–71
53. Lockwood A, Filley T, Rhodes D, Shepson P (2008) Foliar uptake of atmospheric organic nitrates. Geophys Res Lett 35(15):1–5
54. Dobbie KE, Smith KA (2003) Impact of different forms of N fertilizer on N_2O emissions from intensive grassland. Nutr Cycl Agroecosyst 67(1):37–46
55. Forum for the Future (2009) Climate change, foreign policy, and higher education by Levi M, Mullinix J. Forum for the future of higher education. Cambridge, MA. http://forum.mit.edu/articles/%EF%BB%BFclimate-change-foreign-policy-and-higher-education/
56. Singh SK, Rao DN, Agrawal M, Pandey J, Naryan D (1991) Air pollution tolerance index of plants. J Environ Manage 32(1):45–55
57. Awad YM, Blagodatskaya E, Ok Y, Kuzyakov Y (2013) Effects of polyacrylamide, biopolymer and biochar on the decomposition of 14C-labelled maize residues and on their stabilization in soil aggregates. Eur J Soil Sci 64(4):488–499. https://doi.org/10.1111/ejss.12034
58. Yousaf B, Liu G, Wang R, Abbas Q, Imtiaz M, Liu R (2017) Investigating the biochar effects on C-mineralization and sequestration of carbon in soil compared with conventional amendments using the stable isotope (δ13C) approach. Gcb Bioenergy 9(6):1085–1099
59. Wu J, Guo W, Feng J, Li L, Yang H, Wang X, Bian X (2014) Greenhouse gas emissions from cotton field under different irrigation methods and fertilization regimes in arid northwestern China. Sci World J. 407832–407832
60. Glaser B, Lehmann J, Zech W (2002) Ameliorating physical and chemical properties of highly weathered soils in the tropics withcharcoal—a review. Biol Fertil Soils 35:219–230
61. Lehmann J, Gaunt J, Rondon M (2006) Bio-char sequestration in terrestrial ecosystems—a review. Mitig Adapt Strat Glob Change 11:403–427
62. Oertela C, Matschullat J, Zurba K, Zimmermann F, Erasmi S (2016) Greenhouse gas emissions from soils—a review. Chem der Erde-Geochem 76(3):327–352
63. Anonymous (2013) Surface runoff, percolation, total runoff and evaporation from precipitation. Published by: Senate Department for Urban Development and the Environment. No. 02.13, (2013 edn.). Available at: www.stadtentwicklung.berlin.de/umwelt/umweltatlas/e…/ekc213.doc
64. ENPAA, English National Park Authorities Association (2009) Climate change mitigation and adaptation in national parks, London, EC2A 3NW, England. http://www.nationalparksengland.org.uk/__data/assets/pdf_file/0010/767467/ENPAA-Statement-on-Climate-Change-in-National-Parks.pdf
65. Sanderson K (2008) Plants that reduce air nitrous oxide: trees eat pollution products. Nature. https://doi.org/10.1038/news.2008.1046
66. Pouyat RV, Yesilonis ID, Nowak DJ (2006) Carbon storage by urban soils in the United States. J Environ Qual 35(4):1566–1575
67. Hopkin M (2005) Tree planting not always green. Forests can suck up water and change the soil. Nat Mag. Published online 22 Dec 2005. Available at: https://doi.org/10.1038/news051219-14
68. Hunsberger A, Alison F, Angela M, Barbra L, Bart S, Niemann B (2010) The Florida-friendly landscaping guide to plant selection & landscape design. University of Florida publications. Available at: https://fyn.ifas.ufl.edu/pdf/FYN_Plant_Selection_Guide_2015.pdf
69. Boone L (2017) These 7 fire-retardant plants may help save your home (Yucca, Lemonade berry, Oak, Woolly blue curl, Catalina cherry). Available at: http://www.latimes.com/home/la-hm-firescaping-20170930-htmlstory.html
70. Dickie G (2015) Plants that burn fastest in a wildfire. Available at: https://www.hcn.org/issues/47.13/after-a-record-setting-wildfire-a-washington-county-prepares-for-the-next-one/plants-that-burn-fastest-in-a-wildfire
71. Cameron RW, Blanuša T, Taylor JE, Salisbury A, Halstead AJ, Henricot B, Thompson K (2012) The domestic garden—its contribution to urban green infrastructure. Urban For Urban Green 11(2):129–137

72. Wilson M (2007) New gardening: how to garden in a changing climate. Mitchell Beazley and the Royal Horticultural Society. MITCH Publisher, London
73. Cross R, Spencer R (2009) Sustainable gardens. CSIRO, Collingwood. ISBN 9780643094222
74. Walker J (2011) How to create an eco garden: the practical guide to greener, planet-friendly gardening?. Aquamarine, Wigston
75. Clifford DP, Eckbo G (2017) Garden and landscape design. Encyclopædia Britannica, inc. https://www.britannica.com/art/garden-and-landscape-design
76. Haynes J, McLaughlin J, Vasquez L, Hunsberger A (2001) Low-maintenance landscape plants for South Florida. Environmental Horticulture Department, Florida Cooperative Extension Service, University of Florida-IFAS Publication ENH854, Florida, USA
77. Hussein M (2018) Sustainable regeneration of urban green areas in Egypt's desert cities adopting green infrastructure strategies in New Borg El-Arab City. M.Sc. Master of Science in Resource Efficiency in Architecture and Planning Hafen City University, Hamburg, Germany February, available at: http://edoc.sub.uni-hamburg.de/hcu/volltexte/2018/428/pdf/Hussein_Mahmoud_Master_thesis_2018.pdf
78. Prieur D (2016) Environment, topics. Perennial plants can fight climate change, says Dr. Jerry Glover, Medill New Service. Available at: http://dc.medill.northwestern.edu/blog/2016/08/04/perennials-fighting-climate-change/#sthash.lZXd0d2L.DtR9oOsK.dpbs
79. SENES Consultants Limited (2005) NWT environmental audit status of the environment report air quality, climate and climate change. Available at: https://www.aadncaandc.gc.ca/DAM/DAM-INTER-NWT/STAGING/textetext/soeaq_1316619536106_eng.pdf
80. Sawidis T (1988) Uptake of radionuclides by plants after the Chernobyl accident. Environ Pollut 50(4):317–324
81. Itoh S, Eguchi T, Kato N, Takahashi S (2014) Radioactive particles in soil, plant, and dust samples after the Fukushima nuclear accident. Soil Sci Plant Nutr 60(4):540–550
82. Gupta DK, Chatterjee S, Datta S, Voronina AV, Walther C (2016) Radionuclides: accumulation and transport in plants. In: Reviews of environmental contamination and toxicology, vol 241. Springer, Cham, pp 139–160. Gupta DK, Walther C (eds) (2014) Radionuclide contamination and remediation through plants. Springer, Berlin
83. Mimura T, Mimura M, Kobayashi D, Komiyama C, Sekimoto H, Miyamoto M (2014) Radioactive pollution and accumulation of radionuclides in wild plants in Fukushima. J Plant Res 127:5–10
84. Hu N, Ding D, Li G (2014) Natural plant selection for radioactive waste remediation. Radionuclide contamination and remediation through plants. Springer, Cham, International Publishing Switzerland, pp 33–53
85. Pulhani VA, Dafauti S, Hegde AG, Sharma RM, Mishra UC (2005) Uptake and distribution of natural radioactivity in wheat plants from soil. J Environ Radioact 79(3):331–346
86. Li G, Hu N, Ding DX, Zheng JF, Liu YL, Wang YD, Nie XQ (2011) Screening of plant species for phytoremediation of uranium, thorium, barium, nickel, strontium and lead contaminated soils from a uranium mill tailings repository in South China. Bull Environ Cont Toxicol 86:646–652
87. Juszak I, Schaepman-Strub G (2015) The impact of vegetation type on the shortwave radiation balance of the Arctic tundra. In: EGU general assembly conference abstracts, vol 17. April, Bibliographic Code: 2015EGUGA.17.9051J. Available at: http://adsabs.harvard.edu/abs/2015EGUGA..17.9051J
88. Donohoea A, Armour KC, Pendergrass AG, Battisti DS (2014) Shortwave and longwave radiative contributions to global warming under increasing CO_2. Proc Natl Acad Sci 111(47):16700–16705
89. UNFCCC (2012) United Nations framework convention on climate change handbook. https://unfccc.int/resource/docs/publications/handbook.pdf
90. Wang K, Liang S (2009) Daytime net radiation from shortwave radiation measurements and meteorological observations. J Appl Meteorol Climat 48:643–648

91. Heo HK, Park CY, Yang B, Lee DK (2018) Heat stress mitigation of urban trees—fine scale assessment based on terrestrial LiDAR data. Geophysical Research Abstracts 20, EGU2018-11878. Available at: https://meetingorganizer.copernicus.org/EGU2018/EGU2018-11878.pdf
92. Kuusk A (1992) Absorption profiles of shortwave radiation in a vegetation canopy. Agric For Meteorol 62(3–4):191–204
93. Galluzzo G (2012) Particulate matter and climate change. University of Iowa. Available at: https://now.uiowa.edu/2012/03/particulate-matter-and-climate-change
94. Shi J, Zhang G, An H, Yin W (2017) Quantifying the particulate matter accumulation on leaf surfaces of urban plants in Beijing, China. Atmos Pollut Res 2017:1042–1309
95. Chen X, Pei T, Zhou Z, Teng M, He L, Luo M, Liu X (2015) Efficiency differences of roadside greenbelts with three configurations in removing coarse particles (PM10): a street scale investigation in Wuhan, China. Urban Forestry and Urban Greening 14(2):354–360
96. Chen X, Zhou Z, Teng M, Wang P, Zhou L (2015) Accumulation of three different sizes of particulate matter on plant leaf surfaces: effect on leaf traits. Arch Biol Sci 67(4):1257–1267
97. Walling DE, Moorehead PW (1989) The particle size characteristics of fluvial suspended sediment: an overview. Sediment/water interactions. Springer, Dordrecht, pp 125–149
98. Ankers C, Walling D, Smith R (2003) The influence of catchment characteristics on suspended sediment properties. Hydrobiologia 494(1–3):159–1616
99. Krein A, Petticrew E, Udelhoven T (2003) The use of fine sediment fractal dimensions and colour to determine sediment sources in a small watershed. CATENA 53(2):165–179
100. Ingold CT (1971) Fungal spores. Clarendon Press, Oxford, p 302
101. Tewari DN (1994) Urban forestry. Indian For 120(8):647–657
102. Bond TC, Doherty SJ, Fahey DW, Forster PM, Berntsen T, DeAngelo BJ, Flanner MG, Ghan S, Kärcher B, Koch D, Kinne S, Kondo Y, Quinn PK, Sarofi MC, Schultz MG, Schulz M, Venkataraman C, Zhang H, Zhang S, Bellouin N, Guttikunda SK, Hopke PK, Jacobson MZ, Kaiser JW, Klimont Z, Lohmann U, Schwarz JP, Shindell D, Storelvmo T, Warren SG, Zender CS (2013) Bounding the role of black carbon in the climate system: A scientific assessment. J Geophys Res Atmos 118:5380–5552
103. Silva RA, West JJ, Zhang Y, Anenberg SC, Lamarque JF, Shindell DT, Collins WJ, Dalsoren S, Faluvegi G, Folberth G, Horowitz LW, Nagashima T, Naik V, Rumbold S, Skeie R, Sudo K, Takemura T, Bergmann D, Cameron-Smith P, Cionni I, Doherty RM, Eyring V, Josse B, MacKenzie IA, Plummer D, Righi M, Stevenson DS, Strode S, Szopa S, Zeng G (2013) Global premature mortality due to anthropogenic outdoor air pollution and the contribution of past climate change. Environ Res Lett 8(3):1–11
104. EEA (European Environmental Agency) (2007) Air pollution in Europe 1990–2004. Report No 2. European Environmental Agency Report No. 2/2007. Available from: http://www.eea.europa.eu
105. Volk HE, Lurman F, Penfold B, Herz-Picciotto I, McConnell R (2013) Traffic-related air pollution, particulate matter and autism. JAMA Psychiatry 70:71–76
106. Gawrońska H, Bakera B (2015) Phytoremediation of particulate matter from indoor air by *Chlorophytum comosum* L. plants. Air Qual Atmos Health 8(3):265–272.
107. Beckett KP, Freer-Smith P, Taylor G (1998) Urban woodlands: their role in reducing the effects of particulate pollution. Environ Pollut 99:347–360
108. Fuzzi S, Baltensperger U, Carslaw K, Decesari S, Denier Van Der Gon H, Facchini MC, Fowler D, Koren I, Langford B, Lohmann U, Nemitz E, Pandis S, Riipinen I, Rudich Y, Schaap M, Slowik JG, Spracklen DV, Vignati E, Wild M, Williams M, Gilardoni S (2015) Particulate matter, air quality and climate: lessons learned and future needs. Atmos Chem Phys 15(14):8217–8299
109. Ban-Weiss GA, Collins W (2015) Aerosols: role in radiative transfer. Encyclopedia of atmospheric sciences, 2nd edn, pp 66–75. https://doi.org/10.1016/b978-0-12-382225-3.00053-0
110. Lagzi I, Mészáros R, Gelybó G, Leelőssy Á (2014) Atmospheric chemistry. Eötvös Loránd University

111. Kim MJ, Yeh SW, Park RJ (2016) Effects of sulfate aerosol forcing on East Asian summer monsoon for 1985–2010. Geophys Res Lett 43(3):1364–1372
112. Ji J, Wang G, Du X, Jin C, Yang H, Liu J, Yang Q, Si N, Li J, Chang C (2013) Evaluation of adsorbing haze PM2. 5 fine particulate matters with plants in Beijing-Tianjin-Hebei region in China. Scientia Sinica Vitae, 43(8):694–699
113. Wang L, Gao S, Liu L, Ha S (2006) Atmospheric particle-retaining capability of eleven garden plant species in Beijing. Chin J Appl Ecol 17:597–601
114. Zhang ZD, Xi BY, Cao ZG, Jia LM (2014) Exploration of a quantitative methodology to characterize the retention of PM2.5 and other atmospheric particulate matter by plant leaves: taking populous tomentosa leaves as an example. J Appl Ecol 25:1–5
115. Prusty BA, Mishra PC, Azeez PA (2005) Dust accumulation and leaf pigment content in vegetation near the national highway at Sambalpur, Orissa, India. Ecotoxicol Environ Safety 60:228–235
116. Freer-Smith PH, El-Khatib AA, Taylor G (2004) Capture of particulate pollution by trees: a comparison of species typical of semi-arid areas (fichus nitride and Eucalyptus globules) with European and North American species. Water Air Soil Pollut 155:173–187
117. Zhang W, Wang B, Niu X (2015) Study on the adsorption capacities for airborne particulates of landscape plants in different polluted regions in Beijing (China). In: Bhamidiammarri R, Tota-Maharaj K (eds) Int J Environ Res Public Health 12(8):9623–9638
118. Li Y, Zhao M, Mildrexler DJ, Motesharrei S, Mu Q, Kalnay E, Zhao F, Li S, Wang K (2016) Potential and actual impacts of deforestation and afforestation on land surface temperature. J Geophys Res Atmos 121(24):14372–14386
119. Wheeler M (2011). Effects of fertilizer on nitrogen cycling in residential cape cod lawns. M.Sc. thesis, Harvey Mudd College, Brown University. http://www.mbl.edu/ses/files/2014/09/Wheeler-Megan.pdf
120. Nowak DJ, Crane D (2002) Carbon storage and sequestration by urban trees in the USA. Environ Pollut 116(3):381–389
121. Ni Y, Eskeland GS, Giske J, Hansen JP (2016) The global potential for carbon capture and storage from forestry. Carbon Balance Manag 11(1):3–8
122. Akbari H (2002) Shade trees reduce building energy use and CO_2 emissions from power plants. Environ Pollut 116:S119–S126
123. USEPA (2017) United States Environmental Protection Agency. Sulfur Dioxide (SO_2) Pollution. https://www.epa.gov/so2-pollution/sulfur-dioxide-basics
124. EPA (Environmental Protection Agency) (2009) Composting. Online at http://www.epa.gov/waste/conserve/rrr/composting/index.htm
125. Johnson J (2014) Methane's role in climate change. Chem Eng News 92(27):10–15
126. Brassard P, Godbout S, Raghavan V (2016) Soil biochar amendment as a climate change mitigation tool: Key parameters and mechanisms involved. J Environ Manage 181:484–497
127. WCD (Wildlife Center Drive) (2007) Gardeners can play an important role in reducing global warming. Source: National Wildlife Federation. Available at: https://www.sciencedaily.com/releases/2007/05/070519084046.htm
128. Favoino E, Hogg D (2008) The potential role of compost in reducing greenhouse gases. Waste Manag Res 26(1):61–69
129. Termorshuizen AJ, Moolenaar SW, Veeken AHM, Blok WJ (2004) The value of compost. Rev Environ Sci Bio/Technol 3(4):343–347
130. University of Minnesota (2006) Grass plant growth and its relationship to lawncare. Sustainable urban landscape information series. Available at http://www.sustland.umn.edu/maint/grasspla.htm
131. Falloon P, Jones CD, Cerri CE, Al-Adamat R, Kamoni P, Bhattacharyya T, Easter M, Paustian K, Killian K, Coleman K, Milne E (2007) Climate change and its impact on soil and vegetation carbon storage in Kenya, Jordan, India and Brazil. Agr Ecosyst Environ 122(1):114–124
132. Zirkle G, Augustin B, Lal R (2009) The potential for soil organic carbon sequestration in home lawns. Presented at the ASA-CSSA-SSSA International Annual Meetings. Hemenway, Toby. Gaia's garden: a guide to home-scale permaculture. Chelsea Green Publishing

133. Zirkle G, Lal R, Augustin B (2011) Modeling carbon sequestration in home lawns. HortScience 46(5):808–814
134. Palmer C (2012) To what extent could planting trees help solve climate change? This article was written by Dr. Charles Palmer of the Grantham Research Institute on Climate Change and the Environment at LSE in collaboration with the Guardian. https://www.theguardian.com/environment/2012/nov/29/planting-trees-climate-change
135. Hasan R, Othman N, Ismail F (2017) Tree species selection in street planting: it's relationship with issues in urban area. Environ Behav Proc J 2(6):185–194
136. Sjöman H, Gunnarsson A, Pauleit S, Bothmer R (2012) Selection approach of urban trees for inner city environments: larning from nature. Arboricult Urban For 38(5):194–204
137. FAO "Food and Agriculture Organization" (2010) Adaptation of forests to climate change. Global Change 11:403–427
138. Hemenway T (2009) Gias's garden. A guide to home-scale permaculture, 2nd edn. Chelsea Green Publishing Company, USA
139. Koriesh EM (2018b) Studies on air pollution 4: evaluation of some ornamental trees and shrubs for reducing atmosphere carbon dioxide. In: The 8th international Arabic conference, Ismailia, Egypt, Ismailia
140. Doherty C (2012) Ficus religiosa—new crop summary and recommendations. New floricultural crops: formulation of production schedules for wild, non-domesticated species. University of Minnesota. https://conservancy.umn.edu/bitstream/handle/11299/143450/Ficus%20religiosa%20Paper%20Colin%20Doherty.pdf;jsessionid=3B3618BE0BB5D32CB95878FD99DC1DE3?sequence=2
141. Padhy PK, Varshney CK (2005) Emission of volatile organic compounds (VOC) from tropical plant species in India. Chemosphere 59(11):1643–1653
142. El-Hadidi M, Boulos L (1989) Street trees in Egypt. American University Press, Cairo, vol 3, pp 337–349
143. Shaltout K (2004) An updated flora of Egypt. Divers Distrib. https://onlinelibrary.wiley.com/doi/full/10.1111/j.1472-4642.2004.00065.x
144. Dunnett N, Kingsbury N (2008) Planting green roofs and living walls. Timber Press Inc., Oregon, USA
145. Safikhani ST, Abdullah AM, Ossen DR, Baharvand M (2014) Thermal impacts of vertical greenery systems. Environm Climate Technol 14(1):5–11
146. Mwendwa M (2016) Vertical gardens technology to help reduce climate change stress. Clim Change. http://www.talkafrica.co.ke/vertical-gardens-technology-to-reduce-climate-change-stress/
147. Blanc P (2008) The vertical garden. From nature to the city. W. W. Norton and Company Inc., New York
148. Manso M, Castro-Gomes J (2015) Green wall systems: a review of their characteristics. Renew Sustain Energy Rev 41:863–871
149. Susorova I (2015) Green facades and living walls: vertical vegetation as a construction material to reduce building cooling loads. Eco-efficient materials for mitigating building cooling needs, design, properties and applications. Elsevier Ltd., Amsterdam, pp 127–153
150. Koriesh EM, Moghazy EI, Abdel Salam AM, Abd El-Mo'omen A (2018) Raising the efficiency of some ornamental plants to get rid of formaldehyde. In: The 8th international Arabic conference, Ismailia, Egypt
151. Fassden G (2018) Wall greening. Available at: https://www.fassadengruen.de/en/wall-greening.htm
152. Koriesh, EM (1998) Ornamental horticulture (translated), 2nd edn. King Saud University Publications. 19/0818 ISBN:9960-05-759-3. Original text: Prentice-Hall, Inc. Englewood Cliffs
153. Sharp R, Sable J, Bertram F, Mohan E, Peck S (2008) Introduction to Green Walls: technology, benefits & design. Green Roofs for Healthy Cities, Toronto. Available at: http://greenscreen.com/docs/Education/greenscreen_Introduction%20to%20Green%20Walls.pdf

154. Littlewood R (2008) Living roofs and walls-technical report: supporting London plan policy. Published by Greater London Authority, p 12. Available at: https://www.london.gov.uk/sites/default/files/living-roofs.pdf
155. Turgeon N, Le Bihan Y, Buelna G, Bourgault C, Verreault S, Lessard P, Nikiema J, Heitz M (2011) Application of methanotrophic biofilters to reduce GHG generated by landfill in Quebec City (Canada). In: Brebbia CA, Longhurst JWS, Popov V (eds) Air pollution XIX (transactions on ecology and environment), vol. 147. WIT PRESS, UK
156. McPherson EG (1994) Cooling urban heat islands with sustainable landscapes. In: Platt Rutherford H, Rowntree Rowan A, Muick Pamela C (eds) The ecological city: preserving and restoring urban biodiversity. University of Massachusetts Press, Amherst
157. Chavarria G, Santos HP (2012) Plant water relations: absorption, transport and control mechanisms. In: Montanaro G (ed) Advances in selected plant physiology aspects. InTech. Available from: https://www.intechopen.com/books/advances-in-selected-plant-physiology-aspects/plant-water-relations-absorption-transport-and-control-mechanisms
158. Raizada P, Singh A, Raghubanshi AS (2009) Comparative response of seedlings of selected native dry tropical and alien invasive species to CO_2 enrichment. J Plant Ecol 2(2):69–75
159. Dukes JS, Mooney HA (1999) Does global change increase the success of biological invaders? Trends Ecol Evol 14:135–139
160. Keenan TF, Prentice IC, Canadell JG, Williams CA, Wang H, Raupach M, Collatz G (2016) Recent pause in the growth rate of atmospheric CO_2 due to enhanced terrestrial carbon uptake. Nat Commun 7:13428
161. Nowak DJ, Kuroda M, Crane DE (2004) Tree mortality rates and tree population projections in Baltimore, Maryland, USA. Urban For Urban Green 2:139–147
162. Hasegawa S, Macdonald CA, Power SA (2016) Elevated carbon dioxide increases soil nitrogen and phosphorus availability in a phosphorus-limited Eucalyptus woodland. Glob Chang Biol 22(4):1628–1643
163. Lotfiomran N, Köhl M, Fromm J (2016) Interaction effect between elevated CO_2 and fertilization on biomass, gas exchange and C/N ratio of European Beech (*Fagus sylvatica* L.). Plants 5(3):38–50
164. Wang L (2016) Assessment of benefits of urban forests under the ecosystem services framework: a literature review. B.Sc. dissertation. The University of British Columbia, Bachelor of Science in Forest Resource Management, Faculty of Forestry. The University of British Columbia (Vancouver)
165. Whitehead D (2011) Forests as carbon sinks—benefits and consequences. Tree Physiol 31(9):893–902
166. Binkley D, Campoe O, Gspaltl M (2013) Light absorption and use efficiency in forests: why patterns differ for trees and stands. For Ecol Manage 288:5–13
167. Charlton A, Hamilton I (2008) Tree planting: a key weapon against global warming. Carbon Posit. Available at: http://forestindustries.eu/de/content/tree-planting-key-weapon-against-global-warming
168. Koriesh EM (2001) Studies on air pollution. 2. Effects of greenbelts. In: The Fifth Arabian Horticulture Conference, Ismailia, Egypt, vol 2, pp 315–320
169. Koriesh EM, Moghazy EI (2013) Studies on indoor air pollution. 9- air ions, oxygen evolution & air pollution tolerant index (APTI) of some indoor plants. In: The seventh international conference for horticulture, Held at Suez Canal University from 24–26 March 2013. Hortsci J Suez Canal Univ 1:339–34
170. Koriesh EM, Abd El-Fattah YM (2008) Studies on air pollution. 3: Evaluation of some shrubs to their oxygen, negative air ions evolutions and air pollution tolerance index in relation to some fertilization treatments. Agric Res J 8:63–68
171. Khan FI, Abbasi S (2000) Attenuation of gaseous pollutants by greenbelts. Environ Monit Assess 64:457–475
172. Ruth AE, William RE (1994) The encyclopedia of the environment. The Rene Dubor Centre for Human Environments, Houghton Miffin Company, U.S.A

173. Larkham A, Adams R (2011) The post-war reconstruction planning of London: a wider perspective. Planning Birmingham School of the Built Environment, Birmingham City University. ISBN 978-1-904839-54-5. http://www.bcu.ac.uk/cmsproxyimage?path=/_media/docs/cesr_working_paper_8_2011_larkham_adams.pdf
174. Hill AC (1971) Vegetation: a sink for atmospheric pollutants. J Air Pollut Control Assoc 21:341–346
175. Lin DA (1976) Air pollution-threat and responses. Adison Wesley Publishing Company, London
176. Smith WH (1990) Forest as sinks for air contaminants: vegetative compartment. Air pollution and forests: interactions between air contaminants and forest ecosystems, 2nd edn. Series on environmental management. Springer, New York, pp 147–180. https://link.springer.com/content/pdf/10.1007%2F978-1-4684-0104-2.pdf
177. Rawat JS, Banerjee SP (1996) Urban forestry for improvement of environment. J. Energy Environ Monitor 12(2):109–116
178. William HS (1990) Interaction between air contaminants and forest ecology. In: Air pollution and forest, 2nd edn. Springer, New York, pp 618
179. Kabel RL, O'Dell RA, Taheri M, Davis D (1976) A preliminary model of gaseous pollutant uptake by vegetation. Center for Air Environment Studies, Publ. No. 455–76
180. Earth talk, Social Sciences (2018) Which trees offset global warming best? https://www.thoughtco.com/which-trees-offset-global-warming-1204209

Livestock, Fishery, and Aquaculture

Potential Climate Change Impacts on Livestock and Food Security Nexus in Egypt

Tarik S. K. M. Rabie

Abstract Climate on earth has changed ordinarily amid the presence of our planet, going from the ice ages to times of warmth. During the most recent quite a few years increments in average air temperatures have been accounted for and associated consequences for atmosphere have been debated widely in an assortment of discussions. Because of its significance around the world, agriculture was one of the first divisions to be contemplated as far as potential effects of climate change. The agriculture was the key advancement in the ascending of current human civilization, whereby developing of domesticated species (Poultry and ranch creatures) produced sustenance which is fundamental for the practicality of the development. Nowadays, great awareness and concern for problems related to environmental quality, and human welfare, which are propagating at standalone terms subjected every day for debates and discussions such as climate change, biodiversity, soil fertility corrosion and beyond all food quality and pollution. Domesticated animals are the real consumers of water yet additionally support a considerable number of pastoralist and farmers' families. In Egypt where water is a scarce item, there is a requirement for strategies to enhance livestock productivity. The complexity of the problems raised a combined with the heating of the global which called "Heat Stress" as a result of global climate changes, and the unlikelihood about numerous essential information regularly make discourses uncertain; even indication issued by scientific specialists are now and then misleading, and the problems are exacerbated by the continuous influence of ideological positions. A great net interaction gathering impact of climate change on soil, water, population growth, and energy, where more administrative and non-legislative tasks expected to supply farmers with high-quality seeds, high-quality forage with sufficient amount for animals, and veterinary administrations. In this way, thinking about the decent variety of farming frameworks, one rural strategy won't fit all farmers' classifications. Additionally, administrative and non-legislative administrations ought to be contemplated by the district and the season.

Keywords Climate change · Farm animal · Poultry · Genetic resources

---

T. S. K. M. Rabie (✉)
Poultry Breeding & Genetics, Animal Production and Fish Resources Department, Faculty of Agriculture, Suez Canal University, Ismailia 41522, Egypt
e-mail: tarik.rabie@agr.suez.edu.eg

© Springer Nature Switzerland AG 2020
E.-S. E. Omran and A. M. Negm (eds.), *Climate Change Impacts on Agriculture and Food Security in Egypt*, Springer Water, https://doi.org/10.1007/978-3-030-41629-4_17

1 Introduction

Egypt is the second national communication and will be a typical example of a developing country that is very susceptible to global climate change. The final climate of Egypt is dry, hot, with a gentle winter season with rain over the coastal areas, and a hot and dry summer season that exposes it to numerous threats to its economic, social and environmental property. Global climate change has a general trend towards warming of the air temperature, with an increase of humidity in the hazy days, misty days, the turbidity of the atmosphere, regularity of hot days and sand storms [1]. Also, expected to feature to the growing threats to national security fueled by a growing population of regarding 104 million, of that almost 56.7% unit of measurement a rural population, and 43.3% unit of measurement urban population to keep with applied analysis for 2017 [2]. The population is calculated to realize 120 million by 2020, and then the growth rate anticipated to be unchanged. Regarding 97% of the population in Egypt lives on the Nile River depression (11 km^2), and so the Delta which about of 29,000 km^2, and represents regarding 4% of Egypt's total area [3, 4]. Incredible population intensification associated with harboring an enlarged demand on the already irregular resource base. Egypt's security threats which can be extra tormented by the impacts of global temperature change on animal and agriculture in general as well as water security, food security and impoverishment gap, and energy security also. The increasing of human population related to loss of agricultural land because of urbanization processes and alteration water accessibility concomitant with global climate change create quiet opposes to agriculture [5]. Additionally, the drought and salinity are the foremost vital environmental factors that triggered the decreasing growth and productivity of plants [6]. Consequently, increasing water productivity is a fundamental strategy to extend food production underneath conditions of restricted water system [7]. In Egypt, the forage crops play crucial dependability in farm animal production. Wherever, the forage crops are used in different features (Green fodder, hay, and silage) characterized by the high feed conversion ratio to achieve the highest net profit. Consequently, the water use efficiency had the highest correlation with each contemporary and dry forage yields [8]. In this chapter, many attempts that discussed are summarized at Fig. 1. Which illustrate the interactions between and among the energy, water, land, and climate systems take place within a social and economic context.

2 Climate Change Impact

Having explored the term environment and its implications, this part presents the importance of water resources values to externalities related to environmental impacts.

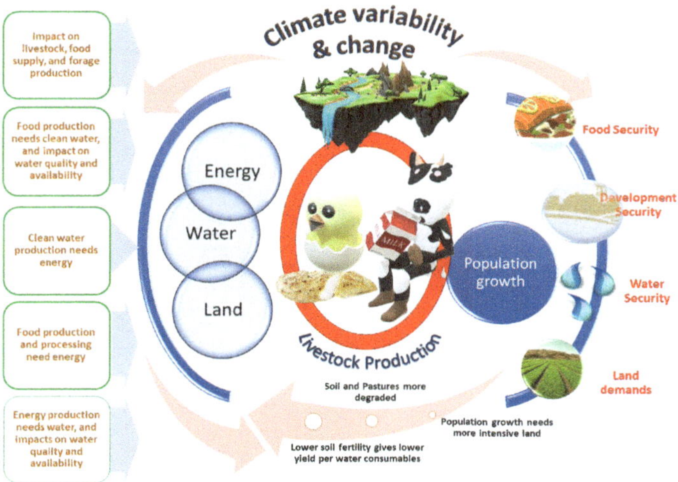

Fig. 1 Livestock production, water, land, energy, and climate change interactions

2.1 *Impact on Water Resources*

Water is equal to lifespan protection. It can be used for several purposes such as agriculture, industry, fisheries forestry, recreation, domestic needs, transport, and so on. Also, alternative uses can become challenging uses if there is a shortage of water to satisfy them all [9]. Competition in water resources increasingly occurs among all sections of agriculture activities, rural, urban, and industrial. Therefore, it is critical to saving and optimizing on-farm water resources.

The Nile River is considering the major source of water supply in Egypt, which provisions over 95% of the country's annual water needs. The changeable climate could cause a significant variability in Nile flood rate. The high growth rate of population size, the increasing of water demands for their life, therefore fresh water per capita is estimated to drop to 550 m^3 by 2020 (well below the water poverty line of 1000 m^3 per capita). Approximately 66% of Africa is classified as arid or semiarid, with extreme variability in precipitation [10]. Over four-hundredth of Africa's population lives within the arid, semiarid, and dry sub humid areas wherever demand for water and different ecosystems services is on the increase [11, 12]. Droughts throughout the past 3 decades and land degradation at the desert margins, significantly the desert, have raised considerations regarding increasing desertification [11]. The entire nature of this drawback and therefore the degree to that human activities and global climate change area unit conducive there to area unit still being determined. In Egypt, the Nile basin is identified as one hot spot where there will be increased water stress in the future. The surface territory of the Nile basin is around 3 million km^2. There are ten riparian countries "Burundi, Vote based Republic of Congo, Egypt, Eritrea, Ethiopia, Kenya, Rwanda, Sudan, Tanzania, and Uganda".

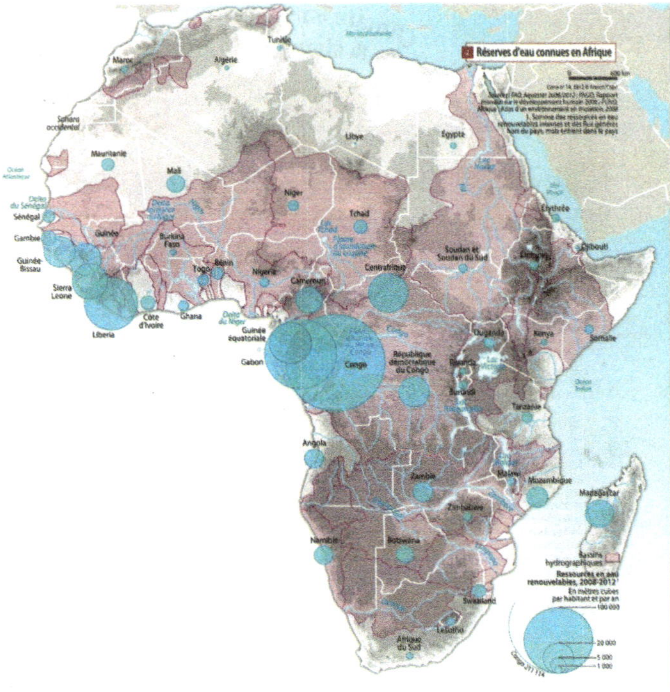

Fig. 2 Renewable water resources in Africa. *Source* Pinterest, Explore water resources and more. https://www.pinterest.com

Add up to rainfall is roughly about 2000 billion m^3 [13]. One of the basic difficulties is that upstream or riparian utilization of water influences downstream users, and how is it conceivable to share a basin's water in an equitable manner (Fig. 2). Egypt is experiencing physical water shortage, i.e., limited water sources accessible within its borders and its dependency on outside water sources [14].

The nonviolent water resource in Egypt is Aswan high dam (111 m high, with a crest length of 3830 m) and a volume of 44,300,000 m^3, impounds a reservoir, Lake Nasser, that has a gross capacity of 169 billion m^3 (BCM). Of the Nile's total annual discharge, 74 BCM of water have been allocated by treaty between Egypt and Sudan[1] while the annual freshwater supply in Egypt is about 64 BCM/year. About 55.5 BCM apportioned to Egypt and the remainder to Sudan[1] (about 85% are received from the Ethiopian Highlands via the Blue Nile and about 15% from the White Nile), in addition to about 1.3, and 7 BCM of rainfall and groundwater respectively [15].

[1]In 1929, there was a treaty between Egypt and Britain securing the water share of Egypt and Sudan (that was part of the Egyptian Kingdom) to be about 84 BCM. In 1959, the 84 BCM were divided between Egypt and Sudan, where Egypt received 55.5 BCM and Sudan 18.5 BCM and about 10 BCM were estimated as the annual water loss due to evaporation.

Table 1 Water resources in Egypt

	1998–2002	2003–2007	2008–2012	2013–2017
Long-term average annual precipitation in volume (10^9 m^3/year)	51.07E (2002)	51.07E (2007)	51.07E (2012)	51.07E (2014)
National Rainfall Index (NRI) (mm/year)	106.6E (2002)			
Surface water produced internally (10^9 m^3/year)	0.5 (2002)	0.5 (2007)	0.5 (2012)	0.5 (2014)
Groundwater produced internally (10^9 m^3/year)	1.3 (2002)	1.3 (2007)	1.3 (2012)	1.3 (2014)
Total internal renewable water resources (IRWR) (10^9 m^3/year)	1.8 (2002)	1.8 (2007)	1.8 (2012)	1.8 (2014)
Total internal renewable water resources per capita (m^3/inhab/year)	25.38 K (2002)	23.19 K (2007)	21.01 K (2012)	19.67 K (2014)
Surface water				
Entering the country (total) (10^9 m^3/year)	84 (2002)	84 (2007)	84 (2012)	84 (2014)
Inflow submitted to treaties (10^9 m^3/year)	84 (2002)	84 (2007)	84 (2012)	84 (2014)
Inflow secured through treaties (10^9 m^3/year)	55.5 (2002)	55.5 (2007)	55.5 (2012)	55.5 (2014)
Accounted inflow (10^9 mm^3/year)	55.5 (2002)	55.5 (2007)	55.5 (2012)	55.5 (2014)
Total external renewable (10^9 m^3/year)	55.5 (2002)	55.5 (2007)	55.5 (2012)	55.5 (2014)
Groundwater				
Entering the country (total) (10^9 m^3/year)	1 (2002)	1 (2007)	1 (2012)	1 (2014)
Accounted inflow (10^9 m^3/year)	1 (2002)	1 (2007)	1 (2012)	1 (2014)
Water resources				
Total external renewable (10^9 m^3/year)	56.5 (2002)	56.5 (2007)	56.5 (2012)	56.5 (2014)
Total renewable surface water (10^9 m^3/year)	56 (2002)	56 (2007)	56 (2012)	56 (2014)

(continued)

Table 1 (continued)

	1998–2002	2003–2007	2008–2012	2013–2017
Total renewable groundwater (10^9 m^3/year)	2.3 (2002)	2.3 (2007)	2.3 (2012)	2.3 (2014)
Total renewable water resources (10^9 m^3/year)	58.3 (2002)	58.3 (2007)	58.3 (2012)	58.3 (2014)
Total renewable water resources per capita (m^3/inhab/year)	822.2 K (2002)	751.2 K (2007)	680.6 K (2012)	637.1 K (2014)
Exploitable				
Irregular renewable surface water (10^9 m^3/year)	49 (2002)	49 (2007)	49 (2012)	
Regular renewable groundwater (10^9 m^3/year)	0.7 (2002)	0.7 (2007)	0.7 (2012)	
Total exploitable water resources (10^9 m^3/year)	49.7 (2002)	49.7 (2007)	49.7 (2012)	
Total dam capacity (km^3)	168.2 L (2000)	168.2 L (2005)	168.2 L (2008)	168.2 L (2015)
Dam capacity per capita (m^3/inhab)	2372 K (2000)	2167 K (2005)	1964 K (2008)	1838 K (2015)

Source AQUUASTAT (http://www.fao.org/nr/water/aquastat/main/index.stm)

In contrast, both of rainfall and groundwater are limited and therefore, cannot be depended upon which makes Egypt highly dependent on the Nile River for all service activities which affect the economy. Moreover, 80% of freshwater from the Nile is used by Agriculture activity only. Table 1 summarize freshwater resources in Egypt. Most of the dams under construction are concentrated in the developing countries [16]. The construction of substantial dams on global rivers does not just influence the country inside which the dam is built yet impacts rise above the limits to downstream countries and their communities likewise, particularly those relying upon freshwater for rural productions.

The development of the Fantastic Ethiopian Renaissance Dam (GERD) with water stockpiling limit of 74 BCM to create 6000 megawatts (MW) of low-cost energy on the Nile River is required to influence Sudan and Egypt as downstream countries and their rural communities. Water diminishment, particularly amid the dam repository filling period, is expected to intensify water deficiency in Egypt and like this influence agricultural production, domesticated animals' production, and the jobs of the livelihoods of the rural poor. Moreover, creating lands in the Nile Delta confront the danger of immersion from sea level rise [17, 18]. This may altogether diminish the agricultural production of Egypt since it is evaluated that 30–40% of Egypt's agricultural production is from the low-lying zones of the Delta and drift that are vulnerable to sea level rise [19]. Likewise, many studies have been launched by the collaborations between Alexandria University (AU), and the Coastal Research Institute of the National Water Research Centre. The general point was to improve Egypt's versatility and diminish weakness to environmental change impacts. A combination of spatial planning and enhanced coastal defenses to manage the risk of sea level rise (SLR) and storm surge is needed. Proactive adaptation requires more effective partnerships among different stakeholders to facilitate the transition towards safe urban development in large port cities and to implement effective disaster management in the event of flooding [20].

2.2 *Agricultural Water Withdrawal*

The annual measure of self-provided water is pulled back for irrigation, domesticated animals and cultivation purposes. It can grasp water from essential inexhaustible and secondary freshwater resources, yet as water from over-deliberation of sustainable groundwater or withdrawal from fossil groundwater, coordinate utilization of agricultural drainage water, coordinate utilization of (treated) sewer water, and desalinated water. Water for the dairy farm and meat industries, and the modern procedure of agribusiness is encased underneath mechanical water withdrawal. Table 2 demonstrate water withdrawal in Egypt in different aspects.

Table 2 Water withdrawal in Egypt

	2008–2012 (2010)
Agricultural water withdrawal (10^9 m^3/year)	67
Industrial water withdrawal (10^9 m^3/year)	2
Municipal water withdrawal (10^9 m^3/year)	9
Total water withdrawal (10^9 m^3/year)	78
Agricultural water withdrawal as % of total water withdrawal (%)	85.9
Industrial water withdrawal as % of total water withdrawal (%)	2.564
Municipal water withdrawal as % of total withdrawal (%)	11.54
Total water withdrawal per capita (m^3/inhab/year)	910.6 K
Fresh groundwater withdrawal (primary and secondary) (10^9 m^3/year)	8.045
Total freshwater withdrawal (primary and secondary) (10^9 m^3/year)	73.8
Desalinated water produced (10^9 m^3/year)	0.2
Direct use of treated municipal wastewater (10^9 m^3/year)	1.3

Source AQUASTAT (http://www.fao.org/nr/water/aquastat/)

2.3 Water Demand for Agriculture, Pasture and Forage Crops

More than 76 billion m^3/year are estimated to be required for all socioeconomic sectors [21]. There was a gap between freshwater resources and Egypt's requirement. In the year 2000, Egypt considered as below the water poverty line. The only solution for this risk is increasing the freshwater resources from conventional sources, which is improbable. In the meantime, nonconventional sources could be a hope such as desalination or cloud seeding which are also infeasible and costly. As the result of incredible increase of the population, the water requirements of different sectors increase rapidly as well, ambitious agricultural expansion, and elevated living standards. In 2014, total water withdrawal was estimated at 76 billion m^3 which include 62.35 billion m^3 for agriculture (83.3%), 9.95 billion m^3 for municipalities (12.8%), 1.2 billion m^3 for industries (1.6%), and 2.5 billion m^3 loss and evaporation from Nile and canals (3.3%). So that the agricultural sector is considered to be the significant high freshwater consumables. The prediction of future water requirements will be accelerated with a combined effect of rapid population growth and an increase of food demand and living standards [22, 23], as the results, increase of meat and milk consumption causing need more water resources to adequate the forage crop cultivation.

Potential negative impact on water availableness and access to and demand for water might occur in most countries attributable to global climate change and its variabilities. Where, global climate change can have comprehensive magnitudes for livestock production, that preponderantly result from its impact on rainfall patterns that confirm each the amount and quality of grassland and land post-productivity

[24]. Overall, the diminished amount and quality of accessible water might occur attributable to adverse global climate change influence on water resources and freshwater ecosystems [25]. Increasing heat stress as a result of temperature change can significantly increase water necessities for a farm animal. Heat stress will ofttimes intensify water issues, in hot dry areas resulting in overgrazing that ultimately culminate in a rapid runoff in these areas resulting in flooding. Frequent droughts may be a reason behind concern in terms of sickness and parasites distribution and transmission, aside from the physical losses to a farm animal. As a result of temperature change, all water resources can dry up as a result of extreme temperatures, and farm animal production is going to be severely hampered in such cases [26]. Further, the drying of water resources can produce a state of affairs wherever livestock have to be compelled to walk long distances in search of water, making extra stress to those animals. Hence, it is attending to be an enormous challenge for livestock researchers across the world to develop appropriate ways to confirm access to water for livestock production.

3 Agriculture and Food Security

The agriculture sector is enormous water-consumer where, 90% of the land currently used for agriculture system which is characterized by sustenance but with low yields of both crops and farm animal's production. Food crops production has been expanding over the last decade, but food production faced of production shortage which doesn't cover local demands in the Nile Basin, so, all net food has to be imported. To enhance provincial vocations and upgrade food security in the area, it will be important to enhance the efficiency of this productivity framework by presenting water and soil conservation strategies, giving quality seeds, and empowering the utilization of composts. Administration objective ought to be to expand both soil infiltration rate and the water-holding ability to increase the accessible water in the soil for crop development. Infiltration will result in improvements in the quality of natural ecosystems and water in aquatic ecosystems [27, 28]. To enhance production, and subsequently rustic employments, interests in water administration are the main point to open the potential in rainfed farming. Also, populace and food securities are the two fundamental components prompting the expansion and the requirement for more developed agriculture in both rainfed and irrigated agribusiness.

Food security, as defined by the United Nations' Committee on World Food Security, is "The condition in which all people, at all times, have physical, social and economic access to sufficient safe and nutritious food that meets their dietary needs and food preferences for an active and healthy life" [29]. Over the coming decades, an evolving atmosphere, increasing of the worldwide populace, rising nourishment costs, and environmental stressors will have extremely unverifiable impacts on sustenance security (see Fig. 1). Rising temperatures and expanded recurrence of a high degree of dry and wet years are relied upon to moderate advance toward expanded profitability of yield and domesticated animal's frameworks and enhanced food security as well. Adjustment procedures and strategy reactions to worldwide climate change, including alternatives for taking care of water portion, land utilize designs, sustenance exchange, post-harvest nourishment handling, and food costs and security are desperately required. These strategy reactions will be fundamental to enhance the living states of smallholders moreover, country populaces over the globe.

3.1 Land Degradation and Desertification

Even though the Aswan high dam is exhaustive control of the Nile's yearly flooding, it has created a few negative impacts, nonetheless, highest of which is a progressive diminishing in the richness and subsequently the efficiency of Egypt's riverside agricultural lands. Aswan high dam is considered as one of the biggest dams on the planet, it is 2142 m in length and is penetrated by 180 floodgates that in the past passed the entire Nile flood, with its overwhelming heap of sediment. In like manner, a significant part of the flood and its heap of rich fertilizing silt is presently impounded in reservoirs and canals; the sediment is consequently never again kept by the Nile's rising waters on farmlands. Egypt's yearly use of around 1 million tons of counterfeit manures is a deficient substitute for the 40 million tons of sediment earlier kept every year by the Nile flood [30].

3.2 Growth and Development

The farming area is an awesome significance to the Nile Bowl countries particularly Egypt since it is a noteworthy supporter of total national output (Gross domestic product, GDP), business, and nourishment security. Gross product per capita is demonstrating that there are expanding of improvement in the Gross domestic product in Egypt from 2011 to 2016 (Fig. 3).

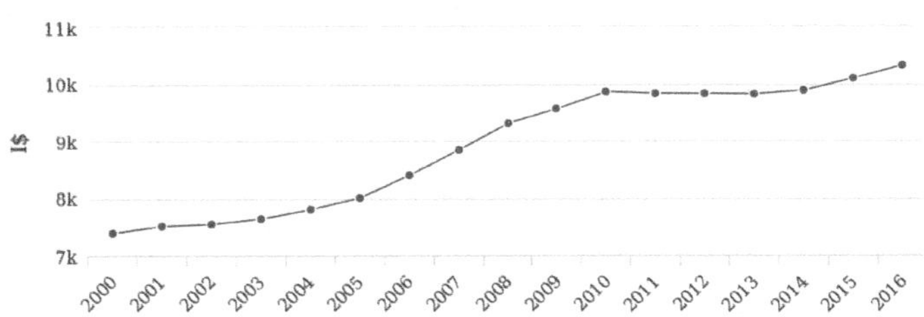

Fig. 3 Gross domestic product (GDP) per capita, PPP (constant 2011 international $). *Source* FAOSTAT [2]

The effects of environmental change at the Nile basin are not uniformly dispersed—the poorest individuals will endure soonest and most. Environmental change is a grave danger to the creating scene and a noteworthy hindrance to proceeded with destitution decrease crosswise over numerous measurements that is claiming the developing regions are as of now hotter, by and large, than developed districts, and they likewise experience the ill effects of high precipitation changeability. Subsequently, developed nations are intensely subject to horticulture, the most atmosphere delicate of every monetary part, and experience the ill effects of lacking wellbeing arrangement and low-quality open administrations. Low salaries and vulnerabilities roll out adjustment to atmosphere improvement remarkably difficult. Because of these resistances, climate change is probably going to lessen advance effectively low wages and increment ailment and passing rates in creating nations. Falling farm livelihoods will build neediness and decrease the capacity of family units to put resources into a superior future, constraining them to go through pitiful reserve funds just to survive. At the national level, environmental change will cut salaries and raise spending needs, disturbing open finances.

3.3 Food Security and Poverty

Extended masses population and upstream country headway will impact the volume of water in the Nile accomplishing Sudan and Egypt and their nourishment security. A severe factor to environmental change is the advancement of the human population, foreseen to be around 35% in the vicinity of 2006 and 2050. In developing countries, the consumption rate of meat and dairy products is brought down per-capita contrasted with the developed countries where consumers regularly watch meat and dairy product as a part of an unrivaled eating regimen and a predominant life and have not yet been instructed about their negative impacts (Fig. 4). In any case, meat and dairy analogs can yield by long shot unrivaled outcomes, particularly if they are publicized with such purpose. To enhance country vocations and upgrade nourishment security in the district, it will be vital, in this way, to enhance the profitability of this cultivating framework through presenting water and soil preservation strategies, giving quality seeds, and empowering the utilization of manures. From the viewpoint of water administration, mediations to increment horticultural profitability ought to incorporate projects to build water collecting, grow flooded territories, enhance the water-maintenance properties of soil in the upstream countries, and enhance profitability and water-utilize efficiency in the downstream countries. Additionally, administration objective ought to be to augment both soil infiltration rate and the water-holding ability to build the accessible water in the soil for crop development. This infiltration will bring about changes ecosystem and of water in aquatic biological systems [27, 28]. Recently, environmental change is a characterizing challenge of our circumstances. Its effect will be worldwide, sweeping and generally irreversible. Atmosphere change is now expanding the risk of presentation to hunger and malnutrition and food insecurity among the poorest and most defenseless individuals.

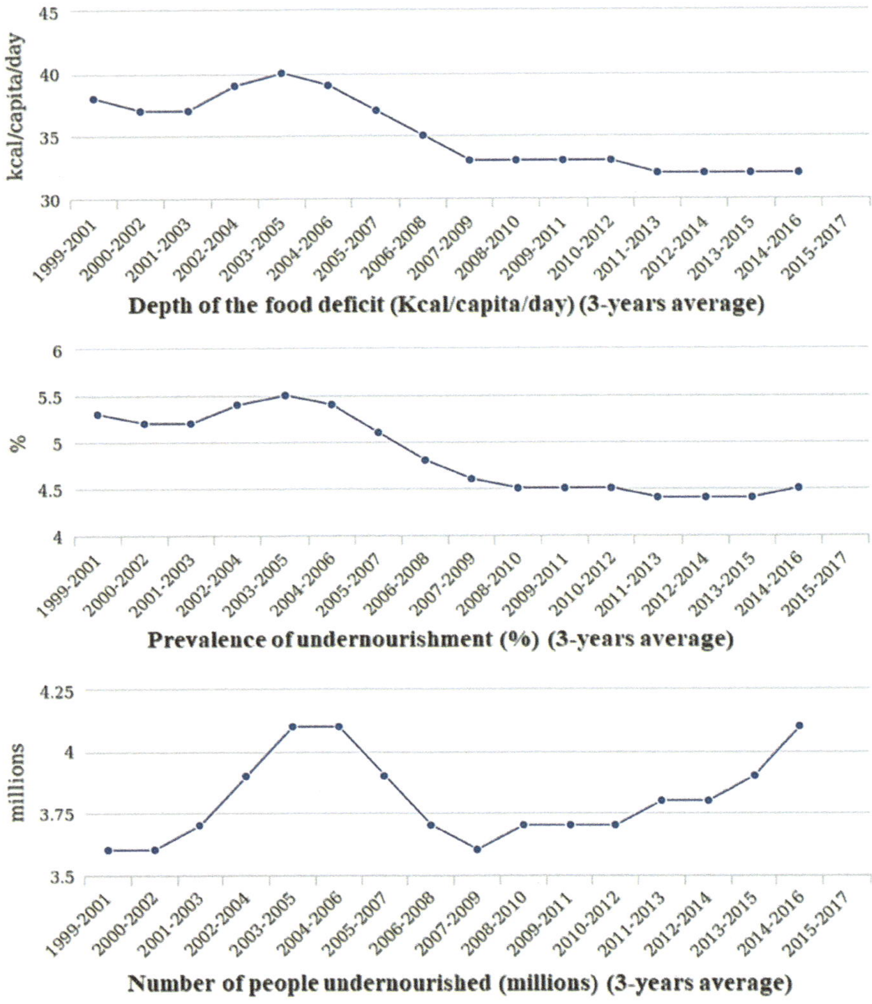

Fig. 4 Effect of climate change nexus on poverty in Egypt. *Source* FAOSTAT [2]

Natural disasters are winding up more regular and extraordinary, land and water are becoming more rare and hard to access and increments in agricultural profitability are ending up harder to accomplish.

4 Livestock Production

In this part addresses in detail the effect of environmental change on domesticated animals' generation. This part involves five principle items covering the immediate effects of environmental change on poultry and farm animal's development, milk and veal production, and its genetic resources in Egypt. This part additionally talks about extravagantly on the aberrant effects of environmental change on animals, water and forage accessibility, and other feed resources. Moreover, this part features the significance of various inborn components by which animals adjust to the evolving the climate variability.

4.1 *Poultry Production Sector*

In value terms, 26% of Egypt's total livestock products came from poultry meat and egg production, and Egypt's livestock sector contributed 27% of total domestic agricultural production in 1999 [31].

In the 1990s, Egypt's 8.6% annual growth rate of poultry meat production exceeded the world average of 5.2% and was much higher than other middle-income countries such as Saudi Arabia (5.2%), Turkey (3%), and Colombia (4.1%). In comparison, the U.S. annual growth rate was 4.3%, and the EU's growth rate was 3.3% during the 1990s. China, Argentina, and Brazil, where poultry meat production grew at an annual rate between 9 and 12.5% in the 1990's, exceeded Egypt's annual growth rates (Fig. 5).

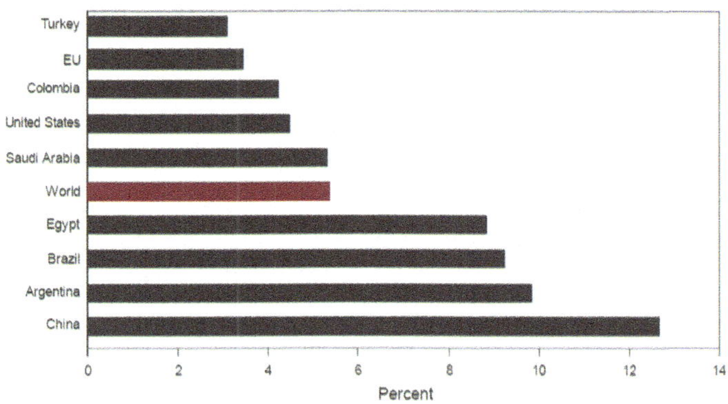

Fig. 5 Average annual poultry production growth rates in selected middle-income countries compared with the United States, EU, and world average; 1990–2000. *Source* Economic Research Service/USDA

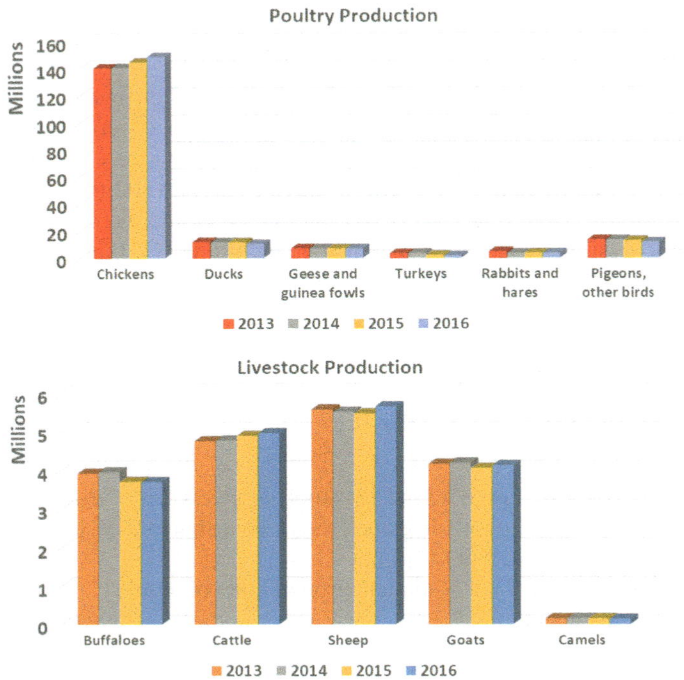

Fig. 6 Egypt's total meat production, by type; 2013–2016. *Source* FAOSTAT [2]

In 2016, poultry meat production, at 148,517 metric tons, exceeded all other meats, including beef and veal, buffalo meat, sheep and goat meat, camel meat, and others (Fig. 6).

Broiler and layer chickens produced in tropical atmospheres battle to perform in high temperature and mugginess conditions. These fowls show outer indications of heat stress; including surprised breathing, wing spreading, execution misfortune and once in while mortality. With these undeniable indications of heat stress come more subtle signs, for example, profound body temperature increment and the biochemical and physiological changes that happen, therefore. At the point when profound body temperature is lifted for prolonged periods, biochemical and physiological changes harm essential body organs, for example, the heart, lung, liver and kidney. The heat stress impacts on poultry production are summarized in Fig. 7.

4.1.1 The Impact of Climate Change on Poultry (Case Study)

The term for climate change includes both low and high temperature, the humidity percentage, Co_2 emissions, oxygen shortage, etc. The description of the case study can be illustrated regarding pulmonary hypertension syndrome in chicken (PHS)

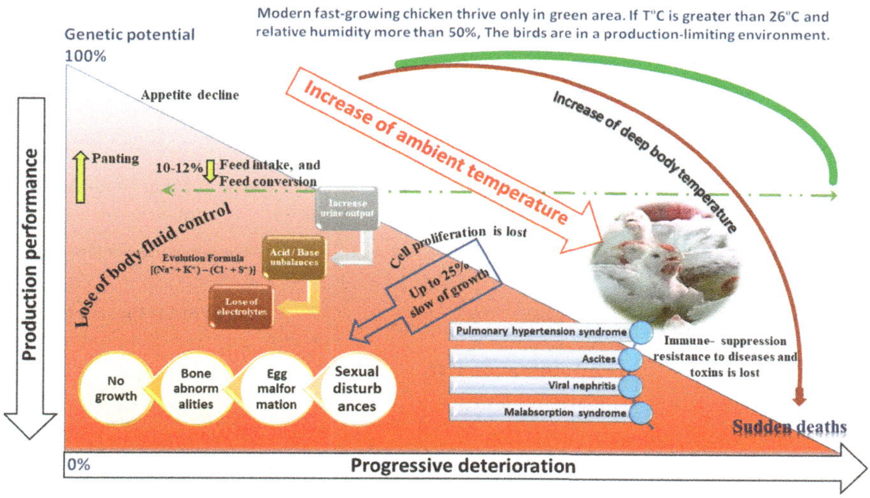

Fig. 7 The effect of heat stress on poultry production performance

which may result from physiological and/or environmental factors [32]. The initial event can come in many forms, but the ultimate factors can be divided to external environmental factors, for example, altitude above sea level, frosty temperature and hatching condition, and internal factors, for example, physiological changes can be separated into numerous general classifications [33]. Figure 8 showed the progression of events and factors contributing to ascites syndrome in broilers. Management also plays a significant role in ascites development. Feed limitation, lighting, air quality and ventilation have all been ensnared in ascites advancement. Also, hereditary inclination assumes an impressive part in the frequency of the PHS [34].

What Increases Oxygen Demand in Broiler Chickens?

The essential internal factors that outcome in an insufficient oxygen supply are rapid growth and proficient feed efficiency. This has brought about broilers with such a high rate of digestion that its heart and lungs are scarcely fit for sufficiently giving oxygen to support the body. The increased body demand for oxygen prompts an increase in cardiac output because the same amount of oxygen is needed by the body. Therefore, the body starts producing more red blood cells to increase the oxygen carrying capacity. This increase in the number of erythrocytes resulted in a concomitate increase in hematocrit [35, 36], and blood viscosity and caused the difficulties to be pumped through the lungs. Also, found that hematocrit in ascitic broilers increased after exposure to cold and that the increase became significant approximately two weeks before death [37].

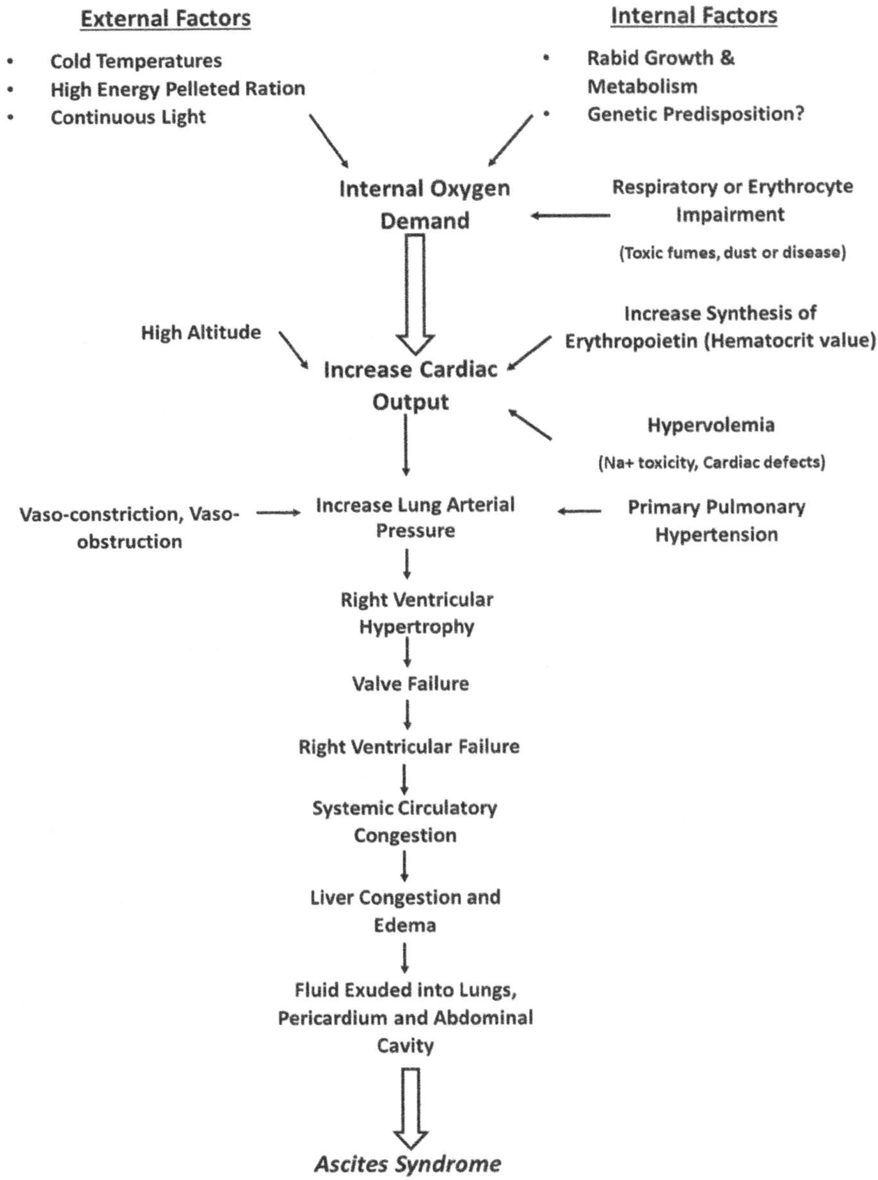

Fig. 8 Progression of events and factors contributing to ascites syndrome in broilers

External cold temperatures, coupled with high energy diets, can result in ascites-related deaths because of insufficient pulmonary vascular capillary capacity for the blood flow necessary to meet a high metabolic oxygen requirement. Because of the right ventricle of birds has developed as a volume pump, not as a pressure pump, since the heart rarely had to respond to changes in pressure [38], hypertrophy of the right ventricle will, in the end, result in diminished effectiveness of the privilege atrio-ventricular valve. Valvular disappointment will cause the reverse of blood from the right ventricle to the atrium, net enlargement of the right ventricle and the right ventricle failure, orderly circulatory congestion, trailed by liver congestion. This has been well documented in humans with liver cirrhosis or renal failure [39]. The link between cold stress and ascites incidence soon led to investigations of exact temperature effect on metabolism and chicken oxygen requirements. The oxygen consumption by broilers strains which selected for either feed efficiency or weight gain was increased 43–59% by lowering the temperature [40].

Environment and Management

Besides the effect of cold temperature to introduce the PHS in chickens, the effect of high altitude (either natural or simulated) is apparent to be the 2nd indirect reason for PHS induced via decreasing in partial pressure of oxygen. The equivalent percentage of oxygen drops approximately 1.0% for every 500 m rise in elevation [41]. Significant microscopic damage to the heart, lungs and kidneys of the birds reared at high altitude have been noticed. These microscopic changes also were seen in one-week-old broilers raised at high altitude [35], and in birds exposed to simulated high altitude. When the PHS was first observed, it was seen mainly in areas of higher altitude and fast-growing broilers although it now occurs at all altitudes. In susceptible lines, hypoxic conditions amid hatching of the chicken egg have been appeared to cause an expansion in the occurrence of the PHS in the grown-up chicken. Unfortunately, good management and environmental practices that encourage high feed intake and rapid growth tend to exacerbate ascites. Although the PHS is not fully understood, this heart failure is thought to be a consequence of cold temperature a combined with a lowered oxygen tension in poultry houses.

4.2 Large Animal Production

In Egypt, foreign agricultural service (FAS-Cairo) figures calf production in 2018 to increment to 1.93 million head, up 4% from 1.85 million head in 2017. The government is seeking after expanded local meat production. The Egyptian crowd's shortage of heifers, including more young female calves, held for multiplication is because of ranchers benefiting from solid local domestic beef demand driving prices. Smallholder ranchers likewise move animals to slaughter rapidly to maintain a strategic distance from exorbitant costs on vaccinations and costly sustains. Post characteristics the expansion in overall calf numbers to the government's enhanced disease control measures and inoculation programs, and it is setting minimum slaughter weights. The legislature is additionally controlling the cost of some animal feed input costs [42]. It's clear that after the animal exposure to heat stress not only decreases the milk yield in dairy animals, but it has dramatic effects on body weight in beef cattle because of decreasing of feed intake and feed conversion and increasing of water consumption. In the following parts, the effect of high temperature on the domestic animal will be discussed.

4.2.1 Livestock Production Chains, Impact on Rural Society

Crop/livestock framework is the fundamental framework at the worldwide scale, gives around 75% of dairy, 60% of meat and up to half of cereals production [43]. Milk production in little scale gives rural work and cash money while in the meantime helping in enhancing, strengthening and balancing out agricultural production framework [44]. In Egypt, crop-livestock production framework is considered as the principal farming system including around 95% of the bovine population (dairy cattle and buffaloes). Crop-livestock production framework gives around 75% of milk production at the national scale, while 25% delivered by business ranches [45]. Moreover, livestock is considered as security assets influencing access to informal credits and loans to aid smallholders in crises and events (like bride price payments). In numerous smallholder frameworks in developing countries, manure is considered as critical as milk and meat. Likewise, buffalo is considered as the backbone of the farmer's financial aspects of in 40 countries, notably Egypt [46]. In Egypt, water buffaloes are the principal source for milk production and contribute over 70% of the yearly milk production [47]. The buffalo population was around 3.5 million head in the year 2016 [2]. Around 42% of the buffalo population is dairy cows, 32% heifers, 6% bulls and 20% male calves. Although the farmers are spending a lot of effort, time and investment to produce milk or meat, and they receive a small proportion of the profit—which considered one of the threats to the whole production system sustainability—,livestock is considered as security assets influencing access to informal credits and loans to aid smallholders in crises and events (like bride price payments).

(a) Dairy Production Chains, and Climate Impacts

Dairying is often part of a mixed farming system in which manure is used for fertilizing cash crop production. Dairy animals are fed on different feed stuff such as grass, crop residues and cultivated grains. Supplementary feeding is practiced only when feasible. In Egypt, the traditional system is characterized by small-herd size less than 50 head of the animal, less than 5 feddans and keep native and crossbred animals. The feeding system relies upon Egyptian clover in winter and maize and sorghum in summer, supplemented with a constrained amount of fixing ingredients such as corn, wheat bran, soybean meal, and so on.) or concentrate feed mixture. This framework gives around 75% of milk production on the national scale [45].

Milk production in small scale farms enhances the sustenance security of family members, as well as makes a few work opportunities along the dairy chain [48]. There were different types of chains, first and regular one called short chain and abundance at towns which there were no merchants for buffalo milk which might be in charge of a couple of buffaloes raised to cover just family consumption, and the rest was sold to neighbors and customers in town without mediators amongst makers and purchasers. On the other hand, the long chain included producers, traders, milk shops, milk processing units and consumers. In this kind of chain, farmers were gotten lower prices in a regular payment (weekly payment). Moreover, more jobs opportunities offered along the chain. Also, the price of milk is increased regarding the distance between the production area and milk marketing area. Both of chains depending on dairy cattle, nature food resource, clean water, and sustainable environments. Opposite of the climate stability effect on production, the main effect of preeminent high temperature on milk production is most likely most injurious for livestock production system where, which animals must be forced to reduce feed intake, resulting in lowered milk yield [49]. The heat stress not just reductions the milk yield yet it remarkably influences the quality of milk moreover. Furthermore, not only a high temperature is the main key, but the humidity is also an important aspect of controlling milk yield in the animals. The reduction of milk production can range from 10 to >25%. As much as 50% lessening in milk yield can be due to reduced feed intake during heat stress, and the other 50% might depend on heat-related lactogenic hormone fluctuations. Besides, the low milk yield production is also dependent upon breed, the age of the animal, stage of lactation, and forage availability [50]. Moreover, the heat stress resulted in decreasing of the percentage of both protein and fat in milk and colostrum. In the hot atmosphere, the milk composition goes to down for fat and protein percentage, solid not fat (SNF), and total solid % in dairy cows [51]. The milk freezing point is also affecting by heat stress because of decreasing of the percentage of lactose and acidity in the milk. Also, milk produced by the animal which exposed to heat stress has lower calcium (Ca), phosphorus (P), and magnesium (Mg) content and higher on chloride [52]. Correspondingly, fatty acid content conveys the heat-stress by a significant deficit of the proportion of short-chain and medium-chain (C4–C10, and C12–C16 respectively), while remaining or increasing the long-chain fatty acid (C17–C18) in milk [52]. The reasons for these changes in the fatty acid chain contents may be due to either by reducing of synthesis of these free fatty acids (FAA) in the mammary glands or by destructive energy of the cow revealed to heat stress. Moreover, heat stress negatively affects α- and β-casein in the milk which will prompt expand the pH of the milk and lower P content [53].

(b) Veal Production Chain, and Climate Impacts

In the old lands (OL) which called Nile Valley, the mixed crop-livestock systems were described as dominance, and in the newly reclaimed lands (NRL) in the west Delta. This system relies on the allocation of lands between fodder and food crops; green fodder (Clover, horse grass and corn) and grain (corn) square measure the most parts of the feed ration of the herd, and crop residues from food and cash crops. Wheat bran as an industrial byproduct is intensively utilized in each OL, and NRL because it is reasonable price and its more substantial volume of the same unit weight of other alternatives like grains. In most cases, manure is directly applied to the land; and veal value used to finance annual expenses like rent land or animal investments, it is easy to recognize that the OL is distinguishable to the NRL. So, the animal charge per ha is between 3 and 11 animal/ha OL, compared to 2–3 animals/ha in the NRL. These variations will affect the needs of each group and each region for services and development. Newly reclaimed land may need balance services for both livestock and other agriculture activities. Old land needs more low-cost services to be sustained and very tight vaccination and biosecurity programs to protect the main source of income and the main protein source for families. Many farmers in new land tended to increase the number of livestock especially cattle to deal with the high fluctuation of vegetables and fruits price in the local market, also to provide daily income and high-quality organic fertilizer for land which will decrease the need for buying more chemical fertilizers. Most of the buffalo breeders sold young calves (veal) to save milk and to get a quick source of money. Cattle breeders, mainly breeders of Baladi (native) cattle, prefer to valorize the veal. Sometimes, certain use of the crossbred or Baladi cows are to suckle buffalo veal, which will decrease the cost of rearing buffalo calves and increasing the revenues of selling more buffalo milk. The marketing decision between veal and milk can vary from one year to another according to family needs. For land renters, they usually rear and sell their calves to cover the cost of renting land. More generally, most of them were depending on milk as family food and as a daily cash income [54].

It is realized that domesticated animals that are presented to high surrounding temperatures enlarge the endeavors to disperse body warm, bringing about the expansion of banting, body temperature, water consumption and a decrease in feed consumption [43]. Aside from feed consumption, sustain transformation additionally significantly diminishes after the presentation to warm pressure. Exposure of the animal to a high natural temperature stimulates the peripheral thermal receptors to transmit suppressive nerve driving forces to the craving focus in the hypothalamus and in this way causes an abatement in feed intake [55]. That is might be due to animals adapt themselves to produce a low amount of body heat. The decrease in feed intake could be due to the adaptive mechanism of the animal to produce less body heat. Animal growth is the expansion in the live weight or cell duplication. It is controlled by both genetically and environmentally [55]. Increased of encompassing temperature is thought to be one of the natural components influencing growth and average daily weight gain in domesticated animals. The explanation behind the impacts of raised surrounding temperature on growth lessening could be because of the reduction in anabolic action and the expansion of tissue catabolism [55]. The expansion in tissue catabolism could be credited to the increment in catecholamines and glucocorticoids after exposure to heat stress in domesticated animals.

5 Livestock Genetic Resources

Among the domesticated avian species, customary breeds developed for meat or egg production are a colossal source of genetic variety. Such breeds result from several times of determination by farmers and breeders, consolidating an assortment of unique genes and gene combinations. In like manner, exceptional selection for odd and interesting traits brought about the variety of favor breeds treasured by specialists. The moderately little size of most poultry species (especially the chicken and Japanese quail), alongside their marvelous regenerative limit and genetic variability, make them especially appropriate to multi-generational examinations with thorough selection or extensive levels of inbreeding. This has fortified the improvement of numerous imperative genetic resources of chicken, turkey, and quail by scientists in agricultural departments at public universities or in agricultural research institutions. These particular genetic stocks have contributed broadly to explore in fundamental, biomedical, and rural sciences. To date, no formal preservation plan exists for these stocks at any authoritative level. Rather, protection of these imperative avian genetic resources is generally individually dependent, here and there associated through breed affiliations or scholarly circles, yet to a great extent as free endeavors. This absence of a formal conservation design is especially risky to inquiring about genetic resources. Not at all like the more broadly scattered conventional or interest breeds, particular research stocks are regularly kept at just a single location, which makes them significantly more powerless against extinction.

Specifically, when such stocks can never again be kept up by their curating specialist, they are regularly seen as a superfluous deplete on decreasing financial assets at that institution and might be dispensed with at short notice. So that, it is vital to discuss around the avian and livestock genetic diversity in Egypt serving many individual attempts to reach the goal of keeping indigenous animal far from the risk especially with the circumstances of severe climate change.

5.1 Avian Genetic Diversity

Poultry is an important enterprise in that it contributes to farm families' income, nutrition and social obligations. In Egypt, indigenous chicken produces 44% of meat products [56]. These indigenous chickens have not been characterized, and therefore long-term breeding plans become difficult to organize. Through evolution and long-term exposure, some indigenous chicken may be suited to the tropical conditions and local management practices. Tropically relevant genes, e.g. the necked neck gene in heat stress can be an important contribution of the indigenous chicken. Some indigenous chicken in Africa is genetically tolerant to some endemic diseases and parasites, e.g. Marek's disease, coccidiosis, and climatic stresses. Moreover, indigenous chickens are generally better in disease resistance and could maintain a higher level of performance under poor nutrition and high environmental temperatures compared to commercial strains under village systems. An increasing loss of genetic diversity has been observed for all agriculturally used species, and more than half of common livestock breeds especially poultry are now endangered or at risk of extinction [57]. Agriculturally important genetic stocks include those selected for various production-related characteristics. These include egg production, body shape, feed-use efficiency, leg strength, and disease resistance. Ironically, these selected stocks that are particularly vulnerable when funds become limiting, because improvements in production-related traits require many years and a relatively large number of birds each year. With the increasing attention given to the characterization of the African genetic resource [58]. The first phase of characterization involves the identification of populations based on morphological evaluations. Morphometric measurements have been used to evaluate the characteristics of various Egyptian local strains of chickens [59], and production and some physiological aspects as well. Recently, Molecular markers have proven to be superior to the conventional methods for determining gene flow in chicken stocks. Also, information on hereditary variety can give essential contribution to design successful strategies for preservation and restoration of characteristic local breeds. The generation of the apparatuses in advance of sequencing (molecular markers, linkage maps, BAC libraries, and so on.) was being started to preserve the accessible unrivaled germplasm of the Egyptian indigenous chickens utilizing different molecular methods [60–62]. Even though the local breeds contain genes germane to their adaptation to specific environments and local breeding objectives, the local chickens confront genetic disintegration which may prompt the loss of profitable genetic variability in particular traits. Like this, indigenous chicken has continuously been replaced and substituted by intensive industrial stocks, undermining genetic power and diversity. Besides, Policies to protect the local species will require more scientific contributions on the genetic structure in its local dissemination extend. The assumption is that exotic breeds are better. On the other hand, this is not generally the case.

5.2 Animal Genetic Diversity

The primary main purpose in domesticated animals' developments in Egypt is to give animal protein to take care of the growing demand in the short run. This has been done regularly with little thought for environment and supportability factors. In this setting, local breeds are infrequently wrongly thought of as something of the past. In Egypt, the normal lactation drain yield in Friesian dairy animals is less of what 33% of its genetic potential. Exotic breeds have an important role in expanding national production, yet their utilization must be supported, and the expansion of their genetics must be under control. The Egyptian dairy cattle had eight breeds: two of them were vanished or reduced (African aurochs and Hamitic longhorn) while the other six breeds are Baladi, Domiette, Egyptian, Maryuti, Menufi, and Saidi, are yet present. Conversely, the Egyptian animal breeders mentioned that the Egyptian local cattle are just a one breed called Baladi. Indigenous population may have distinctive names, however without evident contrasts in phenotype; an adjustment in phenotype may happen without change in name, or all population may have only one name and be exact phenotypically. The outcomes of molecular research gave a confirmation that all Baladi cattle brought up in Egypt have genetic similarity and could be considered as one breed [63], these results comparable with which detailed that Egyptian cows have numerous breeds [2]. The strategy of cattle genetic improvements in Egypt have included crossbreeding with exotic high production dairy cattle breeds, for example, Holstein, Brown Swiss, Friesian and Simmental [64], notwithstanding Abondance and Tarentaise as double reason French breeds. Tarentaise dairy cattle, plunges from the French Savoie cows, was decided for its flexibility to intense conditions. It makes great feed conversion, withstands temperature variety, good adaptation to hot climate, and classified as an easy-calving breed. Finally, no genuine or virtual institutional structures exist in Egypt whose primary concern is animal genetic resource administration responsible for breed characterization, development or improvement, and finally breed conservation.

6 Summary

Regardless of the way that the phenomenon of worldwide atmosphere changes, and the local effects, it is normal that Egypt would be one of the countries most influenced by the impacts of climate change. These impacts are reflected in high temperatures, changing rainfall pattern, rising sea levels, and the expanding recurrence of atmosphere related disasters, which posture dangers to farming, agriculture land, water supply, and food security. Agriculture land is a constrained that ought to be utilized effectively and in a way that keeps up its esteem with the goal that it can produce an adequate amount of food, fuel and forage in both the short and long term. To enhance rural life and enhance the food security in the district, it will be fundamental, subsequently, to enhance the efficiency of this agricultural framework through presenting water and soil preservation procedures, providing quality seeds, and empowering the utilization of manures. From the viewpoint of water administration, intercessions to increment agricultural profitability ought to incorporate projects to extend irrigated areas, enhance the water-maintenance properties of soil, and enhance profitability and water-utilize efficiency. To enhance production, and accordingly, farmers, population and food security are the two primary components prompting the expansion and the requirement for more developed agribusiness in both rainfed and irrigated farming. To avoid poverty, a mix of animal product from animals that have a good feed conversion (e.g. cattle and chickens) which is great for saving the land's esteem and adds to effective land utilize. The benefit of choosing local and regional animal products is that they contribute to supporting local farming and keeping the agricultural landscape. This can be a particularly strong argument in regions with a small share of agricultural land. Another benefit is that transport for e.g. the distribution of meat and milk can be minimized. Transport can also be limited through a more

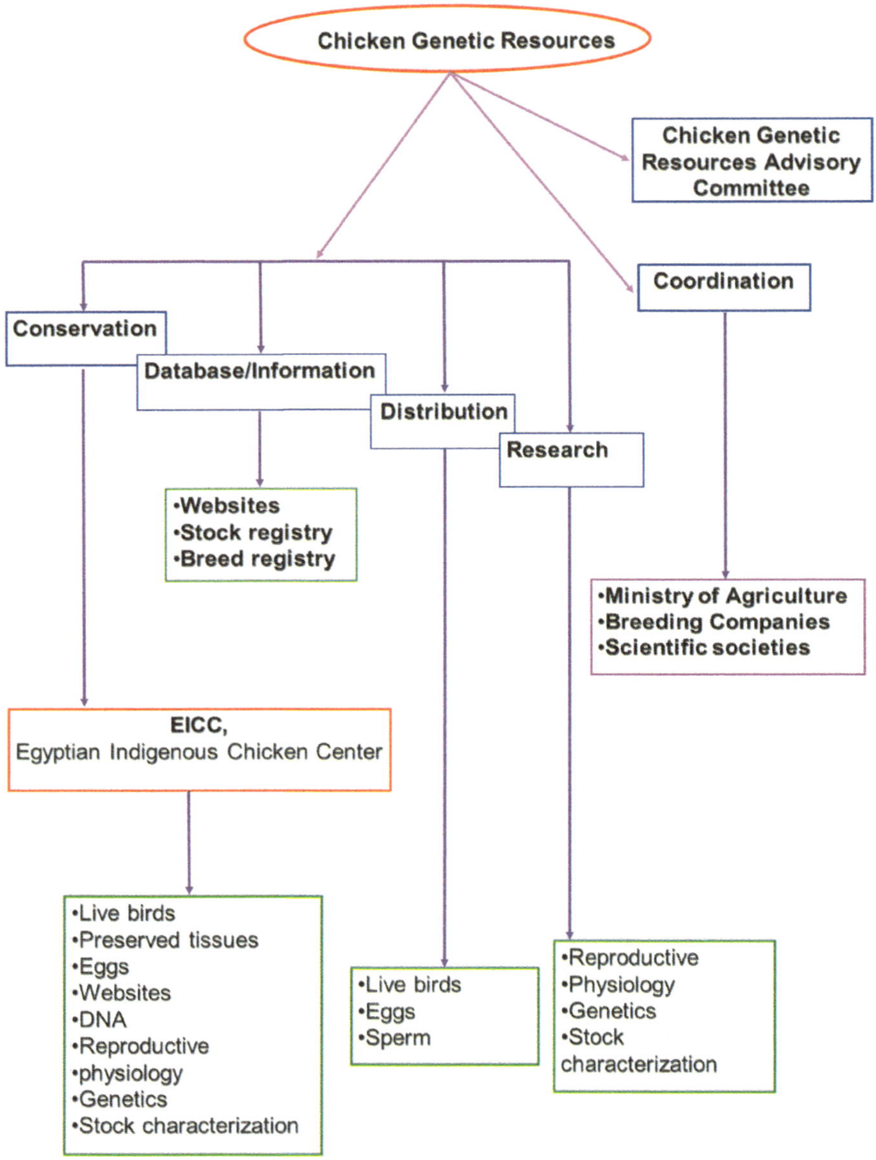

Fig. 9 Components of the Egyptian chicken genetic resources system

significant proportion of locally and regionally produced feed and more even distribution of animal and feed production and favors a more even balance between farm animals and poultry production within domestic agriculture.

7 Recommendations

Although the local breeds contain genes and alleles pertinent to their adaptation to a particular environments and local breeding goals, the local chickens face genetic erosion which may lead to the loss of valuable genetic variability in specific characteristics. Furthermore, policies to protect the species will need more scientific inputs on the genetic structure of the species in its native distribution range. Selection and production of disease and parasite resistant chicken, for an example, will assist to decrease the dependency on drugs and medication which will reduce chemicals in animal products and the environment and good animal welfare. Creating of the Egyptian Indigenous Chicken Center "EICC" will be recommended, which will help the rural community to be able to buy cockerels from superior indigenous strains for upgrading their stock. Furthermore, it is expected that breeding and multiplication center of breed stock "EICC" will produce seed stock for different species of birds that will be distributed to farmers in the villages. This will directly contribute to improved village poultry production and diversity, resulting in increased animal protein intake among smallholders. The incorporates poultry production within existing farming systems, then ensure a sustainable contribution to stable food security. The smallholder will subsequently profit by the combination of various species nutritiously, socially and economically, this will also lead to sustainable conservation, management, improvement and utilization of indigenous poultry genetic resources (Fig. 9).

References

1. Egyptian Environmental Affairs Agency (EEAA). http://www.eeaa.gov.eg/. Accessed on 20 Dec 2017
2. Faostat. http://www.fao.org. Accessed on 6 Feb 2018
3. Aly AA (2007) The suggested cropping patterns for desert lands on the context of groundwater limitation. In: Eleventh international water technology conference, IWTC11. Sharm El-Sheikh, Egypt
4. El Arabi NE (2012) Management of coastal aquifers and groundwater. The legal and institutional framework for coastal aquifers in Egypt. UNESCO-IHP-Sub-component 1.1
5. Mittler R, Blumwald E (2010) Genetic engineering for modern agriculture: challenges and perspectives. Annu Rev Plant Biol 61:443–462
6. Naz N, Hameed M, Ahmed MSA, Ashraf M, Arshad M (2010) Is soil salinity one of the major determinants of community structure under arid environments? Community Ecol 11:84–90
7. Rosegrant MW, Cai X, Cline SA (2002) World water and food to 2025: dealing with scarcity. International Food Policy Research Institute, Washington, DC

8. Mohamed MM, El-Nahrawy MA, Abdu MA, Shams SA (2013) Development of fodder resources in Sinai: The role of forage crops in agriculture development. North Sinai-Governorate, Egypt
9. Chevassus-Au-Louis B, Salles J-M, Pujol J-L, Bielsa S, Richard D, Martin G (2009) An economic approach to biodiversity and ecosystem services—contribution to public decision making # La Documentation francaise—Paris, June 2009. Water resource and challenges
10. UNEP (2002) Africa environment outlook: past, present and future perspectives. United Nations Environment Programme. UK, Earthprint
11. IFOAM Website. http://www.ifoam.org/organic_facts/food/. Accessed on 27 Jan 2018
12. IFOAM, International Federation of Organic Agriculture Movement (2004) Organic farming what is it? http://www.ifoam.bio/. Accessed on 17 Dec 2017
13. New M, Lister D, Hulme M, Makin I (2002) A high-resolution data set of surface climate over global land areas. Clim Res 21(1):1–25
14. Karajeh F, Oweis T, Swelam A, El-Gindy A, El-Quosy D, Khalifa H, El-Kholy M, El-Hafez SA (2011) Water and agriculture in Egypt. Technical paper based on the Egypt-Australia. ICARDA workshop on on-farm water-use efficiency. ICARDA, Egypt
15. CAPMAS (2015) Statistical year book. CAPMAS, Cairo
16. Zarfl C, Lumsdon AE, Berlekamp J, Tydecks L, Tockner K (2015) A global boom in hydropower dam construction. Aquat Sci 77(1):161–170
17. Intergovernmental Panel on Climate Change (IPCC) (2007) Climate change 2007: the physical science basis. Cambridge University Press, Cambridge
18. Loutfy NM (2010) Reuse of wastewater in Mediterranean region, Egyptian experience. In: Barcelo D, Petrovic M (eds) Waste water treatment and reuse in the Mediterranean region. Springer, Berlin
19. Darwish K, Safaa M, Momou A, Saleh SA (2013) Egypt: land degradation issues with special reference to the impact of climate change. Combating desertification in Asia, Africa and the Middle East. Springer, Netherlands, pp 113–136
20. Elshinnawy IE, Borhan M, ElRaey M, Dougherty B, Fencl A (2010) Climate change risks to coastal development and adaptation options in the Nile Delta. Retrieved 2018 from http://www.nile-delta-adapt.org/index.php?view=DownLoadAct&id=4
21. Hoff H, Falkenmark M, Gerten D, Gordon L, Karlberg L, Rockström J (2009) Greening the global water system. J Hydrol 56:199–211
22. El Bedawy R (2013) Water resources management: alarming crisis for Egypt. J Manag Sustain 4(3):2014. ISSN 1925-4725 E-ISSN 1925-4733
23. Zaghloul SS (2013) Consideration of the agricultural problems as a base of water resource management in Egypt. In: Seventeenth international water technology conference, IWTC17. Istanbul, 5–7 Nov 2013
24. Assan N (2014) Possible impact and adaptation to climate change in livestock production in Southern Africa. IOSR J Environ Sci Toxicol Food Technol 8(2):104–112
25. IPCC (Intergovernmental Panel on Climate Change) (2013) Impacts, adaptation and vulnerability. http://ipcc-wg2.gov/AR5/images/uploads/WGIIAR5-TS_FGDall.pdf
26. Rust JM, Rust T (2013) Climate change and livestock production: a review with emphasis on Africa. S Afr J Anim Sci 43(3):255–267
27. African Water Atlas (2010) United Nation Environment Programme. ISBN:978-92-807-3110-1
28. IWMI 2009. http://www.iwmi.cgiar.org/Publications/CABI_Publications/CA_CABI_Series/Rainfed_Agriculture/Protected/Rainfed_Agriculture_Unlocking_the_Potential.pdf
29. FAO World Food Summit [Internet] Rome (Italy) FAO (1996) Available from: http://www.fao.org/docrep/003/w3613e/w3613e00.htm
30. Britannica (2018) https://www.britannica.com/topic/Aswan-High-Dam. Accessed on 6 Feb 2018
31. CAPMS (Central Agency for Public Mobilization and Statistics) (2000) Statistical yearbook. Egypt, Cairo
32. Currie RJW (1999) Ascites in poultry: recent investigations. Avian Pathol 28:313–326

33. Balog JM (2003) Ascites syndrome (pulmonary hypertension syndrome) in broiler chickens: are we seeing the light at the end of the tunnel? Avian Poult Biol Rev 14(3):99–126
34. Decuypere E, Buyse J, Buys N (2000) Ascites in broiler chickens: exogenous and endogenous structural and functional causal factors. World Polt Sci J 56:367–377
35. Maxwell MH, Spence S, Robertson GW, Mitchell MA (1990) Haematological and morphological response of broiler chickens to hypoxia. Avian Pathol 19:23–40
36. Wideman RF Jr, Wing T, Kirby YK, Forman MF, Marson N, Tackett CD, Ruiz-Feria CA (1998) Evaluation of minimally invasive indices for predicting ascites susceptibility in three successive hatches of broilers exposed to cool temperatures. Poult Sci 11:565–1573
37. Luger D, Shinder D, Rzepakovsky V, Rusal M, Yahav S (2001) Association between weight gain, blood parameters, and thyroid hormones and the development of ascites syndrome in broiler chickens. Poult Sci 80:965–971
38. Akester AR (1984) The cardiovascular system. In: Freeman BM (ed) Physiology and biochemistry of the domestic fowl, vol 5. Academic Press, London, p 215
39. Henriksen JH, Siemssen O, Krintel JJ, Malchow-Moller A, Bendsten F, Ring-Larsen H (2001) Dynamics of albumin in plasma and ascitic fluid in patients with cirrhosis. J Hepatol 34:53–60
40. Jones GPD (1994) Energy and nitrogen metabolism and oxygen use by broilers susceptible to ascites and grown at three environmental temperatures
41. Julian RJ (2000) Physiological, management nad environmental triggers of the ascites syndrome: a review. Avian Pathol 29:519–527
42. Global Agriculture information network (GAIN) Report (2017) Ali abdi 2017: strengthening Egyptian pound to ease beef prices. Consumption and imports to rise in 2018
43. Herrero M, Thornton PK, Notenbaert AM, Wood S, Msangi S, Freeman HA et al (2010) Smart investments in sustainable food production: revisiting mixed crop-livestock systems. Science 327:822–825
44. Hahlani CD, Garwi J (2014) Operational challenges to smallholder dairy farming: the case of Mayfield dairy settlement scheme in Chipinge district of Zimbabwe. IOSR J Human Soc Sci (IOSR-JHSS) 19:87–94
45. Abdel Aziz A, Sadek RR (1999) Policy issues in the dairy sub-sector. Workshop on production, processing and marketing policy issues in the dairy sub-sector, 46 p
46. Singh B, Gauhan SK, Chauhan MS, Singla SK (2015) Textbook of animal biotechnology. The Energy and Resources Institute (TERI), New Delhi, India, 678 p
47. Arefaine H, Kashwa M (2015) A review on strategies for sustainable buffalo milk production in Egypt. J Biol Agricult Healthcare 5:63–67
48. Hemme T, Otte J (2010) Status of and prospects for smallholder milk production—a global perspective. FAO, Rome, p 185
49. Dunn RJH, Mead NE, Willett KM, Parker DE (2014) Analysis of heat stress in UK dairy cattle and impact on milk yields. Environ Res Lett 9:064006, 11 pp
50. Ernabucci U, Calamari L (1998) Effect of heat stress on bovine milk yield and composition. Zootec Nutr Anim 24:247–257
51. Ozrenk E, Inci SS (2008) The effect of seasonal variation on the composition of cow milk in Van Province. Pak J Nutr 7:161–164
52. Bernabucci U, Basiricò L, Morera P (2013) Impact of hot environment on colostrum and milk composition. Cell Mol Biol 59(1):67–83
53. Kume S, Takahashi S, Kurihara M, Aii T (1989) The effects of a hot environment on the major mineral content in milk. Jpn J Zootech Sci 60:341–345
54. Ngongoni NT, Mapiye C, Mwale M, Mupeta B (2006) Factors affecting milk production in the smallholder dairy sector in Zimbabwe. Livestock Res Rural Dev 18(05). http://www.cipav.org.co/lrrd/lrrd18/5/ngon18072.htm
55. Marai IFM, El-Darawany AA, Fadiel A, Abdel-Hafez MAM (2007) Physiological traits as affected by heat stress in sheep—a review. Small Rumin Res 71:1–12
56. Economic Research Service/USDA website. www.ers.usda.gov. Accessed on 17 Feb 2018
57. Hoffmann I (2005) Research and investment in poultry genetic resource—challenges and options for sustainable use. World's Poult Sci J 61(10):57–70

58. Osei-Amponsah R, Kayang BB, Naazie A (2012) Age, genotype and sex effects on growth performance of local chickens kept under improved management in Ghana. Trop Anim Health Prod 44:29–34
59. Kosba MA, Abd El-Halim HAH (2008) Evaluation of the Egyption local strains of chickens. Egypt Poult Sci 28:1239–1251
60. Rabie TSK (2016) Conducting and interpretation a cluster analysis for enhancing phylogenetic trees using chicken taxa. In: 9th International Poultry Conference Proceedings, Hurghada, Egypt
61. Rabie TSK (2012) Protein pattern divergence between three of Egyptian endogenous chicken strains and Rohde Island Rod. In: 3rd Medetranian poultry summit and 6th International conference for poultry sciences. PortoMarina, Egypt Porto Marina, Egypt
62. Rabie TSK, Abdou A (2010) Genetic diversity and relationship among Egyptian indigenous chicken strains using random amplification of polymorphic DNA. Egypt Poult Sci 30:473–482
63. Hassanane MS, El-Ashmaoui HM, Abd El-Baset S, Mabrouk MMS, Abdel-Hameid ZG (2006) Preliminary results on genetic diversity in Egyptian cattle using microsatellite markers. Egyptian J Anim Prod 43(1):11–24
64. MoDAD (2004) Secondary guidelines for development of national farm animal genetic resources management plans, measurement of domestic animal diversity: recommended microsatellite markers. Retrieved from http://www.fao.org/3/a-aq569e.pdf. Accessed on 20 Dec 2017

Influence of Climate Changes on Animal Feed Production, the Problems and the Suggested Solutions

Salah A. Attia-Ismail

Abstract Climate change is a long-term change in the earth's climate, especially a change due to an increase in the average atmospheric temperature. Climate change demonstrates itself through the noticeable variation in the weather, including temperature, rainfall, and wind. Agriculture in the arid land is vulnerable to climate changes since it depends on water resources and prevailing atmospheric conditions. Influences and consequences of these variables on agriculture might operate negatively or positively. There is a reciprocal effect between climate change and animal production. The production of livestock contributes about 18% of global greenhouse gas emissions from all human activities. The farm animals are prone to the adverse impacts of the changes in climate. The changes in climate will impact both of the quantity and quality of forage production as well as their reliability. Major impacts of climatic changes on feed crops and grazing systems would be changes in herbage growth, changes in the composition of pastures, changes in herbage quality, and the offset of biomass yield increases. Since pastures depend on rainfall, any changes in rainfall patterns will affect the plants on pasture. Climate Change Convention in Paris and declared an agreement to: "*hold the increase in the global average temperature to well below 2 °C above pre-industrial levels.... recognizing that this would significantly reduce the risks and impacts of climate change*". Also to pursue efforts to limit the temperature increase even further to 1.5 °C. It targeted to undertake and communicate ambitious efforts to contribute to the global response to climate change by strengthening the mobility of countries to deal with the impacts of climate change. The efforts of some selected countries across the world have been explained as examples and experience to get the benefit of as learnt lessons.

Keywords Climate change · Impacts · Mitigation · Animals · Forages · Pasture · Feed production · World experience

---

S. A. Attia-Ismail (✉)
Department of Animal and Poultry Nutrition, Desert Research Center, Cairo, Egypt
e-mail: saai54@hotmail.com

© Springer Nature Switzerland AG 2020

E.-S. E. Omran and A. M. Negm (eds.), *Climate Change Impacts on Agriculture and Food Security in Egypt*, Springer Water, https://doi.org/10.1007/978-3-030-41629-4_18

1 Introduction

1.1 Definition

1. Climate change in a dictionary "is a long-term change in the earth's climate, especially a change due to an increase in the average atmospheric temperature" (http://www.dictionary.com/browse/climate-change). Therefore, the climate of a region is its typical or average weather (e.g. annual rainfall, temperature, etc.).

1.2 Global Demonstrations of Climate Change

Climate change demonstrates itself through the noticeable variation in the weather, including temperature, rainfall, and wind. Climate change is foreseen to aggravate in the coming decades [1].

There are several demonstrations (natural phenomena) that proof the climate changes.

1. Global sea level: has risen about 8 inches in the last century. The rate in the last two decades, however, has become twice as much that of the last century (https://www.ipcc.ch/pdf/assessment-eport/ar5/syr/AR5_SYR_FINAL_SPM.pdf).
2. Global temperature rise: the planet's average surface temperature has risen about 1.1 °C along the 19th century. The increased carbon dioxide and other human-made emissions (greenhouse gases like methane, nitrous oxide, etc.) into the atmosphere (https://www.giss.nasa.gov/research/news/20170118/).
3. Warming oceans: much of the increased heat is absorbed by oceans. Top 700 m of ocean show an increase of 0.302° Fahrenheit since 1969 [2].
4. Shrinking ice sheets: the loss of 150–250 km^3 (in the Greenland) of ice per year between 2002 and 2006, and about 152 km^3 (in the Antarctica) between 2002 and 2005 are the best demonstration for the effect of climate change (https://climate.nasa.gov/evidence/).
5. Climate changes (e.g. increases in air temperatures and decreases in precipitation) will make drought conditions to aggravate and prevent the sustainability of the growth of vegetation. Increased global temperature leads to desertification, especially in dry areas. Desertification has been identified in multiple landscapes all over the map of the world.

However, different weather and climate patterns may occur as the climate changes. Major impacts will take place, then, not only on physical processes (like the water cycle), but also on the function of the ecosystem. In dry-land areas, the climate becomes severer and, as a result, rivers, lakes and underground water sources will dry up. These changes mean that feed-production zones are declining, and in many regions crops and livestock are deteriorating. Figure 1 shows the effect of drought disasters in the world.

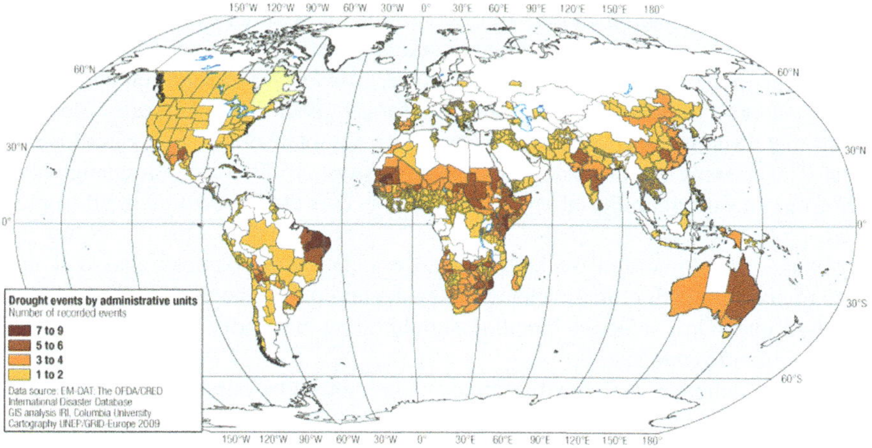

Fig. 1 Number of disasters as recorded by EMDAT (1974–2004). *Source* www.emdat.net

2 Effect of Climate Change on Agriculture

Agriculture in the arid land is vulnerable to climate changes since it depends on water resources and prevailing atmospheric conditions. The biomass in these areas is water stressed since the water resources are limited and subject to variable climatic changes that are global in range and impacts.

Climatic variables such as increasing temperature, changes in precipitation rate and pattern, human activity induced elevation of carbon dioxide concentrations, play solo or in concert to affect agriculture.

Influences and consequences of these variables might operate negatively or positively while positive impacts are advantageous, the negative consequences present true challenges to the future of agriculture.

The greenhouse gases such as carbon dioxide, methane, and nitrous oxide concentrations have increased in the atmosphere over the past centuries [3]. Agriculture is recognized as the first element in nature which is most affected by climate change. The impacts of climate change on vegetation are due to the rising temperature, changes in precipitation, and climatic patterns. One function of carbon dioxide, which is present naturally in the atmosphere, is to partially keep the earth's temperature [4]. The increase in the concentration of carbon dioxide will lead to global atmospheric changes through increasing temperature and its resulting changes in other climate parameters and thereby, affects vegetation.

The multitudinous potential impacts of climate change on agriculture may include consequences on crop growth, the pattern and quality, livestock health, and pest control. Climate change may adversely affect farming practices and management. While the long-term effects of climate change on agriculture have been modeled in some areas of the world, yet still ambiguous, the short term ones have been identified. Even though, the effect of extreme climate weather changes such as heat waves, droughts, strong wind, and heavy rains is still debatable among scientists. On the other hand, with increasing temperature, growing seasons will elongate, thus, opening doors for the use of potential new forage species and cultivars. This also means that a significant decline, and sometimes, complete elimination of crops may occur. The growing seasons will get longer because of expected climatic changes and, therefore, the sustainability of crop production is expected to decline in all tropical areas.

Changes in agricultural techniques and management practices also will take place to cope up with climatic change. New technologies, the introduction of varieties/hybrids and/or selective breeding should take top priority as a means to cope up with climatic changes.

However, whereas the growth of plants is influenced by climate change impacts, the plants at the same time have their influence on the development of climate change. Plants absorb atmospheric carbon dioxide, use it in photosynthesis and finally release carbon dioxide into the atmosphere through the process of respiration.

2.1 Effect of Climate Changes on Animal and Feed Production

2.1.1 Effect on Animals

There is a reciprocal effect between climate change and animal production. The production of livestock contributes about 37% of total methane, 65% nitrous oxide and 9% of carbon dioxide emissions, all of which makes up to about 18% of global greenhouse gas emissions from all human activities. In addition, 64% of ammonia emissions come from livestock production and contaminate air, soil, and water, and cause acid rain and damage to the ozone layer [5].

On the other hand, the farm animals are prone to the adverse impacts of the changes in climate. For instance, the increments in air temperature and humidity can cause stress to the animals, therefore, decrease their productivity. In periods of heat increment, the reproduction, and appetite of animals declines. Appetite decrement may elongate the time required to reach a targeted weight because animals consume almost half of their feed intake. Decreased feed intake is a major cause of deteriorated animal production and health. The heat waves usually happen to occur during the night—in some areas—which may lead to increasing mortality rate among animals.

However, animals have mechanisms to adapt to the hot climate. Sometimes the adverse impacts of climate change might be disadvantageous. The effects of climate change on animals are either direct or indirect. The direct effects manifestations come from the increased atmospheric temperature (Fig. 2).

Temperatures changes (either very cold or very hot) may result in increased heat production by the animal (that is more loss of energy), and consequently less

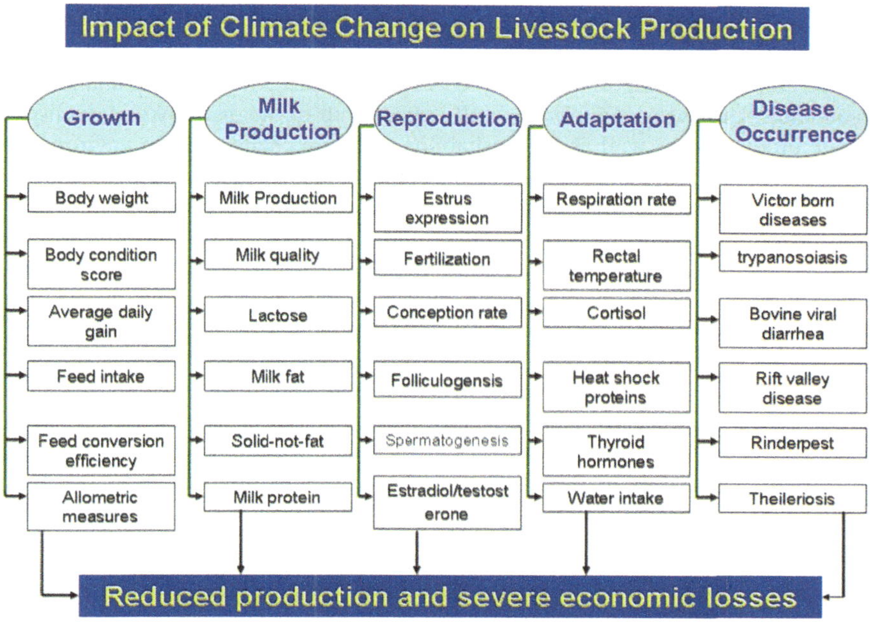

Fig. 2 Effect of climate changes on livestock production [6]

energy remains for production, and the efficiency of energy utilization deteriorates. Although other factors may intervene, the critical temperatures that the animals may feel comfortable in are shown in Table 1.

Table 1 Critical temperatures for animals [7]

Animal species	Critical temperature	
	Lower	Higher
Dairy cows	−12/−1[a]	24
New born dairy calf	8–10	35
One day old chicken	32	35
Finishing broiler	16	26
One day old turkey	35	38
Finishing turkey	16	26
Laying hens	16	27–29

[a] −12 °C for Holstein and Brown Swiss, −1 °C for Jersey

2.1.2 Effect on Feed Production

Forage crops are an important part of agriculture because they provide quality feed to animals. The changes in climate will impact both of the quantity and quality of forage production as well as their reliability. Climate changes cause environmental stresses (e.g. increasing temperature, elevated carbon dioxide concentrations) that affect pasture and forage production. Availability of irrigation water will be affected due to change precipitation patterns. These changes in climate may reduce the harvested forage yield, alter its nutritive value. Societal pressure and emphasis on the impacts of climate changes and the feed production problem have aroused and become more important. Therefore, the knowledge of how abiotic or environmental stresses limit forage production must increase. Taking into account that the available agricultural area is limited, pasture productivity and forage quality are critical issues. Researchers, then, are required to address such problems including botanical changes in vegetation.

However, depending on the climate changes affected region, the impacts will be either positive or negative. Increasing the length of the growing season, and increased plant growth and reducing frost damage are typical impacts of climate change in boreal climates. Increased plant growth may cause depletion to the soil moisture. This will be at the expense of subsequent pasture growth in the spring. Therefore, changes in seasonal patterns of forage availability may take place and could impose additional provocation for grazing management. On the other hand, in regions with warmer climates the negative impacts of climate changes are clear. Increased evaporation as a result of rising atmospheric temperature along with environmental stresses like drought will lead to reducing forage production and quality.

2.1.3 Effect on Forage Crop Production

Forage crops are an important part of agriculture. They provide high-quality feed to enable high production from farm animals. Climate change will offer significant impacts in the productivity of forage crops. As mentioned earlier in this chapter, effects of climate change on agriculture may include impacts on crop growth, pattern, quality, and diversification.

Major impacts of climatic changes on feed crops and grazing systems may include changes in herbage growth, composition and quality. Changes in the composition of pastures, include changing grasses/legumes ratio. Changes in herbage quality are due to changing concentrations of water-soluble carbohydrates and nitrogen compositions. The increased incidences of drought would offset any biomass yield increases. Changed rainfall pattern and intensity will lead to, increased nitrogen leaching in certain systems [8].

The changes in the climate may result in severe deterioration in crop production and to some extent the disappearance of certain crops. Quantity and quality of forage feeds will be mainly affected because of the increase in atmospheric carbon dioxide levels and temperature [9]. The Intergovernmental Panel on Climate Change (http://www.ipcc.ch/publications_and_data/ar4/wg2/en/ch5s5-4-3-1.html), reviewed the impacts of elevated carbon dioxide in the atmosphere and concluded that nutritional content of forage may decrease when carbon dioxide concentration increases. That is because of elevated carbon to nitrogen ratios in the plants themselves and because of the increase in the dominance of unpalatable grassland species.

These impacts are displayed in several manifestations such as grain yield and so does the herbage growth [10]. These are positive effects because of the reduced transpiration, induction of partial closure of stomata, and improved water-use efficiency of some plants [11]. In this case, the more intense management would increase operating costs and therefore may negate any benefits in forage production.

The increased temperature and carbon dioxide will affect the biodiversity of pasture composition by altering species competition dynamics through the changes that take place in growth rates [12]. These competitive changes are due to change the timing and rates of germination and subsequent vegetative and reproductive development.

The changes in chemical compositions of forage (like nitrogen and water-soluble carbohydrates) as induced by climatic changes in temperature and dry conditions will affect the quality of forage to be changed [13]. This author also mentioned that lignin and cell wall constituents might be altered as the climate temperature increase. Form and structure of roots, change in leaf growth rate may occur due to extreme climatic changes like a flood [14]. On the other hand, Craine et al. [15] showed that measures of forage quality (crude protein and digestible organic matter) declined with temperature increases and increased when precipitation is increasing.

The changes in forage quantity and quality depend on the region [13, 16]. The increment of 2 °C will negatively affect pasture and livestock production in arid and semiarid regions and positively impact the humid temperate regions. However, the warming has been shown to increase soil water content by accelerating plant senescence [17].

2.1.4 Effect on Pastures

Pastures depend on rainfall. Any changes in rainfall patterns will affect the plants on pasture. The other changes in climate will also exert their effects on pastures. Changes in temporal rainfall distribution will reduce the effectiveness of rain through the increased variation within and between seasons [18]. As mentioned earlier, the increased atmospheric temperature will increase the length of the growing season and, thus, exhaust soil moisture affecting the subsequent pasture growth, especially in cold areas. In warmer areas, the effect of rising temperature is negative through the increased evaporation.

In hotter and drier climates, pasture composition may shift to species that could be less suitable for grazing and so does grazing management to suit the new prevailing conditions. Increased carbon dioxide concentrations will change soil moisture availability by delaying soil moisture depletion.

3 World Mitigating Experiences and Strategies

3.1 Worldwide Climate Change Agreement

The effect of climate change on feed crop production was investigated in several climatic areas of the world. Since the concentration of atmospheric carbon dioxide in 2005 exceeded the natural range, the (IPCC, https://www.ipcc.ch/pdf/assessment-eport/ar5/syr/AR5_SYR_FINAL_SPM.pdf) started a campaign to reduce/eliminate the reasons that cause this increment in carbon dioxide and to reduce global temperature. As a result, in 2015, representatives of 197 nation [19], held a meeting called Climate Change Convention in Paris. They announced that they have signed an agreement to: "*hold the increase in the global average temperature to well below 2 °C above pre-industrial levels.... recognizing that this would significantly reduce the risks and impacts of climate change*". The agreement also aimed at the pursuit of efforts to limit the temperature increase even further to 1.5 °C. They decided to undertake and communicate ambitious efforts to contribute to the global response to climate change by strengthening the mobility of countries to deal with the impacts of climate change. Later on in 2016, only 94 out of the 197 nations have agreed to enforce the Paris Agreement. The developed country partners assured the goal of the mobilization of US $100 billion to support climate projects.

3.2 Country Actions

In New Zealand (a country where the economy is based on agriculture and forestry) Trolove et al. [20] carried out a research program to (identify the major opportunities that climate change brings to extend the growing season and achieve a step change in cumulative forage crop yields via radical changes in forage germ plasm and management). They used existing crop models and simulated weather data to predict likely changes in forage crop production in 2040 and 2090. They grew maize silage (summer season) and wheat (winter season) at six regions to represent weather changes and summed the biomass to give total annual biomass production. The researchers concluded that climate change might slightly decline the maize yields in the northern regions of New Zealand, and increase as it is grown in southern New Zealand. The predicted warmer winters resulting from climate change would increase the biomass of wheat by about 13–19% by 2040, and 17–38% by 2090. Generally, they found that there was little difference in total annual biomass yields for different sowing dates of maize. The report recommended the use of tropical or late-maturing maize hybrids to make benefit of the expected extension to the growing season.

In Australia Howden et al. [21] recommended the creation of linkages between industrial and political. These linkages aim to take measures to adapt to climate changes in taking into consideration the current governmental policies and initiatives.

One of the goals of the created linkages is to manage a transition period in order to support newly devised systems. Other goals are to ensure the smooth flow of information on climate change, breed and select for plants to have suitable options for higher temperature regimes and changed moisture availability. It also aims to develop grazing systems models that are capable to link with meteorological data to provide quantitative approaches to risk management. On farm level adaptations to climate change, they recommended managing pasture productivity and grazing pressure by the selection of sown pastures better adapted to climate changes, the introduction of adaptive stocking rate strategies.

In Ethiopia Feyssa and Gemeda [22] concluded that Semi-arid areas are more fragile to the adverse effect of climate change because they already suffer from erratic rainfall and rising temperatures. The author suggested ways to overcome the unfavorable impacts of climate change such as community adaptation to these climatic hazards by employing certain mechanisms (e.g. the use of developed rainwater-harvesting techniques, diversifying crop varieties, and improved agro forestry practices).

In Zimbabwe Mubaya and Mafongoya [23] reviewed the role of institutions in managing local level climate change adaptation. They noted that the growing role of the public and private institutions in facilitating adaptation to climate changes through institutional frameworks falls far from the requisite capacity. They attributed this shortage to being under-resourced and fragmented. The authors also noted the fuzziness and fluidity of institutional categorization on the ground. They highlighted the Agrawal and Perrin [24] remarks on ongoing research opportunities to mitigate climate change impacts that this type of research provides options tend to focus more on technology and infrastructure alternatives for reducing vulnerabilities, with little focus on social or institutional alternatives.

Indians Tripathi and Mishra [25] took measures to mitigate and climate change impact like improved crop seeds, livestock, and fish culture; water use efficiency; pest management; improved farm practices; nutrient management; agriculture insurance; credit support; markets; access to information; and livelihood diversification strategies.

In Bangladesh Bangladesh is affected by pollution emitted from India's (surrounding country) power plants (3% of global emission). India is the 4th largest polluter in the world [4]. This author expected that agricultural activities (the main source of the Bangladesh economy), would receive a severe setback along with forest and general ecological destruction and anticipated that there will be no vegetation left into receive the benefit of enhanced growth. There have been efforts by the Bangladesh government to combat pollution and emitted greenhouse gas to alleviate the impacts of climatic changes like the promulgation of the Environment Pollution Control Ordinance [4]. This ordinance was meant to deal with the regulations of industrial discharge. Also, [26] showed the commitment of the government to (achieve carbon dioxide neutrality). Also, Bangladesh is recognized as the most vulnerable country in terms of the cyclone and the 6th most vulnerable country in terms of floods [27]. Frequency and severity of climatic change events are increasing day by day [28].

In China's arid lands which are located in the central part of Eurasian continent, the vegetation growth has shown an overall increase over the last two decades [29]. This may suggest that part of Northern China is undergoing reversal of desertification. Therefore, decreased desertification rate went from 1.2% at the 1950's to 0.2% in 2003 [30]. Air temperature and precipitation rate show elevated trends in the arid region of China [29]. These trends are expected to increase. These climatic changes will have their impact on vegetation in the area and will lead to more vulnerable ecosystems. The forages and feed crops will get affected in a matter that shifts their production in both quantity and quality. Managing water resources, adopting adaptive measures along with rational planning will all lead to mitigating the impacts of climate changes on vegetation.

In the Mediterranean area Pastoralism is a major agricultural activity, supplying about 30% of agricultural income according to [31]. These authors showed that climate change impacted the area of the eastern Mediterranean basin negatively as showed in their Normalized Difference Vegetation Index (NDVI). In North Africa, the erratic rain and high evaporation rate, due to increased temperature, had their effect on crops. Christensen et al. [32] predicted that precipitation would decrease and air temperature will increase in the future. This climatic variability may change land use and, therefore, feed crop production. So far, no models to predict the impact of climate changes in the area of the Mediterranean basin.

In California, USA (a Mediterranean climate type region) Shaw et al. [33] predicted that climate changes (e.g. precipitation patterns) might cause the areas of the state suitable for cattle grazing to change. This might due to that some areas become wetter and others become drier and that the range forage production would decline between 14 and 58 percent. In the San Francisco Bay Area of this state, they suggested the impacts would be more positive, with production increases for the change in precipitation only not involving warming effect. The California Energy Commission [34] predicted that (in a future with higher temperatures and altered precipitation patterns, ranchers will need to consider management options for grazing shorter or less reliable seasons and for forage of questionable nutritional content).

In Argentina The country has a variety of climatic ecosystems from sub-tropical to sub-Antarctic [35] leading to a nice combination of forage diversity. These authors concluded that the climate change in rangelands of Argentina may impose certain changes in decomposition rates, aboveground net primary production, evapotranspiration and runoff. The changes include forage quality; grassland storage of soil carbon, soil quality, ecosystem structure, water storage, and habitat suitability. Additionally, there is an alteration in plant cover; and in biodiversity; as well as shifts in C3/C4 species mix in grasslands/rangelands, fluxes of greenhouse gases. They warned that the ecosystem could change to an irreversible stage of desertification. Unless decisive measures are taken, the situation there may worsen.

In Egypt Egypt has high vulnerability to climate change because depends to a large extent on natural resources. Symptoms of the climate change in Egypt are evidently the:

1. Decreasing precipitation rates,
2. Changing weather patterns,
3. Increased CO_2 emission
4. Increased methane and nitrous emission and
5. Sea level rise.

These changes are slow but steady. This makes a difficult situation worse. Since agriculture in Egypt is dependant on natural resources like rain, Nile river, atmospheric temperature, etc., it is dramatically affected by climate change. A study by the World Bank [36] showed that high temperatures will constrain agricultural production and that warming may also affect water resources. According to the agricultural value added decreased by 5.65% from 2000 to 2014 as a result of climate change [37].

To address these climate changes, Egypt has adopted a (National Strategy for Adaptation to Climate Change and Disaster Risk Reduction) announced in 2011 [38]. In the context of the (National Strategy for Adaptation to Climate Change) [38], it was stated that the agriculture sector including feed and forage production, *agriculture is particularly sensitive to climate change.*

The national strategy aims at the following:

1. *Increasing the flexibility of the Egyptian community in dealing with the risks and disasters caused by climate change and its impact on different sectors.*
2. *Enhancing the capacity to absorb and contain climate-related risks and disasters.*
3. *Reduction of climate change – related disasters.*

The strategy depends on several determinants as follows:

1. *Political will at All Levels*
2. *Availability of Human, Financial and Natural Resources*
3. *Reform and Adjustment of Institutional Frameworks*
4. *Amendment of Legislations and Laws*
5. *Strengthening the National Information Exchange System*
6. *Monitoring, Assessment and Follow-up, and Identification of Performance Indicators*
7. *Development of a National Model for Social and Economic Analysis and Projections.*

4 Conclusions

Climate change affects almost every aspect in our life. It affects animal feed production, forage production, plant feed production. The above discussion provides a view of the world as well as individual countries experience to adapt to climate change. On world wide scale, it is recommended to follow the official adaptation measures of Intergovernmental Panel on Climate Change (IPCC). On individual country scale, the adaptation and mitigation measures should take into considerations the socio-economic dimensions aspects.

5 Recommendations

To adapt to climate changes in the field of agriculture, the national strategy for climate change adaptation stated different measures in two fields. First are the water resources and second is the agriculture. In water resources the national strategy recommended specific measures for adaptation to uncertainty, increased flow of the River Nile, low flows of the River Nile. It also recommended the maximum utilization of rainfall and flash flood water.

In the field of agriculture, the national strategy recommended the followings:

1. Building an Effective Institutional System for Crisis and Disaster Management
2. Biodiversity:
3. Plant Production
4. Soil and Agricultural Land Management
5. Management of Water Resources and Farm Irrigation
6. Livestock Production
7. Fish Wealth
8. Modification and improvement of agricultural economic systems
9. Rural Community.

Acknowledgement I would like to thank Dr. Veerasamy Sejian for making available the drawing scheme on the effect of climate changes on livestock production.

References

1. Stern N (2006) What is the economics of climate change? World Economics-Henley on Thames 7(2):1
2. Levitus S, Antonov JI, Boyer TP, Locarnini RA, Garcia HE, Mishonov AV (2009) Global ocean heat content 1955–2008 in light of recently revealed instrumentation problems. Geophys Res Lett 36:L07608
3. Cubasch U, Wuebbles D, Chen D, Facchini MC, Frame D, Mahowald N, Winthe J-G (2013) The physical science basis. In: Stocker TF, Qin D, Plattner G-K, Tignor M, Allen SK, Boschung J, Nauels A, Xia Y, Bex V, Midgley PM (eds) Contribution of working group I to the fifth assessment report of the intergovernmental panel on climate change. Cambridge University Press, Cambridge/New York
4. Ali M (2013) Climate change impacts on plant biomass growth. Springer, Dordrecht, Heidelberg, New York, London, p P29
5. www.ciwf.org, (2008)
6. Sejian V, Gaughan JB, Bhatta R, Naqvi SMK (2016) Impact of climate change on livestock productivity. Feedipedia (www.feedipedia.org) No. 26 Feb 2016
7. FASS (Federation of Animal Science Societies) (2010) Guide for the care and use of agricultural animals in research and teaching, (January 2010), ISBN: 978-1-884706-11-0, Champaign, Illinois, USA. www.fass.org/docs/agguide3rd/Ag_Guide_3rd_ed.pdf
8. Hopkins A, Del Prado A (2007) Implications of climate change for grassland in Europe: impacts, adaptations and mitigation options: a review. Grass Forage Sci 62:118–126
9. Chapman SC, Chakraborty S, Dreccer MF, Howden SM (2012) Plant adaptation to climate change: opportunities and priorities in breeding. Crop Pasture Sci 63:251–268
10. Thornton PK, Boone RB, Ramirez-Villegas J (2015) Climate change impacts on livestock. CGIAR Research program on Climate Change, Agriculture and Food Security (CCAFS), Working Paper No. 120
11. Wand SJ, Midgley GF, Jones MH, Curtis PS (1999) Responses of wild C4 and C3 grass (Poaceae) species to elevated atmospheric CO_2 concentrations: a meta-analytic test of current theories and perceptions. Glob Change Biol 5:141–723
12. IFAD (International Fund for Agricultural Development) (2010) Livestock and climate change. http://www.ifad.org/lrkm/events/cops/papers/climate.pdf. Iglesias A, Avis K, Benzie M, Fisher P, Harley M, Hodgson N, Horrocks L, Moneo M, Webb J (2007) Adaptation to climate change in the agricultural sector. AEA Energy & Environment and Universidad de Politécnica de Madrid
13. Polley HW, Briske DD, Morgan JA, Wolter K, Bailey DW, Brown JR (2013) Climate change and North American rangelands: trends, projections, and implications. Rangeland Ecol Manage 66:493–511
14. Baruch Z, Mérida T (1995) Effects of drought and flooding on root anatomy in four tropical forage grasses. Int J Plant Sci 156:514–521
15. Craine JM, Elmore AJ, Olson KC, Tolleson D (2010) Climate change and cattle nutritional stress. Glob Change Biol 16:2901–2911
16. Thornton PK, Van de Steeg J, Notenbaert A, Herrrero M (2009) The impacts of climate change on livestock and livestock systems in developing countries: a review of what we know and what we need to know. Agric Syst 101:113–127
17. Zavaleta ES, Thomas BD, Chiariello NR, Asner GP, Shaw MR (2003) Plants reverse warming effect on ecosystem water balance. Proc Nat Acad of Sci 100:9892–9893
18. Stokes AC, Ash J, Howden SM (2008) Climate change impacts on Australian rangelands. Society for range management. June 2008, 40:45
19. UNFCCC, Conference of parties 21 (COP 21) (2015). https://unfccc.int/files/meetings/paris_nov_2015/application/pdf/paris_agreement_english_.pdf
20. Trolove S, Kerckhoffs H, Zyskowski R, Brown H, Searle B, Tait A, Pearson A, Reid J (2008) Forage crop opportunities as a result of climate change. Climate change—'Plan of action' research programme 2007–08. A response to MAF-POL/CP03 RESEARCH-01Crop & Food Research Confidential Report No 2161

21. Howden SM, Ash A, Barlow S, Booth T, Charles S, Cechet R, Crimp S, Gifford R, Hennessy K, Jones R, Kirschbaum R, McKeon GM, Meinke H, Park S, Sutherst B, Webb L, Whetton P (2003) An overview of the adaptive capacity of the Australian agricultural sector to climate change: options, costs and benefits. Canberra, Australia: CSIRO sustainable ecosystems. Available at: http://www.cse. csiro.au/publications/2003/AGO Ag Climate Adaptation Report
22. Feyssa DH, Gemeda DO (2015) Impact of climate change on production systems in Semiarid Regions of Ethiopia. J Biol Chem Res 32(2):755–764
23. Mubaya CP, Mafongoya P (2017) The role of institutions in managing local level climate change adaptation in semi-arid Zimbabwe. Clim Risk Manag 16: 93–105. http://creativecommons.org/licenses/by-nc-nd/4.0/
24. Agrawal A, Perrin N (2008) Climate Adaptation, Local Institutions, and Rural Livelihoods. IFRI Working Paper # W08I–6. Accessed at http://www.umich. edu/~ ifri/on 20 Sept 2015
25. Tripathi A, Mishra AK (2017) Knowledge and passive adaptation to climate change: an example from Indian farmers. Clim Risk Manag 16:195–207
26. Rahman AA (2009) Seal the deal in Copenhagen: the most vulnerable communities demand climate justice. Clime Asia. COP 15/MOP5 Special Issue, 1–8
27. Garai J (2014) The impacts of climate change on the livelihoods of coastal people in Bangladesh: a sociological study. In: Filho WL et al. (eds) International perspectives on climate change, climate change management, https://doi.org/10.1007/978-3-319-04489-7_11, Springer International, Switzerland
28. Siddiqui T, Abrar CR, Alamgir M, Ali MS (2010) Climate change and community adaptation: case study from Bangladesh. Monash University, Monash Sustainable Institute, Australia
29. Ma S (2018) Drylands of China: problems and prospects. In: Gaur MK, Squires VR (eds) Climate variability impacts on land use and livelihoods in drylands. Springer International AG
30. Piao S, Fang J, Liu H, Zhu B, (2005) NDVI-indicated decline in desertification in China in the past two decades. Geophys Res Lett 32: L06402
31. Mohamed AH, Squires VR (2018) Drylands of the Mediterranean Basin: challenges, problems and prospects. In: Gaur MK, Squires VR (eds) Climate variability impacts on land use and livelihoods in drylands. Springer International AG
32. Christensen JH, Hewitson B, Busuioc A, Chen A, Gao X, Held I, Jones R, Kolli RK, Kwon WK, Laprise R, Magaña Rueda V, Mearns L, Menéndez CG, Räisänen J, Rinke A, Sarr A, Whetton P (2007) Regional climate projections, in climate change: The physical science basis. In: Solomon S, Qin D, Manning M, Chen Z, Marquis M, Averyt KB, Tignor M, Miller HL (eds) Contribution of working group I to the fourth assessment report of the intergovernmental panel on climate change, Cambridge University Press, Cambridge
33. Shaw MR, Pendleton L, Cameron DR, Morris B, Bachelet D (2011) The impact of climate change on California's ecosystem services. Clim Change 109:465–484
34. Chaplin-Kramer R (2012) Climate change and the agricultural sector in the San Francisco Bay Area: changes in viticulture and rangeland forage production due to altered temperature and precipitation patterns. California Energy Commission. Publication number: CEC-500-2012-033
35. Busso CA, Fernńdez OA (2018) Arid and semiarid rangelands of Argentina. In: Gaur MK, Squires VR (eds) Climate variability impacts on land use and livelihoods in drylands. Springer International AG, Gewerbestrasse 11, 6330 Cham, Switzerland
36. Eid HM, El-Marsafawy SM, Ouda SA (2007) Assessing the economic impacts of climate change on agriculture in Egypt a Ricardian approach. The World Bank Development Research Group Sustainable Rural and Urban Development Team. Policy Research Working Paper 4293
37. Fawaz MM, Soliman SA (2016) Climate change and its impacts on Egyptian agricultural plant sector. Agricultural Economy Dep., Faculty of Agriculture, Kafr El-Sheikh University
38. IDSC The Egyptian cabinet (Information and Decision Support Center) and the UNDP (2011) National strategy for adaptation to climate change and disaster risk reduction

Algae and Fishes: Benefits and Hazards

Nermin Adel Hussein El Semary

Abstract Aquatic animals such as fish and crustaceans provide human beings with high-quality protein diets. Meals of aquatic animals are based partially or even entirely on algae. Marine algae can be grown as maricultures in marine environments in various forms. Also, algae can be used for fish diets in inland aquacultures. Algae can also be grown not only for their nutritional value but also for their pigments that can affect the colour and health of some aquatic animals thereby affecting their quality and marketability. However, algal phytoplanktons are not massive in large areas of major oceans. The reason behind that lies in the deficiency of iron despite the presence of other major nutrients such as phosphorus and nitrogen in high concentrations. Recently this iron limitation in marine environments attracted the attention of scientists. Thirteen iron fertilisation experiments in those oceans were performed to enrich them with iron which in turn would lead to the increase of the phytoplankton and consequently aquatic animals and fishes. Scientists propose that this would cause mitigation of global warming as algae take in carbon dioxide, the main factor responsible for global warming, and release oxygen during photosynthesis. So double benefits are expected in terms of increasing aquatic animals and decreasing global warming. However, the short and long-term overall effects of those experiments are yet to be validated and evaluated. On the other hand, some algae can have a negative impact on fishes. Toxic algae can cause massive fish mortalities. The two main algal groups responsible for toxins production are cyanobacteria and dinoflagellates. They can spread from one water body to another through ballast water or via biofouling aquatic vessels such as ships and boats. This, in turn, would be harmful and even lethal for aquatic animals and would jeopardize food security. Thereby, surveillance of aquatic vessels must be performed.

Keywords Algae · Fish · Fisheries · Crustaceans · Iron fertilisation · Mariculture

---

N. A. H. El Semary (✉)
Biological Sciences Department, College of Science, King Faisal University, 31982 Al Ahsa, Kingdom of Saudi Arabia
e-mail: nelsemary@kfu.edu.sa

Botany and Microbiology Department, Faculty of Science, Helwan University, 11795 Cairo, Egypt

© Springer Nature Switzerland AG 2020
E.-S. E. Omran and A. M. Negm (eds.), *Climate Change Impacts on Agriculture and Food Security in Egypt*, Springer Water, https://doi.org/10.1007/978-3-030-41629-4_19

1 Introduction

1.1 Importance of Algae

Algae are photosynthetic plant-like organisms that have chlorophyll a in common as well as having no embryo, no vascular tissues, no roots, stems or leaves. Their size varies from unicellular microscopic forms to multi-cellular macroscopic organisms [1]. Their diverse cellular organization, sizes and structure indicate that they are not closely-related phylogenetic group but in fact a diverse group. They occupy almost all habitats even the most extreme. Marine/brackish water algae are being the focus of this chapter as they represent the base of food chain for edible aquatic animals such as fish and crustaceans. Based on their sizes, marine algae can be classified into macroscopic and microscopic forms. The macroscopic forms are usually called seaweeds. They are found in supra-littoral, littoral and sub-littoral zones of marine environments and are mostly benthic. In their comprehensive review, [2] enumerated the advantage of seaweeds where most of the marine seaweeds are green algae, brown algae and red algae. The latter represents an important protein-rich dietary component for animals and humans. However, algae represent a characteristic direct human food, especially *nori* (*Porphyra* spp.), *wakame* (*Undaria pinnatifida*), and *kombu (Laminaria japonica)*. Securing marine food sources would primarily depend on securing dietary supply that is mainly based on algae. According to [2], seaweeds represent primary producers of oxygen and organic matter through photosynthesis and represent food for heterotrophs. They also act in nutrient recycling and waste treatment and can be used as biomonitors of eutrophication as well as indicators of natural and/or artificial changes in biodiversity. They are of huge economic importance in different aspects such as food, industry, agriculture and aquaculture [2]. The continuous production of those algae year round is necessitated by their essence as a feed for aquaculture. Most important algae that are cultivated are *(Laminaria, Undaria,* and the red algae *Porphyra* and *Eucheuma/Kappaphycus.* These red seaweeds are widely cultivated as a source of carrageenan, which is used in the manufacture of food, both for humans and other animals [2].

With regard to algal nutritional value and according to FOA report, the protein content of some filamentous cyanobacteria or bluegreen algae ranged between 60–74% and for filamentous green algae (16–32%). The protein contents of green and red seaweeds ranged between 6–26% and 3–29%, respectively. The lipid levels for filamentous green algae ranged between 2–7%. The lipid contents of both green (0.3–3.2%) and red seaweeds (0.1–1.8%) were less than those values in case of filamentous algae. The ash content of filamentous blue-green algae ranged from 3–11%. The ash contents of green seaweeds ranged from 12–31%. Red seaweeds had an extremely wide range of ash contents (4–47%) (http://www.fao.org/docrep/012/i1141e/i1141e01.pdf).

The chapter here focuses on the benefits that algae offer aquatic animals whether as a feed or as a colourant and methods of cultivating algae to allow continuous year-round biomass supply. The chapter also discusses ways of maximizing algal numbers in the open oceans through iron fertilization with a subsequent increase in phytoplanktons leading to the increase in the number of aquatic animals. However, this scenario has not been fully achieved and results so far are hereby discussed.

On the other hand, not all algae are beneficial. Some algae are toxic as they produce toxins that cause massive fish mortalities. They can be transferred from one water body to another and from a country to another through ballast water and/or biofouling. These overlooked hazards can jeopardize marine food security and endanger local environments. Thereby, surveillance and tight monitoring are needed. The current chapter raises concerns about those overlooked hazards.

1.2 Microalgae in Aquaculture

According to FAO report on the use of microalgae in aquaculture, all stages of marine bivalve molluscs such as clams, oysters, scallops are feeding on live algae as well as the larval stages of some marine gastropods (abalone, conch), larvae of several marine fish species and penaeid shrimp, and zooplankton hatchery. Eight algal species (*Isochrysis* sp.; *Chaetoceros gracilis*; *Chaetoceros calcitrans*; *Tetraselmis suecica*; *Thalassiosiera pseudonana*, clone 3H; *Pavlova lutheri*; *Isochrysis galbana*; *Skeletonema costatum* are usually used for feeding through active uptake by the larvae of polysaccharides present in the algal cell walls. Algae can also play a role in stabilizing the water quality via removal of wastes and production of oxygen as well as the possible stimulation of the non-specific immune system in the larvae (http://www.fao.org/docrep/003/W3732E/w3732e08.htm).

1.3 Description of the System of Inland Aquaculture

The use of underground water for culturing aquatic animals in desert areas is a relatively new approach. The underground water receives no domestic or industrial wastewater, thereby providing a fairly unpolluted thermal water source suitable for their growth. The desert provides a geographical barrier that protects aquacultures against disease and pathogens. The water is either freshwater or slightly brackish. The wastewater of the fish culture is recycled to grow plants as it is rich in phosphate and ammonia. This maximizes water usage efficiency and decreases the usage of hazardous chemical fertilisers. Algae can be grown in man-made ponds to provide feed for aquatic animals grown. A small-scale study was carried out to validate the use of algae grown on brackish water collected from Wadi El Natroun, Western desert, Egypt as fish feed and the use of fish waste water for growing plants and the results were highly encouraging [3]. The algae to be grown are to be fed to fishes as a dietary supplement or main meal for larvae.

The following is a simplified description of the inland aquaculture used for growing fishes with some details on components and units of the system (translated from Arabic and summarized from [4]).

1.3.1 Fish Hatchery

Cement basins are built of red brick and are cemented with waterproof cement. During this stage the basins are inoculated with fish hatchery with an average weight of 1 gm until it reaches 40 gm in average.

1.3.2 The Rearing Basins

Cement basins are built by red bricks with water proof cement. During this stage, the basins are inoculated with fish fingerlings of the average weight of 45 gm until it reaches 400–450 gm.

1.3.3 Units Associated with the System

1. **Water storage reservoir**

This is a floor basin that can take 100 m^3 of water volume to guarantee sustainable water supply in the culture. The basin will be lined with an insulating plastic to detain water within the basin.

2. **Sedimentation basin**

This basin is built to allow sedimentation where it has a small tilt with its two thirds in the direction of waste opening, and the last third is separated with a short blocking wall for water to rise above it to move to that third. This specific third is the one that contains clean water. The culture water is dispensed after forcing it through biological filter.

3. **Biological filter**
4. It is opaque and well-aerated tank that functions in purifying water. The water tank is of 50 m^3 and empty area of 50 m^3
5. **A headquarters**

One floor designed and furnished to meet the needs of the project team.

6. **Fodder store room**

This is used for preserving fodder enough for at least one month. Storage conditions are met, and the humidity must be kept to the minimum and ventilation openings are provided.

7. **Maintenance workshop**

It is used for keeping all the necessary tools/instruments and hardware used in the project such as nettings, wood, supporters, robes, chains, and equipment.

8. **Area of fish assortment and marketing preparations**

It is used in fish sorting and preservation in ice. Floors are to be lined with ceramics.

2 Mariculture of Seaweeds

According to [5], algae whose life cycle does not involve alternation of generation can be cultivated vegetatively. The vegetative culturing involves the excision of small parts of the seaweed and cultivating them in either their natural marine environment or in a suitable environment until reaching a suitable size. If placed in their natural environment, the algal parts are tied to ropes or nets. Then either the whole alga is removed or most of it, leaving a small part that will regrow into a whole thallus once again. For hydrocolloid industry (agar and carrageenan), the vegetative method is mostly used. On the other hand, cultivation of algae that involves alternation of generations is difficult as it involves the transition between completely two different phases such as *Laminaria* species. For this alga, the diploid sporophyte that is harvested as seaweed must first develop into a haploid microscopic gametophyte then to a new sporophyte. The transition is performed in land-based facilities. The growth rate can be increased by bubbling of air, CO_2 or pumping seawater. The latter can be useful also for manipulating temperature and salinity. The pH of the tanks should be in the range of 7.9–8.3 [5].

3 Natural Carotenoids as Nutrients and Colourants for Fishes: Astaxanthin as an Example

According to [6] and references therein, aquatic animals cannot synthesise carotenoids de novo whereas algae can. The latter represents main meal of those animals. Carotenoids are fat-soluble natural pigments whose colour is attributed to the conjugated double bonds in the hydrogen backbone [7]. The carotenoids have several biological roles where they function as.

Vitamin A precursor [8]; antioxidants [9]; reproduction [10] and immune system enhancers [11]. Their overall effects would lead to growth promotion [12]. Different carotenoids have different colours: astaxanthin (red), tunaxanthin (Yellow), lutein (Greenish-yellow), beta-carotene (orange), alpha and beta-doradexathins (Yellow), canthaxanthin (orange-red), xeaxanthin (yellow-orange), eichinenone (red). Carotenoids are being utilized in the ornamental fish feed as colourants [11]. The coloured fishes as salmonids are considered quality fishes among the consumers.

Faded colours or discolouration of fishes would reduce their market value [6]. According to [13], the application of carotenoids is important to tiger shrimps as their deficiency can lead to discoloration called "Blue disease".

3.1 The Use of **Haematococcus** *Alga for Astaxanthin Production*

In nature, algae such as *Haematococcus pulivaris* synthesizes the carotenoid pigment astaxanthin which is transferred through the food web to zooplankton and crustaceans, which in turn are prey for salmonoids and trout [14]. Astaxanthin is biosynthesized through the isoprenoid pathway which proceed from acetyl-Co-A, phytoene, lycopene, carotene, and canthaxanthin and then astaxanthin. According to [15], *Haematococcus pluvialis* is a green alga belonging to family Haematococcaceae, order Volvocales. This alga predominates in temporary water bodies near marine water bodies via its ability to cope under conditions of extreme fluctuations in light, temperature and salinity levels due to its rapid ability to encyst [16]. There are four types of cells in the life cycle of this alga: microzooids, large flagellated macrozooids, non-motile palmella forms; and haematocysts, which are large red cells with a heavy resistant cell wall. When environmental conditions become unfavorable the palmella stage results, followed by the resistant haematocysts and the accumulation of astaxanthin where same conditions apply for encystment and carotenogenesis as the two phenomena usually occur together. This alga does not produce toxin. Its mean protein content is 23.62%, carbohydrate 38%, fats 13%, and its mineral composition includes Fe, Mg, Ca as well as vitamins belonging to vitamin B complex, vitamin C, vitamin E and its astxanthin content is 1%.

[17] reported that under light conditions astaxanthin can be produced whereas in the dark it can be produced only if an organic carbon source is present. Carotenogensis and encystment can be induced by low nitrate or phosphate, high temperature or light, or the addition of sodium chloride in the culture medium [18, 19]. When *Haematococcus* cultures are exposed to continuous light ~350 μmol m^{-2}, astaxanthin formation is increased [20]. Other studies support the major role of astaxanthin accumulation in *Haematococcus* as a form of protection against high light and oxygen radicals [21]. Carotenoid bio-synthesis in cystsis also enhanced by oxidative stresses [22]. According to [15] the industrial production of *Haematococcus* alga involves growing pure cultures in closed culture container with regulated pH and temperature. After suitable growth, cultures are transferred to open culture system ponds under the proper stress conditions such as high light intensity to allow carotenogenesis and encystment. Algal cysts then open to allow standardized production of Astaxanthin.

4 Iron Fertilization

Climatic changes exemplified in ocean acidification and global warming mainly caused by increased CO_2 concentrations urged the international community to tackle those critical environmental issues [23]. There are two main proposed scenarios which include the control of aerosols emissions and to increase oxygen input and decrease the release of carbon dioxide through inducing massive increases in algal phytoplankton as a result of ocean iron fertilization (OIF) experiments or biological pump [24].

Iron fertilization is achieved either naturally by gyres or high dust input or artificially by sinking iron powder/flakes or even iron old vehicles to induce a biomass increase in algal phytoplankton. This would increase in aquatic animals whose basic diet is algae [25]. The increase in iron which is an essential micro-nutrient necessary for chlorophyll formation, electron acceptors such as ferredoxin, flavodoxin and cytochromes and as a cofactor of many enzymes in the algal cell would thereby increase growth and the uptake of macro-nutrients for phytoplankton growth [26]. According to [23], iron is known to be limiting in the subarctic Pacific, equatorial Pacific, and the Southern Ocean, which are known as high-nutrient and low-chlorophyll (HNLC) regions. Phytoplankton cannot completely utilize the available macro-nutrients (especially NO_3) during photosynthesis due to a lack of iron. For this reason, primary production in these HNLC regions is relatively low despite the availability of other nutrients. However, the biological pump would only act efficiently if there is enough supply of macro-nutrients (i.e., nitrate, phosphate, and silicate) into the euphotic zone [27]. To test this hypothesis, a total of 13 OIF experiments have been performed. These OIF field experiments demonstrated that primary production could be significantly increased after artificial iron addition [28]. However, export production efficiency revealed by the OIF experiments was unexpectedly low compared to production from natural processes in all, except one of the experiments [23]. The overall conclusion so far is that even though there was an increase in phytoplankton due to iron fertilisation, there was not any pronounced decrease in CO_2 concentrations which continue to rise [29].

4.1 Side Effects of Iron Fertilization

Side effects such as production of green house gases (e.g., N_2O and CH_4) [30] development of hypoxia/anoxia [31], and toxic algal blooms such as those of the diatom *Pseudo-nitzschia* [32] and the dinoflagellate *Alexandrinum* sp. [33] have been reported.

[23] reviewed some of those investigations. For methane it is used as a source of energy for microorganisms and is converted to CO_2 [34, 35]. For N_2O, it is already produced in the ocean, so any enhancement of biological production would increase

N_2O hence increasing atmospheric greenhouse gas levels rather than decreasing them [36].

Degradation of iron-enhanced biomass may cause decrease oxygen concentrations in the subsurface waters [35]. Dimethylsulfide (DMS) the possible precursor of sulfate aerosols that cause cloud formation and climate cooling was estimated during the experiments [30]. [23] reviewed the results of the different experiments and concluded that significant increases in DMS production were found in some experiments but not in all. Also, iron fertilization experiments resulted in stimulated growth and even dominance in phytoplankton community of the neurotoxic domoic acid producer the diatom '*Pseudo-nitzschia*' in some experiments [32].

5 Toxic Algae

Microalgae are microscopic forms of algae which play an important role in primary production in vast areas of fresh and marine water bodies [37]. However, some microalgae are toxic such as cyanobacteria and dinoflagellates. Cyanobacteria are prokaryotic oxygenic phototrophs, whereas dinoflagellates are eukaryotic algae that are major toxin producers in marine environments [38]. The latter group of flagellated phytoplankton largely proliferate during summer time as well as other microalgal groups such as diatoms, coccolithophorids and cyanobacteria [38] in marine water bodies. Those microalgae owe their success in dominating their niches to their remarkable metabolic ability that enabled them to efficiently utilize all available resources such as nutrients and light. However, more information is needed to investigate and clarify the factors that increase their proliferative capacity especially for dinoflagellates, which causes red tide phenomenon during summer. The interest in these algae stems from the fact that 75% of them are toxin-producers and can profoundly increase fish mortalities and poison the people who eat them as well as decreasing water quality thereby causing tremendous economic loss and imposing health threats. For example, cyanobacteria and dinoflagellates sometimes bloom during summer in concentrations of more than a million cells/ml. Some species produce toxins, which in such quantities kill fish. Moreover, their toxins can accumulate in aquatic animals which in turn may pass them on to people who eat them (http://www.issha.org/Welcome-to-ISSHA/Harmful-algae).

5.1 Cyanobacterial Toxins

According to [39], cyanobacteria are prokaryotic oxygenic phototrophs that are of wide occurrence worldwide. They are unicellular, colonial or filamentous. The latter can be heterocystous or non-heterocystous. Heterocystous cyanobacteria are capable of nitrogen fixation under conditions of limited available combined nitrogen. Several cyanobacterial strains produce toxins of variable toxigenic activities. Some strains are freshwater while others are marine. Cyanobacteria are responsible for the formation of harmful blooms, in nearly all water bodies.

5.1.1 Cyclic Peptides

Genera like *Microcystis*, *Anabaena*, *Planktothrix*, *Oscillatoria*, *Nostoc*, *Haplosiphon* and *Anabaenopsis* are responsible for the production of cyclic peptides. Hepatotoxins produced in freshwater (microcystins) and marine water (nodularins) operate through their strong covalent binding to hepatic enzymes called protein phosphatases thereby causing their inactivation, depolymerization of hepatic cells, hemorrhage and eventually death.

5.1.2 Neurotoxic Alkaloids—Anatoxins and Saxitoxins [39]

These neurotoxins produce rapid respiratory arrest. Three groups of cyanobacterial neurotoxins are found:

- anatoxin-a and homoanatoxin-a, similar to acetyl choline in effect,
- anatoxin-a(S), which is an anticholinesterase, and
- saxitoxins, or paralytic shellfish poisons (PSPs) which block nerve cell sodium channels. Anatoxin-a was found in *Anabaena*, *Oscillatoria* and *Aphanizomenon* whereas homoanatoxin-a from *Oscillatoria*, anatoxin-a(S) from *Anabaena*, and saxitoxins from *Aphanizomenon*, *Anabaena*, *Lyngbya* and *Cylindrospermopsis*.

The alkaloid toxins are diverse heterocyclic nitrogenous compounds. The non-sulphated alkaloid toxins of freshwater cyanobacteria (anatoxins and saxitoxin) are all neurotoxins. The sulphated PSPs, C-toxins and gonyautoxins (sulphated derivatives of saxitoxin) are also neurotoxins, but the sulphated alkaloid cylindrospermopsin blocks protein synthesis with a major impact on liver cells. Some marine cyanobacteria also contain alkaloids (lyngbyatoxins, aplysiatoxins) which are dermatoxins [39].

5.2 *Toxins Produced by Dinoflagellates*

These eukaryotic unicellular protists possess two unequal flagella: one smooth longitudinal & one hairy transverse. If two flagella arise from anterior end they are called desmokont and if they arise in grooves they are called dinokont. There are also two depressions one transverse called cingulum and one longitudinal called sulcus. The nucleus is dinokryon. They are either thecate or naked. The emphysema vesicles are found beneath plasma membrane. It can form cellulose plates in thecate species. They can be photo-or heterotrophs. They are classified according to the International Botanical code as well as the International Zoological code. Seventy five percent of dinoflagellates are toxin- and endocrine disturbing chemicals-producers with serious health effects on other organisms and they are the major causal agents for Red-tide phenomenon. They are major killers of fish through poisoning and blocking of gills. On the other hand, they are major contributors to the food web as primary producers the base of food chain. Major contributor to several marine coral communities in symbiosis with other organisms such as zoothanthalae.

According to [40], the extent of biodiversity within dinophytes is largely underrepresented and new species/genera are being added annually. Areas like Middle East and Africa are profoundly underexplored and only limited records, if any, are available. Culture collections to study their characteristics are not available for most of heterotrophic species simply because their living prey is unknown. Common taxonomic treatment is largely morphology-based. The information from light and electron microscopy provides the basis for morphological description. Taxonomic criteria: cingulum position and sulcus width, tabulation pattern & hyposome and episome size, nucleus position & presence of lists. For heterotrophic sp. we refer to its parts as epi- and hypocone unlike phototrophic sp. we refer to them as epi- and hypotheca. [41] listed some examples of toxin-producing dinoflagellates which include *Gymnodinium* spp., *Alexandrium* spp. and *Pyrodinium* spp. that are responsible for Shellfish Poisoning caused by Saxitoxins (heterocyclic neurotoxic Guanidine). The symptoms include Tingling and numbness of the perioral area and extremities, loss of motor control, drowsiness, in severe cases respiratory paralysis. Also *Karenia brevis* toxins are responsible for neurotoxic shellfish poisoning caused by brevetoxins (lipid soluble cyclic polyethers). They cause gastrointestinal symptoms like nausea and diarrhea as well as neurological symptoms. Other genera such as *Dinophysis* spp. and *Procentrum* spp. cause Diarrhetic Shellfish Poisoning caused by okadaic acid. The symptoms include gastrointestinal symptoms like vomiting and diarrhea. Nevertheless, okadaic acids are carcinogenic.

6 Overlooked Hazards: Toxic and Harmful Invasive Species Introduction by Ballast Water and Biofouling

Biofouling means the colonization of aquatic structures/vehicles such as ships with algae and other organisms. With regard to biofouling microalgae, the following website (http://www.gisp.org/publications/toolkit/BiofoulingGuidelines.pdf) reported that they can cling to other fouling species and to aquatic vehicles such as ships and boats as well as residing in microspaces and pipes. Biofouling not only causes a negative ecological impact but also negative economic and health impacts. For example, biofouling can introduce harmful invasive species into national waters. This has many negative impacts that may include: Outcompeting indigenous species with regard to space and food; preying upon indigenous species; altering environmental conditions and reducing indigenous biodiversity through causing extinctions [42].

Invasive species can harm the ecological balance and cause the death of important local organisms especially if those invasive species include toxic algae that overgrow and outcompete native species and cause massive fish mortality in addition to lowering the quality of recreational and drinking water. On the economic level, biofouling can cause loss of productivity in aquaculture.

Therefore, there is a necessity for early surveillance of water bodies and complementing this with environmental data that include all relevant information including substrate nature, community structure, interrelations between different community members and ecological parameters associated with this phenomenon. Alarmingly, the continual rise of sea levels together with recurrent catastrophic natural disaster of tsunami has exposed wide terrestrial coastal areas, objects and vehicles to invasion by marine microalgae, some of which are toxic which in turn lead to contamination of local water bodies as recorded in villages in Southern India [43].

Another overlooked hazard is ballast water which is the water used by oil and cargo tankers to maintain its balance until being fully loaded with its cargo. The water is then disposed of in the hosting port. Because of the richness of this water by non-indigenous living organisms of algae and other biological forms, they pose an environmental danger and can cause serious ecological damages. Invasive harmful or toxic species can cause environmental imbalance where these organisms can be spread and compete with local organisms for their food resources.

They may also devour many of them or produce toxins causing disease or death, which poses a huge threat to the local environment. In addition, discharge of ballast water may cause sediments in those waters to increase the acidity and mortality of marine larvae (http://main.omanobserver.om/?p=57413).

In 2014, Sultanate Oman held an International conference on the management of ballast water discharge and marine invasive species. One of the key topics addressed was "Biofouling on ship hulls: The vector for moving marine invasive species". Light was also shed on the intensity & frequency of Harmful Algal Blooms (HABs) in the coastal water of Oman and a look at harmful algal blooms (including invasive species) and their effect in Oman waters and regional water as well with experts attending from all over the world and recognizing the magnitude of the problem. Quite recently, The International Convention for the Control and Management of Ships' Ballast Water and Sediments (BWM) has come to force by the end of 2017 (https://www.imo.org/en/About/Conventions/ListOfConventions/Pages/International-Convention-for-the-Control-and-Management-of-Ships'-Ballast-Water-and-Sediments-(BWM).aspx). This convention requires aquatic vessels to keep records of sources of ballast water and dates of renewal. Samples are to be taken regularly to check for any harmful species present.

Therefore, it is important to both recognize the factors leading to biofouling as well as the types of algae able to biofoul aquatic structures. It is worthy noting that some of the most important factors that are most likely to affect the algal community include temperature, light intensity, pH, nutrient status (mainly P and N), dissolved oxygen [44] and substrate nature. For example, In the Foa reviews, [45, 46], http://www.fao.org/docrep/012/i1141e/i1141e01.pdf and http://www.fao.org/docrep/003/W3732E/w3732e08.htm, listed some of the growth condition requirements of several algal taxa that are used in aquaculture and showed the physical and chemical conditions that affect them.

Blooms of phytoplanktonic algae can flourish and can decline according to several factors including for example the presence of iron in marine habitats [47]. However, toxic and harmful algal blooms should particuarly be monitored [48] as they can biofoul ships and spread from one country to another (http://hab.ioc-unesco.org/index.php?option=com_content&view=article&id=5&Itemid=16, http://www.whoi.edu/redtide/, http://www.issha.org/Welcome-to-ISSHA/Harmful-algae). According to [49], biofouling guidelines can be useful in pointinng out risk sources and control measures (http://www.gisp.org/publications/toolkit/BiofoulingGuidelines.pdf). On the biological front, biofouling algal species are mostly microscopic and lack enough diagnostic phenotypic features. Misidentification or overlooking some harmful algal species due to the sole use of traditional methods in the examination can lead to erroneous conclusions. Similarly, ballast water can pose a risk of transferring harmful algal blooms [50], http://main.omanobserver.om/?p=57413 and therefore must be controlled through conventions among nations [51], http://www.imo.org/en/About/Conventions/ListOfConventions/Pages/International-Convention-for-the-Control-and-Management-of-Ships'-Ballast-Water-and-Sediments-(BWM).aspx.

7 Conclusion

Algae are primary producers that form the base of the food web and represent the basic meal of aquatic animals such as fishes and crustaceans upon which the human beings feed. Therefore and in order to secure food resources from aquatic water bodies, the beneficial and nutrient-rich algae whether macroscopic or microscopic can be cultured and grown in both marine environments and inland. The algae can also be grown for their carotenoids that play many functions in aquatic animals. However, some oceans are high in their nutrients and low in their iron content. This causes a reduction in phytoplankton densities. One solution for this and for global warming issue is iron fertilization of those oceans that would increase phytoplankton, increase aquatic animals and decrease global warming as a result of carbon dioxide consumption and oxygen release. However, toxic algae can harm and kill edible aquatic animals thereby causing massive aquatic animals die off Toxins can even be transferred to human beings leading to health risks. Therefore, toxic algae and factors leading to their proliferation must be studied extensively and thoroughly monitored as they can be transferred through biofouling and ballast water.

8 Recommendations

1. The expansion of non-toxic algal mariculture whether for micro and macroalgae.

2. The monitoring of ballast water, especially after the International Convention for the Control and Management of Ships' Ballast Water and Sediments (BWM), has come to force by the end of 2017.
3. Studies should be conducted to analyse the correlation between the diversity of biofouling algal community and the physicochemical factors associated with them which will help not only in gaining a better understanding about the harmful species present but also will highlight the factors that lead to their growth. Generating better surveillance and detection policy of the biofouling algal communities could contribute to our ability to better control the development of harmful and toxic invasive species.

References

1. Van den Hoek C, Mann DG, Jahns HM (1995) Algae: an introduction to phycology. Chapter 1 Cyanobacteria and prochlorophyte
2. Chopin T, Sawhney M (2009) Seaweeds and their mariculture. Encycl Ocean Sci. https://doi.org/10.1016/B978-012374473-9.00757-8
3. Mahmoud RM (2017) Sustainable aquaculture chain for growing algae, fishes and plants. M Sc Thesis, Department of Botany and Microbiology, Faculty of Science, Helwan University, Cairo, Egypt
4. Hussein M (2007) The technology of aquaculture and oysters cultivation in desert and coastal areas, Dr Al Fekr Ararabi
5. McHugh DJ (2003) Fao (food and agricultural organization of the United Nations, Rome) fisheries technical paper 441 http://www.fao.org/docrep/006/y4765e/y4765e00.htm
6. Daniel N, Sivaramakrishnan T, Subramaniyan S, Faizullah MM, Fernando H (2017) Application of carotenoids on coloration of aquatic animals. Int J Fish Aquat Res 2(1):01–07
7. Bendich A, Olson JA (1989) Biological actions of carotenoids. FASEB J. 3(8):1927–1932
8. Guillou A, Choubert G, De La No J (1992) Absorption and blood clearance of labelled carotenoids ([14C] astaxanthin, [3H] Canthaxanthin and [3H] zeaxanthin) in mature female rainbow trout (*Oncorhynchusmykiss*). Comp Biochem Physiol Part A Physiol. 103(2):301–306. https://doi.org/10.1016/0300-9629(92)90584-D
9. Bjerkeng B (2000) Carotenoid pigmentation of salmonid fishes. Carotenoid pigmentation of salmonid fishes–recent progress. In: CruzSuárez LE, Ricque-Marie D, Tapia-Salazar M, Olvera-Novoa MA, Civera-Cerecedo R (eds), AvancesenNutrici6nAcuícola V. Memorias del V Simposium. Internacional de Nutrici6nAcuícola. 19–22 Noviembre, M6rida
10. Chou YH, Chien YH (2001) Effects of astaxanthin and vitamin E supplement in Japanese sea bass Lateolabrax japonicas brood stock diet on their fecundity and egg quality. 6th Asian Fisheries Forum, Book of Abstract, p 60
11. Shahidi F, Metusalach, Brown JA (1998) Carotenoid pigments in seafoods and aquaculture. Crit Rev Food Sci 38:1–67
12. Christiansen R, Glette J, Lie Ø, Torrissen OJ, Waagbø R (1995) Antioxidant status and immunity in Atlantic salmon, Salmo salar L., fed semi-purified diets with and without astaxanthin supplementation. J Fish Diseases 18:317–328
13. Howell BK, Matthews AD (1991) The carotenoids of wild and blue disease affected farmed tiger shrimp (*Penaeusmonodon, Fabricus*). Comp Biochem Physiol B 98(2–3):375–379
14. Terpstra AHM (2015) De Piscium Colore Carotenoidibus The Role of Carotenoids in Coloring Fish An Overview of Data from the Literature Composit et scripsit: Antonius H.M. Terpstra Philosophiae Doctor Universitate Vadensi

15. Lorenz RT, Cysewski GR (2000) Commercial potential for Haematococcus microalga as a natural source of astaxanthin 2000. Trends Biotechnol 18:160–167
16. Proctor VW (1957) Some controlling factors in the distribution of *Haematococcuspluvialis*. Ecology 38(3):457–462
17. Droop MR (1955) Carotenogenesis in *Haematococcuspluvialis*. Nature 175(42). https://doi.org/10.1038/175042a0
18. Boussiba S, Vonshak A (1991) Astaxanthin accumulation in the green alga *Haematococcus-pluvialis*. Plant Cell Physiol 32:1077–1082
19. Kobayashi M, Kakizono T, Nishio N, Nagai S (1992) Effects of light intensity, light quality, and illumination cycle on astaxanthin formation in green alga *Haematococcus pluvialis*. J Ferm Bioeng 74(1):61–63
20. Domínguez-Bocanegra AR, Legarreta IG, Jeronimo FM, Campocosio AT (2004) Influence of environmental and nutritional factors in the production of astaxanthin from *Haematococcus pluvialis*. Bioresour Technol 92:209–214. https://doi.org/10.1016/j.biortech.2003.04.001
21. Kobayashi M, Kakizono T, Yamaguchi K, Nishio N, Nagai S (1992) Growth and astaxanthin formation *Haematococcus pluvialis* in heterotrophic and mixotrophic conditions. J Ferm Bioeng 74(1):17–20
22. Kobayashi M, Kakizono T, Nagai S (1993) Enhanced carotenoid biosynthesis by oxidative stress in acetate-induced cyst cells of a green unicellular alga *Haematococcus pluvialis*. Appl Environ Microbiol 59(3):867–873
23. Yoon J-E, Yoo K-C, Macdonald AM, Yoon HI, Park K-T, Yang E-J, Kim H-C, Lee JH, Lee MK, Jung J, Park JS, Choi J-M, Kim T-J, Kitae, Kim I-N (2016) Ocean iron fertilization experiments: past–present–future with introduction to Korean iron fertilization experiment in the Southern Ocean (KIFES) project. Biogeosci Disc 1–41. https://doi.org/10.5194/bg-2016-472
24. IPCC: Climate Change (2014) Mitigation of climate change, contribution of working group III to the Fifth Assessment Report of the Intergovernmental Panel on Climate Change. Eds.: Edenhofer O, Pichs-Madruga R, Sokona Y, Farahani E, Kadner S, Seyboth K, Adler A, Baum I, Brunner S, Eickemeier P, Kriemann B, Savolainen J, Schlömer S, vonStechow C, Zwickel T, Minx JC, Cambridge University Press, Cambridge
25. De La Rocha CL (2007) The biological pump. In: Holland HD, Turekian KK (eds) Treatise on geochemistry update, Elsevier Pergamon, Oxford, pp 1–29
26. Martin JH (1990) Glacial-interglacial CO_2 change: the iron hypothesis. Paleoceanography 5:1–13
27. Sarmiento JL, Gruber N (2006) Ocean biogeochemical dynamics, Princeton University Press
28. de Baar HJW, Boyd PW, Coale KH, Landry MR, Tsuda A, Assmy P, Bakker DCE, Bozec Y, Barber RT, Brzezinski MA, Buesseler KO, Boyé M, Croot PL, Gervais F, Gorbunov MY, Harrison PJ, Hiscock WT, Laan P, Lancelot C, Law CS, Levasseur M, Marchetti A, Millero FJ, Nishioka J, Nojiri Y, van Oijen T, Riebesell U, Rijkenberg MJA, Saito H, Takeda S, Timmermans KR, Veldhuis MJW, Waite AM, Wong C-S (2005). Synthesis of iron fertilization experiments: from the iron age in the age of enlightenment. J Geophys Res 110, C09S16
29. Hugh P (2008) Dumping iron and trading carbon. Profits, pollution, and policies all will play roles in ocean iron fertilization. Climate Mitigation Research 46(1)
30. Lawrence MG (2002) Side effects of oceanic iron fertilization, science, 297, 1993
31. Sarmiento JL, Orr JC (1991) Three-dimensional simulations of the impact of Southern Ocean nutrient depletion on atmospheric CO_2 and ocean chemistry. Limnol Oceanogr 36:1928–1950
32. Trick CG, Bill BD, Cochlan WP, Wells ML, Trainer VL, Pickell LD (2010) Iron enrichment stimulates toxic diatom production in high-nitrate, low-chlorophyll areas. Proc Natl Acad Sci USA 107:5887–5892
33. Moore SK, Johnstone JA, Banas NS, Salathé EP (2015) Present-day and future climate pathways affecting *Alexandrium* blooms in Puget sound, WA, USA. Harmful Algae 48:1–11
34. Smetacek V, Naqvi SWA (2008) The next generation of iron fertilization experiments in the Southern Ocean. Philos Trans A Math Phys Eng Sci 366:3947–3967
35. Williamson P, Wallace DWR, Law CS, Boyd PW, Collos Y, Croot P, Denman K, Riebesell U, Takeda S, Vivian C (2012) Ocean fertilization for geoengineering: a review of effectiveness, environmental impacts and emerging 35 governance. Process Saf Environ Protect 90:475–488

36. Bange HW (2006) New directions: the importance of oceanic nitrous oxide emissions. Atmos Environ 40:198–199
37. Charpy-Roubaud C, Sournia A (1990) The comparative estimation of phytoplanktonic, microphytobenthic and macrophytobenthic primary production in the oceans. Marine Microbial Food Webs, 1990, 4(1): 31–57. In: Microalgal growth: inputs and losses, practical approaches. Maestrini SY, Rassoulzadegan F (eds)
38. Thornton DCO (2012) Chapter 27 primary production in the ocean. In: Advances in photosynthesis–fundamental aspects. pp 503–588. http://www.Intechopen
39. Chorus I, Bartram J (1999) Toxic cyanobacteria in water: a guide to their public health consequences, monitoring and management. In: Chorus I, Bertram J (eds) Who (world health organisation). Publisher E & FN Spon, Geneva
40. El Semary NA (2016) Benthic dinoflagellates from Red Sea, Egypt: early records. Egypt J Aquat Res 42:177–184
41. Whittle K, Gallacher S (2000) Marine toxins. Br Med Bull 56(1):236–253
42. Muthukumaravel S, Padmanabhan V, Boopathidoss PS, Sadanandane C, Srinivasan R, Gunasekaran K, Sabesan S, Balaraman K (2010) Invasion of toxic marine cyanobacteria in to the tsunami affected coastal villages of southern India. J Commun Dis 42(2):135–138
43. Buck EH (2012) Ballast water management to combat invasive species. U.S. Congressional Research Service. Report No. RL32344
44. El Semary NA (2005) *Anabaena* and associated bacteria: molecular approaches to studying community structure and taxonomy. Ph D thesis, Faculty of Science, University of Bristol, UK
45. http://www.fao.org/docrep/012/i1141e/i1141e01.pdf. Accessed on 10 Feb 2018
46. http://www.fao.org/docrep/003/W3732E/w3732e08.htm. Accessed on 10 Feb 2018
47. Boyd PW, Law CS, Wong CS, Nojiri Y, Tsuda A, Levasseur M, Takeda S, Rivkin R, Harrison PJ, Strzepek R, Gower J, McKay RM, Abraham E, Arychuk M, Barwell-Clarke J, Crawford W, Crawford D, Hale M, Harada K, Johnson K, Kiyosawa H, Kudo I, Marchetti A, Miller W, Needoba J, Nishioka J, Ogawa H, Page J, Robert M, Saito H, Sastri A, Sherry N, Soutar T, Sutherland N, Taira Y, Whitney F, Wong S.-KE, Yoshimura T: The decline and fate of an iron-induced subarctic phytoplankton bloom. Nature 428: 549–553
48. http://hab.ioc-unesco.org/index.php?option=com_content&view=article&id=5&Itemid=16, http://www.whoi.edu/redtide/, http://www.issha.org/Welcome-to-ISSHA/Harmful-algae
49. http://www.gisp.org/publications/toolkit/BiofoulingGuidelines.pdf
50. http://main.omanobserver.om/?p=57413
51. http://www.imo.org/en/About/Conventions/ListOfConventions/Pages/International-Convention-for-the-Control-and-Management-of-Ships'-Ballast-Water-and-Sediments-(BWM).aspx

Climate Change Impact on Immune Status and Productivity of Poultry as Well as the Quality of Meat and Egg Products

Ahmed Soliman and Assem Mohamed Safwat

Abstract The potential impacts of climate change on poultry may include changes in production, reproduction, quality of their products (meat or eggs) and diseases. It is noted that high temperature during the summer season or in regions with hot weather and high relative humidity prevent broiler chicks and laying hens to express their high genotypes, especially when they are raised in an open production system. Current poultry production systems comprise large numbers of birds being housed together making them more susceptible to heat stress. Heat stress not only causes inconvenience and high mortality rate for birds, but results also in lower or lost production which therefore reduces the profitability. Both of production performance and feed conversion ratio are affected by heat stress conditions, which affect the production rate. Other effects associated with heat stress include immunity reduction and weak immune response to vaccines that decrease the resistance of birds to many infectious diseases. In laying hens the production is markedly decrease and does not reach to the peak with declining egg quality (e.g. thin and breakable eggshell) in addition to lower egg weight with small size. This chapter discusses some of the key principles (nutritional or managerial practices) that could be used in order to alleviate the adverse effects resulting from heat stress.

Keywords Climate change · Heat stress · Poultry production · Feeding strategies

1 Introduction

In the current decades, many countries of the world (mainly in tropical and sub-tropical regions) are affected by climate change and global worming which resulted from emitting greenhouse gases (GHG) such as water vapor (H_2O), Carbon dioxide (CO_2), Methane (CH_4) and Nitrous oxide (N_2O). The negative effects of increased

A. Soliman (✉) · A. M. Safwat
Department of Poultry Production, Faculty of Agriculture (El-Shatby), Alexandria University, Alexandria, Egypt
e-mail: ahmed_soliman919@yahoo.com

A. M. Safwat
e-mail: assemsafwat2004@yahoo.com

© Springer Nature Switzerland AG 2020
E.-S. E. Omran and A. M. Negm (eds.), *Climate Change Impacts on Agriculture and Food Security in Egypt*, Springer Water, https://doi.org/10.1007/978-3-030-41629-4_20

environmental temperature due to climate change on poultry production representing lower feed intake, reduction of weight gain and meat quality in broilers, in addition to low egg production rate and egg quality, increasing feed conversion ratio and high mortality rate in laying hens.

An increase in global average surface temperature by 2100, between 0.3 and 4.8 °C, is reported in the IPCC Fifth Assessment Report [1]. That could have several impacts on livestock production including variation in production and quality of feed crops and forage, and biodiversity, water availability and animal health, growth, production and reproduction. Therefore, effective procedures must be taken to challenge this potential risk of high temperature and its impact on livestock in general and on poultry in particular through the implementation of effective strategies in order to protect poultry survival and increase their productivity during hot weather.

2 Response of Heat Stress in Poultry Herds

2.1 Maintenance of Body Temperature

Chicken is a homeothermic animal, which has the ability to maintain a relatively constant temperature of its internal organs. But this ability is highly efficient only when the thermoneutral zone about 21–28 °C which called the temperature range or comfortable zone, through balancing between the body amounts of heat produced and the amount of heat lose. The large part of the heat produced through the body metabolism, which increases the need to increase the elimination of heat to maintain the body temperature at the normal limit of 41.4–42.9 °C. Therefore, the high temperature of the surrounding environment to the direction of a large part of the production capacity in order to maintain the body temperature at the natural limits through a series of physiological changes that work on the stability of body temperature. These changes have direct and indirect negative effects in efficient bird production.

The loss of heat in poultry when exposed to high environment temperature is not through sweating due to lack of sweat glands in the body of poultry and the process of controlling the mechanism of regulating the temperature of the birds by hypothalamus in two ways: The first is sensible heat loss through conduction, convection, fecal excretion, egg production. This system works efficiently in the thermal range, here the size of the comb and wattles plays an important role by providing the body with sufficient space bare skin which provides blood, which transfers the heat from the body to the head to be lost in concrete form, we found that the bird loses about 40% of the heat who wants to get rid by this way. The second is an insensible heat loss, which includes exhaustion and increased evaporative cooling due to increased heart rate, blood flow to the skin and expansion of peripheral blood vessels. Evaporation occurs mainly in the lungs, air sacs, and mucous area between the nasal openings and the tracheal base.

2.2 Behavioral Response

The birds begin to change their behavior to facilitate heat exchange or to get rid of excess heat, which affects the productivity. The birds are exhausted and lose appetite and begin to remove their wings from the body, and increase the speed of breath and exhaustion to get rid of excess heat through water vapor, look for isolated places with the lowest temperature as they stick to cold joints. This is worse if the temperature increases with the increase in humidity as in our country now. Therefore, the ability of the bird to evaporate is reduced, thus increasing the risk of warming. The birds also take some physiological pathways, such as conversion and flow of blood from internal organs to skin [2].

When the temperature rises below 29 °C the thermal capacity of the bird decreases, leading to deterioration of the feed conversion efficiency. This results in a negative effect on production. Large and overweight birds are more prone to stress and mortality than small ones. The ideal range of the bird begins to perform several mechanical procedures to make its temperature constant can be explained as follows:

- Birds try to move away from each other.
- Move to cold surfaces, such as walls or places with air masses and cold currents.
- Remove the wings from the body to reduce the insulation and stripping any areas of the skin without feathers.
- Reduce the temperature of the fatty cover under the skin.
- The flow of blood to the skin of the limbs, especially comb and wattles.
- Reduce movement and reduce food consumption.
- Increasing water consumption.
- Eliminate excess heat through intense panting.

2.3 Physiological Response

The optimum temperature degree of production performance ranging between 20 and 24 °C for laying hens and 18–20 °C for broilers. Body temperature of chickens began increasing when the ambient temperature rose above 30 °C if the rate of increase in ambient temperature was rapid. Physiological response to heat stress in birds involve the functional integration of several organs to meet the metabolic needs of birds that are trying to dissipate heat and maintain homeostasis. The chicken response occurs through increased activity of hypothalamus, pituitary gland and adrenal gland. A series of physiological measures to counter this stress begin by increasing the rate of respiration from 25 times/min. to 260 times/min., resulting in large amounts of carbon dioxide and thus bicarbonate, in blood plasma [3]. In turn, the lowered concentration of hydrogen ions causes a rise in plasma pH, a condition generally referred to as alkalosis.

The hypothalamus starts to stimulate the secretion of hormones that inhibit urinary incontinence, becoming more watery. Corticosteroids are also released, increasing losses in electrolytes, sodium, potassium and bicarbonate. Some organs, such as the brain and circulatory system, are also affected and the burden on the heart and blood vessels increases. The rate of urinary incontinence is the result of water consumption. In the end, the immunity of birds is significantly reduced, especially the reduction of immunoglobulin in the blood, and the herds become more susceptible to diseases. All previous manifestations of high temperature affect the herds produced in the form of symptoms such as bird swings with peripheral spasms, low body weight, increased predation disease, decreased egg production, decreased cortex quality, decreased fertility and poor semen quality, and finally in severe cases to death.

Many studies have shown the possibility of adapting birds to the conditions of heat stress and their tolerance to this burden and low mortality if exposed early in the age to high temperature, and this adaptation is reflected in increased rates of fatigue and low body weight and the lack of water lost by evaporation, high on several factors, including: type of strain, average body weight, quantity of drinking water available and quantity of production.

2.4 Immune Response

The immune system functions can be broadly classified into: (i) innate immunity and (ii) adaptive immunity. Innate immunity is the germ line encoded, non-specific, preliminary line of defense against the invading pathogens. The first entry of the pathogens is prevented at body's entry site through antimicrobial components in mucosa, sweat, tears, saliva, etc. Heat stress directly or indirectly favours disease occurrence in animal host. Directly, high temperature favours the survival of organisms outside the host for a long time. Indirectly, chronic heat stress causes immune suppression in birds and makes them susceptible to diseases. Heat stress reduces the relative weights of lymphoid organs like spleen, thymus and cloacal bursa. This might be due to glucocorticoid induced lympholysis and redistribution of lymphocytes from systemic circulation to other organs NK cells are important components of the innate immune system present in systemic circulation and also in lymphoid organs like lymph nodes, spleen and bone marrow. They are involved in destruction of tumour cells and infectious agents like bacteria, fungi and viruses. Chronic heat stress reduced the splenic NK cell cytotoxic functions. The inhibition may be due to increased glucocorticoid influence in the immune cell [4].

Cytokine interactions are responsible for altered immune functions during heat stress. Stress induced glucocorticoids act to inhibit the pro-inflammatory cytokines, which are required to initiate an innate immune response through inhibition pathway.

Heat stress leads to activation of hypothalamic-pituitary adrenal axis and ultimately the release of glucocorticoids. Glucocorticoids, in normal pulsatile release, enhance the pro-inflammatory cytokine release. However, chronic rise in glucocorticoid levels is inhibitory to majority of immune cytokines. The stress related immune responses in poultry species revealed that acute stress is beneficial to the bird as it the immune system. On the other hand, chronic stress shifts the T helper cell response to T regulatory cell and also TGF-β, a regulatory cytokine production; thereby it suppresses immune response [5].

3 Effect of Heat Stress on Poultry Production

The rapid development of poultry production in many countries of the world with warm climates, such as most countries of the Arab global has, the interest for studying the impact of environment temperature increased as one of the consequences of climate change on poultry production, which is one of the biggest problems can be encountered, because it is the most important needs for poultry. The growth achieved for the quality and quantity production of the bird depends not only on the capabilities of the genetic and the food quality provided to him, but also depends on the environmental factors surrounding it, which negatively affect the ability of productivity and the economic benefits envisaged. Therefore, air temperature is considered one of the most important external environment factors, which directly affect the physiological and productive status on the birds causing constant stress.

Heat stress is the most stressful factor that causes severe losses in poultry farms, especially when high temperatures are accompanied by high humidity, such as some countries due to climatic changes like Egypt. Also, heat stress severely affects the productivity of poultry directly on egg production and growth rates of broilers or indirectly by increasing the probability of growth and reproduction of toxin-producing fungi in feed or by reducing the ability of birds to cope with other stressors. In females, heat stress can disrupt the normal status of reproductive hormones at the hypothalamus and at the ovary, leading to reduced systemic levels and functions.

3.1 Egg Production

The optimal care of the laying hens under the conditions of heat stress is one of the most important criteria that affect productivity efficiency. The factors related to the herd care and management is combined with the provision of the optimal ingredients of balanced nutrition during the growth stage to reach the production of the young layers to the optimal weight, good health and physical condition at sexual maturity. It is known that high temperature contributes significantly to the lack of access to optimal weight, which leads to the deterioration of productive efficiency throughout the cycle of egg production, it was found that the care of herds leaves within the temperature of 35 °C reduces the body weight at sexual maturity by 20–30% compared to the same hens rearing at a temperature of 21 °C.

It is preferable for the breeders during the heat stress to reach laying hens at sexual maturity to a weight slightly higher than the standard weight, which increases the consumption of food later, improve produced of the eggs weight and make them more persistent on production [6].

As soon as laying hens reach sexual maturity, ambient temperature becomes an important factor affecting production. The average feed consumption is 1.6% and the energy consumed is 2.3% for every 1° above the ideal temperature. At temperatures above 30 °C, both food consumption and egg production are reduced by 5% and 1.5%, respectively. The decrease in food consumption (50%) was found when the temperature rises from 21 to 38 °C and most of this decrease is due to the lack of the needs of the preservative body. Treating this decrease by increase the rest of the nutrients in the feed to overcome most of the harmful effects of the heat from an improvement in egg production and the egg weight, but he could not overcome the deterioration of the egg shell quality.

3.1.1 Egg Quality

Several studies have indicated that the bad effect of heat stress on the quality of internal and external eggs. This effect is clearly visible on the egg shell and calcium carbonate deposition. This is reflected in the appearance of the eggshell and its characteristics, with insufficient mineral sources and bicarbonate ions. The decrease in egg weight was estimated at 0.07–0.98 g/egg for each 1° rise. The thickness of the eggshell is reduced by high temperature, and it becomes more complicated at the later ages of the egg production cycle. This decrease is attributed to the low intake of calcium because of the low food consumption. The quality of the egg includes visible natural measurements as well as flavor and aroma. The effect of high temperature extends to the liquid components of the eggs, which quickly damage the internal contents of the egg yolk and reduces the quality and grade of the egg. Leaving the eggs for long periods under high temperature conditions lead to its total corruption, and these changes include the rapid erosion of the chemical layer, the number of open gaps, evaporation of large amounts of moisture content of the whiteness and with the persistence of high temperature, the yolk loses part of its moisture, the coagulation of the liquid components with rise. Fresh eggs production from layers is always of high quality, and there are many factors that affect the quality of eggs before produce it.

Changes in environmental conditions, especially the increase in temperature for the laying hens housing, resulting in lower quality of albumen and shell weak strength and small thickness and egg size, may be due to low consumption of food in the hot atmosphere. The importance of the role of management in the production of high egg quality in the case of high temperature, which is the selection of the strain that produces high egg quality, the balanced diet, diseases control, physiological disorders, and the organization of lighting, and food programs to control the beginning of the egg production. The eggs are damaged and reduced quality if not handled correctly. The most changes associated with changes in temperature is to increase the air sac size as the resulting of losing moisture, where the amount of evaporation for stored eggs depends on the environment temperature, humidity and ventilation as well as shell porosity. The signs of low egg quality turn into the water situation, and this shift depends on the heat stress, which is accompanied by natural reactions and chemical result in the degradation of the construction of the albumen thick protein. As the water dissolves, the water leaks to the yolk through the yolk membrane with the toxic properties, causing the yolk to grow and weakening the vitelline membrane and becoming extensible and facilitates the permeability of the albumin to the yolk, which causes its appearance to be spotted. Fresh egg contains 0.5% CO_2, which is lost during the storage period, resulting in a change in the flavor of the eggs, increasing pH alkalinity from 7.6 to 9.5.

There are some recommendations that must be taken into account to maintain the high egg quality after the production such as, collecting eggs constantly 2–3 times a day, separate the eggs clean from the dirty and broken eggs, and mobilization eggs in clean containers, storing the fresh egg in 10–13 °C and humidity ratio about 70–85%. Also, washing eggs with water temperature higher than the temperature of eggs; because the opposite causes shrinkage of the contents of the egg, resulting in the discharge accompanied by the entry of contaminated water into the components of the egg through the internal gaps, causing the contamination of the egg internally bacteria. Finally, it is possible to dip eggs in paraffin oil to close the gaps and reduce evaporation during storage; because the process of washing eggs lead to the removal of the shell cuticle.

Egg Shell Quality

The eggshell formation one of the processes that interferes with many factors such as genetics, nutrition, health status, hormones and environmental factors surrounding the birds these factors must be in a similar case to obtain the high eggshell quality. The amount of calcium element in the shell about 2.2 g Calcium carbonate form (94%), organic matrix (about 4%), the chicken pours about 25 mg of calcium on the egg surface, which is formed every 15 min during the 20 h period needed to form the eggshell. Due to the spawning of laying hens production, it deposits 24 times more calcium in the eggshell than in its bones. The quality of the eggshell including the color, the number of porosity, the shell strength and its durability as well as its tolerance to stress and shock.

The birds during heat stress resort to increase the rate of breathing in the minute, which phenomenon of panting. In this way, the excess heat is eliminated through the surfaces of the mouth and respiratory passageways. In general, the bird begins panting when the temperature reaches 29 °C as a visible response to stress. The increase in the rate of panting, increase losing of CO_2 by the lungs, leading to a decrease in its molecular pressure and consequently lower bicarbonate ions in the blood, which leads to a decrease in hydrogen ions causing a rise in alkalinity of blood [7].

This egg shell forming is reflected process in the uterus, where carbon dioxide combines with water to give carbonic acid and then enters the carbonic anhydrase enzyme in the reaction and produces carbonic acid in the mucosal cell of the secretion of the shell (uterus). Then combines the bicarbonate with the calcium ion to form carbonate calcium. High temperature and imbalance of pH, where the concentration of bicarbonate become low (the second slice necessary to form the egg shell) and do not find enough calcium ions to connect with it for the formation of calcium carbonate, which is the basic component of the shell as mentioned above. As a result of this, eggs are produced with weak shell sickness and the proportion of eggs without shell.

Improve Egg Shell Quality

There are many trends and practices that minimize the harmful effects of exposure to long periods of heat stress, including environmental issues related to management factors and herd care, including the treatment of nutrition [8]. For example, the addition of ascorbic acid at the rate of 250 mg/kg of important treatments that reduce the impact of heat stress on birds and improve the properties of eggshell produced, this vitamin reduces the level of cortisone in blood plasma and helps stimulate the absorption of calcium and vitamin D of the intestine, then increase the calcium level in the plasma to the point where it supports the normal deposition of minerals in the bones [9].

Respiratory alkalosis can also be combated nutritionally by providing a source of anion via feed or water. Supplemental ammonium chloride in drinking water of chronically heat-stressed birds can return blood pH to normal [10]. During acute heat stress, the provision of ammonium chloride or carbonate water has been found to decrease blood pH. Many researchers pointed to the need to add sodium bicarbonate in the drinking water of the laying hens at a rate of 0.5–2% under the conditions of thermal stress, lead to increase the shell strength and improve its quality. The addition of bicarbonate leads to the adjustment of pH, which positively affects the quality of egg shell [11].

Some minerals, such as zinc, manganese and copper, are involved in the metabolism of the eggshell, by acting as catalysts for enzymes involved in the formation and stimulation of calcium carbonate deposition to form the egg shell. The addition of the zinc element of the important points to increase the level of the enzyme carbonic anhydrase, which enters as a component in the composition of the enzyme, with increases its activity in the liquid of egg gland.

3.2 *Meat Production*

The heat produced from broiler is very high, because it consumes high energy in food while keeping 40%. Thus, 60% of food energy is eliminated in the form of heat production. Under heat stress birds reduce feed consumption and thus produce degradation. The performance of birds varies according to the environmental conditions in the breeding area. In areas with high temperatures during the day and low at night, performance is better than those characterized by high temperature. On this basis, the raising of broilers in the open bays gives better results in the temperate seasons (spring and autumn) compared to the other seasons in year, where growth rates are clearly lower during the summer.

The efficiency of the birds is also different depending on the composition of the temperature and humidity. We find a decrease in the weight of the birds raised in areas with high temperature and humidity, as in Egypt, compared to other high temperature and low humidity. Also, the continuity of high and constant temperature for long periods has a serious role in the deterioration of growth rates and low conversion efficiency of food, and increase the mortality rate when compared to fluctuating temperature (cycling). The latter allows the bird to recover and increase its consumption of feed during periods of moderate heat, and the occurrence of hot flashes during the early stages of broiler is less dangerous than in the advanced stages, especially before marketing.

3.2.1 Meat Quality

It is known that the bird's under cold weather conditions needs more energy to maintain its body temperature. In the case of high temperature, this energy decreases and increases in the ratio of solids, fats and energy in the carcass, while the percentage of protein decreases. Unsaturated fatty acids such as oleic, linoleic, linolenic, in the fat of the carcass are high in ambient temperature, which means high saturated fatty acids in the carcasses intended for marketing. This increase was found to be higher in females than in males during high temperature. High temperatures also caused a decrease in the aminoity of glycine and proline in carcass tissues, especially the muscles of the breast. Early finishing of broilers from the viewpoint of fatty acid composition because the polyunsaturated fatty acid to saturated fatty acid ratio declines significantly with age, regardless of temperature.

The symptoms of heat stress on the carcasses in the form of weakness of coloration and roughness of the skin, dry muscles, increase blood density and color density and then the value of carcasses. The heat stress has been found bad effect on the ratio of reflux and the amount of meat eaten and the proportion of breast muscles. Broilers may be exposed to a variety of stressors during transport from the production farms to the processing facilities, including thermal challenges of the transport microenvironment, acceleration, vibration, motion, impacts, fasting, withdrawal of water, social disruption, and noise. As part of this complex combination of factors, thermal stress, in particular heat stress, plays a major role. The adverse effects of these factors and their combinations range from mild discomfort to death. In fact, heat stress during transport has been associated with higher mortality rate, decreased meat quality, and reduced welfare status.

3.2.2 Pathological Phenomena Associated with Heat Stress

Ascites

Ascites is a symptom of metabolic disorder due to several factors affecting the environment, temperature, carbon dioxide/oxygen ratio, rapid growth rates, high energy content, and high basal metabolic rate. As a result, a combination of lymphatic liquids and blood plasma that leaks from the liver accumulates in the body cavity, often leading to death. The symptoms of ascites are associated with an abnormal increase in blood pressure between the heart and lungs, leading to the failure of the right ventricle and increase blood pressure in the veins, and the great acceleration in the production of fluids from the liver that seeps into the body cavity. The symptoms appear on the birds in the form of inactivity and laziness and difficulty breathing, Skin and congestion, decreased growth and fill the abdomen with liquids.

The incidence of ascites has increased throughout the world over the past years. This increase coincides with continuous improvements in the genetic structure and nutritional efficiency of the herds, which caused the presence of strains with a rapid growth rate and food efficiency. Broiler led to a high rate of metabolism and increase in the size of the breast muscles, which did not develop with the internal organs as much as the heart and lungs and increased pressure to meet the needs of the body of chicken overload of blood feeding, unable to provide enough oxygen to maintain these high rates. The number of lighting hours can also be reduced to 18 h/day, which reduces the volume of cellular blood components and reduces the level of T3 hormone, which reduces the mortality rate resulting.

Pendulous Crop

Birds are under the conditions of thermal stress for long periods and in the presence of some genetic factors responsible for this phenomenon, begin to eat large amounts of feed over consuming when the moderate temperature and then drink large amounts of water with high temperature, which leads to fill the bulk of these quantities and cannot easily discharged. The muscles of the compass relax and lie in the shape of a bag filled with the front of the breast. This helps the weakness of elastin fibers forming the membrane of the compass of heat stress [12].

It should be noted that this phenomenon is more common in the strains of broiler chickens and turkeys especially heavy ones, and there are some administrative practices in the farm or food additives that help to the emergence of this phenomenon. Such as bird staging for long periods during the hot weather, containment of the feed on chemicals or disinfectants mycosis is caused by a secondary infection in the membranes and muscles, leading to weakness, and injury to the fungal contaminants or internal worms.

Breast Blisters and Crooked Sternum

This phenomenon occurs in heavy-weight birds which brooding in wire cages or on the floor. As age and weight increase, birds sit on the wired floor or mattress. During this period, this weight is placed on the bone sheath. There are blisters and breast ulcers, especially during high temperature, the birds search for the wet places inside the house. Also, the increase of humidity mattresses with poor specifications, high content of microbes and pathogens. This leads to the occurrence of breast defects and lower the carcass quality during the stages of slaughtering, then reduce economic return. The breeders can reduce the incidence of this phenomenon by work for a program to market heavy birds first to early ages, ensure the cleanliness and quality of the mattress and finally avoid the presence of dark areas in the house so as not to be left to birds.

4 Feeding Strategies to Deal with Heat Stress

4.1 Feed Restriction

Starving the birds during peak heat stress is an effective way to reduce body temperature and increase survival, as feed intake increases heat increment resulting from ingestion, digestion, absorption and metabolism of nutrients. Feed restriction, in the meantime, makes birds quiet and less energy-producers. Metabolic heat increment is a maximum after 3–6 h of feed intake. Therefore, it is necessary to feed the birds early in the morning and late in the afternoon so that the production of metabolic heat will not occur during higher temperature period of the day (09:00 to 16:00 h). By this way birds will be encouraged to eat during cooler periods of the day and help maintain their performance during hot weather.

4.2 Feed Pelleting

The amount of feed intake by chickens is reduced at hot weather [13] and as a result, the amount of feed components consumed decreases. This is one of the main reasons leading to lower productivity of birds and their performance with the high feed conversion ratio to non-economic degree. Therefore, there are many attempts to increase the quantity and/or the concentration of feed ingredients consumed, without increasing internal body heat, in order to improve birds' performance and productivity.

Pelleting process increases the physical density of feeds; the pelleted diet also improves the digestibility, decreasing the production of metabolic heat with respect to the mash feed [14]. Over and above, birds prefer to eat more feed with larger particle size in hot weather.

4.3 Separated Feeding

A strategy of dividing the diet components at different times of the day could be applied during the hot climate. Birds under this regime are provided with their daily requirements of minerals, vitamins and other supplements in hot periods of the day, when on the other hand they are provided with their basic energy and protein requirements during cooler periods in the evening. This strategy is useful for birds to tolerate heat stress; considering that vitamins and minerals reduce the adverse impact of heat stress during the midday, while proteins and energy are important as major nutrients required to maintain good performance and productivity, providing those major nutrients at the evening reduces the excess heat overload resulted from nutrients digestion and metabolism during high temperature.

4.4 Self-selection

This method depends on giving the birds the opportunity to select from various feed ingredients that match their physiological requirements; this method is similar to what wild or even backyard birds naturally obtain their feed. Under heat stress conditions, the bird can adjust its consumption of different feed ingredients so that it consumes less protein and a slightly higher amount of energy (by consuming high caloric feed ingredients) compared to the complete mixture diet, in an attempt to regulate the heat load associated with metabolism during hot cyclic periods.

In this strategy, feeders containing feed ingredients (each separately) should be provided, including: (1) cereal grains such as corn, wheat, barley or other energy source either together or separately, (2) protein sources (26–48% protein content), (3) major minerals containing calcium and phosphorus, and (4) minor minerals and vitamins premix.

However, cautions should be taken with this feeding system in case of laying hens, which may prefer only grains feeders, which will lead to fat birds with low egg production rate. In general, containers of grains should represent only two-thirds of the total containers number, while the protein source and the premix containers represent one third of the total containers.

The results of the choice feeding strategy are variable, as it could be affected by some factors such as birds' age, the initial age from when the selection was offered, the particle size of the feed ingredients, the quality of energy and protein sources, as well as the difference between daily maximum and minimum temperatures.

5 Nutritional Applications for Heat Stress Mitigation

All poultry species are severely affected by heat stress conditions, preventing their access to the maximum biological performance, so it was necessary to search for modern solutions along with the traditional methods to challenge heat stress in order to protect birds and maintain their production. Nutritional solutions have now become more important to mitigate the detrimental effects of heat stress on birds, and this can be achieved by applying one or more nutritional procedures.

5.1 Using a High-Energy Diet

Birds mainly consume food to satisfy their energetic needs required for maintain life and perform various physiological functions. Fats contain 2.25 times the calories per unit of weight than protein or carbohydrates; generally they are used to increase the energy density of poultry diets. Indeed, fat gives less biological heat production compared to either protein or carbohydrates, because dietary fat metabolism has lower heat increment than protein or carbohydrates. Fat is characterized by decreasing the physiological duty inside the body leading to higher utilization of food energy to be used in different production functions, unlike the carbohydrate energy, which is characterized by the high amount of energy loses during digestion and metabolism.

Recently, researches revealed the benefits of supplementing poultry diets with vegetable oils to account for about 33% of the total dietary energy, replacing part of the traditional energy source (carbohydrate). Where it was noted that supplementing oils or fats by 3–5% of the diet increased the appetite and taste of feed, also it increased the utilization of various nutrients in the feed consumed, besides fat contains good amounts of soluble vitamins (A, D3, E, K) and the essential fatty acid (linoleic acid) needed for growth and egg production.

However, caution should be considered when formulating such these feeds since an appropriate antioxidant should be supplemented and not storing the feed for a long time in order to avoid fat rancidity.

5.2 Dietary Protein and Amino Acid

The requirements of protein and amino acids should be met at the recommended levels for growth and egg production, taking into consideration not to excessively increase the dietary protein level especially under heat stress conditions as the heat increment associated with protein catabolism is higher when compared to that of fats or carbohydrates.

Supplementation of essential amino acids is profitable for reducing the internal heat loss which leads to reduce the adverse effect of high temperature, in this regard it was noted that increasing dietary lysine concentration is necessary to compensate partly for the reduced feed intake [15] and improve feed efficiency. There is a positive effect to increase the Arginine: Lysine ratio on feed conversion ratio and growth performance at high temperatures.

The imbalanced amino acids in the diet results in higher excretion of nitrogenous substances in faeces, which produces and accumulates ammonia in the house atmosphere causing harmful effects on the performance and welfare of birds. The accumulation of ammonia may increase the ambient temperature in poultry houses; therefore it makes difficult for birds to regulate their body temperature.

5.3 Vitamins

Supplementing vitamins to poultry diets is highly recommended during the hot weather, such as Vitamin A, E, C, etc., which play an important role in improving the performance and immunity of birds and significantly reduce mortality rates, especially when birds exposed to heat stress. It is suggested to increase the levels of vitamins during hot weather by 10–15% than the normal requirement because of the low feed intake and potential damage due to feed storage or exposure to oxidation factors, such as high temperature and/or humidity and trace elements presented in the diet, unsaturated fatty acids and oxidation of peroxides, which affects the stability of these vitamins.

5.3.1 Vitamin A

The supplementation of vitamin A in the diet reduces the detrimental effect of heat stress on egg production and has a positive effect on the immunity of laying hens. Higher dietary vitamin A intake is required by birds suffering from heat stress after Newcastle vaccination; in order to produce the maximal level of antibody [16]. Also, supplementing vitamin A to broiler diets improves body weight gain, feed efficiency and carcass traits.

5.3.2 Vitamin C

Ascorbic acid has an important role as an antioxidant in the biological system and under the heat stress conditions, it is important in maintaining the oxidation and reduction processes in birds body. Supplementing ascorbic acid to the diets improves egg production and egg shell quality of broiler breeders; additionally it has beneficial effects on the fertility and hatchability of broiler chicks as well as molting process of laying hens.

Furthermore broiler chicks tend to consume more amount of the diet supplemented with ascorbic acid in the hot weather. Ascorbic acid supplementation improves carcass quality, increases the carcass weight associated with higher crude protein retained in the carcass and reduces the carcass content of fat.

5.3.3 Vitamin E

Vitamin E acts as a natural antioxidant that protects the fat from oxidation rancidity and peroxides formation, especially at high temperatures. It also preserves some fat soluble substances such as vitamins A and D_3, carotene from damage.

Supplementing vitamin E to heat stressed laying hen diets enhanced egg production; this beneficial effect is accompanied with an increase in feed intake and the solids of albumen and yolk. The supplementation of vitamin E at high levels in the diet not only increases egg production but also improves the immunity of bird sat hot weather.

5.4 Electrolytic Balance

Birds get rid of the excess body heat through the respiratory system by panting. With the increase of panting, the body loses large quantities of carbon dioxide, leading to imbalance in the acid/base balance of the bird's body causing an imbalance in the electrolyte balance with severe high blood alkalosis that increases blood viscosity and blood pressure which may lead to an explosion in the arteries of the brain, liver or kidneys.

Supplementing electrolytes such as sodium or potassium bicarbonate ($NaHCO_3$, $KHCO_3$) and ammonium, sodium or potassium chloride (NH_4Cl, $NaCl$, KCl) in the diet or in drinking water by 0.5–1.0% stimulates water consumption and increases bird tolerance to heat stress [17], where electrolytes contribute to adjust the imbalance resulted from increasing blood pH (blood alkalosis), which contributes to produce stronger eggshell for laying hens and have a positive impact on broiler performance.

6 Managerial Application for Heat Stress Mitigation

Management of stress is the key to profitable poultry production during the hot summer months where temperature reaches 40 °C or more, moreover some sea dominating regions may have a relative humidity of 80%. Many production systems suffer from low efficiency and the losses are sometimes costly, such as low growth and survival rate. Decision-making on the type of strategy should be applied to manage stress requires an understanding of the physiological effects of heat stress. The profitability of any strategy during the summer months could be increased when using proper managerial applications.

6.1 Drinking Cold Water

The internal environment of the bird is water medium in which food components are exchanged. In addition to its other important functions of heat regulation, lubrication, softening and hydration verify that water as a food component must be considered. The positive relationship between water intake and body temperature indicates that water used plays an important role as a heat discharger and thus makes energy available for growth rather than body cooling.

Water consumption is very important for poultry in a heat stress environment. Birds must drink more water to prevent dehydration, since the primary method of reducing body temperature is by evaporation of water from the lungs. Securing cold drinking water will help to reduce body temperature and thus increase the survival rate.

Under normal conditions water intake is usually 1.5–2 times feed intake, however during hot weather water consumption increases by 3–6 times than the normal consumption, with an increase by 5% per every temperature degree above 22 °C.

6.2 Ventilation

Both of heat production and heat loss of birds are affected by the efficiency of ventilation system, as the thermal loss through heat emission by evaporation depends on the ability of air to carry water vapor from the hot exhalation air resulting from panting. Therefore, it is important to consider the ventilation system to remove the hot air out of poultry house and replace it with dry, fresh and cool air.

Good ventilation reduces the relative humidity at the house environment and reduces harmful gases (ammonia—carbon monoxide—carbon dioxide); it also provides birds with sufficient oxygen supply required for the biological activities. Ventilation should be at a rate of 5–6 m^3 air/h/kg live weight and the air should be changed at a rate of 20–50 times per hour, so that the excess heat inside the house could be eliminated.

6.3 Stocking Density

Increasing the number of birds in poultry house than the recommended number reduces the ability of birds to get rid of heat during hot weather. Therefore, it is necessary to decrease the density of birds in order to reduce; first its requirements of ventilation, oxygen, cooling … etc., second to reduce the complications of excreta, moisture, carbon dioxide, ammonia (resulting from the fermentation of litter material with organic matter) and urea resulting from birds urine. As the less density per square meter of floor the more air available to birds and the lower relative humidity, which means better environment and condition to achieve higher profitability.

The number of birds per square meter of floor should not be increased than 7–8 birds in case of deep litter system, taking into account the strain size.

7 Conclusions

At hot weather, the use of light is important to regulate feeding behavior so that birds eat better during the cold hours of the night than during the hot times of the day. The intermittent lighting system improves feed efficiency and productivity, in addition to the above, has a positive effect in reducing body heat production during both lighting and dark periods.

References

1. IPCC (Intergovernmental Panel on Climate Change) Climate change (2013) The physical science basis. In: Stocker TF, Qin D, Plattner GK, Tignor M, Allen SK, Boschung J, Nauels A, Xia Y, Bex V, Midgley PM (eds) Contribution of working group I to the fifth assessment report of the intergovernmental panel on climate change. Cambridge University Press, Cambridge, United Kingdom and New York, NY, USA, p 1535
2. Rose SP (1997) Principles of poultry science. CAB International, Wallingford, Oxon, UK
3. Wang S, Bottje WG, Kinzler S, Neldon HL, Koike TI (1989) Effect of heat stress on plasma levels of arginine vasotocin and mesotocin in domestic fowl (Gallus domesticus). Comp Biochem Physiol 93A(4):721–724
4. Won SJ, Lin MT (1995) Thermal stresses reduce natural killer cell cytotoxicity. J Appl Physiol 79(3):732–737
5. Shini S, Huff G, Shini A, Kaiser P (2010) Understanding stress-induced immunosuppression: exploration of cytokine and chemokine gene profiles in chicken peripheral leukocytes. Poult Sci 89(4):841–851
6. Daghir NJ (1995) Poultry production in hot climate. CAB International, Wallingford, Oxon, UK
7. Teeter RG, Smith MO, Owens FN, ARP SC (1985) Chronic heat stress and respiratory alkalosis: Occurrence and treatment in broiler chicks. Poult Sci 64:1060–1064
8. Balnave D, Muheereza SK (1997) Improving eggshell quality at high temperatures with dietary sodium bicarbonate. Poult Sci 76:588–593
9. Sahin K, Onderci M, Sahin N, Gursu MF, Kucuk O (2003) Dietary vitamin C and folic acid supplementation ameliorates the detrimental effects of heat stress in Japanese quail. J Nutr 133:1882–1886
10. Teeter RG, Smith MO (1986) High chronic ambient temperature stress effects on broiler acid-base balance and their response to supplemental ammonium chloride, potassium chloride, and potassium carbonate. Poult Sci 65:1777–1781
11. Chen JLI, Balnave D, Barke J (2005) The influence of dietary sodium chloride, arginine: lysine ratio, and methionine source on apparent ileal digestibility arginine and lysine in acutely heat-stressed broilers. Poult Sci 84:294–297
12. Yahav S (2000) Relative humidity at moderate ambient temperature: its effect on male broiler chickens and turkey. Br Poult Sci 41:94–100

13. NRC (1994) Nutrient requirements of poultry, 9th ed. National Research Council
14. Ruhnke I, Röhe I, Krämer C, Boroojeni FG, Knorr F, Mader A, Schulze E, Hafeez Abdul, Neumann K, Löwe R, Zentek J (2015) The effects of particle size, milling method, and thermal treatment of feed on performance, apparent ileal digestibility, and pH of the digesta in laying hens. Poult Sci 94(4):692–699
15. Corzo A, Moran ET, Hoehler D (2003) Lysine needs of summer-reared male broilers from six to eight weeks of age. Poult Sci 82:1602–1607
16. Lin H, Wang LF, Song JL, Xie YM, Yang QM (2002) Effect of dietary supplemental levels of vitamin A on egg production and immune responses of heat-stressed laying hens. Poult Sci 81:458–465
17. Lin H, Jiao HC, Buyse J, Decuypere E (2006) Strategies for preventing heat stress in poultry. World's Poult Sci J 62:71–85

Climatic Change and Chicken Immunity

Akrum M. M. Hamdy

Abstract As the most socio-economic problems in the world, climatic change is an issue that will cause most damaging for those least able to adapt. But nobody is immune to climatic change—making it a truly global problem. Climate change affirm that the pollution of our atmosphere could result in extreme weather events. With intensify in global temperatures, processes such as desertification are transforming once thriving areas into arid environments. Since warm air is capable of stocking huge quantities of water, due to higher evaporation rates, storms and other extreme climate events have become more frequent and extremist. The challenges posed by climatic change fit broadly into one of two categories: loss of productivity or increasing costs. Stress is the nonspecific response of the body to any demand, whereas stressor can be defined as "an agent that produces stress at any time". Therefore, stress represents the reaction of the animal organism to stimuli that disturb its normal physiological equilibrium or homeostasis, due to the mess up of energy and water balances through the evaporative heat loss mechanism. Concerning the effects of heat stress on chickens, the following findings were reported: (1) energizing glucocorticoid-dependent mechanisms and increasing the corticosterone serum levels (2) decreases in the feed consumption, growth rates, the intestinal villi heights, and the wet and dry weights of jejunum, the total white blood cell count; (3) decreases in the number of peripheral blood lymphocytes and induction of an electrolyte imbalance; (4) decreases in the blood lymphocytes and spleen weight; (5) decreases in macrophage activity and (6) decreases in CD4+ and CD8+ lymphocytes and decreases the antibody production against sheep red blood cells (SRBC).

Keywords Climatic change · Heat stress · Heat balance · Chicken production · Chicken immunity

---

A. M. M. Hamdy (✉)
Animal and Poultry Production Department, Faculty of Agriculture, Minia University, Minia, Egypt
e-mail: akrum312@hotmail.com

© Springer Nature Switzerland AG 2020

E.-S. E. Omran and A. M. Negm (eds.), *Climate Change Impacts on Agriculture and Food Security in Egypt*, Springer Water, https://doi.org/10.1007/978-3-030-41629-4_21

1 Introduction

Whether we believe global warming is real or imagined, the atmospheric concentrations of certain gases are rapidly increasing to levels that we have not seen before. While the impacts of these increased concentrations on climate are less certain, it is notarized that these gases will gain heat in the atmosphere and could lead to global warming. The impacts of environmental change have achieved a point where the accentuation has moved from verification of its reality to approaches of moderation. In this way, farming and different divisions are under expanding open strain to diminish outflows of these gases [1].

Chickens, like all animals, have very strong, built-in defenses (immunity) against diseases that are caused by overrunning of the body by various microorganisms and toxins (collectively called 'antigens'). The essential part of the resistant framework is to see life forms and substances that are viewed as outside, or "non-self" (antigens), that can enter the body. A variety of mechanisms are employed to achieve this goal, including inactivation of biological agents, agglutination or precipitation of molecules or cells, or phagocytosis of foreign agents [2].

Under certain circumstances, these normally protective responses can result in significant tissue damage, which leads to immune-mediated diseases. Chick builds up a pressure protection relies upon how well it battles the intrusion that thus relies upon: (a) The chick's condition, condition of prosperity and level of resistance; (b) The quantity of the attacking life forms, called 'the test' and (c) The destructiveness or quality of the attacking living beings [3].

Chicken's scientists are constantly dealing with ways to make chicken houses more effective. Different methods for ending up more proficient are through enhancing development rate, nourish productivity, and malady protection and additionally invulnerable limit. This has just been significantly enhanced in the chicken business through hereditary determination and enhanced sustenance.

At the point when contrasted with other creature generation frameworks, the broiler, layer and parent stocks chickens are thought to be exceptionally proficient. However, the scale of the chicken sector is vast and even small impacts can add up; therefore, we must be prudent and continue handling to diminish the impact of heat stress as a consequence of climatic change and make the industry more sustainable.

The objective of this chapter is to compile the current knowledge about the importance and impact of climatic change, warming the global, on chicken production, focusing on chicken immunity.

2 Chicken Immunity

Immunity is the biological capacity dealing to protect and prevent the chicks against diseases that are caused by foreign antigens [4]. These defenses include: (a) The skin,

(b) Mucous membranes and (c) The immune system. Generally chicken's immune capacity could be classified into:

2.1 Innate Immunity

This alludes to the common or acquired capacity to resist sickness. This type of immunity includes a number of non-specific disease response mechanisms: (a) **Genetic factors**: birds may lack the receptors required by disease organisms in order to be able to infect. (b) **Body temperature**: the high body temperature of the chicken blocks numerous diseases. (c) **Anatomic features**: many disease organisms cannot penetrate intact body coverings (skin and mucous membranes) or are trapped in the mucus secretions. (d) **Normal microflora**: the skin and gut normally maintain a dense, stable microbial population. (e) **Respiratory tract cilia**: parts of the respiratory tract are fixed with fine hair-like bulges, called cilia, which expel illness living beings and spoil. (f) **Immune system cells**: are those cells that assault remote cells in the body and are enacted by the synthetic properties of the antigens, these cells include: (i) **White blood cells (WBC's) or leukocytes**: they move freely in body, interacting with and capturing cellular debris, foreign particles, and invading microorganisms. (ii) **Mast cells**: They are residing in connective tissues and mucous membranes. The substance released by activated mast cells "histamine" also initiate the recruitment of **neutrophils** and **macrophages**. (iii) **Phagocytes**: These immune cells kill the pathogens by engulfing them. (iv) **Natural killer cells (NK)**: The cells of an innate immune system that do not directly attack pathogens are called NK cells. These cells act by invading the infected host cells, such as tumor cells or virus-infected cells [5, 6]. Different elements associated with the viability of inborn protection incorporate sufficient nourishment, stress, age, inflammatory processes and metabolic factors.

2.2 Adaptive Immunity

Adaptive immunity also originates "immunological memory" once an initial response to a specific pathogen has been outranged, which can enhance future responses to that specific pathogen and act more efficiently. The immune cells or antibodies of the adaptive immune system are **T** and **B** lymphocytes [7–9].

The adaptive immunity is specific to an antigen and can take one of two forms [4]: (a) **Passive immunity**: this is achieved when antibodies are transferred to the individual either from the mother (such as during egg formation in the hen or by thrift of antiserum (blood serum containing antibodies) either orally or through injections. (b) **Active immunity**: this alludes to a dynamic invulnerable reaction in a chick because of recuperation from the illness or by reaction to an immunization. The chick delivers its own resistant cells as well as antibodies to give assurance. (c) **The non-cellular**: (or humoral) immune response: The B-cells then respond by producing antibodies specific to the disease organism. (d) **The cellular immune response**: The cellular component of the immune response includes all the cells that react with specificity to antigens, except those associated with antibody production. (e) **Lymphokines**: which are soluble proteins that assist in activating other components of the immune response? The lymphokines act by: Reacting with white blood cells to increase their ability to swoop on the foreign infestation, increasing the rate of lymphocyte production, breaking down cells (damaged body cells, bacteria) and performing other related functions [3, 10].

3 Climate Changes Challenges

Climate change acts as a major threat to climate—sensitive sectors. This change in climate is the greatest challenge to about 1.3 billion population who depends on animal husbandry as their livelihood. "Livestock are called as the living bank for farmers. They contribute about 53% of world agricultural" [11] and also to the economy by means of milk, meat, hide, eggs, drought power, manure etc.

Heat stress is considered as one of the primary factors that imposes negative impacts on chicken production and reproduction. Due to global warming, abrupt climatic conditions like storms, droughts, floods and extreme hot and cold temperatures are predominant around the globe. With the changing climatic scenario, the frequency and duration of exposure of livestock to abiotic and biotic stressors increases.

In the changing climate scenario, other factors like solar radiation, photoperiod and humidity also synergizes with the above stressors. Furthermore, the concept of global warming is alarming the earth planet day by day. According to the latest report released by IPCC [1], there is an average of about 1.53 °F (0.85 °C) rise in global surface temperature from 1880 to 2012. The earth's climate has warmed in the last century with current climate models indicating a 0.28 °C increase per decade for the next two decades and predicted the increase in global average surface temperature by 2100 to be between 1.88 and 4.08 °C [12]. Therefore, the increasing concern on the thermal comfort of agricultural animals is not only justifiable for countries in tropical areas, but also for nations with high ambient temperatures [13]. Latent and sensible heats are types of energy released or absorbed in the atmosphere. Latent heat is related to changes in phase among liquids, gases, and solids. Sensible heat is related to changes in ambient [14].

3.1 Latent Heat

Latent heat plays a very important role in thunderstorms and hurricanes. Warm air rises and the water vapor it contains condenses in particles called the cloud. When the water vapor condenses to form clouds, it releases latent heat into the atmosphere.

The latent heat then warms the surrounding air around the new cloud droplet causing instability. This increases the cloud height and depending on how unstable the atmosphere is, thunderstorms could form from these growing clouds. Thunderstorms release enormous amounts of latent heat which add to the instability in the atmosphere causing some thunderstorms to become severe. In hurricanes, latent heat is released within the clouds of the hurricane, warming the air inside the clouds. The storm will then intensify, or gain strength [14, 15].

3.2 *Sensible Heat*

Sensible heat is the energy required to change the temperature of a substance. The temperature change can come from the absorption of sunlight by the soil or the air itself. Energy moves through the atmosphere using both latent and sensible heat acting on the atmosphere to drive the movement of air molecules which create wind and vertical motions [16].

How Does This Relate to Chicken?

Evaporation of panting in chickens leads to cooling of the body surface on hot days. Evaporation is more effective when there is some wind to carry the moisture away. However, in really hot conditions the body may shut down and stop panting. This is particularly true in areas of bright sunlight, which will increase their body temperature due to the sensible heat released by absorbing the solar radiation. In that case, the body's temperature can rise very quickly and cause hyperthermia or heat stroke, which can lead to death [17].

4 Climate Change and Chicken Production

Chicken flocks are particularly vulnerable to climate change because birds can only tolerate narrow temperature ranges [18]. Regarding productivity, housing systems need to be managed to maintain optimal seasonal temperatures and reduce the embargo of heat stress, and increased investment will be required in ventilation and cooling [19]. Costs are likely to increase as the result of the need to cool buildings especially in summer and reduce house humidity [14]. Building infrastructure and maintenance will have to cope with more intense weather events and increased rainfall. Stocking density in the house may need to be reduced in extremely high temperatures, and actively controlled ventilation could become essential in transportation vehicles [20].

4.1 Adaptation Suggestions

Heat stress begins when the ambient temperature surges above 80 °F and is readily apparent above 85 °F. When a bird begins to pant, physiological changes have already started within its body to dissipate excess heat. Even before the bird reaches this point, anything that you do to help birds remain comfortable will help to maintain optimum growth rates, hatchability, egg size, egg shell quality, and egg production [21, 22] Chicken producers should reconsider building design in new designs to be more effectively cope with new climate and weather extremes, including the installation of more/new equipments to cope with extreme climatic changes [23]. These incorporate the establishment of sustainable power source—, for example, sun oriented or wind control—to control poultry sheds, and utilizing biomass boilers or anaerobic aging of chicken litter. Albeit a portion of the effects may happen to a more noteworthy or lesser degree in the short, medium or longer term, it is critical to think ahead for the future, particularly in connection to issues, for example, building plan.

4.2 Heat Stress

Heat stress (HS) occurs after exposure to high ambient temperatures beyond the thermo-neutral zone (TNZ) for a given species. Both acute, a brief intense HS episode, and chronic pro-longed exposure to high temperatures, or HS, can cause adverse effects on the chicken's well-being. The TNZ is a physiological range with limited variability of biological function [24]. Individual chickens can exhibit a wide range of reactions to HS, showing from the little effect on their health to experiencing mortality. However, even with the determined optimal temperature per species, different biological functions and activities play a role for different individuals. The optimum temperature for hen egg production is estimated to be between 19 and 22 °C, but between 18 and 30 °C for meat producing birds [2, 14]. Chicken farms have various zones of thermal comfort predominantly dependent on the species of chicken and their physiological status (internal factors) and relative humidity, the velocity of ambient air, and the degree of solar radiation (external factors) also contribute to their TNZ [25]. By including these additional factors, a more useful index can be created in determining the optimal temperature and potential for HS [26]. As guidelines, the theory of Livestock and Poultry Heat Stress indices, [27], was developed in order to determine the extent of HS on chickens. Two various indices derived from meteorological measurements have been developed and recently reviewed by Hahn et al. [26]. The most common one is the temperature humidity index (THI) [28]. The following formula for THI takes into account the ambient temperature and the humidity to estimate the magnitude of HS.

$$\mathrm{HS}:\mathrm{THI} = \mathrm{db}^{\circ}\mathrm{F} - \{(0.55 - 0.55\mathrm{RH})(\mathrm{db}^{\circ}\mathrm{F} - 58)\}$$

In the formula, db is defined as the dry bulb temperature in °F and RH is the relative humidity (RH%/100). The obtained values are used to establish the severity of HS: 86 indicates extreme severe HS [27]. THI, however, does not account for important climatic variables such as air movement or solar radiation. In addition, THI does not include management or animal factors, such as the effect of shade or the genotype and phenotypic differences. Several other formulas have been developed for determining the HS potential to account for the additional factors. Tao and Xin [29], for example, made improvements on the THI by creating the temperature-humidity-velocity index (THVI), which includes wind speed to determine the HS level in poultry. Wind speed has been shown to be favorable to chicks exposed to HS conditions [22]. Therefore, the following formula of THVI may be advantageous to improve the method in determining the severity of HS.

$$\text{THVI} = (0.85 \times \text{DBT} + 0.15 \times \text{WBT}) \times \text{V} - 0.058$$

In the formula, DBT is defined as the dry bulb temperature in °C, WBT is the wet bulb temperature in °C, and V represents the air velocity. Similarly to THI, the obtained numerical values are used to establish the negative effects of HS: $\leq$70 being normal, 70–75 indicates alert, 76–81 being in danger, and $\geq$82 indicating emergency [29]. Additional indices have incorporated the previously stated variables to overcome the limitations of different THI [30]. For example, Mader et al. [31] developed a comprehensive climate index applicable under a wide range of environmental conditions by providing an adjustment to ambient temperature, relative humidity, wind speed, and radiation.

4.3 *Impact of Heat Stress*

Stress, a response to adverse stimuli, is difficult to define and understand because of its nebulous perception. Stress is the nonspecific reaction of the body to any request, though stressor can be characterized as "an operator that produces worry". Therefore, stress represents the biological reaction of the chicken organism to stimuli that disturb its normal physiological equilibrium or homeostasis.

Heat stress results from a negative balance between the net amount of energy flowing from the chicken's body to its surrounding environment and the amount of heat energy produced by the chicken. Heat stress is a critical stressor, especially in regions with hot climates and farms unable to control the ambient temperature [14, 32]. HS has been consistently found to prohibit the chicken production. In general, temperature variation affects the endocrine system, which controls chicken metabolism, growth and reproduction, which implies that a decrease or increase in temperature will have an impact on avian production. It includes a decrease in feed intake [33–35], impaired growth performance [36–38], decrease in egg production [39–42], impaired egg qualities [21, 40, 43], and an increase in mortality [41, 44]. Consolidation of these results was speculated by Mack et al. [40]. The egg production is dependent on gonadal steroids, such as progesterone, testosterone, and negatively affects the bioactivity of gonadotropin is less understood. Most likely, it is due to multiple factors affecting animals' physiological homeostasis under HS, such as a reduction in feed intake and an increase in oxidative stress [15].

The variability of the effects of HS on chicken health may be explained by the examined birds, such as health and production status, and/or genetic background. The variation may also be due to the variety of intensity and duration of HS applied to the chickens. Stocking density is also a major factor affecting chicken productivity and physiological homeostasis, as well [20].

4.4 *Heat Regulation*

Chickens produce and dissipate heat to maintain a relatively constant body temperature. When the temperature and relative humidity exceed the comfort level, chickens lose their ability to efficiently dissipate heat, which then leads to behavioral and physiological changes [14, 45–47]. Chickens become heat stressed without the balance between heat production and heat dissipation, or thermoregulation. Chickens do not have sweat glands. Therefore, the primary methods of heat loss include respiratory evaporative cooling, or panting and conductive heat transfer through their feet and wattles [48]. Likewise, panting relies on water loss which may lead to dehydration [49] or respiratory alkalosis [50]. Chickens have air sacs as an additional mechanism to promote heat exchange between their body and the environment. Air sacs are useful during panting, as they promote air circulation on surfaces, increasing heat loss via evaporation along with gas exchanges [51]. Sensible heat is used by chickens to release heat, including radiation, convection, and conduction [52, 53]. However, some methods of heat loss are challenging to the avian species due to their feather coverage trapping heat to prevent hypothermia. During HS, capillary blood flow is redistributed throughout the body to improve sensible heat loss [54]. Mujahid et al. [55] presented evidence that broiler chickens exposed to different durations of acute HS resulted in distinct time-dependent physiological responses.

5 Impact of Heat Stress on Chicken Health

To maintain flocks of chicks in good health managers should ensure that: (1) The challenge is kept as low as possible and (2) The flocks' ability to fight infection is kept as high as possible.

In some cases, the chicks' immune system requires frequent support to be powerful against an antigen, while in different cases, less incessant incitement is required. In a few circumstances, it is conceivable to rely upon the natural infection to trigger the immune system to produce the antibodies [7]. However, in these cases it is necessary to take precautions to ensure that the infection does not get out of control. Stress is always beneficial at it minimal level. A minimal level of stress called as good stress is required by all living organisms to perform a task. Stress becomes a distress when it alters the normal biological functions like production and reproductive performances of the chicken. Furthermore, stress mediated interactions with the immune system suppress its normal functions and lead the chicken to a disease-prone status [24].

In general, chickens encounter various stressors such as thermal stress, production stress, transportation stress, nutritional stress, immune stress, and stress due to crowding, drought and environmental conditions [56]. In the event of a climatic change, stress due to adverse climatic conditions and emerging microbes and vector—borne diseases are some of which the chickens have to encounter [57].

6 Molecular Responses to Heat Stress

High ambient temperature can adversely affect the structure and physiology of cells causing impaired transcription, RNA processing, translation, oxidative metabolism, membrane structure and function [58]. Cells generate small amounts of free radicals or reactive oxygen species (ROS) during their normal metabolism. Although low levels of ROS are essential in many biochemical processes, accumulation of ROS may damage biological macromolecules i.e. lipids, proteins, carbohydrates and DNA [59]. External factors such as heat can lead to increased free radicals and other ROS and may lead to oxidative stress [60].

Increased mitochondrial reactive oxygen species (ROS) production and decreased avian uncoupling protein was confirmed in HS birds [55]. Heat stress could cause oxidative damage and lead to accumulation of free radicals and other reactive oxygen species (ROS). The balance between ROS and antioxidant is retarded in broilers exposed to HS [61], leading to oxidative stress during acute HS [15, 62]. Malondialdehyde is a biomarker applied to measure oxidative stress in chickens by indirectly measuring the level of peroxidation due to ROS [55]. In order to establish this balance during HS, the strategy for increasing antioxidant capability and activity in birds is required [43].

Antioxidants, both enzymatic (viz. superoxide dismutase, glutathione peroxidase and catalase) and non-enzymatic (vitamins C, E and A, glutathione, pyruvate etc.) provide necessary defense against oxidative stress generated due to high ambient temperature. In such conditions, the administration of antioxidants has proved to be useful for improvement of several immune functions [63]. The immune cell functions are associated with the production of ROS such as that involved in the microbial activity of phagocytes or lymph proliferative response to mitogens [64, 65]. However excessive production of ROS due to heat stress renders harmful effect on cells of immune system. When exposed to oxidative stress, polymorph nuclear leukocytes (PMNs) change their pattern of oxygen uptake sharply while releasing large amounts of superoxide anion into the cell environment. PMNs play an important role as mediators of tissue destructive events in inflammatory diseases, ranging from rheumatoid arthritis and myocardial reperfusion injury to respiratory distress syndrome [66].

Whenever a chick encounters a stressor, the central nervous system first begins to respond by sending signals to any of the biological systems via behavioral, autonomic nervous system, neuroendocrine system and immune system in order to alleviate or compensate the impendence. The response of each chick to a particular stress condition varies depending upon its previous exposure to the stressors, genetic makeup [67], age [68], season and physiological state. The immune system does not respond directly to stress but via the neuroendocrine system. The neuroendocrine system responds mainly to the Hypothalamo—pituitary—Axis (HPA) by the release of glucocorticoids that are normally called as stress hormones [69]. They act to adapt the chick to cope up with the long-term stressors. The immune system responds to stress by enhancement or suppression of immune functions [70]. The stress related hormones act on the immune cell receptors to modulate the immune response.

7 Heat Stress Affecting Chicken Immunity

7.1 Heat Stress Axis

The HPA axis is one of the most important systems for the integration of the body and is activated in response to stressful stimuli or homeostatic inconvenience [71]. Increased of plasma corticosterone levels may cause consequent deteriorations in chicken's health status [44]. It is the thoughtful trunk of the autonomic nervous system (ANS) arranged in the adrenal medulla. It acts by the arrival of epinephrine and nor epinephrine which are in charge of the responsible for the characteristic of flight or -fright mechanism. These hormones act to enhance glycogenolysis causing increased glucose levels in circulation. The blood glucose will reach the stressed organ in order to meet the energy requirements and cope up with the stressor [72].

The corticosterone serum levels will be higher when chicks are exposed to heat. Performance indices deficits and macrophage activity are observed for broiler stressed chicks. The intestinal integrity might have been a consequence of modifications in HPA function. As a matter of fact, results from Shini et al. [73] indicated that heat exposed chicks caused release in serum corticosterone level. This in turn increased the heterophil: lymphocyte ratio. Moreover it induced ultrastructural morphological changes in heterophil size, shape, and granulation and lymphocyte cytoplasmatic characteristics.

The HPA axis and sympathetic-adrenal-medullary system (SAM) are the crucial components that receive stress signals and act to relieve the stress mechanism or adopt the chick to the stressful condition [74]. However, a significant increase in T3 but not in T4 level was observed during heat stress [75]. During short—term exposure to high ambient temperature, the concentrations of glucocorticoids and catecholamines were found to be elevated [76]. Moberg and Mench [24] documented the simultaneous relationship among thermal stress, plasma aldosterone level and urine electrolyte concentration. During prolonged heat exposure plasma aldosterone level was reported to decline. Concurrent with this, there was significant fall in serum K+ [77].

7.2 *Hypothalamus—Pituitary—Adrenal Axis (HPA)*

The center of pressure reaction is accepted to be founded on the enactment of HPA. The key segments of this system include: (i) hypothalamus, and (ii) brain stem. The parvocellular neurons of CRH, arginine vasopressin (AVP) neurons of paraventricular cores of the hypothalamus, CRH neurons of paragigantocellular and parabranchial nuclei of medulla and locus ceruleus innervate the system [72]. They sense stressor and get activated to release CRH from the paraventricular nucleus and arginine vasopressin from magnocellular neurons of the hypothalamus. They act on the anterior pituitary cells to release adrenocorticotropic hormone (ACTH). ACTH, in turn, acts on adrenal cortex to stimulate synthesis and release of steroids leading to cholesterol uptake. Cholesterol further gets converted into cortisol and corticosterone. Glucocorticoid receptor exists as a complex in the cytoplasm of most cells including the immune cells. Glucocorticoid Responsive Elements (GRE), located in the promoter region of the target gene, regulates the expression of target genes either positively or negatively. Both conditions are associated with (1) increases in hypothalamic-pituitary-adrenal (**HPA**) axis activity and elevation in serum corticosterone [78], (2) overall leukocytosis, (3) mild reductions in absolute natural killer (**NK**) cell counts and relative T-cell proportions, (4) marginal increases in CD4+ :CD8+ ratios, and (5) moderate decreases in T-cell and NK-cell function [79].

7.3 *The Link Between Stress and the Immune System*

Sympathetic innervation from brain connects the primary lymphoid organs and secondary lymphoid organs [80]; thus, the catecholamine receptors in immune cells also contribute to immune cell activation by stress mechanisms. The end product of HPA axis activation is glucocorticoid release, which has receptors in almost all organs including the immune cells [81].

Glucocorticoids increase the transcription of anti- inflammatory cytokines [interleukin (IL)-4 and IL-10] and decrease that of pro inflammatory cytokines (IL-1, IL-12, IL-6, tumor necrosis factor-α, interferon, and granulocyte macrophage colony-stimulating factor) (i.e., they modulate the T-helper type 1 vs. T-helper type 2 cytokine profile). These hormones decrease the transcription factors for adhesion molecules such as intercellular adhesion molecule-I and vascular cell adhesion molecule-I [14]. Restraint stress decreases NK cell activity in chickens, most probably due to a transient increase in corticosterone serum levels. Heat stress (39 °C for 7 h/d) presents decreased CD4+ and CD8+ cells and SRBC antibody titers in chickens [82]. Injection of corticosterone decreases antibody production against SRBC and impaired bronchitis vaccine response in chickens [83].

7.4 Heat Stress and Innate Immunity

This refers to the natural or inherited ability to resist disease and the involved mechanisms that come into play immediately after a few hours that an antigen has appeared in the body. However, it is possible that specifically through corticosterone release; heat stress could decrease the chickens' innate immune response in commercial industrial production, thus decreasing their resistance to other pathogens, such as coccidia. Altogether, the central nervous system activation by heat stress is responsible for the changes in intestinal immunity, which reinforces the idea of the existence of a link between the nervous and immune systems in maintaining intestinal homeostasis. Included in this type of immunity are a number of non-specific disease response mechanisms. Concerning the effects of heat stress on avian species, the following findings were reported: (1) decreases in the feed consumption, BW gain, as well as the total white blood cell count and antibody production [41]; (2) decreases in the number of peripheral blood lymphocytes and induction of an electrolyte imbalance [48]; (3) decreases in the blood lymphocytes and spleen weight [84]; (4) decreases in CD4+ and CD8+ lymphocytes and antibody production against SRBC [82] and (5) decreases in the wet and dry weights of jejunum [85].

7.5 Heat Stress and Adaptive Immunity

Adaptive immunity creates "immunological memory" once an initial response to a specific pathogen has been triggered, which can enhance future responses to that specific pathogen and act more efficiently. These antigen-specific immune responses are provided by immune cells or antibodies that are produced in response to exposure to an antigen. These antibodies are very small, very special proteins (globulin proteins) that chickens release into their circulatory system, which fight foreign invaders e.g. invading viruses and bacteria [14].

The adaptive immunity is specific to an antigen and can take one of two forms:

a-**Antibodies**: Antibodies do not have the capability to kill disease organisms directly. Antibodies perform their function by attaching to disease organisms and blocking their receptors. The disease organisms are then prevented from attaching to their target cell receptors in the chicken. The attached antibodies also immobilize the disease organism which assists their destruction by macrophages'. b-**The cellular immune response**: The cell segment of the resistant reaction incorporates every one of the cells that respond with specificity to antigens, aside from those related with counter acting agent generation. The cells related with this system, the T-lymphocytes

(T-cells), start as a similar immature microorganisms as the B-cells. A range of T-cells is produced to perform various roles. Some T-cells act by producing chemicals called lymphokines; others directly destroy disease organisms; some T-cells act to enhance the response of B-cells, macrophages, or other T-cells (called helper T-cells); while still others have the opposite effect and act to inhibit the activity of these cells (suppressors) [58, 75, 86].

7.6 Mechanism of Heat Stress Impacting the Immune System

High environmental temperatures activate the hypothalamic-pituitary-adrenal (HPA) axis. The stress-activated HPA axis was found to be responsible for the negative effects of HS on broiler performance and immune function [34]. The HPA axis regulates corticotrophin releasing hormone and adrenocorticotropic hormone are released from hypothalamic and pituitary cells [87]. In response to the physiological disruptions, more glucocorticoids are released. Glucocorticoids participate in the control of body homeostasis and stress response of various organisms [43]. In chickens, HS has been shown to cause elevated corticosterone concentrations [42, 44, 88, 89]. Changes in corticosterone levels also occur due to environmental stimuli [90]. Increased corticosterone, as the final product of the HPA axis, causes numerous effects on behavior, metabolic pathways, and immune functions [91].

7.7 Heat Stress and Heat Shock Protein

Heat-shock proteins (HSPs) are produced in abundance within the cell in response to various stressors. Proteins are an evolutionarily conserved family of proteins whose expression increases in response to a variety of different metabolic insults. Despite their designation, most of the HSPs are constitutively expressed and perform essential functions [92, 93]. The function of the heat shock proteins is to activate ATP hydrolysis and binding Hsc70 (constitutive form) to the aggregated protein [94]. HSP110 combined with HSP70 into an HSP110/HSP70 bichaperone and supported by gene activity, creates a complex that can unfold even stably misfolded and aggregated proteins [95]. As a part of adaptation, cells are able to adjust to the changing environmental conditions by modulating their gene expression. In chickens, transcriptomic approaches have been applied in studying heat stress with both in vivo and in vitro models, which allowed identification of stress-related gene expression changes in liver [96, 97], testes [98], brain [99], heart [97], muscle [97, 100] and hepatocellular carcinoma cell line [101]. However, there has been no attempt to study the molecular responses to heat stress and the accompanying endotoxemia directly in the avian immune system.

Heat shock proteins greatly enhance the efficiency of intracellular protein manufacture and transport and may enhance immunity against pathogens by improving immune surveillance of infected cells [102]. A number of herbs or their derivatives have been showed to induce or to facilitate HSP responses listed such herbs may strengthen the systemic response to antigens. All have been traditionally used as tonics; adapt gens, or immune-modulators.

7.8 Heat Stress and Gut Health

Physiologically, warm pressure can cause multi-organ system disappointment, including the intestinal tract. The tight intersections between the epithelial cells of the guts lose network because of oxidative pressure that takes after overheating. As an outcome, penetrability of the intestinal layer increments and the intestinal substance begins spilling out of the gut, prompting foundational bacterial contamination of the living being and the sudden incendiary reaction of the safe structure [103]. One of the fundamental poisons that is associated with warm cytotoxicity is lipopolysaccharide (LPS), which is a cell divider segment of the Gram-negative microscopic organisms and the endotoxin that triggers resistant reactions. The primary line of resistance against endotoxemia is the intrinsic invulnerable structure, which is involved for the most part of phagocytic cells, for example, monocytes and macrophages [104–106]. These cells respond to stressors by delivering a lot of chemokines, cytokines, and nitric oxide species to draw in other safe cells to the irritation site and in addition to specifically ward off the pathogens. Normal morphology and integrity of the small intestine are important to prevent bacterial translocation from the intestinal tract to the body as for digestion and absorption of nutrients [14]. Moreover, chickens submitted to chronic heat stress presented decreased villus heights (19%) and wet (26%) and dry (31%) intestine weights of the jejunum [39]. However, it is relevant to point out that an increase in the number of heterophils was found in the stressed chickens, a fact that is likely to be indicative of a possible intestinal mucosal barrier dysfunction and, consequently, a bacterial infection. Inflammatory infiltrate contributes to the production of pro-inflammatory cytokines that act in the intestinal epithelium's tight junctions, in turn increasing the mucosa permeability to pathogenic bacteria [107].

7.9 Significance of Optimum Nutrition for Chicken Immune Function

The relationship between nutrition and the immune system has been the Centre of attention in scientific communities in last decade. National Research Council (NRC) recommendation for feeding regimes are usually based on the needs of healthy birds under ideal management, but chickens in commercial systems are normally exposed to various kinds of stresses and diseases. It is well known that birds reduce feed intake as well as feed efficiency [108] under high ambient temperature. This reduction in feed may avoid the thermogenic effect associated with nutrient absorption, assimilation, and utilization [109]. When the environmental temperature increases above 30 °C, several mechanisms are activated to allow the blood flow increase to upper respiratory tract and other active organs for heat dissipation (combs and wattles) at the cost of blood flow to the digestive system [110]. Wolfenson et al. [54] showed a descending gradient of blood flow along the digestive tract from the ileum to proventriculus in response to hyperthermia. This would reduce proteolytic enzymatic activities occurring in the upper part of the digestive tract and decreases subsequent protein digestibility [111]. It is necessary to establish specific digestibility coefficients for high environmental temperature conditions. The reduction of triiodothyronine and changes in bodily hemodynamics has been suggested to be part of the mechanisms associated with improved thermotolerance in acclimated chickens [112]. The high ambient temperature may suppress nitrogen and total amino acids ileal digestibility by almost 4.5 and 5.5%, respectively [77].

It has been reported that chicks fed deficient amino acid had suboptimal interleukin production during immunologic stress. Rama-Rao et al. [113] showed that methionine levels lower than 0.50% in broiler diet generates a poorer immune cell response as compared to higher concentrations. Utilization of immune stimulants is one solution to improve the immunity of chickens and to decrease their susceptibility to infectious diseases [8]. Glucose is the primary energy source for all metabolic reactions. The plasma glucose concentration is primarily regulated by hormones secreted by the pancreas. The hormones glucagon and insulin function to increase and decrease glucose levels, respectively. Disturbances to this balance can result in damage to nerves, blood vessels, and even death. Changes in physiology may occur by alterations in electrolyte balance which is essential for acid-base balance, maintenance of cellular homeostasis, synthesis of tissue protein, electrical potential of cell membranes, enzymatic reactions, and maintaining osmotic pressure [48]. Increased fluid excretion in the urine is accompanied by increased concentrations of electrolytes in the urine [49]. Altering electrolyte amounts in feed partially ameliorates the negative impacts of heat stress in broiler chickens [2].

Antioxidant vitamins are effective ways of alleviating heat stress in poultry. Antioxidant vitamins such as vitamins A (retinol), E (α tocopherol) and C (ascorbic acid) are used in poultry diets because of their anti-stress effects and because their synthesis is reduced during heat stress. Heat stress stimulates the release of corticosterone and catecholamines and initiates lipid peroxidation in cell membranes [18].

Modern herbs research and a new understanding of the immune system have explained many mechanisms by which these herbs work. Instead, they rely on the innate immunity of each cell and on systemic signals emanating from infection sites [114, 115]. Flavonoids and hydroxylase phenols are naturally synthesized by plants in response to infection [116]. Flavones and flavanones, being bitter, also have natural anti-feeding effects. Alkaloids are the most common plant metabolites [117]. More recently, many secondary metabolites have been suggested to have Immunomodulation properties in chickens [118].

7.10 Genetic Impact

Climate change is expected to increase the average global temperature and negatively impact on the food supply. Increasing population and economies increase demand for dietary animal protein. Heat stress in chicken decreases production and increases disease susceptibility. Understanding the genetic control of response to heat in chickens could enable breeding of more climate-adaptable chickens that are heat and/or disease resistant. At the population level, a highly advanced intercross line originating from the broiler (heat susceptible) and Fayoumi (heat resistant) were applied.

Not only, it is desirable to determine the genetic component by estimating heritability and performing association, but also another approach to understanding genetic control is to quantify the expression of transcripts. The most wide transcripts are mRNA which is transcribed from DNA, processed within the nucleus, and subsequently translated into amino acids on ribosomes within the cytoplasm [100]. There are many technologies to quantify mRNA expression including northern blot, RT-PCR, and whole transcriptome profiling using RNA-sequencing methods. RNA-sequencing is currently the preferred method to assay transcriptional levels because it is an a priori approach that doesn't require investigators to choose genes to assay expression but, rather, assays expression for all genes that are transcribed. RNA-sequencing can reveal genes and pathways that are responsive to treatment effects. The transcriptomic response to heat stress in chickens has been documented in several tissues and cell lines. The liver of heat stressed broilers revealed 43 differentially expressed genes using RNA-sequencing technology. In Silkie fowl, the breast muscle revealed 110 genes differentially expressed using microarray technology [100]. In male white leghorn liver cancer cell line, RNA-sequencing revealed 812 transcripts that were responsive to in vitro heat treatment [101]. In meat-type birds exposed to heat stress, the hypothalamus was evaluated using microarray, which showed 1,967 genes to be differentially expressed [101]. No studies, however, have been done to investigate the global transcriptomic response in an immune tissue to the double stressor of heat stress and immune stimulation.

8 Conclusions

Climate change is defined as increased severity and variability of weather patterns. This includes increasing the severity and incidence of droughts, flooding, cold bouts, and heat waves. Weather variability is the overarching theme, but it is expected that the global average temperature will increase by 4 °C by 2100. The changing environmental temperatures require adaptation from all forms of life.

Under stress conditions, avian blood tolerates a change from acid—base balance to alkaline balance. There is a decline in the plasma, a reduce level of vitamin C in the adrenal cortex, a reduction in lymphocytes and a depression of immune response.

As the temperature rises, the birds undergo many changes—increased water consumption, respiration rate, body temperature, inferior egg quality and susceptibility to diseases.

In particular, the highly specific adaptive immune mechanisms are affected by heat stress. In more specific, heat stress deteriorates the cell mediated immune responses. This reveals that heat stress needs to be considered as an important factor compromising chicken's health. As a result of heat conditioning, biochemical and physiological mechanisms were induced to cope with heat stress; this induction may have delay production of additional acute phase proteins to protect the cells from damage. The stress hormones—cytokine interactions are responsible for altered immune functions during heat stress. Modern day molecular biology tools can help in understanding various cellular and molecular mechanisms involved in the production, physiological and immunological aspects of the poultry birds, which in turn can help in the development of breeds more adapted to the climate changes.

Heat stress is one of the most important environmental stressors challenging poultry production worldwide. The negative effects of heat stress on broilers and laying hens range from reduced growth and egg production to decreased poultry and egg quality and safety. However, a major concern should be the negative impact of heat stress on poultry welfare.

Finally, it is important to mention that intervention strategies to deal with heat stress conditions including environmental management (such as facilities design, ventilation, sprinkling, shading, etc.), nutritional manipulation (i.e., diet formulation according to the metabolic condition of the birds), as well as inclusion of feed additives in the diet (e.g., antioxidants, vitamins, minerals, probiotics, prebiotics, essential oils, etc.) and water supplementation with electrolytes. Over the past fifty years the global poultry production experienced leaps and bounces to accommodate rising demand. On the other hand, popular demand and scientific interest for organic poultry production, particularly feeding with medicinal botanicals, has considerably increased in recent years. Many studies have clearly established the fact that herbal plants and their derivatives have potential as immune modulators. Both, the innate and adaptive components of the immune system are stimulated by phylogenic. Moreover, most studies have exercised herbal extracts rather than the purified compounds. Therefore, there is still distrust concerning the efficacy and optimum dosage of herbal plants and their derivatives as immune stimulators. Hence, more research is required for scientific validation of herbal plants as potent chicken immune stimulators. Nevertheless, the effectiveness of most of the interventions has been variable or inconsistent. More recently, two innovative approaches have been explored, including early-life conditioning (i.e., prenatal heat acclimation) and genetic selection of breeds with the increased capacity of coping with heat stress conditions (i.e., increased heat tolerance). However, these potential opportunities, although promising (particularly, for poultry production in hot climatic regions), still require further research and development.

References

1. IPCC (Intergovernmental Panel on Climate Change) (2013) Climate change: the physical science basis. In: Stocker TF, Qin D, Plattner GK, Tignor M, Allen SK, Boschung J, Nauels A, Xia Y, Bex V, Midgley PM (eds) Contribution of working group I to the fifth assessment report of the intergovernmental panel on climate change. United Kingdom and New York, NY, USA: Cambridge University Press, Cambridge, 1535pp
2. Charles DR (2002) Responses to the thermal environment. In: Charles DR, Walker AW (eds) Poult. Environment problems, a guide to solutions, vol 38. Nottingham University Press, UK, pp 1–16
3. Abbas AK, Lichtman AH, Pillai S (2007) Cellular and molecular immunology, 6th edn. Elsevier, Philadelphia, pp 32–34
4. Weir D, Stewart J (1997) Immunology, 8th edn. Churchill Livingstone
5. Basset C, Holton J, Mahony R, Riott I (2003) Innate immunity and pathogen-host interaction. Vaccine 21:2–23
6. Janeway CA, Medzhitov R (2002) Innate immune recognition. Annu Rev Immunol 20(1):197–216
7. Hamdy AMM, Tellez GA (2009) Chicken's lymphatic system and immunity. In: 5th world poultry conference, Taba, Egypt
8. Liu XY (1999) Stress and immunity. In: Yin TB (ed) Poultry immunology. China Agriculture Press, Beijing, China, pp 230–252
9. Roitt IM (1997) Essential immunology, 9th edn. ELBS, Blackwell Scientific Publications, United Kingdom
10. Abbas AK, Murphy KM, Sher A (1996) Functional diversity of helper T lymphocytes. Nature 383:787–793
11. World Bank (2009) Minding the stock: bringing public policy to bear on livestock sector development. Report no. 44010-GLB. Washington
12. Intergovernmental Panel on Climate Change (IPCC: AR4) (2007) The intergovernmental panel on climate change 4th assessment report. http://www.ipcc.ch/publications_and_data/publications_and_data_reports.htm#2
13. Nardone A, Ronchi B, Lacetera N, Ranieri MS, Bernabucci U (2010) Effects of climate changes on animal production and sustainability of livestock systems. Livest Sci 130:57–69
14. Sturkie PD (2005) In: Whittow GC (ed) Avian physiology, 5th edn. Academic Press, USA
15. Lin H, Jiao HC, Buysr J, Decuypere E (2006) Strategies for preventing heat stress in poultry. World's Poult Sci J 62:71–87
16. Yanagi T, Xin H, Gates RS (2001) Modeling partial surface evaporative cooling of chickens. Paper no. 01-3011 presented at American society of agricultural and engineering meeting, Sacramento, CA
17. Welc SS, Clanton TL, Dineen SM, Leon LR (2013) Heat stroke activates a stress induced cytokine response in skeletal muscle. J Appl Physiol 115(8):1126–1137
18. Yahav S (2009) Alleviating heat stress in domestic fowl: different strategies. World Poult Sci J 65(04):719–732
19. Krüger E, González Cruz E, Givoni B (2010) Effectiveness of indirect evaporative cooling and thermal mass in a hot arid climate. Build Environ 45:1422–1433
20. Estevez I (2007) Density allowances for broilers: where to set the limits? Poult Sci 86:1265–1272
21. Ajakaiye JJ, Perez-Bello A, Mollineda-Trujillo A (2011) Impact of heat stress on egg quality in layer hens supplemented with l-ascorbic acid and dl-tocopherol acetate. Vet Archiv 81:119–132
22. Ruzal M, Shinder D, Malka I, Yahav S (2011) Ventilation plays an important role in hens' egg production at high ambient temperature. Poult Sci 90:856–862
23. Lara LJ, Rostagno MH (2013) Impact of heat stress on poultry production. Anim 3:356–369
24. Moberg GP, Mench JA (2000) The biology of animal stress: basic principles and implications for animal welfare. CABI Publishing, USA

25. NRC (1997) Effects of environment on nutrient requirements of domestic animals. National Academic Press, Washington, p 30
26. Hahn G, Gaughan JB, Mader TL, Eigenberg RA (2009) Thermal indices and their applications for livestock environments. In: JA DeShazer (ed) Livestock energetics and thermal environmental management. American Society of Agricultural and Biological Engineers, pp. 113–130
27. LPHSI (1990) Livestock and poultry heat stress indices. Agriculture engineering technology guide. Clemson University, Clemson, SC 29634, USA
28. Gates RS, Zhang H, Colliver DG, Overhults DG (1995) Regional variation temperature humidity index for poultry housing. Trans ASAE 38(1):197–205
29. Tao X, Xin H (2003) Acute synergistic effects of air temperature, humidity, and velocity on homeostasis of market–size broilers. Trans ASAE 46:491–497
30. Mader TL, Davis MS, Brown-Brandl T (2006) Environmental factors influencing heat stress in feedlot cattle. J Anim Sci 84:712–719
31. Mader TL, Johnson LJ, Gaughan JB (2010) A comprehensive index for assessing environmental stress in animals. J Anim Sci 88:2153–2165
32. Kadim IT, Al-Qamshui BHA, Mahgoub O, Al-Marzooqi W, Johnson EH (2008) Effect of seasonal temperatures and ascorbic acid supplementation on performance of broiler chickens maintained in closed and open-sided houses. Int J Poult Sci 7:655–660
33. Habibian M, Ghazi S, Moeini MM, Abdolmohammadi A (2013) Effects of dietary selenium and vitamin E on immune response and biological blood parameters of broilers reared under thermoneutral or heat stress conditions. Int J Biometeorol 10:1007
34. Quinteiro-Filho WM, Gomes AV, Pinheiro ML, Ribeiro A, Ferraz-de-Paula V, Astolfi-Ferreira CS, Ferreira AJ, Palermo-Neto J (2012) Heat stress impairs performance and induces intestinal inflammation in broiler chickens infected with Salmonella Enteritidis. Avian Pathol 41:421–427
35. Sohail MU, Hume ME, Byrd JA, Nisbet DJ, Ijaz A, Sohail A, Shabbir MZ, Rehman H (2012) Effect of supplementation of prebiotic mannanoligosaccharides and probiotic mixture on growth performance of broilers subjected to chronic heat stress. Poult Sci 91:2235–2240
36. Attia YA, Hassan RA, Tag El-Din AE, Abou-Shehema BM (2011) Effect of ascorbic acid or increasing metabolizable energy level with or without supplementation of some essential amino acids on productive and physiological traits of slow-growing chicks exposed to chronic heat stress. J Anim Physiol Anim Nutr 95(744–755):36
37. Ghazi SH, Habibian M, Moeini MM, Abdolmohammadi AR (2012) Effects of different levels of organic and inorganic chromium on growth performance and immunocompetence of broilers under heat stress. Biol Trace Elem Res 146:309–317
38. Imik H, Ozlu H, Gumus R, Atasever MA, Urgar S, Atasever M (2012) Effects of ascorbic acid and alpha-lipoic acid on performance and meat quality of broilers subjected to heat stress. Br Poult Sci 53(800–808):43
39. Deng W, Dong XF, Tong JM, Zhang Q (2012) The probiotic Bacillus licheniformis ameliorates heat stress-induced impairment of egg production, gut morphology, and intestinal mucosal immunity in laying hens. Poult Sci 91:575–582
40. Mack LA, Felver-Gant JN, Dennis RL, Cheng HW (2013) Genetic variation alters production and behavioral responses following heat stress in 2 strains of laying hens. Poult Sci 92:285–294
41. Mashaly MM, Hendricks GL, Kalama MA, Gehad AE, Abbas AO, Patterson PH (2004) Effect of heat stress on production parameters and immune responses of commercial laying hens. Poult Sci 83:889–894
42. Star L, Juul-Madsen HR, Decuypere E, Nieuwland MG, de Vries Reilingh G, van den Brand H, Kemp B, Parmentier HK (2009) Effect of early life thermal conditioning and immune challenge on thermotolerance and humoral immune competence in adult laying hens. Poult Sci 88(2253–2261):52
43. Lin H, Decuypere E, Buyse J (2004) Oxidative stress induced by corticosterone administration in broiler chickens (Gallus gallus domesticus): 1. Chronic exposure. Comp Biochem Physiol B 139:737–744

44. Quinteiro-Filho WM, Ribeiro A, Ferraz-de-Paula V, Pinheiro ML, Sakai M, As LR, Ferreira AJ, Palermo-Neto J (2010) Heat stress impairs performance parameters, induces intestinal injury, and decreases macrophage activity in broiler chickens. Poult Sci 89:1905–1914
45. Gonzalez-Esquerra R, Leeson S (2006) Physiological and metabolic responses of broilers to heat stress—implications for protein and acid nutrition. Worlds Poult Sci J 62:282–295
46. Hamdy AMM (2005) Some immune responses related to feed restriction in laying hens. In: 12th world veterinary poultry conference, Istanbul. Turkey
47. Toyomizu M, Tokuda M, Mujahid A, Akiba Y (2005) Progressive alteration to core temperature, respiration and blood acid-base balance in broiler chickens exposed to acute heat stress. J Poult Sci 42:110–118
48. Borges SA, Fischer da Silva AV, Majorka A, Hooge DM, Cummings KR (2004) Physiological responses of broiler chickens to heat stress and dietary electrolyte balance (sodium plus potassium minus chloride, milliequivalents per kilogram). Poult Sci 83:1551–1558
49. Belay T, Teeter RG (1993) Broiler water balance and thermobalance during thermoneutral and high ambient temperature exposure. Poult Sci 72:116–124
50. Yahav S, Goldfeld S, Plavnik I, Hurwitz S (1995) Physiological responses of chickens and turkeys to relative humidity during exposure to high ambient temperature. J Therm Biol 20:245–253
51. Fedde MR (1998) Relationship of structure and function of the avian respiratory system to disease susceptibility. Poult Sci 77:1130–1138
52. Yahav S, Lugar D, Cahaner A, Dotan M, Ruzan M, Hurwitz S (1998) Thermoregulation in naked neck chickens subjected to different ambient temperatures. Br Poult Sci 39:133–138
53. Yahav S, Straschnow A, Plavnik I, Hurwitz S (1997) Blood system response of chickens to changes in environmental temperature. Poult Sci 76:627–633
54. Wolfenson D, Frei YF, Snapir N, Berman A (1981) Heat stress effects on capillary blood flow and its redistribution in the laying hen. Pflugers Arch 390:86–93
55. Mujahid A, Akiba Y, Toyomizu M (2009) Olive oil-supplemented diet alleviates acute heat stress-induced mitochondrial ROS production in chicken skeletal muscle. Am J Physiol Regul Integr Comp Physiol 297:R690–R698
56. Kelley KW (1980) Stress and immune function: a bibliographic review. Annales de Recherches Veterinaires 11(4):445–478
57. Epstein P, Diaz H, Elias S, Grabherr G, Graham N, Martens W, Thompson EM, Susskind J (1998) Biological and physical signs of climate change, focus on mosquito borne diseases. Bull Am Meteor Soc 79(3):409–417
58. Cheng HW (2010) Breeding of tomorrow's chickens to improve well-being. Poult Sci 89:805–813
59. Fussell LW (1998) Poultry industry strategies for control of Immunosuppressive diseases. Poult Sci 77:1193–1196
60. Hashemi SR, Zulkifli I, Hair Bejo M, Farida A, Somchit MN (2008) Acute toxicity study and phytochemical screening of selected herbal aqueous extract in broiler chickens. Int J Pharmacol 4:352–360
61. Wang L, Piao XL, Kim SW, Piao XS, Shen YB, Lee HS (2008) Effects of Forsythia suspensa extract on growth performance, nutrient digestibility, and antioxidant activities in broiler chickens under high ambient temperature. Poult Sci 87:1287–1294
62. Mujahid A, Sato K, Akiba Y, Toyomizu M (2006) Acute heat stress stimulates mitochondrial superoxide production in broiler skeletal muscle, possibly via down-regulation of uncoupling protein content. Poult Sci 85:1259–1265
63. Peter SF (2015) Antioxidant systems in poultry biology: superoxide dismutase. J Anim Res Nutr 1:8
64. Halliwell B (2012) Free radicals and antioxidants: updating a personal view. Nutr Rev 70:257–265
65. Halliwell B, Gutteridge JMC (1999) Free radicals in biology and medicine, 3rd edn. Oxford University Press, Oxford

66. Niki E (2014) Antioxidants: basic principles, emerging concepts, and problems. Biomed J 37:106–111
67. Kelley KW (2004) From hormones to immunity: the physiology of immunology. Brain Behav Immun 18(2):95–113
68. Surai PF (1999) Vitamin E in avian reproduction. Poult Avian Biol Rev 10:1–60
69. Webster MJI, Glaser R (2008) Stress hormones and immune function. Cell Immunol 252(1):16–26
70. Dhabhar FS (2009) Enhancing versus suppressive effects of stress on immune function: implications for immunoprotection and immunopathology. Neuro- immunomodulation 16(5):300–317
71. Slawinska A, Hsieh JC, Schmidt CJ, Lamont SJ (2016) Heat stress and lipopolysaccharide stimulation of chicken macrophage-like cell line activates expression of distinct sets of genes. PLoS ONE 11(10)
72. Tort L, Teles M (2011) The endocrine response to stress-a comparative view. In: Basic and clinical endocrinology up-to-date. ISBN 978-953
73. Shini S, Huff GR, Shini A, Kaiser P (2010) Understanding stress-induced immunosuppression: exploration of cytokine and chemokine gene profiles in chicken peripheral leukocytes. Poult Sci 89(4):841–851
74. Riedemann T, Patchev AV, Cho K, Almeida OF (2010) Corticosteroids: way upstream. Mol Brain 3(1):2–20
75. Hamdy AMM (2000) Evaluation of some immunological and hormonal variations among several laying strains. Egyptian J Anim Prod 37(2):121–131
76. Shini S, Shini A, Huff GR (2009) Effects of chronic and repeated corticosterone administration in rearing chickens on physiology, the onset of lay and egg production of hens. Physiol Behav 98:73–77
77. Virden WS, Thaxton JP, Corzo A, Dozier WA, Kidd MT (2007) Evaluation of models using corticosterone and adrenocorticotropin to induce conditions mimicking physiological stress in commercial broilers. Poult Sci 86:2485–2491
78. Vahdatpour T, Adl KN, Nezhad YE, Sis NM, Riyazi SR, Vahdatpour S (2009) Effects of corticosterone intake as stress-alternative hormone on broiler chickens: performance and blood parameters. Asian J Anim Vet Adv 4:16–21
79. Song Z, Yuan L, Jiao H, Lin H (2011) Effect of corticosterone on hypothalamic corticotropin-releasing hormone expression in broiler chicks (Gallus gallus domesticus) fed a high energy diet. Asian-Australas J Anim Sci 24:1736–1743
80. Ader R, Cohen N, Felten D (1995) Psychoneuroimmunology: interactions between the nervous system and the immune system. Lancet, 345(8942):99–103
81. Ader R, Felten DL, Cohen N (2001) Psychoneuroimmunology, 3rd edn. Academic Press, USA
82. Koga T, Harada H, Shi TS, Okada S, Suico MA, Shuto T, Kai H (2005) Hyperthermia suppresses the cytotoxicity of NK cells via down-regulation of perforin/granzyme B expression. Biochem Biophys Res Commun 337(4):1319–1323
83. Meng D, Hu Y, Xiao C, Wei T, Zou Q, Wang M (2013) Chronic heat stress inhibits immune responses to H5N1 vaccination through regulating CD4 + CD25 + Foxp3 + Tregs. BioMed Research International. Retrieved from https://www.hindawi.com/journals/bmri/2013/160859/
84. Dohms JE, Metz A (1991) Stress—mechanisms of immunosuppression. Vet Immunol Immunopathol 30:89–109
85. Hu XF, Guo YM, Huang BY, Zhang LB, Bun S, Liu D (2010) Effect of corticosterone administration on small intestinal weight and expression of small intestinal nutrient transporter mRNA of broiler chickens. Asian-Australas Assoc Anim Prod Soc 23:175–181
86. Siegel HS (1995) Stress, strains and resistance. Br Poult Sci 36:3–22
87. Bakshi VP, Kalin NH (2000) Corticotropin-releasing hormone and animal models of anxiety: gene–environment interactions. Biol Psych 48:1175–1198

88. Garriga C, Hunter RR, Amat C, Planas JM, Mitchell MA, Moreto M (2006) Heat stress increases apical glucose transport in the chicken jejunum. Am J Physiol Reg Integ Comp Physiol 290:R195–R201
89. Star L, Decuypere E, Parmentier HK, Kemp B (2008) Effect of single or combined climatic and hygienic stress in four layer lines: 2. Endocrine and oxidative stress responses. Poult Sci 87:1031–1038
90. Korte SM, Koolhaas JM, Wingfield JC, McEwen BS (2005) The Darwinian concept of stress: benefits of allostasis and costs of allostatic load and the tradeoffs in health and disease. Neurosci Biobehav Rev 29:3–38
91. Haller J, Millar S, van de Schraaf J, de Kloet RE, Kruk MR (2000) The active phase-related increase in corticosterone and aggression are linked. J Neuroendocrinol 12:431–436
92. Bukau B, Weissman J, Horwich A (2006) Molecular chaperones and protein quality control. Cell 125:443–451
93. Lee S, Tsai FT (2005) Molecular chaperones in protein quality control. J Biochem Mol Biol 38:259–265
94. Laufen T, Mayer MP, Beisel C, Klostermeier D, Mogk A, Reinstein J (1999) Mechanism of regulation of Hsp70 chaperones by DnaJ cochaperones. P Natl Acad Sci USA 96:5452–5457
95. Mattoo RUH, Sharma SK, Priya S, Finka A, and Goloubinoff P (2013) Hsp110 is a bona fide chaperone using ATP to unfold stable misfolded polypeptides and reciprocally collaborate with Hsp70 to solubilize gene expression in HD11 cell line stimulated with heat stress and LPS. J Biol Chem 288:21399–21411
96. Coble DJ, Fleming D, Persia ME, Ashwell CM, Rothschild MF, Schmidt CJ (2014) RNA-seq analysis of broiler liver transcriptome reveals novel responses to high ambient temperature. BMC Genom 15:1084
97. Xie J, Tang L, Lu L, Zhang L, Xi L, Liu HC (2014) Differential expression of heat shock transcription factors and heat shock proteins after acute and chronic heat stress in laying chickens (Gallus gallus). PLoS ONE 2014(9):e102204
98. Wang SH, Cheng CY, Tang PC, Chen CF, Chen HH, Lee YP (2013) Differential gene expressions in testes of L2 strain Taiwan country chicken in response to acute heat stress. Theriogenology 79:374–382
99. Luo QB, Song XY, Ji CL, Zhang XQ, Zhang DX (2014) Exploring the molecular mechanism of acute heat stress exposure in broiler chickens using gene expression profiling. Gene 546:200–205
100. Li C, Wang X, Wang G, Li N, Wu C (2011) Expression analysis of global gene response to chronic heat exposure in broiler chickens (Gallus gallus) reveals new reactive genes. Poult Sci 90:1028±1036
101. Sun L, Lamont SJ, Cooksey AM, McCarthy F, Tudor CO, Vijay-Shanker K (2015) Transcriptome response to heat stress in a chicken hepatocellular carcinoma cell line. Cell Stress Chaperones 20:939–950
102. Bergner P (2005) Antiviral botanicals in herbal medicine. Med Herbalism 14:1–12
103. Hamdy AMM (2006) Lymphatic system and immunity of chickens. In: 12th European Poultry conference, Verona, Italy
104. Ciraci C, Tuggle CK, Wannemuehler MJ, Nettleton D, Lamont SJ (2010) Unique genome-wide transcriptome profiles of chicken macrophages exposed to Salmonella-derived endotoxin. BMC Genom 11:545
105. Lambert GP, Gisolfi CV, Berg DJ, Moseley PL, Oberley LW, Kregel KC (2002) Selected contribution: hyperthermia-induced intestinal permeability and the role of oxidative and nitrosative stress. J Appl Physiol 92(4):1750–1761
106. Leon LR, Helwig BG (2010) Heat stroke: role of the systemic inflammatory response. J Appl Physiol 109:1980–1988
107. Hassanpour H, Zamani Moghaddamb AK, Khosravi M, Mayahi M (2013) Effects of synbiotic on the intestinal morphology and humoral immune response in broiler chickens. Livest Sci 153(1):116–122

108. Franco- Jimenez DJ, Beck MM (2007) Physiological changes to transient exposure to heat stress observed in laying hens. Poult Sci 86:538–544
109. Koh K, Macleod MG (1999) Effects of ambient temperature on heat increment of feeding and energy retention in growing broilers maintained at different food intakes. Br Poult Sci 40:511–516
110. Wolfenson D (1986) The effect of acclimatization on blood flow and its distribution in normothermic and hyper-thermic domestic fowl. Comp Biochem Physiol 85A:739–742
111. Hai L, Rong D, Zhang ZY (2000) The effect of thermal environment on the digestion of broilers. J Anim Physiol Anim Nutr 83:57–64
112. Yahav S, Hurwitz S (1996) Induction of thermotolerance in male broiler chickens by temperature conditioning at an early age. Poult Sci 75:402–406
113. Rama-Rao SV, Praharaj NK, Ramasubba RV, Panda AK (2003) Interaction between genotype and dietary concentrations of methionine for immune function in commercial broilers. Br Poult Sci 44:104–112
114. Ausubel FM (2005) Are innate immune signaling pathways in plants and animals conserved? Nat Immunol 6:973–979
115. Dangl JL, Jones JDG (2001) Plant pathogens and integrated defence responses to infection. Nature 411:826–833
116. Dixon RA, Dey PM, Lamb CJ (1983) Phytoalexins: enzymology and molecular biology. Adv Enzymol Related Areas Mol Biol 55:1–136
117. Bergner P (2004) Immunity: Herbal medicine and the TH1/TH2 balance. Med Herbalism 14:1–9
118. Hashemi SR, Davoodi H (2010) Phytogenics as new class of feed additive in poultry industry. J Anim Vet Adv 9:2295–2304

Socioeconomic Impacts and Green Sustainability

Climate Change, Agriculture and Rural Communities' Vulnerability in the Nile Delta

Dalia M. Gouda

Abstract Climate change is a global phenomenon that has different regional impacts. Some countries and sectors are to be affected more than others, especially those with inadequate adaptive capacities that would increase their vulnerability. According to the climate change projections of potential impacts, Egypt and the Nile Delta are to be affected. The effects of climatic changes are to be experienced more in agriculture, which is to negatively affect rural communities. Meanwhile, agriculture in Egypt and the Nile Delta is countering immediate socio-economic and natural resources related challenges which continuation would exacerbate the impacts of climate change. Therefore, the impacts of these challenges on agriculture in the Nile Delta governorates, in terms of population growth, agricultural land fragmentation, and urban expansion on agricultural land are reviewed. For the identification of the impacts of climate change: temperature increase and sea-level rise (SLR) on agriculture in the Nile Delta governorates, the vulnerability assessment framework (VAF) is used. Findings show that—immediate challenges—urban expansion on agricultural land if not curbed, even without climate change, would severely impact agriculture and thus the national food security. Nevertheless, climate change would have great impacts on agriculture and farmers in the Nile Delta, whereas some governorates are to be more vulnerable to the climate change stimuli than others depending on their exposure, sensitivity and adaptive capacities. The chapter concludes that more 'context-specific' attention needs to be directed towards agriculture in the Nile Delta to preserve it—from man-made and environmental challenges—for future generations.

Keywords Agriculture · Climate change · Nile Delta · Rural communities · Urbanization · Vulnerability assessment framework (VAF)

D. M. Gouda (✉)
Socio-Economic Expert, Giza, Egypt
e-mail: Dalia.gouda@gmail.com

© Springer Nature Switzerland AG 2020

E.-S. E. Omran and A. M. Negm (eds.), *Climate Change Impacts on Agriculture and Food Security in Egypt*, Springer Water, https://doi.org/10.1007/978-3-030-41629-4_22

1 Introduction

Climate change is a global phenomenon affecting countries and economic sectors differently through the increase or decrease in temperature, decrease or increase in rainfall and water resources availability, sea-level rise and extreme weather events (floods and droughts). Climate models predict that the change in climatic conditions will affect agriculture by reducing land productivity between 10 and 20% by 2050 in many developing countries and affecting about 10% of the agricultural land [1, 2]. Accordingly, many farming households are expected to counter crop deficiencies and yield reductions that would affect their livelihoods and in turn their vulnerability [3, 4].

Farmers in Egypt are equally expected to suffer from climate change impacts, as Egypt has been classified as one of five countries in the world, vulnerable to the negative impacts of climate change [5, 6]. Climate change is to negatively affect prime agricultural land and thus harm agricultural production. A UNDP supported study conducted in 2013 [7], expects a decrease between 4 and 17% of agricultural production by 2030. The Nile Delta in the northern part of Egypt—comprising about 50% of the agricultural land—is identified as one of the world's three "extreme" vulnerability hotspots. It is expected to experience temperature and crop water requirements' increase, and sea-level rise (SLR) that would negatively affect large agricultural land areas and in turn impact agricultural production [8, 9]. This would sequentially have a bearing on farmers' resilience by increasing their vulnerability.

Nevertheless, agriculture in Egypt in general and in the Nile Delta in particular, is currently facing other intense challenges resulting from the rapid population growth that is exceeding 2.5% annually [10], thus augmenting agricultural land and irrigation water scarcities. These challenges can be categorized into socio-economic and natural resources challenges. These challenges, if not immediately countered, are expected to have far-reaching future impacts on agriculture and would exacerbate the impacts of climate change in Egypt and the Nile Delta.

Though aware of the importance of the impacts of the natural resources challenges on agriculture, this chapter is only to address the socio-economic challenges impacts on agriculture in the Nile Delta governorates. It is believed that the socio-economic aspects fuel and reinforce the natural resources challenges, thus, if the former are curbed the negative impacts of the latter could be contained.

To this end, this chapter attempts to explore the change in agriculture in the Nile Delta governorates, in terms of field crops' agricultural land area, production, net value and employment. This attempt includes changes resulting on the one hand from the continuation of the socio-economic challenges (without climate change), and on the other from those caused by the climate change stimuli (without urban expansion), here, temperature increase and sea-level rise (SLR).

The focus is on socio-economic and climatic changes that are expected to take place by the year 2030, which is only 12 years ahead from today that allows the assumption that similar socio-economic conditions in the Nile Delta governorates would continue to exist.

The assessments are mainly based on the 'Population, housing and establishment census of 2017' [11] for population-related data, and on the 'Annual bulletin of the agricultural cooperatives' activities of 2015/2016' [12] for agricultural land area and number of landownerships data. For data on agricultural field crops (the focus of this chapter), the 'Annual bulletin of statistical crop area and plant production of 2014/2015' is used [13]. In addition to climate change-related literature, such as the 'Proposed climate change adaptation strategy for the ministry of water resources and irrigation in Egypt' [9] and the 'Potential impacts of climate change on the Egyptian economy' report [7].

The impacts of the socio-economic challenges in the Nile Delta governorates by 2030, are assessed by extrapolating available statistics on rural population growth, agricultural land fragmentation and urban expansion on agricultural land. For the determination of the climate change stimuli impacts on agriculture in the Nile Delta governorates and the identification of farmers' vulnerability in 2030, the vulnerability assessment framework (VAF) is applied. The VAF is an important tool for supporting policy makers in formulating necessary strategies of climate change adaptation for the enhancement of the adaptive capacities of local farming communities by improving awareness of the causes and the implications of vulnerability [14, 15].

The framework considers the exposure and sensitivity of agriculture and farmers to the climate change stimuli, and their local adaptive capacities [16] that are necessary for coping with climate change adverse impacts on farmers' livelihoods [16, 17]. For the exploration of the adaptive capacities of farmers the livelihood assets—defined by the sustainable livelihood framework—are used. These are composed of human, social, natural, physical, and financial assets. Assets influence farmers' vulnerability and determine how they would deal with the negative impacts of climate change [18, 19]. The chosen indicators for the determination of farmers' adaptive capacities vis-à-vis climate change stimuli in the Nile Delta governorates are concerned with farmers' awareness of climate change hazards and their ability to decide on choosing the suitable adaptation measure [20]. These indicators include farmers' experience, age, access to agricultural information and their level of education, and access to funding [17].

2 Agriculture and Its Challenges in Egypt

The Nile Delta is central for the production of field crops' and the provision of agricultural employment in Egypt. However, there are present challenges that negatively affect agriculture. The agricultural importance of the Nile Delta and the countered challenges are introduced in this section.

2.1 The Nile Delta and Agriculture

Egypt is an agriculture-based country though it only contributed about 11% of total GDP in 2016/2017 [21]. It provides livelihoods for about 50% of the rural population and employs about 30% of the labor force in farms and agriculture-based industries [22]. Agriculture is also an important source of food for most of the Egyptian population, especially those living in rural areas [23]. Over the past decades, agricultural research has allowed Egypt to increase its self-sufficiency in many field crops', such as wheat and sugar crops [24]. In 2015, self-sufficiency in cereals reached about 60%, 100% in sugar crops, and 70% in oil crops [13, 25, 26].

The Nile Delta, located in the north of Egypt, overlooking the Mediterranean Sea with a coastline of 240 km stretched out from Alexandria in the west to Port Said in the east [27], is central for Egypt's agricultural production. Agricultural land in the Nile Delta covers about 2% of Egypt's total area [27], while the total agricultural land in Egypt (Old and New Lands)[1]—about 9 million feddan—constitutes about 4% of Egypt's total area. The major agricultural governorates in the Nile Delta include: Damietta, Dakhahlia, Sharkia, Qalyubia, Kafr El-Sheikh, Gharbia, Menufia, Beheira, Ismailia and Alexandria[2] (see Map 1). Most of the agricultural land in these governorates are Old Lands, except for Alexandria, Ismailia and Beheira where about 60%, 40%, and 30% respectively of their total agricultural land area is New Lands. While 100% of the agricultural land in Daqahlia, Gharbia and Menufia is Old Lands (see Table 8 in Annex for the agricultural land area in the Nile Delta governorates and agricultural land by type: Old and New Land categorized under different agricultural cooperatives: credit, agrarian reform and land reclamation cooperatives).

The Nile Delta governorates are home for about 50% of Egypt's population, for more than 50% of the rural population and about 60% of the farmers (landowners). Moreover, the cultivation of field crops' in the Nile Delta governorates, makes available more than 55% of working days dedicated to field crops production in Egypt.

These governorates encompass about 50% of Egypt's agricultural land and contributed about 50% of the total field crops' net value in 2015. In Table 1, the field crops are grouped into crop groups: cereals, sugar crops, oil crops, pulses, medicinal crops, vegetables, clover, and cotton. Cereals include wheat, barley, rice, and maize; sugar crops include sugarcane and sugar beet; oil crops include peanut, sesame, flax, sunflower, and soya beans; and vegetables include vegetables, potatoes, sweet potatoes, onion, and garlic.

---

[1]The agricultural land in Egypt is composed of Old and New Lands, the former is the alluvial land formed over thousands of years through the silt coming with the annual flood, while the latter refers to the land in the desert that has been reclaimed over the past six to seven decades.

[2]The focus in this chapter is on 10 major agricultural governorates in the Nile Delta. These include: Damietta, Dakhahlia, Sharkia, Qalyubia, Kafr El-Sheikh, Gharbia, Menufia, Beheira, Ismailia and Alexandria. Port Said was excluded due its limited agricultural area accounting for about 100,000 feddan.

Map 1 The Nile Delta governorates. *Source* Schewe [28]

The Nile Delta governorates comprise about 60% of the land cultivated with cereals that contribute more than 60% of the cereal production in Egypt, and they constitute about 50% of the sugar crops' area that yields more than 30% of the national sugar crops' production. These governorates are also home for about 30% of the oil crops area that supplies about 30% of the total produce and includes more than 55% of the pulses area that contributes to Egypt's self-sufficiency by producing more than 50% of the pulses. The Nile Delta governorates are also important for producing vegetables, as they comprise more than 55% of the total vegetables area that produces about 50% of total vegetables in Egypt. The production of clover and cotton is mostly concentrated in the Nile Delta governorates with about 70% of the clover area and more than 60% of its production, and about 90% of the cotton area and 90% of its production in Egypt in 2015. For the cultivation of these field crops,

Table 1 Field crops' groups by area, production, net value and water requirements in the Nile Delta governorates in 2015

Field Crops' Group	% of Nile Delta field crops' area of total field crops area in Egypt	% of Nile Delta field crops' production of total field crops production in Egypt	% of Nile Delta field crops' net value of total field crops net value in Egypt	% of Nile Delta field crops' water requirements of total field crops water requirements in Egypt	% of Nile Delta field crops' working days of total field crops working days in Egypt
Cereals	57.60	62.50	52.40	67.13	56.60
Sugar crops	46.00	31.10	30.90	18.67	34.10
Oil crops	31.20	28.60	30.50	29.59	31.30
Pulses	55.80	53.60	23.50	54.67	55.70
Vegetables	55.40	49.40	39.60	55.47	56.10
Clover (berseem)	68.90	61.00	82.50	81.01	68.90
Medicinal crops	9.20	2.50	10.30	9.16	9.20
Cotton	87.80	85.80	79.30	87.82	87.80
% of total Egypt	57.40	51.70	50.20	59.30	55.30

Source Calculated by the Author based on Capmas [13] and Mohamed et al. [29]

about 60% of the total field crops' water in Egypt is used for irrigation[3] (see Table 1) during the Summer, Winter and Nili seasons.[4]

Despite its importance, agriculture in the Nile Delta is subject to challenges that need immediate attention and intervention to limit their negative impacts. These are addressed below.

2.2 Challenges Countering Agriculture in the Nile Delta

Agriculture in the Nile Delta is countering socio-economic and natural resources challenges that affect agricultural production and farmers' livelihoods. The socio-economic challenges are concerned with population growth related problems, while the natural resources challenges involve water and soil-related difficulties. These challenges are briefly framed in the following paragraphs.

[3]Fresh water used for the cultivation of field crops was about 40 BCM in 2015, accounting for 72% of Egypt total share from the Nile River (55.5 BCM) excluding horticulture and permanent trees, see [26].

[4]Agricultural land in Egypt is cultivated at least twice a year: during the Winter and Summer seasons in addition to the Nili season.

2.2.1 Socio-economic Challenges in the Nile Delta

The socio-economic challenges are composed of population growth, agricultural land fragmentation and urban expansion on agricultural land. These are three interlinked aspects that impact agricultural production and food self-sufficiency; however, for analytical purposes, they are addressed separately below.

a. Population Growth

According to the population and establishment census of 2017 [11], the population in Egypt has reached about 95 million in 2017, revealing an increase of about 10 million people over the past 3 years [30]. The population of the Nile Delta governorates has also increased by more than 4 million people, from about 42 million in 2015 to 46 million people in 2017 [11, 30].

The increase in the Nile Delta population in general and that of the rural population—accounting for about 64% of the population—contributes to the decrease of the per capita agricultural land. Furthermore, population growth coupled with the limited land area of the Nile Delta have made it among the most densely populated areas in the world [7]. The average density in the Nile Delta governorates, is about 1800 per km^2, it was highest in Qalyubia with about 5000 people per km^2 and lowest in Ismailia with about 260 people per km^2 in 2017 (see Table 9 in Annex for Population density by governorate) [31].

Population growth and increased densities, amplify the pressure on agricultural land causing its fragmentation.

b. Agricultural Land Fragmentation

Rapid population growth has contributed to the drop of the per capita share in agricultural land from 0.29 feddan in 1950 to about 0.10 feddan since the 1990s in Egypt [11, 13, 30], while the average landownership on the village level is 2 feddan. In the Nile Delta governorates, the average landownership is 2.3 feddan; however, when excluding Alexandria and Ismailia governorates—due to their relatively sizeable landownerships owed to the added New Lands areas—the average land ownership drops to 1.46 feddan. It is lowest in Gharbia with 0.9 feddan, and Damietta and Qalyubia with 1.2 feddan respectively, and highest in Ismailia and Alexandria with 6.3 feddan and 4.9 feddan respectively (see Table 9 in Annex for Average landownership by governorate in 2017).

Land fragmentation resulting from demographic changes and inheritance reduces the financial returns of agricultural land, by making it inadequate for supporting a farming household, which forces farmers to seek to increase their non-farm income sources. One of these sources is the transformation of their agricultural land into urban land [32].

c. Urban Expansion on Agricultural Land

According to the Sustainable Agricultural Development Strategy of 2030 [24], the annual loss of agricultural land to urbanization is about 20,000 feddan in the Nile Delta and Valley. In contrast, Tate [33] states that the annual urban growth on agricultural land was about 30,000 feddan, but after 2011 increased to about 50,000 feddan annually, which means that urban expansion on agricultural land has increased by 40% after 2011 in Egypt.

Moreover, Alfiky et al. [34] utter that most of the agricultural land lost to urbanization is concentrated in the Nile Delta, especially in the Middle Delta governorates that lack desert hinterland for urban expansion. Thus, urban expansion on agricultural land in the Nile Delta between 1952 and 2002 has consumed about 300,000 feddan of the agricultural land [35], i.e. about 6000 feddan were lost per year, while El-Kholei [36] states that the annual loss of agricultural land between 1984 and 2007 was about 13,000 feddan. To adjust the agricultural land lost to urbanization in the Nile Delta for the after 2011 period, the 13,000 feddan are increased by 40%, resulting in an annual loss of 18,200 feddan. Therefore, it is sometimes argued that if the current urbanization trends continue, the Nile Delta will soon be totally transformed into urban areas [37].

2.2.2 Natural Resources Challenges

The natural resources challenges are composed of water and soil-related impediments. These affect agricultural land and land productivity in the Nile Delta and are addressed below.

a. Water Scarcity

Egypt is a water scarce country depending on an outside source for about 90% of its uses [38]. Egypt's annual freshwater supply is about 64 BCM/year, of which 55.5 BCM are received via the Nile River, in addition to about 1.3 BCM of rainfall and about 7 BCM of groundwater [30]. About 80% of the fresh water is used for agricultural production.

Agricultural production in the Nile Delta governorates is dependent on the Nile water. Nearly 60% of fresh water used for the irrigation of field crops in Egypt is used in the Nile Delta governorates because precipitation during December to March (winter season) on the coastal governorates is only between 100 and 200 mm [39].

Therefore, any change in the flow of the Nile, which supplies irrigation water for agriculture, can have large impacts on agricultural production in the Nile Delta [40]. Consequently, the construction of the Grand Ethiopian Renaissance Dam (GERD) with a water storage capacity of 74 BCM might have negative impacts on water availability for Egypt, especially during the dam reservoir filling period. The reduction in the available fresh water would further reduce the already below—less than 700 cubic meters per capita per year (m^3/year)—'water poverty line' (1000 m^3/year) water per capita share in Egypt [41]. This, in turn, would affect agricultural production and rural livelihoods [26].

b. Soil Salinity

In Egypt, the agricultural land is cultivated at least twice a year, in the summer and winter in addition to the Nili season [13], which depletes the soils from their nutrients that necessitates the use of fertilizers. The use of fertilizers is also accelerated due to land fragmentation that urges landowners to overuse nutrients [42] to increase their yields to be able to support their families.

Furthermore, areas in the Nile Delta that are suffering from water scarcity—e.g. located at the end of the irrigation system—do not have access to adequate fresh water to leach the soil from excessive salt. A study found that about 2 million feddan in the northern parts of the Nile Delta are suffering from salt-affected soils [27]. Moreover, during the summer season, some farmers irrigate from drains to counter water shortage [43], which further deteriorates the fertility of the soil.

Accordingly, the excessive use of fertilizers and the irrigation from drains [9] negatively affect the soils, especially if coupled with fresh water scarcity that harms soil fertility, crop yields, and livestock and human health [42, 44].

In the following section, the impacts of the socio-economic challenges on agriculture in the Nile Delta governorates by 2030 are explored.

3 Impacts of Socio-economic Challenges on Agriculture in the Nile Delta Governorates by 2030

In this section, the socio-economic challenges in the absence of climate change by 2030 are explored, while assuming the natural resources related challenges remain unchanged. These challenges include population growth, agricultural land fragmentation and urban expansion on agricultural land. Then, the impact of urban expansion on agricultural land, in terms of field crops' agricultural land area, production, net value and employment is assessed.

3.1 Population Growth

According to the population projections of the UN Population Division in 2008 the population of Egypt might reach, 104 million people in case of a slow growth rate and 117 million people in case of a fast growth rate by 2030 [43]. In this section, the fast growth rate scenario is used since in 2017 the population has already reached about 95 million from about 85 million in 2014, which refers to a 10 million people increase in just 3 years. Such an increase will have serious implications for food security, poverty and social stability [45].

For the estimation of the population of the Nile Delta governorates by 2030, the percentage share of every governorate's population of total Egypt in 2017 is multiplied by the total projected population, of 117 million people. The calculation results in an increase of about 11 million people in the Nile Delta governorates, from about 46 million in 2017 to 57 million people by 2030, of which about 60%

would live in rural areas. This, in turn, would increase the average population density in the Nile Delta governorates.

By 2030 the average population density in the Nile Delta governorates would reach about 2300 per km^2. The highest density is to be in Qalyubia with about 6000 inhabitants per km^2 and lowest in Ismailia with about 300 inhabitants per km^2 (see Table 10 in Annex for Population and population densities by governorate by 2030).

Population growth and high population densities would put pressure on available land in the Nile Delta. It is also to reduce the per capita agricultural land as well as the average size of agricultural land owned by farmers, i.e. it would further fragment agricultural land till 2030.

3.2 Agricultural Land Fragmentation

Population growth increases the number of landowners—due to demographic growth and inheritance—that causes land fragmentation. The percentage change of landownerships by governorate between 1999/2000 and 2015/16 is used to assess the growth in the number of landowners between 2015/16 and 2030 [30]. The result shows an increase of about 0.7 million landowners.

The increase in the number of landowners would contribute to the shrinking of the average landownership in the Nile Delta governorates to 2 feddan—assuming no increase in the agricultural land area through land reclamation (see Table 10 in Annex for Average landownership due to population growth by 2030). However, when excluding Alexandria and Ismailia—that have relatively large landownerships—the average landownership would be 1.2 feddan by 2030. The governorates with the lowest landownership average would be Gharbia and Damietta, with 0.6 feddan and 0.8 feddan respectively. The shrinking agricultural landownerships, low returns and population growth are expected to accelerate the transformation of agricultural land into urban land.

3.3 Urban Expansion on Agricultural Land

This section relies on the literature discussed under 2.2.1. It assumes that the annual loss of agricultural land to urbanization in the Nile Delta governorates is 18,200 feddan. This means that by 2030 (i.e. 12 years from now) the loss of agricultural land to urbanization would amount 218,400 feddan.

The 218,400 feddan account for about 4.5% of the total agricultural land area (field crops area) in the Nile Delta governorates. For the determination of the lost cropping area by 2030, the field crops area by governorate is reduced by 4.5% (based on 2015 data). Adding up the reduced areas would equal a reduction of about 330,000 feddan cropping area in the Nile Delta governorates. This reduction would have local and national implications.

Table 2 Field crops' groups and the impact of urban expansion on agricultural land by 2030

Field crops' group	% reduction of field crops' agricultural land area in the Nile Delta due to urbanization of total Egypt	% reduction of field crops' production in the Nile Delta due to urbanization of total Egypt	% reduction of field crops' net value in the Nile Delta due to urbanization of total Egypt	% reduction of field crops' water requirements in the Nile Delta due to urbanization of total Egypt	% reduction of field crops' working days in the Nile Delta due to urbanization of total Egypt
Cereals	2.59	2.81	2.40	3.02	2.55
Sugar crops	2.07	1.40	1.39	0.84	1.53
Oil crops	1.41	1.29	1.37	1.33	1.41
Pulses	2.51	2.41	1.06	2.46	2.52
Vegetables	2.49	2.22	1.78	2.50	2.52
Clover (berseem)	3.10	2.75	3.71	3.65	3.10
Medicinal crops	0.41	0.11	4.41	0.41	0.41
Cotton	3.95	3.86	3.57	3.95	3.95
% reduction of total Egypt	2.58	2.33	2.30	2.67	2.49

Source Calculated by the author based on Capmas [13] and Mohamed et al. [29]

3.3.1 Impact of Urban Expansion on Agricultural Land

The reduction of the field crops cropping area by 330,000 feddan would affect agriculture in the Nile Delta governorates. The impact on the field crops' land area, production, net value and employment is addressed below.

Impact on Field Crops' Agricultural Area

The loss of 218,400 feddan would reduce the cropping area by 330,000 feddan. This loss would reduce the field crops' agricultural land area in Egypt by about 2.6% (see Table 2).

Impact on Field Crops' Production

The reduction of the agricultural area by about 2.6% would cause a decrease in the total production by about 2.3% compared to the field crops' situation in Egypt in

2015 (without urban expansion). The cotton production is to be affected the most by a decrease of about 4% followed by cereals and clover with about 3% each. The total reduction in production accounts for about 3 million tons of total field crops.

Impact on Net Field Crops' Value

The reduction of field crops' production would lead to a decrease of 2.3% of the total net field crops value compared to the situation in Egypt in 2015 (without urban expansion). The medicinal crops net value is to be affected the most by a decrease of 4.4%, followed by clover with about 4% and cotton with 3.6%. This loss amounts about 2 billion Egyptian pounds of the total net field crops value in Egypt based on 2015 data that are equivalent to about US$270 million.[5]

Impact on Agricultural Employment

Dividing the loss of agricultural land per governorate by its average landownership in 2030—accounting for population increase and land fragmentation—shows that the loss of 218,400 feddan to urbanization would mean that about 240 thousand landowners in the Nile Delta governorates have decided to transform their land into urban land. More than 40 thousand landowners in each of Daqahlia, Gharbia, Kafr El-Sheikh and Beheira would have decided to urbanize their land (see Table 10 in Annex for Number of agricultural landowners and average landownership by governorate by 2030). These landowners would have to look for other sources of income.

Furthermore, the reduction of the field crops' cropping area by 330,000 feddan would affect agricultural employment by reducing agricultural working days by about 9 million days.[6] This would equal a total decrease of 2.5% of working days compared to the situation in Egypt in 2015 (without urban expansion). Moreover, the loss of about 218,400 feddan would cause a loss of about 360,000 direct and indirect jobs in the Nile Delta governorates [46].[7]

Population growth and land fragmentation are expected to constrain farmers' resilience to averse the negative impacts of climate change, while urbanization on agricultural land would increase the effects of climate change on the national level. Nevertheless, for simplification, the future impacts of the climate change stimuli on agriculture in the Nile Delta governorates are analyzed, while assuming the agricultural land area remains unchanged.

---

[5]Calculations are based on the 7.5 exchange rate of 2015.

[6]Dividing the 9 million working days by 180 days (the average annual working days per agricultural laborer) shows that 50,000 laborers will be affected.

[7]Multiplying the number of land left fallow by 1.7 (1 direct job and 0.7 indirect job lost by the loss of one feddan), see [46].

4 Impacts of Climate Change Stimuli on Agriculture in the Nile Delta Governorates in 2030

This section focuses on the impacts of temperature increase and sea-level rise on agriculture to elucidate the vulnerability of agricultural communities in the Nile Delta governorates in 2030. This is done by (1) identifying the potential future impact of temperature increase and sea-level-rise (SLR) on field crops' agricultural land, production, net value and agricultural employment; then (2) assessing the vulnerability of agricultural communities in the Nile Delta governorates. To this end, this section is structured in line with the vulnerability assessment framework (VAF) into exposure, sensitivity and adaptive capacity based on which the vulnerability of the Nile Delta governorates is determined.

4.1 Exposure

Agriculture in the Nile Delta is highly exposed to climate change effects via temperature increase that is further expected to cause field crops' deficiencies, and sea-level rise that might lead to the inundation of large areas of agricultural land in the coastal governorates of the Nile Delta (the impact of increased salinity due to inundation is not addressed in this chapter).

Increased temperature threatens agriculture in the Nile Delta, and the condition would exacerbate if coupled with water shortage [7, 47]. Temperature increase affects field crops by lowering their yields as well as through expanding crop diseases[8] [48]. According to the proposed climate change adaptation strategy for the Ministry of Water Resources and Irrigation in Egypt [8], the temperature is expected to increase by 1 °C by 2025, while the Potential impact of climate change on the Egyptian economy report [7] expects an increase between 0.9 and 1 °C by 2030. The temperature increase is to affect field crops' yields differently, i.e. through different percentage reductions.

The Egyptian Third National Communication in 2016 [49] listed percentages for field crops' yields expected reductions in case of a temperature increase of 1.5 °C, while the Potential impact of climate change on the Egyptian economy used lower percentages (see Table 11 in Annex for Percentages of yield reduction and water requirements increase) under the B1 marker scenario. The analysis in this section uses the percentages of field crops' reductions stated in the latter report.

Climate change impacts on irrigation water requirements may be great, in the case of upstream precipitation decrease and/or temperature increase in Egypt. The former would affect the quantity of fresh water received by Egypt via the Nile, while the latter would increase the evapotranspiration (ET) of field crops [7, 9], thus increasing

[8]Climate change and heat waves might also affect human health [8]; however, this topic is beyond the focus of this chapter.

field crops water requirements. Therefore, the impact of climate change on water is a serious concern for Egypt [40]; however, it is not within the focus of this chapter.

Studies predict that agricultural activity in the Nile Delta could be substantially affected by SLR through inundation and salinization [47, 50, 51]. The Potential impact of climate change on the Egyptian economy report [7] used the IPCC's (B1) and (A1F1) scenarios for the determination of the impact of SLR on the agricultural land in the Nile Delta regions in 2030. The impact on governorates is grouped by region into Northeast, North middle, and West Delta region. The middle impact scenario (B1) would affect the Northeast and North middle Delta regions and would have the same impact on agricultural land in both cases, protected or unprotected. In the Northeast Delta, the agricultural land loss would be 0.0% (620 feddan), while in the North Middle Delta 0.2% (1860 feddan) would be lost. The agricultural land in the West Delta would not be affected—though the land elevation would be below the sea level in 2025—because of the Mohamed Ali Wall [52].

The high impact scenario (A1F1) in case of protection would cause a loss of 0.7% (2700 feddan) of the agricultural land in the Northeast Delta and 0.2% (3200 feddan) of the North middle Delta. In contrast, the unprotected case would affect the agricultural land by a loss of 25.7% (90,310 feddan) in the Northeast Delta, 1.6% (20,070 feddan) in the North middle Delta and by 0.1% (1430 feddan) in the West Delta. For the analysis in this section, the middle impact scenario (B1) is used. Thus, the focus is on the loss of agricultural land in the Northeast and Middle Delta regions. This scenario predicted an approximate total loss of 10.4 km^2 of the valuable agricultural land in the Nile Delta in both cases, the protected and unprotected one.

Temperature increase and SLR that are defined as exposure are expected to impact agricultural land and farmers (landowners and agricultural laborers) in the Nile Delta governorates.

4.2 *Sensitivity*

Sensitivity here refers to the degree to which farmers and agriculture are affected, either adversely or beneficially, by climate change [53]. The sensitivity of agriculture and farmers in the Nile Delta governorates to temperature increase and SLR (assuming no change of agricultural land area) is explored in the following part.

4.2.1 Impact of Temperature Increase

The impact of temperature increase on the field crops' production, field crops' net value and employment in the Nile Delta governorates (see Table 11 in Annex for Percentage reduction of field crops' yields) is assessed below.

Table 3 Impact of temperature increase on agriculture in the Nile Delta governorates in 2030

Field crops' group	% reduction/increase of Nile Delta field crops production of total field crops production in Egypt in 2015	% reduction/increase of Nile Delta field crops net value of total field crops net value in Egypt in 2015	% increase of Nile Delta field crops water requirements of total water requirements in Egypt in 2015
Cereals	5.00	4.29	2.02
Sugar crops	0.33	0.30	0.56
Oil crops	1.08	1.07	0.89
Pulses	9.01	3.95	1.57
Vegetables	7.19	6.01	1.42
Clover (berseem)	5.13	6.93	2.53
Medicinal crops	0.21	4.66	0.29
Cotton	(+8.75)	(+10.59)	2.32
% Reduction of total Egypt	4.39	4.95	1.75

Source Calculated by the Author based on ETNC [49], Smith et al. [7], Capmas [13]

Impact on Field Crops' Agricultural Land Area

Temperature increase is expected to have a bearing on the agricultural land area in the Nile Delta governorates. Remote and marginal land areas could be spoiled due to their insufficient access to fresh water that would increase desertification [54]. Nevertheless, the potential area to be lost due to temperature increase is not quantified in this chapter.

Impact on Field Crops' Production

Temperature increase would affect field crops' production in the Nile Delta governorates. Table 3, shows that the temperature increase would decrease the total field crops' production in Egypt by 4.4% compared to 2015. However, the impact on the governorates is expected to differ, depending on their cropping pattern. For example, pulses are to be affected the most with a 9% reduction of total production in Egypt. The production of pulses is concentrated in Alexandria, Damietta and Kafr El-Sheikh where it covers 2.1%, 1.9%, and 1.2% of their total field crops' area respectively, followed by vegetables' production that would suffer from a reduction of 7.2%. However, the impact on governorates cultivating vegetables is expected to be greater, since vegetables cover larger agricultural areas in Alexandria, Ismailia and Beheira, amounting about 43%, 35% and 22% respectively. In contrast, the production of cotton is to increase by about 9% of total cotton production in Egypt (based on 2015 data). This increase is to mainly benefit Kafr El-Sheikh, and Beheira governorates as cotton covers 7.1% and 3.5% of their respective agricultural land areas (see Table 12 in Annex for Percentage of field crops' groups area by governorate).

Moreover, the production of strategic field crops[9] is to suffer as the cereals production is to decrease by 5%, sugar crops production by 0.3%, and oil crops production by about 1.1% of their total production in Egypt based on 2015 data. This is to affect almost all Nile Delta governorates; however, Daqahlia, Sharkia and Qalyubia governorates are to be affected the most as about 70% of their agricultural land is covered with strategic crops (see Table 13 in Annex for Percentage of field crops' groups production by governorate).

This percentage reduction in production assumes the availability of the necessary water requirements as the temperature increase would increase water requirements per field crop per feddan. Water shortage would exacerbate the field crops' yield deficiency. The cultivation of the field crops' in the Nile Delta governorates in 2030 (under B1 scenario), would increase the field crops' water requirements by about 2% of the total field crops water requirements in Egypt based on 2015 data that is equivalent to about 0.7 billion m^3.

Impact on Field Crops' Net Value

The total field crops' net value in the Nile Delta governorates is expected to decrease by about 5% of the total field crops' net value in Egypt based on 2015 data. The net value of clover is to be affected, with a decrease of about 7%, followed by vegetables with a net value reduction of 6%, and medicinal crops with a reduction of about 5%, while the net value of cotton is to increase by about 11% of the total net value of cotton in Egypt based on 2015 data. The governorates to be affected the most due to these reductions are: Damietta and Menufia as about 50% and 40% of their field crops' net value respectively comes from clover. Alexandria and Ismailia as about 70% and 65% of their total field crops' net value comes from vegetables. Daqahlia and Sharkia as their net value from cereals is about 50% and 40% respectively (see Table 14 in Annex for Percentage of net value per field crop by governorate).

Based on the average yield per field crop per feddan per governorate the farm level income would differ from one governorate to the other, which would consequently affect the livelihood of farmers (see Tables 15 and 16 in Annex for Average yield per field crop per feddan per governorate without and with temperature increase).

[9]For the Egyptian government, wheat, sugar crops, and oil crops are considered strategic crops that are important for the Egyptian populations' diet and clover for livestock breeding (meat production), see [24].

Impact on Employment (Landowners and Laborers)

Temperature increase is to affect all agricultural landowners cultivating field crops in the Nile Delta governorates. For the determination of the number of landowners to be affected, the agricultural land is divided by the average landownership by governorate. Accordingly, about 3 million landowners (when accounting for population growth the number of affected landowners would be about 3.5 million by 2030) are to be affected in the Nile Delta governorates.

To compensate their loss, landowners could resort to intensification, either by increasing the use of labor and/or fertilizers that in turn would increase the cost they would incur and reduce the net income they receive [55]. On the other hand, labor days per feddan are expected to increase because of intensification as well as due to the potential migration of pests to Egypt that would require fighting to prevent them from further harming the field crops' yields (resulting from increased temperature) [47, 56]. Thus, labor intensification strategy and pest-fighting might contribute to the increase in the number of agricultural working days per governorate.

4.2.2 Impact of Sea-Level-Rise (SLR)

According to scenario (B1) the Nile Delta regions to be affected by SLR are the Northeast and the North Middle Delta regions. These are composed of four governorates. The Northeast Delta region is composed of Damietta, Daqahlia and Sharkia; while the North middle Delta region is composed of Kafr El-Sheikh governorate. The impact of SLR in these governorates is explored on the field crops' agricultural land area, production, net value and employment.

Impact on Field Crops' Agricultural Land Area

The middle impact scenario (B1) would reduce the field crops' agricultural land area by 0.04% of total field crops' area in Egypt. Damietta, Daqahlia and Sharkia governorates constituting the Northeast Delta Region would lose a total of about 620 feddan in both cases; protected and unprotected. For the analysis, it is assumed that the loss of agricultural land would be equally distributed between the governorates. Accordingly, every governorate would lose about 207 feddan. This area equals different cropping areas in the governorates based on their respective average cropping intensity (see Table 9 in Annex for Land intensity by governorate). Multiplying the cropping intensity by governorate by 207 feddan shows that Damietta would lose a cropping area of 341.5 feddan, in Daqahlia the lost cropping area would equal 364.3 feddan, while in Sharkia it would equal 372.6 feddan. Thus, the 620 feddan would equal a total of 1078.4 feddan cropping area that is subtracted from the field crops area based on their respective share of the total agricultural cropping area per governorate.

Kafr El-Sheikh governorate constituting the North Middle Delta Region would alone lose about 1860 feddan in both cases; protected and unprotected. The average cropping intensity in Kafr El-Sheikh is 1.82. Therefore, the cropping area to be lost would be 3385.2 feddan. This area is deducted from the field crops area based on every field crops share of the total agricultural cropping area in the governorate.

The reduction of the agricultural area due to SLR is to reduce the field crops' water requirements by 0.04% of the field crops' water requirements of total Egypt based on 2015 data that is equivalent to about 16 million m^3.

Impact on Field Crops' Production

The agricultural land reduction would affect field crops' production by reducing the total field crops' production by about 3200 tons in Damietta, 3000 tons in Daqahlia, 2800 tons in Sharkia, and 34,200 tons in Kafr El-Sheikh. The strategic crops would experience reductions, as the cereals production would decrease by 478 tons in Damietta, 868 tons in Daqahlia and 777 tons in Sharkia and about 5995 tons in Kafr El-Sheikh. Sugar crops production losses would be 232 tons in Damietta, 711 tons in Daqahlia, 351 tons in Sharkia, and about 10,000 tons in Kafr El-Sheikh due to the large areas cultivated with sugar beet. The loss of oil crops' production would be 0.5 tons in Damietta, 0.7 in Daqahlia, 11.3 tons in Sharkia and 7.1 tons in Kafr EL-Sheikh.

The total loss of field crops' production would amount about 0.2% in Damietta, 0.03% in Daqahlia, 0.03% in Sharkia, and 0.34% for Kafr El-Sheikh. Though, the loss may appear small as a percentage; for the single farmer, the loss would be severe.

Impact on Field Crops' Net Value

SLR would cause a decrease in the field crops net value of about LE2 million in Damietta, Daqahlia and Sharkia collectively that is equivalent to about US$280 thousand, while the field crops' net value loss in Kafr El-Sheikh would be about LE20 million that equals US$2.5 million.[10]

However, the average annual net income loss for the regular farmer would range between 6 and 8 thousand Egyptian pounds[11] (see Table 4 for selected field crops and their average net income per feddan by governorate depending on yield per feddan). The average net income of selected field crops in Damietta would be LE6.5 thousand, in Daqahlia LE6.4 thousand, in Sharkia LE6.8 thousand and LE7.9 thousand in Kafr El-Shekih.

[10]Calculations are based on the 7.5 exchange rate of 2015.

[11]The result of the average net income of the 7 selected field crops in Table 4. These calculations are based on prices and costs of 2015 before the devaluation of the Egyptian pound in 2017. The crop net value after reduction is determined as follows: The average LE income per feddan is divided by average yield per crop in tons per feddan to determine the unit price of the ton. Then the field crop yield per feddan per governorate in tons is multiplied by the price per ton.

Table 4 Farmers average net income per feddan of selected field crops by governorate

Field crop/governorate	Wheat	Rice	Sugar beet	Vegetables	Onions and garlic	Clover	Cotton
Damietta	3794.10	2809.42	5408.77	11,735.92	10,221.68	8789.45	2787.79
Daqahlia	4166.96	4027.76	6468.52	9402.27	11,800.77	8303.56	2316.58
Sharkia	4357.75	3573.15	5217.50	9813.11	10,425.42	12,268.06	1966.45
Kafr El-Sheikh	3887.29	3482.24	4959.00	6684.41	16,961.44	16,828.81	2536.96

Source Calculated by the Author based on Capmas [13], Nour El-Din [9], and Smith et al. [7]

Accordingly, Kafr El-Sheikh farmers are to experience the highest net income loss per feddan. Furthermore, the loss would be greater for large farms that cultivate vegetables for local and international markets; however more severe for small farmers that depend on agriculture as the main income source and on the field crops production to feed their families and livestock.

Impact on Employment (Landowners and Laborers)

The SLR would impact both landowners and laborers. The total number of landowners to be affected in the four governorates is 1572 landowners and their respective families.[12] The landowners to be affected are determined by dividing the reduced area by the average landownership per governorate. Thus, in Damietta, 173 landowners would be affected, in Daqahlia 138 landowners, in Sharkia 99 landowners and 1163 landowners in Kafr El-Sheikh (when accounting for population growth, the number of affected landowners in the four governorates would increase by about 500 landowners: 259 in Damietta, 173 in Daqahlia, 109 in Sharkia and 1550 landowners in Kafr El-Sheikh, totaling 2090 landowners).

Moreover, the reduction of the cropping area would decrease the number of agricultural working days. In Damietta, about 8240 working days would be lost, 9026 working days in Daqahlia, while in Sharkia 10,280 working days, i.e. a total loss of 27,550 working days[13] in the Northeast Delta. Furthermore, the permanent loss of 620 feddan would cause a loss of 1054 direct and indirect jobs in the Northeast Delta governorates. In Kafr El-Sheikh, the working days would decrease by 89,241 working days,[14] in addition to the loss of 3162 direct and indirect jobs.

[12]The average family size in the Nile Delta governorates is 4.12 members per family.

[13]Dividing 27,550 working days by 180 days—average working days per agricultural laborer per year- equals 153, i.e. 153 laborers would lose their jobs.

[14]Dividing 89,241 working days by 180 days, equals 500, i.e. 500 laborers would lose their jobs.

4.2.3 Combined Climate Change Stimuli Impact on Agriculture in the Nile Delta Governorates in 2030

The combined impact of the climate change stimuli of temperature increase and SLR on agriculture in the Nile Delta governorates in 2030 is presented in Table 5.

The loss of field crops' area would be 0.04% of the total field crops' area in Egypt. This accounts for about 2480 feddan and 4463 feddan cropping area in the four governorates (Damietta, Daqahlia, Sharkia and Kafr El-Sheikh) caused by SLR, as the potential impact of temperature increase on marginal agricultural land is not quantified.

The total impact of temperature increase and SLR on agricultural field crops' production is a decrease of about 4.4% of total field crops' production in Egypt based on 2015 data. This is to reduce the total field crops' net value by about 5%, despite the increase in cotton production by about 9% and its net value by about 10.5% of total cotton production and net value in Egypt. The small area cultivated with cotton mainly in Kafr El-Sheikh, Beheira and Damietta governorates with about

Table 5 Impact of temperature increase and SLR on agriculture in the Nile Delta governorates in 2030

Field Crops' Group	% reduction of field crops' agricultural land area in the Nile Delta of total Egypt	% reduction/increase of field crops' production in the Nile Delta of total Egypt	% reduction/increase of field crops' net value in the Nile Delta of total Egypt	% increase of field crops' water requirements in the Nile Delta of total Egypt	% reduction of field crops' working days in the Nile Delta of total Egypt
Cereals	0.03	5.04	4.32	1.98	0.03
Sugar crops	0.06	0.37	0.34	0.54	0.05
Oil crops	0.01	1.09	1.08	0.89	0.01
Pulses	0.06	9.06	3.97	1.51	0.07
Vegetables	0.02	7.21	6.02	1.40	0.02
Clover (berseem)	0.04	5.17	6.99	2.48	0.04
Medicinal crops	0.01	2.54	4.66	0.28	0.01
Cotton	0.11	(+8.64)	(+10.48)	2.21	0.11
% Reduction of total Egypt	0.04	4.42	4.95	1.71	0.03

Source Calculated by the Author based on ETNC [49], Smith et al. [7], Capmas [13]

7%, 3.5% and 3% respectively would be insufficient to compensate for the losses in other field crops yields and net value.

Temperature increase would increase field crops' water requirements by 1.75%, while the SLR would decrease the field crops' water requirements by 0.04% of total field crops' water requirements in Egypt. Thus, Egypt's total field crops' water requirements is to increase by 1.71% based on 2015 data. Freshwater shortage due to climate change and/or the construction of GERD in Ethiopia would further increase the severity of field crops' deficiencies in Egypt. Moreover, the inundation of agricultural land would reduce the working days by about 0.03% of the total working days in Egypt, while temperature increase might increase working days per crop per feddan to counter yield deficiencies through labor intensification, and crop diseases fighting. The negative impact of yield reductions on landowners might partially be offset by the price increase of agricultural products [43, 57].

The negative impacts of the climate change stimuli on the field crops' area, production, net value and employment would affect Egypt's food self-sufficiency as well as the national budget by increasing imports. These impacts could also contribute to the eruption of social tensions/instabilities.

After having discussed, the exposure of the Nile Delta governorates to the climate change stimuli and the sensitivity of agriculture and farmers, the following section is dedicated to the identification of the adaptive capacities of farmers in the Nile Delta governorates.

4.3 Adaptive Capacity

The loss of agricultural field crops' land, production, and net value would negatively affect farmers (landowners and laborers) in the Nile Delta governorates and increase their vulnerability. However, the impact on governorates and farmers would differ depending on their adaptive capacities. The adaptive capacity refers to available resources that help farmers cope with the impact of climate change on agricultural land and livelihoods [6]. To establish the vulnerability of the Nile Delta governorates and its farmers, their adaptive capacities are explored in this section.

According to Maddison [20] the adaptation of local communities to climate change involves a two-stage process: first being aware of the change and then taking the suitable decision regarding the adoption of a certain climate change adaptation measure. Farmers' awareness is dependent on their farming experience and access to climate change related information. The selection of a suitable adaptation measure to the climate change stimuli by farmers is supported by their education and other socio-economic aspects that include the size of landownership, and access to funding [17]. These stages are investigated in the following paragraphs.

Egyptian farmers have a long farming history and experience that enabled them over the years to cope with scarcities, water shortages and unpredictable events [58].

A recent example of farmers' experience in Mahmoudia district in Beheira governorate shows that farmers[15] have perceived the change in water availability, especially during the summer season. This made them irrigate from private and public drains, and cultivate crops requiring less irrigation water as well as shifting their cultivation and harvesting dates. Moreover, a survey of 900 households in 20 governorates in Egypt revealed that 85% of the households noticed an increase in temperature (heat waves) in the summer, as well as an increase in the winter temperature. Accordingly, farmers resorted to water-use efficient crop varieties and/or early maturing ones, as well as to using wells and drainage water for irrigation [59].

Nevertheless, most farmers' experience is related to the cultivation of traditional crops (cereals, sugar cane, clover and cotton). Therefore, the average cropping area cultivated with traditional crops in the Nile Delta governorates is about 70%. The cultivation of traditional crops is an indication of subsistence farming, as farmers mainly produce to eat, not for the market as villages have become central consumers of agricultural products [60].

In Egypt, the elderly population possesses the farming knowledge and experience, while the younger population is busy in education or absent due to migration [61, 62]. In the Nile Delta governorates, farmers older than 45 years, accounted only for about 14% of the rural population in 2017 [11]. Rapid population growth has caused that about 60% of the rural population in the Nile Delta governorates was less than 25 years old and about 90% were less than 45 years old in 2017. This means that by 2030 the youngest of the elderly population would be more than 57 years old, while those who were less than 45 years old in 2017 would be between 25 and 57 years old. The latter age categories are expected to be the ones to take responsibility for agricultural production by 2030 and, thus, they would require support through access to information and to new technologies to adapt to climate change [62].

They would need to have access to information on various adaptation measures, in terms of costs and benefits of such measures, and their impact on crops as well as on their livelihoods. A study conducted in the Nile Delta found that farmers are willing to adapt to climate change if they have access to reliable information, e.g. on heat-tolerant crop varieties, use of pesticides and fertilizers, and efficient water use that are expected to substantially reduce the adverse effects of climatic impacts on farmers' livelihoods [63].

The public agricultural extension sector could play an important role in providing climate change related information as it is supposed to be the link between research and farmers [17]. For the public extension sector to carry out its tasks, it relies on research findings, e.g. of alternative crops testing that is suitable for different agro-climatic conditions [64].

Nevertheless, agricultural research in Egypt is currently experiencing budget and staffing weaknesses that affect its performance, and in turn that of the extension sector.

[15]A pilot area of the GIZ Programme "The Adaptation to Climate Change in the Water Sector in the MENA Region Programme" (ACCWaM) a project I was personally involved in, in 2013.

The Agricultural Research Center (ARC),[16] although its budget has increased from 0.42 to 0.44 of agricultural GDP between 2009 and 2012, the increase was mostly due to the increase in staffing (i.e. salaries). Furthermore, the hired researchers are not efficiently distributed across disciplines and/or across the governorates [65].

The public agricultural extension sector in Egypt is composed of three levels: i) the central level extension staff at the Ministry of Agriculture and Land Reclamation (MALR) with their subject matter specialists who provide the technical support to the ii) district level extension staff, who in turn support iii) the village extension workers (VEW). The latter is the closest to farmers—most of them reside in the villages—and are affiliated to village agricultural cooperatives. The VEW can be affiliated to one of two types of agricultural cooperatives: credit, and agrarian reform cooperatives[17] (see Table 8 in Annex for Agricultural Cooperatives' Types: inlcuding land area and farmers/ landowners' numbers). The credit cooperatives are present in almost all villages in the Nile Delta governorates. These oversee about 70% of the agricultural land, and comprise about 80% of the farmers in the Nile Delta governorates as members, while the agrarian reform and land reclamation cooperatives encompass the remaining 20% of farmers and 30% of the land [12, 66].

In the Nile Delta governorates, the average agricultural land area by credit cooperative is about 1900 feddan. The largest area per credit cooperative is in Ismailia and Alexandria with about 4300 feddan and 3800 feddan respectively, while the smallest average area is in Qalyubia and Gharbia with about 900 and 1000 feddan respectively. On average one VEW is responsible for about 800 feddan [12]. Dividing the average area per cooperative per governorate by 800 feddan would mean that the average number of VEW per credit cooperative in the Nile Delta governorates is 2 workers. Dividing 800 feddan by the average landownership per governorate shows that on average, one VEW is responsible for about 510 farmers in the Nile Delta governorates.

Agricultural cooperatives were established as an arm of the Ministry of Agriculture to guide and supervise farmers' agricultural production [67].[18] However, since the late 1980s, the agricultural liberalization has stripped agricultural cooperatives of most of their functions, by the abolishment of the compulsory crop rotation and its replacement with the guided crop rotation that enables every farmer to cultivate what he wants, and leaving small individual farmers subject to market forces. This has weakened the cooperatives' role as a means of enabling small farmers to benefit from the economies of scale [68, 69].

In the absence of strong associative structures, the VEW would have to provide individual farmers with information on climate change adaptation measures, which

[16]The Agricultural Research Center (ARC) was established in the 1970s—as part of the organigram of the Ministry of Agriculture and Land Reclamation—to be responsible for increasing agricultural productivity (crop yields).

[17]The agrarian reform cooperatives were formed to supervise the land that was confiscated from big landlords and distributed to small farmers in the 1950s and 1960 s, while the land reclamation cooperatives were formed to support farmers in the reclaimed areas.

[18]This is supposedly done through providing farmers with agriculture inputs, such as seeds, fertilizers and pesticides, machines, access to loans as well as market farmers crops.

is challenging due to their lack of capacity. In 2015, a study conducted in Sharkia, Qalyubia and Beheira governorates, to analyze the extension staff performance in supporting farmers by providing them with climate change adaptation related information found that about 40% of the extension staff had a low performance, 40% middle performance and about 20% high performance. These results are referred to the lack of a clear climate change extension policy and the lack of funds, in addition to the VEW unclear climate change related responsibilities [70]. Furthermore, the extension workers suffer from the aging of their staff, poor training, low salaries that distract them from doing their job in a proper way, and lack of operational funds that hinder them from carrying out their activities, as well as moving around villages due to the lack of transportation facilities that limit their ability to communicate with farmers [24, 71, 72, 73]. These challenges hamper the public extension sector from playing a significant role in enhancing agricultural production in Egypt [60].

Besides, more and more women are playing a greater role in the farming operations through their participation in crop cultivation, pest control and harvesting activities due to men's migration to the Gulf countries or Cairo and Alexandria in search for additional income sources [74, 75]. In the Nile Delta governorates, the average women's engagement in agriculture of total agricultural engagement is about 30%. Nevertheless, the public extension service has not taken account for such a change, because there is a lack of women extension workers on the village level as they do not exceed 4% of total VEW [60, 73].

Even if all farmers become aware of the climate change impacts and VEW have the necessary means to communicate with farmers in the Nile Delta governorates—though farmers do not always follow the advice of the VEW—[76], there are other hurdles that would affect farmers' decisions to adapt to climate change.

According to Capmas [77], the highest percentage of the poor was among the agriculture and fishing workers with 36.3% in 2012/2013. The average poverty rate in the Nile Delta governorates was 15% in 2012/2013. It was highest in Qalyubia and Beheira with 21% and 20% respectively and lowest in Gharbia and Alexandria with 11% and 12% respectively in 2012/2013. Farmers with large agricultural landownerships, such as in Alexandria and Ismailia with 4.9 and 6.3 feddan, are expected to be more willing to adapt to climate change, while small hold farmers might be hampered by the low profitability of their land and their fear of taking risks [20].

In Egypt, statistics of 2012/2013, show that the highest percentage of the poor was among the illiterate and accounted for about 40%, while the lowest percentage was among university graduates and post-graduates accounting for about 15% [77]. In the Nile Delta governorates, the average illiteracy rate was about 20% of the population older than 10 years in 2017. It was highest in Alexandria with about 30%, and lowest in Damietta with about 20% (see Table 9 in Annex for Nile Delta governorates' and farmers' assets). Illiteracy coupled with poverty could reduce farmers' chances for increasing their knowledge and thus, their ability to prepare for future climate change impacts [78].

Furthermore, farmers' dependence on agriculture as the main source of income and their access to funds would affect their decisions regarding investing in climate change adaptation measures. The average percentage of employment in agriculture of those older than 15 years is about 25% in the Nile Delta governorates (men and women). Nevertheless, over the past decades, many farmers in Egypt have become part-time farmers—due to the low returns of the shrinking agricultural landownerships—seeking to diversify and/or complement their agriculture-based income through nonfarm sources. In 2015, the nonfarm sources accounted for about 55% of farmers' income, while the agricultural income accounted for 45% [26, 79].

For financing agriculture-related activities, farmers either resort to personal funds or loans. Farmers' personal funds originate from personal savings and/or from inheritance. To maximize the use of their funds, they participate in informal saving and lending mechanisms, the rotating savings and credit association (ROSCA) that is known in Egypt as 'Gamiaa'[19] [80–82]. Farmers also access additional funds through borrowing from the Agricultural Bank of Egypt.[20] The bank is the main source of agricultural financing for small farmers and provides them with short, medium and long-term loans using their agricultural landownership as collateral. These loans cover agricultural activities, by financing inputs and equipment, animal breeding, as well as the provision of personal loans.

Farmers might be willing to decide to use their limited funds to apply inexpensive adaptation measures by, e.g. changing their planting dates and resorting to other heat tolerant crop varieties [76]. However, the adoption of adaptation measures by individual farmers—seeking to maximize their private interest—will be inadequate to counter climate impacts on agriculture. Individual adaptations might lead to damages that would affect other farmers or the environment, and in turn agricultural production on the national level [83]. Furthermore, adaptation measures with greater advantages would require concerted efforts and sizeable funds to be incurred by farmers, the government and development organizations [76, 84].

4.4 *Vulnerability*

The interplay between the socio-economic, natural resources, institutional and financial assets—including access to funding, poverty, landownership, education, experience and access to information—on the one hand, and exposure and sensitivity to climate change stimuli impacts on the other determines farmers' and governorates' vulnerability.

Thus, the Nile Delta governorates and their farmers are not expected to be equally vulnerable to climatic changes, due to varied assets' possession and because vulnerability is 'context-specific' [85]. In this section, the vulnerability of the Nile Delta governorates is explored.

Table 6 summarizes the climate change impacts on agriculture and farmers in the

[19]ROSCA operates as follows: each month members pay in a certain amount of cash, and each month, on a rotating basis one of the members receives the total amount paid in by all members.

[20]Before 2016 it was known as the Principal Bank for Development and Agricultural Credit (PBDAC).

Table 6 Summary of climate change impacts on agriculture in the Nile Delta governorates

Climate change impact (% impact on Nile Delta governorates based on 2015 data)	Alexandria	Damietta	Daqahlia	Sharkia	Qalyubia	Kafr El-Sheikh	Gharbia	Menufia	Beheira	Ismailia
Field crops' area	–	0.19	0.03	0.03	–	0.34	–	–	–	–
Field crops' production	13.15	9.17	6.89	8.73	9.40	6.85	7.58	9.37	9.36	12.22
Field crops' net value	14.49	10.28	8.30	9.89	10.07	8.04	8.19	10.25	10.59	13.38
Field crops' employment (landowners)	100	100	100	100	100	100	100	100	100	100
Field crops' employment (laborers)	–	0.19	0.03	0.03	–	0.34	–	–	–	–

Source Calculated by the Author based on ETNC [49], Smith et al. [7], Capmas [13]

Nile Delta governorates. It shows that the field crops' cropping area in four Nile Delta governorates is to decrease, whereas Kafr El-Sheikh and Damietta are to experience the largest reduction of 0.34% and 0.19% respectively of their cropping area. The field crops' production is to decrease in all governorates; however, the decrease is to be highest in Alexandria and Ismailia with about 13.2% and 12.2% respectively of their production. The reduction in production leads to a reduction in the field crops' net value. Thus, the field crops net value reduction is also to be highest in Alexandria and Ismailia with about 14.5% and 13.4% of their respective field crops' net value. In contrast, the lowest percentage of field crops' production reduction is to be in Kafr El-Sheikh and Daqahlia with about 7% each, and 8% and 8.3% respective reductions of their field crops' net value.

Climate change is to affect all landowners in the Nile Delta governorates. However, Kafr El-Sheikh and Damietta governorates are to experience the highest reduction of working days by 0.34% and 0.19% respectively of their total working days.

For the determination of the vulnerability of the Nile Delta governorates and their farmers, the climate change impacts are reviewed against the governorates' and farmers' adaptive capacities (see Table 9 in Annex for a summary of assets in percent, and Table 17 in Annex for a summary of assests in numbers). The following paragraphs review governorates and farmers' assets:

- Rural Population: Beheira, Menufia and Kafr El-Sheikh have the highest percentage of the rural population of their total population with about 82%, 79% and 76% respectively, accounting for about 5 million, 3.4 million, and 2.6 million people in 2017. In contrast, Alexandria and Ismailia have the smallest rural population accounting for 1.3 and 55.5% of their total population, of about 70 and 700 thousand people respectively in 2017.
- Engagement of Men and Women in Agriculture: A large percentage of the employed population—over 15 years—in Beheira, Kafr El-Sheikh and Menufia governorates are dependent on agriculture as a main source of income. About 47%, 40% and 33% of the total employed in these governorates are engaged in agricultural activities that make up about 960 thousand, 400 thousand and 400 thousand farmers respectively, versus about 4.5% and 6.7% engaged in agriculture in Ismailia and Alexandria accounting for 106 thousand and 92 thousand farmers of the total employed in the respective governorate in 2016. Furthermore, Menufia, Beheira and Qalyubia have the highest percentage of women's employment in agriculture of their total agricultural engagement with about 51%, 45% and 43%, amounting 200 thousand, 430 thousand and 100 thousand women respectively. The lowest percentage of women's engagement is in Alexandria and Ismailia with about 14.5% and 28% in 2016 that equal 13 thousand and 29 thousand women respectively.
- Age of Rural Population: The percentage of those less than 45 years in the Nile Delta governorates is almost the same in all governorates. However, it is highest in Alexandria and Ismailia with about 90% and 88% respectively accounting for about 60 thousand and 62 thousand people, while lowest in Qalyubia, Sharkia and Beheira with about 87.5%, 87.4%, and 87%, nevertheless, accounting for about 3

million, 4.7 million and 4.4 million people respectively in 2017. If coupled with unemployment—e.g. due to climate change—the large number of the young rural population could lead to tension and unrest. At the same time, the young rural population would constitute a sufficient labor force in case of resorting to a labor-intensive climate change adaptation strategy. The average percentage of those over 45 years in the Nile Delta governorates, is highest in Daqahlia and Damietta with 20% and about 19% respectively, accounting for 930 thousand and 170 thousand, while it is lowest in Alexandria and Ismailia with about 10% and 11%, accounting for about 7 thousand and 85 thousand. This could be an indication that farmers in Alexandria and Ismailia rely less on the elderly's' agricultural experience and more on formal sources of information.

- Rural Illiteracy: Rural illiteracy is highest in Alexandria, Beheira and Kafr El-Sheikh with about 28%, 27% and 23% respectively, amounting about 20 thousand people in Alexandria, and about 1.3 million people in Beheira and 600 thousand people in Kafr El-Sheikh. Illiteracy rates are lowest in Damietta, Menufia and Gharbia with about 18%, equalling about 160 thousand, 620 thousand and 660 thousand people respectively in 2017.
- Poverty: Poverty rates are highest in Qalyubia, Beheira, and Kafr El-Sheikh with 21%, 20%, and 18% in 2012/2013, equalling about 1.2 million, 1.2 million, and 600 thousand people respectively, while Damietta, Gharbia and Alexandria have the lowest poverty rates with 10%, 11% and 12% accounting for about 150 thousand, 550 thousand and 620 thousand people.[21]
- Type of cultivated Field Crops: In Menufia and Daqahlia, about 80% of the area is cultivated with traditional crops (cereals, sugar cane, clover and cotton) that account for about 500 thousand feddan cropping area and 1 million feddan cropping area respectively in 2015. Much of the traditional crops are consumed locally by farmers, i.e. for their subsistence use [80, 86]. In Alexandria and Ismailia, the areas cultivated with traditional crops account for about 52% and 54% respectively that equal about 150 and 140 thousand feddan cropping area, while the areas with nontraditional crops account for 48% and 46% of their cropping area that is equivalent to about 130 and 90 thousand feddan respectively. The nontraditional crops—mainly vegetables accounting for 43% and 35% of total cultivated field crops area—are produced for the local and international markets. The cultivation of traditional crops—in the Nile Delta governorates—is mostly connected with small size agricultural land plots.
- Average Agricultural landownership: The average agricultural landownership is smallest in Gharbia and Qalyubia with about 0.9 and 1.2 feddan respectively, while it is 4.9 and 6.3 feddan in Alexandria and Ismailia. The number of landowners is highest in Beheira, Daqahlia and Gharbia with 560 thousand, 450 thousand and 430 thousand landowners, and lowest in Alexandria and Ismailia with 36 thousand and 45 thousand landowners in 2015/2016.

[21] No recent poverty rates per governorate were available, only those of 2012/13, multiplied by population by governorate in 2017.

- Access to credit cooperative VEW: Alexandria and Ismailia have access to the highest average number of VEW per credit cooperative, about 5 VEW,[22] whereas it is lowest in Qalyubia and Gharbia with 1 VEW per cooperative. Dividing 800 feddan—average area of responsibility per VEW—by average landownership by governorate, shows that one VEW in Alexandria and Ismailia must deal with about 160 and 130 farmers, while in Gharbia and Qalyubia one VEW deals with about 889 and 667 farmers respectively.
- Access to Formal Funds: The percentage of agricultural land benefitting from loans from the Agricultural Bank is highest in Sharkia, Kafr El-Sheikh, and Alexandria with about 25%, 20% and 16%, constituting about 200 thousand feddan, 115 thousand feddan, and 30 thousand feddan, while the lowest percentage is in Ismailia, Qalyubia and Gharbia with about 4%, 5% and 7% respectively accounting for about 12 thousand feddan, 10 thousand feddan and 28 thousand feddan in 2015/16.

Though Kafr El-Sheikh and Damietta are to experience the highest area and working days' reductions and Alexandria and Ismailia are to experience the largest field crop's production and net value reductions. However, the most vulnerable governorate is determined based on the above review of governorates and farmers' assets. Thus, the most vulnerable governorate to climate change impacts in the Nile Delta would be Beheira as it has the lowest adaptive capacity compared to the rest of the governorates in the Nile Delta. Beheira has the highest percentage of rural population, high illiteracy and poverty rates, in addition to the high total and female agricultural engagement, which indicates that climate change impacts—reduction of field crops' production and net value—on agriculture would have severe impacts on rural livelihoods.

Furthermore, on the national level, the impact of climate change on Beheira governorate is also to be higher compared to Alexandria and Ismailia. The rural population in Beheira represents 9.2% of Egypt's rural population, while in Alexandria and Ismailia the rural population accounts for 0.12% and 1.32% of Egypt's rural population respectively. The percentage of landowners in Beheira accounts for 11% of total landowners in Egypt owning 12% of the agricultural land in Egypt, while landowners in Alexandria and Ismailia account only for about 0.8 and 1% of total landowners in Egypt owning about 2% and 3% of the agricultural land in Egypt.

Moreover, the total loss of field crops production in Alexandria and Ismailia would be about 340 thousand and 250 thousand tons, while in Beheira the loss would equal about 1 million tons. Consequently, the decrease of the field crops net value in Alexandria and Ismailia would be about LE370 million and LE260 million (US$84 million for Alexandria and Ismailia), while in Beheira the reduction would be about LE920 million (about US$123 million).[23] Therefore, the impact of climate change

---

[22]However, large farms would depend more on private extension workers—to access agricultural information and technologies for the cultivation of high value crops for local and international markets—that provide the service against a fee (GFAAS; [87] since the public extension workers do not have the necessary advanced expertize for more specialized production [24].

[23]Calculation are based on the 7.5 exchange rate of 2015.

on Beheira is expected to have greater impacts on local food self-sufficiency, on the national budget, and in turn on GDP as well as on social stability.

5 Conclusions

The conclusion presents a brief comparison between the impact of urban expansion on agricultural land and that of climate change on agriculture. The separation between the impacts of the socio-economic challenges and that of climate change stimuli on agriculture in 2030 is for analytical purposes; however, in reality, change in climatic conditions and urban expansion on agricultural land, are expected to take place in conjunction, and even reinforce one another.

The assessment of the impact of urban expansion and that of climate change on agriculture in the Nile Delta governorates shows that both urbanization and climate change would have severe impacts on agriculture by 2030. The impact of urbanization would be higher on the field crops' cropping area, as it would reduce the area by about 2.6% of the total field crops' cropping area in Egypt based on 2015 data, while climate change would only reduce the area by 0.04%. In contrast, the impact of urban expansion on field crops' production and net value in the Nile Delta governorates would account only for about 50% of the impact of climate change (see Table 7). Urbanization would reduce the field crops' production by 2.33% of total field crops' production in Egypt, while climate change would reduce it by 4.42%. This would be reflected in the field crops' net value, as urbanization is to decrease the net value by 2.29%, whereas climate change is to reduce it by 4.95% of total field crops' net value in Egypt based on 2015 data.

Table 7 Impacts of climate change and urbanization on agriculture in the Nile Delta governorates by 2030

Stimuli	% reduction of field crops' agricultural land area in the Nile Delta of total Egypt	% reduction of field crops' production in the Nile Delta of total Egypt	% reduction of field crops' net value in the Nile Delta of total Egypt	% increase/reduction of field crops' water requirements in the Nile Delta of total Egypt	% reduction of field crops' working days in the Nile Delta of total Egypt
Climate change	0.04	4.42	4.95	1.71	0.03
Urban expansion on agricultural land	2.58	2.33	2.29	2.67	2.49

Source Calculated by the Author

Climate change would increase the field crops' water requirements by 2%, while urbanization would decrease it by about 2.7%. However, climate change impacts are dependent on the availability of adequate irrigation water. Any reduction in Egypt's water supply—due to environmental and physical causes, such as rain reduction on the Ethiopian highlands and/or the potential water shortage due to the GERD construction, and increased ET of field crops in Egypt—would increase the severity of annual crop yield deficiencies affecting farmers (landowners and laborers) [26]. This might induce tension among farmers competing over irrigation water and in turn harm the social stability [9]. All landowners and some laborers (about 0.03%) are to be affected by climate change, while the impact of urbanization is to be more severe on laborers as it would reduce the working days by about 2.5% of total field crops' working days in Egypt.

Furthermore, agricultural land reduction and increased agricultural land fragmentation would increase the intensity by which farmers cultivate and service their small land plots that make it prone for other environmental damages. At the same time, climate change impacts on agriculture, in terms of decrease of field crops' production and net value, might accelerate urbanization on agricultural land, especially in marginal areas [9].

Finally, the review of farmers' and Nile Delta governorates' adaptive capacities shows that climate change is to significantly affect Alexandria and Ismailia by reducing their field crops' production and net value. Nevertheless, the most vulnerable governorate to climate change due to its poor assets is to be Beheira. Its adaptation process would be hampered by the large number of the rural population to be affected, small size of landownership, illiteracy and poverty of farmers [20] and their inadequate access to VEW and funding sources. Additionally, the losses of Beheira governorate would have a large impact on the national level due to its share of field crops' agricultural land area, production in tons, net value and farmers' (women and men) percentage of total Egypt. The vulnerability of Beheira might further be affected by potential food price increases and lack of job opportunities, both in Egypt and abroad [88].

6 Recommendations

Agriculture in Egypt is a main source of food, income and employment for about 50% of the Egyptian population, thus, it should receive appropriate attention from the policy makers to preserve it for future generations, either from man-made and/or environmental challenges.

Urban expansion on agricultural land is a current challenge that requires immediate attention to contain its growth—through law enforcement and the provision of alternative affordable housing [32]—or it would severely reduce agricultural land in the Nile Delta and thus could exacerbate the projected impacts of climate change.

To counter land fragmentation which is a main cause of urban expansion on agricultural land, the role of the agricultural cooperatives should be enhanced. Agricultural cooperatives should make sure that agricultural land—especially small land plots—is cultivated as one big plot to allow farmers to benefit from the economies of scale. This, in turn, would improve farmers' well-being and reduce the need to transform agricultural land into urban land [89]. However, the pressure on the land by the population as well as by developers who seek to make profits in the absence of law enforcement—due to the inadequate capacity of national and local institutional structures—increase land prices and make it more profitable for farmers to sell the land and/or use it for urban purposes [32]. This situation might aggravate with the parliament's ratification of the new law resolution that is to allow the reconciliation with those who have built on agricultural land [90].

As for climate change, there is a need to prepare farmers to deal with climate change to reduce its potential negative impacts on agriculture through the application of adaptation measures that would improve farmers' resilience. To this end, many government strategies, such as the Sustainable Agricultural Development Strategy 2030 [24], and Egypt's National Strategy for Adaptation to Climate Change and Disaster Risk Reduction [91] have addressed the need to adapt to the potential impacts of climate change, in terms of temperature increase and sea level rise.

Strategies have also acknowledged the importance of improving the institutional structures of research and extension services in Egypt as well as the importance of allocating sufficient funds to agricultural research and public extension services [92]. However, there are shortcomings related to funding, implementation and time frames, lack of responsibility division between ministries and/or organizations for the implementation of strategies [93], as well as the inadequate attention to farmers on the local level [9].

Therefore, the capacity building of public extension workers and of farmers is expected to have great benefits [76], in addition to enhancing the structures of cooperatives through the introduction of necessary changes to guarantee a more participatory involvement of farmers in the decision-making processes [94, 95] and the regulation of the involvement of the private extension services to complement governmental efforts [96]. However, the provision of support to governorates and farmers should be based on their exposure, sensitivity and adaptive capacity that determine their vulnerability, as governorates and farmers are not homogeneous [97].

Annex

See Tables 8, 9, 10, 11, 12, 13, 14, 15, 16 and 17.

Table 8 Agricultural cooperatives in the Nile Delta governorates

Governorate	Average agricultural land per credit cooperative in feddan in 2015/16	Average agricultural land per agrarian reform cooperative in feddan in 2015/16	Average agricultural land per land reclamation cooperative in feddan in 2015/16	% of credit cooperatives area of total agricultural land per governorate in 2015/16	% of agrarian reform cooperatives area of total agricultural land per governorate in 2015/16	% of land reclamation cooperatives area of total agricultural land per governorate in 2015/16	% of credit cooperative membership (farmers) of total farmers per governorate in 2015/16	% of agrarian reform cooperative membership (farmers) of total farmers per governorate in 2015/16	% of land reclamation cooperative membership (farmers) of total farmers per governorate in 2015/16	Average landownership in feddan per governorate in 2015/16
Alexandria	3802.7	774.9	1727.4	32.17	7.43	60.40	26.13	11.52	62.35	4.9
Damietta	1284.6	709.4		89.94	10.06		93.56	6.44		1.2
Daqahlia	1214.7	788.8	2858.7	76.08	11.46	12.46	87.77	8.34	3.89	1.5
Sharkia	1358.6	752.1	2555.2	79.91	8.62	11.46	85.89	10.84	3.27	2.1
Qalyubia	868.8	544.7	2948.3	88.84	4.91	6.25	92.75	6.17	1.08	1.2
Kafr El-Sheikh	1545.4	1000.5	2032.0	67.78	11.10	21.12	81.15	9.34	9.51	1.6
Gharbia	1055.1	843.7		90.09	9.91		93.06	6.94		0.9
Menufia	1859.8	531.9		136.23	217.23		96.59	3.41		1.4
Beheira	1574.9	912.3	2479.4	56.67	12.19	31.14	70.47	17.58	11.95	2.1
Ismailia	4265.2	1017.3	3283.9	55.53	2.86	41.60	67.59	7.68	24.73	6.3

Source Capmas [12]

Table 9 Nile Delta governorate's and farmers' assets

Farmers' and governorates' assets (%)	% of population of total Egypt population in 2017	% of rural population of total rural Egypt population in 2017	% of rural population of governorate population in 2017	Population density (Inhabitants/km^2) in 2017	% of rural population less than 45 years in 2017	% of rural population over 45 years in 2017	% of rural illiteracy (+10 years) in 2017	% of poverty in 2012/2013	Average family size in rural areas in 2017	% of employed in agriculture of total employed (15 + years) per governorate in 2016
Alexandria	5.45	0.12	1.32	2245.11	90.07	9.93	27.64	12	4.91	6.70
Damietta	1.58	1.66	60.63	1644.25	81.15	18.85	17.79	10	3.99	19.90
Daqahlia	6.85	8.50	71.72	1834.92	79.98	20.02	19.32	14	3.89	25.70
Sharkia	7.56	9.82	75.06	1458.73	87.38	12.62	21.38	14	4.16	30.70
Qalyubia	5.94	5.89	57.31	5005.35	87.47	12.53	20.24	21	3.94	14.00
Kafr El-Sheikh	3.55	4.67	76.05	969.85	85.72	14.28	23.16	18	4.06	40.80
Gharbia	5.27	6.56	71.89	2574.03	84.61	15.39	18.27	11	3.87	22.10
Menufia	4.54	6.23	79.29	1721.33	85.74	14.26	18.12	15	4.12	32.49
Beheira	6.51	9.22	81.84	628.09	86.86	13.14	26.53	20	4.03	46.90
Ismailia	1.38	1.32	55.53	257.30	88.24	11.76	20.86	15	4.18	4.50

(continued)

Table 9 (continued)

Farmers' and governorates' assets (%)	% women agricultural employment in 2016	% male agricultural employment in 2016	Average agricultural land ownership in 2015/16	% of traditional crops area in 2015	% of nontraditional crops area in 2015	% of area benefitting from loans of total governorate area 2015/16	Average VEW per credit cooperative 2015/16	Land intensification in 2015
Alexandria	14.47	85.53	4.90	51.61	48.39	16.37	4.75	1.55
Damietta	29.50	70.50	1.23	78.10	21.90	10.66	1.61	1.65
Daqahlia	21.90	78.10	1.52	80.43	19.57	14.26	1.52	1.76
Sharkia	33.40	66.60	2.10	76.03	23.97	24.57	1.70	1.80
Qalyubia	43.30	56.70	1.19	80.13	19.87	5.16	1.09	1.22
Kafr El-Sheikh	29.30	70.70	1.64	72.92	27.08	20.17	1.93	1.82
Gharbia	34.80	65.20	0.89	76.81	23.19	7.35	1.32	1.56
Menufia	51.10	48.90	1.40	81.29	18.71	8.22	2.32	1.45
Beheira	44.60	55.40	2.15	72.45	27.55	14.02	1.97	1.36
Ismailia	27.60	72.40	6.27	54.36	45.64	4.18	5.33	0.80

Source Calculated by the Author based on Capmas [11, 22, 30, 77]; ESIS

Table 10 Nile Delta governorate's population and agricultural landownership in 2030

Governorate	Population in 2017	% of total population in Egypt in 2017 (94,798,827)	Population in 2030 (% * by 117,000,000)	Population density (Inhabitants/km^2) in 2030	Number of agricultural landownerships in 1999/2000	Number of agricultural landownerships in 2015/16	Number of agricultural landownerships in 2030	Average landownership in feddan in 2030
Alexandria	5,163,750	0.05	6,373,061.45	2770.9	29,007	36,204	43,401	4.1
Damietta	1,496,765	0.02	1,847,296.12	2,029.4	47,590	91,789	135,988	0.8
Daqahlia	6,492,381	0.07	8,012,847.85	2264.6	346,820	451,859	556,898	1.2
Sharkia	7,163,824	0.08	8,841,537.75	1800.4	414,717	382,621	350,525	1.9
Qalyubia	5,627,420	0.06	6,945,319.48	6177.6	151,785	158,401	165,017	1.1
Kafr El-Sheikh	3,362,185	0.04	4,149,583.46	1197.0	204,104	347,108	490,112	1.2
Gharbia	4,999,633	0.05	6,170,509.48	3176.8	255,036	429,766	604,496	0.6
Menufia	4,301,601	0.05	5,309,003.63	2124.5	266,505	287,464	308,423	1.3
Beheira	6,171,613	0.07	7,616,958.39	775.2	339,003	523,181	707,359	1.6
Ismailia	1,303,993	0.01	1,614,600.00	318.6	44,671	45,327	45,983	6.2

Source Calculated by the Author based on Capmas [17, 22, 25], ESIS

Table 11 Percentage change in field crops yield and water requirements due to climate change in Egypt

Field crops	% of yield reduction due to 1.5 °C (1)	% of yield reduction due to temperature increase (B1 Scenario) (2)	% of water requirements increase due to climate change (B1 scenario)
Wheat	(−11) (−12)	−9	2.64
Barley		−12	2.64
Rice	(−26) (−47)	−6.6	3.12
White maize	(−40) (−47)	−8.4	3.12
Yellow corn		−8.4	3.12
Sugar beet		−0.96	3
Sugarcane	−15.2	−8.4	3.12
Flax		−9	2.64
Soya beans	−28	−16.8	2.52
Peanuts		−0.96	3
Sesame		−8.4	3.12
Sunflower	−29	−16.8	3.12
Broad beans	(−8.4) (−38.35)	−16.8	2.88
Lentils	−28	−16.8	2.52
Chickpeas		−16.8	2.52
Lupine		−16.8	2.52
Vegetables	−28	−16.8	2.52
Onion and garlic	−1.53	−0.96	3
Clover		−8.4	3.12
Medicinal crops		−8.4	3.12
Cotton	19.8	10.2	2.64

Source (1) ETNC [49] and (2) Smith et al. [7]

Temperature and water requirements' change for sunflower are assumed to be the same as that for sesame

Temperature and water requirements' change for chickpeas and lupine are assumed to be the same as that for lentils

Temperature and water requirements' change for garlic are assumed to be the same as that for onions

Temperature and water requirements' change for medicinal crops are assumed to be the same as that for clover

Table 12 Percentage of field crops' group area by governorate

Traditional crops	Alexandria	Damietta	Daqahlia	Sharkia	Qalyubia	Kafr El-Sheikh	Gharbia	Menufia	Beheira	Ismailia
Cereals	43.21	47.16	62.62	62.90	61.90	53.34	58.20	60.31	54.39	42.37
Sugar cane	0.00	0.00	0.02	0.00	0.15	0.04	0.24	–	0.01	–
Clover	7.81	27.75	15.20	10.86	18.05	12.40	16.68	20.67	14.59	11.44
Cotton	0.59	3.19	2.59	2.26	0.03	7.14	1.69	0.31	3.47	0.55
%	51.61	78.10	80.43	76.03	80.13	72.92	76.81	81.29	72.45	54.36
Nontraditional crops										
Sugarbeet	2.87	3.18	7.62	4.58	0.26	15.07	2.50	0.20	3.36	3.14
Oil crops	0.55	0.47	0.22	2.88	0.35	0.22	0.38	0.09	1.39	7.02
Pulses	2.13	1.89	0.53	0.57	0.18	1.20	0.16	0.01	0.69	0.76
Vegetables	42.80	16.36	11.16	15.92	18.91	10.48	19.93	18.33	21.96	34.56
Medicinal crops	0.04	–	0.05	0.01	0.18	0.11	0.23	0.08	0.16	0.16
%	48.39	21.90	19.57	23.97	19.87	27.08	23.19	18.71	27.55	45.64

Source Calculated by the Author based on Capmas [13]

Table 13 Percentage of field crops' group production by governorate

Traditional crops	Alexandria	Damietta	Daqahlia	Sharkia	Qalyubia	Kafr El-Sheikh	Gharbia	Menufia	Beheira	Ismailia
Cereals	11.76	15.09	28.98	28.02	20.59	17.53	20.63	18.66	23.25	13.66
Sugar cane	0.01	0.01	0.11	0.01	0.54	0.16	1.11	–	0.04	–
Clover	20.49	58.94	34.41	40.12	51.02	46.30	45.95	68.96	44.26	24.83
Cotton	0.04	0.26	0.20	0.16	0.00	0.49	0.21	0.02	0.27	0.04
%	32.31	74.30	63.71	68.31	72.15	64.48	67.90	87.64	67.81	38.53
Nontraditional crops										
Sugarbeet	5.77	7.31	23.63	12.64	0.64	29.15	7.05	0.41	8.42	6.84
Oil crops	0.03	0.02	0.02	0.41	0.05	0.02	0.03	0.01	0.19	0.85
Pulses	0.47	0.27	0.10	0.12	0.14	0.19	0.05	0.00	0.12	0.26
Vegetables	61.43	18.10	12.54	18.51	26.87	6.16	24.91	11.93	23.42	53.47
Medicinal crops				0.01	0.15	0.01	0.07	0.01	0.03	0.04
%	67.69	25.70	36.29	31.69	27.85	35.52	32.10	12.36	32.19	61.47

Source Calculated by the Author based on Capmas [13]

Table 14 Percentage of field crops' group net value by governorate

Traditional crops	Alexandria	Damietta	Daqahlia	Sharkia	Qalyubia	Kafr El-Sheikh	Gharbia	Menufia	Beheira	Ismailia
Cereals	15.16	23.95	45.90	40.95	29.59	33.85	32.42	30.56	33.21	16.32
Sugar cane	0.00	0.00	0.01	0.00	0.05	0.02	0.10	–	0.00	–
Clover	9.40	39.97	23.32	23.46	30.78	36.77	29.32	47.72	26.13	11.67
Cotton	0.15	1.42	1.08	0.76	0.01	3.10	1.06	0.11	1.26	0.15
%	24.71	65.34	70.31	65.18	60.43	73.75	62.90	78.39	60.60	28.14
Nontraditional crops										
Sugarbeet	1.50	2.82	9.10	4.21	0.22	13.17	2.56	0.16	2.83	1.83
Oil crops	0.19	0.15	0.17	2.83	0.30	0.15	0.19	0.08	1.22	4.68
Pulses	0.50	0.46	0.18	0.18	0.20	0.35	0.07	0.00	0.17	0.30
Vegetables	73.09	31.22	20.24	27.59	38.42	12.56	34.09	21.34	35.10	64.96
Medicinal crops	0.00	–	–	0.02	0.44	0.03	0.20	0.02	0.09	0.10
%	75.29	34.66	29.69	34.82	39.57	26.25	37.10	21.61	39.40	71.86

Source Calculated by the Author based on Capmas [13]

Table 15 Average yield per field crop per feddan per governorate in 2017

Governorate	Alexandria	Damietta	Daqahlia	Sharkia	Qalyubia	Kafr El-Sheikh	Gharbia	Menufia	Beheira	Ismailia	Average yield per feddan in the Nile Delta governorates
Wheat	2.5	2.6	2.9	3	2.9	2.69	2.9	3	2.8	2.67	2.80
Barley	1.3	2	1.8	1.8		1.79	2.2		1.6	1.48	1.75
Rice	3.6	3.1	4.4	3.9	3.7	3.83	3.5	3.8	3.8	3.6	3.72
White maize	2.8	3.4	4	3.4	3.2	3.63	3.3	3.5	3.5	3.12	3.39
Yellow corn	2.6	3.6	3.8	3	3.3	3.65	3.3	3.4	3.2	3.04	3.29
Sugarbeet	18.9	21.3	25.5	20.6	23.5	19.54	25.4	21.4	18.8	19.32	21.43
Sugarcane	30	27.5	38	28.4	33.8	40.17	41.1		28		33.37
Flax		0.3	0.8	0.7	0.8	1.11	0.5	0.4	0.5		0.64
Soya Beans			1.3	1.3	1.5	0.91	1.2	2.4	1.3	8	2.24
Peanuts				1.4	1.4		1.4	1.4	1.6	1.45	1.44
Sesame	0.5		0.6	0.5	0.56	0.48	0.5	0.4	0.5	0.41	0.49
Sunflower	1.1		0.7	1.5	1.7	0.78			1.1		1.15
Broad beans	2.1	1.3	1.6	1.6	7.4	1.56	2.5	1.1	1.3	3.43	2.39
Lentils		1.1	0.8	0.9						0.5	0.83
Lupine				0.7						0.89	0.80
Vegetables	13.6	10.1	8.1	8.5	13	5.76	7.3	6.9	7.7	14.09	9.51
Onion and garlic	8.9	12.6	14.6	12.9	14.3	20.97	18.4	10.1	17.3	5.89	13.60
Clover (berseem)	24.7	19.7	18.6	27.5	26.6	37.71	24.7	35.7	22.8	19.26	25.73

(continued)

Table 15 (continued)

Governorate	Alexandria	Damietta	Daqahlia	Sharkia	Qalyubia	Kafr El-Sheikh	Gharbia	Menufia	Beheira	Ismailia	Average yield per feddan in the Nile Delta governorates
Medicinal crops	0		–	3.7	8.2	0.7	2.6	1	1.5	2.39	2.51
Cotton	0.7	0.76	0.6	0.5	0.84	0.69	1.1	0.7	0.6	0.66	0.72

Source Calculated by the Author based on Capmas [13]

Table 16 Average yield per field crop per feddan per governorate in 2030 (accounting for yield deficiency due to temperature increase)

Governorate	Alexandria	Damietta	Daqahlia	Sharkia	Qalyubia	Kafr El-Sheikh	Gharbia	Menufia	Beheira	Ismailia	Average yield per feddan in the Nile Delta governorates
Wheat	2.29	2.39	2.62	2.74	2.65	2.45	2.63	2.77	2.56	2.43	2.55
Barley	1.19	1.76	1.58	1.58		1.58	1.91		1.44	1.30	1.54
Rice	3.36	2.89	4.14	3.67	3.48	3.58	3.26	3.51	3.58	3.36	3.48
White maize	2.59	3.15	3.66	3.10	2.95	3.33	3.02	3.20	3.16	2.85	3.10
Yellow corn	2.41	3.28	3.49	2.73	2.99	3.34	2.98	3.14	2.95	2.78	3.01
Sugarbeet	18.77	21.11	25.24	20.36	23.28	19.35	25.11	21.18	18.64	19.14	21.22
Sugarcane	27.48	25.19	34.84	25.99	30.93	36.79	37.68		25.65		30.57
Flax		0.31	0.70	0.61	0.74	1.01	0.48	0.40	0.44		0.59
Soya beans			1.07	1.04	1.29	0.76	1.03	2.01	1.10	6.66	1.87
Peanuts				1.35	1.36		1.34	1.43	1.55	1.43	1.41
Sesame	0.44		0.55	0.50	0.51	0.44	0.42	0.39	0.43	0.37	0.45
Sunflower	0.92		0.55	1.23	1.44	0.65			0.88		0.94
Broad beans	1.71	1.10	1.37	1.31	6.13	1.30	2.07	0.90	1.11	2.86	1.99
Lentils		0.89	0.65	0.76						0.42	0.68
Lupine				0.54						0.74	0.64
Vegetables	11.30	8.42	6.74	7.04	10.85	4.79	6.09	5.76	6.38	11.72	7.91
Onion and garlic	8.85	12.52	14.45	12.77	14.19	20.77	18.22	9.97	17.08	5.83	13.47
Clover (berseem)	22.67	18.04	17.04	25.18	24.39	34.54	22.66	32.71	20.87	17.64	23.57

(continued)

Table 16 (continued)

Governorate	Alexandria	Damietta	Daqahlia	Sharkia	Qalyubia	Kafr El-Sheikh	Gharbia	Menufia	Beheira	Ismailia	Average yield per feddan in the Nile Delta governorates
Medicinal crops			–	3.42	7.49	0.64	2.42	0.91	1.39	2.19	2.05
Cotton	0.73	0.84	0.69	0.59	0.92	0.76	1.22	0.78	0.64	0.73	0.79

Source Calculated by the Author based on Capmas [13], Smith et al. [7]

Table 17 Nile Delta governorates and their farmers

Farmers and governorates' assets	Alexandria	Damietta	Daqahlia	Sharkia	Qalyubia	Kafr El-Sheikh	Gharbia	Menufia	Beheira	Ismailia
Number of rural population of governorate population	68,293	907,542	4,656,592	5,377,062	3,224,929	2,557,058	3,594,336	3,410,855	5,050,630	724,046
Number of rural population less than 45 years	61,513	736,497	3,724,327	4,698,561	2,820,808	2,191,825	3,041,145	2,924,387	4,386,988	61,513
Number of rural population over 45 years	6780	171,045	932,265	678,501	404,121	365,233	553,191	486,468	663,642	662,533
Number of rural illiterate population	18,876	161,496	899,806	1,149,609	652,730	592,279	656,797	618,183	1,339,788	151,057
Number of poor population	619,650	149,677	908,933	1,002,935	1,181,758	605,193	549,960	645,240	1,234,323	195,599
Number of employment in agriculture of total governorate employment	91,900	84,600	436,900	586,500	215,200	391,600	302,900	413,700	958,900	106,100

(continued)

Table 17 (continued)

Farmers and governorates' assets	Alexandria	Damietta	Daqahlia	Sharkia	Qalyubia	Kafr El-Sheikh	Gharbia	Menufia	Beheira	Ismailia
Number of women employment in agriculture by governorate	13,300	24,900	95,900	196,000	93,300	114,900	105,500	211,600	429,100	29,300
Number of male employment in agriculture by governorate	78,600	59,700	341,000	390,500	121,900	276,700	197,400	202,100	529,800	76,800
Number of landowners	36,204	91,789	451,859	382,621	158,401	347,108	429,766	394,530	523,181	45,327
Area of traditional crops in feddan	143,656	145,967	979,166	1,143,019	185,008	754,894	462,069	466,511	1,127,269	139,008
Area of nontraditional crops in feddan	131,747	39,808	234,910	305,378	44,871	277,223	136,586	106,753	399,379	87,477
Area benefitting from loans of total governorate in feddan	29,035	12,024	98,110	197,192	9,738	114,515	28,135	32,446	157,450	11,866
Number of farmers per VEW	163	651	525	381	671	489	898	571	373	128

Source Calculated by the Author based on Capmas [11, 22, 30, 77]

References

1. FAO (2007) Adaptation to climate change in agriculture, forestry and fisheries: perspective, framework and priorities. Food and Agriculture Organization of the United Nations Rome, Italy
2. Nelson GC, Rosegrant MW, Koo J, Robertson R, Sulser T, Zhu T, Ringler C (2009) Climate change: impact on agriculture and costs of adaptation. IFPRI, Washington
3. Care (2016) Fleeing climate change: impacts on migration and displacement. Care Danemark. http://careclimatechange.org/wp-content/uploads/2016/11/FleeingClimateChange_report.pdf. Accessed in Mar 2018
4. Thorlakson T, Neufeldt H (2012) Reducing subsistence farmers' vulnerability to climate change: evaluating the potential contributions of agroforestry in western. Kenya Agric Food Secur 1:1–13. http://www.worldagroforestry.org/downloads/Publications/PDFS/OP11183.pdf. Accessed Feb 2018
5. Elbehri A, Sadiddin A (2016) Climate change adaptation solutions for the green sectors of selected zones in the MENA region. Futur Food J Food Agric Soc 4(3):39–54. http://futureoffoodjournal.org/index.php/journal/article/viewFile/231/pdf. Accessed Feb 2018
6. Fawaz MM, Soliman SA (2016) The potential scenarios of the impacts of climate change on Egyptian resources and agricultural plant production. Open J Appl Sci 6:270–286
7. Smith J, Deck L, McCarl B, Kirshen P, Malley J, Abdrabo M (2013) Potential impacts of climate change on the Egyptian economy. United Nations Development Programme (UNDP), Cairo, Egypt
8. IPCC (2007) Climate change 2007: impacts, adaptation and vulnerability. In: Parry ML, Canziani OF, Palutikof JP, van der Linden PJ, Hanson CE (eds) Contribution of working group II to the fourth assessment report of the intergovernmental panel on climate change. Cambridge University Press, Cambridge, UK, 976pp
9. Nour El-Din MM (2013) Proposed climate change adaptation strategy for the ministry of water resources and irrigation in Egypt. Joint programme for climate change risk management in Egypt
10. Capmas (2017e) Egypt 2018 statistical abstract. CAPMAS, Cairo, Egypt
11. Capmas (2017f) Population, housing and establishment census. CAPMAS, Cairo, Egypt
12. Capmas (2017c) Annual bulletin of the agricultural cooperatives' activities 2015/2016. CAPMAS, Cairo, Egypt
13. Capmas (2017b) Annual bulletin of statistical crop area and plant production 2014/2015. CAPMAS, Cairo, Egypt
14. Fuessel HM (2009) Review and quantitative analysis of indices of climate change exposure, adaptive capacity, sensitivity, and impacts. World Development Report, World Bank. http://siteresources.worldbank.org/INTWDR2010/Resources/5287678-1255547194560/WDR2010_BG_Note_Fussel.pdf. Accessed July 2018
15. Panpeng J, Ahmad MM (2017) Vulnerability of fishing communities from sea-level change: a study of Laemsing district in Chanthaburi province, Thailand. http://www.mdpi.com/2071-1050/9/8/1388/htm#B9-sustainability-09-01388. Accessed Feb 2018
16. IPCC (Intergovernmental Panel on Climate Change) (2014) Annex II: glossary. In: Mach KJ, Planton S, von Stechow C (eds) Climate change 2014: synthesis report. Contribution of working groups I, II and III to the fifth assessment report of the intergovernmental panel on climate change. IPCC, Geneva, Switzerland, pp 117–130. [Core Writing Team, Pachauri RK, Meyer LA (eds)]
17. Brussel SEC (2009) Adapting to climate changes: the challenge for European agriculture and rural areas. Commission of the European communities. Commission working staff working document accompanying the white paper no 147. http://www.europarl.europa.eu/meetdocs/2009_2014/documents/com/com_com%282009%290147_/com_com%282009%290147_en.pdf. Accessed Apr 2018

18. UNFCCC—United Nations Framework Convention on Climate Change. Background paper: integrating socio-economic information in assessments of impact, vulnerability and adaptation to climate change. https://unfccc.int/sites/default/files/background_paper_tt_web.pdf. Accessed June 2018
19. Abdul-Razak M, Kruse S (2017) The adaptive capacity of smallholder farmers to climate change in the northern region of Ghana. Clim Risk Manag 17:104–122. https://www.sciencedirect.com/science/article/pii/S2212096316300419#b0020. Accessed June 2018
20. Maddison D (2007) The perception of and adaptation to climate change in Africa. Policy research working paper policy 4308. Development Research Group, Sustainable Rural and Urban Development Team, The World Bank. http://documents.worldbank.org/curated/en/479641468193774164/pdf/wps4308.pdf. Accessed Apr 2018
21. CBE—Central Bank of Egypt (2018) GDP at factor cost (constant prices). http://www.cbe.org.eg/_layouts/xlviewer.aspx?id=/Time%20Series%20Documents/GDP/GDP%20at%20factor%20cost%20(constant)/GDP_factor%20cost_Constant%20prices_Annual.xls.xlsx&DefaultItemOpen=1. Accessed Mar 2018
22. Capmas (2017a) Annual Bulletin of Labor Force for 2016
23. Sholkamy M (2015) Reviving agricultural research and development in Egypt. Egypt Network for Integrated Development, Policy Brief 032. http://www.enid.org.eg/Uploads/PDF/PB32_reviving_agric.pdf. Accessed Feb 2018
24. SADS 2030 (2009) Sustainable agricultural development strategy towards 2030. Ministry of Agriculture and Land Reclamation (MALR), Cairo, Egypt
25. Capmas (2017d) Annual bulletin of the movement of production, foreign trade and available for consumption of agricultural commodities 2015. CAPMAS, Cairo, Egypt
26. Gouda DM (2017) The Grand Ethiopian Renaissance Dam, agriculture, and the rural poor in Egypt. In: The handbook of environmental chemistry. Springer, Berlin, Heidelberg
27. Negm AM, Saavedra O, El-Adawy A (2016) Nile Delta biography: challenges and opportunities. In: Negm A (eds) The Nile Delta. The handbook of environmental chemistry, vol 55. Springer, Cham
28. Schewe E (2012). District map of the presidential election in lower Egypt: an environmental history. Blog by Eric Schewe. https://ericschewe.wordpress.com/2012/07/07/district-map-of-the-presidential-election-in-lower-egypt-an-environmental-history/. Accessed Apr 2018
29. Mohamed M, Sakr Y, Shaaban S, Abdelwahab R, El-Gizawy G (2015) Economic impacts of expected water resources decrease on the agricultural sector. Agricultural Research Center (ARC), MALR, Egypt
30. Capmas (2015) Statistical year book 2015. CAPMAS, Cairo, Egypt
31. ESIS (Egypt State Information Service) website Local Administration. http://www.sis.gov.eg/section/210/16?lang=en-us. Accessed Mar 2018
32. Masoumi HE, Hosseini M, Gouda AA (2018) Drivers of urban sprawl in two large Middle-eastern countries: literature on Iran and Egypt. Mitig Adapt Strat Glob Chang 12(1). https://humangeographies.org.ro/articles/121/a_121_4_masoumi.pdf. Accessed July 2018
33. Tate B (2016) Egyptian land reclamation efforts. USDA foreign agricultural service. Global agricultural information network. https://gain.fas.usda.gov/Recent%20GAIN%20Publications/Egyptian%20Land%20Reclamation%20Efforts_Cairo_Egypt_5-16-2016.pdf. Accessed in Dec 2017
34. Alfiky A, Kauleb G, Salheena M (2012) Agricultural fragmentation of the Nile Delta; a modeling approach to measuring agricultural land deterioration in Egyptian Nile Delta. Procedia Environ Sci 14:79–97
35. Noseir D (2014) Assessing the potentials of multi-functional urban agriculture in Egypt. Towards cultivating the new urban settlements: the case of Al Sadat city. A thesis submitted in the partial fulfillment for the requirement of the degree of Master of Science in integrated urbanism and sustainable design. Ain Shams University, Egypt and Stuttgart University, Germany. https://iusd.asu.edu.eg/wp-content/uploads/2015/11/2ndInt_Nossier1.pdf

36. El-Kholei A (2012) Country study on status of land tenure, planning and management in oriental near east countries: case of Egypt. Food and Agriculture Organization of the United Nations Regional Office for the Near East (FAO-RNE), Cairo, Egypt. http://www.fao.org/3/a-as710e.pdf. Accessed Jan 2018
37. Wolters W, Smit R, Nour El-Din M, Ahmed E, Froebrich J, Ritzema H (2016) Issues and challenges in spatial and temporal water allocation in the Nile Delta. Sustain-Basel 8:383. https://doi.org/10.3390/su8040383
38. Karajeh F, Oweis T, Swelam A, El-Gindy A, El-Quosy D, Khalifa H, El-Kholy M, El-Hafez SA (2011) Water and agriculture in Egypt. Technical paper based on the Egypt-Australia-ICARDA workshop on on-farm water-use efficiency. ICARDA, Egypt
39. Shalaby A, Moghanm FS (2015) Assessment of urban sprawl on agricultural soil of northern Nile Delta of Egypt using RS and GIS. Chin Geogr Sci 25(3):274–282. Link: https://link.springer.com/content/pdf/10.1007%2Fs11769-015-0748-z.pdf. Accessed Jan 2018
40. Agrawala S, Möhner A, El Raey M, Conway D, van Aalst M, Hagenstad M, Smith J (2004) Development and climate change in Egypt: focus on coastal resources and the Nile, Organisation for Economic Co-operation and Development, Paris, France. http://fs.wa4.lucklaboratories.com/placemarks/files/435/33330510.pdf. Accessed Jan 2018
41. El-Gammal H (2012) Water quality protection in rural areas of Egypt. Int Water Technol J (IWTJ) 1(3):230–238
42. Zeidan BA (2016) Groundwater monitoring, modeling and assessment in the Nile Delta. Unpublished Chapter. https://www.researchgate.net/publication/305109484_Groundwater_
43. McCarl BA, Musumba M, Smith JB, Kirshen P, Jones R, El-Ganzori A, Hynninen R (2013) Climate change vulnerability and adaptation strategies in Egypt's agricultural sector. Mitig Adapt Strat Glob Chang 20(7):1097–1109
44. Khater AE, Kitamura Y, Shimizu K, Somura H, Abou El Hassan WH (2014) Improving water quality in the Nile Delta irrigation network by regulating reuse of agricultural drainage water. J Food Agr Environ 12(3&4):329–337
45. UNFPA—United Nations Population Fund (2016) Population situation analysis Egypt. http://egypt.unfpa.org/sites/default/files/pub-pdf/PSA%20Final.pdf. Accessed in Mar 2018
46. El-Hefnawi A (2005) Protecting agricultural land from urbanization or managing the conflict between informal urban growth while meeting the demands of the communities. In: 3rd urban research symposium on land development, urban policy and poverty reduction, Brasilia, Brazil, World Bank Institute of Applied Economic Research (IPEA) Brasilia, DF, Brazil, Apr 2005
47. Abdrabo MA, Hassaan MA, Selmy ARN (2015) Economic valuation of sea level rise impacts on agricultural sector in northern governorates of the Nile Delta. Low Carbon Econ 6:51–63. https://file.scirp.org/pdf/LCE_2015062315324024.pdf. Accessed Feb 2018
48. Khalil I, Milliken K, Gaad M, Ekdahl O, Yousri A, Pergolini G (2017) Engaging stakeholders and building ownership for climate adaptation: best practice from Egypt. World Food Programme (WFP), Egypt
49. ETNC (2016) Egypt third national communication: under the united nations framework convention on climate change Egyptian Environment Affairs Agency (EEAA), UNDP, Egypt
50. El-Nahry AH, Doluschitz R (2010) Environ Earth Sci 59:1497. https://link.springer.com/article/10.1007/s12665-009-0135-0#citeas. Accessed Mar 2018
51. Sušnik J, Vamvakeridou-Lyroudia LS, Gebert N, Kloos J, Renaud F, La Jeunesse I, Mabrouk B, Savić DA, Kapelan Z, Ludwig R, Fischer G, Roson R, Zografos C (2014) Interdisciplinary assessment of sea- level rise and climate change impacts on the lower Nile Delta, Egypt. Sci Total Environ 503:279–288. https://doi.org/10.1016/j.scitotenv.2014.06.111. https://ore.exeter.ac.uk/repository/bitstream/handle/10871/16667/J%20Susnik%20et%20al%20CLIWASEC%20Paper%20STOTEN%20-%20Nile.pdf?sequence=1&isAllowed=y. Accessed in Mar 2018
52. Elshinnawy I, Borhan M, ElRaey M, Dougherty B, Fencl A (2010) Climate change risks to coastal development and adaptation options in the Nile Delta. Coastal Research Institute, Stockholm Environment Institute, UNDP, Egypt, Jan 2010. http://www.eg.undp.org/content/dam/egypt/docs/Environment%20and%20Energy/Climate%20Change%20Project%20Document.pdf. Accessed in Jan 2018

53. GIZ (Deutsche Gesellschaft für Internationale Zusammenarbeit) (2014) A framework for climate change vulnerability assessments. Satapathy S, Porsché I, Rolker D, Bhatt S, Tomar S, Nair S (eds) Deutsche Gesellschaft für Internationale Zusammenarbeit (GIZ) GmbH, India. https://www.adaptationcommunity.net/?wpfb_dl=236. Accessed May 2018
54. El-Raey M (1999) Impact of climate change on Egypt. Environmental Software and Services. http://www.ess.co.at/GAIA/CASES/EGY/impact.html. Accessed Mar 2018
55. Hazell P (1995) Managing agricultural intensification. International Food Policy Research Institute, 2020 Brief, Washington, Feb 1995
56. Jaggard K, Qi A, Ober E (2010) Possible changes to arable crop yields by 2050. Philos Trans R Soc B 365:2835–51. https://www.ncbi.nlm.nih.gov/pmc/articles/PMC2935124/. Accessed June 2018
57. Jarvis A, Jamirez R, Anderson B, Leibing C, Aggarwal P (2010) Scenarios of climate change within the context of agriculture. In: Reynolds MP (ed) Climate change and crop production, CABI series in climate change, vol 1. CAB International, Wallingford, Oxfordshire/Cambridge
58. Layton JJ, Santopolo FA, Naguib M (1993) Social power, water control and irrigation systems: the Egyptian farmers' ability to obtain irrigation water. Irrig Drain Syst 7:291–304
59. Eid H, El-Marsafawy SM, Ouda SA (2007) Assessing the economic impacts of climate change on agriculture in Egypt. Pol Res Work Pap, The World Bank, Washington
60. Shalaby MY, Al-Zahrani KH, Baig MB, Straquadine GS, Aldosari F (2011) Threats and challenges to sustainable agriculture and rural development in Egypt: implications for agricultural extension. J Anim Plant Sci 21(3):581–588. http://thejaps.org.pk/docs/21-3/25.pdf. Accessed in May 2018
61. Gouda D (2016) Social capital and local water management in Egypt. AUC Press, Cairo, Egypt
62. Lewis LL (2009) Egypt's future depends on agriculture and wisdom. http://aic.ucdavis.edu/calmed/August%2008,%20final%202d%20Egypt.pdf. Accessed Aug 2018
63. Attaher SM, Medany MA, Abou-Hadid AF (2009) Possible adaptation measures of agriculture sector in the Nile Delta to climate change impacts. Adv Sci Res 3:123–126. https://www.adv-sci-res.net/3/123/2009/asr-3-123-2009.pdf. Accessed Apr 2018
64. Albehri A, Sadiddin A (2016) Sustainable agriculture and climate change: challenges and solutions for adaptation in selected zones in the MENA region. In: 3rd SITES-IDEAs annual conference on development topics. Florence, Italy. http://www.sitesideas.org/public/ideas/files/F30M1_1_Sadiddin_et_al.pdf. Accessed Jan 2018
65. Stads GJ, Moussa H, Badwan R (2015) Egypt: agricultural R&D indicators factsheet. Agricultural Research Center (ARC), International Food Policy Research Institute (IFPRI), Cairo, Egypt
66. Abdel Al M (2007) Rural responses to globalization in Egypt. Ford Found. http://dar.aucegypt.edu/bitstream/handle/10526/2875/M%20%20Aal%20report.pdf?sequence=2. Accessed Apr 2018
67. Richards A (1980) Egypt's agriculture in trouble. MERIP Rep 84:3–13
68. ILO (2013) Egypt's farmers: sowing the seeds of an agricultural revolution. http://www.ilo.org/global/about-the-ilo/newsroom/features/WCMS_204313/lang–en/index.htm. Accessed June 2018
69. Rivera WM, Elshafie EM, Aboul-Seoud KH (1997) The public sector agricultural extension system in Egypt: a pluralistic complex in transition. J Asso Inter Agric Ext Edu 4:67–74
70. Abdel Halim HK, Aboelmatty M, Hikal SA (2015) The role of agricultural extension agents to reduce the effects of the phenomenon of climate change in some governorates. Egypt J Agric Econom Social Sci Mansoura Univ 6(12):2295–2306
71. Abdel-Ghany MM, Diab AM (2013). Reforming agricultural extension in Egypt from the viewpoint of central level extension employees. Arab Univ J Agric Sci 21(2):143–154. https://www.researchgate.net/profile/Ahmed_Diab13/publication/258282975_Reforming_Agricultural_Extension_in_Egypt_From_The_Viewpoint_Of_Central_Level_Extension_Employees/links/577a713008ae213761c9b8a7.pdf. Accessed Apr 2018
72. Mamdouh A (2017, July 10) Agriculture: insufficient staff and decreasing the budget paralyzes public extension in Egypt. Veto online Newspaper (Arabic). https://www.vetogate.com/2784340. Accessed Jan 2018

73. Global Forum for Agricultural Advisory Services website. http://www.g-fras.org/en/world-wide-extension-study/africa/northern-africa/egypt.html. Accessed Mar 2018
74. Abdou DK, El Eraki M, Al Nagar K, Sholkamy M (2014) Reforming the agricultural extension services in Egypt: the case of the farmers field schools in Qena. Egypt Network for Integrated Development, Policy Brief 029
75. Radwan H (2010) Women and water management in Egypt: an empirical study. Egyptian-German Water Resources Management Reform Programme, GIZ, Egypt
76. Abou Hadid AF (2006) Assessment of impacts, adaptation and vulnerability to climate change in North Africa: Food production and water resources. A final report submitted to assessments of impacts and adaptations to climate change (AIACC), project no AF90
77. Capmas (2014) Biannual bulletin on population research and studies, Egypt. CAPMAS, Cairo, Egypt
78. Wall E, Marzall K (2006) Adaptive capacity for climate change in canadian rural communities. Local Environ 11:373–397. http://citeseerx.ist.psu.edu/viewdoc/download?doi=10.1.1.551.4289&rep=rep1&type=pdf. Accessed Feb 2018
79. Capmas (2016) Income, expenditure and consumption survey 2015, vol V, CAPMAS, Cairo, Egypt
80. Aboulnaga A, Siddik I, Megahed W, Salah E, Ahmed S, Nageeb R, Yassin D, Abdelzaher M (2017) Study on small-scale family farming in the Near East and North Africa Region. Focus Country: Egypt. FAO, Center de Cooperation Internationale en Recherche Agronomique pour le Development International Centre for Advanced Mediterranean Agronomic Studies, Cairo, Egypt
81. Achy L, Selim R (2017) Micro and small enterprises in Egypt: access to finance and job creation dynamics. Paper for the 23rd ERF annual conference, March 2017. http://erf.org.eg/wp-content/uploads/2017/03/Mic_ERF23AC_AchySelim.pdf. Accessed Aug 2018
82. Helmy IAF (2015) Understanding the saving behavior of poor households: evidence from Egypt. Master of Arts in Economics dissertation, The American University in Cairo. http://dar.aucegypt.edu/bitstream/handle/10526/4548/2015%20ECON%20Imane%20Abdel%20Fattah%20Helmy.pdf?sequence1/41
83. Fahim MA, Hassanein MK, Khalil AA, Abou Hadid AF (2013) Climate change adaptation needs for food security in Egypt. Nat Sci 11(12)
84. Kurukulasuriya P, Rosenthal S (2003) Climate change and agriculture: a review of impacts and adaptations. The World Bank Environment Department, World Bank. http://documents.worldbank.org/curated/en/757601468332407727/pdf/787390WP0Clima0ure0377348B00PUBLIC0.pdf. Accessed Aug 2018
85. Wolf S, Hinkel J, Hofmann M, Bisaro A, Lincke D, Ionescu C, Klein RJT (2012) Clarifying vulnerability definitions and assessments using formalisation. Int J Clim Change Strateg Manag 5(1):54–70. https://doi.org/10.1002/wcc.314
86. Metz HC (ed) (1990) Egypt: a country study. GPO for the Library of Congress, Washington. http://countrystudies.us/egypt/5
87. Fleischer G, Waibel H, Walter-Echols G (2002) Transforming top-down agricultural extension to a participatory system: a study of costs and prospective benefits in Egypt. Public Adm Dev 22(4):309–322. https://onlinelibrary.wiley.com/doi/pdf/10.1002/pad.233. Accessed June 2018
88. Ludena CE, Yoon SW (2015) Local vulnerability indicators and adaptation to climate change: a survey. Inter-American Development Bank, Technical Note no 857 (IDB-TN- 857), Washington. https://www.uncclearn.org/sites/default/files/inventory/idb01062016_local_vulnerability_indicators_and_adaptation_to_climate_change_a_survey.pdf. Accessed Dec 2017
89. Mohamed NN (2018) Egyptian food insecurity under water shortage and its socioeconomic impacts. In: The handbook of environmental chemistry. Springer, Berlin, Heidelberg. https://link.springer.com/chapter/10.1007/698_2018_239. Accessed June 2018

90. Hussein M (2018) Amid the refusal of the government and the division of parliamentary members due to the reconciliation of building on agricultural land. Parliament is close to the resolution of the law of reconciliation in the violations of construction. Housing Committee: this concerns 40 million citizens and will be temporary. And would require the modification of the "unified construction" as well. Elyoum7 Newspaper, 05 February 2018 (Arabic). https://www.youm7.com/story/2018/2/5/-وسط-رفض-حكومى-وانقسام-نيابى-بسبب-التصالح-بشان-الأراضى-الزراعية/3634625 Accessed July 2018
91. UNDP (2011) Egypt's national strategy for adaptation to climate change and disaster risk reduction. United Nations Development Programme (UNDP) and Information and Decision Support Centre (IDSC), Egypt
92. Siam G (2013) The agricultural sector: policy brief to climate change adaptation of agricultural sector. Climate Change Risk Management Programme in Egypt. http://www.mdgfund.org/sites/default/files/ENV_POLICY%20BRIEF_%20Egypt_%20%20Agr%20Sector%20(MALR).pdf. Accessed Feb 2018
93. Abutaleb KAA, El-Sayed Mohammed AH, Ahmed MHM (2018) Climate change impacts, vulnerabilities and adaption measures for Egypt's Nile Delta. Earth Syst Environ. (Springer). https://www.researchgate.net/profile/Khaled_Abutaleb2/publication/324434143_Climate_Change_Impacts_Vulnerabilities_and_Adaption_Measures_for_Egypt%27s_Nile_Delta/links/5ace37934585154f3f42f9f2/Climate-Change-Impacts-Vulnerabilities-and-Adaption-Measures-for-Egypts-Nile-Delta.pdf. Accessed July 2018
94. Elmenofi GAG, El Bilali H, Berjan S (2014) Governance of rural development in Egypt. Ann Agric Sci 59:285–296. https://www.sciencedirect.com/science/article/pii/S0570178314000566. Accessed in May 2018
95. Haars C, Lönsjö EM, Mogos B (2016) The uncertain future of the Nile Delta. https://www.researchgate.net/profile/Christian_Haars/publication/301549227_The_uncertain_future_of_the_Nile_Delta/links/5718b94d08ae30c3f9f188f8/The-uncertain-future-of-the-Nile-Delta.pdf. Accessed July 2018
96. Enid (2011) Reforming the agricultural extension services in Egypt: the case of the farmers field schools in Qena. Egypt network for integrated development, Policy Brief 029. http://www.enid.org.eg/Uploads/PDF/PB29_agric_extension.pdf. Accessed Feb 2018
97. Veland S, Howitt R, Dominey-Howes D, Thomalla F, Houston D (2013) Procedural vulnerability: understanding environmental change in a remote indigenous community. Glob Environ Change Human Policy Dimens 23:314–326. https://s3.amazonaws.com/academia.edu.documents/46590794/Procedural_vulnerability_Understanding_e20160618-16257-1gbnb9w.pdf?AWSAccessKeyId=AKIAIWOWYYGZ2Y53UL3A&Expires=1530884943&Signature=tLhFAd4zYu96GL4nJNft9TqPdN4%3D&response-content-disposition=inline%3B%20filename%3DProcedural_vulnerability_understanding_e.pdf. Accessed May 2018

Water Climate Food Nexus for Green Sustainability

Sherien A. Zahran

Abstract Egypt is characterized as arid and semi-arid country with very limited water resources, population growth, and proving with climate changes such reasons grow pressures on the environment and natural resources, and consequently affecting on per capita sharing for water and land, human securities and eventually political stability. There are interrelationships between food, water and climate that affect in sustainable development generally and especially for green sustainable. Green sustainable could support the human being, environment, and prevent poverty in terms of sufficient water and food security. Whenever Climate change is considered as cross cutting issue with food and water security and consequentially for green sustainable Development. Applying Water-climate-Food Security (WCF) Nexus approach, could be save time and efforts need to cope with the risk of climate at these crosscutting areas. It is highly appreciated if there are a coordination between governmental sectors, stakeholders and different beneficiaries. By identifying entry points, processes and partners for WCF nexus mainstreaming, it could be add value to implementation of sectorial strategies and to contribute to cross-sectorial policy goals such as the national vision for sustainable development 2030 and climate change adaptation and mitigation strategies to meet Sustainable Development Goals (SDGs). The next sections identify the interrelationships between Climate water food nexus and discuss the national and international mechanisms that need to conduct this approach efficiently. In addition to drawing attention to some related national strategies in order to meet the green sustainable development.

Keywords Climate · Water · Food nexus · Sustainable development goals · SDGs

S. A. Zahran (✉)
Environment and Climate Change Research Institute (ECRI), National Water Research Center (NWRC), NWRC Building, Delta Barrage, Cairo 13621/5, Egypt
e-mail: sherienzahran@gmail.com

Canadian International Collage (CIC), Land#6, Central Services Area, Fifth Settlement, New Cairo, Egypt

© Springer Nature Switzerland AG 2020
E.-S. E. Omran and A. M. Negm (eds.), *Climate Change Impacts on Agriculture and Food Security in Egypt*, Springer Water, https://doi.org/10.1007/978-3-030-41629-4_23

List of abbreviations

FAO	Food and agriculture organization
(GSD)	Green sustainable development
IPCC	Inter-governmental panel of climate change
RCPs	Representatives concentration pathways
(SDGs)	Sustainable development goals
WFP	World food program

1 Introduction

Increasing water and food demands in Egypt is very fast, which resulting mainly from many challenges such as population growth and climate change. These challenges add more stress on gross demotic product (GDP) and national policy. Such stresses need better management and conservation of our natural resources to fulfill our requirement.

Egypt is very limited water resources which characterized as arid and semi-arid country. It depend on the River Nile that comes from outside the country and shared with other riparian basin countries, with limited amount of precipitation at coastal area and Sinai, and some confined and non-confined aquifer. All these resources, doesn't satisfy the demand especially with increase of population growth rate. Whatever the agriculture is an important sector for food security and the biggest user of water. Rather than the other demands for the development requirement which also is rising fast which should be has a green sustainably sound.

In this context and especially for the rainfall agriculture which faced particular challenges of unpredicted rainfall due to climate change and low productivities and because of the generally arid climate, there are a high level of development of water resources for irrigation and relatively high levels of performances in agricultural water management. In addition to that the Groundwater as well has become a significant source of agriculture water across Egypt.

Green Sustainable Development (GSD) is that meets the needs of the present without compromising the ability of future generations to meet their own needs. At the core of sustainable development is the need to consider "three pillars" together: society, the economy and the environment. No matter the context, the basic idea remains the same—people, habitats and economic systems are inter-related.

These may be conducted by applying the new approach of nexus especially between water, climate and foot security (WCF), which are interrelated with each other and facing from the same challenges with cross cutting strategies and plan for green sustainable development.

Achieving for the "GSD" green strategies and plans for WCF nexus need special institutional inter sectorial mechanism for cooperation and formulating throw the mainstream national strategy.

This chapter discus and present some of the WCF nexus interrelation; impact and strategies, an overview of related sustainable development goals (SDGs), indicators for Egyptian sustainable development 2030 SDGs, some idea for interrelated sectorial mechanism, and system analysis for this mechanism, the Egyptian efforts at national and international legislations and committed and finally Egyptian entity working in those areas, then conclusions and recommendation reflecting the progresses and steps forward.

2 Sustainable Development Goals (SDGs) Overview

Green Sustainable Development is that meets the needs of the present without compromising the ability of future generations to meet their own needs. At the core of sustainable development is the need to consider "three pillars" together: society, the economy and the environment. No matter the context, the basic idea remains the same—people, habitats and economic systems are inter-related as shown in Fig. 1 [1]. http://www.un.org/sustainabledevelopment/news/communications-material.

2.1 *Sustainable Development Goals*

The **Sustainable Development Goals** (**SDGs**) are interlinked through their target and indicators which are covered three dimension of social, economic and environment. Those are 17 SDGs include poverty, hungry, health, education, climate change,

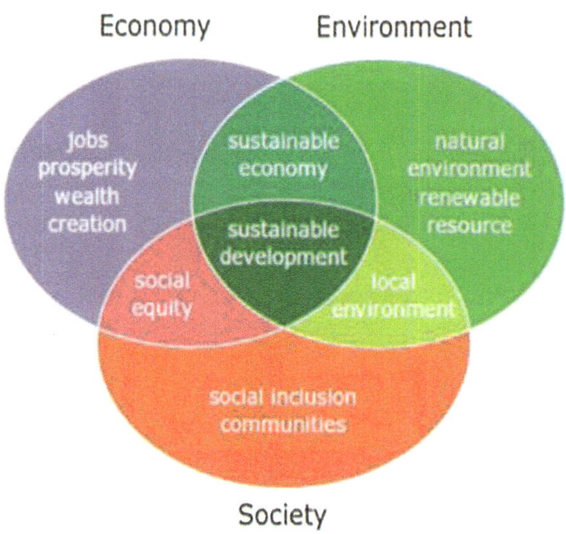

Fig. 1 Socio-economic and environment pillars for sustainable development. http://www.un.org/sustainabledevelopment/news/communications-material/

gender equity, water and sanitation, energy, environment and social justice. It were developed by United Nation to replaced the Millennium, Development Goals which end at 2015 as shown in Fig. 2 [2].

Among the SDGs 17 goals; Achieving SDG 13 on climate Action required a close nexus with other key SDGs on the social agenda, helping local communities for building resilience against now a day risk posed by climate change. This includes climate risk for achieving SDGs 1 reducing poverty, SDG 2 for food security, SDG 3 improving health, SDG5 Gender equity, SDG6 for water Access, SDG 10 for inequality and SDG 15 for land and Ecosystems, and finally SDG 17 for Partnerships. Synergic between the WCF nexus could be helping specially for:

- Foster engagement with policy makers across traditional sectorial line
- Provide partner and entity with capacity development and policy tools that reduce the risk for SDG achievement and move quickly towered the climate resilience pathways
- Quantify emerging climate risks within SDG social sector lines

The UN discussions and negotiations identified the links between the post-2015 SDG process and the Financing for Development process that concluded in Addis Ababa in July 2015 and the COP 21 Climate Change conference in Paris in December 2015 [3, 4].

In May 2015, a report concluded that only a very ambitious climate deal in Paris in 2015 could enable countries to reach the sustainable development goals and targets [5]. The report also states that tackling climate change will only be possible if the SDGs are met. Further, economic development and climate are inextricably linked, particularly around poverty, gender equality, and energy. The UN encourages the public sector to take initiative in this effort to minimize negative impacts on the environment as well [6].

Fig. 2 Sustainable development goals (SDGs) for transforming the world. http://www.un.org/sustainabledevelopment/news/communications-material/

Achieving the SDGs goals need cooperation between different parties; to exchange knowledge, expertise, technology and financial support. These patties should include governmental and non-governmental agencies, privet sectors and donors as well.

2.2 *Implementation*

Since 2016, implementation of SDGs started at national levels by moving it to national implementation plans with mobilize all required resources, monitoring plane, researches, interlinked stakeholders responsible and tools. The UNFCCC "United Nation framework convention of climate change" provides all support for building capacity to formulate national planes in this regards especially for less development countries (LDC).

2.3 *Egypt Vision for Sustainable Development 2030 SDGs*

Egypt has taken steps forward to achieving SDGs by 2030 in all relevant sectors with economic environment, and political dimensions.

The steps that were taken toward SDGs related WCF nexus approach include socio-economic and environmental dimensions. Some indicator will be achieving by the year 2030 in that dimensions. Such indicators are decreasing water consumptions ratio, increasing of Egyptian Per capita share of fresh water, enlarging of using Non-conventional Water Resources Ratio with respect to the total water resources, increasing of Treatment drainage water ratio related to total drainage, decreasing of Industrial waste water ratio deliver to River Nile without treatment, treat the Domestics water delivered to River Nile, reducing the ratio of water network transmitters losses. The trend of these indicators shown in Figs. 3, 4, 5, 6, 7, 8 and 9 [7]. http://sdsegypt2030.com.

3 Water–Climates–Foot Nexus

3.1 *Climate Change and Food Security Link*

During the last two decades, 200 million have been lifted out of hunger and the prevalence of chronic malnutrition in children has decreased from 40 to 26%. In spite of this progress, according to the World Bank, 702 million people still live in extreme poverty and, according to this year's report on the State of Food Insecurity

Fig. 3 Egypt water consumptions ratio. *Source* http://sdsegypt2030.com

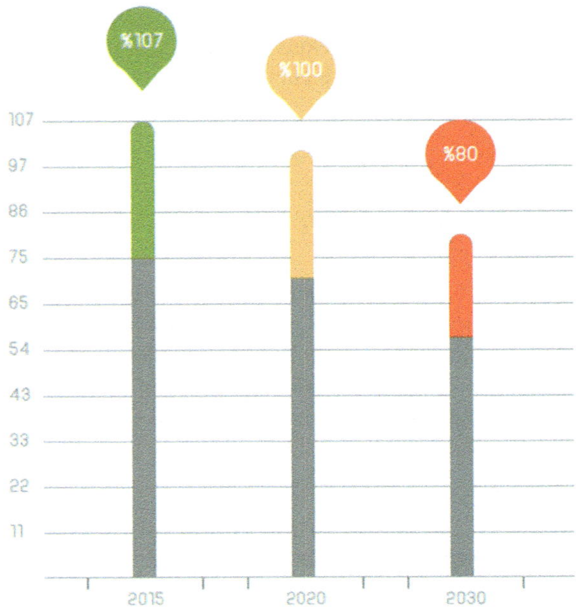

Fig. 4 Egyptian Per capita share of fresh water. *Source* http://sdsegypt2030.com

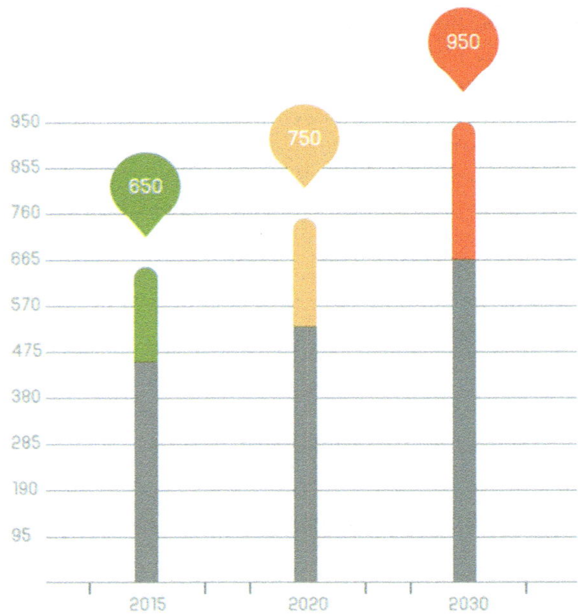

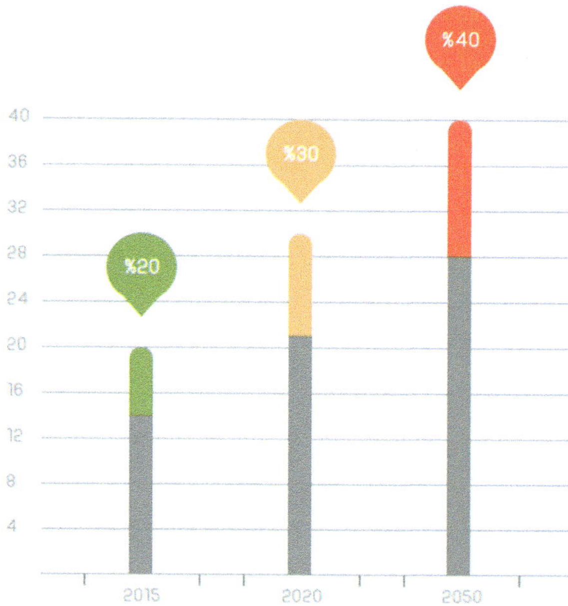

Fig. 5 Non-conventional water resources ratio with respect to the total water resources. *Source* http://sdsegypt2030.com

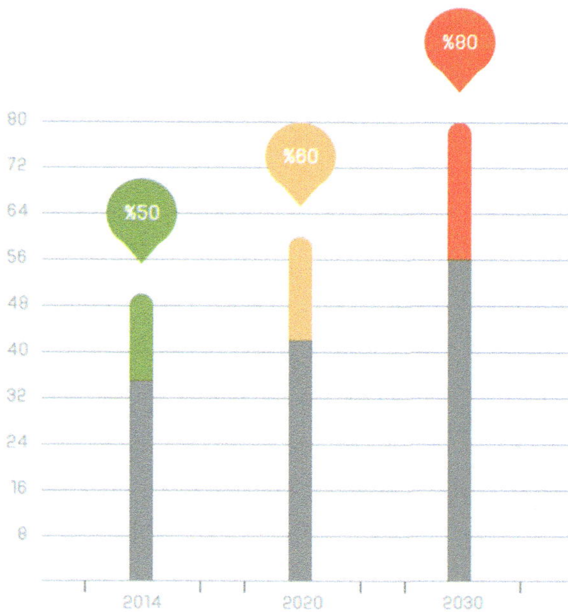

Fig. 6 Treatment drainage water ratio related to total drainage. *Source* http://sdsegypt2030.com

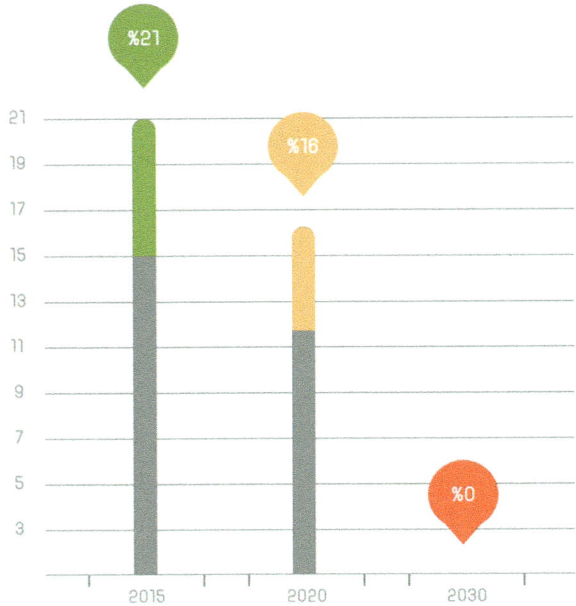

Fig. 7 Industrial waste water ratio delivers to River Nile without treatment. *Source* http://sdsegypt2030.com

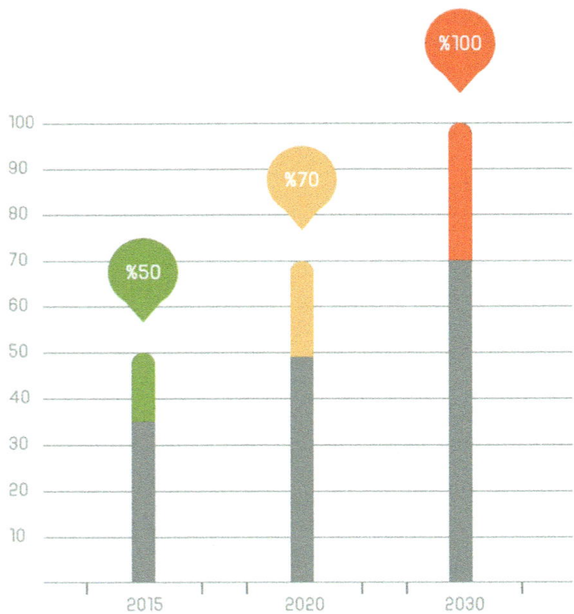

Fig. 8 Egyptian domestics treated water delivered to river Nile. *Source* http://sdsegypt2030.com/

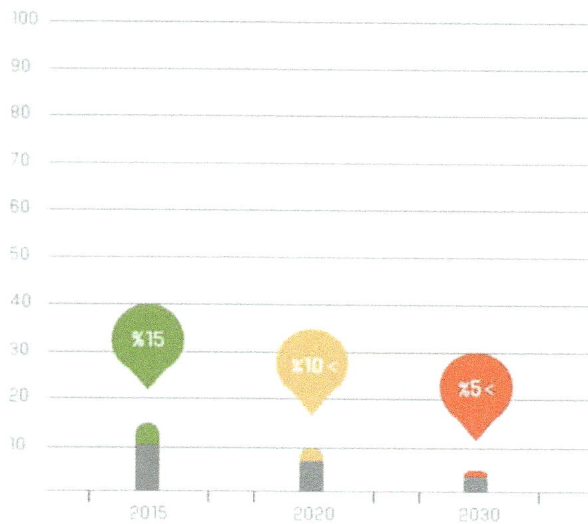

Fig. 9 Egyptian ratio of water network transmitters' losses. *Source* http://sdsegypt2030.com

in the World [8], 793 million people are undernourished [9]. Among the most significant impacts of climate change is the potential increase of food insecurity and malnutrition. https://m.wfp.org/climate-change/climate-impacts.

3.1.1 Climate Change and Risk of Hunger

Many factors increase the risk of climate change on hunger and mal nutrition. These factors include:

- Extreme weather event (intensity and frequency); throw adverse impact on livelihood and food security by climate related disaster to destroy crop and infrastructure.
- Gradual and long term climate risk; as increase in sea level rise with on sequently impact on all the live and activity at costal area and river delta with change the pattern of flooding and drought as well.
- Food availability, access, stability and utilization; change in climate variables (temperature and rainfall) affected on crop productivity and quality which sacrificed the poor farmer and vulnerable people with increase the prices of main and major crops. In addition to the impacts on chronic food insecurity and create a vicious cycle of disease and hunger. All these factor could be create the state of instability of the strategic plans of food security.

3.1.2 Egypt Food Insecurity and Climate Change Vulnerability Map

UN World Food Programmed (WFP) and the Met Office Hadley Centre produced The Food Insecurity and Climate Change Vulnerability map. It illustrates index for how strong adaptation and mitigation efforts will prevent the worst impacts of climate change on hunger globally and help make people less vulnerable to food insecurity.

This index is calculated based on exposure to climate change related hazard, sensitivity of agriculture production to these hazards, and the capacity of country with climate change-related food shocks.

- Index is calculated from a measure of the average length flood and drought events in each country as the exposure component for the baseline period (1981–2010). The adaptive capacity component of the index is calculated using socio-economic indicators relevant to the food security system. Criteria for inclusion, all data included in the vulnerability to food insecurity index correlates with the Food and Agriculture Organization (FAO) measure of under nutrition. Three representative concentration pathways (RCPs) scenario are considered in this map; low emission with temperature around 2 °C. "RCP 2.6", intermediate emission with temperatures between 2.5 and 3 °C above pre-industrial levels "RCP 4.5", and high emission with temperature rise in global average of 4 °C or more above pre-industrial levels by the end of the 21st century. This scenario is also known as RCP 8.5.

The projections of the climate model are used to calculate future projections of the exposure component of the index. Twelve climate models from the latest generation of climate models used to inform the most recent Inter-governmental Panel on Climate Change report (IPCC AR5) were used, and the average value of the index for each country across the models is shown

There are three level of different adaptation; low; high and none as following (Fig. 10):

- The 'high' adaptation scenario corresponds to a change of approximately 10–15% in the 2050s compared to the present-day, and a further change of approximately 10–15% in the 2080s compared to the 2050s.
- The 'low' adaptation scenario corresponds to a change of approximately 5–10% in the 2050s compared to the present-day, and a further change of approximately 5–10% in the 2080s compared to the 2050s.
- The no adaptation scenario maintains the sensitivity and adaptive capacity components of the index at the present-day level.

As illustrates in the Fig. 11, and related to vulnerability map. It should be noted that by 2050 and 2080 with high emission scenario for climate change and with conducting a High level of adaptation, Egypt vulnerability to food insecurity increase with 35 and 62% from the present day respectively. However if there are maladaptation and with high scenario for emission the Egypt vulnerability to food insecurity will be increase with 47 and 85% from the present day by 2050 and 2080 respectively. Table 1 shows the percentages of increasing the vulnerability of Egyptian

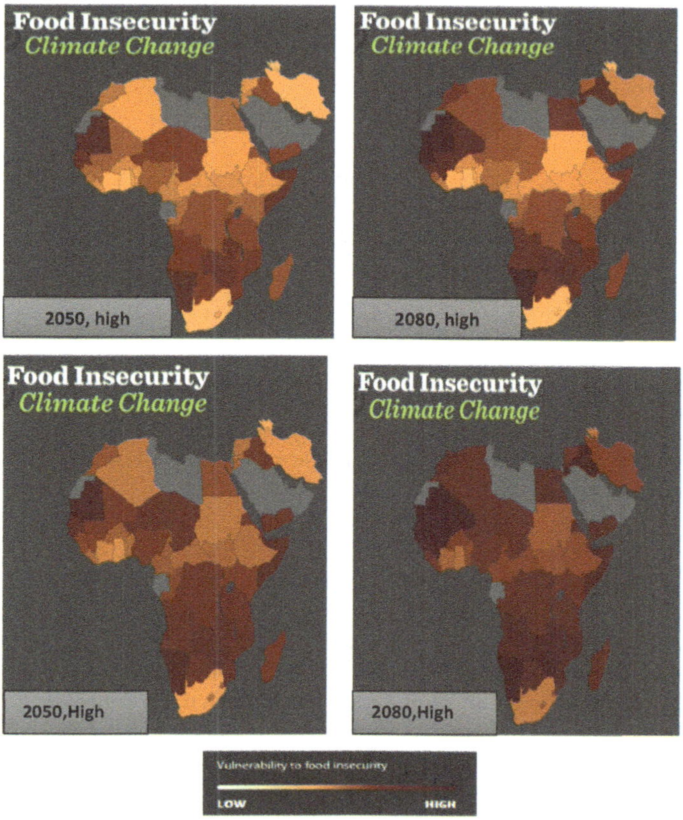

Fig. 10 Africa and middle east vulnerability to food insecurity for 2050 and 2080. *Source* WFP & UK Mett office, https://www.metoffice.gov.uk/food-insecurity-index

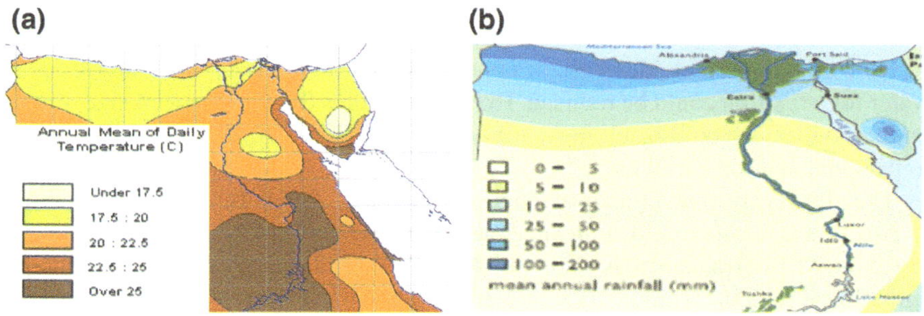

Fig. 11 **a** The average annual temperature across Egypt from 1961 to 2000 and **b** the average annual precipitation (NWRP 2005)

Table 1 Percentages of increasing the vulnerability of Egyptian food insecurity from present day due to climate change

Scenarios	Emission scenarios, RCPs (W/m^2)					
Time period	2050			2080		
Adaptation level	Low (RCP 2,6)	Medium (RCP 4.5)	High (RCP 8.5	Low (RCP 2.6)	Medium (RCP 4.5	High (RCP 8.5)
High	18	20	35	6	11	62
Low	24	26	41	17	22	73
None	30	32	47	29	34	85

Source FAO & Mett office

food insecurity from present day due to climate change for different future projection of emission scenarios and different level or adaptation. https://m.wfp.org/climate-change/climate-impacts.

3.1.3 Climate–Water–Crop Production

Crop production in Egypt will be affected by three ways by climate change. First, higher temperatures will change yields and water demand. EEAA report at [10] displays projected changes in crop yields, all of which are projected to decrease except cotton. This is mainly the result of higher temperatures because all crops are irrigated. The IPCC stated that by 2050, rice yields in Egypt could decrease by 11% and soybean yields by 28% [10].

Change in irrigation is the second way with climate change may affect crop yields. Higher temperatures will likely increase demand for water by crops (although higher atmospheric concentrations of carbon dioxide (CO_2) without a change in climate will reduce water demand by crops). The TNC cited several studies that project a 5–13% increase in irrigation needs by Egyptian crops [10].

If flows into the Nile decrease, it is possible that, in the long-term, deliveries of water for irrigation could be reduced. Even if the Nile's flow does not change, higher population levels could result in a shift of available water supplies from agriculture to personal and industrial uses.

The third way that climate change could affect agriculture is through SLR. A rise in sea levels could inundate low-lying and unprotected agricultural lands along the Mediterranean coast. The low-lying Nile Delta is Egypt's most productive agricultural region.

There are other direct and indirect ways that climate change could affect crop production in Egypt. A change in climate could affect pests and disease. For example, a warmer climate may enable some pests and diseases to migrate into Egypt, but a warmer climate may also make it too warm for some pests and diseases to survive. Wetter conditions could enhance migration of pests and disease, whereas drier conditions could limit migration of some pests and disease but perhaps make it possible for others to migrate into agricultural areas in Egypt [10].

Change in global supply and demand for certain crops could also affect production in Egypt. On average, warmer temperatures will increase relative yields of grain crops such as wheat in higher latitudes and will decrease relative yields in lower latitudes. This should shift the competitive advantage to growing areas at higher latitude [10], putting Egypt at a competitive disadvantage and decreasing exports and increasing imports.

3.1.4 Livestock—Climate Impacts

There is limited information on how livestock in Egypt could be affected by climate change. Higher temperatures can decrease livestock productivity, and extreme hot and dry conditions can be fatal to livestock. The SNC notes that bluetongue and Rift Valley fever have recently emerged in Egypt and this emergence may be related to climate factors. Another risk to livestock from climate change is decreased fodder production.

3.1.5 Fisheries—Climate Impacts

In general, higher water temperatures will affect fish production. In addition, higher salinity levels could limit the production of freshwater fish.

3.2 Climate and Water Nexus

Egypt's climate is hot, dry, deserted and is getting warmer. During the winter season (December–February), Lower Egypt's climate is mild with some rain, primarily over the coastal areas, while Upper Egypt's climate is practically rainless with warm sunny days and cool nights. During the summer season (June–August), the climate is hot and dry all over Egypt. The temperature in summer ranged between 38 and 43 °C with extremes 49 °C at the desert and south part but it is cooler at mediterranean with max. 32 °C as Fig. 1a [11].

Table 2 shows water budget in Egypt. The conventional water resources in Egypt are the Nile water, rainfall, and deep groundwater. The non-conventional water resources are the shallow groundwater in the Nile delta, the re-used agricultural drainage water, and re-used treated waste water. The Nile water supplies are extremely limited by the 55.5 billion m^3 at High Aswan Dam (HAD), and projected to become even more limited due to the increased competition on water resources among the Nile basin countries. Annual water share of Nile water per capita in Egypt has decreased from

Table 2 Egypt water budget [12]

	Water source	Volume in billion (m^3/year)
Conventional	Nile	55.5
	Groundwater	2.0
	Rainfall and flash flood	1.30
	Desalination	0.2
	Total	59.0
Non-conventional	Shallow GW	6.2
	Reuse of drainage water	16.0
	Total	22.2

2500 m^3/capita/year in the 1950s to about 680 m^3/capita/year in 2012, and is projected to drop to about 350 m^3/capita/year in 2050 (www.mwri.gov.eg).

Climate change and climate variability, especially regarding to temperature and precipitation have a great impacts on the water resources. Many studies investigated the climate historical trends and future climate projections for the Rive Nile basin using various GCMs and its impacts as shown in the Table 3.

The water demands in Egypt lie into two categories; agricultural water demand and municipal/industrial water demand. Low irrigation efficiencies and inadequate irrigation management practices have made the irrigation water demand highly vulnerable to climate change. It is expected that the climate change would lead to future temperature increase associated with the increase in crop-water requirement and thus decrease in crop use efficiency.

They concluded that for many strategic crops in Egypt, the irrigation water demands shall increase by 13% by the end of 2100, considering the IPCC emission and socioeconomic scenarios. They also discussed the crop evapo-transpiration (ET) and its expected impact on irrigation demands by 12% by the end of 2100. The [14] report cited several studies that project a 5–13% increase in irrigation requirements by Egyptian crops [10]. Table 4 shows the impacts of climate change on water demands up to the year 2100. Assuming constant water consumption rates, the holding company for water and wastewater projected water demand to 2050, rough estimates were used afterwards [11]. The allocation of future municipal and industrial water shall depend on the agricultural future water allocations.

Table 3 Studies of climate historical trends and future climate projections for the Rive Nile basin using various GCMs [13]

Author	Projection	Brief
Elshamy et al. (2000)	IPCC projections (2050)	2–4.3 °C increase over Nile Basin; 3–4 °C increase in Northern Sudan and Egypt −22 to +18% change in precipitation
Conway (2005)	Historical trends	No precipitation trend over Blue Nile
Hulme et al. (2001)	Historical trends (20th Century)	0.5 °C increase in Africa, 0.6 °C in Ethiopia
Nyssen et al. (2004)	Historical trends	No precipitation trend over highlands in Ethiopia/Eritrea
Sayed and Nour (2002)	IPCC projections	−2 to +11% change in Blue Nile precipitation; −1 to +10% change in White Nile precipitation; −14 to +32% inflows to Lake Nasser
SNC-Lavalin (2006)	IPCC projections (2050)	+7.4% mean increase in precipitation in Equatorial Lakes; +23% change in inflows to Southern Nile
IPCC (2007)	IPCC projections	Increased rainfall over Nile Equatorial Lakes Region, GCMs inconsistent over Ethiopia and Sahel
Conway et al. (2004)	IPCC projections (2050)	+2.2 °C mean increase in Ethiopia +1 to 6% mean increase in precipitation in Ethiopia
Elshamy (2008)	IPCC projections (2081–2099)	2–5 °C increase over Nile Basin +2.4% change in precipitation +2–14% increase in potential evapotranspiration −15% mean change in runoff

4 International, Regional, and National Efforts

4.1 Water Climate Food (WCF) Nexus System Analysis and Coordination Mechanism

A nexus approach can increase resource efficiency and support decoupling of economic development from use of resources and environmental pressure. These can be done by fostering coordination and cooperation among the relevant sectors and cross

Table 4 Impacts of climate change on water demands [11]

Year	2010	2025	2050	2075	2100
Population (million)	79	104	146	191	237
Mean air temperature increase (°C)	–	1.0	1.7	2.5	3.6
Evapotranspiration and irrigation water requirements (%)	–	2	4.5	8	12
Municipal water (million m^3/year)	9	9.6	12.55	14.75	17.2
Industrial water (million m^3/year)	2	2.2	3.4	4	4.9

Source Practice Assessment and Recommendations Report/REC-WEE pros/WPOrt EG [13]

cutting policy. The decision maker and all stakeholders should be engaged and work closely to fill the gaps and construct the bridge linked between all sides in the Crosse areas to get the benefit from inter-relationship for the natural resources in terms of water management, climate change and food security with effective way. Such way need form the most pressing Nexus problem, that require system analysis, develop storyline for plausible futures, and co-design polices and investment strategies based on modeling input to caches SDGs (Figs. 12, 13 and 14).

Development of scenarios and PATHWAYS needs to be interactive between science, policy, investors and others to establish priorities and ownership. This joining venture could be study the current situation, externalities with different scenarios and proposed pathway strategies. Under the good selection of tools, data, information and use suitable technologies, one can predicts, simulate, trade off and differentiate between many sustainable alternatives or synergies between those to reduce the future challenges. See Fig. [7].

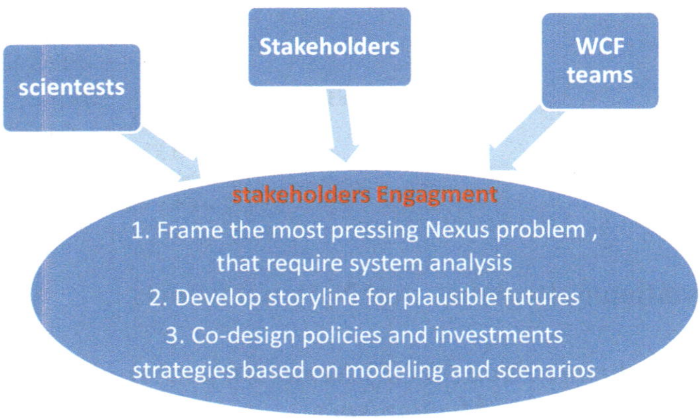

Fig. 12 Stakeholder engagement

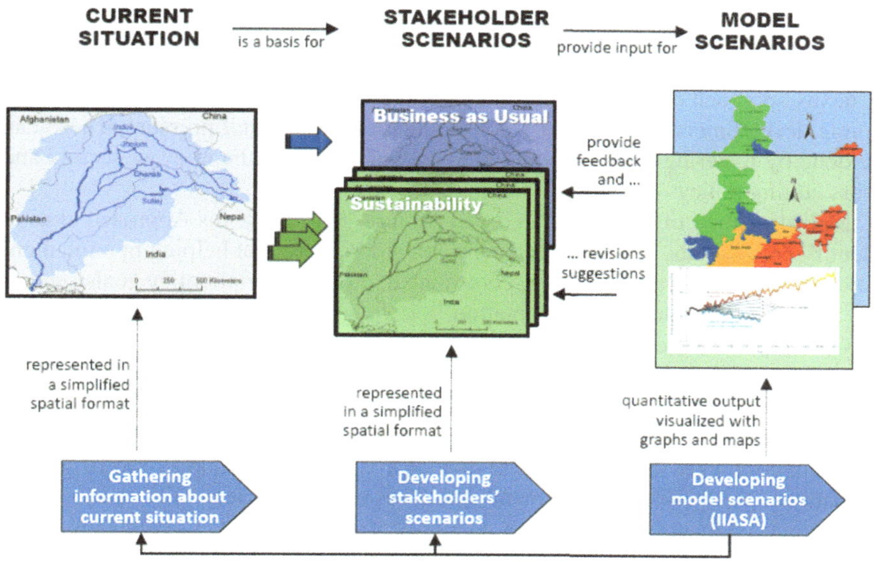

Fig. 13 Scenarios for system analysis. www.iiasa.ac.at

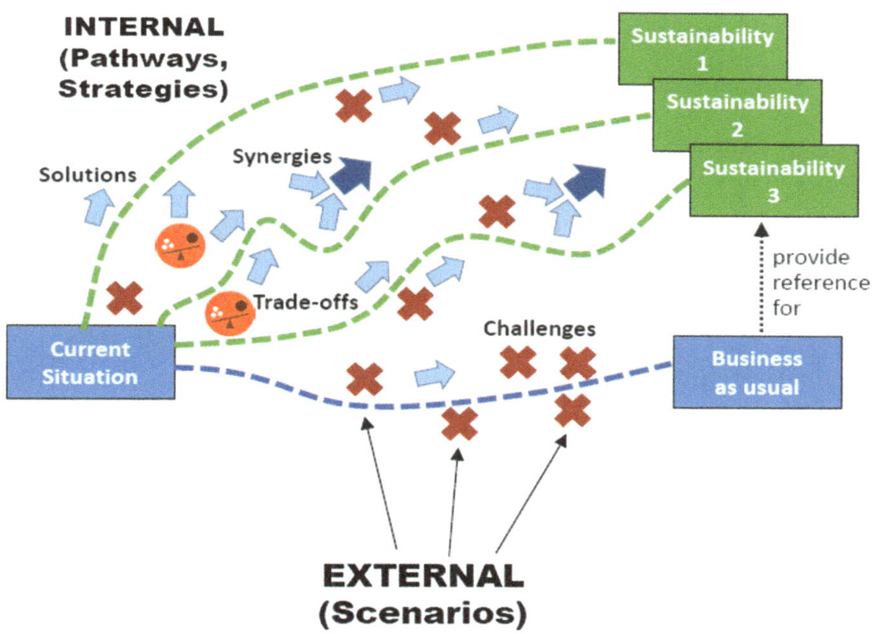

Fig. 14 Synergies between internal and external pathway and strategies. www.iiasa.ac.at

4.2 WCF Nexus Coordination Mechanism

A nexus approach can increase resource efficiency and support decoupling of economic development from use of resources and environmental pressure. These can be done by fostering coordination and cooperation among the relevant sectors and cross cutting policy.

Mainstreaming of water climate Food Security (WCF) Nexus Approach aims to identify entry points, processes and partners into Policies that helping in implementation of sectorial strategies and to contribute to cross-sectorial policy goals such as the SDGs and climate change adaptation.

One important step in the process of WCF Nexus mainstreaming policies and institutions is to improve coordination and collaboration across sectorial activities, initiatives, policies and strategies. This will contribute to merge cross-sectorial synchronization and minimize trade-offs.

The main player sectors are the bodies related to water resources and irrigation, agriculture production, environment and ecosystem and energy resources; rather than which is governmental or non-governmental. Some related organization may supporting the process and roles of emphasis the cross sectorial coordination. There are, three types of instruments to support policy integration [15]: (a) Communicative instruments; (b) Organizational instruments and (c) Procedural instruments.

4.2.1 Communicative Instruments

Set longer-term objectives to guide strategy development in the sectors concerned. They could involve, for example:

- including WCF-nexus-related provisions in overall (sustainable-development) strategies, or adopting a specific WCF-nexus strategy/action plan;
- Obligations for sectorial departments to report on performance with regard to specific nexus indicators, etc.

4.2.2 Organizational Instruments

Encompass institutional coordination mechanisms at high political levels or at the administrative level. They could involve, for example:

- Amalgamation of departments responsible for water, and food security;
- Setting-up of interdepartmental coordination committees or institutionalizing WCF-nexus working groups at various political or administrative levels, etc.

4.2.3 Procedural Instruments

Procedural instruments could involve on

- Bureaucratic rules, such as the obligation to inform and consult other sectors on policy developments;
- Instruments that aim to influence the direction of decision-making, e.g. by making a nexus perspective a prerequisite for funding of sectorial programs;
- Obligations to carry out assessments of the impact of WCF nexus within project or program planning (similar to Strategic Environmental Assessments);
- Requirements to include a WCF-nexus perspective in the assessment of new policies and strategies, etc.

The importance of policy learning and awareness for successful cross-sectorial integration also implies that, besides conductive administrative set-ups for WCF nexus coordination, increasing knowledge and public awareness can be powerful instruments in support of policy change towards a more integrated WCF-nexus perspective. Consequently, coordination of sectorial strategies can also be supported by expert advisory bodies or targeted research programs that provide expertise and recommendations on how synergies from WCF nexus can best be exploited and trade-offs be minimized. Moreover, public opinion, non-governmental organizations or civil society organizations often play an important role in promoting sustainable development. Raising their awareness of the issue and providing opportunities for their engagement in strategy development processes can thus strengthen integration of the WCF-nexus approach to sectorial strategies [16].

In the end, the question will be whether expert recommendations are converted into political decisions, whether sectorial administrative bodies are encouraged to coordinate across sectors, and whether WCF nexus requirements will be enforced. These questions boil down to the crucial role of high level political will in mainstreaming a WCF-nexus approach [15]. www.accwam.org, www.giz.de.

4.3 Institutional and Legal Status for WFC Nexus

International climate policy, grounded in the United Nations Framework Convention on Climate Change (UNFCCC), recognizes this discrepancy. Within the UNFCCC process, the international community now faces the challenge of negotiating a post-2020 agreement. The parties to the Convention have set the goal of limiting global warming to 2 °C relative to the pre-industrial level by using some instruments of adaptation and mitigation strategies of climate change These strategies sometime depend on financial support from industrialized nations in terms of some mechanism such Green climate fund (GCF).

In this context, there are National and international climate change legislative at Egypt supporting water Climate food nexus that can be summarized in the following sections.

4.3.1 International Level

- Egypt signed the UFCCC on 10 November 1994. Egypt signed the Kyoto Protocol, on 15 March 1999.
- Egypt prepared and submitted three communications under the UFCCC, which triggered various useful studies and strategies relating to climate change; first in 1999, second in 2010 and third is in 2013.
- In 2008, Egypt partnered with the Millennium Development Goals Achievement Fund (MDG-F) to establish and implement a Climate Change Risk Management Program.
- In Dec 2015, Egypt participated in Paris international conference for climate change with highest level officials.
- In April 2016, Egypt signed the Paris convention for combating climate change.
- In May 2017, the Egyptian Parliament approved the Egyptian involvement in Paris convention for combating climate change.
- In Dec 2018, Egypt participated in Paris international conference for climate change with highest level officials.

4.3.2 National Level

- Egypt formed an inter-ministerial National Climate Change Committee in October 1997 and restructured it in 2007.
- On 14 March 2005, the Minister of Environmental Affairs established the Designated National Authority of Clean Development Mechanism by Decree no. 42 for the year 2005. Egypt is deemed to be 'among the leading countries in Arab States region in terms of the number of registered Clean Development Mechanism CDM projects' [17].
- In 2010, Egypt published a National Environmental, Economic and Development Study (NEEDS) for Climate Change to outline the financial and institutional needs for adaptation measures.
- In 2013, the Ministry of Environmental Affairs placed a strategy for renewable energy adoption in Egypt to 2035.
- In 2014, the Ministry of Environmental Affairs signed an agreement in January 2014 with Italian Minister of Environment to transform El-Gouna City into the first carbon-neutral city in Africa.
- In 2016, Egypt launched 2030 national strategy towards sustainable development, with the fourth strategic goal of combating the effects of climate change (http://sdsegypt2030.com).
- In 2017, the Agriculture research center of the Ministry of agriculture and land reclamation prepared the Egyptian agricultural strategy 2030 including a complete chapter on climate change and related adaptation measures.

- In Sept 2017, the Ministry of Water Resources and Irrigation obtained official approval of the national water resources plan 2017–2037 including climate change adaptation (www.mwri.gov.eg).
- In 2017, the Ministry of Environmental affairs is currently finalizing project report for climate change adaptation to be submitted to national committee for green climate.
- In 2017, the Ministry of Environmental affairs has formulated stronger strategies for mitigating the effects of climate change and included in a new law, which has been submitted to the parliament for discussion.

4.4 Existing Climate Change Institutional Setup

This section gives an overview of the Existing climate change institutional setup [18, 19, 11].

- In 1992, a Climate Change Unit (CCU) was established within The Egyptian Environmental Affairs Agency (EEAA). An inter-ministerial National Committee of Climate Change (NCCC) was established in 1997 and reformed at 2007 by decree of prime-minister and headed by the Minister of State for Environmental Affairs. The major responsibility for NCCC is: Development of a national policy for mitigating effects of climate change, Organizing the participation of Egypt in the United Nations Framework Convention on Climate Change (UNFCCC), and Reviewing developed national climate change plans.
- The Coastal Research Institute (CORI) since 1972, The Nile Forecast Center (NFC) since mid-1990s as well as the Environmental and Climate Research Institute (ECRI) of the National Water Research Center (NWRC) since 1997, have been established by The Ministry of Water Resources and Irrigation (MWRI). There are studying the impacts of climate change on water resources project, hydrology especially for Nile basin and Egyptian coasts as well with relevant strategies for adaptations.
- Climate and Environment Risk Assessment Laboratory (CERAL) that was establish under ECRI is deal with general and regional climate model and impact of climate change at water resources for national and regional dimension (Africa, Nile Basin, and Arab Region …). In addition to establishing the real time monitoring network for climate data.
- The ministry of Agriculture and land reclamation (MALR) also established the Central Laboratory for Agricultural Climate (CLAC) since mid-1990s. The CLAC aimed at estimating and managing crop water requirement and effect of weather on them.
- The National Authority of Remote Sensing and Space Science (NARSSS) had also created a climate change sector in the water resources department (WRD). The WRD has a specific plan targeting the global climate change effects on national water resources.

4.5 *Implementation Plans for Water Climate Food Nexus*

4.5.1 Indicators for Monitoring and Evaluations

Climate protection and adaptation to the impacts of climate change in water and agricultural sectors are among the biggest challenges for sustainable development. Industrialized countries have therefore committed themselves to provide USD 100 billion per year for mitigation and adaptation measures in developing countries, starting in 2020. With increasing volumes of adaptation finance, and growing numbers of adaptation activities and strategies at the national level, the need for robust monitoring and evaluation (M&E) approaches is growing alike. Indeed, it is important not just to secure sufficient financial resources but to ensure that investment in adaptation to climate change at these sectors actually contributes to climate resilient sustainable development.

Therefore Conducting the WCF nexus approach need to monitor and evaluate some interlinked Indicators based related nexus sectors and their plans. Table 5 illustrates some indicators to be monitor and evaluate for Climate impact, climate parameters, adaptive action and their results in Water resources and agricultural sectors regarding to the possible climate impacts to meet the target result.

These items are considered the highest important socio-economic and environmental indicators which are direct related to Agriculture, climate and water projects.

5 Conclusion and Recommendation

5.1 *Conclusions*

Green sustainable development (GSD) is one of the most valuable goal that everybody looking forward. It is clear that the Nexus approach is very important issue toward meeting this goal. Meanwhile Water–Climate–food nexus approach is supporting GSD. This will happen by applying sectorial interrelationships for these areas. whatever, it is essential need to defined the measures, instruments, tools, communicate, coordinate and implementing plan between those sectors and put it throw the national mainstream policy for sustainable development. In this context, Egypt has taken many steps as presented in this chapter at international and National levels for achieving these goals but there are still some challenges. Lack of public awareness and communication between private, governmental, and non-governmental sectors are major challenges. Meanwhile increasing knowledge and public awareness could be powerful instruments in supporting policy change towards a more integrated WCF-nexus perspective.

Table 5 Monitoring indicators for CWF nexus

	Water resources sector	Agriculture sector
Climate parameter	1. Change in annual temperature 2. Number of hot days 3. Monthly precipitation 4. Change in annual precipitation 5. Change in wind speed and direction 7. Change in relative humidity 8 Extremes precipitation events	
Climate impact	1. Number of household affected by drought 2. Percentage of total livestock killed by drought 3. Number of surface water areas subjected to declining water quality due to extreme temperatures 4. Loss of GDP in percentage per year due to extreme rainfall 5. Number of people permanently displaced from home as a result of flood, drought or sea—level sea	1. Number of hectares of productive land lost to soil erosion 2. Percentage of area of ecosystem that has been disturbed or damage 3. Area covered by vegetation affected by plague of fires 4. Shift of agrophenological phases of cultivated plane 5. Decreased annual average fish catch as resulted of temperature change 6. Decline in fish habitats due to temperature change
Adaptation action	1. Number of public awareness campaign on water efficiently 2. Number of communication tools that incorporate climate change adaptations 3. Number of vulnerable stakeholder using climate responsive tools to response to climate variability and climate change 4. Degree of adaptation of climate change into development planning 5. Percentage of treated wastewater 6. Early warning system 7. Number of metrological stations per territorial unite 8. Climate change vulnerability maps 9. Number of financial mechanisms identified to support climate change adaptation 10. Existing of inter-ministerial/inter-sectorial commissions working on adaption	
		1. Uptake of soil conservation measures 2. Percentage of Agricultural land with improved irrigation
Adaptation results	1. Number of governmental staff that have received training on adaptation 2. Number of inventories of climate change impacts on biodiversity 3. Number of farmers involved in pilot irrigation message projects 4. Percentage of poor people in drought—prone areas with access to safe and reliable water 5. Number of cubic meters of water conserved 6. Percentage of water demand being met by existing supply	1. Increase in agriculture productivity through irrigation of harvested land 2. Increase in the percentage of climate resilient crops being used 3. Percentage of farmer and fisher folk with access to financial services 4. Number of people supported to cope with the effects of climate change through the availability of service facility 5. Number of women organized in agriculture cooperatives

In addition to how to quantified the impacts and vulnerable sectors and consequently communicate and preparing a good portfolio for climate change mitigation and adaptation projects to be submitted to international agency is another problem.

In the end, the question will be whether expert recommendations are converted into political decisions, whether sectorial administrative bodies are encouraged to coordinate across sectors, and whether WCF nexus requirements will be enforced. These questions boil down to the crucial role of high level political will in mainstreaming a WCF-nexus approach.

5.2 Recommendations

Lunched an integrated initiative for water and food security (WFSI), could be helped in conduct collaborative strategies on sustainable agricultural water management that helping the countries in identifying and mainstreaming policy, governance and practices toward improve agricultural productivity and food security with saving water and applying with sustainable manner.

WFSI, can Promote the implementation of cost-effective water investment and management practices to fulfill the food security requirement under the future projections climate change conditions.

Building capacity for integrating between the different new techniques and approaches for climate science such using numerical forecasting model and regional climate model. Such model could reduce the uncertainties for future predations whereby we can draw accurate adaptive strategies in the field of water management and food security using optimums coast recovery with high efficiency.

References

1. UNSC (2015) Technical report by the Bureau of the United Nations Statistical Commission (UNSC) on the process of the development of an indicator framework for the goals and targets of the post-2015 development agenda—working draft (PDF), Mar 2015. Retrieved 1 May 2015
2. United Nations General Assembly (2015) Draft outcome document of the United Nations summit for the adoption of the post-2015 development agenda. UN. Retrieved 25 Sept 2015. http://www.un.org/sustainabledevelopment/news/communications-material
3. Anzoátegui A, Greco P, Houlden V et al (2015) The impact of climate change on the achievement of the post-2015 sustainable development goals (PDF). CDKN & HR Wallingford. Retrieved 20 May 2015
4. Paris Climate Change Conference: COP21 (2015) United Nations Development Programme. Retrieved 25 Sept 2015
5. UNDP (2018) Goal 15: life on land. Retrieved 12 Mar 2018
6. UNDP, Goal 12 (2017) Sustainable Development UN, Sustainable Development Goal 12. Sustainable Development UN, 16 Nov 2017. Retrieved 16 Nov 2017
7. Sustainable Development Strategies Egypt' Vision 2030 (2016). http://sdsegypt2030.com

8. SOFI (2018) The state of food security and nutrition in the world, building climate resilience for food security and nutrition. FAO, IFAD, UNICEF, WFP and WHO, 2018. https://www.wfp.org/climate-change/climate-impacts
9. World Food program (WFP) and UK (2018) Mett office Haddely center. Food insecurity and climate change technical report, Nov 2015. Food insecurity and climate change vulnerability map. https://www.metoffice.gov.uk/food-insecurity-index
10. EEAA (2010) Egypt state of the environment report 2009. Egypt Environmental Affairs Agency, Cairo, Egypt. Available: http://www.eeaa.gov.eg/english/info/report_search.asp#soe
11. NSACC (2011) Egypt's National Strategy for adaptation to climate change and disaster risk reduction. Egyptian Cabinet/Information and Decision Support Center/UNDP, Dec 2011
12. MWRI (2017) National Water Resources plan in Egypt. Ministry of Water Resources and Irrigation, Cairo, Egypt. www.mwri.gov.eg
13. Practice Assessment and Recommendations Report (2017) REC-WEE pros/Port EG. Water Sum Project
14. SNC (2010) Egypt's Second National Communication. Egyptian Environmental Affairs Agency (EEAA-May 2010), under the United Nations Framework Convention on Climate Change on Climate Change
15. ACCWaM (2017) Adaptation to climate change in the Water Sector in the MENA Region. Deutsche Gesellschaft für Internationale Zusammenarbeit (GIZ) GmbH "Options for a Nexus Coordination Mechanism for the League of Arab States Mainstreaming the Water-Energy-Food Security Nexus into Sectoral Policies and Institutions in the Arab Region, Sept 2017
16. ACCWaM (2013) Adaptation to climate change in the Water Sector in the MENA Region. Deutsche Gesellschaft für Internationale Zusammenarbeit (GIZ) GmbH, Mainstream Climate Change Adaptation in National policy str Adin the Water Sector in the MENA Region (Strategy and Action Plan Summery, 2013
17. Lord RQ, Goldberg S, Rajamani L, Brunnee J (2012) In: Lord R (ed) Climate change liability. Cambridge University Pres
18. Blanken J (2012) Mainstreaming of MDGF projects (FAO component) and provision of tools for strategic decisions for the adaptation to Climate Change in the Egyptian Agricultural Sector. Final mission report—UNJP/EGY/022/SPA, Climate Change Risk Management Program
19. El-Din MN (2013) Proposed climate change adaptation strategy for the ministry of water resources & irrigation in Egypt. Joint programme for climate change risk management in Egypt
20. International Institute for Applied Systems Analysis (IIASA). www.iiasa.ac.at

Climate Impact on Egyptian Agriculture: An Efficiency Analysis Approach

Fadi Abdelradi and Dalia Yassin

Abstract The Intergovernmental Panel on Climate Change (IPCC) 2018 special report on the implications of 1.5° warming (SR 1.5) concludes that the expected impacts of the 2° increase are more critical than expected leading to imminent critical tipping points. Additionally, the report indicates that there is a window of only 12 years to take actions to achieve the 1.5° target. While Africa is contributing the least to global warming by 4% of global greenhouse-gas emissions, it is one of the vulnerable regions to Climate Change (CC) effects. The objective of this study is to assess different climatic variables on the technical efficiency of the cereal crops production in the largest five producing governorates in Egypt. This objective will be achieved through using panel frontier model to assess the impact of average maximum temperature, humidity and solar radiation on the wheat, rice and maize producing governorates. Results of the analysis show that the average maximum temperature and humidity contribute to worsening the technical efficiency of the cereals producers.

Keywords Climate change · Frontier · Efficiency analysis · Egypt · Panel data

1 Introduction

Both agriculture and CC affect each other. CC affects agriculture in different ways through changing the intensity, the frequency, and occurrence of events (rainfall, heat waves, changes in pests and diseases, changes in carbon dioxide, droughts) that are not common in the location's history. For example, the increased temperature has contributed to changes to growing seasons affecting the flowering and harvesting of crops. Agriculture produces Greenhouse Gases emissions (GHGs) due to the

F. Abdelradi (✉)
Department of Agricultural Economics, Faculty of Agriculture, Cairo University, 3 Gamaa Street, Giza 12613, Egypt
e-mail: fadi.abdelradi@agr.cu.edu.eg

D. Yassin
Agricultural Research Centre, Agricultural Economics Research Institute, Giza, Egypt
e-mail: dalia.yassin@hotmail.com

© Springer Nature Switzerland AG 2020

E.-S. E. Omran and A. M. Negm (eds.), *Climate Change Impacts on Agriculture and Food Security in Egypt*, Springer Water, https://doi.org/10.1007/978-3-030-41629-4_24

practices of conventional food production systems [1]. Additionally, according to the World Bank report [2]. CC and poverty reduction are major issues to the international society; furthermore, the challenges of food and nutrition security in developing countries are increasing [3]. As a result, the need for innovative and resilient agricultural production practices to cope with these factors is essential. Climate Smart Agriculture (CSA) is an approach that deals with restructuring agricultural systems to cope with CC through three main objectives, which are sustainability increasing productivity and incomes, mitigating GHG emissions, adaptation and building resilience.

Egypt faces many food and nutrition security challenges despite the rich agricultural resources, for example, the high levels of food losses experienced in all value chain with an average of 35% based on the FAOSTAT food balance sheet [4]. These, in part, due to the low capability for efficient use of these resources, inefficient policies, poor labor skills, poor market dynamics, inefficient market information systems, increasing intensity of climate change, poverty, dependence on rainfall for irrigation, poor infrastructure [5, 6]. Historically, Egyptian agro-ecological systems recognized the importance of trees within and outside the farm for climate and livelihood improvement and as such endeavored to preserve a specific type of species through cultivation to harvest. However, present stagnation and/or reduction in the number of trees on and around farmland [7], as well as water scarcity in Egypt [8], contribute to low agricultural production and increasing environmental degradation which is detrimental to agrarian livelihoods (food, feed, energy, and material demand). Furthermore, faced with growing local demand and competition for resources, Egypt's food production needs to analyze in a wider context, affecting 50% of the population lives in rural areas, through finding synergies among agriculture, energy, water and ecosystem.

According to the [3] the highest GHG emissions contributor is Eastern and South-Eastern Asia with 1.2 billion tons of Carbon Dioxide Equivalent (CO_2e), Latin America and the Caribbean 0.909 billion tons of CO_2e, Sub-Saharan Africa 0.768 billion tons of CO_2e, and the lowest Northern Africa and Western Asia 0.156 million CO_2e. The highest contributors in Sub-Saharan Africa are Ethiopia (96 million tons of CO_2e) Sudan (72.5 million tons of CO_2e), Nigeria (64.2 million tons of CO_2e), South Sudan (43 million tons of CO_2e) and Kenya (37.1 million tons of CO_2e). The main sources of Sub-Saharan Africa's CO_2e are burning Savana (lowest in Kenya), enteric fermentation, manure left on pastors and rice cultivation (highest in Nigeria). The highest contributors in Northern Africa and Western Asia countries are Turkey with around 43 million tons of CO_2e and Egypt with 31 million tons of CO_2e. The sources of Egypt's CO_2e are enteric fermentation with 10.1 million tons of CO_2e, production of synthetic fertilizers with 8.4 million tons of CO_2e, manures left on pastors 6.6 million tons of CO_2e and rice cultivation contributing with 3.7 million tons of CO_2e.

Egypt is considered one of the countries that are vulnerable to CC impact because it is located in a region that falls between semi-arid and arid conditions, where 94.5% of Egypt total area is desert. The 5.5% of the area is populated with 95 million inhabitants that are located along the River Nile and the coastal area. In Egypt, current annual average temperatures increase from about 20 °C on the Mediterranean coastline to 24 °C. Normal maximum daytime temperature range from 30 °C in Alexandria on the Mediterranean coastline to 41 °C at Aswan near the borders with Sudan. However, there is a general warming trend since 1960 with warmer summers. Additionally, during the period 1960–2003, a general increase in the number of warning nights and a decrease in the frequency of cold nights. The projected increase towards 2030 there will be an increase by 1.0°–1.27 °C using the forecast models RCP4.5 and RCP8.5 median ensemble, while the projected 2050 increase range from 1.64 to 2.33 °C using the forecast models RCP4.5 and RCP8.5 median ensemble [9].

The objective of this study is to assess the impact of climatic variables on the technical efficiency of the cereal crops production for the most significant five producing governorates in Egypt for the period of 2000–2009. This objective will be achieved through using panel frontier model to assess the impact of average maximum temperature, humidity and solar radiation on the technical efficiency of the highest cereals producing governorates (Behera, Gharbia, Kafr-Elsheikh, Dakahlia, and Sharkia). The contribution of this work is twofold: first, it focuses on the Egyptian agricultural, in contrast to the predominant literature, developing countries have not received much attention. Second, we compare efficiency scores among different governorates to propose suitable policies that match their needs.

2 The State of Climate Change in Egypt

Agricultural production is hugely reliant on climate and profoundly impacted by climate variability. For this reason, the FAO Global Early Warning System FAO-GIEWS is dedicated to monitor food demand and supply globally together with providing indicators measuring climate-induced impacts through the Earth Observation for Crop Monitoring [10]. An example of these indicators are the annual Agricultural Stress Index (ASI) that represent the arable land percentage that was affected by drought conditions over an entire cropping season was explicitly designed to meet the need for information like the drought that will be provided only timely basis about affected agricultural zones. Drought Intensity (DI) represents the higher the drought severity. The lower the vegetation health over the entire cropping season. The Vegetation Health Index (VHI) is a composite index that combines both the Vegetation Condition Index (VCI) and the Temperature Condition Index (TCI). The TCI relates the current temperature to the long-term maximum and minimum, based on the assumption that higher temperatures cause negative impacts on the vegetation conditions. The VHI relates the vegetation conditions with temperatures, for example, a decrease in VHI means poorer vegetation condition and higher temperatures signifying stressed vegetation conditions, and an indication of drought over a more extended period.

Graphical representations (Figs. 1 and 2) of ASI and VHI indicators developed by FAO based on two growing seasons per year for analyzed Egyptian governorates during the period 1984–2017. Egypt can be categorized into three areas that each has its unique climatic characteristics. Upper Egypt that and contains five governorates, Middle Egypt includes four governorates, Lower Egypt that represent thirteen governorates, and five governorates are located outside the Nile valley that has a different type of soils compared inside the Nile valley. We have selected five governorates that are located in Lower Egypt, those governorates were selected based on the level of production for strategic agricultural crops as staple food and export cash crops. The graphical analysis shows that the mean VHI exhibits a positive trend over time

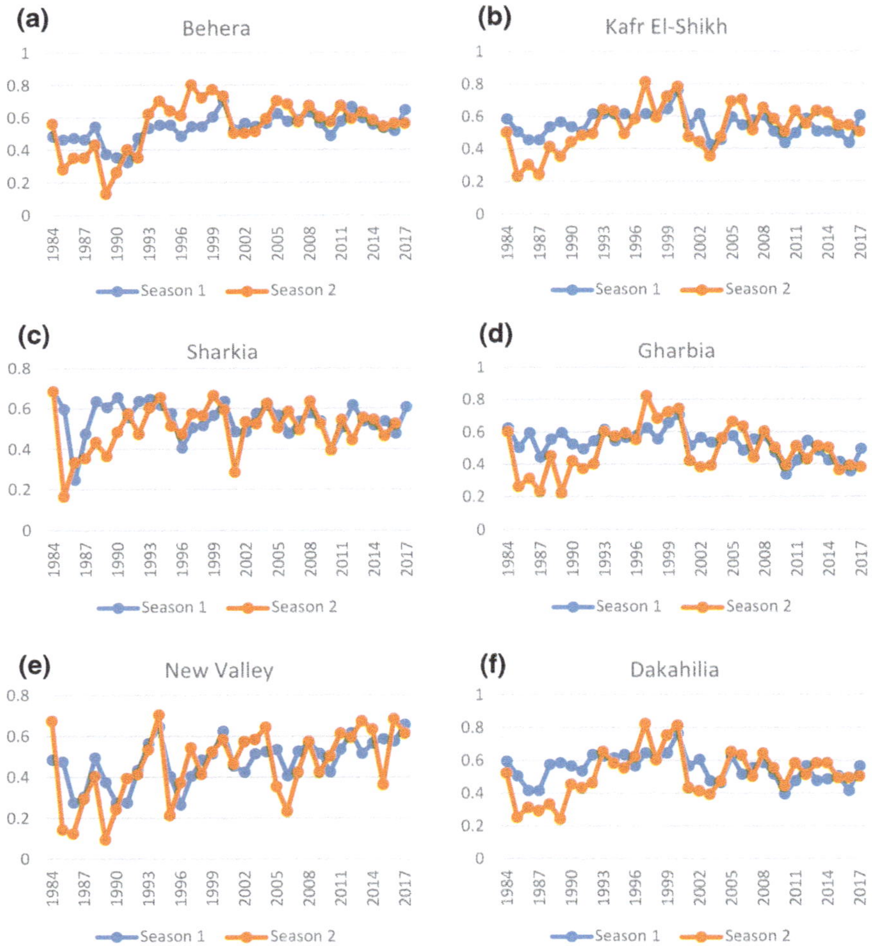

Fig. 1 Graphical representation for the mean VHI indicator for seasons 1, 2

Fig. 2 Graphical representation for the ASI indicator for seasons 1, 2

for the analyzed governorates and the ASI indicator exhibits high fluctuations for season 2 compared with season 1 considering the higher the index value the more likelihood of facing drought. The increase in the ASI was more intense for Behera, Gharbia, Sharkia, which represent the old lands and New Valley that represent the new lands.

3 Initiatives and Policies

The actual movement towards dealing with CC issues began in 1972 in Stockholm and advocate government, and then the first international forum on CC that was held in 1988. Since CC is considered as an international phenomenon and an issue that has cross-border impacts. The international community formed workgroups to investigate the type and magnitude these impact, these workgroups took its final legal shape that is called the Intergovernmental Panel on CC (IPCC). The IPCC was established by World Meteorological Organization (WMO) in 1988 that has published the first report on CC in 1990. Based on this report the UNFCCC agreement, which Egypt was one of the members was signed in 1992 committing industrial countries to keep the levels of CO_2 emissions to allow the ecosystems to recover without human intervention. In 1997, a conference in Japan was held on CC to sign the Kyoto protocol that commits industrial countries to reduce their CO_2 emissions by 5.2% of the 1990 levels. The IPCC has released a special report on 2018 on the impact of global warming of 1.5 °C and GHG emission pathways. Additionally, the report has shown that most climate models showed robust results of regional climate characteristics differences [6]. A number of initiatives taken to deal with CC through adaptation in Sub-Saharan Africa for the period 2000–2015 [11]. The report concludes that there are three pillars to reach sustainable adaptation benefits: first, adopt collaborative management practices involve affected parties in the center of the innovation process together with involvement in the decision-making over the adaptation process. Second, develop sustainable programs that identify and deal with barriers that can be replicated for adaptation interventions. Third, capacity building for evidence-based policy analysis and evaluation.

In Africa, a 4 years program developed by the UK Strategy for the Global Challenges Research Fund GCRF-AFRICAP that promotes for a more productive, resilient to CC and sustainable agriculture production in Sub-Saharan Africa (Malawi South Africa, Tanzania, Zambia) through a top-down approach from national governments and bottom-up approach with local organizations in the selected countries. The objective of the GCRF-AFRICAP program is to support climate-smart solutions and sustainable development for agriculture that seeks reduced poverty and improves livelihood considering the Sustainable Development Goals and the Malabo Declaration in Africa. These initiative and policies impact will be evaluated in different locations for an increased application that requires a capacity building that depends on evidence-based policy development. In this line, GCRF-AFRICAP is carrying out the necessary research in selected countries that cover a wide range of areas of crop and animal sciences to soil, ecosystems and social and political sciences. The GCRF-AFRICAP program assesses existing farmers' practices and their impact on yields, incomes and compare it with predicted CC induced events. The GCRF-AFRICAP program aim is to put the results obtained from research into empirical work. This empirical output will be translated into policy into action.

The Regional Center for Renewable Energy and Energy Efficiency (RCREEE) is an organization that coordinate efforts among governments that advocate for improving energy efficiency practices and use of renewables in the Arab region through collaboration among policymakers, industry and research for key areas like policies, regulation, technical assistance. The RCREEE is a non-profit and is located in Egypt. Additionally, a set of projects is launched in Egypt to tackle CC is running, for example, "Enhancing Climate Change Adaptation in the North Coast of Egypt" (ECCANCE) project managed by the Ministry of Water Resources and Irrigation with a budget of US$31.4 million over seven years. The project works on helping the highly populated areas in the Nile Delta that represent 25% of the Egyptian population, which have been considered as extremely at risk to CC prompted sea-level rise [12].

Egypt has developed the National Strategy for Adaptation to CC and Disaster Risk Reduction in 2011. The strategy is a framework of adaptation policy, which aims at increasing the flexibility of the Egyptian society in relation to dealing with CC-induced risks and its impact on different activities and sectors and reinforcing the capacity to adapt and reduce the risks caused by such changes [13]. An Egyptian Climate Change Risk Management program is developed to incorporating GHG emissions mitigation into the national policy and investment frameworks, including Clean Development Mechanism financing opportunities, Enhancing the national capacity to CC adaptation through synergies among water, energy and food sectors, and finally advocacy and awareness-raising [14]. Furthermore, work on reducing CO_2 emissions, the Egyptian government is advocating a higher energy efficiency in the industrial sectors and improved practices in the agricultural sector. For example, the use of drought-resistant crops and improved irrigation systems for water saving, and encouraging citizens to adopt alternative technologies of energy consumption like using generating electricity from solar panels, increasing awareness of using water resources [15]. In relation to the adoption of post-2015 Sustainable Development Goals (SDGs), Egypt has established the "Sustainable Development Strategy; Egypt's Vision 2030" that represent a roadmap for the country to achieve its desired SDGs during the next 15 years. This strategy is having three dimensions focusing on economic, environmental and social sides. The three dimensions consist of a set of pillars that target for improvement, for example in the case of economic dimension it focuses on the improvement of Egypt's economic development, energy efficiency, and knowledge and innovation. While the environmental dimension is focused on improving ecosystems and improving environmental efficiency and optimal use of available resources, moreover, an action plan to be implemented to promote resilience in different sectors [15].

At the industry level, Agribusiness corporates are taking actions towards CC induced events adaptation and mitigation through their Corporate Social Responsibility (CSR) programs is SEKEM that is focused on organic agricultural production. SEKEM applies biodynamic agricultural methods, a specialized method of organic agriculture which has lower GHG emissions compared to conventional practices. With regard to adaptation, biodynamic agriculture has a high impact on strengthening CC adaptation through plant diversification, increased water efficiency, using resilient crop varieties towards drought and heat. SEKEM relies on Key Performance Indicators to evaluate their actions in adaptation and mitigation towards CC through using Sustainability Balanced Scorecard (SBSC) system which follows up on the practices carried out to face the challenges identified and to keep track of sustainable development. Furthermore, Sustainability Flower (SF) technique is adopted, SF is a tool for information and communication management, that represent the concept of sustainable development in its four dimensions, which are ecological, social, cultural, and economic components. Each component contains many performance indicators with specific targets that meet the Global Reporting Initiative standard.

4 Literature Review

CC impacts in Africa has become a reality [16]. The warming of Africa is very likely to be severe than in other regions [17, 18]. Studies assessing CC impacts found that Africa is the most vulnerable in the world due to poor adaptation practices and high vulnerability [19–21]. The low adaptive capacity at the financial, technical and institutional level caused that African countries are the most affected [22–25]. Taking into account that agriculture is the most significant economic sector in Africa with the largest employment share of 60% [26] besides land degradation and desertification [27]. The more prolonged and severe droughts in Eastern Africa; increased floods in Western Africa; deforestation in equatorial Africa; and increased acidity across Africa's southern coast together with climate extreme events will have severe impacts on food security, health, water and energy security, which at the end will weaken African's potential to grow and develop [16]. Furthermore, the population in Africa exhibits a growth trend which puts more pressure of natural resources, the increase was estimated for the 1980s to be 478 million and has experienced an increase to nearly 1.2 billion and is expected to increase to 1.5 billion by 2025 and rise to 2.4 billion by 2050 [28]. Besada and Sewankambo [16] investigated CC impacts in Ethiopia, Kenya, and Somalia and they have shown that these countries were influenced by persistent droughts and losses of livestock where the majority of people are dependent on livestock for their livelihood. For instance, Collomb [29] predicted that by 2050, food production needs to be increased by quintuple in Africa, double in Asia, and almost double in Latin America.

In the northern part of the Mediterranean, a decrease in rainfall during the summer that is partially recovered during the winter by precipitation increase. In the central Mediterranean, periods of droughts are increased for one month, beginning one week earlier and finishing 3 weeks after. These changes have impacts on all economic activities such as agriculture, energy, tourism and environmentally through affecting the natural ecosystems. The results of the analysis show that in the case of agriculture, cultivated autumn and winter seasons are not affected or an increase in production. While in the case of summer crops a significant fall of yield was indicated [30].

Evidence of CC impact is found in other areas of the world. For example, in India, Aggarwal [31] has shown a potential decrease from 3 to 7% decrease in the productivity of field crops like wheat, soybean, groundnut, and potato because of 1 °C increase in the temperature. Kumar and Parikh [32] indicated that CC would affect significantly rice and wheat production and wheat by 2060, which are a staple food necessary for the livelihood and food security of the Indian population. Geethalakshmi et al. [33] have provided evidence of a 41% decline in the production of production due to the increase in the temperature by 4 °C in Tamil Nadu.

Berg et al. [34] used agro-DGVM model (Dynamical Global Vegetation Model) driven by projections from several climate models and two (Special Report on Emissions) SRES scenarios to assess CC influences on possible C4 crop productivity in Africa and India for the period 1960–2100. The authors distinguish the impact of rising CO_2 emissions. They adopted transient simulations by the climate model outputs: to ensure the analysis consistency regardless of regional biases, investigate yield change using bioclimatic basis rather than geographical basis. The results of the analysis show a decreased average productivity of millet crop, which is considered an important staple crops in those regions, by −6% (projections range from −29 to +11%). The overall effect is yield decrease over the Equatorial and Temperate Köppen zones; significant inconsistencies change of yields that arise in Arid Köppen zones because of uncertainty in precipitation projections compared with climate models. Finally, the authors have concluded that climatic impacts put more pressure to achieve food security in developing countries in the Tropics, and these impacts can be alleviated through adopting adaptation practices and best farming practices. Deryng et al. [35] used a global crop model, PEGASUS 1.0 (Predicting Ecosystem Goods And Services Using Scenarios) that incorporates the different indicators of planting dates, cultivar choices, type of irrigation and fertilization of the crop yield for maize, soybean, and spring wheat, in addition to climate indicators. The results of the model are consistent with global data. Additionally, the used planting and harvesting dates are found to be within crop calendar observations of almost 75% of the total crop-harvested areas. The authors found that variations in the temperature and rainfall lead to yield reduction, which as predicted by global climate models for the 2050s considering planting and harvesting dates remain unchanged. While adopting adaptation strategies for planting dates and choosing suitable cultivars can help increase yield in temperate regions and evades 7–18% of global losses.

CC influences different aspects of the economy from infrastructure, health, ecosystem, food and water security, to displacement. Logistical infrastructure like roads and bridges and services infrastructures like power and water stations are impacted due to flood, landslides and heavy rainfalls. The previous literature has shown evidence on the consequences of extreme climate change-induced events that influence road networks in Africa in for food security purposes [36]. In different African countries like those that Ethiopia, Uganda, Ghana, and Tanzania have gone through drought events which affected their dams and caused a reduction in the hydropower generation [37]. For the case of human health, CC increases the occurrence and gravity of disease outbreaks. The literature has shown that hard climate conditions like increased temperature and high rainfall levels are enabling causes for malaria epidemics in Kenya, Uganda, Ethiopia, Tanzania [38].

CC impact ecosystems and as a result cause more reduced ecosystem services like lower production of oxygen, lower biodiversity, loss of soil and trees. There is a growing empirical work on the impact of CC on forests ecosystems in Africa. As well as the livelihoods of the forest-dependent communities' activities that rely on vegetation services [39]. On the other hand, food security is impacted by CC in the form of reduced food production due to changes in rainfall and temperatures, which has led to increased demand for groundwater and decreased groundwater recharge. For example, Funk et al. [40] have shown that for the period 1996–2003, a reduction in rainfall of 50–150 mm that has caused a reduction in maize and sorghum production across most of the eastern African countries. Correspondingly, for energy consumption, it is expected to decrease because of warmer winters periods. However, there is a substantial increase in the temperature during the summer together with better climate circumstances in northern Europe, may cause a decrease in tourism during the summer and an increase in spring and autumn.

5 Methodology

There are two main approaches extensively used in the literature to measure technical efficiency: parametric SFA and non-parametric DEA. To take care of the identification problem caused by non-parametric models, SFA is used as an alternative. The SFA approach, was commenced simultaneously by Aigner et al. [41]; Meeusen et al. [42], makes a distinction between exogenous shocks outside the control and inefficiency of the firm. In contrast to DEA, SFA takes into account random noise and allows for conventional tests of hypotheses. Additionally, SFA requires the specification of a functional form for the production function and a distributional assumption for the inefficiency term. Agricultural production is stochastically determined since it is affected by climatic changes, and the agricultural production literature is likely affected by measurement and variable omission errors [43–45], choosing a robust model is vital to consider these problems.

The literature extended the efficiency analysis from cross-section to panel estimation, which has helped to relax some of the assumptions and consider closer realistic characterization technical efficiencies. Pitt and Lee [46] were the first to extend the cross-section specification to panel data. They have developed ML estimation approach using Normal-Half Normal stochastic frontier model. Battese and Coelli [47] have extended Pitt and Lee [46] model to a Normal-Truncated Normal specification. Schmidt and Sickles [48] have shown that the estimation of a time-invariant inefficiency frontier model could be carried out through conventional fixed-effects estimation approaches. However, invariant inefficiency has been questioned especially with empirical applications [49]. Cornwell et al. [50] have introduced a solution to this issue by proposing a quadratic specification model with individual-specific slope parameters. This quadratic specification permits a unit specific pattern over time of inefficiency. However, it needs a large number of parameters to be estimated. Lee and Schmidt [51] proposed a different specification that is more parsimonious than the one proposed by Cornwell et al. [50]. Additionally, it does not enforce any parametric specification, yet it restricts the temporal pattern of inefficiency to be the same for all productive units. Ahn et al. [52, 53] have proposed a different estimation methodology using Generalized Methods of Moments (GMM) to the [50, 51] developed models, respectively. The authors have shown that GMM is performing better because it is asymptotically efficient. Kumbhakar [54] proposed a Maximum Likelihood (ML) estimation of a time-varying stochastic frontier model. A similar model proposed by Battese and Coelli [55] termed as a time decay model.

The common characteristic among all the frontier models is that the intercept has not been changed over productive units, which leads to misspecification in the existence time-invariant unobservable factors, which cause that these factors may be captured by the inefficiency term, resulting in producing biased estimates. Greene [56] has introduced a model to solve this issue using a time-varying stochastic frontier model with unit-specific intercepts. In contrast with the previous models helps to distinguish time-varying inefficiency from unit specific time-invariant unit heterogeneity. For this reason, Greene named these models as True Fixed Effects (TFE) or True random effects (TRE). While the estimation of the TRE specification is computationally simple using simulation-based techniques, the ML estimation of the TFE model needs to solve two main challenges related to nonlinear panel data models estimation. The first is related to a computational challenge because of the large size of the parameters space. However, Greene [56, 57] has provided evidence that a Maximum Likelihood Dummy Variable (MLDV) methodology is feasible in the case of a large number of nuisance parameters α_i ($N > 1000$). The second challenge is the incidental parameters problem, which is an inferential problem that is a result of a larger number of units, compared to the length of the panel. We adopt a True Fixed Effects (TFE) stochastic production function:

$$y_{it} = \alpha_i + \boldsymbol{X}'_{it}\beta + \varepsilon_{it} \tag{1}$$

$$\varepsilon_{it} = v_{it} - u_{it} \tag{2}$$

where y_{it} denotes the level of output of governorate i at time t, $\boldsymbol{X}_{it}$ is a vector of input costs of the ith governorate at time t. β is a vector of parameters to be estimated. Component α_i in Eq. (1) is the random governorate specific effect, this specification allows to disentangle time-varying inefficiency from unit specific time invariant unobserved heterogeneity. The error term ε_{it} in Eq. (2) can be decomposed into two components. The first component, v_{it} is a normally distributed disturbance capturing the random variation in output due to statistical noise that occurs due to the unintended omission of relevant variables from vector $\boldsymbol{X}_{it}$; errors (measurement and approximation) associated with the choice of the functional form; unexpected stochastic the control of the industry. Component v_{it} is usually assumed to be symmetric, independent and identically distributed as $N(0, \sigma_v^2)$. The second component u_{it} is a one-sided, non-negative random variable, representing the stochastic shortfall of the i-th governorate output from its production frontier, assumed to be independently distributed following a

Table 1 Frontier models specification

Wheat	$y_{it} = \alpha_i + \beta_1 AREA_{it} + \beta_2 fertilizers_{it} + \beta_3 Pesticides_{it} + \beta_4 labor_{it} + \beta_5 y_{it-1}$ $u_{it} = \delta_1 MaxTemp_{it} + \delta_2 Humidity_{it} + \delta_3 SolarRad_{it} + \xi_{it}$
Rice	$y_{it} = \alpha_i + \beta_1 AREA_{it} + \beta_2 fertilizers_{it} + \beta_3 Pesticides_{it} + \beta_4 labor_{it} + \beta_5 y_{it-1}$ $u_{it} = \delta_1 MaxTemp_{it} + \delta_2 Humidity_{it} + \delta_3 SolarRad_{it} + \xi_{it}$
Maize	$y_{it} = \alpha_i + \beta_1 AREA_{it} + \beta_2 fertilizers_{it} + \beta_3 Pesticides_{it} + \beta_4 labor_{it} + \beta_5 y_{it-1}$ $u_{it} = \delta_1 MaxTemp_{it} + \delta_2 Humidity_{it} + \delta_3 SolarRad_{it} + \xi_{it}$

truncated normal with mean ($Z_{it}\delta$) and variance σ^2, Z_{it} is a vector of exogenous variables associated with the inefficiency of production of governorates over time distribution (3). Compared with previous literature on panel stochastic frontiers, this specification can differentiate unit specific time invariant unobserved heterogeneity and time varying inefficiency. To this end, TFE model is specified and is estimated using a Simulated Maximum Likelihood (SML). The technical inefficiency effects model:

$$u_{it} = \delta Z_{it} + \xi_{it} \tag{3}$$

where u_{it} is the technical inefficiency for governorate i at time t, obtained from the stochastic frontier model and ξ_{it} is the error term. Selection of Z_{it} variables is based on previous literature and data availability. The method of estimation is the simulated maximum likelihood for simultaneous estimation of the stochastic frontier model and the technical inefficiency model following [56]. We have specified a dynamic panel frontier models for cereals crops which are (wheat, maize, rice) for the ith governorates (Behera, Gharbia, Kafr-Elsheikh, Dakahlia, and Sharkia) for specified T time periods based on data availability. The frontier and efficiency models specifications for the selected cereal crops are presented in Table 1. Logarithmic transformations of variables are used and we followed a one-step estimation of the likelihood function.

6 Results

The present empirical application aims at studying the impact of climate conditions on the technical efficiencies of producing cereal crops in Egypt using SFA approach applied on a panel dataset obtained from Ministry of Agriculture and the Central Laboratory for Agricultural Climate. Our sample consists of five governorates for the period 2000–2009 yielding 50 observations. The analysis was carried out using the econometric software Stata 11.2. The data are transferred to logs to have a more straightforward interpretation of the estimated parameters as elasticities. The SFA model estimates are obtained with a Cobb-Douglass functional form as presented in Table 1. The dependent variable (y_{it}) is the production quantity in tons per year of the identified crop. The crops used in the analysis are necessary for food security in Egypt since they are considered a staple food. The inputs includes: (x_{1it}) area of the selected crops in the targeted governorates measured in acres; (x_{2it}) is the cost of fertilizers measured as Egyptian pounds per acre; (x_{3it}) is the cost of pesticides measured as Egyptian pounds per acre; (x_{4it}) represent the cost of labor hours per acre; (y_{it-1}) is the lagged production quantity in terms of tons per lagged year. Summary statistics for the variables by groups used in the analysis are presented in Table 2.

The summary statistics show the relative similarity among the four groups in terms of output, land, livestock and labor, and heterogeneity in Machinery and fertilizer. For the efficiency equations, The dependent variable (u_{it}) is the technical inefficiency obtained from stochastic frontier model; $(MaxTemp_{it})$ is average maximum temperature for the governorates under analysis measured as Celsius degrees; $(Humidity_{it})$ is the water vapor present in the air and this indicator could have a reason for potential plant infections; $(SolarRad_{it})$ is the power received per meter squared from the sun measured as Mega Joules per square meter (MJ/m^2).

Results of the estimated stochastic frontier models are presented in Table 3. Results of the production elasticity estimates indicate that, for all governorates, the cultivated area has the largest positive effect on the quantity produced. In the case of the rice model, cultivated land, labor and fertilizers are found to have statistically significant influence; lagged rice production has a negative impact on current production because of the government policy to minimize water resources consumption. In the case of the Maize model, cultivated area and fertilizers are found to be relevant. For the wheat model, the cultivated area and lagged production are the relevant factors affecting production. Other factors in the three models were not statistically significant because these input factors are subsidized since the government provides cereal farmers, especially wheat producers, with subsidized inputs quantities. These results are compatible with [58] that identified these factors of production are the most productive in Egypt's agriculture sector. For the technical inefficiency equations for the three models, the climatic indicators were found to be relevant in causing a problem of increasing the technical inefficiency and as a result preventing achieving the maximum potential of producing these crops.

According to Table 4, TE scores for the three cereal crops in the selected governorates are calculated as an output-oriented measure following [59]. The average technical efficiency score is for wheat and maize is above 90% in Kafr-Elsheikh, Beher, and Gharbia, while Dakahlia is more technically efficient in Maize production only, and this is expected since these governorates are the biggest wheat producers. The average technical efficiency score is for rice range from 80 to 85% along with the four governorates yet Dakahlia governorate minimum technical efficiency is 80%, which is higher than the minimum levels in other governorates. These results indicate that if these governorates effectively use available resources and at the current technology, it will be able to increase the output by 10% on average. Improving technical efficiency allows reducing production costs and increases competitiveness, which can help farmers face changing conditions and unfortunate economic circumstances. Climatic variables are found to be relevant in affecting the technical efficiency and this call for adopting new agricultural practices that are adaptive to these climate induced-events. As in the case of poor climatic circumstances, farm margins may be

Table 2 Summary statistics for variables in the analysis

Variable		Production	Cultivated area	Fertilizers	Pesticides	Labor	Max. Temp.	Humidity	Solar radiation
Behera governorate									
Maize	Mean	480,035	125,391	293	43	375	27.7	52	20
	Std. Dev.	70,657	17,713	167	29	113	1.4	4.1	0.5
	Normality	−0.948 (0.828)	0.362 (0.358)	0.808 (0.209)	−0.558 (0.711)	0.533 (0.297)	1.113 (0.132)	−1.461 (0.928)	−4.639 (1.000)
Wheat	Mean	317,502	6,077,439	753	102	1315	29.6	48.2	20.5
	Std. Dev.	37,863	696,170	1142.7	36.6	2001	0.8	1.6	0.8
	Normality	−0.945 (0.827)	0.160 (0.436)	4.343 (0.000)	0.002 (0.499)	4.126 (0.000)	1.960 (0.025)	−0.212 (0.583)	1.425 (0.077)
Rice	Mean	908,280	219,068	157	44	364	28.5	50	20
	Std. Dev.	82,455	19,417	86	14	141	1.08	3.4	0.57
	Normality	−0.759 (0.775)	0.836 (0.201)	2.318 (0.010)	−2.197 (0.985)	0.966 (0.167)	−3.910 (0.999)	−0.203 (0.580)	1.876 (0.030)
Gharbia governorate									
Maize	Mean	261,494	73,754	225	67	495	28.4	49.8	19.6
	Std. Dev.	53,071	15,286	154	14	81	1.8	4.7	0.84
	Normality	0.164 (0.434)	−0.369 (0.644)	3.221 (0.000)	1.364 (0.086)	3.433 (0.000)	0.373 (0.354)	−0.415 (0.661)	0.910 (0.181)
Wheat	Mean	150,796	2,891,901	786	112	1406	30.3	45.9	20.3
	Std. Dev.	9686	267,846	1371	30.	2113	1.5	2.4	0.78
	Normality	1.467 (0.071)	– 0.746 (0.772)	4.496 (0.000)	−0.115 (0.545)	4.168 (0.000)	−0.334 (0.630)	−1.278 (0.899)	1.313 (0.094)

(continued)

Table 2 (continued)

Variable		Production	Cultivated area	Fertilizers	Pesticides	Labor	Max. Temp.	Humidity	Solar radiation
Rice	Mean	653,953	161,027	168	80	430	28.4	49.8	19.6
	Std. Dev.	103,239	22,140	131	18	95	1.8	4.7	0.84
	Normality	0.431 (0.333)	0.242 (0.404)	1.689 (0.045)	−1.630 (0.948)	2.444 (0.007)	0.373 (0.354)	−0.415 (0.661)	0.910 (0.181)
Kafr-Elsheikh governorate									
Maize	Mean	182,373	50,960	309	93.7	537	25.9	61.4	20.08
	Std. Dev.	27,708	6825	116	40	201	3.08	6.34	0.66
	Normality	−0.781 (0.782)	−0.111 (0.544)	0.850 (0.197)	−1.051 (0.853)	1.633 (0.051)	2.338 (0.009)	1.946 (0.025)	−4.162 (0.999)
Wheat	Mean	233,887	4,205,080	699	369	1312	27.7	57.5	19.78
	Std. Dev.	14,640	216,522	1457	844	1770	2.3	4.1	0.89
	Normality	0.524 (0.300)	−0.038 (0.515)	4.085 (0.000)	4.560 (0.000)	4.483 (0.000)	0.537 (0.295)	0.321 (0.374)	0.399 (0.344)
Rice	Mean	1,177,232	285,148	181	127	496	25.1	62.8	20
	Std. Dev.	126,709	33,039	120	28	132	2.6	5.99	0.67
	Normality	0.979 (0.163)	1.685 (0.046)	2.237 (0.012)	1.185 (0.118)	1.052 (0.146)	2.502 (0.006)	2.608 (0.004)	−4.265 (0.999)
Dakahlia governorate									
Maize	Mean	210,351	54,210	387	115	553	24.7	63.9	20.5
	Std. Dev.	50,031	12,879	156	32.6	215	1.33	3.5	0.52
	Normality	0.610 (0.270)	0.370 (0.355)	1.555 (0.059)	1.188 (0.117)	0.587 (0.278)	1.171 (0.120)	1.494 (0.067)	−4.262 (0.999)

(continued)

Table 2 (continued)

Variable		Production	Cultivated area	Fertilizers	Pesticides	Labor	Max. Temp.	Humidity	Solar radiation
Wheat	Mean	295,195	5,600,073	417	141	556	26.2	61.5	20
	Std. Dev.	17,756	337,746	125.8	40.6	268	1.2	1.9	0.91
	Normality	0.290 (0.385)	−0.825 (0.795)	3.086 (0.001)	2.504 (0.006)	1.403 (0.080)	−1.505 (0.933)	1.401 (0.080)	1.536 (0.062)
Rice	Mean	1,691,729	403,732	234	106	365	24.7	63.9	20.5
	Std. Dev.	539,775	124,181	126	36	105	1.33	3.54	0.52
	Normality	3.271 (0.000)	3.858 (0.000)	2.963 (0.001)	−0.669 (0.748)	2.627 (0.004)	1.171 (0.120)	1.494 (0.067)	−4.262 (0.999)
Sharkia governorate									
Maize	Mean	210,351	54,210	311	89	614	28	51.3	20.1
	Std. Dev.	50,031	12,879	172	38	171	1.6	4.5	0.57
	Normality	0.610 (0.270)	0.370 (0.355)	1.380 (0.083)	−0.515 (0.696)	−0.648 (0.741)	1.558 (0.059)	−0.703 (0.758)	1.876 (0.030)
Wheat	Mean	395,447	7,054,729	464	148	646	29.3	48.3	20.7
	Std. Dev.	32,326	759,039	166	34.5	251.4	1.5	2.2	0.82
	Normality	1.767 (0.038)	−1.271 (0.898)	0.794 (0.213)	−1.986 (0.976)	0.781 (0.217)	0.899 (0.184)	1.412 (0.0780	2.180 (0.014)
Rice	Mean	1,132,717	287,202	182	93	609	27.7	52	20
	Std. Dev.	112,544	30,097	74	30	173	1.4	4.1	0.5
	Normality	−0.310 (0.621)	−1.152 (0.875)	2.083 (0.018)	−0.373 (0.645)	−0.738 (0.769)	1.113 (0.132)	−1.461 (0.928)	−4.639 (1.000)

Table 3 Parameter estimates of the stochastic frontier models

	Wheat	Maize	Rice
Cultivated area	0.726*** (0.050)	0.973*** (0.025)	1.074*** (0.008)
Fertilizers	0.020 (0.020)	0.054* (0.030)	0.049*** (0.009)
Pesticides	0.010 (0.010)	−0.025 (0.027)	0.036 (0.032)
Capital	0.010 (0.023)	–	0.026 (0.074)
Labor	−0.024 (0.020)	0.011 (0.073)	0.081** (0.033)
Seeds	−0.012 (0.23)	0.038 (0.035)	−0.075*** (0.014)
Lagged production	0.311*** (0.061)	0.029 (0.026)	−0.022*** (0.007)
Returns to scale	0.726	1.027	1.129
Parameter estimates of the inefficiency model			
Average Max. Temperature	–	2.341*** (0.221)	1.017*** (0.202)
Humidity	0.406*** (0.104)	0.973*** (0.191)	0.331** (0.155)
Log pseudo likelihood	100.464	108.689	108.672

*** (**) indicate significance level 1% (5%)
Standard errors in parenthesis

squeezed due to the loss in production or affect crop quality. In this line, improving technical efficiency can help farmers endure times of hardship. A strategy based on increasing the skills of farmers is relevant as well as improving local and regional markets infrastructure and transportation networks in Egypt. However, improving technical efficiency is not sufficient to be more competitive and economically efficient. Farmers also need to improve their allocative efficiency, an essential component of economic efficiency. Future extension of this work will concentrate on the analysis of allocative efficiency for agricultural subsectors, the second component of economic efficiency and the study total factor productivity and its components for each group.

7 Conclusions

The agricultural sector constitutes a large portion of the economies in Africa, which can help in solving major issues like malnutrition and hunger. CC has been identified as a significant threat especially to vulnerable regions especially Africa. Two main

Table 4 Summary statistics for technical efficiency scores

	Mean	Std. Dev.	Min.	Max.
Behera governorate				
Wheat model	0.967	0.013	0.950	0.989
Maize model	0.955	0.047	0.878	0.999
Rice model	0.810	0.018	0.779	0.835
Gharbia governorate				
Wheat model	0.985	0.015	0.955	0.999
Maize model	0.956	0.045	0.862	0.999
Rice model	0.814	0.030	0.753	0.846
Kafr-Elsheikh governorate				
Wheat model	0.905	0.025	0.861	0.937
Maize model	0.934	0.043	0.872	0.999
Rice model	0.853	0.051	0.764	0.894
Dakahlia governorate				
Wheat model	0.877	0.012	0.861	0.898
Maize model	0.989	0.008	0.972	0.999
Rice model	0.860	0.027	0.809	0.888
Sharkia governorate				
Wheat model	0.967	0.019	0.947	0.999
Maize model	0.988	0.012	0.972	0.999
Rice model	0.831	0.024	0.776	0.851

factors affect agricultural growth namely labor and agricultural land together with other factors that can help to boost food production; the development of secure market infrastructure and a transportation network that facilitates the supply of food. All these factors are impacted by climate change, which leads to a negative impact on food security. Since CC may cause a reduction in agricultural production, impact negatively logistical infrastructure which may reduce access to food, a loss in natural resources impacting the sustainability of using these resources, and finally changing the cropping pattern of a region affecting the needed nutrition requirements of these regions.

The objective of this study is to assess climatic conditions on the technical inefficiency of the agricultural production of cereal crops for the largest five producing governorates in Egypt. To address this objective, a TFE model is applied following Greene [56]. The method of estimation is a maximum likelihood for simultaneous estimation of the parameters of the stochastic frontier model and the technical inefficiency model. The results obtained from the stochastic frontier analysis using 5 governorates for 10 years for the period 2000–2009. We provide information on the output elasticity of different inputs for the different crop; we find that agricultural land, labor and are the most productive inputs. Maize and rice production exhibit increasing rate return to scale meaning that maize and rice producers could become more profitable with larger operations. The empirical findings in the inefficiency equation indicate that average maximum temperature and humidity have a significant positive effect on technical inefficiency. Results of the efficiency scores show, on average, that Behera, Gharbia, Kafr El-Shiek and Sharkia are more efficient in producing wheat and maize compared to rice, while Dakahlia is more technically efficient in Maize compared with rice and wheat.

8 Recommendations

Based on the results, this information could help policymakers understand how to apply any new policies. For example, making new investment plans for introducing climate-smart technologies and adopting new production practices that may help increase farms' technical efficiencies that in turn will help increasing agricultural output and reduce the variability of yields. Additionally, improving local markets infrastructure and logistics, which will help to smooth the supply of food and facing international markets volatilities. Considering that family farming which represents the majority of all farms in Egypt, improving their technical efficiency allows reducing production costs and improving competitiveness that can help them face fluctuating market conditions and economic hardships. As climate impact on agriculture is intensifying, Egyptian agricultural policies needs to be developed that technically support farmers through training to improve their skills to face these effects, minimizing carbon emission from agriculture the recycling of agricultural waste instead of burning it. Use new varieties of cereal crops that are more resistant to climate impact. Develop markets for carbon credits to regulate the right to pollute. Promote sustainable use of natural resources like the current government initiatives of reducing rice and sugar cane crops cultivated areas to reduce water consumption and adoption of advanced irrigation systems that efficiently use water.

Introducing significant changes in current Egyptian agricultural policies focused through giving main attention to achieving the higher growth rates in agricultural production. The current growth rate does not exceed 3% annually for the past decade; this rate should be doubled to at least 5% through increasing agricultural investments from 8 billion Egyptian pounds at the present time to 16 billion annually at constant prices over the next decade. Increase agricultural research budget, especially research related to the production and development of adaptable varieties to climate change, like salt-resistant varieties, heat-resistant varieties, short-lived varieties, and water-saving varieties and spending on improving animal breeds. It should be noted that the current share of agricultural research does not exceed 25 million pounds per year. This is incompatible with the challenges facing Egypt related to food security. In the future, this budget needs to increase to at least 500 million pounds annually, which does not exceed 0.5% of the total agricultural output. Follow appropriate strategies to resist rising sea levels to prevent erosion of the northern shores of the Delta. Future research will not only focus technical efficiency, which is the first component of economic efficiency but will investigate the allocative efficiency as well that represents the second component of economic efficiency and total factor productivity for each farming system. Since investigating only on tuning the technical efficiency is not sufficient to be more competitive.

References

1. Olila DO, Wasonga VV (2016) Nexus between climate change and food security in the East Africa Region: an application of autoregressive modelling approach. Invited paper presented at the fifth African higher education week and Ruforum Biennial conference, 17–21 Oct 2016, Cape Town, South Africa
2. World Bank (2016) Shock waves: managing the impacts of climate change on poverty
3. Food and Agricultural Organization (FAO) report (2016) The state of food and agriculture: climate change, agriculture, and food security. Available at: http://www.fao.org/3/a-i6030e.pdf
4. Food and Agricultural Organization (FAOSTAT) (2018) Available at: http://www.fao.org/faostat/en/
5. International Food Policy Research Institute and World Food Program, (IFPRI-WFP) (2013) Tackling Egypt's Rising food insecurity in a time of transition. Joint IFPRI-WFP country policy note. Available at: http://www.ifpri.org/sites/default/files/publications/ifpriwfppn_egypt.pdf
6. IPCC (2018) Summary for policymakers. In: Global Warming of 1.5°C. An IPCC Special Report on the impacts of global warming of 1.5°C above pre-industrial levels and related global greenhouse gas emission pathways, in the context of strengthening the global response to the threat of climate change, sustainable development, and efforts to eradicate. World Meteorological Organization, Geneva, Switzerland, 32 pp
7. Zomer RJ, Neufeldt H, Xu J, Ahrends A, Bossio D, Trabucco A, Noordwijk Mv, Wang M (2016) Global tree cover and biomass carbon on agricultural land: the contribution of agroforestry to global and national carbon budgets. Sci Rep 6:29987. https://doi.org/10.1038/srep29987
8. MWRI (2015) Egypt 2012 state of the water report. Monitoring and evaluation for water in North Africa (MEWINA) project. Ministry of water resources and irrigation (MWRI)—Egypt water resources management program—CEDARE. Available at: http://web.cedare.org/wp-content/uploads/2005/05/Egypt-2012-State-of-the-Water-Report.pdf
9. USAID fact sheet (2015) Climate change information fact sheet Egypt. Available at: https://www.climatelinks.org/sites/default/files/asset/document/Egypt%20Climate%20Info%20Fact%20Sheet_FINAL.pdf
10. Food and Agricultural Organization, Global Early Warning System, (FAO-GIEWS) (2018). Available at: http://www.fao.org/giews/earthobservation/index.jsp?lang=en
11. Synthesis of Experiences and Recommendations. UNDP report (2018) Climate change adaptation in Africa. Available at: http://www.fao.org/giews/earthobservation/asis/index_1.jsp?lang=en
12. Enhancing Climate Change Adaptation in the North Coast of Egypt (ECCANCE) project overview (2018). Available at: https://www.adaptation-undp.org/projects/enhancing-climate-change-adaptation-north-coast-egypt
13. Egypt's National Strategy for Adaptation to Climate Change and Disaster Risk Reduction, (NSACCDRR) strategy (2011) Available at: http://www.climasouth.eu/docs/Adaptation011%20StrategyEgypt.pdf
14. Climate Change Risk Management in Egypt (CCRM) (2013). Available at: http://www.mdgfund.org/sites/default/files/Egypt%20Joint%20Programmes%20Fact%20Sheet.pdf
15. Egypt NDC (2015) Egyptian intended nationally Intended Nationally Determined Contributions as per United Nation Framework Convention on Climate Change. Available at: https://www4.unfccc.int/sites/ndcstaging/PublishedDocuments/Egypt%20First/Egyptian%20INDC.pdf

16. Besada H, Sewankambo N (2009) CIGI special report: climate change in Africa: adaptation, mitigation and governance challenges. Available at: https://www.cigionline.org/sites/default/files/climate_change_in_africa_3.pdf
17. Collier P, Conway G, Venables T (2008) Climate change and Africa. Oxford Rev Econ Policy 24(2):337–353. https://doi.org/10.1093/oxrep/grn019
18. Gemeda DO, Sima AD (2015) The impacts of climate change on African continent and the way forward. J Ecol Nat Environ 7(10):256–262
19. Clements R (2009) Economic-cost-of-climate-change-in-Africa
20. Niang I, Ruppel O, Abdrabo M, Essel A, Lennard C, Padgham J, Urquhart P (2014) Africa. In: Climate change 2014: impacts, adaptation and vulnerability. Contribution of working group II to the fifth assessment report of the intergovernmental panel on climate change. Cambridge University
21. World Wide Fund For Nature (2006) Climate change impacts on East Africa. World wide fund for nature. Retrieved from https://repositories.tdl.org/tamugir/bitstream/handle/1969.3/28826/east_africa_climate_change_impacts_final[1].pdf?sequence=1
22. Huq S, Reid H, Konate M, Rahman A, Sokona Y, Crick F (2004) Mainstreaming adaptation to climate change in Least Developed Countries (LDCs). Clim Policy 4(1):25–43. https://doi.org/10.1080/14693062.2004.9685508
23. Nordhaus WD (2006) Geography and macroeconomics: new data and new findings. Proc Natl Acad Sci USA 103(10):3510–3517. https://doi.org/10.1073/pnas.0509842103
24. Rose RM (2015) The impact of climate change on human security in the Sahel region of Africa. Donnish J Afr Stud Dev 1(2):9–14
25. Singh A, Purohit BM (2014) Public health impacts of global warming and climate change. Peace Rev 26(1):112–120. https://doi.org/10.1080/10402659.2014.876326
26. Awojobi O, Tetteh J (2017) The impacts of climate change in Africa: a review of the scientific literature. J Int Acad Res Multi 5(11):39–52
27. Hummel D (2015) Climate change, land degradation and migration in Mali and Senegal—some policy implications. Migr Dev 5(2):211–233
28. UNEP report (2011) Livelihood security: climate change, migration and conflict in the Sahel. United Nations Environment Programme, Châtelaine, Geneva
29. Collomb P (1999) Une voie étroite pour la sécurité alimentaire d'ici a 2050. FAO Economica, Paris
30. Giannakopoulos C, Le Sager P, BIndi M, Moriondo M, Kostopoulou E, Goodess CM (2009) Climatic changes and associated impacts in the Mediterranean resulting from a 2°C global warming. Glob Planet Change 68:209–224
31. Aggarwal, PK (2008) Global climate change and Indian agriculture: impacts, adaptation and mitigation. Indian J Agric Sci 78(10):911–919
32. Kavi Kumar KS, Parikh J (2001) Indian agriculture and climate sensitivity. Global Environ Change 11(2):147–154
33. Geethalakshmi V, Lakshmanan A, Rajalakshmi D, Jagannathan R, Gummidi S, Ramaraj AP, Bhuvaneswari KL Gurusamy, Anbhazhagan R (2011) Climate change impact assessment and adaptation strategies to sustain rice production in Cauvery basin of Tamil Nadu. Curr Sci 101(3):342–347
34. Berg A, de Noblet-Ducoudre N, Sultan B, Lengaigne M, Guimberteau M (2013) Projections of climate change impacts on potential C4 crop productivity over tropical regions. Agric For Meteorol 170:89–102
35. Deryng D, Sacks WJ, Barford CC, Ramankutty N (2011) Simulating the effects of climate and agricultural management practices on global crop yield. Glob Biogeochem Cycles 25:GB2006. https://doi.org/10.1029/2009gb003765
36. Cervigni R, Losos AM, Chinowsky P, Neumann JL. (2016) Enhancing the climate resilience of Africa's Infrastructure: the roads and bridges sector (No. 110137). The World Bank, pp 1–0. Available at: http://documents.worldbank.org/curated/en/270671478809724744/Enhancing-theclimate-resilience-of-Africa-s-Infrastructure-the-roads-and-bridges-sector

37. Karekezi S, Kamani J, Onguru O, Kithyoma W (2012) Energy security and adaptation to climate change in East Africa and the horn of Africa: large scale hydropower vs. decentralized renewables. Available at: https://ke.boell.org/2012/08/19/energy-security-andadaptation-climate-change-east-africa-and-horn-africa-large-scale
38. Zhou G, Minakawa N, Githeko AK, Yan G (2004) Association between climate variability and malaria epidemics in the East African highlands. Proc Natl Acad Sci USA 101(8):2375–2380
39. Chidumayo E, Okali D, Kowero G, Larwanou M (2011) Climate change and African forest and wildlife resources. African Forest Forum, Kenya, Nairobi, Nairobi
40. Funk C, Senay G, Asfaw A, Verdin J, Rowland J, Michaclson J, Choularton R (2005) Recent drought tendencies in Ethiopia and equatorial-subtropical eastern Africa. FEWS-NET, Washington DC. Available at: http://pdf.usaid.gov/pdf_docs/Pnadh997.pdf
41. Aigner D, Lovell CAK, Schmidt P (1977) Formulation and estimation of stochastic frontier production functions models. J Econometrics 6:21–37
42. Meeusen W, Van Den Broek J (1977) Efficiency estimation from Cobb-Douglas production functions with composed error. Int Econ Rev 18:435–444
43. Chakraborty K, Misra S, Johnson P (2002) Cotton farmers' technical efficiency: stochastic and non-stochastic production function approaches. Agric Resour Econ Rev 31:211–220
44. Coelli TJ (1995) Recent developments in frontier modeling and efficiency measurement. Aust J Agric Res Econ 39:219–245
45. Oude Lansink A, Pietola K, Bäckman S (2002) Efficiency and productivity of conventional and organic farms in Finland 1994–1997. Eur Rev Agric Econ 29:51–65
46. Pitt M, Lee L (1981) The measurement and sources of technical inefficiency in the Indonesian weaving industry. J Dev Econ 9:43–64
47. Battese G, Coelli T (1988) Prediction of firm-level technical efficiencies with a generalized frontier production function and panel data. J Econometrics 38:387–399
48. Schmidt P, Sickles R (1984) Production frontiers and panel data. J Bus Econ Stat 2(4):367–374
49. Belotti F, Daidone S, Ilardi G, Atella V (2012) Stochastic frontier analysis using Stata. Centre Econ Int Stud 10(12):251
50. Cornwell C, Schmidt P, Sickles R (1990) Production frontiers with cross-sectional and time-series variation in efficiency levels. J Econometrics 46:185–200
51. Lee Y, Schmidt P (1993) The measurement of productive efficiency: techniques and applications, chap. A production frontier model with flexible temporal variation in technical inefficiency. Oxford University Press
52. Ahn SC, Hoon Lee Y, Schmidt P (2001) GMM estimation of linear panel data models with time-varying individual effects. J Econometrics 101(2):219–255
53. Ahn SC, Orea L, Schmidt P (2005) Estimation of a panel data model with parametric temporal variation in individual effects. J Econometrics 126(2):241–267
54. Kumbhakar S (1990) Production frontiers, panel data and time-varying technical inefficiency. J Econometrics 46:201–212
55. Battese G, Coelli T (1992) Frontier production functions, technical efficiency and panel data: with application to paddy farmers in India. J Prod Anal 3(1/2):153–169
56. Greene W (2005) Reconsidering heterogeneity in panel data estimators of the Stochastic frontier model. J Econometrics 126(2):269–303
57. Greene W (2005) Fixed and random effects in Stochastic frontier models. J Prod Anal 23:7–32
58. Elasraag Y (2015) Economics of wheat in Egypt. Published Ph.D. dissertation. Technical University of Madrid. Available at: http://oa.upm.es/39024/1/YAHIA_HAMID_AMIN_ELASRAAG.pdf
59. Battese G, Coelli T (1995) A model for technical inefficiency effects in a stochastic frontier production function for panel data. Empirical Econ 20:325–332

Conclusions

Update, Conclusions, and Recommendations to "Climate Change Impacts on Agriculture and Food Security in Egypt"

El-Sayed Ewis Omran and Abdelazim M. Negm

Abstract This chapter encapsulates the essential climate change challenges (in terms of conclusions and recommendations) of the agricultural environment in Egypt and presents insights derived from the cases in the volume. In addition, some (update) findings from a few recently published research work related to the climate change covered themes. This chapter focuses on climate change impact on agriculture in Egypt that was documented during the book project. To this end, we identify five main contribution areas, which include an overview of climate change in Egypt, land and water resources, smart farming, livestock, fishery, and aquaculture, and socioeconomic impacts. Therefore, conclusions will be built on researcher vision gained concerning study findings and limitations. In addition, this chapter encompasses evidence on a set of recommendations to direct future research towards climate change impact on agriculture, which is a main strategic theme of the Egyptian Government. The set of recommendations is presented for specialists involved in following additional research to exceed the scope and findings of this book.

Keywords Climate change · Egypt · Climate variability · Land and water resources · Smart farming · Livestock · Fishery · Aquaculture · Socioeconomic impacts

---

E.-S. E. Omran (✉)
Soil and Water Department, Faculty of Agriculture, Suez Canal University, Ismailia 41522, Egypt
e-mail: ee.omran@gmail.com

Department of Natural Resources, Institute of African Research and Studies, Aswan University, Aswan, Egypt

A. M. Negm
Water and Water Structures Engineering Department, Faculty of Engineering, Zagazig University, Zagazig 44519, Egypt
e-mail: amnegm85@yahoo.com; amnegm@zu.edu.eg

© Springer Nature Switzerland AG 2020

E.-S. E. Omran and A. M. Negm (eds.), *Climate Change Impacts on Agriculture and Food Security in Egypt*, Springer Water, https://doi.org/10.1007/978-3-030-41629-4_25

1 Introduction

While the world is facing unprecedented transitions and threats, we need to deeply rethink the relationships between agriculture, food production, and climate change. Climate change is the goal number 13 in the 2030 agenda for sustainable development adopted by the United Nations, which addresses the target action to combat climate. Agriculture is one of the most vulnerable economic sectors to climate change, mainly due to the limited availability of water and land resources. There is a future risk of higher skills shortages related to the impact of climate change on agricultural sectors and food production. Particularly, there is a need for highly specialized scientists in the field of agriculture and food security who want to combine scientific and social or policy skills to better understand and make significant contributions to climate adaptation and mitigation in agriculture and food security. Therefore, it is critical to integrate agricultural science with related subjects that impact on sustainability and food security such as geo-politics, legislation, and regulation, consumer pressures, economics, and agro-ecology.

While urgent action to combat climate change and its impacts can be taken through Sustainable Development Goal 13, this book tries to answer the question "How will climate change affect food security and sustainable agriculture?" that will help promote sustainable agricultural development and food security goals. This book takes stock of current knowledge and analyses the most recent trends in climate change impact on agriculture.

Therefore, this chapter will present general conclusions of the climate change impacts on agriculture and food security in Egypt. It is necessary to give due consideration to the characteristics of various climate change impacts, which render the resultant system unsustainable. So, the intention of the book is to address the following main theme.

- An overview of climate change in Egypt.
- Land and water resources.
- Smart farming.
- Livestock, fishery, and aquaculture.
- Socioeconomic impacts.

The next section presents a brief of the important findings of some of the recent (updated) published studies on the climate change impacts on agriculture and food security in Egypt, then the main conclusions of the book chapters in addition to the main recommendations for researchers and decision makers. The update, conclusions, and recommendations presented in this chapter come from an investigation of data collection presented in this book.

2 Update

The following are the major update for the book project based on the main book theme:

2.1 Update on an Overview of Climate Change in Egypt

Three ways were identified to give us an overall impression of the climate change in Egypt. The first way through of Paleo-climate evidence in Egypt. Two main processes can yield the climatic changes; the earth's internal processes and the extraterrestrial impacts. Both of them have a strong effect on the earth's system. The paleoclimatic change is well preserved in the earth's sedimentary record and can be reviled by using multidisciplinary studies including mineralogy, geochemistry, and the fossil contents. Egypt is a key area of one of the most pronounced climatic changes that occurred in the earth's geologic history; the Paleocene Eocene thermal maximum that used recently as analog for the current warming.

Many hypotheses have been proposed to explain the climate variations through the geologic time such as changes in solar radiance due to the instability of earth's Orbital and Milankovitch cycles, change in the greenhouse gases (carbon dioxide or methane) level in the atmosphere [1], or change in the Oceanic thermohaline circulation. The climate can be affected by two different major processes; internal processes such as changes in the atmosphere, ice, vegetation, and changes in the ocean and land surface; and external processes like changes in plate tectonics, change in the Earth's orbit, and changes in the intensity of the Sun [2]. The main control on the paleoclimate was the paleoposition due to the continental drift.

The second way is linked to governance and institutional structure of climate change in Egypt. There is still a marked lack of support for developing the institutional capacity and the corresponding mechanisms necessary to address climate variation at the local level [3, 4]. A cross-cutting issue like climate change, cannot be managed by one government department or a single ministry. Mainstreaming climate change into policy decisions requires the consideration of climate change causes and consequences into non-climate sectors [5]. "Integrated strategies are a relatively novel approach to govern highly complex issues that involve several sectors and levels of governance. They are a key tool to foster not only sustainable development but also environmental policy integration or ecological modernization" [3]. Both of which are concepts that promote the integration of environmental concerns into sectors with significant environmental ramifications long before sustainable development dominated the environmental discourse [6]. In Egypt, despite a stated commitment to grassroots involvement in climate change adaptation processes and actions, the actual focus in national adaptation plans is on technical and infrastructure options for adaptation, with little attention given to the social and institutional context in which these options operate [4]. Future national-level efforts to develop adaptation plans need to consider the role of local institutions more centrally if they seek to serve the needs and interests of the most vulnerable populations better.

The third way is the role of science, technology, and innovation in addressing climate change challenges in Egypt. Climate change presents opportunities for development. Linking climate change and development became important to ensure an increasing resilience of the poorest and most vulnerable communities. It provides them with the needed tools and technologies, which should be the main task of the international communities to approach the impact of climate change [7]. Climate change phenomena are based originally on scientific facts where research tools are being used to confirm the data and information related to the cause of climate change as well as its impact. Extensive scientific evidence made it clear that the main cause of the change in the climate of the past half century is human-induced increases for greenhouse gases. This scientific opinion is expressed in technical reports and scientific articles by scientific national and international entities and by conducting surveys of opinion among climate scientists. Complex, integrated mathematical models are being used for projecting future climate change. Climate models are based on well-established physical principles and have been demonstrated to reproduce observed features of recent climate and past climate changes. Models can simulate the impacts at the local and global scale, e.g., crop yields, hydrological, and water quality models [8].

2.2 *Update on Land and Water Resources*

Five potential impacts were identified to study the effect of climate change on land and water resources in Egypt. First is the effect of climate change on agriculture, which has specifically become a global issue due to the ever-growing need to provide food security and end hunger. The impact is on the whole agro-ecosystem from crops and livestock to fisheries, which in turn can affect economic stability [9]. Implications of climate change on agriculture are not only due to the effects on plants but also due to effects on biogeochemical properties of the soil on which they grow. Not only does the climate-mediated change in microbiota and their behaviors reduce food productivity but also it extends to food safety, human health, and economy when soil microbiota turns toxigenic under the environmental influence. Fungi are the primary decomposers in the soil. It is implicative that changes in temperature and precipitation will change the diversity and functionality of fungal communities [10, 11]. Hence, to find a solution, one has to step back and study the impact. It should address the recent initiative by the Egyptian Ministry of Higher Education and Scientific Research (MHESR) to encourage researchers to find solutions to overcome climate change impacts such as forecasting of future changes in diversity and food productivity, detecting potential natural disasters, changing building codes, protecting coasts and delta region.

Second effect is the impacts of climate change on microbial activity in agricultural Egypt's soil. Climate change is a forward direct and indirect effect on speed up or slows down terrestrial microbial community composition and their functions. The effect of climate change on microorganisms are listed: death and disturbance, metabolic activity is direct and indirect highly influenced, reduction/stimulation of biomass, diversity, and composition lead to extinct, having a negative or positive result on its physiology and greenhouse gas emission. As the temperature increases, microbial community structures are altered and processes climate change is a forward direct and indirect effect on speed up or slows down terrestrial microbial community composition and their functions. Soil-dwelling bacteria and fungi control the breakdown of organic matter in soil and its release as carbon dioxide to the atmosphere [12]. Climate warming is expected to accelerate the activity of soil microbes, stimulating further CO_2 release, and positive feedback to climate change. A better understanding of microbial processes is likely to improve climate change predictions [13]. Recent research has thus sought to quantify the consequences of warming for soil microbes and the carbon cycle processes they govern and to describe this using metrics such as microbial carbon use efficiency [12]. However, the relationships between temperature and soil microbes remain inconsistent in both space and time, preventing consensus on the severity of feedback from microbial activity to future climate change.

The third influence is through the mapping of Egypt vulnerability to Sea Level Rise (SLR). The coastal area of the Nile Delta is one of the areas that are most prone to flooding because of the expected sea level rise. This may be accompanied by soil subsidence at varying rates, depending on the conditions of each region, the topographical and geological characteristics, as well as the current protection means of each area. Most previous studies of the effect of SLR on the Nile Delta region examine only the assumed SLR range [14]. The potential threats and impacts of SLR on the southern Levantine sub-basin, including the Nile Delta coast, therefore deserve greater attention. Moreover, local and global SLR can differ significantly [15, 16] so regional assessments are needed for risk management, adaptation planning, and integrated coastal zone management (e.g., BACC [17]).

Fourth inspiration is soil itself as driver and victim of climate change. Egypt is a country that foresees to face severe effects owing to climate change, although its share in the global greenhouse gas (GHG) emissions is only 0.57%, and despite the fact that it is one of the countries that is not required to manage or reduce its emissions according to Kyoto protocol [18]. Nevertheless, Egypt has considered a national plan that includes mitigation measures to manage and reduce GHG flux from the major sectors contributing to the projected climate change which is agriculture, industry and energy wastes [18]. Soil may consider an important source of greenhouse gas emissions (i.e., carbon dioxide, methane, and nitrous oxides). Soil is also considered as a victim of climate change. Global warming may induce, depletion of soil organic matter that causes the decline of soil fertility, poor soil water regime, shifting of soil microbiome and soil compaction (i.e., increase soil compaction, surface sealing, and crust formation).

The final control is through soil carbon sequestration for climate change mitigation. An increasing trend of the atmospheric temperature and global changes in the weather conditions is indicated as climate change [19]. There are two anthropogenic processes that contribute greenhouse gases to the atmosphere; fossil fuels burning and changes in land use. Climate change can significantly affect soil carbon (C), since changes in temperature, rainfall patterns, and CO_2 concentrations influence C input to the soil, and soil C decomposition. Recently, the impact of climate change on global SOC stocks has been widely recognized [19]. There are many different available options, terrestrial carbon sequestration, particularly in the soil is count as a 'win–win' strategy because generally management practices that lead to increased carbon in the soil also enhance soil fertility. Understanding the effect of land cover/land use and expected climate change on SOC stocks provides valuable information to improve approaches to land planning. However, uncertainties in climate change studies are high, so investigating several scenarios is essential.

2.3 Update on Smart Farming

Seven approaches were acknowledged for using smart farming as a way to mitigate climate change impacts on agriculture. The first is assessing vulnerability and enhancing adaptive capacity and resilience to climate change. In order to implement the 2030 Agenda, the international community has endorsed five principles of sustainable food and agriculture providing a framework for policy dialogue and for developing appropriate policies, strategies, regulations, and incentives [20]. More than 40% of the people in the Arab region have already been exposed to drought and other climatic disasters [21]. Nevertheless, analysis of the impacts of climate change at the farming system level regarding crop and livestock production and productivity, as well as the implications that this has for the livelihoods of those dependent on them will elucidate some of the common trends and challenges. Most of the farmers live in the areas of mixed cereals-livestock farming systems keep goats, which are the most vulnerable to climate change after cattle. Their vulnerability comes from the fact that while the potential impact of climate change on livestock is less than that on crops, there is limited adaptive capacity for any of them [20].

The second approach used the pollination of flowering plants is one of the most important ecological services in natural and agricultural ecosystems. The majority of angiosperm plants rely on animals for pollination. Bees and their life-sustaining relationships with flowering plants occupy keystone in both natural and agricultural ecosystems. Given their ecological services, native bee faunas are an important natural resource that can complement honey bees and other managed pollinators in their ecosystem services [22]. The decline of some species is well documented globally due to several reasons such as climate change, lack of natural foraging resources, an epidemic of some pests and diseases and elimination of nesting habitats; these declines have been reported worldwide [23].

The third approach utilized the change of agricultural calendar as a response to climate variability. Climate change often impacts agricultural production indirectly by affecting the development and distribution of crop pests and animal diseases, increasing the rate and dissemination of harmful diseases, weather conditions, reducing water supplies and irrigation; and enhancing the severity of soil erosion [24, 25]. Such climatic hazards became the driving forces, which challenge many farmers' livelihoods. The rural population, whose main source of food, direct and/or indirect employment and income is agriculture, will be most influenced because of the susceptibility of agriculture to climate change. Adaptation to climate change refers to modification in natural or human systems in response to actual or predictable climatic stimuli [26]. It enables farmers to produce their food, income, and livelihood protection goals in light of climate and socio-economic change [27]. Common methods of adaptation in agriculture involve new crop varieties and animal species that are suitable to drier situations, irrigation, divarication of crops, implementation of mixed crop and livestock farming systems, and changes in the dates of agricultural activity [28, 29]. Some of these methods (e.g., changes in agricultural activity dates) undertaken in response to short-term climate variability are classified as coping responses. Several factors and critical information are considered amongst them crop calendars [30]. The chronological sequence of the occurrence of different phenological stages of a crop in its growth cycle defines the so-called crop calendar [31]. Crop calendars include information on the timing of periods of crop sowing, growing, and harvesting. A huge number of studies take into consideration the techniques established by farmers to deal with climate variability [28, 29] have reported changes in sowing dates.

The fourth approach study the projected crop coefficients under climate change in Egypt. "Climate change is expected to have many adverse impacts on various sectors, yet agriculture is considered to be the most tangible affected sector, as any alteration in the prevailing temperature or precipitation patterns will disturb the agricultural sector as a whole, including crop yields, crops water requirements, and soil fertility" [32]. Climate change has already caused significant impacts on water resources and food security. Land ecosystems would require more water to match increased water demand, and consequently to prevent drought [33]. It was reported that climate change is expected to negatively affect crops productivity and cause increases in water requirements for crops in Egypt [34]. Because climatic parameters are the only factors affecting ETo, it is the most important hydrological and meteorological variable to reflect climate change [35].

The fifth approach make use of the rice production in Egypt: the challenges of climate change and water deficiency. Rice paddies are estimated by a large portion of the wetland ecosystem; mainly in Asian countries [36]. El-Shahawy et al. [37] stated that in Egypt and most other developing countries that produce rice, it is commonly cultivated under continuous flooding with about 5 cm depth of standing water during the growing season. Rice cultivation is known as an important emitter of greenhouse gases emission, especially methane due to rice management practices and burning of rice straw after harvesting.

The six is the smart farming approach, which is using nanotechnology for real-time control of the Red Palm Weevil under climate change. "The red palm weevil has become a global threat and demands a global strategy to eradicate it. One of the biggest threats to global agriculture these days is a tiny, bright red weevil. Because of climate change, the pest can spread to countries where before the weather would not have allowed them to survive" (https://www.npr.org/sections/thesalt/2017/06/22/532619019/red-palm-weevils-wreak). Climate change will profoundly affect the distribution and abundance of all species, including insects, and promote ecology-driven changes in phenology [38]. Climate change may result in a potential expansion of insect populations to polar regions and higher elevations [39]. Increases in global temperatures will expand or reduce the range of insects by converting climatically unsuitable habitats into suitable ones or vice versa [40].

Finally, how to face climate change: Urban gardening and sustainable agriculture. Agriculture is threatened by climate change, responsible for food supply and employs about 30% of Egypt's labor force. Thus, a new criterion, "sustainable agriculture", is needed, which addresses all these issues simultaneously! "Sustainable Agriculture" and climate change adaptation potential adaptation by adopting cropping patterns" (https://www.longdom.org/proceedings/impact-of-climate-change-on-agriculture-sect). Trying to save our global, urban, or city gardening as a sort of horticultural strategies can dealing with climate change. We want to begin thinking now not just about what used to be, but what may be, generally contributing the city regions to be sustainable socially, economically, and environmentally. Climate change goes to pressure us to paintings difficult at something Homo sapiens has never been good. An urban garden like our own families is a great location to begin.

2.4 Update on Livestock, Fishery, and Aquaculture

Five different ways are recognized for potential climate change impacts on livestock, fishery, and aquaculture. First is the potential climate change impacts on livestock and food security nexus in Egypt. Where, global climate change can have comprehensive magnitudes for livestock production, that preponderantly result from its impact on rainfall patterns that confirm each the amount and quality of grassland and land post-productivity [41]. Further, the drying of water resources can produce a state of affairs wherever livestock has to be compelled to walk long distances in search of water, making extra stress to those animals. Hence, it is attending to be an enormous challenge for livestock researchers across the world to develop appropriate ways to confirm access to water for livestock production.

Second is the influence of climate changes on animal feed production. There is a reciprocal effect between climate change and animal production. On the one hand, the production of livestock contributes about 37% of total methane, 65% nitrous oxide and 9% of carbon dioxide emissions, all of which makes up to about 18% of global greenhouse gas emissions from all human activities. Also, 64% of ammonia emissions come from livestock production and contaminate air, soil, and water, and cause acid rain and damage to the ozone layer. On the other hand, the farm animals are prone to the adverse impacts of the changes in climate. Forage crops are an important part of agriculture because they provide quality feed to animals. The changes in climate will influence both the quantity and of quality of forage production as well as their reliability. Climate changes cause environmental stresses (e.g., increasing temperature, elevated carbon dioxide concentrations) that affect pasture and forage production.

The third way is through algae and fish. Aquatic animals such as fish and crustaceans provide human beings with high-quality protein diets. Scientists propose that this would cause mitigation of global warming as algae take in carbon dioxide, the main factor responsible for global warming, and release oxygen during photosynthesis. So double benefits are expected in terms of increasing aquatic animals and decreasing global warming.

The fourth way is climate change impact on immune status and productivity of poultry as well as the quality of meat and egg products. The negative effects of increased environmental temperature due to climate change on poultry production representing lower feed intake, reduction of weight gain and meat quality in broilers, in addition to low egg production rate and egg quality, increasing feed conversion ratio and high mortality rate in laying hens. Effective procedures must be taken to challenge this potential risk of high temperature and its impact on livestock in general and on poultry in particular through the implementation of effective strategies in order to protect poultry survival and increase their productivity during hot weather.

The final way presented in this book is climatic change and chicken immunity. Heat stress is considered as one of the primary factors that impose negative impacts on chicken production and reproduction. Due to global warming, abrupt climatic conditions like storms, droughts, floods, and extreme hot and cold temperatures are predominant around the globe. With the changing climatic scenario, the frequency and duration of exposure of livestock to abiotic and biotic stressors are expected to increase. Chicken flocks are particularly vulnerable to climate change because birds can only tolerate a narrow temperature ranges [42].

2.5 Socioeconomic Impacts

Two scenarios have been identified to study the socioeconomic impacts of climate change. The first scenario deals with agriculture and rural communities' vulnerability in the Nile Delta. Climate models predict that the change in climatic conditions will affect agriculture by reducing land productivity between 10 and 20% by 2050 in many developing countries and affecting about 10% of the agricultural land [43]. Accordingly, many farming households are expected to counter crop deficiencies and yield reductions that would affect their livelihoods and in turn, their vulnerability [44]. Nevertheless, agriculture in Egypt in general and in the Nile Delta in particular, is currently facing other intense challenges resulting from the rapid population growth that is exceeding 2.5% annually [45], thus augmenting agricultural land and irrigation water scarcities. For the determination of the climate change stimuli impacts on agriculture in the Nile Delta governorates, and the identification of farmers' vulnerability in 2030, the vulnerability assessment framework (VAF) is applied. The VAF is an important tool for supporting policy makers in formulating the necessary strategies of climate change adaptation for the enhancement of the adaptive capacities of local farming communities by improving awareness of the causes and the implications of vulnerability [46]. The framework considers the exposure and sensitivity of agriculture and farmers to the climate change stimuli, and their local adaptive capacities that are necessary for coping with climate change adverse impacts on farmers' livelihoods [47].

The second scenario is an efficiency analysis approach. Climate Smart Agriculture (CSA) is an approach that deals with restructuring agricultural systems to cope with climate change through three main objectives, which are sustainability increasing productivity and incomes, mitigating GHG emissions, and adaptation and building resilience. The Intergovernmental Panel on Climate Change (IPCC) has released a special report on 2018 on the impact of global warming of 1.5 °C and GHG emission pathways. Additionally, the report has shown that most climate models showed robust results of regional climate characteristics differences [48]. An Egyptian Climate Change Risk Management program is developed to incorporating GHG emissions mitigation into the national policy and investment frameworks, including Clean Development Mechanism financing opportunities, Enhancing the national capacity to climate change adaptation through synergies among water, energy and food sectors, and finally advocacy and awareness-raising.

Finally, Green Sustainable Development (GSD) is that meets the needs of the present without compromising the ability of future generations to meet their own needs. At the core of sustainable development is the need to consider "three pillars" together: society, the economy and the environment. No matter the context, the basic idea remains the same—people, habitats and economic systems are inter-related. These may be conducted by applying the new approach of nexus especially between water, climate and foot security (WCF), which are interrelated with each other and facing from the same challenges with cross cutting strategies and plan for green sustainable development. Achieving for the "GSD" green strategies and plans for WCF nexus need special institutional inter sectorial mechanism for cooperation and formulating throw the mainstream national strategy. This chapter discus and present some of the WCF nexus interrelation; impact and strategies as well.

3 Conclusions

Throughout the present book project, the editorial' teams were able to draw several conclusions. Besides methodological insights, the chapter originates key lessons from the cases in the book, in particular, the promising characteristics of both the Paleo and current climate impacts on climate change. These conclusions are important to understand the impact of climate change in Egypt. These are discussed in the following in no particular order.

3.1 An Overview of Climate Change in Egypt Was Concluded in Three Approaches

The first approach is an overview of Paleo-climate evidence in Egypt. Study of the different impacts of the climatic changes occurred in the Earth's geologic history can give a clue to predict the consequence of the current warming. The natural warming events took several hundreds of kilo years led to increase the earth temperatures by a few degrees, whereas the human impact accelerates the process by around 250 times. Therefor the human activities should be controlled by lows to stop and/or decrease the emission of the greenhouse gases to slow there steps toward the sex mass extinction event.

The second approach is the governance and institutional structure of climate change in Egypt. Egypt's National Strategy for Adaptation to Climate Change and Disaster Risk Reduction stressed the importance of reforming and modifying the institutional frameworks for climate change adaptation in Egypt. Such reformation would not necessarily entail establishing new institutions, but rather ensuring best practices in existing ones. While a framework for climate change adaptation exists in Egypt, local level policies and actions need to be strengthened to ensure effective implementation of adaptation and mitigation effects. There is a need to formulate integrated climate change strategies, to enable access to climate finance mechanisms, and to develop and strengthen the policies, institutions, capacities, knowledge and transformative change required to build climate-resilient communities (http://www.climasouth.eu/docs/Adaptation011%20StrategyEgypt.pdf).

The third approach is the role of science, technology, and innovation in addressing climate change challenges in Egypt. Science policy interface is vital in approaching the climate change, which challenges globally and specifically in Egypt. Innovations and new technologies have the major role of developing new crops varieties that are resistant to possible impacts of climate change (heat, drought, and salinity stress), disease, and producing higher yields. These technologies should be simplified to be close to the policymakers, who need to understand the nature and processes of scientific innovation and to enhance an environment in which it can flourish in responding to the emerging challenges. Water, food energy, and climate change nexus can be the major concern of the scientific research plans, programs, and projects in Egypt. In addition, Policy time scales are generally shorter than scientific developments, so policy needs to demonstrate more continuity for fostering innovation. Because of the cost of research, funding for scientific research and development needs long-term commitments, which are not affected by rapid policy changes. There is also a need for more international cooperation.

3.2 Land and Water Resources

First is the impact of climate change on Plant-associated fungi. Whether we like it or not, climate change will continue to occur in the future, laying its heavy impact on every agricultural-related element. Finding solutions to shifts in soil microbial community and deterioration of plant produce under extreme climate conditions can be addressed using biotechnology tools such as genetic modification, stress tolerance induction, and synthetic biology to obtain new breeds that tolerate climate change increase productivity or enhance carbon sequestration. Molecular biology is expected to play another role, which is developing new tools for solar energy conversion because a huge percentage of CO_2 emissions are attributed to agricultural emissions. Harnessing energy from photosynthesis is also another biotechnology application that will add an ecological dimension through mitigating greenhouse gas produced from conventional energy production; this will consequently aid in preserving the environment and biodiversity.

Second is the impact of climate change on microbial activity in agricultural Egyptian soil. Climate change has an impact on Egyptian soil microbial community and activities both directly or indirectly. The fluctuation of temperature and elevated CO_2 are the most climatic changes in Egypt that positively, negatively, and neutrally affect soil microbial activity, community, and enzyme activities. However, most of researchers or studies focused on climate change impacts as a single factor or by separating the different climate change components (altered temperature and moisture or elevated CO_2) and manipulating their responses on soil microbial activity independently. The behavior and response of plant growth promoting microorganisms to climate change are very variedly and complicated, especially in saline habitats. The structure and functions of soil microorganisms are tremendously complex and can be impacted by climate changes in several ways that prevent the ability to draw reliable conclusions. Long-term experiments are critical to more illustrate the interactions between biotic and abiotic factors and better predict soil microbial responses to future climate changes. On the other hand, the use of modern molecular techniques such as metagenomics will help and gave more details about soil microbial communities that would otherwise be undetectable.

Third is mapping Egypt vulnerability to sea level rise scenarios. Nile Delta contains high population densities, significant economic activities, and ecosystem services. This area is already subject to climate change has the potential to pose increasing risks in the future. Consequently, a range of negative socio-economic impacts is expected to follow such as food insecurity, loss of income sources and unemployment of small farmers as well as the loss of livelihoods of other people who are currently dependant on the agricultural activities carried out in the Nile Delta. These predictions assure the need for the agricultural sector in Egypt to adapt—in the long term—to the potential threat of SLR as part of its adaptation to climate change in general. The SLRS analysis can be used to identify priority areas for adaptation based on the vulnerability of cities and districts.

Fourth consider the soil as driver and victim of climate change in Egypt. The soil in Egypt may consider an important source of greenhouse gas emissions (i.e., carbon dioxide, methane, and nitrous oxides). The drivers of soil GHG emissions are soil type and composition (i.e., soil texture, pH, soil organic matter (SOM), etc.), soil temperature, moisture, fertilization, soil miss-management (Tillage), rice cultivation and burning of Crop residues. Soil is also considered as a victim of climate change. Global warming may induce, depletion of soil organic matter that causes the decline of soil fertility, poor soil water regime, shifting of soil microbiome and soil compaction (i.e., increase soil compaction, surface sealing, and crust formation). Global warming also induces sea level rise (SLR) on the soil of Egypt, which increases the area of submerged land in northern Nile Delta and consequently soil salinization. With climate change, more frequent extreme precipitation and drought events are projected which may exacerbate the rate and soil susceptibility to accelerated erosion, salinization and other degradation processes, leading to further carbon losses.

Finally, soil carbon sequestration for climate change mitigation: some implications to Egypt. Soil C sequestration is a very important tool for C storage, atmospheric CO_2 reduction, and climate change mitigation. As well, it is critical for enhancing soil fertility, agricultural productivity, and sustainability. Adoption of good management practices may make agricultural soil an effective C sink because of their content of C content is below the C saturation point, which means that an additional C can be sequestered with time. Soil C and N are low in Egyptian soils and Delta's soils, although the Nile Delta is one the most fertile area in Egypt. Although North Nile Delta in Egypt produces a very small percentage of global GHGs emissions, it is one of the most vulnerable areas to the expected impacts and risks of climate change. In addition, although, Nile Delta in Egypt is one of the most fertile areas in the world, it contains a considerable area of degraded and unreclaimed soil. Therefore, the reclamation process in such areas is very important not only for agricultural extension in Egypt but also it may have a significant effect on SOC by alternation of soil condition and inputs and outputs of organic C. Thus, such soil has huge potential to sequester more C and N, subsequently reducing atmospheric CO_2 if improved management practices are adopted.

3.3 Smart Farming

Agricultural production in Egypt: Assessing vulnerability and enhancing adaptive capacity and resilience to climate change. Climate change adaptation is of paramount importance to agriculture, given the reliance of the sector on climate. Climate change adaptation policies should be based on science, and incorporate knowledge of indigenous people and traditional practices. Adaptation considerations should be mainstreamed into sectoral and cross-sectoral policymaking, and promote good adaptation practices to confront the heterogeneity and uncertainty of climate change impacts. Ultimately, adaptation efforts should contribute towards sustainable food production and food security for all.

The vulnerability of crop pollination ecosystem services to climate change. Pollinators play a key role in food security due to their substantial intrinsic value by providing pollination services for agriculture ecosystems. *Bees* are considered the most *important pollinators.* Managed honeybees and other wild bee species have an enormous, unchallenged impact as plant pollinators, both for wild plants and crops. The value of annual global food production relies on direct contributions by pollinators is substantial, with estimates between 235 billion and 577 *billion dollars*. Recently, there is growing *concern* about the decline in insect pollinators across the world. *Pollinator decline* has become a worldwide issue as its potential *consequences* to global food production. The decline in bees is caused by multiple stresses, including habitat fragmentation and loss, *exposure* to pathogens, parasites and *pesticides,* shortage of food sources, and the impacts of climate change. Climate change is considered the most relevant factor responsible for the decline of insect pollinators. It is likely to affect the activity of insect pollinators, thereby influencing its role in crop pollination. Climate change also is likely to alter the pollinator lifecycle, distribution and their interaction with the prevalence of pests and diseases, affect the bee-plant interaction, the mismatch between the flowering time and seasonal activity of bees, lack of floral resources, biological invasion, habitat loss, degradation, and fragmentation.

Change of agricultural calendar as a response to climate variability. Climate and weather have different influences on crop area, intensity, and yield. These influences are modified by farmer decision-making and innovations. Farmers are developing coping strategies, such as adjusting some of their farming practices, given climate variability. Enhancing our understanding of climate impact and management contributions to crop area and intensity as well as yield is crucial for lowering uncertainty regarding long-term effects of climate change on agricultural production and creating more targeted responses to climate adaptation.

Crop calendars are a core component of monitoring systems for agricultural production as they help experts focus on the seasons when different crop types grow in the field. Land surface phenology (LSP) metrics have been used as a proxy for crop calendars and apply criteria such as the start and end of the season (SOS and EOS respectively) to identify the period of pixel-level growth of active agricultural vegetation. This information, indeed, is not crop-specific and so it remains relevant to use crop calendars from independent sources, which provide crop-specific phenological timing like those of sowing, growing and harvesting.

Projected crop coefficients under climate change in Egypt. Because only monthly air temperature and solar radiation are available in the RCP6.0 climate change scenario, it is impossible to use the P-M equation. Instead, monthly ETo can be calculated using H-S equation, and the developed prediction equations for ET(P-M) can be used to calculate the values of monthly ETo using the developed calibration coefficients for each agro-climatic zone. Our results showed that this method was accurate, and the predicted ETo values were close to the calculated ETo values by the P-M equation. Thus, it is recommended to use this procedure in case of unavailability of wind speed and dew point temperature values. The BISm model was used to project Kc values in 2030. The results indicated that Kc_{ini} values for field and vegetable crops were lower in 2030, compared to its counterpart values in 2016. The values of Kc_{med} were higher in 2030, and the value of Kc_{end} was similar or higher in 2030, compared to its counterpart values in 2016. This increase in the Kc values, especially in the middle of the growing season, where maximum growth existed, results in increases in the required irrigation amounts to satisfy the needs of these crops. Whereas, there was no change in the values of Kc for fruit crops between 2016 and 2030. This could be attributed to that fact that fruit trees established ground cover all year long, which makes it less responsive to weather variation between growing seasons. However, the projected values of ETc will increase. This increase can be attributed to the expected rise in weather elements that could lead to a net deficit in atmospheric water content, consequently excessive evaporation from the soil, water, and plant surfaces.

Rice production in Egypt: The challenges of climate change and water deficiency. Agricultural soils in North Nile Delta Egypt are vulnerable to water shortage and climate change. Rice system in this area has higher C and N pools. However, the area of rice production is expected to decline because of many reasons, such as establishing Grand Renaissance Ethiopian Dam and climate change. Management practices should be taken into consideration to avoid such decline in those soils and consequently decline in SOC, which considers one of the most important methods to mitigate climate change if sequestered in the soil. Sequestering SOC decline CO_2 in the atmosphere where CO_2 is one of the most important GHG that cause climate change. Therefore, the attention should be paid for soil C sequestration in the North Nile Delta Egypt for balancing the excepted decline in C and N, due to reducing rice cultivation area because of many challenges such as climate change, water deficiency, and sea level rise, as well for climate change mitigation.

Nanotechnology for real-time control of the Red Palm Weevil (RPW) under climate change. Control of the RPW is difficult due to the concealed nature of the life cycle of the pest since palm trees do not show visual evidence of infection until it is too late for them to recover. The early detection system proposed is based on acoustic monitoring, as the activity of RPW larvae inside the palm, trunk is audible for human operators under acceptable environmental noise levels. The sensor can be installed in a palm tree and can analyze the caught audio signal during large periods. The results of sound analysis would be accounted wirelessly to a control station, to be subsequently processed and conveniently stored. It is modified to send warning messages when predefined alarm thresholds are achieved, thereby enabling managers to check online the status and evolution of the palm trees. The results imply the efficiency of the developed system to detect the existence of the RPW through its feeding sound. Nano-sensors for RPW thermal analysis is efficient to detect the existence of the RPW through its thermal properties. Nano-minerals were used and prepared in natural leaves extract to control RPW. Nano-particles of TiO_2 could be used along with *Moringa oleifera* extract against *Rhynchophorus ferrugineus* larvae. The present study revealed that NME extract drastically decreased developmental stages of *R. ferrugineus*. Unlike traditional methods, which are time-consuming, and labor intensive, this system offers the advantages of keeping the palm trees intact; reduce costs, as well as saving time and money in the process of pest's infestation detection. The compound is a 100% natural solution. It is derived from naturally plants and minerals that can significantly aid in controlling RPW. In addition, it is safe to use on all plants and crops, including natives.

Finally, facing climate change: urban gardening and sustainable agriculture. Trying to save our global, urban, or city gardening as a sort of horticultural strategies can deal with climate change. We want to begin thinking now not just about what used to be, but what may be, generally contributing the city regions to be sustainable socially, economically, and environmentally. Climate change goes to pressure us to paintings difficult at something *Homo sapiens* has never been good. An urban garden like our own families is a great location to begin. Measures to combat climate change encompass 2 primary titles:

1. Removal of the maximum greenhouse effective gases, specifically methane, carbon dioxide, sulfur oxides, nitrogen oxides, particulates, and nucleoids
2. Decreasing causes of greenhouse gas emissions.

Urban gardens, including street trees, public gardens, recreation parks, and greenbelts, especially that designed around the industrial areas can help mitigation of climate change.

3.4 Livestock, Fishery, and Aquaculture

Potential climate change impacts on livestock and food security nexus in Egypt. Regardless of the way that the phenomenon of worldwide atmosphere changes, and the local effects, it is normal, that Egypt would be one of the countries most influenced by the impacts of climate change. These impacts are reflected in high temperatures, changing rainfall pattern, rising sea levels, and the expanding recurrence of atmosphere related disasters, which posture danger to farm, agriculture land, water supply, and food security. Agriculture land is a constrained that ought to be utilized effectively and in a way that keeps up its esteem with the goal that it can produce an adequate amount of food, fuel, and forage in both the short and long term. From the viewpoint of water administration, intercessions to increment agricultural profitability ought to incorporate projects to extend irrigated areas, enhance the water-maintenance properties of soil, and enhance profitability and water-utilize efficiency. To enhance production, and accordingly, farmers, population, and food security are the two primary components prompting the expansion and the requirement for more developed agribusiness in both rainfed and irrigated farming. To avoid poverty, a mix of animal product from animals that have a good feed conversion (e.g., cattle and chickens) which is great for saving the land's esteem and add to effective land utilize.

Influence of climate changes on animal feed production: The problems and the suggested solutions. Climate change affects almost every aspect of our life. It affects animal feed production, forage production, plant feed production. The above discussion provides the view of the world as well as individual countries experiences to adapt to climate change. On a worldwide scale, it is recommended to follow the official adaptation measures of Intergovernmental Panel on Climate Change (IPCC). On an individual country scale, the adaptation and mitigation measures should take into considerations the socio-economic dimensions aspects.

Algae and fish: Benefits and hazards. In order to secure food resources from aquatic water bodies, the beneficial and nutrient-rich algae, whether macroscopic or microscopic can be cultured and grown in both marine environments and inland. The algae can be also be grown for their carotenoids that play many functions in aquatic animals. However, some oceans are high in their nutrients and low in their iron content. This causes a reduction in phytoplankton densities. One solution for this and the global warming issue is iron fertilization of those oceans that would increase phytoplankton, increase aquatic animals, and decrease global warming as a result of carbon dioxide consumption and oxygen release. However, toxic algae can harm and kill edible aquatic animals, thereby causing the massive die off, and toxins can even be transferred to human beings leading to health risks. Therefore, toxic algae and factors leading to their proliferation must be studied extensively and thoroughly monitored as they can be transferred through biofouling and ballast water.

Climatic change and chicken immunity. Climate variability is the overarching theme, but it is expected that the global average temperature will increase by 4 °C by 2100. The changing environmental temperatures require adaptation from all forms of life. Under stress conditions, avian blood tolerates a change from acid-base balance to alkaline balance. There is a decline in the plasma, a reduced level of vitamin C in the adrenal cortex, a reduction in lymphocytes, and a depression of the immune response. As the temperature rises, the birds undergo many changes—increased water consumption, respiration rate, body temperature, inferior egg quality, and susceptibility to diseases. In particular, the highly specific adaptive immune mechanisms are affected by heat stress. In more specific, heat stress deteriorates the cell-mediated immune responses. This reveals that heat stress needs to be considered as an important factor compromising chicken's health. Because of heat conditioning, biochemical and physiological mechanisms were induced to cope with heat stress; this induction may have delayed production of additional acute phase proteins to protect the cells from damage. The stress hormones—cytokine interactions are responsible for altered immune functions during heat stress. Modern-day molecular biology tools can help in understanding various cellular and molecular mechanisms involved in the production, physiological and immunological aspects of the poultry birds, which in turn can help in the development of breeds more adapted to the climate changes.

Over the past fifty years, global poultry production experienced leaps and bounces to accommodate rising demand. However, popular demand and scientific interest for organic poultry production, particularly feeding with medicinal botanicals, has considerably increased in recent years. Many studies have established the fact that herbal plants and their derivatives have the potential as immune modulators. Both, the innate and adaptive components of the immune system are stimulated by phylogenic. Moreover, most studies have exercised herbal extracts rather than purified compounds. Therefore, there is still distrust concerning the efficacy and optimum dosage of herbal plants and their derivatives as immune stimulators.

More recently, two innovative approaches have been explored, including early-life conditioning (i.e., prenatal heat acclimation) and genetic selection of breeds with the increased capacity of coping with heat stress conditions (i.e., increased heat tolerance). However, these potential opportunities, although promising (particularly, for poultry production in hot climatic regions), still require further research and development.

3.5 Socioeconomic Impacts

Climate change, agriculture, and rural communities' vulnerability in the Nile Delta. The conclusion presents a brief comparison between the impact of urban expansion on agricultural land and that of climate change on agriculture. The impact of urbanization would be higher on the field crops' cropping area, as it would reduce the area by about 2.6% of the total field crops' cropping area in Egypt based on 2015 data, while climate change would only reduce the area by 0.04%. In contrast, the impact of urban expansion on field crops' production and the net value in the Nile Delta governorates accounts only for about 50% of the impact of climate change. Urbanization would reduce the field crops' production by 2.33% of total field crops' production in Egypt, while climate change would reduce it by 4.42%. This would be reflected in the field crops' net value, as urbanization is to decrease the net value by 2.29%, whereas climate change is to reduce it by 4.95% of total field crops' net value in Egypt based on 2015 data.

Climate change would increase the field crops' water requirement by 2%, while urbanization would decrease it by about 2.7%. However, climate change impacts are dependent on the availability of adequate irrigation water. Any reduction in Egypt's water supply—due to environmental and physical causes, such as rain reduction on the Ethiopian highlands and/or the building of the GERD, and increased ET of field crops—would increase the severity of annual crop yield deficiencies affecting farmers (landowners and laborers). All landowners and some laborers (about 0.03%) are to be affected by climate change, while the impact of urbanization is to be more severe on laborers as it would reduce the working days by about 2.5% of total field crops' working days in Egypt.

Climate change impacts on agriculture, in terms of decrease of field crops' production and net value, could accelerate urbanization on agricultural land, especially in marginal areas. Finally, the review of farmers' and Nile Delta governorates' adaptive capacities showed that climate change is to significantly affect Alexandria and Ismailia by reducing their field crops' production and net value. Nevertheless, the most vulnerable governorate due to its assets would be Beheira. As the adaptation process would be hampered by the number of the rural population, the size of land ownership, illiteracy, and poverty of farmers [21] and inadequate access to VEW and funding source. Additionally, the losses of Beheira governorate would also have a large impact on the national level due to the size of its field crops' agricultural land, production in tons , net value and farmers' (women and men) percentage of total

Egypt. Potential food price increases and lack of job opportunities, both in Egypt and abroad could further affect the vulnerability of Beheira.

Climate impact on Egyptian agriculture: An efficiency analysis approach. Climate change has been identified as a major threat, especially to vulnerable regions, especially Africa. Two main factors affect agricultural growth that are labor and agricultural land together with other factors that can help to boost food production; the development of strong market infrastructure and a transportation network that facilitate the supply of food. All these factors are impacted by climate change, which leads to a negative impact on food security. Since climate change may cause a reduction in agricultural production, impact negatively logistical infrastructure which may reduce access to food, a loss in natural resources impacting the sustainability of using these resources, and finally changing the cropping pattern of a region affecting the needed nutrition requirements of these regions. Agricultural land, labor, and are the most productive input. Maize and rice production exhibit increasing rate return to scale, meaning that maize and rice producers could become more profitable with larger operations. Our empirical findings in the inefficiency equation indicate that average maximum temperature and humidity have a significant positive effect on technical inefficiency. Results of the efficiency scores show, on average, that Behera, Gharbia, Kafr El-Shiek and Sharkia are more efficient in producing wheat and maize compared to rice, while Dakahlia is more technically efficient in Maize compared with rice and wheat.

Green sustainable development (GSD) is one of the most valuable goal that everybody looking forward. It is clear that the Nexus approach is very important issue toward meeting this goal. Meanwhile Water—Climate-food nexus is always supporting GSD. This will happen by applying sectorial interrelationships for these areas. whatever, it is essential need to defined the measures, instatements tools, communicate, coordinate and implementing plan between those sectors and put it throw the national mainstream policy for sustainable development. In this context, Egypt has taken many steps as presented in this chapter at international and National levels for achieving these goals but there is still some challenges. Lack of public awareness and communication between private, governmental, and non-governmental sectors are major challenges. Meanwhile, increasing knowledge and public awareness can be powerful instruments in support of policy change towards a more integrated WCF-nexus perspective.

4 Recommendations

A key aspect of climate change impacts is the ability to adapt to future challenges. We argue that mitigation and adaptation need built-in flexibility to achieve this goal. Throughout this book project, the editorial' teams noted some areas that could be explored to further improvement. Based on the authors' findings and conclusions, this section offers a set of recommendations providing suggestions for future researchers, stakeholders and decision making regrouped by subjects:

- **An overview of climate change in Egypt**.
  - The awareness rising by the government among businesses and local authorities about the risks and vulnerabilities created by climate change to minimize impacts and help limit the cost of damages.
  - The provision to the EEAA of the full responsibility for the monitoring and management of a national carbon market, including, but not limited to allocation of emissions, development of appropriate standards, regulations and laws, establishment of appropriate bodies and mechanisms for the effective functioning of the market, oversight of development of derivative markets, and provision of guidance to concerned institutions.
  - Guide the research agenda in Egypt to provide scientific facts to inform policy makers with information and knowledge related to climate change and its impact on the agriculture sector.
  - Enhance the role of the scientific research entities in presenting the knowledge in a simple way for the decision makers and the public.
- **Land and water resources**.
  - Mainstream climate change assessment, impact, and measures of adaptation in the research programs of the research institutions.
  - Establish strong linkages between the local civil society organizations and the agricultural research bodies in the country to empower these organizations with the necessary tools and gaudiness to approach climate change challenges at the levels of local communities.
  - Beneficial plant growth promoting microorganisms could be an important biotechnological tool for sustainable agriculture by their positive effect on soil fertility and crop production and quality and could reduce the costs for chemical fertilization and pest control.
  - Staying abreast of technological innovations is vital to the future efficiency and competence of the agriculture system in the region.
  - Determining the best approaches to observe and quantify the response of plant-microbe interactions or microbe-microbe interactions to climate change to predict future ecosystem function.
  - Temporal and seasonal variation and plant root exudates play a vital role in the response of soil microbial activity to climate change, subsequently should be incorporated into future climate change experimental manipulations.
  - Research around institutional reform regarding the distribution of power and the flow of knowledge in the policymaking chain with respect to SLR adaptation is needed.
- **Smart farming**.
  - From an agricultural perspective, in-depth studies are needed to identify the different physical, social, economic, and institutional barriers that stand in the way of agricultural adaptation in the Nile Delta region.

- Regarding farm level adaptation and the resilience of the Nile Delta communities facing potential SLR are also worth studying.
- Follow appropriate strategies to resist rising sea levels to prevent erosion of the northern shores of the Delta.
- Strengthening the climate change institutionally at the country level to include environmental institutions in addition to those of agriculture and water.
- Mainstream issues related to impact of climate change in the national policies and strategies of the agricultural and water sector.
- Enhance the scientific research sector to address climate change concerns.
- Improve irrigation water productivity for sustainable agricultural production and food security.
- Raising awareness on the adverse impacts of climate change on ecosystem services and food security.
- Enhance the public appreciation of the importance of natural pollination by bees as one of the most substantial ecosystem services.
- Fill knowledge gaps in pollinator ecology, declines, and intact management.
- Encourage public engagement in the protection and conservation of honeybees and other wild pollinators.
- Conserving and restoring natural nesting habitats of solitary bees and enhancing nesting materials and nesting sites by creating bee hotels (artificial nests) for the propagation of the most promising solitary bees.
- Encourage farm owners for cultivation bee-friendly plants and other crops with a continuous succession of flowering.
- The policymakers should take into account the expected consequences of climate change in their future plans, regulate the amount of available water for agriculture, and distribute it on the basis of crops need in each agro-climatic zone.
- Policymakers should make efforts to improve irrigation water transport, distribution, and application efficiency to reduce water losses through evaporation and deep percolation.
- Raising awareness on Red Palm Weevil measures among all the stakeholders.
- Urgent need to develop a fast and reliable, cost-effective, and easy to handle early detection device for Red Palm Weevil control.
- Implement a GIS aided monitoring system for efficient mapping, data collection, analysis, and management for climate change control.
- Use remote sensing imagery to geo-reference palm trees in countries to be used as a primary base map in the GIS to control and mitigate climate change.
- Municipality planning and landscape gardeners in conjunction with the Ministry of the Environmental Affairs have to review planning methods and undertake scientific foundations while choosing garden plants and verifying non-emission farming structures that may be harmful to the environment when dealing with soil during soil preparing for agriculture.
- Look for ways or means to change the plants that are not environmentally friendly plants, with no harm to the soil.

- Climate change affects and is affected by all countries, but its treatment must begin with the actions of individuals, which require the provision of appropriate means of awareness in urban communities.
- Gardeners should not lose sight of the importance of reducing water consumption, at least through selecting C4 and CAM plants for their low water needs. Taking into consideration their environmental importance and their efficiency in eliminating greenhouse gases.
- Reduces climate change as urban agriculture contributes to the prevention of the over-heating of city environments (Urban Heat Islands).
- Gardens are one of the maximum valuable legacies that we can go away for future generations.

• **Livestock, fishery, and aquaculture.**

- Selection and production of disease and parasite resistant chicken, for an example, will assist in decreasing the dependency on drugs and medication, which will reduce chemicals in animal products and the environment and good animal welfare.
- Creating of the Egyptian Indigenous Chicken Center "EICC" will be recommended, which will help the rural community to be able to buy cockerels from superior indigenous strains for upgrading their stock.
- EICC will produce seed stock for different species of birds that will be distributed to farmers in the villages. This will directly contribute to improved village poultry production and diversity, resulting in increased animal protein intake among smallholders.
- The incorporates poultry production within existing farming systems, and then ensure a sustainable contribution to stable food security.
- The smallholder will subsequently profit by the combination of various species nutritiously, socially and economically, this will lead to sustainable conservation, management, improvement, and utilization of indigenous poultry genetic resources.
- Increase agricultural research budget, especially research related to the production and development of adaptable varieties to climate change, like salt-resistant varieties, heat-resistant varieties, short-lived varieties, and water-saving varieties and spending on improving animal breeds.

• **Socioeconomic impacts**.

- To counter land fragmentation, which is a main cause of urban expansion on agricultural land, the role of the agricultural cooperatives should be enhanced. Agricultural cooperatives should make sure that agricultural land—especially small land plots—is cultivated as one big plot to allow farmers to benefit from the economies of scale. This, in turn, would improve farmers' well-being and reduce the need to transform agricultural land into urban land.
- There is a need to prepare farmers to deal with climate change to reduce its potential negative impacts on agriculture through the application of adaptation measures that would improve farmers' resilience.

- Strategies have also acknowledged the importance of improving the institutional structures of research and extension services in Egypt as well as the importance of allocating sufficient funds to agricultural research and public extension services.
- The capacity building of public extension workers and of farmers is expected to have great benefits, in addition to enhancing the structures of cooperatives through the introduction of necessary changes to guarantee a more participatory involvement of farmers in the decision-making processes and the regulation of the involvement of the private extension services to complement governmental efforts.
- Making new investment plans for introducing climate-smart technologies and adopting new production practices that may help increase farms' technical efficiency that in turn, will help to increase agricultural output and reduce the variability of yields.
- Improving local markets infrastructure and logistics, this will help to smooth the supply of food and facing international markets volatilities. Considering that family farming, which represents the majority of all farms in Egypt, improving their technical efficiency, allows reducing production costs and improving competitiveness that can help them face fluctuating market conditions and economic hardships.
- As climate impact on agriculture is intensifying, Egyptian agricultural policies needs to be developed that technically support farmers through training to improve their skills to face these effects, minimizing carbon emission from agriculture the recycling of agricultural waste instead of burning it.
- Use new varieties of cereal crops that are more resistant to climate impact. Develop markets for carbon credits to regulate the right to pollute.
- Promote sustainable use of natural resources like the current government initiatives of reducing rice and sugar cane crops cultivated areas to reduce water consumption and adoption of advanced irrigation systems that efficiently use water.
- Future research will not only focus technical efficiency, which is the first component of economic efficiency but as well will investigate the allocative efficiency that represents the second component of economic efficiency and total factor productivity for each farming system. Since investigating only on tuning the technical efficiency is not sufficient to be more competitive.
- More research needs to be done to enhance our understanding the responses of farmers to climate shocks under different economic conditions and different technology access. To develop more climate-resilient crop production systems, a global analysis of these issues is crucial.
- Given global climate change, food trade globalization, and the growing importance of food imports to maintain national food balance in many countries, we claim that addressing these knowledge gaps globally is crucial.
- The establishment of an analysis and reporting mechanism by the NCCC and the Ministry of Finance for direct climate finance to cover the disbursement use and effectiveness of funding.

- The incorporation of carbon reduction and adaptation into climate change elements in the national development strategy and management system for international cooperation and use of these elements as a key criterion for project selection in national and international investments.
- The elevation of climate change collaboration to the level of international relations strategy to facilitate climate funding into Egypt.
- The establishment of an international development administration with responsibility for climate-related international cooperation (including finance) also under the direct supervision of the Ministry of International Cooperation.
- Lunched an integrated initiative for water and food security (WFSI), could be helped in conduct a collaborative strategies on sustainable agricultural water management. WFSI, can Promote the implementation of cost—effective water investment and management practices to fulfill the food security requirement under the future projections climate change conditions.

References

1. Solanki SK (2002) Solar variability and climate change: is there a link? Sol Phys 43:59–513
2. Pfeilsticker K (2006) Paleo-climate. Institut für Umweltphysik, Universität Heidelberg, INF 229, 69120 Heidelberg
3. Mubaya C, Mafongoya P (2017) The role of institutions in managing local level climate change adaptation in semi-arid Zimbabwe. Clim Risk Manag 16:93–105
4. United Nations (2018) Implementation of the framework for capacity-building in developing countries. Framework convention on climate change subsidiary body for implementation forty-eighth session (Bonn, 30 Apr to 10 May 2018)
5. UNEP (2018) Decision adopted by the conference of the parties to the convention on biological diversity. convention on biological diversity United Nations, Treaty Series, Registration No. I-54113. Fourteenth meeting. Sharm El-Sheikh, Egypt, 17–29 Nov 2018, Agenda item 21
6. Kleinschmit D, Arts B, Giurca A, Mustalaht I, Sergent A, Pülzl H (2017) Environmental concerns in political bioeconomy discourses. Int For Rev 19(1):41
7. Hiwasaki L, Hill L (2018) Building resilience through socially equitable climate action. International Development Research Centre
8. Pulatov B, Jönsson AM, Wilcke RAI, Linderson M-L, Hall K, Bärring L (2016) Evaluation of the phenological synchrony between potato crop and Colorado potato beetle under future climate in Europe. Agr Ecosyst Environ 224:39–49. https://doi.org/10.1016/j.agee.2016.03.02
9. FAO (2016) Climate change and food security: risks and responses
10. A'Bear A, Boddy L, Jones T (2012) Impacts of elevated temperature on the growth and functioning of decomposer fungi are influenced by grazing collembola. Glob Change Biol 18:1823–1832. https://doi.org/10.1111/j.1365-2486.2012.02637.x
11. Nottingham AT, Fierer N, Turner BL, Whitaker J, Ostle NJ, McNamara NP, Bardgett RD, Leff JW, Salinas N, Silman M, Kruuk L, Meir P (2018) Microbes follow Humboldt: temperature drives plant and soil microbial diversity patterns from the Amazon to the Andes. bioRxiv:079996. https://doi.org/10.1101/079996
12. Walker WN, Kaiser C, Strasser F, Herbold C, Leblans N, Woebken D, Janssens I, Sigurdsson B, Richter A (2018) Microbial temperature sensitivity and biomass change explain soil carbon loss with warming. Nat Clim Change 8(10):8. https://doi.org/10.1038/s41558-018-0259-x
13. Bradford MAea (2016) Managing uncertainty in soil carbon feedbacks to climate change. Nat Clim Change 6:751–758

14. Hassaan MA, Abdrabo MA (2013) Vulnerability of the Nile Delta coastal areas to inundation by sea level rise. Environ Monit Assess 185:6607–6616
15. Kopp RE, Horton RM, Little CM, Mitrovica JX, Oppenheimer M, Rasmussen DJ, Strauss BH, Tebaldi C (2014) Probabilistic 21st and 22nd century sea-level projections at a global network of tide-gauge sites. Earth's Future 2:383–406. https://doi.org/10.1002/2014EF000239
16. Stammer D, Cazenave A, Ponte RM, Tamisiea ME (2013) Causes for contemporary regional sea level changes. Annu Rev Mar Sci 5:21–46. https://doi.org/10.1146/annurev-marine-121211-172406
17. BACC (2015) Second assessment of climate change for the baltic sea basin. Springer International, Cham, Switzerland, p 501
18. UNFCCC (2016) United Nations framework convention on climate change Egypt third national communication
19. Soleimani A, Hosseini SM, Bavani ARM, Jafari M, Francaviglia R (2017) Simulating soil organic carbon stock as affected by land cover change and climate change, Hyrcanian forests (northern Iran). Sci Total Environ 599–600:1646–1657
20. FAO (2017) The state of food security and nutrition in the world. Food and Agriculture Organization of the United Nations, Rome, Italy
21. Koday K (2017) The international policy centre for inclusive growth. Soc Protect After Arab Spring 14(3)
22. Garibaldi L, Dewenter IS, Winfree R, Aizen MA, Bommarco R, Cunningham SA, Kremen C, Carvalheiro LG, Harder LD, Afik O, Bartomeus I, Benjamin F, Boreux V, Cariveau D, Chacoff NP, Freitas BM, Ghazoul J, Greenleaf S, Hipólito J, Holzschuh A, Howlett B, Isaacs R, Javorek SK, Kennedy CM, Krewenka KM, Krishnan S, Mandelik Y, Mayfield MM, Motzke I, Munyuli T, Nault BA, Otieno M, Petersen J, Pisanty G, Potts SG, Rader R, Ricketts TH, Rundlöf M, Seymour CL, Schüepp C, Szentgyörgyi H, Taki H, Tscharntke T, Vergara CH, Viana BF, Wanger TC, Westphal C, Williams N, Klein AM (2013) Wild pollinators enhance fruit set of crops regardless of honey bee abundance. Sci Technol 339:1608–1611
23. Dicks LV, Abrahams A, Atkinson J, Biesmeijer J, Bourn N, Brown C, Brown JFM, Carvell C, Connolly C, Cresswell EJ, Croft PBD, De Zylva P, Effingham P, Fountain M, Goggin A, Harding D, Harding T, Hartfield C, Heard SM, Heathcote R, Heaver D, Holland J, Howe M, Hughes B, Huxley T, Kunin W, Little J, Mason C, Memmott J, Osborne J, Pankhurst T, Paxton JR, Pocock JOM, Potts GS, Power FE, Raine EN, Ranelagh E, Roberts S, Saunders R, Smith K, Smith MR, Sutton P, Tilley ANL, Tinsley A, Tonhasca A, Vanbergen JA, Webster S, Wilson A, Sutherland JW (2013) Identifying key knowledge needs for evidence-based conservation of wild insect pollinators: a collaborative cross-sectoral exercise. Insect Conserv Divers 6:339–353
24. Intercontinental Panel on Climate Change (2001) Climate Change 2001: impacts, adaptation, and vulnerability. In: McCarthy JJ et al (eds) Contribution of working group II to the third assessment report of the intergovernmental panel on climate change. Cambridge University Press, Cambridge, UK
25. Watson R, Zinyoera M, Moss R (1998) The regional impacts of climate change: an assessment of vulnerability. Cambridge University Press, Cambridge
26. IPCC IPoCC (2001) Climate change 2001: impacts, adaptation, and vulnerability contribution of working group II to the third assessment report of the intergovernmental panel on climate change. Cambridge University Press, Cambridge
27. Kandlinkar M, Risbey J (2000) Agricultural impacts of climate change: if adaptation is the answer, what is the question? Clim Change 45:529–539
28. Nhemachena C, Hassan R (2007) Micro-level analysis of farmers' adaptation to climate change in Southern Africa. IFPRI Discussion Paper No 00714 International Food Policy Research Institute, Washington, DC
29. Yegbemey RN, Yabi JA, Tovignan DS, Gantoli G, Kokoye SEH (2013) Farmers' decisions to adapt to climate change under various property rights: a case study of maize farming in Northern Benin (West Africa). Land Use Policy 34:168–175. https://doi.org/10.1016/j.landusepol.2013.03.001

30. Meroni M, Rembold F, Urbano F, Csak G, Lemoine G, Kerdiles H, Perez-Hoyoz A (2018) The warning classification scheme of ASAP—anomaly hot spots of agricultural production. 2.1. https://doi.org/10.2760/798528
31. Patel JH, Oza MP (2014) Deriving crop calendar using NDVI time-series. https://doi.org/10.5194/isprsarchives-XL-8-869-2014
32. El-Massah S, Omran G (2014) Would climate change affect the imports of cereals? The case of Egypt. In: Handbook of climate change adaptation. Springer, Berlin
33. Karmakar R, Das I, Dutta D, Rakshit A (2016) Potential effects of climate change on soil properties: a review. Sci Int 4(2):51–73
34. Ouda S, Noreldin T, Abd El-Latif K (2015) Water requirements for wheat and maize under climate change in North Nile Delta. Span J Agric Res 13(1):1–7
35. Rahman A, Karim N, Kadir N, Naher T (2015) Impacts of climate change on crop coefficient and reference crop evapotranspiration of boro rice in north-east hydrological region of Bangladesh. In: 5th international conference on water and flood management, pp 345–350
36. Shahid M, Nayak AK, Puree C, Tripathi R, Lal B, Gautam P, Bhattacharyya P, Mohanty S, Kumar A, Panda BB, Kumar U, Shukla AK (2017) Carbon and nitrogen fractions and stocks under 41 years of chemical and organic fertilization in a sub-humid tropical rice soil. Soil Tillage Res 170:136–146. https://doi.org/10.1016/j.still.2017.03.008
37. El-Shahway AS, Mahmoud MMA, Udeigwe TK (2016) Alterations in soil chemical properties induced by continuous rice cultivation: a study on the arid Nile delta soils of Egypt. Land Degrad Develop 27:231–238
38. Nooten S, Andrew N, Hughes L (2014) Potential impacts of climate change on insect communities: a transplant experiment. Plos One 9(1):e85987. https://doi.org/10.1371/journal.pone.0085987
39. Chen Y, Ma C (2010) Effect of global warming on insect: a literature review. Acta Ecol Sin 30(8):2159–2172
40. Sambaraju K, Carroll A, Zhu J, Stahl K, Moore R, Aukema B (2012) Climate change could alter the distribution of mountain pine beetle outbreaks in western Canada. Ecography 35(3):211–223
41. Assan N (2014) Possible impact and adaptation to climate change in livestock production in Southern Africa. IOSR J Environ Sci Toxicol Food Technol 8(2):104–112
42. Yahav S (2009) Alleviating heat stress in domestic fowl: different strategies. World Poult Sci J 65(04):719–732
43. Nelson G, Rosegrant M, Koo J, Robertson R, Sulser T, Zhu T, Ringler C (2009) Climate change: impact on agriculture and costs of adaptation. IFPRI, Washington DC
44. Care (2016) Fleeing climate change: impacts on migration and displacement. Care Danemark http://careclimatechange.org/wp-content/uploads/2016/11/FleeingClimateChange_report.pdf. Accessed in May 2018
45. Capmas (2017) Egypt 2018 statistical abstract. Capmas, Cairo, Egypt
46. Panpeng J, Ahmad M (2017) Vulnerability of fishing communities from sea-level change: a study of Laemsing district in Chanthaburi Province, Thailand. http://www.mdpi.com/2071-1050/9/8/1388/htm#B9-sustainability-09-01388. Accessed Feb 2019
47. IPCC IPoCC (2014) Annex II: glossary. In: Mach KJ, Planton S, von Stechow C (eds) Climate change 2014: synthesis report contribution of working groups I, II and III to the fifth assessment report of the intergovernmental panel on climate change [Core Writing Team, Pachauri RK, Meyer LA (eds)]. IPCC, Geneva, Switzerland, pp 117–130
48. IPCC (2018) Summary for policymakers. In: Global warming of 15°C an IPCC special report on the impacts of global warming of 15°C above pre-industrial levels and related global greenhouse gas emission pathways, in the context of strengthening the global response to the threat of climate change, sustainable development, and efforts to eradicate. World Meteorological Organization, Geneva, Switzerland, 32 pp

The manufacturer's authorised representative in the EU is Springer Nature Customer Service Centre GmbH, Europaplatz 3, 69115 Heidelberg, Germany. If you have any concerns regarding our products, please contact ProductSafety@springernature.com

Printed and bound by CPI Group (UK) Ltd, Croydon, CR0 4YY
30/09/2026
02225331-0007